Grzimek's
Animal Life Encyclopedia

Second Edition

••••

Grzimek's
Animal Life Encyclopedia

Second Edition

●●●●

Volume 7
Reptiles

James B. Murphy, Advisory Editor
Neil Schlager, Editor

Joseph E. Trumpey, Chief Scientific Illustrator

Michael Hutchins, Series Editor
In association with the American Zoo and Aquarium Association

GALE®

TM

GALE

Detroit • New York • San Diego • San Francisco • Cleveland • New Haven, Conn. • Waterville, Maine • London • Munich

THOMSON

GALE

Grzimek's Animal Life Encyclopedia, Second Edition
Volume 7: Reptiles
Produced by Schlager Group Inc.
Neil Schlager, Editor
Vanessa Torrado-Caputo, Assistant Editor

Project Editor
Melissa C. McDade

Editorial
Stacey Blachford, Deirdre S. Blanchfield, Madeline Harris, Christine Jeryan, Kate Kretschmann, Mark Springer

Permissions
Margaret Chamberlain

Imaging and Multimedia
Randy Bassett, Mary K. Grimes, Lezlie Light, Christine O'Bryan, Barbara Yarrow, Robyn V. Young

Product Design
Tracey Rowens, Jennifer Wahi

Manufacturing
Wendy Blurton, Dorothy Maki, Evi Seoud, Mary Beth Trimper

For permission to use material from this product, submit your request via Web at http://www.gale-edit.com/permissions, or you may download our Permissions Request form and submit your request by fax or mail to: The Gale Group, Inc., Permissions Department, 27500 Drake Road, Farmington Hills, MI, 48331-3535, Permissions hotline: 248-699-8074 or 800-877-4253, ext. 8006, Fax: 248-699-8074 or 800-762-4058.

Cover photo of green python by JLM Visuals. Back cover photos of sea anemone by AP/Wide World Photos/University of Wisconsin-Superior; land snail, lionfish, golden frog, and green python by JLM Visuals; red-legged locust © 2001 Susan Sam; hornbill by Margaret F. Kinnaird; and tiger by Jeff Lepore/Photo Researchers. All reproduced by permission.

While every effort has been made to ensure the reliability of the information presented in this publication, The Gale Group, Inc. does not guarantee the accuracy of the data contained herein. The Gale Group, Inc. accepts no payment for listing; and inclusion in the publication of any organization, agency, institution, publication, service, or individual does not imply endorsement of the editors and publisher. Errors brought to the attention of the publisher and verified to the satisfaction of the publisher will be corrected in future editions.

ISBN   0-7876-5362-4 (vols. 1–17 set)
0-7876-5783-2 (vol. 7)

LIBRARY OF CONGRESS CATALOGING-IN-PUBLICATION DATA

Grzimek, Bernhard.
[Tierleben. English]
Grzimek's animal life encyclopedia.— 2nd ed.
v. cm.
Includes bibliographical references.
Contents: v. 1. Lower metazoans and lesser deuterosomes / Neil Schlager, editor — v. 2. Protostomes / Neil Schlager, editor — v. 3. Insects / Neil Schlager, editor — v. 4-5. Fishes I-II / Neil Schlager, editor — v. 6. Amphibians / Neil Schlager, editor — v. 7. Reptiles / Neil Schlager, editor — v. 8-11. Birds I-IV / Donna Olendorf, editor — v. 12-16. Mammals I-V / Melissa C. McDade, editor — v. 17. Cumulative index / Melissa C. McDade, editor.
ISBN 0-7876-5362-4 (set hardcover : alk. paper)
1. Zoology—Encyclopedias. I. Title: Animal life encyclopedia. II. Schlager, Neil, 1966- III. Olendorf, Donna IV. McDade, Melissa C. V. American Zoo and Aquarium Association. VI. Title.
QL7 .G7813 2004

590'.3—dc21
2002003351

Printed in Canada
10 9 8 7 6 5 4 3 2 1

Recommended citation: *Grzimek's Animal Life Encyclopedia*, 2nd edition. Volume 7, *Reptiles*, edited by Michael Hutchins, James B. Murphy, and Neil Schlager. Farmington Hills, MI: Gale Group, 2003.

Contents

Foreword... vii
How to use this book.............................. x
Advisory boards.................................... xiii
Contributing writers............................... xv
Contributing illustrators........................... xvi

Volume 7: Reptiles
What is a reptile? 3
Evolution of the reptiles........................... 12
Structure and function............................. 23
Behavior.. 34
Reptiles and humans 47
Conservation 59

Order TESTUDINES
Turtles and tortoises 65
    Family: Pig-nose turtles....................... 75
    Family: Australo-American sideneck turtles....... 77
    Family: Seaturtles 85
    Family: Snapping turtles 93
    Family: Central American river turtles 99
    Family: Leatherback seaturtles 101
    Family: New World pond turtles............... 105
    Family: Eurasian pond and river turtles, and
        Neotropical wood turtles 115
    Family: American mud and musk turtles 121
    Family: African sideneck turtles 129
    Family: Big-headed turtles 135
    Family: Afro-American river turtles............. 137
    Family: Tortoises 143
    Family: Softshell turtles...................... 151

Order CROCODILIANS
Crocodiles, alligators, caimans, and gharials 157
    Family: Gharials 167
    Family: Alligators and caimans 171
    Family: Crocodiles and false gharials........... 179

Order SPHENODONTIA
Tuatara
    Family: Tuatara 189

Order SQUAMATA
Lizards and snakes 195

Family: Angleheads, calotes, dragon lizards, and
    relatives 209
Family: Chameleons............................. 223
Family: Anoles, iguanas, and relatives 243
Family: Geckos and pygopods.................... 259
Family: Blindskinks............................. 271
Family: Wormlizards 273
Family: Mole-limbed wormlizards 279
Family: Florida wormlizards..................... 283
Family: Spade-headed wormlizards 287
Family: Night lizards 291
Family: Wall lizards, rock lizards, and relatives 297
Family: Microteiids............................. 303
Family: Whiptail lizards, tegus, and relatives 309
Family: Girdled and plated lizards 319
Family: Skinks 327
Family: Alligator lizards, galliwasps, glass lizards,
    and relatives 339
Family: Knob-scaled lizards 347
Family: Gila monsters and Mexican beaded lizards.. 353
Family: Monitors, goannas, and earless monitors ... 315
Family: Early blindsnakes 369
Family: Slender blindsnakes..................... 373
Family: Blindsnakes 379
Family: False blindsnakes 387
Family: Shieldtail snakes........................ 391
Family: Pipe snakes............................ 395
Family: False coral snakes....................... 399
Family: Sunbeam snakes 401
Family: Neotropical sunbeam snakes............. 405
Family: Boas 409
Family: Pythons................................ 419
Family: Splitjaw snakes 429
Family: Woodsnakes and spinejaw snakes 433
Family: File snakes 439
Family: Vipers and pitvipers..................... 445
Family: African burrowing snakes................ 461
Family: Colubrids 465
Family: Cobras, kraits, seasnakes, death adders,
    and relatives 483
For further reading................................ 501

Contents

Organizations 507
Contributors to the first edition 509
Glossary 516

Reptiles species list 520
Geologic time scale 571
Index 573

Foreword

Earth is teeming with life. No one knows exactly how many distinct organisms inhabit our planet, but more than 5 million different species of animals and plants could exist, ranging from microscopic algae and bacteria to gigantic elephants, redwood trees and blue whales. Yet, throughout this wonderful tapestry of living creatures, there runs a single thread: Deoxyribonucleic acid or DNA. The existence of DNA, an elegant, twisted organic molecule that is the building block of all life, is perhaps the best evidence that all living organisms on this planet share a common ancestry. Our ancient connection to the living world may drive our curiosity, and perhaps also explain our seemingly insatiable desire for information about animals and nature. Noted zoologist, E.O. Wilson, recently coined the term "biophilia" to describe this phenomenon. The term is derived from the Greek *bios* meaning "life" and *philos* meaning "love." Wilson argues that we are human because of our innate affinity to and interest in the other organisms with which we share our planet. They are, as he says, "the matrix in which the human mind originated and is permanently rooted." To put it simply and metaphorically, our love for nature flows in our blood and is deeply engrained in both our psyche and cultural traditions.

Our own personal awakenings to the natural world are as diverse as humanity itself. I spent my early childhood in rural Iowa where nature was an integral part of my life. My father and I spent many hours collecting, identifying and studying local insects, amphibians and reptiles. These experiences had a significant impact on my early intellectual and even spiritual development. One event I can recall most vividly. I had collected a cocoon in a field near my home in early spring. The large, silky capsule was attached to a stick. I brought the cocoon back to my room and placed it in a jar on top of my dresser. I remember waking one morning and, there, perched on the tip of the stick was a large moth, slowly moving its delicate, light green wings in the early morning sunlight. It took my breath away. To my inexperienced eyes, it was one of the most beautiful things I had ever seen. I knew it was a moth, but did not know which species. Upon closer examination, I noticed two moon-like markings on the wings and also noted that the wings had long "tails", much like the ubiquitous tiger swallow-tail butterflies that visited the lilac bush in our backyard. Not wanting to suffer my ignorance any longer, I reached immediately for my *Golden Guide to North American Insects* and searched through the section on moths and butterflies. It was a luna moth! My heart was pounding with the excitement of new knowledge as I ran to share the discovery with my parents.

I consider myself very fortunate to have made a living as a professional biologist and conservationist for the past 20 years. I've traveled to over 30 countries and six continents to study and photograph wildlife or to attend related conferences and meetings. Yet, each time I encounter a new and unusual animal or habitat my heart still races with the same excitement of my youth. If this is biophilia, then I certainly possess it, and it is my hope that others will experience it too. I am therefore extremely proud to have served as the series editor for the Gale Group's rewrite of *Grzimek's Animal Life Encyclopedia*, one of the best known and widely used reference works on the animal world. *Grzimek's* is a celebration of animals, a snapshot of our current knowledge of the Earth's incredible range of biological diversity. Although many other animal encyclopedias exist, *Grzimek's Animal Life Encyclopedia* remains unparalleled in its size and in the breadth of topics and organisms it covers.

The revision of these volumes could not come at a more opportune time. In fact, there is a desperate need for a deeper understanding and appreciation of our natural world. Many species are classified as threatened or endangered, and the situation is expected to get much worse before it gets better. Species extinction has always been part of the evolutionary history of life; some organisms adapt to changing circumstances and some do not. However, the current rate of species loss is now estimated to be 1,000–10,000 times the normal "background" rate of extinction since life began on Earth some 4 billion years ago. The primary factor responsible for this decline in biological diversity is the exponential growth of human populations, combined with peoples' unsustainable appetite for natural resources, such as land, water, minerals, oil, and timber. The world's human population now exceeds 6 billion, and even though the average birth rate has begun to decline, most demographers believe that the global human population will reach 8–10 billion in the next 50 years. Much of this projected growth will occur in developing countries in Central and South America, Asia and Africa-regions that are rich in unique biological diversity.

Finding solutions to conservation challenges will not be easy in today's human-dominated world. A growing number of people live in urban settings and are becoming increasingly isolated from nature. They "hunt" in super markets and malls, live in apartments and houses, spend their time watching television and searching the World Wide Web. Children and adults must be taught to value biological diversity and the habitats that support it. Education is of prime importance now while we still have time to respond to the impending crisis. There still exist in many parts of the world large numbers of biological "hotspots"-places that are relatively unaffected by humans and which still contain a rich store of their original animal and plant life. These living repositories, along with selected populations of animals and plants held in professionally managed zoos, aquariums and botanical gardens, could provide the basis for restoring the planet's biological wealth and ecological health. This encyclopedia and the collective knowledge it represents can assist in educating people about animals and their ecological and cultural significance. Perhaps it will also assist others in making deeper connections to nature and spreading biophilia. Information on the conservation status, threats and efforts to preserve various species have been integrated into this revision. We have also included information on the cultural significance of animals, including their roles in art and religion.

It was over 30 years ago that Dr. Bernhard Grzimek, then director of the Frankfurt Zoo in Frankfurt, Germany, edited the first edition of *Grzimek's Animal Life Encyclopedia*. Dr. Grzimek was among the world's best known zoo directors and conservationists. He was a prolific author, publishing nine books. Among his contributions were: *Serengeti Shall Not Die, Rhinos Belong to Everybody* and *He and I and the Elephants*. Dr. Grzimek's career was remarkable. He was one of the first modern zoo or aquarium directors to understand the importance of zoo involvement in *in situ* conservation, that is, of their role in preserving wildlife in nature. During his tenure, Frankfurt Zoo became one of the leading western advocates and supporters of wildlife conservation in East Africa. Dr. Grzimek served as a Trustee of the National Parks Board of Uganda and Tanzania and assisted in the development of several protected areas. The film he made with his son Michael, *Serengeti Shall Not Die*, won the 1959 Oscar for best documentary.

Professor Grzimek has recently been criticized by some for his failure to consider the human element in wildlife conservation. He once wrote: "A national park must remain a primordial wilderness to be effective. No men, not even native ones, should live inside its borders." Such ideas, although considered politically incorrect by many, may in retrospect actually prove to be true. Human populations throughout Africa continue to grow exponentially, forcing wildlife into small islands of natural habitat surrounded by a sea of humanity. The illegal commercial bushmeat trade-the hunting of endangered wild animals for large scale human consumption-is pushing many species, including our closest relatives, the gorillas, bonobos and chimpanzees, to the brink of extinction. The trade is driven by widespread poverty and lack of economic alternatives. In order for some species to survive it will be necessary, as Grzimek suggested, to establish and enforce a system of protected areas where wildlife can roam free from exploitation of any kind.

While it is clear that modern conservation must take the needs of both wildlife and people into consideration, what will the quality of human life be if the collective impact of short-term economic decisions is allowed to drive wildlife populations into irreversible extinction? Many rural populations living in areas of high biodiversity are dependent on wild animals as their major source of protein. In addition, wildlife tourism is the primary source of foreign currency in many developing countries and is critical to their financial and social stability. When this source of protein and income is gone, what will become of the local people? The loss of species is not only a conservation disaster; it also has the potential to be a human tragedy of immense proportions. Protected areas, such as national parks, and regulated hunting in areas outside of parks are the only solutions. What critics do not realize is that the fate of wildlife and people in developing countries is closely intertwined. Forests and savannas emptied of wildlife will result in hungry, desperate people, and will, in the long-term lead to extreme poverty and social instability. Dr. Grzimek's early contributions to conservation should be recognized, not only as benefiting wildlife, but as benefiting local people as well.

Dr. Grzimek's hope in publishing his *Animal Life Encyclopedia* was that it would "...disseminate knowledge of the animals and love for them", so that future generations would "...have an opportunity to live together with the great diversity of these magnificent creatures." As stated above, our goals in producing this updated and revised edition are similar. However, our challenges in producing this encyclopedia were more formidable. The volume of knowledge to be summarized is certainly much greater in the twenty-first century than it was in the 1970's and 80's. Scientists, both professional and amateur, have learned and published a great deal about the animal kingdom in the past three decades, and our understanding of biological and ecological theory has also progressed. Perhaps our greatest hurdle in producing this revision was to include the new information, while at the same time retaining some of the characteristics that have made *Grzimek's Animal Life Encyclopedia* so popular. We have therefore strived to retain the series' narrative style, while giving the information more organizational structure. Unlike the original *Grzimek's*, this updated version organizes information under specific topic areas, such as reproduction, behavior, ecology and so forth. In addition, the basic organizational structure is generally consistent from one volume to the next, regardless of the animal groups covered. This should make it easier for users to locate information more quickly and efficiently. Like the original Grzimek's, we have done our best to avoid any overly technical language that would make the work difficult to understand by non-biologists. When certain technical expressions were necessary, we have included explanations or clarifications.

Considering the vast array of knowledge that such a work represents, it would be impossible for any one zoologist to have completed these volumes. We have therefore sought specialists from various disciplines to write the sections with which they are most familiar. As with the original *Grzimek's*, we have engaged the best scholars available to serve as topic editors, writers, and consultants. There were some complaints about inaccuracies in the original English version that may have been due to mistakes or misinterpretation during the complicated translation process. However, unlike the original *Grzimek's*, which was translated from German, this revision has been completely re-written by English-speaking scientists. This work was truly a cooperative endeavor, and I thank all of those dedicated individuals who have written, edited, consulted, drawn, photographed, or contributed to its production in any way. The names of the topic editors, authors, and illustrators are presented in the list of contributors in each individual volume.

The overall structure of this reference work is based on the classification of animals into naturally related groups, a discipline known as taxonomy or biosystematics. Taxonomy is the science through which various organisms are discovered, identified, described, named, classified and catalogued. It should be noted that in preparing this volume we adopted what might be termed a conservative approach, relying primarily on traditional animal classification schemes. Taxonomy has always been a volatile field, with frequent arguments over the naming of or evolutionary relationships between various organisms. The advent of DNA fingerprinting and other advanced biochemical techniques has revolutionized the field and, not unexpectedly, has produced both advances and confusion. In producing these volumes, we have consulted with specialists to obtain the most up-to-date information possible, but knowing that new findings may result in changes at any time. When scientific controversy over the classification of a particular animal or group of animals existed, we did our best to point this out in the text.

Readers should note that it was impossible to include as much detail on some animal groups as was provided on others. For example, the marine and freshwater fish, with vast numbers of orders, families, and species, did not receive as detailed a treatment as did the birds and mammals. Due to practical and financial considerations, the publishers could provide only so much space for each animal group. In such cases, it was impossible to provide more than a broad overview and to feature a few selected examples for the purposes of illustration. To help compensate, we have provided a few key bibliographic references in each section to aid those interested in learning more. This is a common limitation in all reference works, but *Grzimek's Encyclopedia of Animal Life* is still the most comprehensive work of its kind.

I am indebted to the Gale Group, Inc. and Senior Editor Donna Olendorf for selecting me as Series Editor for this project. It was an honor to follow in the footsteps of Dr. Grzimek and to play a key role in the revision that still bears his name. *Grzimek's Animal Life Encyclopedia* is being published by the Gale Group, Inc. in affiliation with my employer, the American Zoo and Aquarium Association (AZA), and I would like to thank AZA Executive Director, Sydney J. Butler; AZA Past-President Ted Beattie (John G. Shedd Aquarium, Chicago, IL); and current AZA President, John Lewis (John Ball Zoological Garden, Grand Rapids, MI), for approving my participation. I would also like to thank AZA Conservation and Science Department Program Assistant, Michael Souza, for his assistance during the project. The AZA is a professional membership association, representing 205 accredited zoological parks and aquariums in North America. As Director/William Conway Chair, AZA Department of Conservation and Science, I feel that I am a philosophical descendant of Dr. Grzimek, whose many works I have collected and read. The zoo and aquarium profession has come a long way since the 1970s, due, in part, to innovative thinkers such as Dr. Grzimek. I hope this latest revision of his work will continue his extraordinary legacy.

Silver Spring, Maryland, 2001
Michael Hutchins
Series Editor

· · · · ·

How to use this book

Grzimek's Animal Life Encyclopedia is an internationally prominent scientific reference compilation, first published in German in the late 1960s, under the editorship of zoologist Bernhard Grzimek (1909–1987). In a cooperative effort between Gale and the American Zoo and Aquarium Association, the series has been completely revised and updated for the first time in over 30 years. Gale expanded the series from 13 to 17 volumes, commissioned new color paintings, and updated the information so as to make the set easier to use. The order of revisions is:

Volumes 8–11: Birds I–IV
Volume 6: Amphibians
Volume 7: Reptiles
Volumes 4–5: Fishes I–II
Volumes 12–16: Mammals I–V
Volume 3: Insects
Volume 2: Protostomes
Volume 1: Lower Metazoans and Lesser Deuterostomes
Volume 17: Cumulative Index

Organized by taxonomy

The overall structure of this reference work is based on the classification of animals into naturally related groups, a discipline known as taxonomy—the science in which various organisms are discovered, identified, described, named, classified, and catalogued. Starting with the simplest life forms, the lower metazoans and lesser deuterostomes, in Volume 1, the series progresses through the more advanced classes of classes, culminating with the mammals in Volumes 12–16. Volume 17 is a stand-alone cumulative index.

Organization of chapters within each volume reinforces the taxonomic hierarchy. In the case of the volume on Reptiles, introductory chapters describe general characteristics of the class Reptilia, followed by taxonomic chapters dedicated to order and family. Species accounts appear at the end of family chapters. To help the reader grasp the scientific arrangement, each type of taxonomic chapter has a distinctive color and symbol:

● = Order Chapter (blue background)

▲ = Family Chapter (yellow background)

⬭ = Monotypic Order Chapter (green background)

As chapters narrow in focus, they become more tightly formatted. Introductory chapters have a loose structure, reminiscent of the first edition. Although not strictly formatted, chapters on orders are carefully structured to cover basic information about the group. Chapters on families are the most tightly structured, following a prescribed format of standard rubrics that make information easy to find. These chapters typically include:

Thumbnail introduction
    Common name
    Scientific name
    Class
    Order
    Suborder
    Family
    Thumbnail description
    Size
    Number of genera, species
    Habitat
    Conservation status
Main chapter
    Evolution and systematics
    Physical characteristics
    Distribution
    Habitat
    Behavior
    Feeding ecology and diet
    Reproductive biology
    Conservation status
    Significance to humans
Species accounts
    Common name
    Scientific name
    Subfamily
    Taxonomy
    Other common names
    Physical characteristics
    Distribution
    Habitat
    Behavior
    Feeding ecology and diet
    Reproductive biology
    Conservation status
    Significance to humans

Resources
  Books
  Periodicals
  Organizations
  Other

Color graphics enhance understanding

Grzimek's features approximately 3,500 color photos, including nearly 130 in the Reptiles volume; 3,500 total color maps, including more than 160 in the Reptiles volume; and approximately 5,500 total color illustrations, including approximately 300 in the Reptiles volume. Each featured species of animal is accompanied by both a distribution map and an illustration.

All maps in *Grzimek's* were created specifically for the project by XNR Productions. Distribution information was provided by expert contributors and, if necessary, further researched at the University of Michigan Zoological Museum library. Maps are intended to show broad distribution, not definitive ranges.

All the color illustrations in *Grzimek's* were created specifically for the project by Michigan Science Art. Expert contributors recommended the species to be illustrated and provided feedback to the artists, who supplemented this information with authoritative references, skins, and specimens from University of Michigan Zoological Museum library. In addition to illustrations of species, *Grzimek's* features drawings that illustrate characteristic traits and behaviors.

About the contributors

All of the chapters were written by herpetologists who are specialists on specific subjects and/or families. Topic editor James B. Murphy reviewed the completed chapters to insure consistency and accuracy.

Standards employed

In preparing the volume on Reptiles, the editors relied primarily on the taxonomic structure outlined in *Herpetology: An Introductory Biology of Amphibians and Reptiles*, 2nd edition, edited by George R. Zug, Laurie J. Vitt, and Janalee P. Caldwell (2001). Systematics is a dynamic discipline in that new species are being discovered continuously, and new techniques (e.g., DNA sequencing) frequently result in changes in the hypothesized evolutionary relationships among various organisms. Consequently, controversy often exists regarding classification of a particular animal or group of animals; such differences are mentioned in the text.

Grzimek's has been designed with ready reference in mind, and the editors have standardized information wherever feasible. For **Conservation Status**, *Grzimek's* follows the IUCN Red List system, developed by its Species Survival Commission. The Red List provides the world's most comprehensive inventory of the global conservation status of plants and an-

imals. Using a set of criteria to evaluate extinction risk, the IUCN recognizes the following categories: Extinct, Extinct in the Wild, Critically Endangered, Endangered, Vulnerable, Conservation Dependent, Near Threatened, Least Concern, and Data Deficient. For a complete explanation of each category, visit the IUCN web page at <http://www.iucn.org/themes/ssc/redlists/categor.htm>.

In addition to IUCN ratings, chapters may contain other conservation information, such as a species' inclusion on one of three Convention on International Trade in Endangered Species (CITES) appendices. Adopted in 1975, CITES is a global treaty whose focus is the protection of plant and animal species from unregulated international trade.

In the Species accounts throughout the volume, the editors have attempted to provide common names not only in English but also in French, German, Spanish, and local dialects. Unlike for birds, there is no official list of common names for reptiles of the world, but for species in North America an official list does exist: *Scientific and Standard English Names of Amphibians and Reptiles of North America, North of Mexico, with Comments Regarding Confidence in our Understanding*, edited by Brian I. Crother (2000). A consensus of acceptable common names in English, French, German, Portuguese, and Spanish for European species exists in the *Atlas of Amphibians and Reptiles in Europe*, edited by Jean-Pierre Gasc, et al. (1997). Two books purportedly contain common names of reptiles worldwide, but these are names mostly coined by the authors and do not necessarily reflect what the species are called in their native countries. The first of these books, *Dictionary of Animal Names in Five Languages. Amphibians and Reptiles*, by Natalia B. Anajeva, et al. (1988), contains names in Latin, Russian, English, German, and French. The second is *A Complete Guide to Scientific Names of Reptiles and Amphibians of the World*, by Norman Frank and Erica Ramus (1995); for those species for which no commonly accepted common name exists, the name proposed in this book has been used in the volume on Reptiles.

Grzimek's provides the following standard information on lineage in the ***Taxonomy*** rubric of each Species account: [First described as] *Atractaspis bibroni* [by] A. Smith, [in] 1849, [based on a specimen from] eastern districts of the Cape Colony, South Africa. The person's name and date refer to earliest identification of a species, although the species name may have changed since first identification. However, the entity of reptile is the same.

Readers should note that within chapters, species accounts are organized alphabetically by subfamily name and then alphabetically by scientific name.

Anatomical illustrations

While the encyclopedia attempts to minimize scientific jargon, readers will encounter numerous technical terms related to anatomy and physiology throughout the volume. To assist readers in placing physiological terms in their proper context, we have created a number of detailed anatomical drawings. These can be found on pages 65–70, 159–161, 191, and 199–201. Readers are urged to make heavy use of these drawings. In addition, terms are defined in the *Glossary* at the back of the book.

Appendices and index

In addition to the main text and the aforementioned *Glossary*, the volume contains numerous other elements. *For further reading* directs readers to additional sources of information about reptiles. Valuable contact information for *Organizations* is also included in an appendix. An exhaustive *Reptiles species list* records all known species of amphibians as of November 2002, based on information in the EMBL Reptile Database (http://www.reptiliaweb.org) and organized according to *Herpetology*, 2nd edition, by Zug, Vitt, and Caldwell; the section of turtle species was supplemented with information obtained from the World Turtle Database, EMYSystem (http://emys.geo.orst.edu/). And a full-color *Geologic time scale* helps readers understand pre-historic time periods. Additionally, the volume contains a *Subject index.*

Acknowledgements

Gale would like to thank several individuals for their important contributions to the volume. Dr. James B. Murphy, topic editor for the Reptiles volume, oversaw all phases of the volume, including creation of the topic list, chapter review, and compilation of the appendices. Neil Schlager, project manager for the Reptiles volume, coordinated the writing and editing of the text. Dr. Michael Hutchins, chief consulting editor for the series, and Michael Souza, program assistant, Department of Conservation and Science at the American Zoo and Aquarium Association, provided valuable input and research support. Judith A. Block, registrar at the Smithsonian National Zoological Park, assisted with manuscript review. Finally, George R. Zug provided helpful advice regarding taxonomic issues.

Advisory boards

Series advisor

Michael Hutchins, PhD
Director of Conservation and Science/William Conway Chair
American Zoo and Aquarium Association
Silver Spring, Maryland

Subject advisors

Volume 1: Lower Metazoans and Lesser Deuterostomes
Dennis Thoney, PhD
Director, Marine Laboratory & Facilities
Humboldt State University
Arcata, California

Volume 2: Protostomes
Dennis Thoney, PhD
Director, Marine Laboratory & Facilities
Humboldt State University
Arcata, California

Sean F. Craig, PhD
Assistant Professor, Department of Biological Sciences
Humboldt State University
Arcata, California

Volume 3: Insects
Art Evans, PhD
Entomologist
Richmond, Virginia

Rosser W. Garrison, PhD
Systematic Entomologist, Los Angeles County
Los Angeles, California

Volumes 4–5: Fishes I– II
Paul Loiselle, PhD
Curator, Freshwater Fishes
New York Aquarium
Brooklyn, New York

Dennis Thoney, PhD
Director, Marine Laboratory & Facilities

Humboldt State University
Arcata, California

Volume 6: Amphibians
William E. Duellman, PhD
Curator of Herpetology Emeritus
Natural History Museum and Biodiversity Research Center
University of Kansas
Lawrence, Kansas

Volume 7: Reptiles
James B. Murphy, DSc
Smithsonian Research Associate
Department of Herpetology
National Zoological Park
Washington, DC

Volumes 8–11: Birds I–IV
Walter J. Bock, PhD
Permanent secretary, International Ornithological Congress
Professor of Evolutionary Biology
Department of Biological Sciences,
Columbia University
New York, New York

Jerome A. Jackson, PhD
Program Director, Whitaker Center for Science, Mathematics, and Technology Education
Florida Gulf Coast University
Ft. Myers, Florida

Volumes 12–16: Mammals I–V
Valerius Geist, PhD
Professor Emeritus of Environmental Science
University of Calgary
Calgary, Alberta
Canada

Devra Gail Kleiman, PhD
Smithsonian Research Associate
National Zoological Park
Washington, DC

Contributing writers

Ardith L. Abate
Chameleon Information Network
San Diego, California

Patrick J. Baker, MS
Miami University
Oxford, Ohio

David G. Barker, MS
Vida Preciosa International
Boerne, Texas

Tracy M. Barker, MS
Vida Preciosa International
Boerne, Texas

Aaron Matthew Bauer, PhD
Villanova University
Villanova, Pennsylvania

Daniel D. Beck, PhD
Central Washington University
Ellensburg, Washington

Robert L. Bezy, PhD
Natural History Museum of Los Angeles County
Los Angeles, California

Bill Branch, PhD
Port Elizabeth Museum
Port Elizabeth, South Africa

Adam R. C. Britton, PhD
Wildlife Management International
Darwin, Northern Territory
Australia

David Chiszar, PhD
University of Colorado
Boulder, Colorado

David Cundall, PhD
Lehigh University
Bethlehem, Pennsylvania

C. Kenneth Dodd, Jr, PhD
U.S. Geological Survey
Gainesville, Florida

Lee A. Fitzgerald, PhD
Texas A&M University
College Station, Texas

L. Lee Grismer, PhD
La Sierra University
Riverside, California

Ronald L. Gutberlet, Jr., PhD
The University of Texas
Tyler, Texas

J. Alan Holman, PhD
Michigan State University Museum
East Lansing, Michigan

John B. Iverson, PhD
Earlham College
Richmond, Indiana

Maureen Kearney, PhD
The Field Museum
Chicago, Illinois

J. Scott Keogh, PhD
Australian National University
Canberra, Australia

Nathan J. Kley, PhD
Field Museum of Natural History
Chicago, Illinois

Harvey B. Lillywhite, PhD
University of Florida
Gainesville, Florida

Leslie Ann Mertz, PhD
Wayne State University
Detroit, Michigan

Göran Nilson, PhD
Göteburg Natural History Museum
Göteburg, Sweden

Javier A. Rodríguez-Robles, PhD
University of Nevada
Las Vegas, Nevada

Manny Rubio
Acworth, Georgia

Eric R. Pianka, PhD
University of Texas
Austin, Texas

Alan H. Savitzky, PhD
Old Dominion University
Norfolk, Virginia

Geoffrey R. Smith, PhD
Denison University
Granville, Ohio

Hobart M. Smith, PhD
University of Colorado
Boulder, Colorado

David Robert Towns, PhD
Department of Conservation
Auckland, New Zealand

Nikhil Whitaker
Madras Crocodile Bank/Centre for Herpetology
India

Romulus Earl Whitaker III, BSc
Madras Crocodile Bank/Centre for Herpetology
India

Contributing illustrators

Drawings by Michigan Science Art

Joseph E. Trumpey, Director, AB, MFA
Science Illustration, School of Art and Design, University
of Michigan

Wendy Baker, ADN, BFA

Brian Cressman, BFA, MFA

Emily S. Damstra, BFA, MFA

Maggie Dongvillo, BFA

Barbara Duperron, BFA, MFA

Dan Erickson, BA, MS

Patricia Ferrer, AB, BFA, MFA

Gillian Harris, BA

Jonathan Higgins, BFA, MFA

Amanda Humphrey, BFA

Jacqueline Mahannah, BFA, MFA

John Megahan, BA, BS, MS

Michelle L. Meneghini, BFA, MFA

Bruce D. Worden, BFA

Thanks are due to the University of Michigan, Museum
of Zoology, which provided specimens that served as models
for the images.

Maps by XNR Productions

Paul Exner, Chief cartographer
XNR Productions, Madison, WI

Tanya Buckingham

Jon Daugherity

Laura Exner

Andy Grosvold

Cory Johnson

Paula Robbins

• • • • •

Topic overviews

What is a reptile?

Evolution of reptiles

Structure and function

Behavior

Reptiles and humans

Conservation

What is a reptile?

The reptiles

The difference between amphibians and reptiles is that reptiles exhibit a suite of characteristics understandable as adaptations to life on land at increasing distance from water. Although many species of amphibians live on land in adulthood, most have an aquatic larval stage, and few can exist for long without moisture even during their terrestrial stages of life. Amphibians are tied to water—most species are not found more than a few meters from water or from moist soil, humus, or vegetation. Reptiles of many species are relatively liberated from water and can inhabit both mesic (moist) and xeric (dry) environments. Reptiles need water for various physiological processes, as do all living things, but some reptiles can obtain the water they need from the foods they eat and through conservative metabolic processes without drinking or by drinking only infrequently. Understanding the nature of reptiles requires focus on their techniques for maintaining favorable water balance in habitats where water may not be readily available and where moist microniches may be uncommon.

Characteristics

Most reptiles have horny skin, almost always cornified as scales or larger structures called scutes or plates. Such integuments resist osmotic movement of water from body compartments or tissues into the surrounding air or soil, thus minimizing desiccation. There are times in the lives of snakes and lizards when their skin becomes permeable to water, as when the animals are preparing to shed their old skin. During such times they seek out favorable hiding places that protect them not only from predators but also from water loss. The combination of integumentary impermeability (most of the time) and innate preferences for favorable microclimates during vulnerable periods allows reptiles to retain body water rather than to lose it to arid surroundings. Some reptiles are known to drink water that condenses on their scales when they reside in cool burrows.

Added to the mechanisms for retaining body water is an excretory system that is considerably advanced over those in fishes and many amphibians. The kidneys are integral components of the circulatory system. They allow constant, efficient filtration of blood. Most aquatic organisms excrete

nitrogenous waste as ammonia. Ammonia readily diffuses across skin or gills, provided plenty of water is present, but is not efficiently excreted by the kidneys. Ammonia is highly toxic, and animals cannot survive if this substance accumulates in their bodies. Terrestrial organisms excrete nitrogenous waste in the form of urea or uric acid, which are less toxic and which require less water than does excretion of ammonia. Urea is the main nitrogenous waste in terrestrial amphibians, whereas uric acid (which requires very little water) is the main nitrogenous effluent in reptiles. Finally, some desert-dwelling reptiles have a remarkable ability to tolerate high plasma urea concentrations during drought. This characteristic allows the animals to minimize water loss that would be coincident with excretion. Rather than being excreted, nitrogenous waste is simply retained as urea, and water is conserved. When a rainfall finally occurs, reptiles (e.g., the desert tortoise *Gopherus agassizii*) drink copiously, eliminate wastes stored in the bladder, and begin filtering urea from the plasma. Within days their systems return to normal, and the tortoises store a large volume of freshwater in their bladders to deal with the next drought.

Feeding

Feeding in a water medium among vertebrates can take several forms ranging from detritus feeding (ingestion of decaying organic matter on the substrate) to neuston feeding (ingestion of tiny organisms residing in the surface film). Probably the most common mechanism of obtaining food is suction feeding, whereby the predator creates a current by sucking water into the expanded buccal cavity and out through gills, causing prey to be captured in the mouth. Most fish rely on suction feeding, and this mechanism contributes to the effectiveness of detritivores, neustonivores, and aquatic predators. As a consequence, most fish have relatively weak mouths and low bite strength. There are exceptions, such as sharks, but the general rule is that fish depend on suction more than on biting, a circumstance that works effectively because of the liquid nature of the water medium and the associated friction arising between the medium and objects suspended in it. Aquatic amphibians also use suction feeding, although some species have lingual and jaw prehension, particularly during terrestrial stages. The transition to land dwelling among most reptiles has necessitated a revolution in oral structures and

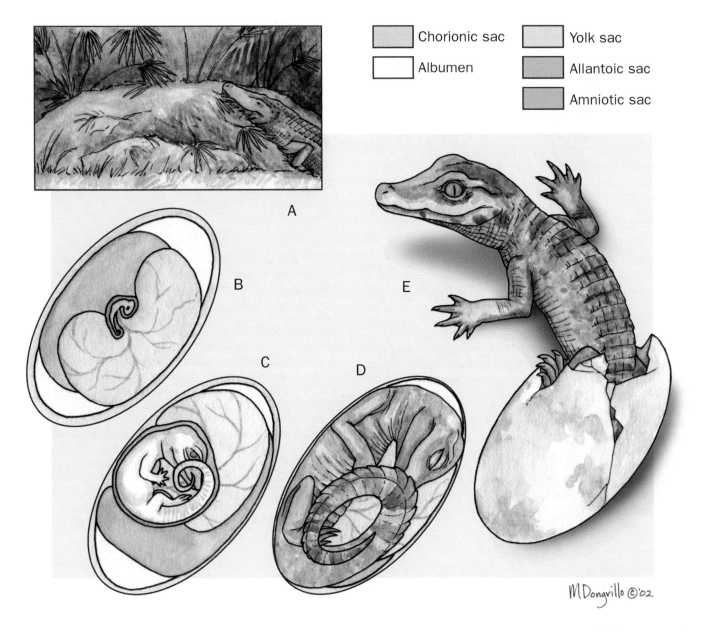

Chorionic sac	Yolk sac
Albumen	Allantoic sac
	Amniotic sac

A. The female American alligator (*Alligator mississippiensis*) buries her eggs under a pile of vegetation. As the plant material decomposes, the heat produced incubates the eggs while the female stays near the nest to guard against predators; B. Embryo development at day 12 after laying; C. Day 30; D. Day 50; E. Eggs hatch at day 65. (Illustration by Marguette Dongvillo)

kinematics to cope with the less dense medium of air. Because suction feeding does not work effectively in air, jaw prehension with consequent increases in bite strength has been emphasized in the evolution of most reptiles. Jaw prehension involves increased number and volume of the jaw-suspending muscles and increased surface area of muscle origins. Associated with this development was the appearance of temporal openings in the dermal bone surrounding the brain, because these openings allowed some of the jaw-suspending muscles to escape from the constraints of the dermal-chondral fossae and to attach at origin sites on the lateral and dorsal surfaces of the skull.

Skulls

The number and position of temporal openings have been used to classify reptiles into taxonomic groups, and the highlights of this classification system are reviewed here. Reptile skulls lacking temporal vacuities are said to be anapsid (without openings). This group includes the fossil order Cotylosauria, also called stem reptiles because of their ancestral position to all higher reptiles and hence to birds and mammals. The turtles, order Testudines, also are anapsid. Synapsid skulls have a single temporal opening on each side. The opening is positioned relatively low along the lateral surface of the skull, within the squamosal and postorbital bones. All synapsid reptiles (orders Pelycosauria, Therapsida, and Mesosauria) are extinct, but they are of great interest because of their ancestral position relative to the mammals. The parapsid condition also has a single vacuity on each side, but it is located rather high on the dorsolateral surface of the skull, within the supratemporal and postfrontal bones. Extinct, fishlike members of the order Ichthyosauria constitute the single order of parapsid reptiles, but these animals were probably closely related to euryapsid reptiles that had a single vacuity in much the same position except that it also invaded the dorsal aspects of the squamosal and postorbital bones. Orders of euryapsids were Placodontia and Sauropterygia, both marine and extinct, in the Triassic and Cretaceous periods, respectively. The diapsid condition is characterized by two temporal vacuities on each side of the skull. Major orders include Thecodontia (small crocodilian-like reptiles ancestral to birds and to all of the archosaurs), Crocodylia, Saurischia (dinosaurs with ordinary reptile-type hips), Ornithischia (dinosaurs with bird-type hips), Pterosauria (flying reptiles), Squamata (lizards, snakes, and several extinct groups), Eosuchia (extinct transitional forms that led to squamates), and Rhynchocephalia (mostly extinct, lizard-like diapsids with one surviving lineage, the tuatara [*Sphenodon punctatus*] on islands associated with New Zealand; *S. punctatus* may be a superspecies containing two or more separable species).

The order Testudines, which contains all living and extinct turtles, has traditionally been grouped with the primitive cotylosaurs because of common possession of the anapsid condition. Most herpetologists and paleontologists have agreed on this matter for many years. Molecular geneticists, however, have found evidence that turtles may actually be closely related to diapsid reptiles. This finding suggests that the anapsid condition of turtles may be secondary. That is, turtles may have evolved from ancestors that possessed two temporal vacuities on each side of their skulls, but in the course of evolution, turtles lost these openings. Essentially the same idea was proposed early in the twentieth century, not on the basis of genetic evidence but on the basis of a paleontological scenario involving a series of extinct but turtle-like diapsid fossils. Few at that time could accept the possibility that temporal vacuities once evolved would ever be abandoned, so this notion was dismissed and has resided in scientific limbo ever since. It has been revived on the strength of genetic data, and this much derided "preposterous idea" may become accepted.

It appears as if there is a contradiction associated with the anapsid status of turtles. Whereas some species are suction feeders with relatively weak mouths, others, such as snapping turtles, have profound bite strength. How is this strength produced, given the absence of temporal openings that would allow large jaw-suspending muscles to anchor (originate) on the dorsal surface of the skull? It turns out that many species of turtles have an analogous adaptation in which sections of dermal bone on the side and back of the skull have become emarginated or notched. Temporal openings are holes surrounded by bone. Emarginations are missing sections of the edges of the flat bones that form the ventral or pleural borders of the skull. With substantial sections of these bones missing, jaw-

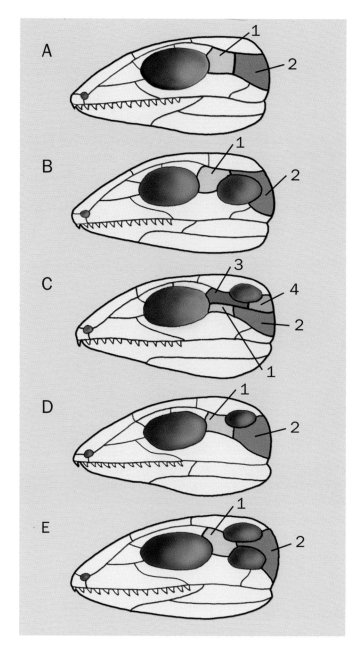

Structural types of the reptilian skull: A. Anapsid (no temporal openings); B. Synapsid (lower temporal opening); C. Parapsid (highest temporal opening); D. Euryapsid (upper temporal opening); E. Diapsid (two temporal openings). Bones shown: 1. Postorbital bone; 2. Squamosal bone; 3. Postfrontal bone; 4. Supratemporal bone. (Illustration by Gillian Harris)

suspending muscles have the same opportunity to escape from the dermal-chondral fossae as is made possible by vacuities. Although turtles are, strictly speaking, anapsid, some have taken an alternative pathway that leads to the bite strength necessary for effective jaw prehension of substantial prey or for tearing vegetation. If the anapsid condition is secondary, turtles have substituted an analogous trait that accomplished much the same biophysical effect as did the former temporal vacuities.

Snapping turtle (*Chelydra serpentina*) embryo in egg. Most reptiles hatch from eggs, although some snakes and lizards are live-bearers. (Photo by Gary Meszaros/Photo Researchers, Inc. Reproduced by permission.)

Reproduction

The earliest reptile fossils known are from the Upper Carboniferous period, approximately 270 million years ago, but by this time several of the reptilian orders were already in evidence, including both anapsid cotylosaurs and synapsid pelycosaurs. This finding implies that reptile evolution began much earlier. Another implication is that temporal vacuities (empty spaces) and emarginations (notches), although widely distributed in reptiles, are not defining characteristics of this class of vertebrates, because several groups do not have them. The earliest defining characteristics may never be known unless some very early fossils in good condition are found. It is likely that a desiccation-resistant integument was present. Another area on which to focus is the egg and the reproductive process. The egg is macrolecithal (contains much yolk) and is surrounded by a hard shell in turtles, crocodilians, and geckos and a soft or parchment-like shell in the other squamates. In either case, a shelled egg requires that fertilization occur before shell formation. This means that fertilization must take place within the female's body (i.e., in her oviducts) rather than externally as is typical of fishes and amphibians. Consequently, most male reptiles possess copulatory organs that deposit sperm into the cloaca of the female. From the cloaca the sperm cells migrate up the oviduct guided by chemical stimuli. Male turtles and crocodilians have a single penis homologous to the penis of mammals. This organ develops during embryogenesis from the medial aspect of the embryonic cloaca. Male lizards and snakes have paired hemipenes, which develop during embryogenesis from the right and left lateral aspects of the embryonic cloaca. Some male snakes have bifurcated hemipenes, so the males appear to have four copulatory organs. Thus internal fertilization is the rule among extant reptiles. Even tuatara, the males of which lack copulatory organs, transfer sperm in the manner of most birds with a so-called cloacal kiss involving apposition of male and female cloacae and then forceful expulsion of seminal fluid directly into the female's cloaca. Internal fertilization is necessary because of shell formation around eggs. Many reptiles live far from standing or running water, thus external fertilization in the manner of most fishes or amphibians would be associated with risk of desiccating both sperm and eggs.

The oviducts of some female reptiles are capable of storing sperm in viable condition for months or even years. In some turtles and snakes, fertilization can occur three years after insemination. Theoretically, a female need not mate each year, but she might nevertheless produce young each year using sperm stored from an earlier copulation. Although this interesting possibility has been known from observation of captive reptiles for approximately five decades, we still do not know whether or how often female reptiles use it under natural conditions. Another curiosity of reptile reproduction is that the females of some species of lizards and snakes are capable of reproducing parthenogenetically, even though reproduction in these species normally occurs sexually. (These species should not be confused with others that only reproduce parthenogenetically. This is not a widespread mode of reproduction in reptiles, but it is known to occur in several species of lizards and at least one snake.) Facultative parthenogenesis has only recently been discovered among captive reptiles, and there is as yet no information on whether it occurs in nature.

Macrolecithal eggs allow embryos to complete development within the egg or within the mother in the case of viviparity, such that the neonate is essentially a miniature version of its parents rather than a larva that must complete development during an initial period of posthatching life, as is common among amphibians. The reptilian embryo lies at the top of the large supply of yolk, and cell division does not involve the yolk, which becomes an extra embryonic source of nourishment for the growing embryo. A disk called the vitelline plexus surrounds the embryo and is the source of the three membranes (chorion, amnion, and allantois) that form a soft "shell" within the outer shell of the reptilian egg. Together these structures defend the water balance of the developing embryo and store waste products. Although reptile eggs absorb water from the substrate in which they are deposited, these eggs do not have to be immersed in water as is required for the eggs of most amphibians. Immersion of most reptile eggs results in suffocation of the embryos. Female reptiles deposit their eggs in carefully selected terrestrial sites that provide adequate soil moisture and protect the eggs from extremes of temperature.

Some species have another strategy for protecting embryos from abiotic and biotic exigencies. These reptiles retain the embryos and incubate them within the maternal body. The mother's thermoregulatory and osmoregulatory behaviors contribute to the embryos' welfare and to the mother's welfare. The mother's predator-avoidance behaviors can enhance the fitness of embryos exposed to greater predation elsewhere. In view of these potential advantages, which in some habitats might be considerable, it is not surprising that live-bearing has evolved many times in reptiles, although it is quite rare in amphibians. All crocodilians, turtles, and tuatara are egg layers. At least 19% of lizard species and 20% of snakes are live-bearers. Cladistic studies have shown that viviparity has evolved independently many times within squamates, in at least 45 lineages of lizards and 35 lineages of snakes. It also appears that viviparity is an irreversible trait and that once viviparity evolves, oviparous descendants rarely occur. The term embryo retention is used for species in which females retain embryos until very near the completion of embryogenesis when shells are added. The eggs are deposited and then hatch within 72–96 hours. Examples include the North American smooth green snake *Liochlorophis vernalis*, and the European sand lizard *Lacerta agilis*. Most important to understand is that the embryos are lecithinotrophic (nourishment of the embryos comes entirely from the yolk) with no additional postovulatory contribution from the mother. The mother, however, may play a role in gas exchange of the embryos. This process can involve proliferation of maternal capillaries in the vicinity of the embryos, a form of rudimentary placentation. Some species that give birth to live young also have lecithinotrophic embryos that undergo rudimentary placentation. Some embryo-retaining species eventually add a shell to their eggs and oviposit them within a few days of hatching. Others never add a shell, and the young are simply born alive, although they need to extricate themselves from the extraembryonic membranes that surround them. Many herpetologists prefer to abandon the term ovoviviparous because this word connotes that shelled eggs hatch in the maternal oviduct. No species is known in which this occurs. Accordingly, the term viviparous is used for all live-bearers, and herpetologists recognize that considerable variation exists in the degree to which viviparous embryos are matrotrophic (supported by maternal resources through a placenta).

Although females of oviparous species deposit their eggs in sheltered positions, the vagaries of climate can result in relative cooling or heating of oviposition sites with associated changes in moisture. This realization has led to considerable research on the effects of these abiotic factors on embryonic development. It is now known that within the range of 68–90°F (20–32°C), incubation time can vary as much as fivefold, and that neonatal viability is inversely related to incubation time. Hatchlings from rapidly developing embryos at high temperatures perform poorly on tests of speed and endurance relative to hatchlings from slower-developing embryos at lower temperatures. The slower-developing embryos typically give rise to larger hatchlings than do their rapidly

The black-breasted leaf turtle (*Geoemyda spengleri*) lives in the mountainous regions of northern Vietnam and southern China. (Photo by Henri Janssen. Reproduced by permission.)

developing counterparts. In the context of this work, it was found that the sex ratio of hatchling turtles varied depending on incubation temperature. In several species of tortoise (*Gopherus* and *Testudo*), for example, almost all embryos became males at low incubation temperatures (77–86°F [25–30°C]), and most became females between 88°F and 93°F (31–34°C). Temperature-dependent sex determination (TSD) is known to be widespread, occurring in 12 families of turtles, all crocodilians, the tuatara, and in at least three families of lizards. However, the effect of temperature differs in the various groups. Most turtles exhibit the pattern described, whereas most crocodilians and lizards exhibit the opposite pattern, females being produced at low incubation temperatures and males at higher ones. In a few crocodilians, turtles, and lizards females are produced at high and low incubation temperatures and males at intermediate temperatures. It is possible that some viviparous species experience TSD, in which case the thermoregulatory behavior of the mother would determine the sex of the embryos, but this phenomenon has not been observed.

The effect of the discovery of TSD has been enormous. Almost all developmental biologists previously believed that sex in higher vertebrates was genetically determined. This phenomenon has important implications for the management of threatened or endangered populations, especially if the program contains a captive propagation component. Unless care is taken to incubate eggs at a variety of temperatures, the program could end up with a strongly biased sex ratio. Reflection on the effects of global warming on reptiles exhibiting TSD generates the worry that extinction could be brought about from widely skewed sex ratios.

Diversity of reptiles

Reptiles range in body form from crocodilians to squamates, tuatara, and turtles. This diversity borders on trivial, however, in comparison with the range of forms and lifestyles that existed during the Jurassic and Cretaceous periods. This point can be further appreciated by considering locomotion

Parson's chameleon (*Calumma parsonii parsonii*) has a prehensile tail as long as its body length that can be used for climbing, grasping, or perching. At rest or during sleep the tail is coiled, as shown here. The specialized feet are divided into two bundles of fused toes, consisting of three on the outside and two on the inside of the rear feet. This is reversed on the front feet, giving them the ability to grasp, perch, and climb, and facilitates their largely arboreal existence. They are also able to use their highly dextrous feet to remove shed skin and put food into or take objects out of their mouths. (Photo by Ardith Abate. Reproduced by permission.)

among lizards with well-developed legs. Although some species are capable of quick movement, the gait of all lizards is basically the same as that of salamanders. The legs extend from the sides and must support the body through right angles, greatly limiting body mass and speed. Within the context of these constraints, lizards do quite well, but their locomotion remains relatively primitive. Truly advanced locomotion, with the legs directly under the body, occurs among mammals, but this pattern of limb suspension evolved in dinosaurs and was clearly a part of their long period of success. All extant reptiles are ectotherms, deriving their body heat from radiation, conduction, or convection, whereas mammals and birds are endotherms, producing body heat by energy-consuming metabolic activity. Thus we see the primitive condition in the reptiles and the advanced condition in the birds and mammals. There is now good reason to believe that at least some dinosaurs were endotherms. Accordingly, it is important to keep in mind that the diversity of extant reptiles is but a fraction of the diversity exhibited by this class of vertebrates during earlier phases of its natural history.

Locomotion

The basic pattern of the tetrapod limbs of amphibians is preserved in reptiles: a single proximal bone is followed dis-

tally by paired bones. In the fore limb is the humerus followed by the radius and ulna. In the hind limb is the femur followed by the tibia and fibula. The wrist and hand are formed from the carpal and metacarpal bones, and the ankle and foot are formed from the tarsals and metatarsals, five or fewer digits bearing horny claws distal to both wrist and ankle. Reptile orders show enormous variation in the precise form and arrangement of these basic elements and in their behavioral deployment. In squamates these elements are abandoned in favor of serpentine locomotion, which requires an elongate body and therefore an increased number of vertebrae, more than 400 in some snakes. Serpentine locomotion depends on friction between the animal and the substrate, which in some animals is accomplished by pressing the posterior edges of the belly scales against stationary objects so that Newton's third law (for every action there is an equal and opposite reaction) can operate. Some lizards have lost their limbs and use serpentine movement. Others with perfectly fine legs will, in bunch grass habitats, fold the limbs against the body and exhibit facultative serpentine movement, presumably because this type of movement produces faster escape behavior than does ordinary running in tangled vegetation. The twisting and bending of the trunk required in serpentine movement enhance the danger of vertebral dislocation. This selective pressure has been answered by the development of an extra pair of contact points between adjacent vertebrae in snakes, bringing the total number of articular points to five per vertebra. The result is that each vertebra is essentially locked to the next and resists dislocating forces arising from roll, pitch, and yaw.

Brain

The brain and spinal cord exhibit several advanced characteristics in reptiles relative to amphibians, including larger size and greater definition of structural divisions and greater development of the cerebral cortex. Neural connections between the olfactory bulbs, the corpus striatum, and several other subcortical structures have become clearly established in reptiles, and these connections have been conserved in subsequent evolution such that they are present in mammals, including humans. This set of connections is sometimes referred to as the "reptilian brain" or "R-complex" and is thought to represent a neural circuit necessary for the mediation of basic functions such as predation and mating as well as the affective concomitants associated with social behaviors ranging from cooperation to aggression. In the study of mammals, we speak of the regulation of emotion by components of the reptilian brain. Herpetologists are generally reluctant to speak of emotion in their animals, but they have no difficulty recognizing the existence of the neural circuit in question and in understanding that it contributes to social and reproductive activities. Whether this contribution is limited to the organization of motor patterns or whether emotion also is involved remains an open question.

Eyes

Sensory structures of reptiles exhibit variations in size and complexity that are roughly correlated with ecological variation and phylogeny. For example, lizards considered to be primitive, such as those of the family Chamaeleonidae, are primarily visually guided in the context of predation as well as in the contexts of social and reproductive behavior. This reliance on vision is reflected in the wonderful mobility of the eyes, the size of the optic lobes, and in the brilliant color patterns in the family. The phenomenon of "excited coloration" (color changes reflecting emotional or motivational states) involves socially important signals that can only be appreciated with vision. More advanced lizards, such as those in the family Varanidae, place greater emphasis on their nasal and vomeronasal chemosensory systems. Associated with this characteristic is a shift in the morphology and deployment of the tongue, which in varanids is used mainly to pick up nonvolatile molecules and to convey them to the vomeronasal organs. There is an associated shift from insectivory to carnivory. In snakes, which may be derived from a varanid-like ancestor, these shifts have been carried to an even greater extreme.

Ears

Audition presents an interesting problem in reptiles. Snakes and some lizards have no external ear, although the middle and inner ears are present. In species with a distinct external auditory meatus, there is little doubt about the existence of a sense of hearing, although it is generally thought that only sounds of low frequency are detected. In species lacking an external ear, seismic sounds are probably conducted by the appendicular and cranial skeletons to the inner ear. It has been suggested, however, that the lungs might respond to airborne sounds and transmit them to the inner ear via the pharynx and eustachian tube. Although no reptiles are known for having beautiful voices, many generate sounds. For example, male alligators bellow, and this sound undoubtedly serves social functions. Many snakes hiss, some growl, and a fair number issue sounds with their tails either with a rattle or by lashing the tail against the substrate. Such sounds are generally aimed at predators or other heterospecific intruders, and herpetologists have believed that the issuing organism was deaf to its own sound, unless the sound had a seismic component. Perhaps this view can be altered if the concept of pneumatic reception of airborne sound is corroborated.

Other senses

Cutaneous sense organs are common, including those sensitive to pain, temperature, pressure, and stretching of the skin. Although pain and temperature receptors are best known on the heads of reptiles, these receptors are not confined there. The mechanoreceptors that detect touch, pressure, and stretch are present over the body, especially within the hinges of scales. Receptors that detect infrared radiation (heat) are also of dermal and epidermal origin. In boas and pythons, these receptors are associated with the lips. In pitvipers such as rattlesnakes, a membrane containing heat receptors is stretched across the inside of each pit approximately 0.04–0.08 in (1–2 mm) below the external meatus. The geometry of the bilateral pits is such that their receptive fields overlap, allowing stereoscopic infrared detection. The nerves of the pits project to the same brain areas as do the eyes, giving rise to images containing elements from the visible part of the spectrum as well as the infrared part. When a pitviper is in the process of striking a mouse, the snake's mouth is wide open with fangs erect, so that the pits and eyes are oriented up rather than straight ahead toward the prey. It turns out that in the roof of the mouth near the fangs are additional infrared sensitive receptors that appear to take over guidance of the strike during these final moments.

Reptiles also possess proprioceptors associated with muscles, tendons, ligaments, and joints. Proprioceptors report the positions of body components to the brain, allowing the brain to orchestrate posture and movement. Another class of internal receptor contains taste buds, which are located in the lining of the mouth and on the tongue. In reptiles with slender, forked tongues specialized for conveying nonvolatile chemicals to the vomeronasal organs, lingual taste buds are generally absent, but taste buds may be present elsewhere in the mouth.

Teeth

With a few notable exceptions, the teeth of extant reptiles are unspecialized; that is, most teeth look alike, and the dentition is called homodont (Latin for "alike teeth"). The teeth may vary considerably in size along the length of the tooth-bearing bones, especially in snakes, because the teeth are deciduous and are replaced regularly. This type of dentition is called polyodont. Teeth are present on the bones of the upper and lower jaw and on other bones forming the roof of the mouth (palatine and pterygoid). If teeth are ankylosed (cemented by calcification) to the inside of jawbones, the dentition is pleurodont. This is the condition of all snakes and most lizards. If the teeth are ankylosed to a bony ridge along the jawbones, as in some lizards, the dentition is acrodont. Crocodilian teeth are situated in sockets, as are the teeth of mammals, and this dentition is called thecodont. The most spectacular type of tooth specialization in extant reptiles involves the fangs of venomous snakes. These fangs are hollow, elongated teeth on each side of the front of the upper jaw, although some species have solid, grooved fangs on each side of the rear of the upper jaw. In front-fanged snakes, venom is forcefully injected through the fangs and exits into the prey through slitlike openings on the lower anterior face of each fang. In rear-fanged snakes, venom runs under little pressure along the grooves and enters prey as the rear fangs successively embed themselves into prey during swallowing. Among the front-fanged species are those with folding fangs that are normally held parallel to the roof of the mouth and rotated down into position as needed. Other front-fanged snakes have less mobility associated with their fangs, which are therefore always in the biting position. The fangs typically are much longer in species with folding fangs than in species with fixed front fangs. With the exception of fangs, most teeth in extant reptiles are used to grip prey, although some lizards have specialized, blunt teeth that crush snail shells. Some extinct reptiles had far more specialized tooth patterns than do the surviving groups.

Venom

All reptiles possess salivary glands that lubricate food and begin the process of digestion. Saliva also cleans the teeth by digesting pieces of organic matter that might adhere to the teeth or be stuck between adjacent teeth. The venom that has evolved in snakes undoubtedly arose from salivary glands, and it has retained its original digestive function. Venom contains elements that immobilize and kill prey, and it facilitates digestion. It has been conclusively demonstrated in force-feeding experiments in which rattlesnakes fed envenomated mice completed the digestion process significantly quicker than did conspecifics fed identical euthanized mice that had not been envenomated. Similar studies have been completed with comparable results for a variety of species, including rear-fanged snakes. In some rear-fanged snakes, venom is apparently used only for digestion and not for subduing prey or for defense. In the Mexican beaded lizard (*Heloderma horridum*) and the Gila monster (*H. suspectum*), the only venomous lizards, venom is apparently used strictly for defense and not for acquisition or digestion of prey.

Energy

In some snakes and lizards, very long periods of time can occur between successive meals, and the reptiles exhibit an interesting form of physiological economy by down-regulating their digestive machinery. This process saves energy, because maintaining functional digestive tissue in the absence of food would require considerable caloric costs. Reptiles retain this down-regulated condition until the next meal has been secured, at which time the gut is up-regulated.

Exercise

Gas exchange occurs through lungs. Most snakes have only one lung (on the left). The heart has three chambers, two atria and one ventricle, except in crocodilians, in which a second ventricle is present, producing a four-chambered heart much like that of mammals. Even in reptiles with a three-chambered heart, a septum exists within the ventricle and minimizes mixing of oxygenated and nonoxygenated blood. Researchers have studied the physiological mechanisms associated with exhaustive locomotion and have found interesting parallels between reptiles and mammals in the rapidity of recovery from exhaustion. A major difference, however, is that mammals exhibit a so-called exercise effect (exercise-induced ability to mobilize greater levels of oxygen and, hence, to work harder than was possible before exercise), whereas no reptile has yet been shown to do this.

Conservation

New species of reptiles continue to be discovered. This is especially true of lizards. Hence the numbers that follow are approximations subject to change. We currently recognize 285 species of turtles, 23 crocodilians, two tuatara, 4,450 lizards, and 2,900 snakes. One of the authors of this chapter (H. M. S.) has named approximately 300 species in his career and is working on projects that will almost certainly add species to the list. In countries such as the United States, where numerous herpetologists have studied the fauna thoroughly, it is relatively unlikely that new species will be dis-

covered. Nevertheless, herpetologists sometimes find reasons to justify the splitting of previously recognized species into two or more species. Third World countries present an entirely different situation because they possess few indigenous herpetologists, and some of these countries have only rarely been visited by herpetologists. Consequently, new species are quite likely to be found in these lands, especially those in the tropics and subtropics. It has been estimated that in most such countries, approximately 30% of the reptile fauna remains to be discovered. Thus much basic work remains to be done. At the same time, we must be mindful of the rate at which species are currently being lost to deforestation, habitat fragmentation, pollution, overharvesting, invasion of harmful exotic species, and other anthropogenic causes. We are now facing a situation in which we are losing species to extinction before they have been given proper scientific names. During the past decade, amphibian biologists have justifiably called attention to the worldwide decline of many salamanders and anurans. Without doubt this is a serious problem, but it has overshadowed the fact that reptiles have been suffering the same fate.

Many of the same factors responsible for amphibian declines have been insidiously working their decimating effects on reptiles. At the heart of the problem is the human population, now much more than six billion, and a drastically uneven distribution of resources. Many people living in areas of high reptile diversity are unable to eke out a living and are therefore tempted to exploit their native fauna, legally or illegally, and to engage in other economic activities that eventually have negative repercussions on the fauna. Hunting of reptiles occurs for local consumption, sale of hides or shells, sale of live animals to the pet trade, and sale of meat or other body parts as exotic food or medicines. China has almost extinguished its turtle fauna, for example, and has put catastrophic pressure on the turtle population in the rest of Southeast Asia. Chinese dealers also purchase several species of turtles during their active seasons in North America, particularly snapping turtles and softshells, for shipment to Asia. A team of biologists conducting a survey of tortoises in Madagascar found hundreds of dead animals, all with their livers removed. Local rumor revealed that these organs are made into an exotic pâté that is shipped to Asia. Although the mathematics of sustainable harvesting have been well worked out and can provide the basis for enlightened commercial practices and population management, the rate at which turtles have been harvested in China, Southeast Asia, Madagascar, and elsewhere is greatly exceeding the rate required for sustainable yields.

A similar situation developed in connection with hides of various reptiles, including crocodilians and several large lizards and snakes. In the case of crocodilians, management programs aimed at providing sustainable yields were developed in several countries, and these measures proved successful, so much so that the species involved recovered from endangered status. This experience indicates that the conservation strategy of management for sustainable yield can work if it is carefully implemented on the basis of good ecological and demographic data and if the harvest is carefully monitored. Enthusiastic participation of local people is an important element of the success of such programs as they have been carried out in Africa, Asia, and South America. It may not be too late to put these ideas into practice to save the turtle fauna of Asia. In the case of the crocodilians, declining populations quickly allowed several secondary events, such as explosive growth in populations of fish that were prey of crocodilians and reductions in populations of fish that depended on the deep holes made by crocodilians. An added benefit of sustainable yield programs was that these pertur-

bations were reversed as the crocodilian populations were restored. It is probable that secondary effects of Asian turtle harvesting will make themselves known in the near future because turtle burrows are homes for a variety of other creatures. Eliminating turtles makes the ecosystem inhospitable for animals that depend on turtles. In short, enlightened management may be a tool for creating sustainable yield and for habitat restoration.

Resources

Books

Auffenberg, Walter. *The Behavioral Ecology of the Komodo Monitor.* Gainesville, FL: University Presses of Florida, 1981.

Bennett, A. F. "The Energetics of Reptilian Activity." In *Biology of the Reptilia.* Vol. 13, *Physiology,* edited by C. Gans and F. H. Pough. New York: Academic Press, 1982.

Carroll, R. L. "The Origin of Reptiles." In *Origins of the Higher Groups of Tetrapods: Controversy and Consensus,* edited by H. P. Schultze and L. Trueb. Ithaca, NY: Comstock, 1991.

Fitch, H. S. *Reproductive Cycles in Lizards and Snakes.* Lawrence, KS: University of Kansas Natural History Museum, 1970.

Garland, T., Jr. "Phylogenetic Analyses of Lizard Endurance Capacity in Relation to Body Size and Body Temperature." In *Lizard Ecology: Historical and Experimental Perspectives,* edited by L. T. Vitt and E. R. Pianka. Princeton, NJ: Princeton University Press, 1994.

Greenberg, N., and MacLean, P. D., eds. *Behavior and Neurology of Lizards.* Rockville, MD: National Institute of Mental Health, 1978.

Pieau, C. "Temperature and Sex Differentiation in Embryos of Two Chelonians, *Emys orbicularis* L. and *Testudo graeca* L." In *Intersexuality in the Animal Kingdom,* edited by R. Reinboth. New York: Springer-Verlag, 1975.

Pough, F. H., R. M. Andrews, J. E. Cadle, M. L. Crump, A. H. Savitzky, and K. D. Wells. *Herpetology.* Upper Saddle River, NJ: Prentice Hall, 1998.

Zug, G. R., L. J. Vitt, and J. P. Caldwell. *Herpetology: An Introductory Biology of Amphibians and Reptiles.* New York: Academic Press, 2001.

Periodicals

de Cock Buning, T. "Thermal Sensitivity as a Specialization for Prey Capture and Feeding in Snakes." *American Zoologist* 23 (1983): 363–75.

Gans, C., and P. F. A. Maderson. "Sound-Producing Mechanisms in Recent Reptiles: A Review and Comment." *American Zoologist* 13 (1973): 1195–203.

Guillette, L. J., Jr., R. E. Jones, K. T. Fitzgerald, and H. M. Smith. "Evolution of Viviparity in the Lizard Genus *Sceloporus.*" *Herpetologica* 36 (1980): 201–15.

Hedges, S. B., and L. L. Poling. "A Molecular Phylogeny of Reptiles." *Science* 283 (1999): 998–1001.

Packard, G. C., and M. J. Packard. "Evolution of the Cleidoic Egg among Reptilian Antecedents of Birds." *American Zoologist* 20 (1980): 351–62.

Schuett, G. W., P. J. Fernandez, W. F. Gergits, N. J. Casna, D. Chiszar, H. M. Smith, J. G. Mitton, S. P. Mackessy, R. A. Odum, and M. J. Demlong. "Production of Offspring in the Absence of Males: Evidence for Facultative Parthenogenesis in Bisexual Snakes." *Herpetological Natural History* 5 (1997): 1–10.

Schwenk, K. "The Evolution of Chemoreception in Squamate Reptiles: A Phylogenetic Approach." *Brain Behavior and Evolution* 41 (1993): 124–37.

Secor, S. M., and J. Diamond. "Adaptive Responses to Feeding in Burmese Pythons: Pay before Pumping." *Journal of Experimental Biology* 198 (1995): 1313–25.

David Chiszar, PhD
Hobart M. Smith, PhD

Evolution of the reptiles

The reptiles make up a huge group of fossil and living vertebrates, ranging in size from tiny thread snakes to sauropod dinosaurs, which are the largest animals ever to have lived on land. Through time reptiles have evolved into unique forms, such as turtles, snakes, and dinosaurs, but they also have taken on the appearance and habits of other vertebrate groups, such as sharks and dolphins. As with other animal classes, reptile groups that are thought to share a common ancestor are known as clades. The application of cladistics has changed ideas about how organisms should be classified. For instance, because they are thought to be descended from small bipedal dinosaurs, birds are now included with the reptiles. But synapsids (once called mammal-like reptiles) are classified with the mammals.

Because so many diverse animals are included under the term reptiles, they are difficult to define as a single group. Reptiles are amniotes, that is, they are tetrapods (four-legged vertebrates) with an amnion that surrounds and protects the developing embryo. Reptiles other than birds and their immediate ancestors lack true feathers, and all of them lack true hair. Common (though varying) characteristics among reptiles include the fact that they cannot regulate their temperature internally (with the possible exceptions of some dinosaurs and all birds), that they have an extensive covering of scales or bony plates or both (individual exceptions occur in many major groups), that they have a three-chambered heart (with the exceptions of crocodilians and, possibly, other archosaurs), and that they have 12 pairs of cranial nerves.

The major reptile groups considered here are Anapsida ("stem reptiles," turtles, and other primitive groups), Euryapsida (the marine nothosaurs, plesiosaurs, placodonts, and ichthyosaurs), and Diapsida. The last group includes the Lepidosauria (sphenodontians, such as tuatara and its fossil relatives; lizards; and snakes) and the Archosauria (pseudosuchians; crocodilians; pterosaurs, also known as "flying reptiles"; and dinosaurs). Each of the three major reptile groups are defined on the basis of the number and position of large openings in the temporal region of the skull behind the eyes (relative to other skull bones). Anapsid reptiles have no large openings in the temporal region of the skull and were the first stem to branch off the reptilian lineage. Euryapsid reptiles have a single temporal opening in the upper part of the skull. Diapsid reptiles have two large temporal openings, one above and one below a horizontal bony bridge.

Anapsida

The earliest reptiles are known from the early Pennsylvanian (323–317 million years ago, or mya). They were quite small and lizardlike in appearance, and their skulls, jaws, and tooth structures strongly indicate that they were insectivorous. In fact, it is thought that they evolved in tandem with insect groups that were beginning to colonize the land. Some Pennsylvanian (323–290 mya) amphibians of the microsaur group also evolved into insectivores that were so superficially similar to early reptiles that, for a time, they were classified as such.

The amniote egg evolved in the earliest reptiles. This allowed for the first true occupation of the land by tetrapods, for the amniote egg allowed the embryo to develop in an aquatic microcosm until it was ready for terrestrial life; this paved the way for the huge adaptive radiation that eventually took place among the reptiles. Robert Carroll of the Redpath Museum in Montreal, Canada, has pointed out that the

Malayan box turtle (*Cuora amboinensis*) hatchling. (Photo by Henri Janssen. Reproduced by permission.)

earliest reptiles probably occupied the land before the amniote egg was developed fully. An analogy may be found in a few modern salamanders (small, somewhat lizardlike amphibians) that lay tiny non-amniote eggs in moist terrestrial places, such as under logs or in piles of damp leaves. These eggs hatch into tiny replicas of the adults rather than going through a larval stage, such as occurs in frogs and other salamanders. The evolution of the amniote egg took place when the membranes within the eggs of the earliest reptiles became rearranged in the form of various sacs and linings and the outer membrane incorporated calcium into its structure to form a shell.

The calcareous (limey) shell afforded protection for the developing embryo and was porous enough to allow for the entrance and exit of essential elements, such as oxygen and carbon dioxide. Food for the developing embryo was supplied in the form of an extra-embryonic sac full of yolk and a system of blood vessels that allowed the yolk to be transferred to the embryo. The amnion itself formed a sac that contained a fluid within which the embryo was suspended. This provided a "private pond" (a term used by the late Harvard University scientist Alfred S. Romer, the world leader in vertebrate paleontology from the late 1940s until his death in 1973) for its occupant and kept the fragile embryonic parts from sticking together. A unit composed of a part of the allantois and the membranous chorion next to the shell allowed for the absorption of oxygen and the excretion of carbon dioxide. A sac formed by the allantois stored the nitrogenous wastes excreted by the embryo. The origin of the amniote egg was one of the most important evolutionary events that ever occurred.

Turning to other anapsid reptiles, turtles were one of the first reptiles to branch off the amniote stem. Turtles are first known from the late Triassic (227–206 mya) but probably evolved in the Permian (290–248 mya). Aside from protection, the turtle shell (which is essentially a portion of its skeleton turned inside out) has other critical functions. A large percentage of the red blood cells of most land vertebrates are formed in the marrow of the long bones. Turtles, however, need sturdy legs to support their shells; thus, their limb bones are very dense, with little space, if any, for red blood cell–producing marrow. It has been shown that the turtle shell is filled with canals and cavities where red blood cells are produced in quantity. Moreover, rather than being just an inert shield, the turtle shell is the site of calcium metabolism and is important in the process of temperature regulation, by absorbing heat during the basking (sunning) process.

Other attributes of turtles include the ability of some of them to absorb oxygen in the water through patches of thin skin on the body, in the lining of the mouth, or within the cloaca, a terminal extension of the gut wall. Some turtles can freeze solid in the winter and thaw out in the spring with no harmful effects—a process that also takes place in various frog species. The ability of turtles to survive severe injuries is well known, and many species can exist in the absence of oxygen for long periods of time. Leatherback turtles have a current-countercurrent blood flow similar to that of deep-sea mammals and can dive in the sea at great depths and remain active

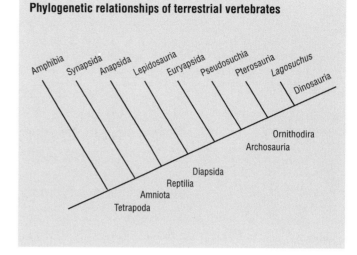

Phylogenetic relationships of terrestrial vertebrates

(Illustration by Argosy. Courtesy of Gale.)

in very cold temperatures. The incubation temperature of the eggs of many species of turtle determines the sex of the hatchlings. In some cases "cold nests" produce females and "warm nests" produce males, but the opposite can also occur. Turtle eggs have large amounts of yolk compared with those of many other vertebrates. This "egg food" sustains the young during long incubation periods.

The origin of turtles is somewhat in doubt. An early anapsid with expanded ribs, *Eunotosaurus*, once was proposed as the ancestral form, but the skull of *Eunotosaurus* was not turtle-like. On the other hand, the body skeleton of early reptiles called procolophonids had a shell, and the skull was somewhat turtlelike. *Owenetta*, a procolophonid found in the Upper Permian (256–248 mya) of South Africa has nine advanced characters in the skull and one in the humerus that are shared with *Proganochelys*, an unquestionable Triassic (248–206 mya) turtle. *Owenetta* lacks a shell, however; thus it has been suggested that the skull changes in turtle ancestors preceded those that led to the origin of the shell.

True turtles (order Testudines) are composed of three major suborders: the proganochelydians, the pleurodirans, and the cryptodirans. The proganochelydians are the most primitive and are known from the late Triassic to the early Jurassic (206–180 mya). The shell is similar to that of modern turtles, except that it has extra bones and the head and limbs cannot be retracted effectively into it. The skull lacks teeth except for a few on the palate. The pleurodirans and cryptodirans have no teeth in the skull and can retract the head, neck, and tail into the shell.

In the pleurodires the neck swings sideways when it is retracted, so that the turtle looks out with only one eye. In the cryptodires the neck folds over itself when it is retracted, so that the turtle gazes with both eyes. Oddly, the neck differentiation in these turtles did not occur until the late Cretaceous (99–65 mya). Reflecting their Gondwana origin, one finds only cryptodires in the Northern Hemisphere, whereas

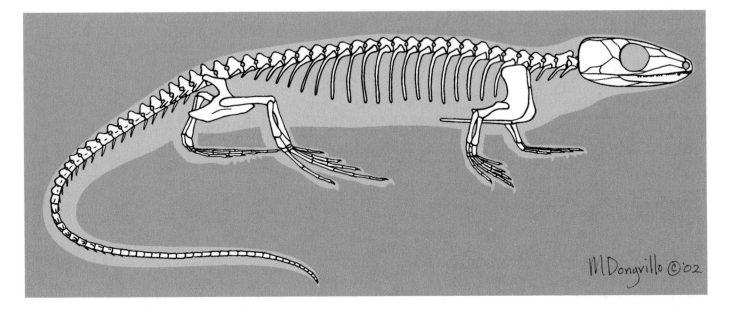

Skeleton of one of the earliest known amniotes, *Hylonomus lyelli*, from the early Pennsylvanian of Joggins, Nova Scotia. Remains were found within the upright stump of the giant lycopod *Sigillaria*. (Illustration by Marguette Dongvillo.)

a significant number of pleurodires occur in the Southern Hemisphere. Seaturtles and soft-shelled turtles, familiar animals throughout much of the world today, appeared in the Upper Jurassic (159–144 mya); by the time the neck specializations came about in the late Cretaceous, all turtles were essentially modern. A few very big seaturtles lived in the Mesozoic (248–65 mya) seas during the time when dinosaurs dominated the land. *Archelon* of the Cretaceous (144–65 mya) had a shell length of 6.3 ft (1.9 m). But the largest known turtle is a freshwater pleurodire (side-neck group) turtle with a shell length of 7.6 ft (2.3 m). Aptly named *Stupendemys*, this turtle was collected on a Harvard field trip to Venezuela. The animal came from Pliocene (5.3–1.8 mya) sediments, not long ago at all in terms of geologic time.

Several odd reptile clades branched off the anapsid stem, including the elephantine, terrestrial pareiasaurs, and the slim, marine mesosaurs. The pareiasaurs (among them, the well-known genus *Scutosaurus*) have been found in the Middle and Upper Permian (269–248 mya) of Africa, western Europe, Russia, and China. They were up to 10 ft (3 m) long and had an upright stance (unlike that of other amniotes of the time) and stout limbs that supported the massive body. The head of pareisaurs was short and thick, with heavily sculptured bones protecting the eyes and tiny brain. The teeth had compressed, leaf-shaped crowns like those of modern leaf-eating herbivorus lizards; thus, pareiasaurs were probably the first large land herbivores.

Mesosaurs, on the other hand, were the earliest truly aquatic reptiles. These gracile animals were close to 1 yd (0.9 m) long, about a third made up of the tail. Fossil mesosaurs are known only from the adjacent coasts of southern Africa and eastern South America, reflecting the fact that Gondwana split into the two continents. In the mesosaurs the snout was

very long, and the mouth was filled with long, needlelike teeth. These teeth apparently formed a specialized straining devise that allowed the animals to feed upon small crustaceans, possibly those found in the same fossil beds as the mesosaurs. The long, compressed tail probably was used for propulsion in swimming. The posterior tail vertebrae have fracture plains, indicating that caudal autonomy (voluntary shedding of the end of the tail during stress) could have occurred; this feature may have allowed them to escape from the grasp of predators. The skull of mesosaurs originally was thought to have a temporal opening, but this was later disproved.

Euryapsida

Turning to the euryapsid reptiles, we find four highly adapted groups of sea reptiles, some of which reached ponderous proportions. Euryapsids are believed to have evolved from diapsids, having lost the lower temporal opening in the process. They were a very important part of the marine environment during the Mesozoic and, in a sense, were as dominant in that setting as the dinosaurs were on the land. A subject that often is neglected in the course of discussions about the reasons for the demise of the dinosaurs at the end of the Cretaceous concerns the reasons behind the extinction all of the euryapsid sea reptiles at the end of the Cretaceous.

Nothosaurs lived from early to late Triassic times but were most common in the Middle Triassic (242–227 mya). They were relatively small compared with the plesiosaurs that followed them. Nothosaurs had moderately long necks and limbs modified as flippers. They are thought to have been possible ecological equivalents of modern seals and otters. Nothosaurs had sharp, conical teeth modified for catching fish. The structure of the nothosaur shoulder girdle is unique among reptiles; it provides little space for the attachment of trunk-supporting body muscles, such as occurs in land reptiles.

Plesiosaurs are thought to have evolved from nothosaurs in the Middle Triassic. They persisted until almost the end of the Cretaceous but were most abundant in the Middle Jurassic (180–159 mya) and slowly dwindled in numbers until their extinction. Plesiosaurs had legs modified as paddles that flapped up and down like aquatic wings. This action not only pushed them through the water but also gave them a lift that allowed them to "fly" through the sea in the same manner as seaturtles and penguins. The bones of the shoulder and hip girdles of plesiosaurs were expanded greatly below, forming an armor on the bottom of the animal that left no "soft underbelly" for attack by such predators as sharks. Two general kinds of body types were prominent in the plesiosaurs, long-necked forms and short-necked forms.

In the first group, a small head was positioned at the end of a very long neck. The body was heavy and bulbous. The teeth were conical and sharp; for this reason it is assumed that these plesiosaurs fed mostly on fishes. The elasmosaurid clade of long-necked plesiosaurs reached a length of more than 40 ft (12.2 m) and had enormous paddles and very small heads. Some researchers have suggested that the Loch Ness monsters (if they actually existed) were long-necked plesiosaurs. If this is true, remarkable physiological changes that allowed them to adapt to icy waters must have occurred since the Mesozoic.

Short-necked plesiosaurs have practically no neck at all and a massive head. It has been suggested that they were the ecological equivalents of the killer whales of present times, as they were consummate carnivores. *Kronosaurus*, the largest marine reptile that ever lived, was a massive animal that reached a length of at least 42 ft (12.8 m). This animal was found on the property of a rancher in Australia and ended up in the Museum of Comparative Zoology at Harvard. One of the problems of collecting such a large animal is how and where to exhibit it. As frequently happens, a room at the museum had to be remodeled to put this "reptilian killer whale" on exhibit.

Placodonts were marine reptiles with short, stout bodies that lived from Middle to Upper Triassic times (242–206 mya). The limbs were only moderately paddlelike. It once was suggested that placodonts were related rather closely to nothosaurs, but there is really no good evidence to support this hypothesis. The most distinguishing feature of placodonts is the form of their teeth, which are flattened rather than pointed. In the well-known genus *Placodus*, the teeth along the margin of the cheek region and on the palate are large and very flat, whereas the long, narrow front teeth protrude from the end of the somewhat narrowed snout. It is thought that *Placodus* used the front teeth to grasp mollusks and the hind ones to crush them. Some placodonts, such as *Henodus*, were superficially like turtles, in that they had an upper shell composed of numerous small polygonal bones. A lower shell was not present.

The most highly specialized marine reptiles were the ichthyosaurs, whose body took on the appearance of modern tunas, sharks, and porpoises. Ichthyosaurs lived from early Triassic to Middle Cretaceous times (248–112 mya). The skull of ichthyosaurs is streamlined, with a long snout; the eyes are very large. The body is spindle-shaped, and the limbs are reduced to fins. The tail fin is fishlike. The individual vertebrae in the spinal cord are in the form of very short and compact biconcave discs, very similar in appearance to those of modern sharks. It is estimated that some ichthyosaurs were very active and could swim 30–40 mph (48.3–64.4 km/h). Why the ichthyosaurs became extinct in the Middle Cretaceous (ca. 121–99 mya), long before the dinosaurs and other marine reptiles, remains a mystery.

Diapsida

Aside from having two openings in the temporal region of the skull, diapsids typically have hind limbs that are longer than the forelimbs. The oldest known diapsid was a small, lizardlike animal with a body length (minus the long tail) of about 8 in (20.3 cm). This slender animal, named *Petrolacosaurus*, was collected from the late Pennsylvanian (ca. 303–290 mya) of Kansas.

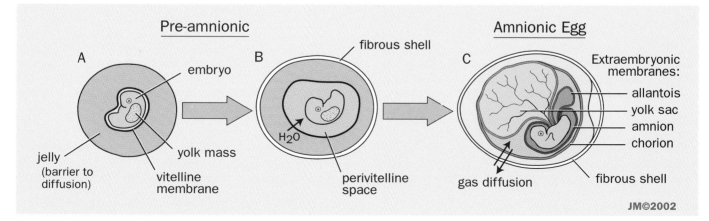

Evolution of the amniotoic egg. (Illustration by Jacqueline Mahannah)

Two distinct clades, the Lepidosauria and the Archosauria, branched off early from the diapsid trunk. These two groups are characterized by their contrasting patterns in locomotion and posture. The lepidosaurs retained the sprawling posture and laterally directed movement of the limbs found in primitive tetrapods. Lateral undulation of the vertebral column was also an important method of locomotion for most lepidosaurs, reaching its highest degree of development in snakes. Loosely separated skull bones, which allowed prey to be swallowed whole, was another important lepidosaur feature. On the other hand, archosaurs had limited or absent lateral undulation in the vertebral column, and the limbs were brought into a position more directly under the trunk. These modifications reached their highest degree of development in the dinosaurs and pterosaurs.

Representative species of the three groups of the Lepidosauria addressed here, the sphenodontids, lizards, and snakes, are presently alive. Turning to the sphenodontids, the tuatara (*Sphenodon*) of New Zealand are the only living members of this once large group. A newspaper article dating to the 1940s explained that the tuatara looked something like a lizard but really was a "living fossil." There are several differences between the sphenodontids and lizards, which split off from each other in the Triassic, possibly early in the epoch. In the sphenodontids, the jaw muscles are massive, which allows them to have a stronger but slower bite than lizards. Sphenodontid teeth are fused to the jaw so firmly (acrodont condition) that the jaw has a sawtooth appearance, as if the teeth are merely serrations of the bone itself. Jaw muscles are less massive in lizards and are located farther back in the mouth, producing a weaker but faster bite. Most lizards have teeth somewhat loosely attached to the inside of the jaw (pleurodont condition), and these teeth are replaced frequently in most species.

Sphenodontids were the dominant lepidosaurs of the Jurassic (206–144 mya), but they sharply declined in the Cretaceous as the lizards began to diversify broadly. Only the two species of New Zealand tuatara have survived to the present, others having died out at the end of the Cretaceous.

Fossil of a mosasaur—a giant sea lizard from the Cretaceous period. (Photo by R. T. Nowitz Photos/Photo Researchers, Inc. Reproduced by permission.)

The tuatara are active at much lower temperatures than most lizards, and the eggs have a gestation period of about nine months before being laid. The incubation period for the eggs is about 15 months, the longest of any known living reptile. Growth in the young is slow, and the animals do not reach sexual maturity until they are about 20 years old. Slow growth then continues until the animal is 50 to 60 years of age. Rather than having hemipenes (double penis) like lizards and snakes, male tuatara transfer sperm to the female by an extension of the gut called the cloaca. Whereas most lizards seem to look right through a person, tuatara have a direct gaze, with big brown eyes that seem more mammalian than reptilian.

Lizards and snakes are considered to be a single clade, the Squamata (scaled reptiles). Both lizards and snakes have legless forms with a jaw structure that allows them to swallow prey whole. Snakes, however, have carried these tendencies to the extreme. The first lizards are represented by an animal known as *Paliguana*, from the late Permian of South Africa. The fact that lizards had a more effective jaw structure, better hearing, and improved locomotion probably allowed them to exploit the habitats occupied by other lizardlike tetrapods, such as the sphenodontids. Most modern lizards, with the exception of the Komodo dragon (a monitor lizard that can take down deer), have not achieved large size. But in the Cretaceous, ancestral monitor lizards evolved into the ecologically important mosasaur, marine lizards that grew to 30 ft (9.1 m) in length. One giant terrestrial lizard of the past was the monitor lizard *Megalania* (probably 20 ft or 6.1 m long), the top predator in the Pleistocene (1.8–0.1 mya) of Australia. Fossils of *Megalania* at first were thought to be those of dinosaurs, but Max K. Hecht of the American Museum of Natural History proved that they were, in fact, giant lizards.

Snakes originated much later in the fossil record than lizards, at some time during the Middle to Upper Cretaceous. The four fossils that bear most closely on the ancestry of snakes are *Pachyrhachis*, *Podophis*, *Lapparentophis*, and *Dinilysia*. The first three are from the Middle Cretaceous, but *Dinilysia*, the most complete and well-studied of the four, is from the late Cretaceous. The marine squamates *Pachyrhachis* and *Podophis* have been considered the most primitive snakes by some researchers, because the configuration of the skull bones resembles that of living snakes, but they have a well-developed hind-limb skeleton. The terrestrial snake *Lapparentophis*, often called the "oldest snake," is represented by vertebrae only, but they are certainly snake vertebrae, with all of the unique modifications found in generalized living snakes. *Dinilysia* has a skull that is a mosaic of lizard and primitive snake characters, but its vertebrae are clearly like those of a boa-like snake. Unfortunately, the whole picture of early snake evolution has been muddled by jargon-filled, convoluted arguments, the problem, as always, being the basic similarity of snakes and lizards.

Primitive snakes were dominant in the world until the Miocene, when modern snakes quickly replaced the less-advanced types. Three factors probably played a part: the return of warm and equable climates in the higher latitudes following the climatic deterioration in the Oligocene (ca. 33–23 mya), the striking spread of grassland habitats worldwide, and the evolution of many rodent groups that could be exploited by snakes as food. The largest modern snakes include the boas of the New World and the pythons of the Old World. The giant python-like snake *Wonambi* lived in the Pleistocene of Australia, along with the giant lizard *Megalania*.

Archosauria

The archosaurs, or "ruling reptiles," branched into an impressive array of important groups, including the pseudo-suchians, crocodilians, pterosaurs, and dinosaurs. The archosaur clade may be defined by two temporal openings in the skull, one (antorbital fenestra) between the eye and the nostril and another (mandibular fenestra) in the hind part of the lower jaw. Many of the other archosaur characters reflect skeletal changes associated with a more upright posture and front-to-back motion in the limbs. It has been pointed out by Robert Carroll that the lineages within the archosaur assemblage were all distinct from one another when they first appeared as fossils.

The living crocodilians are related to the Pseudosuchia, a diverse group of early, sometimes crocodile-like archosaurs that are linked to the true crocodiles mainly on the basis of having the so-called crocodile-normal structure of the tarsus (ankle). The Pseudosuchia is one of the two major clades that branched from the early Archosauria. Pseudosuchians all had extensive external armor composed of bony plates. The group includes the rauisuchids, phytosaurs, and aetosaurs.

Rauisuchids are Middle and late Triassic reptiles that had a more or less upright stance and grew up to 20 ft (6.1 m) in length. *Ticinosuchus*, from the early part of the Middle Triassic of Switzerland, is one of the best-known forms. In the limb skeleton, the ankle and foot advanced to the level of those of modern crocodiles. Supposedly, the upright stance developed independently of the lineage that led to the dinosaurs. *Ticinosuchus* had an armor of two rows of small, bony plates that extended along the trunk and a single row at the top and bottom of the tail. Sharp, piercing teeth set in sockets indicate that this animal was a carnivore.

Phytosaurs occurred abundantly in the late Triassic of North America, India, and Europe. Although they were not true crocodiles, they resembled them in body form and probably had a similar lifestyle. Phytosaurs had a very long snout, with sharp, piercing teeth in sockets along the margin of the jaws. Although the nasal openings were on top of the head, they were set far back on the snout rather than on the tip, as in true crocodiles. Both the trunk and the tail had an extensive covering of dermal armor. A variety of other contemporaneous reptiles have been found in the stomach contents of phytosaurs, thus documenting their carnivorous habits. The fossil record of phytosaurs is confined to the late Triassic.

The aetosaurs formed a distinct group also known only from the late Triassic. Rather than sharply pointed teeth, aetosaurs had small, leaf-shaped teeth that suggest a herbivorous diet. The well-known aetosaur *Stagnolepus*, was a bizarre

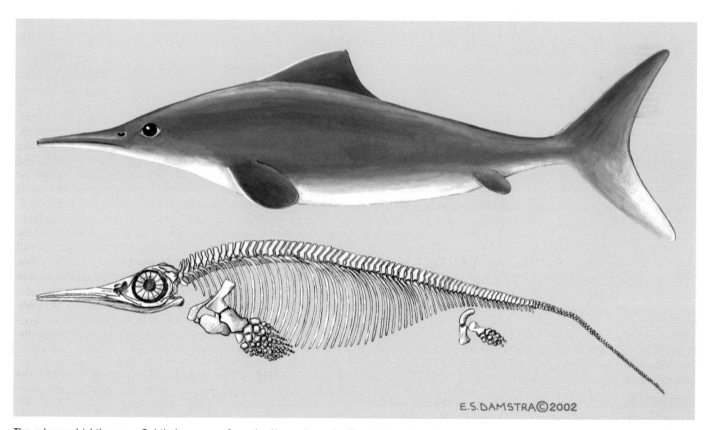

The advanced ichthyosaur *Ophthalmosaurus* from the Upper Jurassic. The skeleton is approximately 11.5 ft (3.5 m) long; the top image shows a restoration. (Illustration by Emily Damstra)

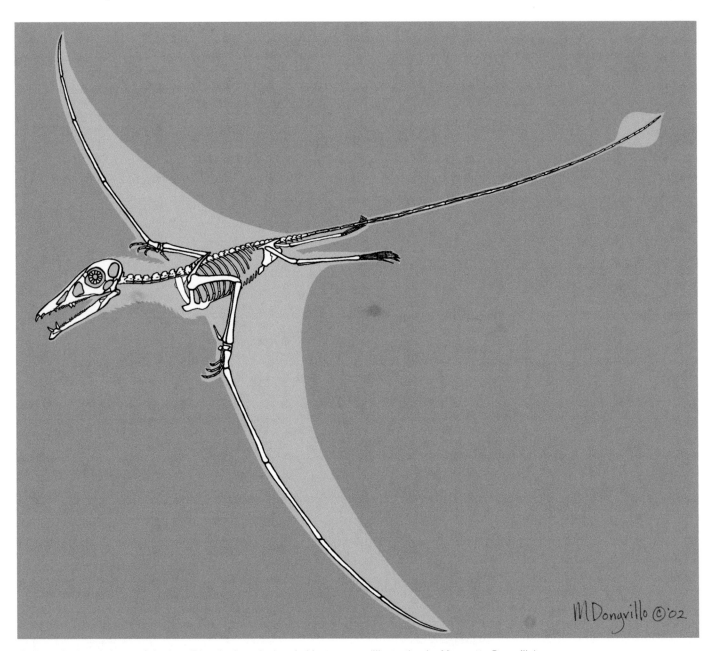

Eudimorphodon skeleton of the late Triassic rhamphorhynchoid pterosaur. (Illustration by Marguette Dongvillo)

beast with short legs and an upright posture. The body was rotund and the tail massive. The body was covered with large quadrangular plates that formed extensive armor along the back, extended down the sides, covered the belly, and surrounded the tail. The head was small in relation to the body and narrowed into a short rostrum anterior to the teeth, capped by a swollen area that indicated that the animal had a piglike nose. The relationship of aetosaurs to other archosaurian clades is poorly understood.

Crocodiles are the only surviving giant archosaurs and the top predators in many aquatic environments throughout the world today. A recent survey in Florida reported a male alligator that was 14 ft (4.3 m) long and weighed 946 lb (429 kg).

Much longer crocodiles have been found in the tropics of the Southern Hemisphere. Why crocodiles survived the rounds of extinction that occurred among the other large archosaurs is a matter of conjecture. The earliest crocodilians were terrestrial and included such forms as *Gracilisuchus* of the Middle Triassic of South America, which walked on two legs, and the four-legged *Terrestrisuchus* of the late Triassic of Europe, which had an extremely gracile body skeleton and must have been a fast runner. *Terrestrisuchus* was about the size of a rabbit but was a carnivore that probably scurried in and out of Triassic hiding places looking for mouse-size prey.

The mesosuchians of the Jurassic looked much more like modern crocodiles than the Triassic forms. Several of them became semiaquatic, and some invaded the marine environment. The process of leaving land to become semiaquatic was important in terms of the body changes that took place in these transitional forms. Modern families of crocodiles, Alligatoridae (alligators and caimans), Crocodylidae (modern crocodiles), and Gavialidae (the gavial), are known to date back to the late Cretaceous. Alligatorids have a broad snout, with a relatively large number of somewhat blunt teeth. Crocodylids have a longer, thinner snout, with a significant number of pointed teeth. Gavialids have an elongated snout and needlelike teeth, as do garfish; gavialids are consummate fish eaters. Some Cretaceous crocodiles grew to very large size and probably preyed upon dinosaurs. *Sarcosuchus imperator* from Africa possibly reached a length of 40 ft (12.2 m), and some from Texas were about that long as well.

Modern crocodilians were much more widespread in the world in the early Tertiary (ca. 65–34 mya) than they are today, their decline probably resulting from climatic deterioration in the Cenozoic era. Crocodilians are the only living reptiles that give true parental care to their young, including nest guarding, helping the young exit eggs by cracking the eggs with their jaws, carrying the young from the nest to the water in their mouths, and protecting them in the water for a time. Vocal communication between the parents and the young also occurs.

The Ornithodiran evolved a neck that could be folded into an S shape and a narrow, compressed foot. One branch of the Ornithodira, the Pterosauria (flying reptiles) of the Mesozoic, had a very long, much enlarged fourth finger on the hand, which supported a wing membrane. The other branch of the Ornithodira, which included *Lagosuchus* and the dinosaurs, had elongated lower legs (tibiae and metatarsals) as well as a thigh bone (femur) with the head turned inward so that it could fit into a deep socket in the hip girdle (pelvic girdle). This socket allowed the legs to support the large hip girdle, which in turn supported the body.

Pterosaurs were the first vertebrates to evolve true wing-flapping flight. They are known from the late Triassic and had become quite diverse by this time. The fact that pterosaurs had a foramen in the skull in front of the eyes (antorbital foramen), legs arranged mainly straight under the body, and a dinosaur-like ankle joint indicates their close relationship to *Lagosuchus* and the dinosaurs. But the fact that they had a hooked fifth metatarsal bone and an long fifth finger suggests that they diverged from the early archosaurs before the dinosaurs (which had the fifth finger reduced or lost) had split off.

Pterosaurs ranged from very small forms to by far the largest creatures that have ever flown. They are divided into two major groups, the primitive rhamphorhynchoids and the more advanced pterodactyloids. Rhamphorhyncoids are first known from the late Triassic and were dominant throughout the Jurassic. They had a short face, a short neck, and a long tail. Some rhamphorhyncoids were as small as English sparrows. From the beginning, rhamphorhyncoids had evolved various specialized characters for flight, including a sternum (breastbone) that had a boatlike keel that supported the wing-flapping muscles, as in birds. Other

birdlike bones included the scapula (shoulder blade), coracoid (upright shoulder girdle bone), and humerus (upper arm bone), all of which were specialized to contribute to active, flapping flight.

The pterodactyloids appeared in the late Jurassic and lasted until the end of the Cretaceous. These reptiles had a much longer face, longer neck, and shorter tail than the rhamphorhynchoids. The skull was highly modified in the Cretaceous genus *Pteranodon*, which had a long, bladelike rostrum (snout) in front of the eye and an almost equally long bladelike projection in back of the eye. On the other hand, *Pterodaustro*, from the Upper Cretaceous of Argentina, had practically no skull at all behind the eye, but it did have large, elongate, upwardly curved jaws. The lower jaw had very long, closely packed teeth that are thought to have strained small invertebrate animals from the water in the same manner as baleen in whales. *Quetzalcoatlas*, from the late Cretaceous of Texas, was the largest animal ever to fly, with a wing span of more than 35 ft (10.7 m). Before the discovery of *Quetzalcoatlas*, British zoologist J. Z. Young suggested that *Pteranodon*, with its 23-ft (7-m) wing span, was probably the largest animal that could possibly fly. Robert Carroll pointed out that *Quetzalcoatlas* was obviously far heavier than *Pteranodon*. It has been suggested that *Quetzalcoatlas*, might have fed upon the dead bodies of dinosaurs, like some gigantic vulture. Pterosaurs were quite different from birds, in that the wing was composed of a membrane, something like the one in bats. The difference is that in pterosaurs the membrane was supported entirely by a long, robust fourth finger. Unlike pterosaurs, bats have a wider wing membrane, and it attaches to the rear limbs.

Lagosuchus, from the Middle Triassic of South America, provides a structural link between early archosaurs and dinosaurs. *Lagosuchus* was only about 1 ft (0.3 m) long and had a very lightly built skeleton, with long delicate limbs; this was a humble ancestor of the gigantic animals to come. In *Lagosuchus* the posterior limbs were much longer than the anterior ones. Moreover, the tibia (lower leg bone) was much longer than the femur (upper leg or thigh bone). The pelvic girdle (hip girdle) was composed of three bones forming a triradiate (three-pronged) structure. Thus, the long ilium was directed forward, the long ischium was directed backward, and the short pubis sat on top of the other two pelvic girdle bones. In the tarsus (ankle) there was a hinge between the upper and lower tarsal bones. All of these features (reflecting changes in the limbs and limb girdles, mainly the pelvic girdle and hind limbs) made *Lagosuchus* the most dinosaur-like of any of the primitive archosaurs.

Dinosauria

Dinosaurs fascinate more people than any other reptile group. The achievement of great size and diversity and the long domination of the earth by dinosaurs form a large part of this fascination. As is true for the great gray apparition in Mozart's *Don Giovanni* or the Frankenstein monster, people love things that are terrible and wonderful at the same time. In fact, the name *dinosaur* comes from roots meaning "terrible lizard."

Unique evolutionary features are evident in the hands of some dinosaurs. The joint between the thumb and the palm was structured so that when the hand was closed, the thumb bent toward the palm, indicating that the hand was used for grasping and holding objects. In effect, the bipedal gait in early archosaurs set the hands free for other functions in both pterosaurs and dinosaurs. In the dinosaur pelvic girdle (hip girdle), the acetabulum (hip socket) either was open or was composed of cartilage in primitive forms, probably increasing the rate and range of leg motion from front to back. Also, one or more vertebrae were incorporated into the pelvic girdle, giving it more strength for vigorous locomotion.

Three main dinosaur clades are recognized by most paleontologists: the Theropoda, which were bipedal and mainly carnivorous; the Sauropodomorpha, which had long necks and were herbivorous; and the Ornithischia, which encompassed a diverse assemblage of beaked herbivores. The earliest members of all three groups are first known at about the same time in the late Triassic. During the early part of the late Triassic, dinosaurs were an important emerging group, but they actually were dominated by pseudosuchian archosaurs and advanced synapsids (ancestral mammals with reptilian characteristics). It was only after the great extinction that took place at the end of the Triassic that dinosaurs became the dominant group of large terrestrial vertebrates. This dominance lasted until dinosaurs became extinct at the end of the Cretaceous.

Theropods had more primitive dinosaur characters than the other two groups. For instance, theropods retained their bladelike, serrated teeth; by this feature it is known that they were carnivores. All known theropods were bipedal, and many retained grasping hands. The most primitive known theropods were the herrerasaurids of the late Triassic of Argentina, Brazil, and North America. One specialization that these animals shared with later theropods was a joint in the lower jaw between the tooth-bearing and non-tooth-bearing portions. This innovation probably counteracted the stress associated with biting relatively large prey.

The two main clades of Theropoda are the Neotheropoda and the Coelurosauria. An early neotheropod, *Coelophysis* of the late Triassic, was gracile in build and had a kink in the upper jaw believed to be an adaption for holding on to small prey. *Ceratosaurus* of the late Jurassic was a robust form, with hornlike knobs on top of the front of its skull. *Spinosaurus* of the Cretaceous had a very long snout and large, conical (rather than bladelike) teeth; the teeth at the end of the snout were larger than the rest. It has been suggested that *Spinosaurus* ate large fish, because some fossil localities where spinosaurs were collected contained abundant fish remains. *Allosaurus* of the Jurassic, a huge bipedal carnivore with a compressed head, is featured in many museums and represents a significant branch of the theropod stem called carnosaurs.

The Coelurosauria, the other main branch of the theropod stem, differed from the neotheropods in numerous ways. The brain cavity was relatively larger, and the hands were more slender. The tail tended to be very stiff. Elements thought to be homologous to the feathers of birds, called protofeathers by some researchers, have been found in both

primitive and advanced coelurosaurs, and it has been suggested that all primitive coelurosaurs may have had them. Several Jurassic and Cretaceous coelurosaurs, both small and large, were not ancestral to more advanced clades, but a clade called maniraptors became increasingly specialized.

Maniraptors are characterized by the development of a secondary palate and several changes in the structure of the brain case. They also had very slender hands and fewer tail vertebrae. Birds are a by-product of maniraptor evolution. There are several important maniraptor groups; among them, ornithomimosaurs, such as *Struthiomimus*, are not bird ancestors, but they show convergence with both modern ostriches and other flightless birds. Early examples had tiny teeth, but the teeth were lost in later forms. The fingers of the hands formed a hook-and-clamp structure that may have been used to grasp branches as the animals searched for food.

Tyrannosaurids are well known in the form of *Tyrannosaurus rex*, popularly known as *T. rex*, as well as other large carnivores. It should be pointed out that tyrannosaurids evolved independently of the narrow-faced *Allosaurus* and its kin. Tyarannosaurids are characterized by massive, rather than narrow heads; nipping teeth (incisors) at the front of the jaws; thickened, rather than compressed lateral teeth on the sides of the jaws; and very minimal forelimbs bearing only two claws. These animals were fast runners for their size and by means of their jaws alone could kill their prey and render it into portions small enough to be swallowed. Early Cretaceous tyrannnosaurids were only about 10 ft (3 m) long, but later ones such as *Tyrannosaurus rex* were 40–50 ft (12.2–15.2 m) long with a weight of up to seven tons or more.

Other maniraptors include such groups as the oviraptorosaurs, of which *Oviraptor* is a well-known genus. Many species of oviraptosaurs are characterized by ornate crests and are thought to have brooded their nests in the manner of modern birds. Dromaeosaurids include the familiar genera *Deinonychus*, *Velociraptor*, and the larger *Utahraptor*. These animals had long, grasping forelimbs and a large, retractable, curved claw on the second digit of the foot. Dromaeosaurids, like birds, had a backward-directed pubis and are thought by some researchers to be near the stem of bird evolution.

Found among the Sauropodomorpha are the largest land animals that ever lived; some reached a length of about 100 ft (30.5 m). All sauropods had long necks and small heads. Some of the primitive sauropodomorphs, known as prosauropods (such as *Plateosaurus*), spent some time on all fours, though they still could grasp objects with the hand. At the end of the Triassic, prosauropods were common large land herbivores. Long necks gave sauropodomorphs access to tree leaves, and their large body size enabled them to have a larger digestive system to process vegetation and served as a defense against small predators.

Sauropodomorphs called the *Neosauropoda* became a diverse group in the Jurassic and Cretaceous. Neosauropods

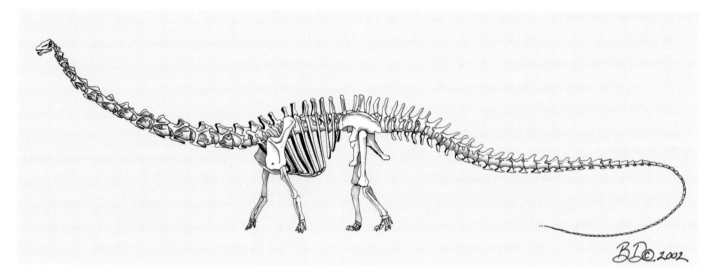

An adult *Diplodocus* was a 89.6-ft (27-m) long, lightly built sauropod, characteristic of the diplodocids. (Illustration by Barbara Duperron)

were characterized by columnar legs, wide hips, and specialized teeth. The two main neosauropod clades are the Diplodocoidea and the Macronaria. Diplodocoideans had thin, pencil-shaped teeth that supposedly were used for harvesting leaves and needles from the branches of trees. The skull was long and sloping, and the nostrils were positioned behind the eyes. The tails of diplodocoideans were very long and whiplike. Familiar examples are *Apatosaurus* (long called *Brontosaurus*) and *Diplodocus*. Shorter-necked forms include *Amargasaurus* of the early Cretaceous of Argentina.

In the Macronaria ("large nostrils") clade are sauropods with enormous nostrils that are larger than the eye sockets. Quite a few members of this group had spatula-shaped teeth as well. Brachiosaurs, including the well-known genus *Brachiosaurus*, make up a clade that had forelimbs much longer than the hind limbs and a very long neck, a condition that brings to mind modern giraffes. The combination of long front legs and a long neck allowed these animals to feed at the very top of trees. *Sauroposeidon* of the Cretaceous of Oklahoma is estimated to have been 80 ft (24.4) long, about 40 ft (12.2) of which consisted of the neck. Titanosaurs were giant macronarian sauropodomorphs that form a clade apart from the brachiosaurs. It has been discovered that advanced forms in this group developed pencil-like teeth similar to those in the diplodocids.

The Ornithischia are the third and last of the three main dinosaur clades. These animals were mainly armored, beaked, four-footed herbivores, but some of the primitive examples were at least partially bipedal. The three major clades presently recognized within the Ornithischia are the Thyreophora (armored dinosaurs), Ornithopoda (beaked, billed, and crested dinosaurs), and Marginocephalia (helmeted and horned dinosaurs). In the diversification of the thyreophorans, one finds that the modification of armored scutes into other defensive structures is a central tendency. The relatively unarmored *Scutellosaurus* of the early Jurassic, thought to be a basal member of the group, had a row

of bony dermal plates from the neck to the tail. It is thought that this feature might have protected it against the small predators of the time but probably not against the large predators that appeared later in the Jurassic. Indeed, later forms modified this dermal armor into plates, spikes, and tail armor. Stegosaurs of the late Jurassic were true armored dinosaurs, in that they had compressed plates, conical spines, or armor of intermediate shape along the middle of the back and tails that were modified as spiked clubs. Some stegosaurs had shoulder spikes, and others had masses of knobs on the throat. Ankylosaurs had an armadillo-like dermal armor fused to the head and protecting most of the body. One group of ankylosaurs had club tails.

Ornithopods may be distinguished from other Ornithischia in that the tooth row in the front portion of the upper jaw (premaxilla) is more depressed than those in the remaining upper jaw. Evolutionary tendencies in this group included an increase in size and changes in the joints of the jaw and teeth that led to a grinding mode of chewing. *Iguanodon*, an early ornithopod first discovered from the early Cretaceous of England, was at one time thought to be a giant lizard, because it had leaflike serrated teeth like the modern iguana lizard. The duckbill dinosaurs of the late Cretaceous, with terminally expanded snouts, have been exceptionally well studied, and some are known literally from the cradle (nest and eggs) to the grave. Each species of lambeosaurs had a unique crest shape and sound-producing tubes within the skull. These features provided both visual and vocal signals that indicate complex social behavior in these animals.

The last group of Ornithischia are the marginocephalians (helmeted and horned dinosaurs), represented by both bipedal and four-footed forms. The main character of this clade was a ridge or shelf of bone that overlapped the back of the skull. This group is divided into two clades, the pachycephalosaurs and the ceratopsians. Pachycephalosaurs were bipedal forms with a thick helmet of bone over the brain case; in the most derived forms, the helmet formed a thickened dome. Many scientists believe that the dome was used for butting clashes, as occur among bighorn sheep today. Others think that because the dome was provided with a large number of blood vessels, that it might have been important in temperature regulation.

Ceratopsians had large beaks at the end of the snout. A primitive ceratopsian, *Psittacosaurus*, of the early Cretaceous had this deep beak but was bipedal and lacked the frill of more advanced forms. Advanced forms, called neoceratopsians, had a shelf modified as a frill at the back of the skull where the jaw muscles were attached. Neoceratopsians had rows of teeth packed together in such a way as to provide a continuous cutting surface, but it is not known what kind of plants these animals fed on. The North American giant ceratopsians had greatly elaborated horns on the skull.

Epilogue

Dinosaurs and vast numbers of other extinct reptiles are represented exclusively by bones. Today we have only turtles, lizards, snakes, crocodiles, and the lonely tuatara of New Zealand to remind us of the glory of the reptiles of the past. Humans have exploited seaturtles to the point of extinction, and in some parts of the world freshwater and land turtles also are being harvested at an alarming rate. Lizards, snakes, and crocodilians likewise are greatly in decline, though, ironically, the rare tuatara appears to be well protected. What a pity if our living reptiles, survivors of the catastrophic extinction of the dinosaurs at the end of the Cretaceous, were to become extinct in a mere instant of geologic time as the result of human exploitation and neglect.

Resources

Books

Benton, Michael J. *Vertebrate Paleontology*. 2nd edition. Oxford: Blackwell Science, 2000.

Benton, Michael J., and D. A. T. Harper. *Basic Palaeontology*. London: Addison Wesley Longman, 1997.

Carroll, Robert L. *Patterns and Processes in Vertebrate Evolution*. Cambridge: Cambridge University Press, 1997.

Colbert, Edwin H., Michael Morales, and Eli C. Minkoff. *Colbert's Evolution of the Vertebrates*. 5th edition. New York: John Wiley and Sons, Inc., 2001.

Cowan, Richard. *History of Life*. 3d edition. Malden, MA: Blackwell Science, 2000.

Farlow, James O., and M. K. Brett-Surman, eds. *The Complete Dinosaur*. Bloomington and Indianapolis: Indiana University Press, 1997.

Hallam, Arthur, and P. P. Wignall. *Mass Extinctions and Their Aftermath*. New York: Oxford University Press, 1997.

Holman, J. Alan. *Vertebrate Life of the Past*. Dubuque, IA: William C. Brown Publishers, 1994.

————. *Fossil Snakes of North America: Origin, Evolution, Distribution, Paleoecology*. Bloomington and Indianapolis: Indiana University Press, 2000.

Lucas, Spencer G. *Dinosaurs, the Textbook*. 3d edition. Dubuque, IA: William C. Brown, Publishers, 2000.

Paul, Gregory S., ed. *The Scientific American Book of Dinosaurs*. New York: St. Martin's Press, 2000.

Pough, F. Harvey, Robin M. Andrews, John E. Cadle, Martha L. Crump, Alan H. Savitzky, and Kentwood D. Wells. *Herpetology*. 2nd edition. Upper Saddle River, NJ: Prentice Hall, 2001.

Sumida, Stuart S., and Karen L. M. Martin. *Amniote Origins, Completing the Transition to Land*. San Diego: Academic Press, 1997.

Weishampel, David B., Peter Dodson, and Halszka Osmólska, eds. *The Dinosauria*. Berkeley: University of California Press, 1990.

J. Alan Holman, PhD

· · · · ·

Structure and function

Structure and function combine anatomy, microstructure, and physiology essentially to describe "how animals work." These subjects are richer and better understood when considered in the contexts of evolution, ecology, and behavior. Thus an integrative discussion of these topics is especially meaningful for understanding the fascinating lives of reptiles.

Skeleton, muscle, and movement

Reptiles evolved from limbed ancestors, and they have an axial skeleton consisting of the vertebral column, limbs, and central nervous system encased in bone. The loss of limbs in snakes and some lizards evolved secondarily from limbed ancestors. Many of the fundamental features of the reptilian skeletal system and its attached musculature reflect adaptations for support and locomotion in terrestrial environments where strong weight-bearing elements are essential for counteracting gravity. The aquatic to terrestrial transition of early vertebrates also required reorganization of many internal organs related to weight bearing, changes in breathing and distribution of body fluids, specializations of sensory systems, and reorganization of the feeding apparatus.

The skeleton of turtles is largely rigid, and aspects of its structure are unique among vertebrates. The shell that encompasses the body, consisting of a dorsal carapace and ventral plastron (joined laterally by a bridge), is composed of fused bone that includes the ribs, vertebrae, and parts of the pectoral girdle. Both the pectoral and pelvic girdles are located within the rib cage. The bony elements of the shell are covered externally by corneous plates or, in fewer species, leathery skin. The overall shape of the shell varies greatly, and the shell is rigid in most species. The enclosure of internal organs by the shell precludes breathing by expanding and contracting the rib cage. In some lineages of turtles, partial flexibility of the body has been achieved by reduction, softening, or hinged articulation of the shell elements. Some species have evolved the ability to close the body within the shell during withdrawal of the head and limbs.

Because the vertebral column is rigidly fused to the carapace, turtles are limited to paddling as a means of locomotion in water. Walking by turtles on land is awkward because the vertebral column cannot be bent to shift the center of gravity, and the limb girdles are enclosed within the shell. A walking turtle tends to pitch and roll, and the forefeet catch the weight of the animal as it falls forward. It has been suggested that slower-acting muscles characteristic of turtles cannot adjust rapidly during movement to eliminate inherent instabilities related to the shifting center of gravity.

Crocodilians are characterized by short, powerful limbs and a muscular tail that is used in swimming. The spine is flexible, and individual vertebrae are strengthened by the addition of bone that is curved in the forward direction. These animals are excellent swimmers, and saltwater crocodiles are known to disperse long distances over the sea. Crocodilians are capable of surprisingly fast movement on land, and one species is known to gallop. The skull has evolved a secondary palate that separates the respiratory passages from the mouth cavity and allows these reptiles, and some turtles, to breathe with only the tip of the snout exposed to air. All crocodilians are aquatic.

Lepidosaurians (the tuatara and squamates—lizards and snakes) are characterized by a flexible vertebral column, reduction and mobility of the skull, in some cases elongated tails, and development of breakage planes in the tails of many species. The reduction of limbs has been a major

Underside of a tail and anal plate of ball python (*Python regius*) where the vestigial legs are located. (Photo by Animals Animals ©Zig Leszczynski. Reproduced by permission.)

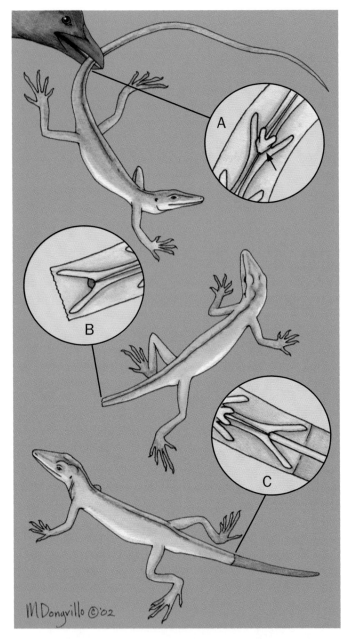

Tail regeneration. A. The autotomic split in the vertebra; B. The vertebra splits and the tail is shed; C. A shorter tail is regrown. Instead of bone, the new tail has a cartilage rod. (Illustration by Marguette Dongvillo)

velopment of some elements of the pelvic girdle and femur. Modifications of *Hox* gene expression appear the likely mechanism for body elongation and a repeated pattern of vertebral development without limbs. These developmental processes produce more than 300 vertebrae in pythons with ribs on all but one of the vertebrae in front of the hind limbs.

The ability to lose the tail easily when seized by a predator is called caudal autotomy and is characteristic of tuatara, many lizards, and some snakes. With few exceptions, autotomy is attributable to fracture planes within the vertebrae. These planes are enveloped by muscle and connective tissue arrangements that allow easy separation of the elements. When the tail is separated, the frayed muscle bundles collapse and seal the end of the broken tail while the lost end reflexly wiggles to attract the attention of would-be predators. Thus autotomy and rapid tail loss are an effective adaptation to confuse or deter predators. The tail regenerates to varying degrees in most lizards, but the regrown tail has an axis of cartilage rather than bone and is not as long as the original structure. Separation of tails in a few species of snakes occurs between the vertebrae, and the broken tail does not regenerate markedly.

The mechanics and energetics of locomotion have been analyzed in some detail for various reptiles, especially snakes and lizards. All reptiles living in water are secondarily aquatic and, with the exception of turtles, swim by means of axial propulsion involving undulating waves that are passed along the body and tail. Such swimming is associated with a large muscle mass, sometimes amounting to more than 50% of the body mass. There is a high conversion of muscle energy to thrust energy, which in some cases may exceed 90%. The more posterior parts of the body contribute most to thrust, and forces are maximized when the greatest depth of the body is posterior. Examples of adaptation to this principle are the paddle-like tails of sea snakes and the vertically flattened tails of crocodilians. Other characteristics that assist swimming are a flexible body to maximize amplitude of undulation, large ratio of muscle mass to body mass, and a relatively large body and caudal area. The energy cost of swimming is generally severalfold more economical than is locomotion on land. Buoyancy contributes to the efficiency of locomotion in water.

Swimming turtles use paired appendages that beat backward and forward, acting like paddles. This mode of locomotion achieves lower power and is less efficient than that involving use of body undulation. The forelimbs of sea turtles can create lift forces similar to those made by an airfoil.

Movement on land includes quadrupedal, bipedal, and limbless locomotion, depending on the species of reptile. The limb or the ventral aspect of the body pushes on the ground at an angle, eliciting an equal and opposite reactive force on the foot or body. The reactive force can be resolved into a forward propulsive component, which generates thrust in the direction of movement. Part of the energy required for locomotion is devoted to support of the body and the maintenance of posture by limbs. When quadrupedal reptiles move at slow to moderate speeds, the body moves in such a manner as to maintain the center of gravity over

evolutionary trend for many lineages within the squamates. Some members of most families exhibit variable degrees of limb reduction, usually the loss of a few phalanges but in numerous lineages the total loss of limbs. All snakes are limbless, and limb reduction has evolved repeatedly among various lineages of lizards. The evolution of limblessness usually is associated with burrowing habits or with life in dense grass or shrubbery.

In pythons, which have vestigial hind limbs, *Hox* genes that specify development of the limbs are differentially expressed. The result is absence of forelimbs and shoulder girdle but de-the shifting base of support during alternating movement of the limbs.

Most lizards have a sprawling limb posture, but some quadrupedal reptiles, such as chameleons and rapidly moving crocodilians, have evolved a more erect or nearly vertical position of the limbs. While the animal is standing, the musculature must support the body between the laterally placed feet. However, most reptiles do not remain standing while motionless but minimize expenditure of energy for postural support by resting the body on the substrate. Sprawling limbs promote stability and are advantageous when lateral undulations of the body are used in locomotion. During walking or running, substantial lateral bending of the body axis promotes longer stride length, which is also increased by specialization of the limb girdles. Progressively more distal elements of the arm or leg describe successively larger horizontal ellipses relative to the shoulder and hip, and the feet may twist in their tracks. The hind limbs are longer and usually more robust than are the forelimbs, and they provide most of the propulsive force.

Other specializations of the hind limbs include elongated toes, fringed scale appendages on toes to assist movement on sandy substrates, and close union of the first four metatarsal bones while the fifth metatarsal functions as a lever to extend the ankle. Several species of lizards are capable of running bipedally on their hind legs, a well-known example being the Neotropical Jesus Christ lizard, the common basilisk, *Basiliscus basiliscus*, which can run over the surface of water.

Snakes and other limbless squamates ordinarily move by one of four locomotion patterns. In lateral undulation, the trunk moves continuously relative to the substrate ("slithering"). Propulsive forces are transmitted by the sides of the trunk as the body slides past sites where resistance forces are exerted. These sites are generally irregularities in the substrate against which the muscles of the body push to move the animal forward. In this process a series of undulant curves pass from front to rear while posteriorly facing surfaces of the body contact and push against the irregularities of the surface on which movement occurs. The propulsive forces are generated entirely by

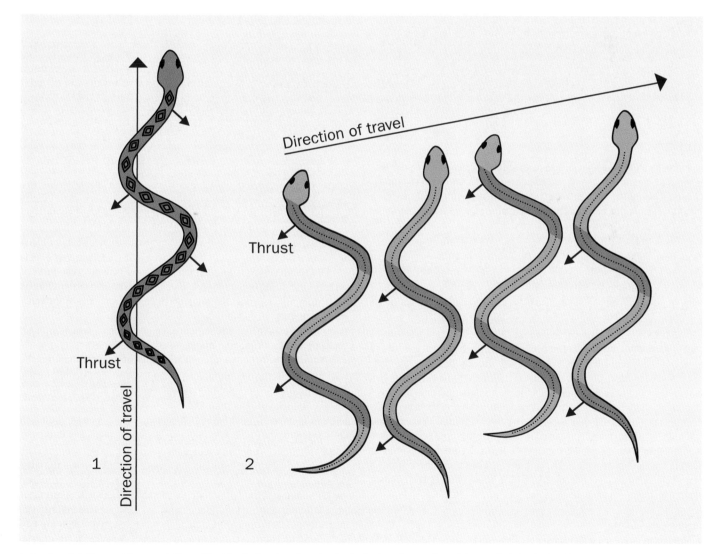

Lateral undulation and its correlation with sidewinding travel: red color indicates the portion of the snake's body that is in contact with the ground. (Illustration by Barbara Duperron)

the axial muscles and are perpendicular to the contact surfaces, so friction is not used. Progression requires at least three posteriorly directed force vectors (contact points) to be stable.

All three of the other modes of locomotion are different from lateral undulation in having propulsive forces that act horizontally in the direction of movement and are transmitted across zones in which the trunk makes static contact with the substrate. In concertina locomotion, progression is accomplished with stationary body parts as an anchor to push or pull the rest of the body forward. The friction associated with static contact is used as a reaction force to allow forward progression. Forward movement involves alternate anchoring, folding, anchoring, and extension of the body as the muscles act somewhat like an accordion between the zones of static contact. Because acceleration alternates with deceleration, this type of movement is comparatively slow and energetically costly.

Sidewinding is a method of locomotion used for progression on relatively flat, low-friction surfaces such as desert sands or mud flats. In this movement, sections of the body are lifted, moved forward, then set down in such a manner that the body is usually in static contact with the ground at two points. Once the neck of a snake contacts the ground at the forward point, the body is essentially "rolled out" on the substrate, producing a series of separate, parallel tracks oriented at an angle to the direction of travel. A variation of sidewinding, termed saltation, occurs when the body is straightened so strongly and rapidly that it lifts completely off the substrate. This mode of locomotion and slide pushing, whereby posterior contact zones of undulating snakes slide backward relative to the ground, are associated with rapid escape movements.

The fourth principal mode of limbless locomotion involves laterally symmetric use of muscles associated with the ribs and the flexible skin. Termed rectilinear locomotion, this mode involves alternate placement of ventral scales in static contact with the ground followed by active stretching to extend the ventral wall (like an inchworm) and produce straight-line progression. The anterior surface of the snake progresses at a more-or-less constant velocity and conserves momentum because the mass of the animal continues to move.

Movements of limbless reptiles such as snakes can entail a mixture of the modes of locomotion, each of which may be influenced by the size of the snake, nature of the substrate, temperature, or stimulus for movement. Both lateral undulation and sidewinding are more rapid modes of progression than are concertina or rectilinear modes of movement. In faster-moving snakes, such as whipsnakes and racers, the locomotor muscles are elongated to span many vertebrae and create long arcs during undulating movements. Yet these muscles are lightweight owing to elongation primarily of tendons rather than of the active muscle mass. In contrast, snakes that constrict their prey, such as boas and pythons, have heavier, shorter muscle units that produce great strength of constriction but slower locomotion.

Feeding and digestion

Many of the prominent and interesting adaptations of reptiles are related to the capture and digestion of food. Most reptiles seize prey as individual items, and feeding strategies can largely determine the shape of the head and characteristics of the skull and jaws.

The reptilian skull varies in relation to feeding requirements. Skulls of turtles and crocodilians are comparatively rigid and compact. Those of lizards and especially snakes exhibit evolutionary reduction of structure and articulation of elements to produce a highly movable skull. The skull of a snake contains eight links that have kinetic joints that allow rotation and an impressive complexity of possible movements. When a snake swallows prey, which is ingested whole, the two sides of the skull alternately "walk" over the prey. Multiple recurved teeth pull the prey into the throat and esophagus. Lizards have mandibles that are joined at the front of the mouth, whereas the mandibles of snakes are attached only by muscles and skin. These elements in snakes can spread apart and move backward and forward independently. The kinetic structure of the skull of snakes allows very large gapes that accommodate prey larger in circumference than the snake's body. Snakes capable of larger gape, such as pit vipers, can swallow larger prey. Prey subdued and swallowed by these snakes can be enormous, sometimes exceeding the body mass of the snake that ate it.

Many snakes seize prey and swallow it as they struggle. However, snakes that consume animals capable of inflicting harm generally subdue them by means of constriction or envenomation. Toxic secretions that immobilize prey are contained in Duvernoy's gland in the upper jaw of many colubrid snakes. These glands are homologous to the venom glands of viperid and elapid snakes. Envenomation of prey occurs with sharp but unmodified teeth or with specialized, enlarged fangs that have evolved from teeth at the front of the maxilla in viperids and elapids or near the rear of the maxilla in colubrids. Fangs are grooved or hollow like a hypodermic needle and inject venom released from the venom glands into the struck prey. Toxic venom immobilizes prey and aids in digestion. Both of these functions are probably more important than is a defensive function.

The teeth of squamate reptiles are generally specialized for seizing and holding prey and have less variability in structure than do teeth needed for mastication in mammals. Reptilian teeth are attached to bone and often undergo replacement several times during the lifetime of individual animals. The teeth of crocodilians are attached inside sockets by ligaments in a manner similar to the attachment of mammalian teeth. One of the remarkable adaptations in a few species of snakes is hinged teeth that facilitate swallowing of hard-bodied prey. Teeth are absent in living turtles, being replaced by a tough, keratinized beak.

The reptilian gut is similar to that of many other vertebrates in being elongated, compartmental, and complex. Digestion depends on gut motility, orchestration of multiple hydrolytic enzymes, and appropriate conditions of temperature and pH. Digestion proceeds most rapidly at warmer body temperatures, and some species of reptiles select body temperatures that are higher during digestion than during nonfeeding periods. On the other hand, foods tend to putrefy and are regurgitated if body temperatures are too low. The importance of mastication or reduction of food to smaller particles varies

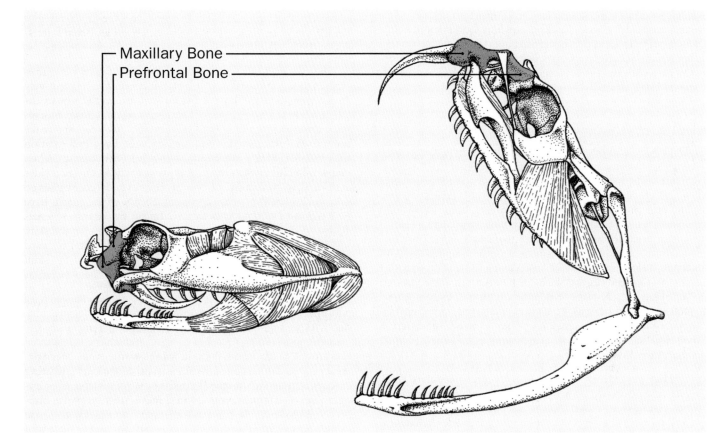

Fang erection by the African puff adder (*Bitis arietans*). The maxillary and prefrontal bones are highlighted. (Illustration by Barbara Duperron)

with species and is generally much less than is required by mammals. Because of low metabolic rates and generally high efficiencies of converting food to assimilated energy, most reptiles do not need food frequently.

Results of studies mainly of snakes indicate that physiological responses of the gut are regulated adaptively in relation to foraging modes and feeding habits. After feeding, snakes that feed infrequently on relatively large prey undergo remarkable increases in metabolic rate, mass of intestinal tissue, and rate of nutrient transport across the gut wall. After digestion and between feedings, the gut atrophies, and digestive functions are down-regulated. These changes are more modest in actively foraging snakes that feed with relatively greater frequency and thus maintain the gut in a more constant state of readiness. The down-regulation responses in snakes that feed less frequently presumably evolved to conserve energy otherwise spent on gut maintenance during extended periods without feeding. Snakes, such as vipers and pythons, that generally capture prey by ambush may feed once a month or less frequently.

Various turtles and lizards are omnivorous, but relatively few lineages have evolved strict herbivory. Herbivorous reptiles exhibit morphological specializations of the gut and depend to varying degrees on fermentation by populations of symbiotic microorganisms that usually reside in enlarged for-

ward portions of the large intestine. Plant material has a lower energy content than does animal tissue, and the high fiber content of plants can be very resistant to digestion. Digestion of plant material requires more time and is less efficient than is generally true of animal tissue. Among lizards, strict herbivory is associated with relatively large body size, possibly as an evolutionary consequence of the energetics involved. Fruits and flowers are common foods that are relatively rich in energy and nutrients, but some lizards and land tortoises have diets composed largely of leaves.

Studies of the Galápagos marine iguana (*Amblyrhynchus cristatus*) have elucidated important couplings between food resources, body size, and survival during periods of food limitation. These lizards are herbivorous reptiles that feed on submerged algae along the rocky shorelines of the Galápagos archipelago of Ecuador. Smaller marine iguanas are prevalent on islands that have lower production of marine algae, apparently because the smaller animals have a higher foraging efficiency than do larger iguanas. Moreover, when food is scarce (e.g., during El Niño events), larger animals suffer higher mortality than do smaller animals. Thus population shifts to smaller body sizes can occur during or after periods of resource limitation. Amazingly, changes in bone metabolism enable marine iguanas to shrink reversibly during times of energy shortages. Thus shrinkage in body size has been shown to coincide with low food availability after El Niño events, whereas body length has been shown to increase during subsequent La Niña conditions, when food is more abundant. Marine iguanas demonstrating greater shrinkage survive longer than do larger iguanas because the efficiency of foraging increases and the total energy expenditure decreases relative to larger animals.

Energy metabolism and ectothermy

Energy is used by organisms to maintain living processes, power locomotion and other activities, and fuel the synthesis of tissue in growth and reproduction. All vertebrates derive their usable energy from the oxidation of carbohydrates, lipids, and proteins in the food they eat. Reptiles do not expend large amounts of energy to maintain high levels of body temperature as do birds and mammals, which are endothermic. The latter maintain high body temperature with internal heat production coupled with insulation—fur or feathers—that minimizes losses of heat to the environment. Reptiles have lower rates of heat production and are not well insulated from thermal exchanges with the environment. They are ectothermic, meaning they depend on external sources of heat in the environment for maintenance of body temperature.

During periods of activity, many reptiles regulate their body temperature within narrow limits by adjusting both position and posture in relation to thermal complexity in the environment. The simplest example of such behavioral thermoregulation is that of many lizards that shuttle between sunlit and shaded microenvironments, maintaining high body temperature, sometimes with remarkable precision. Body temperatures are maintained at levels between 95°F and 108°F (35°C and 42°C) during daylight hours in most diurnal species. Thus ectothermic reptiles can be as warm as many mammals. However, during nighttime hours or colder seasons in the temperate latitudes or high altitude, reptiles usually retreat to locations where temperatures are moderated, such as underground, in rock crevices, or in bodies of water. Unlike mammals, reptiles cannot maintain activity during colder conditions.

The time reptiles spend exposed to solar radiation for thermoregulatory purposes is inversely related to body size and ambient temperature. Thus altitude greatly affects the behaviors required for thermoregulation. Decreasing air temperatures and increasing cloud cover generally accompany increases in altitude. Lizards that have been studied tend to compensate for these conditions by adjusting their exposure to sunlight. Thus at increasingly higher altitudes there is a shift from occupancy of shaded habitats, as at low altitudes, to occupancy of open habitats. Above approximately 9,800 ft (3,000 m), high body temperatures can be attained only by exposure to direct sunlight, a circumstance that also increases exposure of basking lizards to predators.

Because reptiles do not depend on internal metabolism for body heat, their metabolic requirements are sevenfold to tenfold lower than those of most endotherms. Moreover, when the environment cools at night, the decreasing body temperature of reptiles reduces energy expenditure even further, whereas mammals must produce additional heat to maintain warmer bodies. Thus reptiles use far less energy during a day or a season's activity than do birds or mammals of comparable size.

Metabolic rates vary with body size, but the relation usually is not linear; energy expenditure scales typically to the 0.6–0.8 power of body mass. The rate of metabolic energy expenditure per gram of body mass therefore increases at smaller body sizes. Although this relation holds for almost all animals, the rates of energy expenditure in birds and mammals are many-fold higher than those of reptiles of comparable size. All factors that contribute to energy differences considered, a lizard in nature uses only approximately 3% of the energy used by a mammal of similar body size during the course of a day.

Being ectothermic has both advantages and disadvantages for reptiles. The lower metabolic energy requirements of reptiles decrease demand for food resources and increase the efficiency at which energy from food is transformed into production of body tissue. Thus reptiles are better able than endotherms to live in environments of low biological productivity and to withstand periods of food scarcity related to changes in weather or other factors. Ectotherms, however, cannot maintain optimal body temperature in all environmental conditions. The activities of reptiles are limited in time and space by environmental conditions, which also limit geographic distribution and reproductive output. Numerous body functions and processes operate most effectively at higher temperatures, and reptiles are not always in conditions that allow optimal performance, as is the case for endothermic mammals. Exposure to sun or other heat sources also may render reptiles more vulnerable to predation than would be the case if shuttling between environments were not required for thermoregulation.

There are two notable exceptions to the general pattern of ectothermy among reptiles. Females of several species of pythons generate metabolic heat from spasmodic contractions of skeletal muscles during brooding when they coil around their eggs. Thus brooding female pythons are endothermic and transfer body heat to the incubating eggs, which are maintained at approximately 86°F (30°C). Metabolic heat also may accumulate in other reptiles of large body mass. The result is periodic or facultative endothermy. In comparatively cold seawater, body temperatures of leatherback seaturtles (*Dermochelys coriacea*) have been shown to be as much as 32°F (18°C) greater than the surrounding water temperature.

Circulation and respiration

Reptiles have a well-developed blood circulation that plays a central role in transport of respiratory gases. In most species oxygen is acquired from the environment at the internal lung surfaces, which are ventilated by movements that alternately expand and contract the body compartment surrounding the lungs. Lung ventilation is intermittent, and the depth and frequency of breathing vary greatly among species and even among individual animals. Oxygen is transported in blood largely in chemical combination with hemoglobin, which is contained in red blood cells.

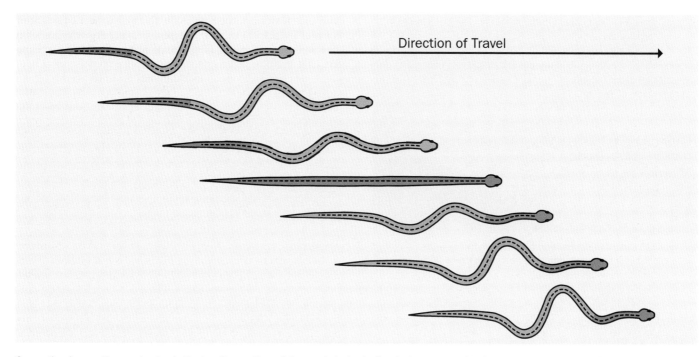

Direction of Travel

Concertina locomotion: red color indicates the portion of the snake's body that is in contact with the ground. (Illustration by Barbara Duperron)

The lungs are paired structures, except in various limbless species in which one of the lungs has become reduced or absent. The functional lung surfaces are complexly honeycombed to increase surface area, and they receive a profuse blood supply. In various reptiles, especially snakes and lizards, the posterior part of the lung is a simple membranous sac that functions to conduct or store air but does not participate directly in gas exchange. The saccular part of the lung may also be used to produce defensive inflation of the body. Among highly aquatic reptiles, the skin and linings of the throat and cloaca serve as accessory respiratory surfaces. However, most oxygen and carbon dioxide are exchanged across the lungs of most species, which require periodic access to air.

The blood is circulated by a central heart connected to distributing arteries, which service the tissue capillaries, where exchange of respiratory gases, nutrients, and other materials take place. Blood leaving the capillaries is returned to the heart in a system of veins, and excess fluid filtered from the blood at the capillaries is returned to the circulation via the lymphatic system. Except for crocodilians, all reptiles have a heart with a single ventricle filled by two atria. Thus in the ventricle there is an admixture of oxygenated blood returning from the lungs with oxygen-depleted blood returning from the tissue capillaries elsewhere in the body. In many species, these two bloodstreams are kept relatively well separated despite the potential for mixture in the single ventricle. On the other hand, the capacity for shunting blood between the lung and body circuits depends on physiological demands. For example, in aquatic reptiles blood flow to the lung is highest during air breathing, when lung and blood oxygen stores are renewed, and declines considerably during diving or submergence as the lung oxygen is used.

Both heart rate and blood pressure are generally lower than those in mammals, but they vary considerably with temperature and activity. The volume of circulating blood in reptiles varies from a few percent to approximately 14% of body mass. The common value is approximately 5–7% of body mass.

Integument and water exchange

As in other vertebrates, the integument of reptiles serves as mechanical protection from the environment, shields the body from unwanted substances or toxins, and largely determines the nature and magnitude of mass and energy transfer between animals and their environment. The skin consists of a fibrous dermis overlain by an epidermis that has multiple layers of living or keratinized cells derived from an active germinative layer beneath. Thin layers of bone called osteoderms are deposited in the dermal layers of many lizards, turtles, and crocodilians. The outermost stratum corneum consists of highly organized layers of keratin derived from dead cells and is periodically shed, either as an entire unit (snakes, some lizards) or in flakes or pieces (turtles, crocodilians, lizards). Periodic replacement of older, keratinized, epidermal generations as a synchronous, whole-body ecdysis (sloughing or shedding) is especially characteristic of snakes.

There are relatively few glands in reptilian skin compared with that of fishes and amphibians. Known glands tend to be concentrated in particular regions such as the femoral and preanal scales of lizards. Glandular secretions mark territories or establish scent trails that are recognized by conspecific individuals. Several species of snakes secrete substances from the skin that contain lipids involved in sexual signaling or

The radiated tortoise (*Geochelone radiata*) is considered to be one of the most beautiful tortoises. (Photo by Animals Animals ©Joyce & Frank Burek. Reproduced by permission.)

contain toxins thought to have defensive functions. These and other glands of reptiles are not well studied.

The external stratum corneum of scales often is highly sculptured with complex patterns of spiny projections or ridges. Patterns of microscopic sculpturing at micrometer and nanometer levels vary among species and may give rise to structural colors exemplified by the iridescence seen in many snakes immediately after they shed their skin and expose a new generation of epidermis. These colors are most likely incidental to other functions of the sculpturing, which remain largely unknown. The color patterns of reptiles are determined principally by pigments present in the skin. Structures called chromatophores bear pigments that reflect light differentially. Nervous or hormonal control of the dispersion of these pigments determines the color seen and gives rise to color changes related to camouflage, excitement, thermoregulation, and defensive and social behaviors. The color patterns of most reptiles probably are functional largely in relation to camouflage and protection from predators.

Water evaporates from the skin and lung and is lost in the urine and feces. Evaporative loss of water from the skin can account for more than half of the total water loss and is extremely important in species with relatively permeable integuments. Reptiles that live in drier environments have been shown to have skin that is more resistant to transepidermal water loss than is skin of reptiles that live in moist environments. The resistance to passage of water is determined largely by a discrete layer of lipids within the keratinized layers of epidermis. These lipids are organized into intercellular layers that surround the keratin filaments to form a complex association that has been likened to bricks and mortar. Thus the barrier to water is a highly ordered, laminated lipid-keratin complex. Variation in the quantity and kinds of lipids presumably accounts for differences in skin permeability among species. Contrary to popular opinion, differences in the thickness of skin, amounts of keratin fibers, or morphology of scales have relatively less influence on skin per-

meability than does the nature of the lipid barrier. Therefore the external appearance of the skin is not a reliable guide to how resistant an animal is to dehydration. Crocodilians have skin that is thick and tough in appearance, yet the lipid barrier is poorly developed, and the skin is quite permeable to water exchange.

Marine reptiles experience loss of water across the integument if the solute level in the external medium exceeds that of the body fluids. For example, 92% of total body water loss of the yellow-bellied sea snake (*Pelamis platurus*) is across the skin. However, the skin of marine reptiles generally has lower permeability than does that of freshwater species. The direction of net water movement is reversed for animals in freshwater, where the solute concentration of the medium is generally less than that of the body fluids, thereby promoting osmotic movement of water from outside to inside the animal. Excess water resulting from osmotic intake is normally eliminated by the kidneys.

A complete picture of water balance must include routes of water gains from food, metabolic water, drinking, and skin as well as losses due to excretion, respiration, and outward diffusion across the skin. The reptilian kidney cannot excrete urine that is more concentrated than the body fluids, although digestive and urinary fluids can sometimes be further concentrated by absorption of water within the cloaca. The cloaca is the terminal region of the hindgut where digestive and urinary contents are joined and temporarily stored before elimination through the anal opening. Fully marine reptiles and some terrestrial ones have excretory organs called salt glands in addition to the kidneys. These structures are located near the tongue (snakes, crocodilians), nasal region (lizards), or eye (turtles), and eliminate concentrated secretions of salts, principally sodium, potassium, and chloride, that might be ingested in excess. The salt gland, however, does not necessarily enable marine reptiles to maintain water balance by drinking seawater. Some marine species have been shown to become dehydrated if kept in saltwater indefinitely. They rely on water contained in their food and drinking of freshwater from rainfall to maintain water balance.

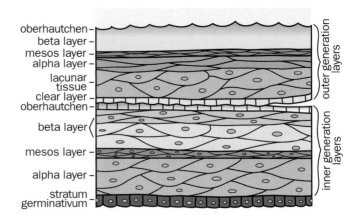

Generalized epidermis of a squamate reptile. (Illustration by Patricia Ferrer)

Nervous system and sensory organs

The reptilian nervous system is organized into identifiable regions based on structure and function. The central nervous system consists of the brain and spinal cord and contains most of the body's nerve cells or neurons. Collections of neurons that have similar function are called nuclei, and the bundles of axons that extend from the cell bodies and transmit messages are called tracts. The peripheral nervous system consists of sensory and motor nerves that communicate between the brain or spinal cord and various other parts of the body. The motor neurons that control contraction of skeletal muscles compose the voluntary nervous system. The autonomic nervous system, consisting of sympathetic and parasympathetic branches, provides unconscious control of the heart and lungs and activity of smooth muscles and various glands.

The spinal cord is enclosed and protected by the vertebral column. It provides reflex actions that occur independently of the brain but also receives input from higher brain centers. The spinal cord has a segmental organization that is largely lost in the brain, although bilaterally paired cranial nerves connect centers of the brain with structures in the head and body and represent a vestige of segmental organization. The cranial nerves carry both sensory and motor information.

Chemical senses

Receptors that sense chemical stimuli are designated gustatory, or taste, receptors if they respond to dissolved molecules and olfactory, or smell, receptors if they respond to airborne particles. Taste receptors typically are present in the mouth, but these vary in occurrence, and many reptiles rely more on olfaction than on taste. Olfactory receptors are located in the nasal passages, where a stream of air flows over them during ventilation of the lungs or during pumping movements of the throat, which may have a "sniffing" function. Olfactory receptors are generally more varied than taste receptors, and many are highly specific for chemical stimuli. A second kind of olfactory chamber, the vomeronasal or Jacobson's organ, is a pair of blind-ended cavities opening into the oral cavity. These are well developed in snakes and some lizards that transfer odor molecules directly from their tongue to the sensory epithelium of the cavity. Jacobson's organs are lacking in crocodilians and are poorly developed in many other reptiles.

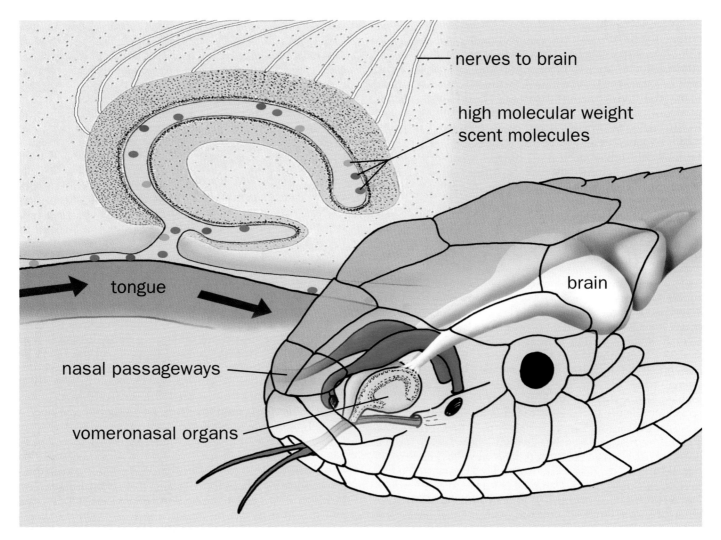

Jacobson's organ, a second kind of olfactory chamber, is well developed in snakes and some lizards. (Illustration by Dan Erickson)

Big-headed turtles (*Platysternon megacephalum*) have such large heads that they do not fit back into their shells. To protect their heads from predators, they have hard horny scutes on the top and sides of their heads and their skulls are solid bone. (Photo by Tom McHugh/Photo Researchers, Inc. Reproduced by permission.)

Mechanoreceptors

A variety of mechanoreceptors are present in reptiles. These structures sense vibrations, sound, and other forms of mechanical stimuli. Specialized structures called muscle spindles are present in the skeletal muscles and detect, as well as control, the stretch of body muscles. Various nerve endings associated with the skin and joints detect a variety of vibrational stimuli. The scales on the body of many lizards and snakes, and on the head of crocodilians, bear small, circular sense organs that are thought to detect vibrations and other mechanical stimuli. Some of these organs appear as flattened plates; others bear complex hairlike structures. The most specialized of mechanoreceptors, however, are those involved in hearing. The inner ear contains collections of specialized receptors called hair cells, named for numerous tiny projections sensitive to movement. These are responsible for mediating the sense of equilibrium and for hearing.

Sound stimuli are conducted through tissues of the head to the inner ear, as in snakes, or impinge on a taut membrane associated with the external ear. In essence, vibrational waves in air bounce off this membrane and set it into vibrational motion in doing so. The vibrations reflect the frequency and intensity of the sound and are transmitted from the external membrane to fluid in the inner ear by means of a small connecting bone. The fluid of the inner ear bathes the hair cells and stimulates their response by means of vibrational waves in liquid. Sounds are an excellent monitor for conditions in the environment and serve primarily as a warning source. Reptiles respond to frequencies from a few tens to several hundred hertz (cycles per second, abbreviated Hz) in tortoises and squamates to more than 10,000 Hz in some species of geckos. Alligators can hear throughout the range of frequencies tested, from approximately 30 Hz to 10,000 Hz, both in air and beneath water. Contrary to popular belief, snakes not only sense vibrations in the substrate on which they rest but also can hear airborne vibrations of low frequencies (especially 150–500 Hz). Relatively few reptiles vocalize or use sounds for communication.

Vision

Most reptiles have well-developed eyes with properties that are generally similar to eyes of other vertebrates. Incoming light is bent by the curvature of a lens that focuses light on photoreceptors arranged on the retina within the eye. An opaque iris controls the entering light by means of a variable aperture called the pupil. Muscles control the diameter of the pupil and help to control the intensity of light entering the eye and the quality of the image on the retina. There are many other structural similarities to a camera.

Photoreceptors consist of rods and cones. These structures bear protein molecules that capture light energy and convert it to nerve signals. The rods function best in dim light, whereas the cones function best in bright light and provide higher resolution. Varying abilities to differentiate color depend on the possession of multiple visual pigments, each of which absorbs maximally at different wavelengths of light. Some reptiles, such as arboreal snakes, have "keyhole" pupils, which enhance binocular vision (similar images are formed simultaneously on both retinas of the two eyes), and a fovea, where high densities of cones on the retina provide high visual acuity. Slender head shape, especially an attenuated snout, confers considerable overlap of vision in the two eyes. The eyes of chameleons are unique among vertebrates in their degree of movement and ability to scan the environment. Each eye is located on a turret and moves independently of the other. The lens of the chameleon's eye focuses very rapidly and produces enlarged images on a comparatively large retina with high densities of cones, acting somewhat like a telephoto lens.

Most reptiles have movable eyelids and a transparent nictitating membrane. Snakes and some lizards lack movable eyelids and have instead a transparent window that covers the eye. Eyes are greatly reduced in snakes and lizards that burrow in soil.

In addition to eyes, most reptiles have a single photoreceptive structure, a parietal organ, on the mid-dorsal aspect of the head and brain and associated with the pineal complex of the brain. Lizards and tuatara have a distinctive parietal organ called a third eye, which is equipped with a lens and a retina. These organs are thought to be ancient structures that evolved as accessory sensory systems sensitive to visible radiation. The parietal organs of living reptiles are photosensitive and appear to be involved with circadian or seasonal cycles and possibly with aspects of thermoregulation. Light-sensitive receptors are thought to be present on the skin of the tails of certain sea snakes.

Thermoreception

Sensory nerve endings in the skin and possibly other tissues are sensitive to temperature change. Extraordinarily sensitive receptors are present in the facial pits of pitvipers and the labial pits of boid snakes. A pit organ consists of a thin membrane stretched across an open cavity. The membrane is richly supplied with nerve endings that have remarkable sensitivity to temperature change. The nerve endings detect infrared or radiant heat energy, which affects the heat-sensitive membrane. The design allows rapid detection of temperature changes as small as 0.005°F (0.003°C). The temperature receptors lie deep within the facial pit and enable a snake to detect the direction as well as the intensity of a radiant heat source. The information is relayed to the brain via nerve tracts and assists the hunting and capture of prey under conditions of dim light.

Resources

Books

Gans, Carl, and Abbott S. Gaunt, eds. *Biology of the Reptilia.* Vol. 19, *Morphology G, Visceral Organs.* Ithaca, NY: Society for the Study of Amphibians and Reptiles, 1998.

Gans, Carl, and David Crews, eds. *Biology of the Reptilia.* Vol. 18, *Physiology E, Hormones, Brain and Behavior.* Chicago: University of Chicago Press, 1992.

Gans, Carl, and F. Harvey Pough, eds. *Biology of the Reptilia.* Vol. 12, *Physiology C, Physiological Ecology.* London: Academic Press, 1982.

———, eds. *Biology of the Reptilia.* Vol. 13, *Physiology D, Physiological Ecology.* London: Academic Press, 1982.

Gans, Carl, and Philip S. Ulinski, eds. *Biology of the Reptilia.* Vol. 17, *Neurology C, Sensorimotor Integration.* Chicago: University of Chicago Press, 1992.

Hildebrand, Milton, et al., eds. *Functional Vertebrate Morphology.* Cambridge: Belknap Press of Harvard University Press, 1985.

Kardong, Kenneth V. *Vertebrates: Comparative Anatomy, Function, Evolution.* 3rd ed. New York: McGraw-Hill, 2002.

Kluge, Arnold G., et al. *Chordate Structure and Function.* 2nd ed. New York: Macmillan Publishing Co., 1977.

McNab, Brian Keith. *The Physiological Ecology of Vertebrates: A View from Energetics.* Ithaca, NY: Cornell University Press, 2002.

Perry, Stephen F. "Structure and Function of the Reptilian Respiratory System." In *Comparative Pulmonary Physiology: Current Concepts.* Vol. 39, edited by Stephen C. Wood. New York and Basel: Marcel Dekker, 1989.

Pough, F. Harvey, et al. *Herpetology.* 2nd ed. Upper Saddle River, NJ: Prentice Hall, 2001.

Periodicals

Andrews, Robin M., and F. Harvey Pough. "Metabolism of Squamate Reptiles: Allometric and Ecological Relationships." *Physiological Zoology* 58 (1985): 214–31.

Angilletta, Michael J., Jr., Peter H. Niewiarowski, and Carlos A. Navas. "The Evolution of Thermal Physiology in Ectotherms." *Journal of Thermal Biology* 27 (2002): 249–68.

Burggren, Warren W. "Form and Function in Reptilian Circulation." *American Zoologist* 27 (1987): 5–20.

Huey, R. B., and M. Slatkin. "Cost and Benefits of Lizard Thermoregulation." *Quarterly Review of Biology* 51 (1976): 363–84.

Lillywhite, H. B. "Gravity, Blood Circulation, and the Adaptation of Form and Function in Lower Vertebrates." *Journal of Experimental Zoology* 275 (1996): 217–25.

Ruben, John A. "Morphological Correlates of Predatory Modes in the Coachwhip (*Masticophis flagellum*) and Rosy Boa (*Lichanura roseofusca*)." *Herpetologica* 33 (1977): 1–6.

Secor, Stephen M., and Jared M. Diamond. "Evolution of Regulatory Responses to Feeding in Snakes." *Physiological and Biochemical Zoology* 73 (2000): 123–41.

Stevenson, R. D., C. R. Peterson, and J. S. Tsuji. "The Thermal Dependence of Locomotion, Tongue Flicking, Digestion, and Oxygen Consumption in the Wandering Garter Snake." *Physiological Zoology* 58 (1989): 240–52.

Wikelski, Martin, Victor Carrillo, and Fritz Trillmich. "Energy Limits to Body Size in a Grazing Reptile, the Galapagos Marine Iguana." *Ecology* 78 (1997): 2204–17.

Wikelski, Martin, and Corinna Thom. "Marine Iguanas Shrink to Survive El Niño." *Nature* 403 (2000): 37.

Harvey B. Lillywhite, PhD

·····

Behavior

Behavior consists of all an animal's reactions to messages received by the central nervous system from any of the several sensory systems of the body. Herpetologists have been most interested in behaviors that have clear functional significance, such as behaviors involved in obtaining food, avoiding danger, finding mates, thermoregulating, and moving between microniches at alternate times of the active season (e.g., migration from hibernacula to feeding grounds and vice versa, migration from feeding grounds to oviposition sites and vice versa, and moving between feeding grounds).

For many vertebrates, vision is the most vital of all the senses, followed generally by hearing. In most reptiles, however, chemical sensitivity is as vital as vision, or more so. As in other vertebrates, olfaction is mediated by the paired nasal organs, which open to the exterior through the nostrils. In reptiles, however, additional mediators of chemical sensitivity are the paired vomeronasal organs, the openings of which are at the anterior extremity of the roof of the mouth, close to the nasal organs. In general, the nasal organs are sensitive to airborne or volatile chemicals, whereas the vomeronasal organs are sensitive to substrate-borne or nonvolatile chemicals. Hearing air vibrations varies in importance across reptilian taxa, although there is a general sensitivity to low-frequency vibrations propagated through the substrate. Touch and taste receptors exist in reptiles, and they play important roles in mediating certain behaviors, but considerable variation exists across species. Sensitivities to polarized light, infrared radiation, and geomagnetism are known to play significant roles in some taxa, but analysis of these sensory processes is in its infancy.

Behaviors guided by chemical senses

Chemical guidance of predatory behavior has been extensively studied in snakes and lizards. Many species exhibit attack and ingestive behaviors on presentation of chemical cues derived from prey. In these experiments, the chemicals typically are presented on cotton-tipped applicators in the absence of any visual or tactile cues associated with prey. Therefore we can be certain that chemicals alone are responsible for triggering the predatory actions. This does not mean that garter snakes (*Thamnophis sirtalis*), for example, pay no attention to other stimuli arising from prey (fish, sala-

manders, frogs, worms) under natural conditions. We know that cues such as movement are vital for attracting the snakes' interest. Yet if investigatory behavior does not bring the snake into contact with appropriate chemical cues, the snake is unlikely to bite the target. Visual cues may command attention and lead to careful inspection, whereas chemical cues trigger the final consummatory acts. Of considerable interest is that neonatal garter snakes exhibit such responses to certain chemical cues before having any experience with prey. This finding leads to the conclusion that responses to these cues are innate. Furthermore, neonatal garter snakes born in different parts of the geographic range respond most strongly to different prey extracts. This finding indicates that subpopulations of snakes have experienced differential selection based on variation in prey abundance. In general, the chemical cues to which snakes respond most strongly are those associated with prey that happen to be available in the snakes' habitat. Changes in prey populations are followed by changes in prey recognition mechanisms within the snake population.

Because reptiles possess nasal and vomeronasal chemosensory systems, herpetologists have been keen to learn the relative contributions of these two systems to predatory behavior. Experimenters have developed various techniques for blocking one or both of the systems to study predatory behavior. Garter snakes have been subjects of most of these investigations, although a few studies have involved western rattlesnakes (*Crotalus viridis*). The common result has been that prey recognition remains undisturbed when the nasal system is blocked but that this behavior almost disappears when the vomeronasal system is blocked. In some particularly elegant experiments, the blocks were reversible, and restoration of the vomeronasal system was followed by a return of the abilities to recognize and respond appropriately to prey. This body of research leaves no doubt about the importance of the vomeronasal organs.

The role of the nasal system and its interaction with the vomeronasal system remains a matter of speculation. Most investigators believe that the nasal system is extremely sensitive but not particularly discriminatory, whereas the reverse is true of the vomeronasal system. The nasal system is thought to serve an alerting function. It informs the snake that something of interest is nearby and sets the vomeronasal system in motion in the form of tongue flicking. Vomeronasal exami-

Reptilian visual displays: 1. Cottonmouth uses gaping mouth as a defensive warning; 2. Frilled lizard looks larger as a defensive display; 3. A ring-necked snake draws attention away from its head and shows its coloration as a defense; 4. The alligator snapping turtle uses a food lure to attract its prey; 5. and 6. Territorial or mating displays for conspecifics—green anole (5) and tuatara (6). (Illustration by Dan Erickson)

nation can identify the molecules in question and activate the appropriate behaviors. In this view, nasal olfaction functions as vision or detection of vibrations does in that ambiguous stimulation of any of these senses can activate tongue flicking so that the animal can conduct a definitive examination. Unambiguous stimulation of any sense, usually by an approaching predator, typically activates immediate escape without the need for cross-modal verification. Detection of potential prey is frequently an ambiguous matter because of prey crypsis (camouflage) and consequent reduction in intensity of stimuli and requires the synergistic action of multiple sensory systems. In the latter context, the vomeronasal system performs the "gold standard" test, and the other senses invite the vomeronasal system to conduct the test.

The vomeronasal system plays an equally important role in the reproductive behavior of garter snakes. In experiments, males with impaired vomeronasal systems did not follow the trails of estrous females and did not attempt to court or copulate with such females when they encountered them. When the vomeronasal systems of the males were restored, normal sexual behavior reappeared, including the ability of the males to follow trails deposited by females. Male garter snakes

Egg laying strategies. 1. Peninsula cooter turtle; 2. American alligator; 3. Python; 4. Copperhead. (Illustration by Dan Erickson)

differentiate the trails of conspecific and heterospecific females only through chemical cues. Males can differentiate large and small (but reproductively mature) conspecific females with chemical information alone. In experiments, males preferred chemical cues from large females over those of smaller females, a fact probably correlated with the greater number of eggs produced by large females. In other words, mating with larger females produces greater fitness benefits for males than does mating with smaller females, and males reflect the effect of this selective pressure in their chemosensory preferences. When the length of females was controlled but mass was varied, males preferred the heavier females and did so when only chemical information was available. Male garter snakes apparently can use chemical information to discriminate females of varying nutritional conditions. This capability probably is associated with an effect of nutritional condition on quantity and quality of eggs produced by females.

Although few other reptile species have been studied in as much detail as have garter snakes, especially *T. sirtalis*, we know that male skinks can do some of the things that male garter snakes can do. In particular, male skinks can differentiate conspecific and heterospecific females with only chemical cues. Likewise, female skinks can differentiate chemical cues of conspecific and heterospecific males. It seems probable that male skinks also are able to differentiate conspecific females of varying size and condition, but these phenomena have not yet been tested. Nor has anyone tested whether fe-

male skinks or garter snakes can differentiate chemical cues derived from males of varying size or condition. We assume the vomeronasal systems of garter snakes and skinks are responsible for discrimination, but this has not yet been demonstrated in experiments that block the vomeronasal system while the olfactory system is unimpaired.

Chemical cues associated with predators are detected by various reptiles. The result is avoidance or other self-protection reactions. We cannot yet be certain that the vomeronasal system mediates these reactions because appropriate blocking experiments have not been conducted, but this is a reasonable hypothesis.

Some reptiles have been shown to discriminate between their own chemical cues and those of conspecifics. Some male reptiles can differentiate male conspecifics on the basis of chemical cues. Experiments along these lines have been done with rattlesnakes, desert iguanas, and sand swimming skinks. It is likely the abilities will be found to be widespread. A number of benefits can be derived from individual recognition, including the interesting social dynamic known as the "dear enemy" phenomenon. Common among territorial birds, the phenomenon consists of individual recognition by neighboring territorial males. These animals respect the boundaries between their territories and do not intrude on each other; thus each animal is allowed to relax the level of vigilance and aggressive behavior that would otherwise be devoted to boundary patrolling and defense. If one of the males is replaced with a new individual, the remaining neighbor exhibits an immediate elevation in vigilance and defense, revealing that he notices that his former dear enemy is no longer present. Eventually the original territory owner and the new neighbor enter a dear enemy relationship, so that each can save energy and avoid injury from fighting. The dear enemies enjoy mutual benefits by respecting each other's property rights. Because individual recognition is clearly an important component of this behavior, the dear enemy phenomenon is generally thought to be an advanced form of social interaction. This does not mean that it is found only in birds and mammals; the behavior has been shown to occur in salamanders and lizards. Whether it occurs in any species of snake, turtle, or crocodilian is conjectural at this time, although a chemosensory basis for the phenomenon clearly exists in at least some species.

Perhaps the most intriguing chemosensory behaviors of reptiles involve venom. Venom immobilizes and kills prey and contains powerful enzymes that greatly facilitate digestion. In experiments, rattlesnakes given a choice between envenomated and nonenvenomated mice otherwise equal in size, age, and sex selected the envenomated prey more frequently than they did the nonenvenomated prey. This remained true when the mice were wrapped in dark nylon mesh that blocked visual or tactile cues; thus we can be certain that chemical cues mediated the selection. When rattlesnakes were given a choice between a trail deposited by an envenomated mouse and one deposited by a nonenvenomated mouse, the snakes reliably selected the former trail, a finding that again indicates chemical cues mediate the behavior.

To understand how sensory bias might contribute to a snake's fitness, it is necessary to understand that rattlesnakes and many other venomous species are ambushers that strike and release adult rodent prey. This style of predation avoids injuries that could occur if struggling rodents were held in the snake's jaws after the strike. Although the venom kills the prey, the process takes as long as several minutes, during which the rodent's teeth, claws, and guard hairs can inflict serious damage. Releasing the rodent after the envenomating strike minimizes the snake's risk of injury, but the snake risks losing the prey, which can travel several meters from the site of attack while the venom takes effect. Recovering the rodent carcass becomes difficult, partly because the immobilized rodent no longer emits motion cues to attract the snake's attention and partly because the rodent may be behind objects or in a burrow, so thermal information (i.e., infrared radiation detectable by pit organs) is obscured or blocked entirely. Recovery of the rodent carcass therefore is mediated by chemical cues detected by the vomeronasal system. A rattlesnake that has delivered a successful predatory strike is capable of following the rodent's chemical trail with exactitude, even though the same snake usually does not follow such a trail in the absence of a predatory strike. The strike is necessary to trigger chemosensory searching and trail-following behavior. This is true for rattlesnakes; cottonmouths (*Agkistrodon piscivorus*), however, follow trails effectively whether or not a predatory strike has been delivered before the snake encounters the trail. Gila monsters (*Heloderma suspectum*) and Australian sand goannas (or monitors) *Varanus gouldii*, behave as cottonmouths do. For these

predators, chemical cues on the substrate are sufficient to induce trail following, and this behavior may be associated with a propensity to search for carrion. For rattlesnakes and various other vipers, chemical cues on the substrate usually are not sufficient, and a successful predatory strike is critical.

In the case of rattlesnakes, two trails always are present in the poststrike environment, one deposited as the prey wanders into striking range (the preenvenomation trail segment) and one deposited as the prey moves away from the site of attack (the postenvenomation trail segment). Following the wrong, or preenvenomation, segment could result in losing the prey, wasted energy, and vulnerability to predators. It is not surprising that rattlesnakes discriminate the two trails and select the postenvenomation segment. What is the chemical basis for this fascinating discrimination?

Venom is a complex material, containing well over 30 identifiable fractions, some of which are lethal to rodents, some not. The latter fractions are thought to be synergizers or amplifiers, not necessarily harmful in themselves but capable of increasing the damaging effects of other fractions. Finding the particular fractions responsible for the snake's discrimination of pre- and postenvenomation trail segments is almost impossible. One reasonable hypothesis is that proteolytic enzymes, major constituents of venom, are responsible. The idea is that such enzymes break down rodent protein and contribute to immobilization, death, and digestion of prey and that the snakes have evolved a perceptual sensitivity to these effects. This sensitivity facilitates the important task of locating and following the postenvenomation trail segment. In other words, proteolytic enzymes have an initial or primary function, and they have acquired a secondary one because of the snake's ability to perceive some of these primary chemical effects. Another hypothesis is that specialized components have been added to venom, not necessarily because of their

A male panther chameleon (*Furcifer pardalis*) behaves defensively by widely gaping its mouth, extending its gular pouch with its tongue, rearing up on its hind legs, laterally compressing its body, and hissing. These displays are common in the genera *Chamaeleo*, *Calumma*, *Bradypodion*, and *Furcifer*. If this display fails, the chameleon must attempt to flee to avoid predation. (Photo by Ardith Abate. Reproduced by permission.)

Green seaturtles (*Chelonia mydas*) mating. (Photo by Animals Animals ©H. Hall. Reproduced by permission.)

lethal effects but because of their perceptual, trail-enhancing effects. The term "trail marker substance" can be used in recognition of the idea that the venom component has acquired its perceptual role not secondarily but as its primary (and perhaps only) function.

Differentiating these hypotheses hinges on the fact that rattlesnake venom passed through gel filtration columns separates into fractions according to molecular weight and that proteolytic enzymes sort into several fractions while other fractions contain little or none of this material. Therefore it is possible to use these various fractions as experimental injectants. That is, mice given injections of each of these fractions suspended in distilled water can be paired with control (nonenvenomated) mice and presented to rattlesnakes. If proteolytic enzymes are the critical elements, then snakes ought to be able to discriminate mice injected with these enzymes from control mice. If such mice cannot be differentiated from controls, one implication would be that some other venom fraction is the critical one. The result in experiments with western diamondback rattlesnakes (*Crotalus atrox*) has been that proteolytic enzymes do not cause a mouse to be discriminated from a control but that another fraction (containing no proteolytic enzymes) has this effect. The chemical composition of this fraction has not yet been identified, but the indications are that the second hypothesis is correct. This work underscores the subtlety of the chemical cues used by reptiles. Reptiles can detect gross cues, such as those associated with rotting carrion, and they can use remarkably subtle cues in remarkably low concentrations.

Ambush predators must first select an appropriate site, a place which prey are likely to visit or to pass through. Several reptiles have been shown to make this selection on the basis of chemical cues deposited by prey that have recently moved through the site. Prairie rattlesnakes making their ver-

nal migration would stop if they were to encounter fresh chemical cues derived from deer mice (*Peromyscus maniculatus*). Once the rattlesnake occupies an ambushing site, however, it usually waits until a successful predatory strike is delivered before paying any further attention to chemical trails. There are exceptions, of course: Some snakes engage in active foraging for burrows containing the neonates of various small mammals, and this search is undoubtedly guided by chemical cues. Thus strikes are not always necessary to activate chemosensory searching. It seems likely that reptiles of many species can use chemical cues to discriminate large versus small populations of prey and to differentiate sites currently occupied by prey from sites previously occupied but now abandoned. Evidence along these lines has been collected, but a great deal of research remains to be done on the topics of habitat selection and chemical ecology.

Although heavy reliance on the nasal and vomeronasal senses is characteristic of many species of reptiles, particularly the most advanced species, numerous species are less reliant on chemical cues than on visual information. Chameleons are examples not only in their use of visual cues in locating and capturing prey but also in their use of visual cues in social and reproductive behavior. Interspecific variation exists among reptiles in the modalities that mediate important behaviors. Therefore statements about the role of vomeronasal stimulation should not be generalized to all reptiles. The relationships between type of food, behaviors involved in acquiring food, and the sensory modalities used are a major area of herpetological investigation. Herbivorous lizards have been found to behave in surprising ways. These animals not only detect edible plants through chemical cues but also detect and reject other plants on the basis of the presence of defensive compounds. Herpetologists have made theoretically important discoveries regarding the chemical senses of reptiles, but we have probably only scratched the surface.

Behaviors guided by vision

Visual cues, usually in the form of prey movement, have long been known to attract a snake's attention and to elicit predatory responses in insectivorous lizards. Equally interesting is the role of visual information in social and reproductive behavior of lizards. Chameleons exhibit "emotional colors" that involve changes in brightness during agonistic interactions and culminate in victory or submission, each with a characteristic pattern. It is partly through these color changes that chameleons have acquired their protean reputation. Iguanian lizards execute a set of movements, including head bobs, pushups, and dewlap extensions, combined in particular sequences that are species specific not only with respect to sequence but also with respect to cadence. Females of sympatric species can discriminate conspecific from heterospecific males on the basis of these display properties. Males use the displays to advertise territories and to settle boundary disputes with other males.

Results of experiments involving presentation of models to territorial males have established that color patterns and postures contribute to agonistic and courtship activities. These behaviors have been found to be innate, and the development of the behaviors is not influenced by social stimulation. Some display elements of the green anole (*Anolis carolinensis*) are present immediately after hatching, whereas other elements emerge later in ontogeny. Even in the later stages, social stimulation appears unnecessary, because the behaviors usually emerge on schedule in animals that live in social settings or in experimental isolation. This characteristic represents a major difference between green anoles and birds and mammals, both of which typically depend on early social stimulation for the proper development of social signals. Because only the green anole has been studied extensively in this regard, we look forward to comparable projects with additional reptile species. Only when such work is completed will we know whether the developmental characteristics of the green anole are common among reptiles. The indications are that the green anole is a reasonable model for reptiles in general, and many herpetologists believe this to be the case, but more taxa must be analyzed.

Reptiles use visual cues as lures for prey. The alligator snapping turtle (*Macrochelys temminckii*) is probably the best known example because of the wriggling wormlike process attached to the floor of the turtle's mouth. The animal remains motionless with jaws agape while the "worm" attracts fish, which are engulfed by a profound oral snap. Several snakes (e.g., the sidewinder rattlesnake [*Crotalus cerastes*] and the massasauga [*Sistrurus catenatus*]) use their tails as lures, in some cases only during the juvenile stage of life. As the snakes grow, they experience an ontogenetic shift in prey preferences such that caudal luring for lizard or frog prey is abandoned in favor of predation on rodents. Ontogenetic shifts in prey preferences are not uncommon, and they are associated with parallel ontogenetic shifts in habitat preferences and even in diel activity patterns. Some snakes rely on caudal luring throughout life. The best example is the death adder (*Acanthophis antarcticus*), which has a highly specialized tail that strongly resembles a wormlike or grublike creature and is quite attractive to lizards.

Play, learning, and plasticity

Although play behavior occurs in a few species at least under captive conditions, most species do not exhibit this phenomenon. Play observed in reptiles has not been a social phenomenon. It has involved the deployment of foraging, feeding, or other behaviors in unusual, nonfunctional contexts, sometimes aimed at inanimate objects, sometimes at humans. The players have been adults, and their playful activities have been idiosyncratic rather than common among conspecifics.

Exhibition of play behavior as a social phenomenon is a major difference between reptiles and mammals, with birds positioned in between. The usual explanation is that most reptiles have been strongly selected for precocity. Neonates are miniature versions of adults and must function effectively as predators, as avoiders of predators, and as competitors. They have no leisure to acquire behavioral skills during playful interactions with peers or by observing parents.

Even the most precocial (capable of independent activity from birth) birds and mammals have opportunities to learn,

A cornsnake (*Elaphe guttata guttata*) kills a rat by constriction. (Photo by Animals Animals ©Carson Baldwin, Jr. Photo reproduced by permission.)

although this process may be accelerated, as it is in filial imprinting in precocial fowl and analogous social phenomena in some precocial mammals. There are important ontogenetic effects of early social stimulation in these birds and mammals. In the more altricial (immature or helpless at birth) species, similar effects occur over a broader span of time, play being an important context for social learning. In reptiles, by contrast, such effects are less common. The usual observation is that neonates are competent at biologically significant tasks the first time they encounter the task.

This does not mean that learning is unimportant in reptiles, only that it is a less conspicuous aspect of reptile ontogeny than is the case for birds and mammals. For example, several species of turtles develop strong preferences for the foods encountered after hatching. The turtles are capable of accepting many different types of prey, but the types available at the time the turtles hatch becomes favored prey as a consequence of an imprinting-like phenomenon that occurs during early feeding experiences. Because piscine prey species can fluctuate in abundance, even replacing each other owing to normal ecological events, hatchling turtles apparently are better off without strong, innate preferences but with the capacity to form preferences after sampling foods that happen to be present in the posthatch habitat. A strong innate preference for a prey type that happens not to be available could have disastrous consequences, whereas experientially induced preferences would simply adjust the predators to prey currently in the food web.

Only when the composition of the food web is predictable or reliable does it make sense for neonates to have innate preferences. This phenomenon is well known in garter snakes.

Adder snake sidewinding in the Namib Desert. (Photo by David Hughes. Bruce Coleman, Inc. Reproduced by permission.)

Neonates exhibit strong preferences, on the basis of chemical cues, for prey normally present in their habitats. Even here two points are of interest. First, results of experiments have shown that neonates typically exhibit several preferences; they do not prefer only a single prey species. If one type of preferred prey is not present, another will probably be available. Second, although they have strong innate preferences, neonates are capable of acquiring new preferences on the basis of early feeding experience. If none of the preferred prey are available, hungry neonates are flexible enough to adjust to new foods.

The flexibility of adult snakes and other adult reptiles has not been studied experimentally, but the husbandry experience of zoo professionals and that of many hobbyists indicates that some species readily adjust to captivity and to the foods normally provided there, whereas other species make this transition only with great difficulty. Most rodent-eating snakes typically do well in captivity, even when captured as adults. The same is true of fish-eating species, especially if rodents are facultatively present in the natural diet, such that the snakes can be switched to rodent prey in captivity. Snakes with highly specialized diets are typically difficult to keep in captivity, but this problem usually is associated with the fact that the keeper has difficulty obtaining the necessary prey in sufficient quantity.

Even notoriously difficult species can be kept in captivity, if specialized habitat factors and required foods can be provided. This judgment is based far less on scientific data than on accumulated experience of gifted, dedicated keepers. Our experience with the viper boas (genus *Candoia*) is consistent with this point of view. Species that take rodents adjust well to captivity, whereas species that take lizards are quite difficult to keep, unless appropriate lizard prey are available. When such prey can be offered, the snakes accept them and remain in good flesh. Lizard prey, however, are not regularly obtainable, and because the lizard-specializing snakes do not readily switch to rodents, the snakes generally become thin and vulnerable to infections that occur as a secondary effect of nutritional compromise. This outcome is a shame because

lizard-eating viper boas have pleasant personalities, being tolerant of handling without offering to bite. They could be easily maintained in captivity if the food problem could be solved. Commercially produced rodent prey are widely available in the United States. The rodents are generally free of infection and relatively free of parasites, especially dangerous ones. Wild-caught lizards are sometimes infested with microbiotic and macrobiotic organisms that can readily colonize snakes with disastrous consequences. This is another reason that rodent-feeding snakes have an advantage in captivity.

Although rodent-feeding snakes usually are easy to keep in captivity, there are exceptions. One of the most famous is the eastern diamondback rattlesnake (*Crotalus adamanteus*). Neonates born in captivity do fairly well, but wild-caught adults rarely thrive, even under the best husbandry conditions. This fact is all the more interesting because western diamondbacks are no less aggressive and no less likely to remain aggressive in captivity, yet they generally do reasonably well, accepting prey readily and breeding with alacrity when given the opportunity under appropriate thermal conditions. Neonatal western diamondbacks born in captivity usually are calmer than wild-caught adults, but the latter usually manage to do well in captivity even with their notorious dispositions. Why wild-caught adult eastern diamondbacks and western diamondbacks differ in their tolerance of captivity remains a mystery that probably involves a difference in behavioral and emotional plasticity. Unraveling this mystery will not only shed light on the dynamics of these species but also contribute to the ability to manage many other species in captivity. Problems similar to those among wild-caught eastern diamondback rattlesnakes have been reported in many other species. Providing appropriate habitat may be part of the solution. Many keepers report success with defensive individuals if hiding places are provided. To our knowledge, experimental tests have not been performed, and we see here an excellent example of the interface between behavioral research and the development of good husbandry techniques.

Behaviors guided by tactile cues

Tactile cues are important contributors to social and reproductive behaviors. Although chemical cues usually guide males to females, once the individuals meet, tactile information comes into play. If several males are simultaneously attracted to the same female, male combat is likely in some species. The winner is the male that eventually mates with the female. In some species, many males are present simultaneously, all competing for one female. This process is called "scramble competition" rather than male combat and leads to formation of the famous mating balls, which can contain dozens or even hundreds of males. Garter snakes (*T. sirtalis*) in some parts of their range form such balls.

Less chaotic mating arrangements are more common, a few males courting one female. In this circumstance, ritualized combat is likely to occur, as in some rattlesnake species. Two males engage each other in a pushing contest involving the anterior parts of their bodies, which usually are raised off the ground, each male pressing against the other. The winner is able to press his opponent down to the ground and hold him there for a few seconds or longer. Losers of such contests typically retreat and are sexually refractory for days afterward. Although the contest requires effort, it seems most unlikely that physical exhaustion can be the explanation for the loser's withdrawal and refractoriness. Instead, a neuroendocrine mechanism must exist by which losing suppresses subsequent aggressive and sexual activities, at least temporarily. This mechanism may benefit both winner and loser in that the winner mates with no further distraction and the loser manages to avoid serious damage. An interesting characteristic of male combat among vipers is that males usually do not bite each other during the ritualized wrestling activity. Because the snakes are not immune to their venom, the vipers could inflict serious damage through fang punctures and venom injection. By replacing such actions with ritualized wrestling, both males benefit. By withdrawing after defeat, the loser spares himself further domination as well as, perhaps, a biting attack by the winner.

Species in which male combat occurs typically exhibit sexual dimorphism in body size, males being larger than females. The idea is that combat creates a selective pressure favoring large males, because size correlates with winning versus losing. On the other hand, females should generally be selected for large body size because this factor determines clutch size. Within the limits of opposing selective pressures, females should generally be large. Males also are large, even larger than females, if male combat is important for mating success. If male combat does not occur in a species, males are smaller than females. These generalizations appear to hold over a broad range of species.

Sexual dimorphism in size is relatively common among reptiles, as is sexual dichromatism. In dichromatic species, males usually are more brightly colored than are females, a fact that correlates with the males' need to defend and advertise territories and to attract females. There are some spectacular cases among the spiny lizards (genus *Sceloporus*) in which both males and females have brilliant (but different) color patterns. These cases are not well understood, but they constitute an extreme form of dichromatism in which both sexes may be sacrificing crypsis to communicate sexual identity. It is possible, however, that future research will reveal that one or both of these patterns, although perceived as brilliant and conspicuous to human observers of preserved specimens, are nevertheless cryptic in the natural habitat.

Because male combat is a form of tactile stimulation, (1) delivery of this stimulation has become highly ritualized in some snakes and lizards, (2) the "meaning" of the stimulation appears to depend on differential neuroendocrine events in winners and losers, (3) delivery of the stimulation appears to inhibit tissue-damaging bites, at least for a while (it is possible that escalated aggression, including bites, could occur if the loser fails to retire), and (4) this form of tactile stimulation has created a selective pressure favoring large body size and increased ability to deliver the tactile information. Females benefit when they mate with males who have won combat encounters and are, by this measure, superior males. Sons of such males are presumed to inherit the size and strength of their fathers. Daughters also may benefit from genes fa-

Bearded dragon lizard (*Pogona barbata*) secures a grasshopper for a meal in Queensland, Australia. (Photo by Jen and Des Bartlett. Bruce Coleman, Inc. Reproduced by permission.)

voring large size. Although females can benefit by doing nothing, that is, by simply letting males fight and then mate with the winner, it has been reported that females sometimes add an element of their own. Females of several pitviper species raise their heads above the ground when a male approaches. This is the same behavior that combating males use at the start of their contests. The female may be making an aggressive gesture that has the effect of differentiating dominant and subordinate males. Males that have recently lost in combat are intimidated by this action on the part of a female, whereas males who have recently won in combat are not intimidated and proceed with courtship. Females can easily discriminate these two types of males on the basis of the males' behavior immediately after the female presents a combat intention display. This is the hypothesis currently held by a number of herpetologists who study the reproductive behavior of vipers.

Courtship and copulation in nonvenomous species have been studied in detail, and a useful terminology has been developed. Precourtship behaviors are those by which the potentially large distances between male and female are reduced. These behaviors usually involve the following of female pheromone trails by males. In some species, males may also use a head-raised posture to search for visual cues arising from females. An important role of visual cues has been established for ratsnakes (genus *Elaphe*) and is believed to exist in other genera as well. Males are attracted by the visual cues, and when the male is relatively near the female, chemical cues become readily available regarding the female's specific and sexual identity and her state of sexual readiness.

On making contact with a receptive conspecific female, the male initiates a series of behaviors typically grouped into three phases: tactile-chase, tactile-alignment, and intromission and coitus. At first the male exhibits a high rate of tongue flicking, during which he is apparently confirming the information he has previously obtained during the trail-following period. Direct contact with sexual pheromone along the female's dorsal surface likely intensifies the male's sexual motivation, leading him to rub his chin in a linear series of jerky movements along the female's back. A female might flee at this point, and the male usually follows, sometimes attempting to mount the female. This tactile-chase phase continues until the female remains stationary and allows the male to mount along part or all of her dorsum. The most conspicuous act usually is chin rubbing by the male along the female's back, advancing toward her anterior, moving posterior, and starting forward again. In the family Boidae, in which males have vestigial pelvic structures in the form of spurs, the tactile-chase phase of courtship entails spurring, which is rapid, oscillating movements of the spurs against the female's body. In some species of rattlesnakes, the male curls his tail around the base of the female's tail and gently massages her by moving his curl over the female's vent and down the length of her tail while repeatedly rubbing his chin along her dorsal, anterior surface. These behaviors may be analogous to the titillation movements by which male pond turtles stroke the heads and necks of their mates with long fore claws.

The tactile-alignment phase contains all of the foregoing behaviors and the juxtaposition of cloacal apertures, a posture that eventually allows copulation. The male and female achieve this important posture entirely with tactile sensation; there is no visual guidance. (Although there is ample evidence of integumentary tactile receptors in snakes and other reptiles, the term "hedonic receptors" is rare or absent in the herpetological literature. This term and various synonyms occur commonly in the mammalian literature, especially the literature on human sexuality, and we suspect it is applicable to the receptors mediating these cloacal alignments during sexual encounters between squamates.) On attaining juxtaposition of the cloacal apertures, the female may exhibit cloacal gaping, and the male may evert one of his hemipenes. At this point, tail-search copulatory attempts occur in which the tails of the partners are engaging each other but the animals' heads are not (i.e., the eyes and other anterior sense organs are not involved). Male garter snakes have been observed to rub their chins on one female while aligning with the cloaca of another female.

Coupling is the eventual outcome of tail-search copulatory attempts. In some colubrid genera, such as *Elaphe*, *Lampropeltis*, and *Pituophis*, it is common for a male to obtain a mouth grip on a female's neck just before penetration and to maintain this grip during coitus. This is a regular feature of mating among lizards (most of the terminology developed by snake researchers can be applied to lizards). One of the most interesting aspects of snake reproduction has to do with the duration of coitus, which ranges from a few minutes in many species to well over 20 hours in others (e.g., vipers and pitvipers). In garter snakes of several species (*T. sirtalis*, *T. sauritus*, *T. butleri*, and probably others), coitus is relatively brief, but a mucoprotein plug forms thereafter in the cloaca of the female that blocks penetration of the female by other males. In vipers and pitvipers, the tendency of the male to remain coupled for extended periods accomplishes the same result as the copulatory plug in garter snakes. On the other hand, the male garter snake is free within 15–30 minutes to seek additional copulation, whereas male vipers and pitvipers are unable to do so for 20–30 hours. Such variation may be associated with several ecological factors, such as the availability of sexually receptive females nearby. If availability is high, it makes sense to uncouple quickly and for the male to initiate courtship with new females. If, however, females are widely spaced and it is likely that several males will be simultaneously attracted to each receptive female, a successful male may benefit more by guarding the inseminated female than by leaving her while he goes on an uncertain hunt for a new female. The density of predators that might take advantage of snakes in copula is another ecological factor that could influence the duration of coitus. A high density of predators leads to selection for brief copulations. These ecological factors have not yet been studied systematically by herpetologists.

Maternal behavior

Little or no interaction occurs between male and female after coitus. With rare exception (male western diamondback rattlesnakes have been observed to remain with females for some days before and after copulation), there have been no scientific reports of long-term pair bonds between males and females in any species of reptile. Females of some species exhibit maternal behavior. Building of nests with attendance of eggs has been reported among king cobras (*Ophiophagus hannah*), Great Plains skinks (*Eumeces obsoletus*), and most crocodilians. Female turtles deposit their eggs in nests, sometimes with elaborate digging and refilling, but no species has been reported to attend the eggs or nest after oviposition and refilling of the nest cavity. Indian pythons (*Python molurus*) have been observed to brood their eggs using shivering movements to generate heat. Although parental (maternal) care of neonates is now well known among crocodilians, early reports of this behavior were doubted until modern researchers developed methods to observe these animals, especially American alligators (*Alligator mississippiensis*) in captivity as well as in the wild. The well-organized maternal behavior of crocodilians, the last of the ruling reptiles, has prompted some paleontologists to speculate about the presence of maternal care in dinosaurs. This topic remains controversial but is gradually being supported by an impressive array of fossil evidence and insightful argument.

New evidence regarding maternal behavior is likely to arise from the study of certain viviparous snakes. Several investigators have reported that female rattlesnakes remain with or very near their litters for several days after parturition. Initially this behavior was considered unimportant, perhaps the result of maternal exhaustion from the act of giving birth. Now it has been hypothesized that maternal attendance of neonates may be more than an artifact of exhaustion. Perhaps the fitness of the neonates is enhanced by the mother's presence, and this means her fitness may be enhanced. The general idea is that the mother's presence may discourage predators who would otherwise take a toll on the neonates. The mother's presence also may allow the neonates to become familiar with her odors and use them to find their way to the hibernaculum. That is, the neonates might locate the den by trailing familiar odors deposited by the mother. None of these ideas has been subjected to rigorous scientific testing, but such tests are likely to be performed.

Activity patterns and thermoregulation

Activity patterns influenced by seasonal, daily, or other rhythms probably have been the targets of more herpetological research than any other aspect of behavior except diet. This is partly a consequence of the development of miniaturized radiotelemetric equipment and partly the result of the simple fact that understanding any animal requires a sense of the actions it performs and when it performs them. Numerous investigators are interested in energy budgets and related phenomena, and activity patterns tell us about most of the energy costs sustained by animals. Feeding success tells us about energy gains. The construction of activity budgets is an important step in the calculation of energy budgets. Many herpetologists consider the study of activity patterns a necessary starting point, once the basic questions of taxonomic allocation are at least tentatively answered. It could be hypothesized that we could predict activity patterns and energy expenditures if we knew the resource needs of a species and the distribution of these resources in the habitat. Although this reasoning is sound, rarely do we know about all the resource needs of a species, and we know even less about their distribution, especially in the perception of the animals. Knowledge of resource needs is an important goal, but instead of using the information to predict activity patterns, we frequently use data on the patterns to infer resource needs and distributions.

The longest migration by any reptile is the oviposition movement of some female green seaturtles (*Chelonia mydas*) from their feeding grounds in the kelp forests off the coast of Brazil to the beaches of Ascension Island in the South Atlantic, about midway between Brazil and the west coast of Africa (approximately 1,600 mi [1,900 km] from Brazil). Other populations of green seaturtles and all other seaturtles make substantial annual movements to nesting beaches, although none of these migrations is as long as the swims of the subpopulation of green seaturtles using Ascension Island. The sensory basis of this remarkable migration involves geomagnetic sensitivity, chemical cues, and probably celestial cues. Even more remarkable is the movement of hatchlings from Ascension Island to the Sargasso Sea and later to the Brazil kelp beds, the turtles again using the same variety of sensory systems. Somewhat less spectacular migrations are known in other reptiles, especially snakes in temperate zones, where hibernation sites and feeding sites are separated by distances on the order of 1–7 mi (1.6–11.2 km). The snakes make an annual vernal migration to the feeding grounds and an autumnal migration back to the hibernacula. Sensory bases for these movements have not been studied sufficiently, but it seems clear that chemical cues are involved, and there is reason to suspect a role for geomagnetic, celestial (especially solar), and visible landmark cues.

Homing has been observed in a variety of reptiles that have returned to sites from which they were experimentally displaced. Alligators, several species of snakes and turtles, and a few species of lizards have been studied, but we know relatively little about the sensory basis of this behavior. Geomagnetic cues were probably used by the alligators, chemical cues were certainly involved in several of the turtle studies, and solar cues were well documented in several of the snake studies. In most other cases, however, our knowledge has not advanced beyond the fact that homing or related behavior (e.g., y-axis orientation) occurred. Some studies have shown no evidence of return to a site occupied before displacement. This lack of evidence gives rise to interesting speculation that the motivation to return depends on the quality of the initial habitat and the quality of the habitat in which the test animals were released. An important implication is that the animals assess their habitats and use this information to decide whether to move toward "home" or to adopt a new home for themselves. If such behaviors occur, the cognitive mechanisms involved in habitat assessment will prove to be even more interesting than the sensory mechanisms mediating orientation and movement toward home. The literature on reptile homing behavior contains many anecdotes that have substantially influenced our collective thinking. For example, rattlesnakes have been removed from a human property (usually a backyard, field, or a common area in a community) and released a distance away (usually on the order of a mile or two [1.6–3.2 km]) only to return to its initial location within a few days. Another interesting case involved a black ratsnake (*Elaphe obsoleta*) removed from a farm several times but always returning within a few days. In cases of this sort, the snakes were not marked at the time of original capture, so it is not absolutely certain that the returning snake and the removed snake were one and the same. On the basis of size, coloration, and unique features such as scars, we can be reasonably confident, but the following case, described by the late Charles M. Bogert, urges caution.

Bogert was collecting reptiles in New Mexico and caught a striped whipsnake (*Masticophis taeniatus*) in a particular bush, which Bogert marked with a golf tee. The next day, while passing the same bush, Bogert caught a second striped whipsnake in it. Later on the second day, Bogert visited the bush again, and, sure enough, a third specimen of the species presented itself. There is no question that the snakes were three different individuals because Bogert caught and preserved each one. He was interested in what might have been special about the bush, but he was able to identify nothing. Perhaps this was pure chance, always a possibility with anecdotes. On the other hand, maybe the bush was in a prey-rich area or in a migration path or in a spot containing an ideal refuge from the sun or from predators. If any of these conditions existed, it would not be surprising that removal of an initial occupant might be followed by the arrival of another. If the first occupant had not been captured and kept out of the habitat, we might easily misinterpret the second and third snakes as being the initial one. Although anecdotal reports of homing should not be discounted, they should be regarded cautiously.

Closely related to homing is territoriality, the defense of a resource, usually in a particular place but sometimes mobile, as in the case of potential mates that might move a considerable distance but are nevertheless defended in each place they occupy. Numerous lizards exhibit unambiguous territoriality, defending feeding areas from conspecifics and sometimes from heterospecifics that compete for the same foods. Clever experiments have shown that adding food to territories results in shrinkage of the area defended, whereas depleting food from territories results in expansion of the area defended. Results of such studies leave no doubt that the territorial behavior of the resident lizard is sensitive to the quantity and quality of the food contained within the territory. The results of the studies also leave no doubt that the lizards assessed their food supplies and responded accordingly. Food, however, is not the only resource that lizards defend. Oviposition sites are important in some species, and refuge is important in others.

Crocodilians are known to behave in a territorial manner, especially mature females who defend their nests. Large males also appear to be territorial, particularly during the breeding season. Turtles and snakes present a far more ambiguous situation, except for the few species in which females attend their eggs or young. Even these cases are not clear examples of territoriality, because there is no evidence that the females defend their eggs or oviposition sites against conspecifics. It is possible that mothers defend their eggs or neonates only against heterospecific oviphages or predators, in which case the term *territoriality* may not be appropriate. Territoriality implies intraspecific interaction. At present we cannot state with certainty that any species of turtle or snake exhibits true territorial behavior. Anecdotal evidence exists for both groups, and behavioral phenomena among captive specimens also are suggestive. The possibility of territoriality among these reptiles should be considered, but caution should be exercised in interpretation of anecdotal evidence.

Another approach to the study of activity patterns of reptiles is to record the number of individuals seen or captured (usually in pitfall traps) during each month of the year. In temperate North America, such projects reveal several annual patterns: a single peak during the warmest months with sharply reduced activity before or after; a single peak during the warmest months but fairly broad activity period such that substantial numbers of individuals are active before and after the warmest months; and two peaks, one in spring and one in autumn. Some species, such as garter snakes, appear to remain active all year if the temperature remains high enough. Other species appear to be endogenously programmed to become inactive during winter months, even if the temperature is artificially elevated. Having maintained many rattlesnakes in captivity, we have found that in species such as the prairie rattlesnake and the western diamondback, some individuals remain active all year if the temperature is kept at 79°F (26°C) or higher. Other individuals "shut down" for several months under the same conditions, refusing food for one to three months each year.

Closely associated with the circannual studies of reptile activity are parallel studies of reproduction. Periods of high activity or capture success can correspond with periods of intense feeding behavior, but they can also correspond with periods of courtship and copulation, particularly because of the likelihood of capturing males active in searching for reproductively motivated females. Pregnant female prairie rattlesnakes, for example, are relatively inactive, perhaps even congregating in birthing rookeries, whereas nonpregnant females have made a vernal migration to hunting grounds. The females that migrate are the ones likely to develop ripe ova and to become sexually motivated. Consequently, they are the ones mature males pursue, making themselves vulnerable to traps or to capture by hand. Although a number of reproductive strategies are observed across the various families of reptiles, most of these strategies involve seasonally increased activity by one sex or the other. It is safe to conclude that circannual activity patterns are partly correlated with ovarian and testicular cycles or with courtship and copulatory seasons in taxa that exhibit a dissociation between reproductive behavior and genital physiology (i.e., some species breed at times of the year different from the times gametes are produced; in such cases, activity is influenced by the breeding season rather than the gamete-producing season).

In addition to circannual studies of reptile activity, there have been many studies of circadian rhythms, that is, changes in activity during the 24-hour period of a typical day in the animal's active season. Some species are strictly diurnal (active only during daylight), others are strictly nocturnal (active only at night), and other species are crepuscular (active at dawn and dusk). Juvenile kingsnakes (*Lampropeltis getula floridana*) exhibit crepuscular and nocturnal patterns, whereas adults are diurnal. The shift toward diurnal behavior occurs when the snakes are approximately 35 in (90 cm) snout-vent length and capable of defending themselves against a variety of predators, particularly birds, that are active during daylight. Nocturnal reptiles exhibit relative cessation of activity during periods of full moon, when ambient illumination at night favors detection by predators.

In some species, daily activity patterns shift with average daily temperature. Among plains garter snakes (*Thamnophis radix*) and western diamondback rattlesnakes, individuals are diurnal at the relatively low temperatures of early spring and late autumn, but during the hottest days of summer, these same individuals become nocturnal. During intermediate parts of the year, such as late spring and early summer, the snakes are crepuscular, because it is too hot during the day and too cold at night. For species with very broad geographical ranges, it is possible that individuals in northern latitudes are diurnal whereas individuals to the south are crepuscular or nocturnal, depending on thermal conditions.

Temperature is a primary modulator of many physiological processes influencing digestion, reproduction, and locomotion in reptiles. Herpetologists have focused on this factor more than any other. Each species has a lethal minimum and maximum temperature, below or above which life ceases immediately. Within this range, which might be 17.6–104°F (−8°C to +40°C), there is a second range from the critical thermal minimum (CT_{Min}) to the critical thermal maximum (CT_{Max}), which might be from 35.6°F to 89.6°F (2°C to 32°C). Survival below the CT_{Min} or above the CT_{Max} is possible for brief periods, longer below CT_{Min} than above CT_{Max}, but death occurs relatively rapidly as the temperature converges with the respective lethal extremes. Yet another range is nested with the boundaries of CT_{Min} to CT_{Max}. This range runs from the voluntary minimum to the voluntary maximum, within which the reptile prefers to remain when choice (i.e., behavioral thermoregulation) is possible. This range defines the animal's normal or preferred activity range, and its midpoint corresponds to the animal's voluntary or preferential mean temperature. The precise temperatures corresponding to these boundaries shift during the course of the year, a phenomenon called acclimation. For example, in early spring, a reptile might prefer to keep its body temperature between 64°F and 82°F (18°C and 28°C), but later in the summer, the voluntary minimum and maximum might shift to 68°F (20°C) and 86°F (30°C), respectively. As this acclimation effect occurs, the lethal minimum, CT_{Min}, lethal maximum, and CT_{Max} all shift upward by 3.6–5.4°F (2–3°C).

One of the dominating necessities of any reptile is to keep its body temperature within the preferred range of its species. Because all extant reptiles are ectotherms, temperature regulation involves a large suite of behaviors, some obvious and some remarkably subtle. Reptiles also exhibit many anatomical and physiological adaptations. Among the more obvious thermoregulatory behaviors are locomotory activities by which the animals shuttle between sunlight and shade as necessary to keep their body temperatures within the optimal range. Basking is another behavior by which body surfaces are exposed to sunlight to increase core body temperature. Reptiles of some species can thermoregulate by altering the amount of surface area exposed to sunlight. They do so with inconspicuous alterations of posture, sometimes associated with color change. Large snakes are known to retain fecal material and to use it for storing heat. If a snake needs to reduce its heat load, one option is to eliminate the feces. Coiling the body and aggregating with other snakes can profoundly reduce the rate of heat loss, such that snakes in winter dens can have higher body temperature than would be indicated by the ambient climate. Because of the many behavioral and other devices they have for regulating body temperature, reptiles can maintain their temperatures within a surprisingly narrow range. The typical standard error of the mean of repeated measurements of an individual reptile's preferred body temperature is less than 1.8°F (1°C). These measurements were made under close to ideal conditions; nevertheless, such studies show how precise reptiles can be when heat sources and refuge are readily available and when no obstructions are placed in the way of locomotion. Although reptiles lack the endothermic mechanisms of birds and mammals and only a few species of reptiles engage in shivering thermogenesis, when solar or other heat sources are available during the active season, many reptiles can maintain high body temperatures with surprisingly little variability.

The temperature at which reptiles maintain their bodies during the active season can be influenced not only by acclimation but also by disease. Infection with pathogenic bacteria can cause the reptile to prefer a higher than usual temperature. This so-called behavioral fever kills the bacteria within a few days. This interesting phenomenon has been found to occur among most ectothermic vertebrates and is straightforwardly analogous to the endogenous fever response of endothermic vertebrates to pathogenic bacteria. Another thermal phenomenon is emotional fever. Lizards handled briefly by humans regulate their body temperature between one and two degrees higher than normal. This elevation may be related to a flight or escape response or to a metabolic response to the immunosuppressive effects of stress. The presence of food in the stomach also induces reptiles to elevate their body temperature and digest the food more rapidly and more thoroughly than would otherwise be the case. Pregnant females prefer higher temperatures, which facilitate gestation. As we discover these fascinating events, our appreciation of the precision of reptile behavior increases, as does our ability to care for these animals in captivity.

Resources

Books

Burghardt, G. M. "Behavioral and Stimulus Correlates of Vomeronasal Functioning in Reptiles: Feeding, Grouping, Sex and Tongue Use." In *Chemical Signals: Vertebrates and Aquatic Invertebrates,* edited by D. Muller-Schwarze and R. M. Silverstein. New York: Plenum Press, 1980.

Gans, C., and F. H. Pough, eds. *Biology of the Reptilia.* New York: Academic Press, 1977.

Halpern, M. "Nasal Chemical Senses in Reptiles: Structure and Function." In *Biology of the Reptilia*, Vol. 18, edited by C. Gans and D. Crews. Chicago: University of Chicago Press, 1992.

Hayes, W. K., I. I. Kaiser, and D. Duvall. "The Mass of Venom Expended by Prairie Rattlesnakes When Feeding on Rodent Prey." In *Biology of the Pit Vipers*, edited by J. A. Campbell and E. D. Brodier. Tyler, TX: Selva, 1992.

Seigel, R. A., J. T. Collins, and S. S. Novak, eds. *Snakes: Ecology and Evolutionary Biology.* New York: McGraw-Hill, 1987.

Periodicals

Aleksiuk, M. "Temperature Dependent Shifts in the Metabolism of a Cool Temperate Reptile, *Thamnophis sirtalis parietalis.*" *Comparative Biochemistry and Physiology* 39A (1971): 495–503.

Andren, C. "The Role of the Vomeronasal Organs in the Reproductive Behavior of the Adder *Vipera berus.*" *Copeia* 1982, no. 1: 148–157.

Bennett, A. F. "The Thermal Dependence of Lizard Behavior." *Animal Behavior* 28 (1980): 752–762.

Brattstrom, B. H. "The Evolution of Reptilian Social Behavior." *American Zoologist* 14 (1974): 35–49.

Brown, W. S., and F. M. MacLean. "Conspecific Scent-trailing by Newborn Timber Rattlesnakes, *Crotalus horridus.*" *Herpetologica* 39 (1983): 430–436.

Carpenter, C. C. "Communication and Displays of Snakes." *American Zoologist* 17 (1977): 217–223.

Resources

Chiszar, D., A. Walters, J. Urbaniak, H. M. Smith, and S. P. Mackessy. "Discrimination between Envenomated and Nonenvenomated Prey by Western Diamondback Rattlesnakes (*Crotalus atrox*): Chemosensory Consequences of Venom." *Copeia* 1999, no. 3: 640–648.

Cooper, W. E., Jr. "Correlated Evolution of Prey Discrimination with Foraging, Lingual Morphology, and Vomeronasal Chemoreceptor Abundance in Lizards." *Behavioral Ecology and Sociobiology* 41 (1997): 257–265.

Devine, M. C. "Copulatory Plugs, Restricted Mating Opportunities and Reproductive Competition among Male Garter Snakes." *Nature* 267 (1977): 345–346.

Dill, C. D. "Reptilian Core Temperatures: Variation within Individuals." *Copeia* 1972, no. 3: 577–579.

Duvall, D., K. M. Scudder, and D. Chiszar. "Rattlesnake Predatory Behaviour: Mediation of Prey Discrimination and Release of Swallowing by Cues Arising from Envenomated Mice." *Animal Behavior* 28 (1980): 674–683.

Ford, N. B., and J. R. Low. "Sex Pheromone Source Location by Garter Snakes: A Mechanism for Detection of Direction in Nonvolatile Trails." *Journal of Chemical Ecology* 10 (1984): 1193–1199.

Garstka, W. R., and D. Crews. "Female Sex Pheromones in the Skin and Circulation of a Garter Snake." *Science* 214 (1981): 681–683.

Gillingham, J. C. "Reproductive Behavior of the Rat Snakes of Eastern North America, Genus *Elaphe*." *Copeia* 1979, no. 2: 319–331.

Greene, H. W. "Dietary Correlates of the Origin and Radiation of Snakes." *American Zoologist* 23 (1983): 431–441.

Hirth, H. F., and A. C. King. "Body Temperatures of Snakes in Different Seasons." *Journal of Herpetology* 3 (1969): 101–102.

Kardong, K. V. "The Strike Behavior of the Rattlesnake, *Crotalus viridis oreganus*." *Journal of Comparative Psychology* 3 (1986): 314–324.

Krysko, K. L. "Seasonal Activity of the Florida Kingsnake, *Lampropeltis getula floridana* (Serpentes: Colubridae), in Southern Florida." *American Midland Naturalist* 148 (2002): 102–114.

Lowe, C. H., P. J. Lardner, and E. A. Halpern. "Supercooling in Reptiles and Other Vertebrates." *Comparative Biochemistry and Physiology* 39A (1971): 125–135.

Mackessy, S. P. "Venom Ontogeny in the Pacific Rattlesnakes, *Crotalus viridis helleri* and *C. v. oreganus*." *Copeia* 1988, no. 1: 92–101.

Naulleau, G. "The Effects of Temperature on Digestion in *Vipera aspis*." *Journal of Herpetology* 17 (1983): 166–170.

Reinert, H. K., D. Cundall, and L. M. Bushar. "Foraging Behavior of the Timber Rattlesnake, *Crotalus horridus*." *Copeia* 1984, no. 4: 976–981.

Rodda, G. H. "The Orientation and Navigation of Juvenile Alligators: Evidence of Magnetic Sensitivity." *Journal of Comparative Physiology A* 154 (1984): 649–658.

Semlitsch, R. D., K. L. Brown, and J. P. Caldwell. "Habitat Utilization, Seasonal Activity, and Population Size Structure of the Southeastern Crowned Snake, *Tantilla coronata*." *Herpetologica* 37 (1981): 40–46.

Shine, R. "Sexual Size Dimorphism and Male Combat in Snakes." *Oecologia* 33 (1978): 269–277.

Shine, R., and J. Covacevich. "Ecology of Highly Venomous Snakes: The Australian Genus *Oxyuranus* (Elapidae)." *Journal of Herpetology* 17 (1983): 60–69.

Yamagishi, M. "Observations on the Nocturnal Activity of the Habu with Special Reference to the Intensity of Illumination." *The Snake* 6 (1974): 37–43.

York, D. "The Combat Ritual of the Malayan Pit Viper (*Calloselasma rhodostoma*)." *Copeia* 1984, no. 3: 770–772.

David Chiszar, PhD
Hobart M. Smith, PhD

·····

Reptiles and humans

From our earliest days on Earth, humans have interacted with nearly all species of reptiles. Some aspects are considered positive, but the relationship has had a largely detrimental impact on reptiles, frequently affecting survival. Today there is a perception that reptiles are an inexhaustible natural resource. When populations are reduced, however, it becomes readily apparent that reptiles have a significant role in the stability of nature's convoluted web of life. It is difficult to comprehend this topic fully without reference to the broader subject of conservation.

Reptiles as food

There is no doubt that many reptiles have the necessary skills and physical characteristics to protect themselves, but generally they are more sedentary and lethargic and less intelligent and aggressive than large birds and mammals. From prehistoric times these qualities have made them vulnerable to human predation. Reptiles remain important food items for isolated tribes in developing countries throughout the world. Human foragers fulfill their need for scarce animal protein with reptiles when the opportunity presents itself, and in certain situations they actively hunt some taxa. In the developed world, turtle, crocodilian, and rattlesnake meats have found their way into a variety of unusual recipes.

Chelonians

Reptile eggs, particularly those of chelonians (turtles and tortoises), provide excellent nourishment and are sought as delicacies nearly worldwide. Although conservation laws protect turtle eggs from harvesting, thousands, perhaps millions, of eggs are dug up and eaten or sold as food annually. When female turtles come onto land to dig nests, they are particularly vulnerable to human predation. Whether they are seaturtles the size of automobile tires or terrapins the size of frying pans, these creatures' graceful and wiry movements, which make them difficult to catch in water, are valueless when they are on the beach. Turtle hunters gather the smaller species and place them in sacks or pens. Seaturtles that have come onto beaches to lay eggs are flipped onto their backs and left to flounder helplessly, unable to right themselves. These massive animals are too big and bulky to be moved, so they are butchered on the beach. Eggs are scooped into buckets, the flesh is cut into chunks, and organs that are thought

to have medicinal value are stored in suitable containers. Large sections of skin are cut out carefully, but smaller pieces, appendages, and any remaining entrails are dumped on the ground for scavengers. Shells are transported carefully to a safe place, where they are air-dried. Then they are cleaned of all remaining skin, polished, and sold as is. Alternatively, pieces may be carved into an array of collectible curio items. Damaged shells are pulverized and used in folk medicines.

Of the Asian countries, China is the biggest consumer of turtle meat; in fact, the Chinese eat more than all other countries combined. Until the 1990s the average Chinese citizen had scant access to this expensive delicacy. Industrialization and an upgrade in the nation's economic structure changed that; turtle meat now has become available to the masses. The Chinese view turtle meat in the same way as Western countries view beefsteak—a delectable, fairly common source of protein. In 1996, 7,716,000 lb (3,499,900 kg) of turtles (roughly three million animals) were imported and consumed in Hong Kong alone.

In the heyday of sailing ships, it was difficult to keep adequate amounts of food and water onboard during long voyages. Sailors learned that live giant tortoises stayed alive for weeks without the need for food or water and yet retained their weight. Thus, tortoises were viewed as an excellent source of fresh meat and could be butchered whenever they were needed as food. The decimation of Galápagos tortoises (*Geochelone nigra*), Aldabra tortoises (*Geochelone gigantea*), and several other large species throughout the world, brought many taxa to the brink of extinction.

In the United States common snapping turtles (*Chelydra serpentina*), soft-shelled turtles (*Apalone* species), and red-eared sliders are farmed and ranched along with the more sought after but protected alligator snapping turtles (*Macrochelys temminckii*). "Snapping turtle soup" is widely available in restaurants in the eastern and southern states.

Lizards and snakes

Eggs of the common iguana, *Iguana iguana*, are a delicacy in Latin America, where they can be bought hard-boiled in markets. The eggs of other lizard species are also available periodically in markets throughout the developing world. The eating of common iguanas and spiny-tailed iguanas (*Ctenosaura similis*) has been traced to prehistoric times, and these animals were common fare for the Mayans in the mid-1500s. They still can be found skinned and dressed in indoor markets or sold by children at bus stops and along roadways in villages throughout Mexico and Central America.

Iguanas, skinned and cleaned, are a main ingredient in many dishes, including casseroles and stews. They also may be baked or grilled. Pregnant females are gutted, leaving the eggs intact, and roasted as a delicacy. The Catholic Church does not consider reptile flesh to be meat, so during the 40 days of Lent, when eating meat is forbidden, Latin Americans often substitute lizards. In Nicaragua enormous numbers of common iguanas are slaughtered during Easter's Holy Week to prepare a traditional soup, *indio viejo*.

Adult iguanas are in such demand that they fetch as much as $10 per animal. The country has attempted to protect the lizard by listing it as Threatened and placing a ban on collecting and eating from December 1 to March 31, the peak of their reproductive cycle. The fine of 50 *cordabas* ($5.50) per animal is mostly a symbolic threat, and it is rarely enforced. Generally, when collectors are caught holding iguanas, the animals are confiscated and released into the wild.

Lizards are eaten nearly everywhere in developing nations. Generally, smaller lizards and snakes require too much effort to catch for the amount of meat they provide; large water monitor lizards, *Varanus salvator*, are the only species that are considered a primary food source, mainly on a few Malaysian islands. Goannas, Australian varieties of monitor lizards, are skinned, gutted, and broiled on a skewer by some aborigines. Several species are eaten throughout their extensive range in Africa, India, Asia, and Australia. The liver and eggs are particular delicacies.

Islamic law considers turtle, crocodilian, and snake meats *haram* (unclean) and forbids eating them. Lizards are labeled *mushbooh* (doubtful or suspect). African and some Arabian Muslims eat monitor lizards (*Varanus griseus* and *V. niloticus*) and dhabb lizards (*Uromastyx* species), which they call "fish of the desert."

Large snakes have substantial flesh, which is palatable when properly prepared and cooked. For a multiplicity of reasons based on fear, religion, and folklore, snakes are rarely eaten by all but the most protein-starved people. They have found a small niche among predominantly North American and European epicureans searching for unusual food items. All rattlesnake meat is from animals taken at rattlesnake roundups or caught by commercial collectors.

Crocodilians

Crocodilian steaks from animals butchered in their second or third year of life are said to be exceptionally tender and succulent, making them a delicacy in posh restaurants throughout the civilized world. Saltwater crocodiles (*Crocodylus porosus*) are eaten in Australia, with a favored dish being crocodile vol-au-vent. Nile crocodiles (*Crocodylus niloticus*) in southern Africa, saltwater crocodiles in Australia and Malaysia, and American alligators (*Alligator mississippiensis*) in the United States are extensively and successfully farmed.

There are several large facilities that produce tons of meat, skins, and other products. These farms meet most commercial needs and have gone a long way toward making the capture of animals in the wild unnecessary.

Reptile farming and ranching

Certain crocodilians (e.g., the American alligator and saltwater crocodile) and turtles (e.g., the common snapping turtle, *Chelydra serpentina*; red-eared slider, *Trachemys scripta*; and Chinese softshell turtle, *Pelodiscus sinensis*) are farm-raised in immense numbers for food and ancillary products. Using crocodilian and turtle farming as a foundation, attempts have been made to breed other reptiles on a large scale in other parts of the world. Specialty breeders supply the needs of the pet trade for some (mostly expensive) animals, and there are carefully monitored endangered species breeding projects undertaken to assure the continuance of a species and possibly providing stock for reintroduction into depleted areas.

Iguana ranches have been established in Honduras, El Salvador, and Belize to provide a renewable natural food supply as well as skins, but the demand for the pet industry has made them more valuable as live exports. It is hoped that eventually the pet requirements will decrease and the original intent of these farms will be realized.

Establishing similar ranches and farms in developing countries throughout the world has been slow. However, by providing jobs and demonstrating the quantity of animals that can be produced with minimal time and work, local people quickly realize the advantages that captive breeding and animal farming have over taking animals from the wild. The amount of space needed is minimal, and because many of the locations are Neotropical or tropical, climates and temperatures provide optimal conditions for inexpensive high yield harvests of many desirable species. Combined, these factors make reptile farming an ideal cottage industry. With such farms, ecologically minded organizations and governmental factions could readily and inexpensively subsidize the reproduction of rare and endangered animals for possible reestablishment in places where collecting has decimated populations.

American alligators: A successful conservation effort

The accomplishment of commercially ranching American alligators, combined with controlled harvesting of animals from the wild, is a classic case of turning a near ecological disaster into a financially rewarding industry that satisfies the demands of the food, fashion, and folk medicine markets while protecting the survival of wild populations and sustaining an excellent natural balance. It could act as a template for all governments facing the unbridled exploitation of their reptile populations.

Habitat destruction, overharvesting, poaching, and wanton killing forced the U.S. government to take measures to protect the American alligator. In 1963 it became illegal to kill alligators in the state of Louisiana, and in 1967 the alligator received federal protection under the U.S. Endangered Species Preservation Act. This protection was strengthened in 1969 with the Endangered Species Conservation Act, which

A view of holding facilities at a reptile importer. The aquarium cages contain medium sized snakes and lizards, and the shoeboxes (in racks along the wall) contain newborn, mostly captive bred snakes. This facility is unusual in that it is carefully maintained and scrupulously clean. The animals are of the highest quality, have an excellent chance for survival, and will be sold at excellent prices. (Photo by Manny Rubio. Reproduced by permission.)

made interstate shipment of illegally taken alligators or hides a federal offense, and again in 1973 with the passage of the Endangered Species Act, which officially listed the alligator as Endangered. As alligator populations recovered in Louisiana, wildlife officials there began promoting heavily controlled ranching and culling of wild animals in three parishes in 1972, an effort that has become a $54 million annual industry in the state. Landowners quickly realized that supporting legal operations was much more productive, so poachers were actively pursued and the illegal taking of animals has been eliminated. All animals are tagged, allowing the skins to be identified easily as being legally produced. The majority of skins are sent overseas, while the meat is shared between domestic and international markets. Skulls, teeth, toes, and small pieces of skin are sold as curios. Tours of the ranches and related swamp tours are estimated to bring more than $5 million into the state's economy.

Controlled ranching and propagation combined with judiciously monitored nationwide protection and reestablishment of populations in areas where they had been eliminated has enabled American alligators to rebound. In 1987 the U.S. Fish and Wildlife Service, which enforces the Endangered Species Act, moved the American alligator from the Endangered list to the Threatened list. This kind of commitment and involvement on the part of federal and state authorities appears to be necessary to aid in the survival of many other

reptiles. Although farming and captive colonies are major components of conservation, restraining habitat destruction and actively enforcing wildlife laws are primary steps to preventing the extinction of innumerable taxa.

Rattlesnake roundups

The "rattlesnake roundup" stands in contrast to these efforts at conservation. Rattlesnake roundups are unique in the United States in that they are permitted to continue regardless of the serious impact they inflict on habitat and snake populations. Although they are widely publicized, rattlesnake roundups are held in very few states. They are run each spring by private organizations in small, otherwise insignificant towns as a way of making money. The most harmful ones are staged by private organizations in Texas, Georgia, Alabama, and Kansas.

Huge numbers of snakes (as many as 5,000 by some estimates) are taken each year at roundups. Visitors leave these events with the impression that killing snakes is good and that doing so should be promoted, because it is a major means of protecting the public from harm from rattlesnakes. Under the guise of education, patently dangerous demonstrations of free handling and other perilous acts are presented. Reckless participants who are bitten in the "quick bagging" competition,

An elderly participant at a Fundamentalist Appalachian Church holds a timber rattlesnake (*Crotalus horridus*) aloft as a sign of his commitment. He stands, nearly entranced, having been consumed in his belief that by holding the venomous snake he is overcoming evil. The intense state and "taking up of serpents" comes after considerable communal praying, chanting, and dancing. It is only after he has been "anointed by the Holy Spirit" that he has the strength for the confrontation. (Photo by Manny Rubio. Reproduced by permission.)

where snakes are grabbed freehand and stuffed into sacks, receive the coveted "White Fang" award and necessary medical attention. This so-called honor has been bestowed on as many as 30 contestants in a single year.

The snakes, nearly all eastern diamondback rattlesnakes (*Crotalus adamanteus*) in Georgia and Alabama, western diamondback rattlesnakes (*Crotalus atrox*) in Texas, and prairie rattlesnakes (*Crotalus viridis*) in Kansas, are mistreated from the moment of their capture. The unspoken rationale is that they will be killed anyway.

At the event, the snakes are weighed, pinned with a hook, stretched to be measured, milked of venom, and placed in a trash can. When the can is filled with snakes, it is taken to a pen, where the snakes are dumped out, piled on one another. They bite each other, and many catch or break their fangs in the wire mesh or strike the transparent plastic sides of the enclosure. A number of them die in the snake pit from this mistreatment. At a few Texas events the snake's future is ended onsite. For $5.00 an attendee can chop the head off a restrained live snake. Hundreds of others are beheaded, strung up, skinned while still writhing, gutted, filleted, and deep-fried. A few more dollars buys a meal of freshly fried rattlesnake,

coleslaw, and cornbread. All the other snakes are sold to one of several specialty butchers (for as much as $12.00 a foot) who slaughter them, producing meat, skins, heads, and rattles. After tanning, the skins are turned into wallets, belts and buckles, boots, vests, skirts, bikinis, and an array of curios. Some snakes are freeze-dried in a defensive position with the mouth agape and fangs erect and sold for $100 or more in gift shops.

Frequently, rattlesnakes destined for these events are taken from dens and hibernacula during the winter. Once a den entrance is located, gasoline is poured through a long flexible tube and forced into the depths of the hibernaculum, creating lethal fumes that drive the snakes to the surface, where they are captured. The 80 or so forms of animal life that have been recorded to cohabit in these retreats likely also become disoriented and suffocate. No one knows how or even if the noxious fumes dissipate, but the niche remains uninhabitable during this time. Rattlesnake roundups are flagrant examples of reptile exploitation at its worst. If any other animals were the brunt of these unconscionable acts (e.g., rabbits, feral dogs or cats), the roundups would be banned without delay.

Folk medicine

Shamans (medicine men or women) in developing countries claim uncountable natural sources for curing myriad maladies and diseases, from common colds to cancers. The efficacy of these "curatives" is suspect, but ethnobotanists and ethnobiologists—often at the behest of pharmaceutical companies—travel to the most remote parts of the world in an attempt to assess such sources as well as identify previously unknown natural chemical compounds. For the most part folk remedies are derived from plants. Nevertheless, animal organs, often from endangered taxa, also find their way into folk prescriptions.

Snakes as major ingredients

Practitioners of Asian medicine use reptiles, mainly snakes, extensively, because they supposedly have wide-ranging medicinal powers. Rare venomous snakes are constantly in demand, making folk medicine a lucrative market, worth many millions of dollars annually. The trend toward holistic remedies has brought worldwide acceptance of traditional oriental medicine. A three-volume pharmacopoeia, *Chinese Materia Medica*, has been the major source of such remedies for centuries. It is updated periodically, and the newest edition shows a trend toward conservation; ingredients from less endangered species are suggested as substitutes for those from seriously endangered animals and plants.

Rubbing snake skin on an affected area is said to treat superficial but chronic problems, such as acne, psoriasis, boils, carbuncles, hemorrhoids, eye infections, and sore throats. A snake's agility and speed are thought to indicate that potions containing snake skin will act rapidly. Snake oil has been sold as a cure-all in Mexico and rural parts of the United States for more than a century. Romans used it as hair restorer, but the Chinese avoid it because it supposedly promotes impotence.

Some Asian cultures believe that eating snake meat improves eyesight. In some places the flesh of venomous species is eaten to treat paralysis, epilepsy, and hypertension. For nearly a thousand years Mexicans have used dried and pulverized rattlesnake for almost every ailment, including terminal diseases like AIDS (acquired immune deficiency syndrome) and cancer. According to Chinese practitioners, gallbladders have been known to have phenomenal medicinal properties as far back as 500 A.D. Crushed gallbladders and bile are taken as a tonic for rheumatism, high fever, whooping cough, infantile convulsions, hemorrhoids, bleeding gums, and skin disorders. Bile is eaten with rice or rice wine as a nourishing, energizing appetite stimulant. It tastes sweet, very much like anise or licorice. Five major Chinese suppliers claim to sell a combined 22,000 lb (10,000 kg) of snake gallbladders each year.

In some Asian cultures, particularly China, mixing two or more ingredients, especially those of venomous snakes, is said to make more powerful remedies. A mixture of skin and gallbladders is considered excellent for skin diseases, internal hemorrhaging, and acute pain. Citrus and gallbladder is combined to treat acute bronchitis. Gallbladders of three different snake taxa (*Naja naja*, *Bungarus fasciatus*, and *Ptyas korros*) combined with the herb *Fritillaria thumbergii* in a powder or liquid is an all-round remedy in Europe and North America. It is said to be particularly effective for burns. Dried snake and snake gallbladders are used to hasten surgical recovery and to alleviate hypertension. Because snakes move so freely and effortlessly, a salve containing snake extracts and organs is thought to work well to treat arthritis.

Worldwide, people go to extremes to attain virility and longevity. Of all the accepted potions used in Asian communities, drinking the blood drained from live venomous snakes ranks high, as does eating their gallbladders, either raw or cooked. Snake gallbladders serve a similar function in Latin America. Soaking rare venomous snakes in alcohol produces a concoction that is the specialty of hundreds of bars in Vietnam and China. The snake may be drowned in the alcohol or killed first, and it remains submerged for hours to weeks before the beverage is consumed as a sexual stimulant. Dropping a still beating snake heart into a glass of rice wine and drinking it makes yet another virility brew.

Lizards are a distant second

Throughout Latin America common iguanas are believed to have medicinal qualities. Eating iguanas is said to transfer their strength to the diner. A paste made by mashing them (*pinol*) is a remedy for many common illnesses. Spiny-tailed iguana flesh is regarded as a cure for a long list of maladies, including impotence. Practitioners of oriental medicine claim that eating live geckos cures tuberculosis and that dried gecko powder in warm rice wine treats ailments ranging from coughs to asthma. Dried monitor lizard gallbladders are believed to heal heart problems, liver failure, and impotence. In Africa the pulverized dried heads of North African monitors are used for numerous diseases, as are tonics derived from various other parts of lizards. Many tribes also claim that eating monitor lizard fat helps deteriorating eyesight, arthritis, rheumatism, hemorrhoids, and muscular pains.

Turtles are not left out

Although their meat is relished throughout Asia, pulverized turtle and tortoise shells are in great demand as folk

The young boys stand alongside a paved road that passes by a mountain village in the Mexican state of Guerrero. The locally captured iguanas, *Ctenosaura* sp., are bound with hemp cord and displayed for passing tourists. Payment is solicited for taking pictures of the quaint scene, or the lizard can be bought. Likely it will end up as a meal. (Photo by Manny Rubio. Reproduced by permission.)

remedies, mostly in the form of drinkable concoctions. Tortoise shell is a general curative; it has been administered for back pain, coughing, dysentery, malaria, rickets, hemorrhaging, and problems associated with birthing and to increase virility and enrich the blood. Freshwater turtle shells treat lethargy, problems with menstruation and menopause, and prostate inflammation.

Venom

The properties of snake venom have given it a place in folk remedies as far back as the seventeenth century, but it has been used in serious medical research only since the latter part of the twentieth century. Of the wide array of components (more than 100), a few enzymes have been singled out by molecular biologists as potentially having a great effect on many serious diseases. Research is being carried out throughout the world.

This chapter is not the place to elaborate on the intricate nature of venoms, but a fundamental understanding can impart their importance. Venoms are divided into two basic categories.

Master Bengali charming a snake in India. (Photo by Jeffrey L. Rotman/Corbis. Reproduced by permission.)

Hemotoxins affect the blood, damaging muscle cells and preventing or, in some cases, causing blood clotting. If the venom is lethal, the victim dies from a painful heart attack by thrombosis. Neurotoxins affect the nervous system by stopping neurons from communicating with one another normally. In this case, death is by relatively painless respiratory paralysis. Snake venoms, which vary widely from species to species, usually attack several organs simultaneously, causing a cascade of complicated physiological responses. To complicate matters, there are snake venoms that are both hemotoxins and neurotoxins.

Ancrod, an enzyme derived from the Malaysian pitviper (*Calloselasma rhodostoma*), is being used successfully in some countries to dissolve fibrinogen, which forms blood clots, a major cause of strokes. Its applications are limited, since the remedy has some serious side effects. Cancer researchers are using a protein from a relative of the Malaysian pitviper, the southern copperhead (*Agkistrodon contortrix*), to prevent the metastasis and growth of cancerous tumors in lab animals. The substance does not kill cancer cells but prevents them from adhering to other healthy cells, essentially producing remission. In the coming years we can look forward to many significant medical applications of snake venoms and their derivatives: as anticoagulants, antiplatelets, and antitumor agents; in the treatment of hypertension and thermal stress; and as anesthetics and analgesics.

Reptile skins

Since the Gilded Age (1890–1915), American alligator skin has been popular in the fashion industry for the strength and lustrous texture and finish of the tanned hide. The numbers of alligators taken from the wild skyrocketed over the years, requiring enactment of a law to protect this denizen of southern swamps in the United States. Farming and ranching brought their numbers back to a sustainable and viable level in the wild.

Other countries have similar projects, but crocodilians continue to be poached and harvested to the point of extirpation in many regions. France and Japan are the biggest importers of crocodilian skins. The plight of crocodilians and chelonians will continue until governments of developing countries accept that there is a serious problem and make a concerted effort to protect and maintain populations. Nonetheless, in many places animal numbers have fallen well below those needed to reestablish sustainable crocodilian populations.

Each year millions of snake and lizard skins are used for a plethora of fashion items. The beautiful skin patterns and textures of monitor lizards such as the water monitor (*Varanus salvator*), Bengal monitor (*V. bengalensis*), and two African species (*V. albigularis* and *V. exanthematicus*), make them very popular. Tanned snake skins are handsome, but they are not as strong or serviceable as those from monitor lizards, seaturtles, and crocodilians. The difficulty in identifying the taxon of the hide animal allows large numbers of endangered species to be shipped posing as legal animals. Even though hundreds of thousands of lizard and snake hides are exported from a host of countries each year, no farming or ranching is undertaken, which could ease the stress placed on wild populations.

Reptiles in the pet trade

Exotic reptiles have long had a special appeal for animal lovers, but only since the 1980s has an amazing diversity of species and a huge number of animals been imported to and exported from the United States by what is known as the "pet trade." At the turn of the twenty-first century, there appeared to be a downward trend in the numbers of reptiles imported to the United States and an increase in those being exported. Two factors are responsible for the increased number of exports: more reptiles are being collected from the wild, and more are being produced through farming, ranching, and captive breeding. This pet trade total is misleading, however, because a substantial number of adult chelonians sent abroad from the United States are used for food or folk medicines, not as pets. The following figures, reported by the Humane Society, are taken from data collected by the United States Fish and Wildlife Service:

	Imports	Exports
1998	2,141,823	10,736,258
1999	1,986,747	10,956,876
2000	1,514,646	13,334,338

(Illustration by GGS. Courtesy of Gale.)

There are literally millions of people keeping reptiles and amphibians as pets in the United States. Some may have one or two animals, but the majority of collectors have a dozen or more. Many commercial breeders retain colonies of hundreds of reptiles, and a few have thousands. An inordinate number of the nine million reptiles kept in the United States as pets do not survive more than two years, mainly because their needs are not properly met and they perish from untreated ailments and diseases.

The situation appears less dire in Europe and Japan. European and Japanese collectors pay higher prices for their animals and tend to have smaller collections than do Americans. To compensate for the limited variety of indigenous animals and the stringent laws protecting those they do have, nearly all pet reptiles and amphibians are imported. Also, pet keepers in these countries are vigilant in their husbandry techniques, taking great pride in the condition and longevity of their captives. Likely because of the expense of housing in Japan, captive breeding is minimal and few animals are shipped out of the country. That appears not to be the situation in parts of Europe, where captive breeding is popular, and certain particularly delicate species are actively bred and exported. However, the number of imports outnumbers those exported.

The exportation of reptiles is fraught with problems. Reptiles frequently spend weeks to months without food and wa-

Detail of a Panamanian cuna mola with a large snake. (Photo by Danny Lehman/Corbis. Reproduced by permission.)

ter in transit from the original captor (frequently a farmer in an isolated region) to the point of shipping. Many are smuggled from countries in which they are protected to those with more liberal laws. Commonly they are stored in crowded, unclean containers and pens that are breeding grounds for disease, while the shipper awaits additional animals to fill an order. There are innumerable reports of suffocated and crushed animals found on the bottom of overstuffed shipping crates and boxes. Malnutrition and dehydration exacerbate the plight of these animals.

Unusual uses for reptiles

In some parts of the world snakes and lizards are tolerated near homes because they control the populations of rats, mice, and other pests. Reptiles may proliferate in these conditions, sometimes reaching impressive densities. Larger lizards, mostly monitors, not only devour rodents but also feed on snakes and small animals that are destructive to crops. They also consume animal and vegetable refuse left in dumps, which helps cut down on vermin and flies and other insects attracted to such sites.

Walter Auffenberg discusses one of the more extraordinary intentional uses of monitor lizards: devouring human corpses. In Bali dead bodies are covered with wicker baskets to deter marauding monkeys and dogs, so that the lizards can feed uninterrupted. He also relates that on the Mergui Archipelago, corpses are placed on platforms in the forest as a feast for lizards. The Hindu ritual of rafting burning corpses down the Ganges River in India provides meals for gavials and Indian crocodiles.

Fear and prejudice

The acceptance of turtles, tortoises, and lizards is widespread, and crocodilians are tolerated from afar. Snakes are an entirely different matter. The mere sight of a snake can elicit repulsion and anxiety from many people, and even genuine terror and aggression. These violent reactions often arise from an aversion to scaly, cold creatures that are capable of biting and, in some cases, injecting their victims with a lethal dose

A woman in Huisache, San Luis Potosí, Mexico, attempts to sell salted snake skins and sun-dried rattlesnake carcasses to passersby. The purchaser will grind them up into fine particles and use it as folk medicine. Natives believe the powder cures a variety of maladies from colds to cancer. (Photo by Manny Rubio. Reproduced by permission.)

of venom. Thousands of snakes are killed annually for no reason other than their being in the proximity of humans; ironically, the encounter is typically in the snake's natural habitat. The evolutionary biologist E. O. Wilson has suggested that a fear of snakes is in our genes and that cultural evolution can be linked genetically to biological evolution over time. There is some justification for this fear, as thousands of people are killed by venomous snakes annually. Many snakes are killed on sight. The weapon of choice may be as unsophisticated as a stick or as refined as a rifle.

Folklore, mythology, religion, and superstition

Legends

People's perception of reptiles has always bordered on fantasy, and most countries in the Old World and New World have stories about these creatures ensconced in their histories. One example is the enchanting myth of Saint Patrick leading all of Ireland's snakes to their demise, to protect the Irish people. (There are no snakes in Ireland.) Druids carried amulets, *gloine nathair* (serpent glass), representing adder eggs, to show respect for the wisdom and cunning of snakes as well as for

their powers of regeneration through molting. Turtles taught the Druids to be methodical, unhurried, and in tune with changes in their environment. From lizards, which represented constant change, they also learned to be aware of what went on around them. A giant sea serpent is referenced in many countries, but it is best known as Scotland's Loch Ness monster. Norway, too, has its serpent monster, Nidhogger, representing volcanic upheavals and decay in the earth.

Reptile associations with dieties

Through equal parts fear, misunderstanding, awe, and respect, reptiles inevitably became an integral part of folklore and religion. They are found throughout Greek and Roman mythology, often in association with deities. Early Greeks and Romans thought that lizards retained divine wisdom and brought prosperity. Their ability to hibernate was seen as the embodiment of resurrection from the dead in early Rome. The Roman goddess Venus (Aphrodite to the Greeks) was said to have descended from the ocean, so the turtle was attributed to her and to fertility. Athena, the Greek goddess of wisdom, carried a shield with the image of a snake, and the Temple of Apollo at Delphi originally was known as Pytho, from which comes our English word "python." The god of medicine was known as Aesculapius in Latin and Asclepius in Greek; his emblem, the caduceus, was two serpents entwined on a winged staff. The entwined snakes originated in Babylonia as representative of healing, wisdom, and fertility. Adding the staff and wings, the Greeks made it the symbol of peace, carried by Hermes, the winged messenger of the gods. The Romans bestowed a similar function on the caduceus for their counterpart to Hermes, the wing-footed god Mercury. In the sixteenth century it became associated with medicine and eventually took its place as the icon of the American Medical Association.

In Egypt, Renenutet (Re), the goddess of fertility and the protector of children, is depicted as a cobra-headed woman capable of destroying enemies by simply looking at them. Like many early peoples living within the range of crocodiles, Egyptians saw them as the epitome of evil, hypocrisy, and treachery. Sobek was a vicious, deceitful, treacherous, crocodile-headed Egyptian idol that was enshrined in the Great Temple of Sobek. Conversely, other Nile peoples viewed them as symbolizing the rising waters of the Nile and the representation of sunrise.

Reptile worship

Reptiles also have been viewed as dieties in their own right. Both early peoples and indigenous peoples of today hold certain reptiles in reverence. The Aztecs and Toltecs saw snakes as the teachers of humankind. Pre-Columbian Mayans named the half-man, half-god incarnation of the serpent sun "Quetzalcoatl," meaning "plumed (feathered) serpent." This god is depicted in stylized art throughout Mexican and Central American ruins of the era. The pattern, shape, and rattle leave no doubt that this is the Neotropical rattlesnake *Crotalus durissus*. Elaborate rituals, including human sacrifice, were carried out to appease the god. Many peoples correlate turtles and tortoises with earth gods, holding the earth atop their shells. In Polynesia lizards are seen as the "god of heaven," and geckos are considered sacred, a manifestation of the mystical, powerful, and terrifying dragon-like monster *mo'o*. Their longevity and apparent intelligence have placed tortoises in an exalted place, as oracles, in Chinese folklore.

Snakes seem to have found the widest place in religion, mythology, and folklore through the ages. In *Serpent Worship in Africa*, Wilfred Hambly lists 16 biologically based reasons why snakes are so important in religions throughout the world. With minor modifications, most of his concepts can be applied to many larger reptiles. Their cryptic coloration, quiet and slow movements followed by rapid and vigorous movements, and appearance after a period of hibernation or estivation (sometimes in great numbers and frequently associated with rains) perpetuate the illusion that snakes can become visible and invisible at will. The large size and strength of some species are viewed with awe. Their ability to inject or spray venom is a unique method of killing or overpowering prey or enemies. Hissing and producing other sounds, along with body inflating or other changes or enlargements in the shape of body parts (e.g., cobra hooding) add to the fear factor. Bright coloration and strange patterns; ridges, spines, and other protuberances of the skin; and a bifurcated penis covered with spines add to their uniqueness. The shedding of their outer layer of skin represents rejuvenation, and the large numbers of young they bear are seen as a bountiful perpetuation of the species. The darting, forked tongue of snakes and their unchanging stare (derived from their lack of eyelids) are perceived as the ability to hypnotize.

The ancient Semites and Mesopotamians believed that snakes were immortal. The figures of cobras were affixed to the crowns of the pharaohs in Egypt. Sects and tribes in India and Africa have worshiped snakes as gods or their messengers since earliest times. In China, snakes, assuming the form of dragons, are seen as fierce protectors of the people. Snake temples can be found throughout Buddhist countries. Tibet, Vietnam, Cambodia, and Laos have many shrines devoted to pit vipers. The snakes roam freely on the altars, where they are carefully fed and tended by priests. Hundreds of myths relating to the powers of snakes have been passed from the earliest generations. Snakes are revered to placate them, so that they will become guardians of the people and not harm them; they are worshiped to encourage a blessing of distinction, prosperity, and contentment. If disrespected, it is thought that they will become angered and place a curse on the offender, which may cause illness, death, or loss of possessions. Additionally, snakes are said to be spirits of the dead and are respected as such. They are given supernatural powers (animism) and, in many cases, humanistic traits.

Mexican anthropologists have suggested that the Virgin of Guadalupe, the patroness of Mexico, is the personification of Cihuacoatl, the Mayan snake goddess. Snake priestesses held a very important place in African snake worship in nearly every one of the nearly thousand known temples. They were considered to be wives of the python god and had sexual relations with the priests. Children born from these unions were considered offspring of the python god.

Snake worship is declining rapidly as remote parts of the world become "westernized," but Africa is apt to retain vestiges for quite some time. Reverence for African rock pythons (*Python sebae*) remains high throughout Africa. In Benin the python god, Danh-gbi, is seen as the great supporter of humankind, god of happiness and wisdom, and overseer of bountiful crops. Elaborate festivals are held, in which large pythons are carried throughout the village. Snake priestesses walk at the front, beating stray dogs, fowl, and pigs to death with clubs, lest they upset the python god. In earlier times, little girls, with the blessing of their parents, frequently were taken to be trained as wives of the python god. In many parts of Africa it is a crime for anyone to mistreat a python. A person that mistakenly or accidentally kills one may be burned to death. It is not uncommon for foreigners to be punished, by flogging or some other harsh physical punishment, for abusing a python. Animal dealers and collectors always have been very careful to avoid taking pythons from snake-worshiping areas.

Indian traditions and offshoots

Nowhere do snakes receive more elaborate adoration than in India. Indian mythology claims eight major snake gods: Shesha, Ananta, Visuki, Manasadevi, Astika, Kaliya, Padmaka, and Kulika. In different regions they may be called by different names, but their physical characteristics and virtues remain constant. Most are depicted with several heads, and each has a specific day dedicated to its worship. The great snake god Vishnu's couch is believed to be the thousand-headed cobra Shesh Nag, who protects him and symbolizes eternity.

Indian cobras (*Naja naja*) and king cobras (*Ophiophagus hannah*) are embedded deeply in India's traditions and venerated by Hindus, especially in the south. Cobra veneration is second only to that of the cow. Cobra symbols, idols, effigies, and temples representing Krishna, Shiva, Rama, and Janardhana, are widespread in the country. Most private homes have a snake idol, usually a cobra, carved into stone and carefully placed in the shade of a tree, where vegetation is permitted to grow freely. In India the cobra is known as Naga, a name derived from the Sanskrit for "serpent."

Naga Panchami is a special all-India festival devoted to worshiping snakes, particularly cobras. It dates back at least a thousand years, and the most elaborate celebrations are held in Baltis Shirale, where the snake goddess Amba Mata is said to have killed the devil. Because of her, snake festivals are seen as praising feminine power. Snakes are caught beginning three weeks earlier, at the Bendur festival that celebrates Amba Mata, and kept in earthen jars at the temple, where they are fed rodents and frogs until Naga Panchami. They are not defanged or mistreated in any way, as such mistreatment is considered sacrilegious.

On Naga Panchami, dancing villagers, accompanied by musicians, lead a procession carrying flowers, milk, and eggs, which are presented to the snakes at the temple. Milk offerings can be traced to the legend of the snake god Visuki, who used his massive tail to churn the oceans of milk in an attempt to raise the elixir of immortality from the submerged city of Atlantis. Unmarried women see the cobra as good luck in marriage, and some will chance kissing the aroused hooded snake on top of its head to solicit additional blessings. Reverence is so pervasive on Naga Panchami that vegetables are not cut, because it is believed that a cobra may have assumed a tiny form and lies hidden in them.

Many Hindu myths involve a semidivine race of snake people also known as Naga. The women, Nagin, are strikingly beautiful serpent princesses capable of changing into cobras or half-snakes, half-humans at will. In the states of Bengal, Orrisa, and Assam, the snake queen Mansadevi (Visuki's sister) is adored as protector of the people. Some snake charmers play lilting and melodious tunes to give her pleasure.

Naga worship started in India and followed trade routes into China and southeastern Asia, where it was accepted and absorbed into Hinduism and Buddhism. This brought the importance of the beautiful Nagin women into predominantly male-dominated beliefs of such countries as China, Laos, Cambodia, Vietnam, Java, and Japan. Frequently, they are seen as having human heads on snake bodies. In parts of China the sister of Emperor Fu Xi, Nu Wa, is said to have created humanity from mud found at the shore. Humans made by hand became aristocracy, whereas those formed from droplets flicked from her tail became peasants.

The best-known Nagin tale is the *Legend of White Snake*, which has been traced to the Tang Dynasty (618 A.D.). All versions are love stories, focusing on a beautiful seductress, a spirit with the form of a white snake. Her friend and hand-maiden is a spirit that takes the form of a green snake. The tale, which has as many as 16 parts, has changed in plot through the centuries and as it passed from country to country.

Snake charming

The followers of a guru named Gorakhnath, said to have been entrusted by the snake god Shiva as the keeper of snakes, began the practice of handling cobras, or "snake charming." It is believed that after being fed cobra flesh and venom at a dinner, he and henceforth all of his disciples became immune to the venom. Because of this supposed immunity, his believers handle snakes that have not been altered in any way; they are fully capable of inflicting deadly bites. Disciples, who are considered holy men, settled in a village near Delhi called Morbandth.

For more than two centuries the majority of India's famed snake charmers learned their craft in Morbandth. Although there are thousands of these snake handlers on the streets of India who employ snakes in the context of their religious beliefs, many others perform in marketplaces as a method of making money by entertaining tourists. Most cobras, vipers (e.g., Russel's vipers, or *Daboia (Vipera) russelii*), and pythons (e.g., Indian and reticulated pythons, or *Python molurus* and *P. reticulatus*) used by bogus snake charmers have had their venom glands removed or their mouths sewn shut to prevent injury to the handler. Although they are in distress in this condition, they still respond to the snake handlers by spreading their hoods and following the swaying movements of the charmer's flute. Thousands of cobras die from mistreatment, starvation, or malnutrition at the hands of these bogus snake charmers each year. Those used by followers of Gorakhnath are set free after a few months or kept and cared for carefully, sometimes for many years.

Under pressure from India's Wildlife Protection Fund, the government is enforcing the 1972 law that makes it illegal to capture snakes. Sadly, most of the snake charmers, both real and fake, are uneducated and have no other method of making a living, so the ban exacerbates the extreme poverty in which the country's population lives.

Hopi snake dance

The most noted Native American snake/human interrelationship is the annual Hopi snake dance, which takes place in the fall. Snakes are used in the elaborate nine-day ceremonies as effigies, to promote spring rain that will ensure strong crops. The snakes are caught and placed in underground rooms known as *kivas*, where they are maintained in earthenware pots, prayed over, and then washed carefully. The event became such a popular attraction that photography was banned in 1915 as a way to maintain decorum and dignity; only Native Americans have been permitted to attend since the latter years of the twentieth century.

The most dramatic part of the ceremony comes at its culmination at sundown on the last day. It is a drama depicting an ancient Hopi myth in which the Corn Maiden confronts the Snake Youth. Masked, bare-chested, bedecked in blue, and elaborately painted in black and white, priests perform a circle dance with snakes, mostly rattlesnakes (*Crotalus viridis nuntius*), dangling from their mouths, while younger members of the tribe accompany them. Whipsnakes (*Masticophis* species) and bullsnakes (*Pituophis catenifer*) also are used. After several minutes of dancing, ending in a prayer by the chief, the priests gather up the snakes, race into the desert, and release them as messengers to the gods.

Judeo-Christian beliefs

The Bible contains many references to snakes, the most well known being the serpent tempting Adam and Eve to eat the apple (forbidden fruit) in Genesis 3:1–24. Out of this temptation, Jews and Christians have been taught that snakes represent evil. Nearly everywhere they are related to Satan, as is evidenced in God's curse of the snake: "Upon thy belly shalt thou go, and dust thou shalt eat all the days of thy life" (Genesis 3:14). In other Old Testament biblical passages, Moses changes rods to serpents (Exodus 7:8–12) and makes fiery serpents (Numbers 21:5–9), signifying God's power to transform an inanimate object into a terrifying one. Also, corrupt judges are likened to snake's venom (Psalms 58:4–5). In the stories of the New Testament, the apostles are called wise as serpents (Matthew 5–7 and 14–16), and Paul is bitten by a viper (Acts, 27:41–44, 28:1–6) but fails to show any signs of sickness.

One unusual Christian religious sect employs venomous snakes in their services. This group is located in very small, poor rural towns scattered in and around the southern Appalachian Mountains of the United States. Following fundamentalist beliefs, the worshipers maintain simple, puritanical lifestyles by strictly adhering to the words of the Bible. Their credo is based on Mark 16:17–18: "And these signs shall follow them that believe; In my name shall they cast out devils; they shall speak with new tongues; they shall take up serpents; and if they drink of any deadly thing, it shall not hurt them."

Participants believe that their personal faith and Christian obedience anoint them with the power to overcome evil, symbolized by the ability to handle venomous snakes with near impunity. "Snake worship" is a misnomer for this practice; the potentially lethal rattlesnakes, copperheads, and cottonmouths they handle are seen not as messengers of God but as the embodiment of Satan. The intensity of this sect's services, which includes praying, chanting, singing, and dancing along with handling snakes and talking in tongues, accompanied by music from electric guitars and keyboards, is quite literally overwhelming.

Voodoo

With its roots in West African beliefs that can be traced back six thousand years, Voudou (anglicized as "Voodoo") and related religions (e.g., Yoruba, Ubanda, Candomble, Quimbada, Lucumi, and Macumba) have strong followings in several parts of the New World, among them, Haiti and Cuba, with isolated groups throughout the West Indies and in some parts of South America and United States. As might be expected for worship so closely entwined with nature and associated with strange rituals, including sacrifice, snakes assume a prominent place. Snakes are seen as symbols of three main deities, called Loas, who parallel Christian saints and manifest themselves by taking over and possessing a human participant involved in a religious ceremony.

Aida-Wedo is the Loa of fertility and new life. She assumes the form of a rainbow snake and is extolled with the sacrifice of white chickens and white eggs. Damballah-Wed (Bon Dieu Loa in Haiti) is known as the Loa Father and represents the ancestral knowledge upon which Voodoo is based. He, too, symbolizes fertility and new life and is said to be a snake living in trees near springs. White chickens and eggs are sacrificed to him as well. Simbi is the Loa of freshwater and rain. He watches over the creation of charms. Symbolized by the water snake, he is one of Haitian Voodoo's three cosmic serpents. He is celebrated by sacrificing spotted roosters.

Superstitions

Some myths are better called superstitions. Some tribes along the Nile River believe that crocodile teeth worn on a cord around the neck protect a person from attack. In many African countries, if a crocodile is killed, its liver must be burned to protect the village from evil. If a python is killed in southern Africa, it must be burned, or, it is believed, it will seek revenge or cause an extended drought. In other African countries, where crocodiles are thought to be reincarnations of the dead, they are fed regularly, to ensure that they will protect the community.

Snakes took such an important place in Asian history that they were chosen as one of the animals in the Chinese Zodiac's 12-year cycle. In India a copper coin is placed in the mouth of a dead snake before it is buried carefully, to forestall evil events. Devout Australian Aborigines believe that killing a goanna will cause the sky to fall. True chameleons are viewed with trepidation in West Africa and believed to have exceptional mystical powers. It is thought that their unusual appearance, independently moveable eyes, and ability to change rapidly into vibrant or dull colors give them the power of deceiving humans. They are said not to eat but rather to take nourishment directly from the air, and they are considered to be directly associated with the sun, from which they can steal fire. Even in the modern world American cowboys think that sleeping encircled with a horsehair rope will keep snakes at bay.

Resources

Books

Anonymous. *Serpent Worship*. Toronto: Tudor Press, 1980.

Auffenberg, Walter. *The Behavioral Ecology of the Komodo Monitor*. Gainesville: University Presses of Florida, 1983.

Aymar, Brandt, ed. *Treasury of Snake Lore*. New York: Greenberg Publishers, 1956.

Bennett, Daniel. *A Little Book of Monitor Lizards*. Aberdeen, United Kingdom: Viper Press, 1995.

Resources

Bjorndal, Karen, ed. *Biology and Conservation of Sea Turtles.* Washington, DC: Smithsonian Institution Press, 1982.

Burghardt, Gordon M., and A. Stanley Rand, eds. *Iguanas of the World: Their Behavior, Ecology, and Conservation.* Park Ridge, NJ: Noyes Publications, 1982.

Burton, Thomas. *Serpent-Handling Believers.* Knoxville: University of Tennessee Press, 1993.

Diaz-Bolio, Jose. *The Geometry of the Maya and Their Rattlesnake Art.* Merida, Mexico: Area Maya-Mayan Area, 1987.

Ernst, Carl H., and Roger W. Barbour. *Turtles of the World.* Washington, DC: Smithsonian Institution Press, 1989.

Fizgerald, Sarah. *International Wildlife Trade: Whose Business Is It?* Washington, DC: World Wildlife Fund, 1989.

Fosdick, Peggy, and Sam Fosdick. *Last Chance Lost?: Can and Should Farming Save the Green Sea Turtle? The Story of Mariculture, Ltd., Cayman Turtle Farm.* York, PA: Irvin S. Naylor, 1994.

Glasgow, Vaughn L. *A Social History of the American Alligator.* New York: St. Martin's Press, 1991.

Guggisberg, C. A. W. *Crocodiles: Their Natural History, Folklore and Conservation.* Harrisburg, PA: Stackpole Books, 1972.

Hambly, Wilfrid D. *Serpent Worship in Africa.* Chicago: Field Museum of Natural History, 1931.

Hoser, Raymond. *Smuggled: The Underground Trade in Australia's Wildlife.* Sydney, Australia: Apollo Books, 1993.

———. *Smuggled-2: Wildlife Trafficking, Crime and Corruption in Australia.* Victoria, Australia: Kotabi, 1996.

Levell, John. P. *A Field Guide to Reptiles and the Law.* 2nd revised edition. Lanesboro, MN: Serpent's Tale Books, 1997.

National Research Council, Committee on Sea Turtle Conservation. *Decline of the Sea Turtles: Causes and Prevention.* Washington, D.C.: National Academy Press, 1990.

Nissenson, Marilyn, and Susan Jonas. *Snake Charm.* New York: Harry Abrams Publishers, 1995.

Rubio, Manny. *Rattlesnake: Portrait of a Predator.* Washington, DC: Smithsonian Institution Press, 1998.

Zhu, Y. P. *Chinese Materia Medica: Chemistry, Pharmacology, and Applications.* Australia: Harwood Academic Publishers, 1998.

Periodicals

Anonymous. "Unto the Church of God." *Foxfire* (spring 1973): 1–96.

Auffenberg, Walter. "Notes on the Feeding Behaviour of *Varanus bengalensis.*" *Journal of the Bombay Natural History Society* 80 (2): 286–302.

Fewkes, J. W. "Tusayan Snake Ceremonies." *Sixteenth Annual Report of the Bureau of American Ethnology, Smithsonian Institution* (1894–1895): 273–312.

———. "Tusayan Flute and Snake Ceremonies: Part 2." *Nineteenth Annual Report of the Bureau of American Ethnology, Smithsonian Institution* (1897–1898): 273–312.

Grove, Noel. "Wild Cargo: The Business of Smuggling Animals." *National Geographic* (March 1981): 287–314.

Speake, D. W., and R. H. Mount. "Some Possible Ecological Effects of 'Rattlesnake Roundups' in the Southeastern Coastal Plain." *Proceedings of the Annual Conference of Southeastern Game and Fish Commissioners* 27 (1973): 267–277.

Manny Rubio

· · · · ·

Conservation

Reptiles are frequently secretive, and knowledge of their biological status often is based on anecdotal information rather than on precise scientific data. Nonetheless, scientists have identified certain characteristics, which make some reptiles particularly vulnerable to changes in their environment. Species that are large; island dwelling; restricted in distribution, habitat, or ecological specialists; require large home ranges; are migratory; or are valued as food or medicine or for their skin are most likely to show population declines when stressed by human activity. In combination, as when a large species used for food lives on an island, these characteristics have made some reptiles extremely susceptible to human influences. In addition to the spatial, size, and behavioral characteristics that make reptiles vulnerable to extinction, certain life history traits, such as delayed maturity, slow growth, a low reproductive output, and high juvenile mortality rate, combine to make recovery difficult for many species. The difficulty arises because these biological traits are the product of a long-term evolutionary history and are slow to respond to rapidly changing conditions brought about by human activity. Because some species have long generation times, they cannot adapt to rapid environmental change or even to well-intentioned conservation management.

It is difficult to determine how many reptiles are endangered worldwide or even to guess at what percentage is threatened. Some groups, such as turtles and tuataras, are declining because they are affected adversely by humans and because they possess the aforementioned life history characteristics that make them exceedingly vulnerable to environmental change. Examples of reptiles in need of the highest-priority conservation efforts include the giant tortoises of the Galápagos and Seychelles, seaturtles, the giant land iguanas of the Caribbean, the Canary Islands giant lizard, several crocodilians (e.g., the Chinese alligator, the Siamese and Philippine crocodiles, the tomistoma, and the gharial), and the two species of tuataras in New Zealand. The following sections provide a broad overview of reptile conservation, the factors that affect reptiles, and the tools available to biologists to reverse declines and to enhance prospects for long-term survival.

Scope of the problem

Of the four reptile lineages, turtles (Chelonia) are the most threatened of the major groups. More than half of the 264 or so recognized species face serious population declines or even extinction. Nearly all tortoises require concerted conservation management, as do the seven species of seaturtles. The regional turtle fauna most endangered is the Southeast Asian turtle fauna: many species are being driven to extinction by the trade for food, traditional medicines, and pets. Turtles everywhere are threatened by habitat loss and the degradation of river and wetland ecosystems.

Although they are feared, crocodilians always have been valued for their skins or as food, and today more than half of the 23 species are endangered or declining. Fortunately, biologists have reversed declines in some of these species through strict legislation, research, and management programs, at least where habitat remains intact. Other species, however, are still extremely vulnerable, because human populations have encroached into their habitats to such a great extent that the crocodilians have no place left to go. Poaching remains a threat, particularly to the rarer species.

The status of squamates (lizards and snakes) generally is less well known than that of turtles and crocodilians, except for some species in commercial trade. Like other reptiles, lizards and snakes are threatened especially by habitat loss. Literally millions of these reptiles are harvested from wild populations for the fashion industry. Some species seem fairly resilient to harvest, whereas other species have declined. Collection for the pet trade and for food is also a source of concern, especially among the chameleons and the attractive and docile snakes. Harvest in certain regions, such as southeastern and eastern Asia, is likewise cause for great concern. Snakes are killed nearly everywhere, even when they pose no threats to humans. Unfortunately, little is known about the basic biology of many squamates that appear to be declining.

There are only two species of tuataras (Rhynchocephalia), both inhabitants of remote New Zealand islands. Although they are lizard-like, they are the sole survivors of an ancient reptile lineage. They are vulnerable to habitat loss, poaching, and especially the introduction of rats onto their small island homes. Both species are strictly protected, monitored, and managed.

Threats to reptiles

The threats that affect reptiles are the same as those that affect biodiversity throughout the world. The primary danger

Turtle eggs, five in each bag, are hanging for sale for about $3 in a village stall by the roadside in Terengganu, eastern Malaysia. Seaturtles have existed virtually unchanged for more than 100 million years. But human activity, including taking eggs, considered an aphrodisiac in some parts of Asia, has helped push several species to the brink of extinction. (Photograph. AP/Wide World Photos. Reproduced by permission.)

a portion of the population may have severe consequences. For example, female seaturtles are particularly vulnerable as they nest. They are easily captured, and the loss of this vital half of the adult population eliminates the chances of recovery. In general, little research has been undertaken on the effects of trade on reptiles, except for seaturtles and crocodilians, because wildlife management agencies traditionally are concerned with popular and charismatic mammals and birds rather than reptiles.

Environments around the world are awash in chemicals whose effects are both direct and indirect. Pollution destroys reptile habitats by changing environmental conditions, by causing direct toxicity to an animal or its prey, or by subtle effects on reptile biophysical or physiological requirements. In contrast, long-term sublethal exposure makes reptiles more vulnerable to predators and disease. For example, such contaminants as PCBs have been shown to mimic the activity of estrogen, an important hormone that plays a part in sex determination. When these chemicals are present, abnormal development takes place, resulting in intersexes and in reduced reproductive ability (by affecting morphologic features), success, and survivorship. The populations of alligators and turtles in certain Florida lakes have declined as a result of endocrine-mimicking chemicals dumped into their habitats years ago.

There are hosts of other factors that affect reptile survivorship, including the overabundance of subsidized predators (e.g., ravens and raccoons), disease, adverse effects from nonindigenous animals and plants (e.g., mongooses, fire ants, and invasive vegetation), malicious killing, and even climate change. For example, certain predators have increased dramatically in proximity to humans. Raccoons are now without natural predators throughout much of their range, and they are subsidized by human garbage and feeding in both urban and rural areas. As their populations increase, they are capa-

to reptiles probably comes from the direct loss of habitats, whether the habitat is a small patch of temperate forest or the vast rainforest of the Amazon basin. Habitat loss is not confined to the surface of the ground but extends both arboreally and deep underground, depending on the life history requirements of the species. Migratory reptiles, such as seaturtles, face threats in different habitats, sometimes located great distances from one another, as the turtles move between natal, breeding, and feeding grounds. If a beach vital to nesting is destroyed, then the life history of the species may be disrupted, even if migratory pathways and feeding grounds are in excellent condition. Complex life cycles make species vulnerable to habitat change.

The alteration of habitats often is more subtle and less dramatic than outright habitat destruction, but it is equally devastating to reptiles. Whereas habitat destruction is immediate, the effects of habitat alteration may take place over a long time period. Thus, the difference between habitat destruction and alteration is often only a matter of scale and time. Perhaps the most common effect of habitat alteration is fragmentation of remaining habitats into smaller patches. These patches may not contain sufficient amounts of habitat to maintain a reptile population, such as a colony of tortoises or a wide-ranging population of indigo snakes. Habitat fragments may be isolated; may require animals to move (if possible) across unfamiliar ground; increase animals' vulnerability to predation and mortality from humans, especially as roads are crossed; and are more susceptible to random environmental events, such as hurricanes. Reptiles living in habitat fragments are exposed to disturbance from predators and competitors living on the edge of the fragment.

Unsustainable use, whether trade for skins, food, or pets, may devastate certain reptiles, because reproduction cannot keep pace with the loss of animals. Selective harvesting of even

Confiscated snakeskin shoe. (Photo by Galen Rowell/Corbis. Reproduced by permission.)

ble of destroying nearly every turtle nest and neonate reptile in their vicinity, and they can seriously threaten adults.

Disease (both viral and bacterial) has become a growing problem, particularly affecting seaturtles and tortoises. Disease outbreaks often are associated with polluted or environmentally stressed habitats, and pathogens may be transmitted through the release of diseased captive animals into the wild. Invasive or nonindigenous species, such as imported fire ants, directly kill reptiles and destroy eggs or modify habitats so that native species can no longer survive there. People everywhere kill harmless snakes and other reptiles for no reason at all, except pure meanness or fear.

Finally, little is known about how reptiles might respond to climate change, although it is certain that barriers between existing fragmented habitats would limit movement to new areas. In addition, global warming might have a more subtle effect. Since the temperature during egg incubation determines the sex of many reptiles, an increase in nest temperatures could produce fewer male and more female turtles and more male and fewer female crocodilians and could have a mixed result among some other species. Changes in rainfall patterns undoubtedly would affect reptiles, although increased ambient temperatures actually might benefit some species.

A KwaZulu-Natal Nature Conservation Service staff member cuts notches into the carapace of a loggerhead turtle hatchling (*Caretta caretta*) as part of a research project. (Photo by Roger De La Harpe: Gallo Images/Corbis. Reproduced by permission.)

Conserving reptiles

The best way to conserve reptiles is to protect their habitats. This does not call solely for the creation of parks or preserves, since such areas are not isolated from the environmental effects surrounding them. Scientists realize that habitat protection is complex, bringing to bear knowledge of the interrelationships of land, water, air, and biotic components. Likewise, people surrounding protected areas must have a stake in the success of the park or preserve, because reptile conservation usually involves "people management" more than "reptile management." Additionally, there are other ways to protect habitats rather than putting them solely in public trust, as through land easements and conservation agreements, tax incentives, land banking, and private acquisition. Habitat protection must encompass the spatial needs of the species or ecosystem to be conserved, whether it is a regional landscape, a linear river or stream, or an underground aquifer.

Knowing the biotic requirements of a species helps in planning reserve design and management. There are no truly pristine areas on Earth, and all ecosystems must be managed at some level to ensure reptile survival. Management must work within the requirements of both the species and the available human resources. If either set of requirements is ignored, efforts to conserve reptiles are unlikely to succeed. Consideration of habitat restoration and manipulation, such as revitalizing ponds for bog turtles or building dens for wintering snakes, increasingly is being included in recovery and conservation plans. Both management and restoration require detailed knowledge of natural history to predict how a species will respond to change and to determine which management approaches benefit the species.

There are several ways to curb habitat fragmentation. Planning could minimize the extent of lands affected, maximize patch size, and reduce the distances between patches. Development plans could allow for corridors between patches, protection of migratory habitats, and construction of ecopassages over or under roads and other transportation routes. For example, barrier walls and culverts have been used successfully in Florida to cut down mortality rates and to facilitate movement under a major highway across a state preserve. Deaths of alligators, turtles, and snakes declined significantly after the ecopassages were built. Effective management of habitat patches might include the removal of subsidized predators and the maintenance of natural disturbances, such as fire.

If the problem affecting reptiles leads to the loss of individuals, such as through habitat destruction or trade, legal protection is appropriate. For example, the American alligator and other crocodilians have made a remarkable comeback after legal protection and vigorous law enforcement prevented unsustainable trade. Protection of individual animals without protection of their habitat or without research designed to understand the cause of their decline will be ineffective, however, and, in some cases, counterproductive. In addition, the presence of an animal on protected lands does not mean that it is protected. For years venomous snakes were killed on sight in some U.S. national parks as part of government policy.

In many countries there are laws that restrict or curtail pollution, but pollution continues to affect ecosystems throughout the world. If reptiles are to be conserved, particularly in aquatic ecosystems, these laws must be enforced and extended vigorously. Measuring sublethal or indirect effects of chemical

A newborn leatherback seaturtle (*Dermochelys coriacea*) makes its way into the ocean in Jupiter, Florida, USA. The nest produced about 200 turtles overall. If the turtle survives, it could grow to a weight of about 350 lb (158 kg); however, only about 10% of the turtles that are born each year survive to adulthood. (Phototograph. AP/World Wide Photos. Reproduced by permission.)

long-term effects on wild populations, and methods of treatment and management.

A few declining reptiles have benefited enormously from advances in husbandry at zoos and aquariums and even by private individuals. Zoos and aquariums serve as refugia for threatened reptiles, allowing scientists to learn much about reptile biology as a prerequisite to the restoration of wild populations. Programs for giant tortoises, crocodilians, and some of the larger lizards, in particular, hold much promise. In addition, zoos and other organizations participate in the formation of "assurance colonies," where animals seized in illegal trade are rehabilitated and held until conservation plans can be developed for their ultimate return, if possible, to natural habitats. Asian turtles currently maintained in assurance colonies offer promise that these species can be saved from extinction.

Ultimately, reptiles and the ecosystems on which they depend can be conserved only via the partnership of research, management, and public support. In this regard, education helps build appreciation for the beauty and functional value of reptiles to ecosystems, whether they control mammal pests or serve as sentinels of environmental health. The public needs to be encouraged to leave reptiles in the wild, to avoid buying products made from declining species, to refrain from keeping as pets any animals caught in the wild, and to support habitat conservation both at home and in exotic lands of wide diversity. Finally, resource managers must rely on proven management techniques, rather than opting for quick-fix "solutions" to complex problems. All conservation efforts must have a solid biological foundation so that self-sustaining and viable populations of reptiles may persist.

New tools for conservation.

At the turn of the twenty-first century exciting methods rapidly were becoming available to assist in the conservation of reptiles, including research techniques that allow for greater knowledge of natural history requirements (such as telemetry and other tracking methods), molecular biology (which helps define populations and measure diversity and relatedness and is critical in the new field of forensic herpetology), landscape ecology (GIS technology, remote sensing and satellite imagery, all of which define large-scale distribution patterns and help scientists understand how land use affects reptiles), and new biometrical research, especially for taking inventory of communities, monitoring populations, and understanding population biology. The wealth of technological and theoretical advances makes reptile conservation a challenging and rewarding field of biology.

contamination is much more difficult, especially with the variety and number of chemicals released each year. Endocrine-mimicking chemicals have the potential to devastate wildlife populations, because they work in trace amounts. The demonstrated effects of some of these chemicals on reptiles, especially as factors affecting development, reproduction, and survival, should bring into question their impacts on humans. In addition, much research needs to be done toward understanding diseases in reptile populations, that is, their causes, the way they spread, the factors that stress immune systems,

---

Resources

Books

Alberts, Allison, ed. *West Indian Iguanas: Status Survey and Conservation Action Plan.* 2nd edition. Gland, Switzerland: IUCN/SSC West Indian Iguana Specialist Group, 2000.

Bambaradeniya, Channa N. B., and Vidhisha N. Samarasekara, eds. *An Overview of the Threatened Herpetofauna of South Asia.* Colombo, Sri Lanka: IUCN Sri Lanka and Asia Regional Biodiversity Programme, 2001.

Resources

Bjorndal, Karen A., ed. *Biology and Conservation of Sea Turtles.* Washington, DC: Smithsonian Institution Press, 1995.

Branch, W. R., ed. *South African Red Data Book: Reptiles and Amphibians.* Report no. 151. Pretoria: South African National Scientific Programmes, 1988.

Cogger, H. G., E. E. Cameron, R. A. Sadlier, and P. Eggler, eds. *The Action Plan for Australian Reptiles.* Endangered Species Unit, project number 124. Sydney: Australian Nature Conservation Agency, 1993.

Committee on Sea Turtle Conservation. *Decline of the Sea Turtles: Causes and Prevention.* Washington, DC: National Academy Press, 1990.

Corbett, Keith. *Conservation of European Reptiles and Amphibians.* London: Christopher Helm, 1989.

Dodd, C. K., Jr., "Status, Conservation and Management." In *Snakes: Ecology and Evolutionary Biology,* edited by R. A. Seigel, J. P. Collins, and S. Novak. New York: Macmillan Publishing Co., 1987.

———. "Strategies for Snake Conservation." In *Snakes: Ecology and Behavior,* edited by R. A. Seigel and J. T. Collins. New York: McGraw-Hill Book Co., 1993.

Klemens, M. W., ed. *Turtle Conservation.* Washington, DC: Smithsonian Institution Press, 2000.

Langton, T., and J. A. Burton. *Amphibians and Reptiles: Conservation Management of Species and Habitats.* Strasbourg, France: Council of Europe Publishing, 1998.

Newman, Don. *Tuatara.* Endangered New Zealand Wildlife Series. Dunedin: John McIndoe, 1987.

Ross, James Perran, ed. *Crocodiles: Status Survey and Conservation Action Plan.* 2nd edition. Gland, Switzerland: IUCN/SSC Crocodile Specialist Group, 1998.

Webb, G., S. Manolis, and P. Whitehead, eds. *Wildlife Management: Crocodiles and Alligators.* Chipping Norton, Australia: Surrey Beatty & Sons, 1987.

Periodicals

Gibbons, J. W., D. E. Scott, T. J. Ryan, et al. "The Global Decline of Reptiles déjà vu Amphibians." *BioScience* 50, no. 8 (2000): 653–666.

C. Kenneth Dodd, Jr., PhD

Testudines

(Turtles and tortoises)

Class Reptilia

Order Testudines

Number of families 14

Number of genera, species About 99 genera; at least 293 species

Photo: Eastern painted turtles (*Chrysemys picta picta*) basking in the sun in New York, USA. (Photo by John M. Burnley/Photo Researchers, Inc. Reproduced by permission.)

Evolution and systematics

Turtles first appeared in the fossil record during the Triassic period, about 220 million years ago. They were originally believed to have evolved from early anapsid reptiles (lanthanosuchids, millerettids, nytiphruretians, pareiasaurs, and the procolophonoids), but recent studies (mostly molecular) argue for a diapsid origin (the group that includes the squamate reptiles, the crocodilians, and the birds). Two mechanisms for retraction of the neck evolved in ancestral turtles. The members of the suborder Pleurodira (or side-necked turtles) retract their necks laterally between the carapace and the plastron, while those in the suborder Cryptodira (or hidden-necked turtles) retract their neck vertically. The pelvic girdle is primitive in shape, and fixed to the plastron in side-necked turtles.

Physical characteristics

These reptiles are easily recognized by the presence of a dorsal bony carapace and a ventral bony plastron, with the limb girdles located inside the ribs. All living forms lack teeth, have internal fertilization, and lay shelled amniotic eggs.

A turtle's shell first and foremost provides protection from predators and serves to buffer harsh environmental conditions. Most species can retract the head and limbs completely within the shell when distressed. The upper shell, known as the carapace, is typically joined to the lower shell or plastron by a bony bridge.

Bony plates, which develop from the dermal (lower) layer of the skin, widen and fuse to one another and with the vertebrae and ribs to form the carapace. Neural bones form along the midline, pleurals from the ribs, as well as the peripherals, which are outermost. The plastron is composed of nine bony elements;

the paired hyo- and hypoplastra, epiplastra, xiphiplastra, and a single entoplastron are derived from the pectoral girdle, the sternum, and abdominal ribs (gastralia). The modified shoulder girdle remains inside the ribs, a remarkable arrangement found in no other vertebrate. The outer surface of the shell is generally covered by horny scutes derived from keratin in the epidermal (upper) layer of skin. The scutes overlap the sutures of the bone, which increases the strength of the shell and protects the growing portions.

Many variations on the basic shell structure have evolved over time. Softshell turtles, which lack horny scutes, have reduced pleurals, and most have completely lost the peripheral bones. Although softshell turtles lack a bony bridge, the carapace is firmly attached to the plastron by connective tissue.

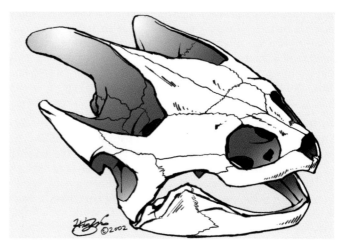

Testudine skull. (Illustration by Jonathan Higgins)

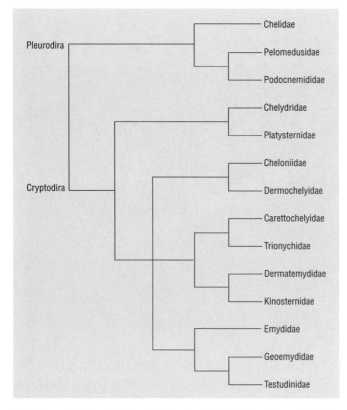

Testudine phylogenetic tree. (Illustration by GGS. Courtesy of Gale.)

The plastron is greatly reduced, and the bones are loosely connected to one another by cartilage. When fully formed, the plastral bones of an adult softshell turtle may be covered by two to nine leathery callosities, thick callous-like layers of epidermis that cover the plastral bones of softshell turtles. Callosities are generally absent in hatchlings, slowly developing with growth and attaining full size when the seaturtle reaches maturity.

An elastic cartilaginous hinge has arisen independently in many lineages. A box turtle can withdraw its head and limbs completely within its shell and draw the anterior and posterior lobes of the plastron tightly against the carapace. Some African tortoises have a carapacial hinge that allows the rear portion to close upon the plastron, thereby protecting the hind limbs. Female semiaquatic and terrestrial species may have a hinge in the posterior lobe of the plastron, providing the flexibility necessary to lay extremely large eggs.

The size and shape of the turtle shell may also be adapted to the environment. The broad, flattened carapace of some aquatic turtles functions like a solar panel. A basking turtle will change its position on a log or rock so that the greatest surface area is exposed to the Sun. In some northern species, the scutes of the carapace are darkened, allowing maximal absorption of solar radiation. Aquatic turtles have lower, more streamlined shells that offer less resistance while swimming. Extreme flattening is found among the softshells, which hide in shallow water beneath a thin layer of sand or mud. The flattened shell of the pancake tortoise (*Malacochersus tornieri*)

in East Africa allows it to squeeze into narrow crevices within its rocky habitat. With the force generated by its legs and the natural elasticity of the shell bones, this tortoise is extremely difficult to extract once it has wedged itself between rocks.

Aquatic species that share their habitat with large crocodilians have traded the advantages of streamlining in favor of high-vaulted, strongly buttressed shells for protection from being crushed. Among Asian river turtles (e.g., river terrapins [*Batagur baska*], crowned river turtles [*Hardella thurjii*], and painted roofed turtles), these buttresses form bony chambers that enclose the lungs and prevent compression during deep dives. In desert-dwelling tortoises, the domed shell reduces surface area relative to volume, while a thickened keratin layer retards evaporative water loss. Some tortoises have even been observed gathering drinking water by angling the carapace

Turtle and tortoise limb differences for living on land and in the sea: a. Great African tortoise (*Geochelone sulcata*) lives on land; b. Red-eared slider (*Trachemys scripta elegans*) is semi-aquatic; c. Hawksbill (*Eretmochelys imbricata*) lives in the sea. (Illustration by Jacqueline Mahannah)

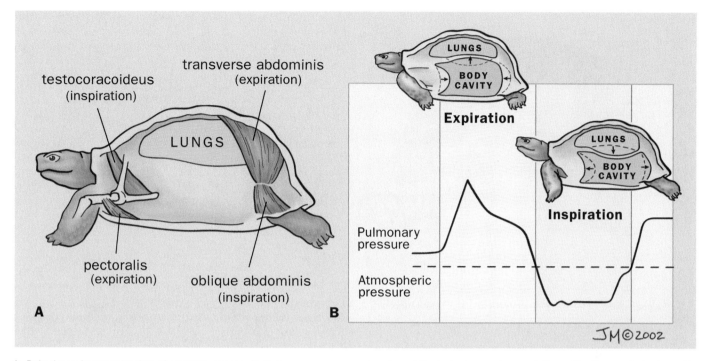

A. Paired respiratory muscles of *Testudo graeca*; B. Pulmonary pressure changes of *Testudo graeca*, in correlation with changes in body cavity volume during expiration and inspiration. (Illustration by Jacqueline Mahannah)

forward during a rainfall to catch water dripping from the shell.

Large openings, or fontanelles, between the pleural bones in the carapace have evolved in several genera. The leatherback seaturtle (*Dermochelys coriacea*) shows the greatest divergence from the characteristic turtle shell; tiny platelets embedded in the leathery carapace are all that remain of the bony shell. A reduced carapace decreases the physiological costs associated with building and maintaining a heavy shell (such as sequestering minerals to a substrate where they are not readily accessible for some physiological processes), as well as the energy cost for locomotion in terrestrial species, and provides greater buoyancy in the aquatic forms.

The turtle skull is also unique among living vertebrates for the absence of temporal fenestra. Formerly believed to be a true anapsid (i.e., without such openings), many researchers now believe that the turtle evolved from diapsid ancestors. In this scenario, the two pairs of temporal openings were secondarily lost, giving an anapsid-like appearance. Vestiges of the temporal openings can be seen in the slightly arched posterior margin of some turtle skulls. These large openings in the back of the skull allow the muscles of the jaw to expand beyond the confines of the adductor chamber.

Dietary preferences range from completely herbivorous to totally carnivorous; however, many turtles consume a mixture of plant and animal matter. In some species there is a dietary shift from the carnivorous diet of hatchlings to the mostly herbivorous adults. Although modern turtles lack teeth, there are many modifications of the maxillary, premaxillary, and dentary bones for feeding. A pronounced beak made of ker-

atin may be used to hold and tear food. The palate of herbivorous turtles contains a series of ridges that assist in the maceration of plant matter. Macrocephaly, characterized by an enlarged head (as in many female map turtles), often develops in mollusk-feeding species. The broad crushing surfaces and powerful musculature allow them to exploit an abundant food item that may be unavailable to turtles that cannot extract this meal from the mollusk's protective shell.

The limbs of most aquatic turtles terminate in five independent digits; however, most terrestrial turtles and tortoises have reduced phalanges. The limbs of freshwater turtles are flattened laterally, and the digits are generally webbed. The sturdy limbs of tortoises are round in cross section. In some highly aquatic species, such as seaturtles and pig-nose turtles, the limbs are paddlelike, and digits are reduced to just a few claws.

The texture of turtle skin ranges from virtually scaleless and smooth in highly aquatic softshell species, to the coarse scaly texture of terrestrial tortoises. Keratin scales of various shapes and sizes are found on the head and limbs of most species. The large, thickened scales of tortoises are adapted to their dry environments. Epidermal appendages such as chin barbels, warts, and the fringe of matamatas provide cryptic camouflage that may assist in prey acquisition and/or provide protection.

The major organs of the turtle circulatory system are similar to those of other reptiles. Although heart rate is largely dependent upon temperature, the three-chambered heart beats slowly, especially in tortoises. The paired lungs are dorsal to the visceral organs and cannot be expanded by the action of

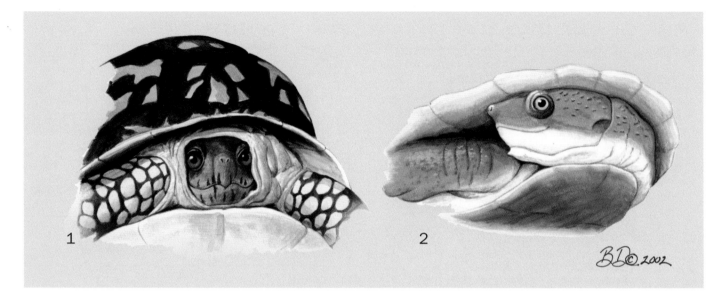

1. A straight necked turtle, the eastern box turtle (*Terrapene carolina*); 2. A sideneck turtle, the New Guinea snapping turtle (*Elseya novaeguineae*). Straight necked and sideneck turtles protect their heads in their shells differently, based on their different anatomy. (Illustration by Barbara Duperron)

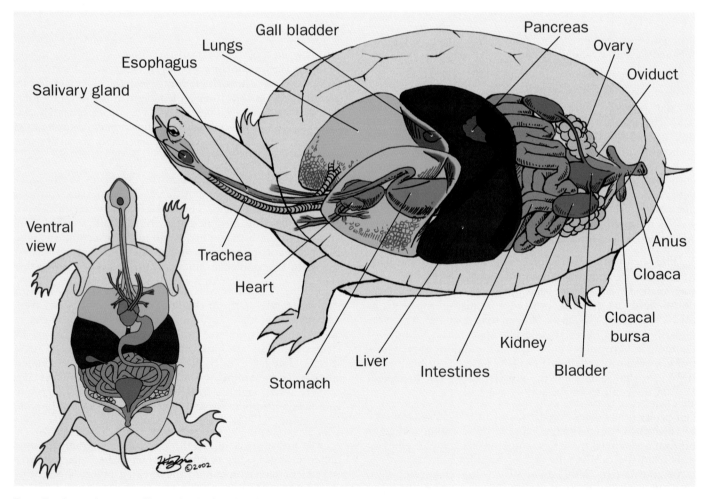

Testudine internal organs. (Illustration by Jonathan Higgins)

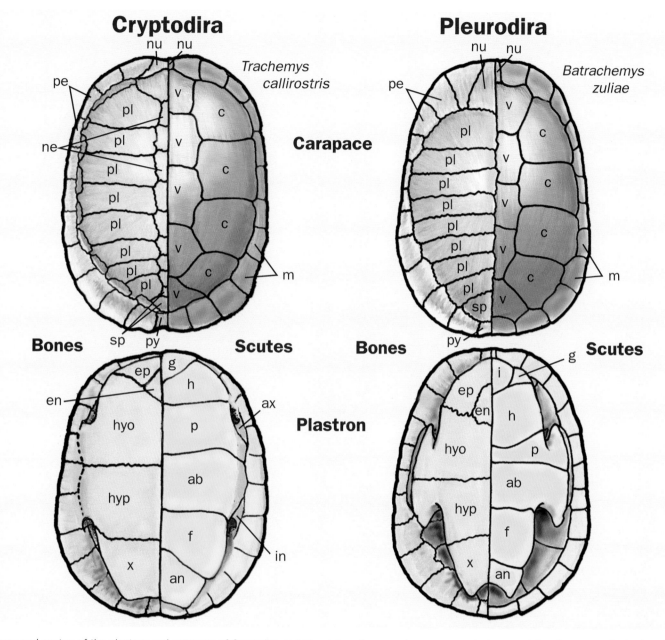

Cryptodira

Pleurodira

Trachemys callirostris

Batrachemys zuliae

Carapace

Bones **Scutes**

Bones **Scutes**

Plastron

Bones and scutes of the plastron and carapace of Cryptodira and Pleurodira. Carapace bones: nu=nuchal, pe=peripheral, ne=neural, pl=pleural, sp=suprapygal, py=pygal. Carapace scutes: nu=nuchal, m=marginal, v=vertebral, c=costal. Plastron bones: ep=epiplastron, en=entoplastron, hyo=hyoplastron, hyp=hypoplastron, x=xiphiplastron. Plastron scutes: i=intergular, g=gular, h=humeral, p=pectoral, ab=abdominal, f=femoral, an=anal, ax=axial, in=inguinal. Some pleurodires have mesoplastrons between the hyoplastron and hypoplastron. (Illustration by Gillian Harris)

the rib muscles. Ventilation of the lungs is controlled by contraction of lung muscles; however, in the relaxed state the lungs are maximally filled with air. By manipulating airflow from one chamber of the lung to another, aquatic turtles can adjust their position in the water much like a fish uses a swim bladder. This ability is impaired in turtles with respiratory ailments and results in a diagnostic lopsided appearance.

Aquatic species may also respire through their skin, the lining of the throat, and through thin-walled sacks, or bursae, in the cloaca. The Fitzroy River turtle (*Rheodytes leukops*),

an Australian sideneck living in well-oxygenated streams, maintains a widely gaping cloacal orifice and rarely surfaces. The turtle pumps water through the cloaca, which gapes in sequence to the pumping. Although common to most aquatic species, cloacal bursae are absent in softshell turtles. In these aquatic turtles, 70% of the submerged oxygen intake is through the skin and 30% is through the lining of the throat. In northern climates, turtles that spend most of the winter trapped below the ice must rely upon submerged oxygen uptake or tolerate long periods without oxygen. The mineralized shell of the painted turtle buffers the accumulation of

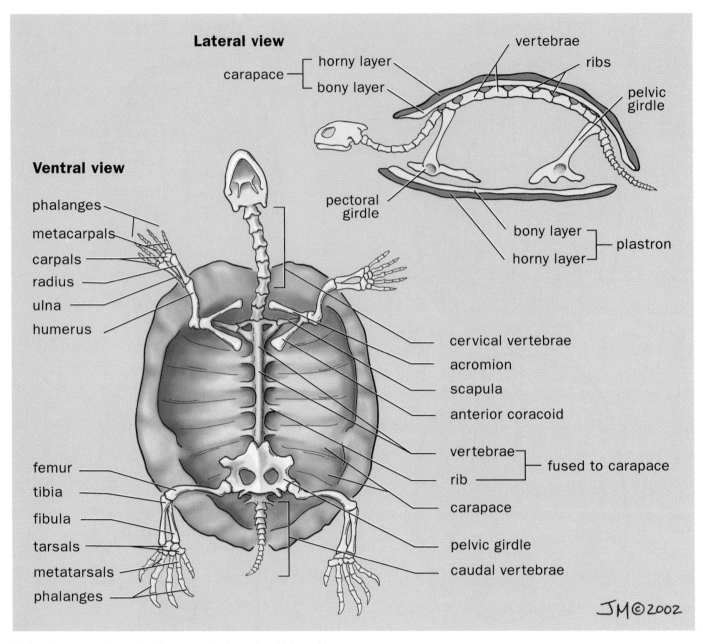

Lateral view

carapace — { horny layer / bony layer

vertebrae

ribs

pelvic girdle

pectoral girdle

bony layer — } plastron
horny layer —

Ventral view

phalanges
metacarpals
carpals
radius
ulna
humerus

cervical vertebrae
acromion
scapula
anterior coracoid

vertebrae — } fused to carapace
rib —

carapace

femur
tibia
fibula
tarsals
metatarsals
phalanges

pelvic girdle
caudal vertebrae

JM©2002

Skeletal structure of a turtle. (Illustration by Jacqueline Mahannah)

lactic acid formed under anaerobic conditions to maintain a stable blood pH through the winter.

The largest extant species is the leatherback seaturtle, which attains a shell length of 96 in (244 cm) and may weigh up to 1,191 lb (867 kg). Of the freshwater species, the alligator snapping turtle (31 in/80 cm; 249 lb/113 kg), the Asian narrow-headed softshell turtle (*Chitra indica*) (47 in/120 cm; 330 lb/150 kg), and the South American river turtle (42 in/107 cm; 198 lb/90 kg) attain impressive sizes. The Aldabra tortoise (55 in/140 cm; 562 lb/255 kg) is the largest living terrestrial species. With maximum shell lengths of less than 4.7 in (12 cm), the speckled cape tortoise, flattened musk turtle, and bog turtle are among the world's smallest turtles.

Distribution

Turtles and tortoises exist on all continents except Antarctica. The diversity of these species allows them to inhabit both temperate and tropical regions, as well as all bodies of water.

Reproductive behavior

Most turtle species exhibit sexual size dimorphism. Among aquatic species, males are generally smaller than females and have elaborate courtship behavior. However, in semiaquatic, bottom-walking species and tortoises, in which males are equal to or larger than females, courtship displays are gener-

Galápagos tortoise conflict. (Photo by Laura Riley. Bruce Coleman Inc. Reproduced by permission.)

nesting beach are overwhelmed by the reproductive output and many nests escape detection.

Turtle eggs are usually deposited in flask-shaped chambers excavated into the ground. However, some turtles may oviposit, or deposit their eggs, in decaying vegetation and litter, in nests of other animals, or even in a nest constructed while the female is completely underwater or underground. Some species quickly cover the eggs and leave the area, while others spend considerable time concealing the nest. Despite their vulnerability on land, leatherback seaturtles obscure the site completely before returning to the sea. Some species may construct a false nest some distance from the first or divide the clutch between two or three nests to confound predators. Although parental care is rare in turtles, the Asian giant tortoise, which nests in mounded vegetation, will defend her eggs from potential predators for several days following oviposition.

Reproductive output is related to body size, both within and across species. Smaller species lay one to four eggs per clutch, whereas large seaturtles regularly lay over 100 eggs at a time. The majority of species lay two or more clutches each nesting season. At higher latitudes there is also a general trend, both within and across species, toward the production of one large clutch of smaller eggs.

Turtle eggs are of two shapes: elongate or spherical. Although egg shape is usually consistent within a genus, members of diverse families such as the tortoises and side-necked turtles may lay eggs of either shape. The spherical shape has the lowest possible surface-to-volume ratio, and therefore is less vulnerable to dehydration. Turtles that produce large clutches (50 eggs or more) have spherical eggs to make efficient use of the limited space available.

ally minimal, and combat for territories and/or mates is common. In temperate climates, courtship and mating may occur in the fall or the spring, but nesting usually occurs in the spring to early summer.

Although individual females may not reproduce every year, nesting in most species is annual and seasonal. Females of many species can store sperm in their oviducts for years and produce fertile eggs without mating annually. In addition, DNA analysis has shown that eggs within the same clutch are sometimes fertilized by more than one male.

The majority of turtles select nest sites from the available upland habitats found in the vicinity of their foraging areas. However, some sea and river turtles make extensive migrations to nesting beaches. Seaturtles, which nest every two to three years, may migrate over 2,796 mi (4,500 km) to nest in a specific location. During the arribada (a massive, coordinated arrival of seaturtles, and some freshwater species, at a nesting beach) of the olive ridley seaturtle, as many as 200,000 females nest on the same small beach over a period of one or two days. The large freshwater river turtles of South America and Asia similarly nest en masse. The predators on the

Pair of European pond turtles (*Emys orbicularis*). (Photo by Jane Burton. Bruce Coleman, Inc. Reproduced by permission.)

Aldabra tortoise (*Geochelone gigantea*) feeding in La Digue, Seychelles. (Photo by K & K Ammann. Bruce Coleman, Inc. Reproduced by permission.)

The eggs of most turtles have flexible, leathery shells, but the shells of other turtles are more inflexible and often brittle. Eggs with brittle shells tend to be more independent of the environment, losing and absorbing less water than eggs with flexible shells. However, those with flexible shells often develop faster. Species that do not dig sophisticated nests, or those that nest in particularly dry or very moist soils, tend to lay eggs with brittle shells. Conversely, turtles nesting on beaches prone to flooding, or in areas with limited growing seasons, situations where rapid egg development is important, are more likely to lay eggs with flexible shells.

In most turtles, the temperature during incubation also determines the sex of the hatchling. In species with "temperature-dependent sex determination" (TSD), the temperature during the middle third of incubation affects the biochemical pathway that determines the sex of the hatchling. Two patterns of TSD have been described for turtles. Type I species have a narrow pivotal temperature range (usually between 80.6–89.6°F/27–32°C) above which only females are produced and below which only males result. Type II species have two pivotal temperature ranges, with males predominating at intermediate temperatures, and females predominating at both extremes. Sex determination appears to be genetically determined (GSD) in the Austro-American side-necked turtles, all softshells, and a few musk and pond turtles. Among species with GSD, only the wood turtle, two species of giant musk turtles, the black marsh turtle, and the brown roofed turtle have dimorphic sex chromosomes; all others have identical chromosome sizes in males and females. The evolutionary advantage conferred by these modes of sex determination remains unknown.

When fully developed, hatchling turtles use their caruncle, a small tubercle on the upper beak, to slice through the embryonic membranes and eggshell. Soon after hatching, most neonates emerge from the nest and head directly for cover of water or vegetation. Vibrational cues, such as movement by hatchlings within the nest, may help neonate sea-turtles to coordinate the intense effort necessary for emergence from their sandy nest chamber. Hatchlings of a

few temperate species (ornate box turtle, yellow mud turtle) delay emergence from the nest. After hatching, they immediately dig downward a few feet or more below the nest, presumably a behavioral adaptation to avoid the impending lethal winter temperatures in shallow water or soil. Hatchlings of a few other temperate species, such as the painted turtle, remain in the nest over the winter where they may be exposed to temperatures of 10.4°F (−12°C) or lower. Although these turtles tolerate freezing at high subzero temperatures (e.g., to 24.8°F/−4°C), they must remain supercooled (i.e., without the tissues freezing) in order to survive colder temperatures. Still other turtles, particularly those in highly seasonal tropical environments, must remain in their nests until rain softens the soil, allowing them to dig out. In dry years, the neonates may remain in the nest chamber for more than a year after hatching.

Growth may vary considerably even within the same clutch. Habitat, temperature, rainfall, sunshine, food type and availability, and sex have each been associated with growth rate in turtles. Growth can be conveniently studied in turtles because many species retain evidence of seasonal growth on their scutes. The rate of growth is also reflected in the ring-like layers of bone deposited on the femur and humerus. In most species, the turtle grows rapidly to sexual maturity; then the growth rate slows markedly. In later years, small species may stop growing completely.

Conservation status

Commonly known as turtles, tortoises, and terrapins, members of the order Testudines are distinguished from all other vertebrates by their bony shell. The protection conveyed by this morphological curiosity has contributed to the persistence of this group through more than 200 million years of evolution. Turtles and their shells have survived the conditions that resulted in the fall of the dinosaurs, the shifting of continents, and the ebb and flow of glaciers with little structural modification. They appear in the folklore, art, and creation myths of many human cultures, but humans are a major reason for the precipitous decline in turtle populations world-

A snapping turtle kills a watersnake. (Photo by Tom Brakefield. Bruce Coleman, Inc. Reproduced by permission.)

wide. The characteristics of turtle life history (e.g., late maturity, extreme longevity, and low adult mortality) make them especially vulnerable to the habitat destruction and deterioration associated with the expansion of human activities. Indeed, nearly 50% of living species are listed as Endangered or Vulnerable.

Resources

Books

Ernst, C. H., and R. W. Barbour. *Turtles of the World.* Washington, DC: Smithsonian Institution Press, 1989.

Ernst, C. H., J. E. Lovich, and R. W. Barbour. *Turtles of the United States and Canada.* Washington, DC: Smithsonian Institution Press, 1994.

Pough, F. H., R. M. Andrews, J. E. Cadle, M. L. Crump, A. H. Savitzky, and K. D. Wells. *Herpetology.* Upper Saddle River, NJ: Prentice Hall, 1998.

Zug, G. R., L. J. Vitt, and J. P. Caldwell. *Herpetology: An Introductory Biology of Amphibians and Reptiles.* San Diego, CA: Academic Press, 2001.

Patrick J. Baker, MS

Pig-nose turtles
(Carettochelyidae)

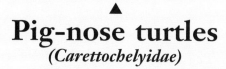

Class Reptilia
Order Testudines
Suborder Cryptodira
Family Carettochelyidae

Thumbnail description
A turtle of moderate size with paddle-like forelimbs, a rough, leathery shell, and a pig-nose snout

Size
Up to 22 in (56 cm) and 50 lb (22.5 kg)

Number of genera, species
1 genus; 1 species (*Carettochelys insculpta*)

Habitat
Rivers and lakes

Conservation status
Vulnerable

Distribution
Southern New Guinea and northern Australia

Evolution and systematics

Although pig-nose turtles represent a very distinctive family, they are most closely related to the softshell turtles (Trionychidae). This turtle is also commonly known as the Fly River turtle and pitted-shelled turtle (New Guinea). No subfamilies are recognized.

The taxonomy for this species is *Carettochelys insculptus* Ramsay, 1886, Fly River, Papua New Guinea.

Physical characteristics

This turtle is readily distinguished by its rough leathery shell (without scutes), its paddle-like forelimbs (each with two well-developed claws), and its blunt, piglike snout. The carapace of the adult has a smooth margin and the keel along the midline of the carapace is only obvious posteriorly; however, small juveniles have a serrated shell margin and a knobby midline keel. The adult bony carapace is well developed, and not reduced as in the softshell turtles.

Distribution

The pig-nose turtle inhabits the river systems of southern New Guinea (Papua and Irian Jaya) as well as along the northern coast of Australia's Northern Territory.

Pig-nose turtle (*Carettochelys insculpta*). (Illustration by Barbara Duperron)

Habitat

These turtles primarily inhabit freshwater ecosystems (rivers, lakes, lagoons, swamps) with slow-moving or still waters and soft bottoms. They also are found in estuaries associated with these systems.

Behavior

Little is known about the behavior of this species, since they emerge from the water only to nest. However, they are known to aggregate around food sources. In northern Australia, the average home ranges for females and males cover about 5 mi (8 km) and 2 mi (3 km) of river length, respectively. Adults will also thermoregulate underwater by lying over small thermal springs on the river bottom.

Feeding ecology and diet

Pig-nose turtles are opportunistic omnivores, with herbivorous tendencies. Their principal food is the fruits of shoreline trees, although they also eat their leaves and flowers as well as algae and other submergent plants. Animal foods include insect larvae, mollusks, and crustaceans. Fish and mammals are also eaten, but probably as carrion.

Reproductive biology

Courtship and mating have not been described. Nesting occurs late in the dry season, from September to December or possibly even January in New Guinea, and from June, July, or August (depending on locality) to October or November in northern Australia. The species produces multiple clutches, but the maximum number per year is uncertain. On the Daly River in northern Australia, females lay two clutches per season, but only nest every other year. Nests are excavated by the hind limbs at sites 20 in (50 cm) to 16 ft (5 m) above the water level. The brittle-shelled eggs are spherical, 1.5–2.1 in (38–53 mm) in diameter, and 1.1–1.6 oz (32–46 g) in mass. Clutch sizes range from 7 to 39 eggs, the number probably related to the female's size. At 86°F (30°C) embryos develop to full term in 64 to 74 days; however, they then begin estivating within the egg until rainy season flooding stimulates their hatching. Thus, the total incubation period for natural nests is 86 to 102 days. Sex is determined by temperature during the middle third of incubation, with only females produced at high temperatures and only males at low ones.

Conservation status

This species is listed as Vulnerable on the IUCN Red List. It may be negatively affected by the trampling of its nests by water buffalo and by fishing, logging, intensive grazing, and agriculture.

Significance to humans

Pig-nose turtles are consumed by people across their range. In New Guinea, indigenous peoples collect the eggs from nesting beaches and capture the turtles in nets, in traps, with spears, and by hand. In Australia, the turtles are speared, fished, and grabbed by hand, although the eggs apparently are not harvested.

Pig-nose turtle (*Carettochelys insculpta*) is a popular food item on Fly River, Papua New Guinea. This close-up shows why it has the common name "pig-nose." (Photo by Jeffrey W. Lang/Photo Researchers, Inc. Reproduced by permission.)

Resources

Books

Cann, John. *Australian Freshwater Turtles.* Singapore: Beaumont Publishing, 1998.

Periodicals

Doody, J. S., R. A. Sims, and A. Georges. "Use of Localized Thermal Springs to Elevate Body Temperatures by the Pig-Nosed Turtle, *Carettochelys insculpta*." *Chelonian Conservation and Biology* 4 (2001): 81–87.

Doody, J. S., J. E. Young, and A. Georges. "Sex Differences in Activity and Movements in the Pig-Nosed Turtle, *Carettochelys insculpta*, in the Wet-Dry Tropics of Australia." *Copeia* 1 (2002): 93–103.

Georges, Arthur, and Mark Rose. "Conservation Biology of the Pig-Nosed Turtle, *Carettochelys insculpta*." *Chelonian Conservation and Biology* 1, no. 1 (1993): 3–12.

John B. Iverson, PhD

Australo-American sideneck turtles
(Chelidae)

Class Reptilia
Order Testudines
Suborder Pleurodira
Family Chelidae

Thumbnail description
Medium to large sideneck turtles with a nuchal scute (absent only in *Elseya*), zero to six neural bones, the pleural bones almost always meeting at the midline behind the neurals, mesoplastral bones absent, and the pelvis fused to the carapace and plastron

Size
6–19 in (14–48 cm)

Number of genera, species
16 genera; 50 species

Habitat
Swamps, marshes, seasonally flooded wetlands, ponds, lakes, streams, and rivers

Conservation status
Critically Endangered: 3 species; Endangered: 4 species; Vulnerable: 6 species; Lower Risk/Near Threatened: 8 species; Data Deficient: 2 species

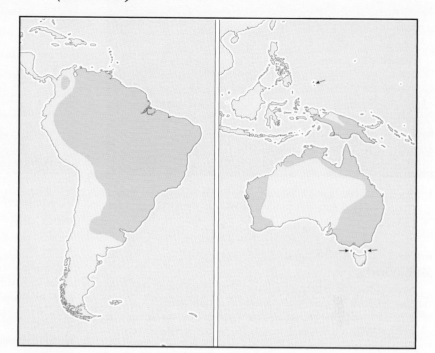

Distribution
South America, Indonesia, New Guinea, and Australia

Evolution and systematics

Fossils of these turtles are known from the Miocene of Australia and South America. They are most closely related to the other pleurodiran families: the Pelomedusidae of Africa and the Podocnemididae of South America and Africa. Their subfamilial relationships remain poorly understood. Molecular evidence suggests that the Chelidae can be further divided into three subfamilies: the Hydromedusinae, containing only the genus *Hydromedusa*; the Chelidinae, containing all other South American chelids; and the Chelodininae, containing both the short- and long-neck species of Australia and New Guinea. This taxonomic relationship has yet to be diagnosed in terms of morphological characters. Genetic and morphological data support the revision of the genus *Phrynops*, which resulted in the erection of two new genera (*Bufocephala* and *Ranacephala*) and the resurrection of three previously recognized genera (*Mesoclemmys*, *Batrachemys*, and *Rhinemys*). The status of several morphologically distinct Australian "species" is debated among taxonomists because the species are virtually indistinguishable using genetic analysis.

Physical characteristics

Sideneck turtles are extremely diverse in size, shape, and coloration. Most species have four claws on the hind feet and five claws on the forelimbs, zero to six neural bones present, the pleural bones almost always meeting at the mid-

line behind the neurals, mesoplastral bones absent, and the pelvis fused to the plastron. Neck elongation is extreme; in some species the length of the neck may exceed that of the carapace. While the carapace tends to be dark and cryptic, there are a few spectacularly colored chelids that may have bright red, pink, orange, and yellow on the plastron and/or soft parts.

Distribution

With the exception of *Batrachemys dahli*, all of the South American chelids are found east of the Andes and as far south as northern Argentina. The remaining species are found in tropical to temperate regions of Australia and New Guinea. This curious distribution may be the result of a common ancestor that dispersed across Antarctica before the southern continents separated and drifted apart. No fossils attributable to this family have been found outside the present-day range.

Habitat

Primarily aquatic, most species inhabit permanent freshwater rivers, streams, lakes, and ponds. Some species are found in seasonal wetlands that are dry for most of the year, while others expand their range into flooded forests during the rainy season. The New Guinea snakeneck turtle (*Chelodina siebenrocki*) may frequent estuaries and coastal waters.

Common snakeneck turtle (*Chelodina longicollis*) in Australia. (Photograph by Tom McHugh. Photo Researchers, Inc. Reproduced by permission.)

Male toad head turtle (*Phrynops hilarii*). (Photograph by Henri Janssen. Reproduced by permission.)

Behavior

Several species estivate by burying themselves in the mud during periods of extreme drought. Steindachner's turtle (*Chelodina steindachneri*), an aquatic species that resides in the deserts of Western Australia, is well adapted to the extremely high temperatures and desiccating conditions prevalent for most of the year. This species is resistant to evaporative water loss and stores fluids in accessory bladders in order to survive while buried for the year or two until the rains return.

Feeding ecology and diet

Most species are omnivorous to totally carnivorous; however, the adult *Elseya dentata* may be completely herbivorous. Many diverse adaptations for specialized feeding are found in this family. The snakeneck turtles of Australia and South America have developed very similar modes of prey capture. Both groups have independently evolved a long neck that is adapted for striking at prey while suspended in the water column. The negative pressure generated by the rapid expansion of the mouth and neck pulls the head toward the prey while simultaneously pulling the prey item into the turtle's open jaws. Matamatas lie in wait on the murky bottom of their aquatic habitat and use a gape-and-suck method similar to that of the snakeneck turtles to capture their prey. Alternatively, some Australian chelids are specialized for feeding on mollusks. As they age, macrocephalic females of the genus *Emydura* develop enormous skulls that provide the force required for crushing larger prey.

Reproductive biology

Females are generally larger than males; however, the reverse is true in *Pseudemydura* and *Elusor*. Males of the latter genus possess an extremely long tail with a unique bony structure that may play a role in copulation. Breeding has been observed in the early spring for many species, but may be year-round in the Tropics. Nests are typically constructed in spring and early summer; however, some tropical species nest during the winter as well. Clutch size ranges from one to 28 and two or more clutches may be produced per year. The brittle-shelled eggs may be elongate or spherical, 1.2–2.4 in (3–6 cm) in greatest diameter. Incubation may last more than 200 days in species that require a period of diapause (early developmental arrest) for proper development.

A variety of nest chambers are constructed. The gibba turtle may lay its elongate eggs in shallow nests beneath vegetation, whereas the western swamp turtle uses her forelimbs to dig a chamber, which she then enters to deposit her eggs. Researchers in Australia have discovered that, while completely submerged, the female northern snakeneck (*Chelodina rugosa*) lays her eggs in the muddy bottom of temporary ponds. Development is arrested until the water subsides during the dry season, but resumes in time for incubation to be completed before the pond fills again. Hatchlings may remain in the nest until rain softens the hardened soil plug with which most species seal the nest chamber. In extreme cases emergence may occur nearly two years after the eggs were laid. All species analyzed thus far exhibit genetic sex determination.

Male northern Australian snakeneck turtle (*Chelodina rugosa*). (Photograph by Henri Janssen. Reproduced by permission.)

Conservation status

Thirteen species are listed as Threatened on the IUCN Red List: six are Vulnerable, four Endangered, and three Critically Endangered. Eight species are listed as Lower Risk/Near Threatened. The western swamp turtle (*Pseudemydura umbrina*), with known populations of fewer than 400 individuals, may be one of the world's most endangered turtles. Similarly, Hoge's sideneck (*Ranacephala hogei*) is known from just a few locales in Brazil, and is thought to be Critically Endangered. The major threat to most species is habitat destruction or degradation. Many of the threatened species are being bred in captivity, offering hope for future repatriations. However, active conservation efforts in nature are lacking for most species.

Significance to humans

Many species are consumed locally.

1. Common snakeneck turtle (*Chelodina longicollis*); 2. Victoria river snapper (*Elseya dentata*); 3. Gibba turtle (*Mesoclemmys gibba*); 4. Western swamp turtle (*Pseudemydura umbrina*); 5. Matamata (*Chelus fimbriatus*). (Illustration by Barbara Duperron)

Species accounts

Gibba turtle
Mesoclemmys gibba

SUBFAMILY
Chelidinae

TAXONOMY
Emys gibba Schweigger, 1812, locality unknown.

OTHER COMMON NAMES
None known.

PHYSICAL CHARACTERISTICS
This is a small to medium (up to 9 in [23 cm] maximum carapace length) sideneck turtle with a prominent medial keel on its oval carapace. The nuchal scute is long and narrow. The carapace and upper surfaces of the soft parts are uniformly dark in color. The bridge and underside of the marginals are mostly yellow with some dark mottling at the seams. With the exception of the gular and anal scutes, which are predominantly yellow, the plastral scutes are mostly dark with yellow markings at the edges. The relatively small head has wrinkles of skin that give a marbled texture and appearance. Two small, fleshy barbels are widely separated on the chin. Up to five rudimentary neural bones may be present.

DISTRIBUTION
This species is found in the Amazon basin in northeastern Peru, eastern Ecuador, southeastern Colombia, northern

Mesoclemmys gibba
Chelus fimbriatus

Brazil, to the Rio Negro of southwestern Venezuela. The distribution is interrupted by the Sierra Nevada de Mérida, but resumes in northeastern Venezuela and continues through Guyana, Suriname, French Guiana, and northeastern Brazil.

HABITAT
The gibba turtle is found in marshy prairies and slow-moving creeks of lowland tropical rainforests.

BEHAVIOR
In Trinidad this species is primarily nocturnal; however, basking may occur in the early morning. Although relatively docile, this species may emit a foul-smelling musk from the inguinal glands when handled.

FEEDING ECOLOGY AND DIET
The natural diet may be omnivorous; however, in captivity it is primarily carnivorous, feeding on small fish, frogs, and worms.

REPRODUCTIVE BIOLOGY
The nesting season lasts from July to November. The female may nest in vegetation, among the roots of trees, or excavate a shallow chamber to deposit two to four elongate eggs. The small clutch size is offset by the relatively large size (up to 1.8 in [4.5 cm] long and 1.3 in [3.2 cm] wide) of the brittle-shelled eggs. A flexible connection between the carapace and plastron may allow the diminutive female to pass these enormous eggs successfully. Incubation may take up to 200 days under natural conditions.

CONSERVATION STATUS
Not threatened. Unfettered habitat loss and degradation, however, may reduce or extirpate populations before they are ever recorded.

SIGNIFICANCE TO HUMANS
They are consumed locally by indigenous peoples, and are also gathered to supply the international pet trade. ◆

Matamata
Chelus fimbriatus

SUBFAMILY
Chelidinae

TAXONOMY
Testudo fimbriata Schneider, 1783, Approuague River, Guisanbourg, French Guiana.

OTHER COMMON NAMES
French: Matamata; German: Fransenschildkröte.

PHYSICAL CHARACTERISTICS
A medium to large turtle (up to 18 in [45 cm] in carapace length), the matamata is probably the most distinctive of all turtles. This species is recognized by its flattened yet rugged shell, and the rough skin with fringelike appendages that gives the head a triangular appearance when viewed from above. The raised conical knobs on each scute form three keels on the dark carapace to enhance the cryptic appearance of this bottom-walking species. Furthermore, the skin on the broad

flat head forms small flaps that are most pronounced above the tympana and waver gently in a slow current. The tiny eyes are set forward and may be of little use in the turbid waters this species inhabits. The tubelike proboscis is used to breathe without fully surfacing.

DISTRIBUTION
Northern South America, including the Amazon and Orinoco Rivers.

HABITAT
This highly aquatic freshwater species prefers the still waters of oxbow lakes and ponds. Anecdotal evidence suggests that this species survives extended periods in brackish water, but it is not known whether it enters these habitats freely or is deposited there by flooding rivers.

BEHAVIOR
These turtles are poor swimmers and spend the majority of their time walking along the bottom. When found in rivers, matamatas avoid the current by moving beneath cut banks and submerged logs.

FEEDING ECOLOGY AND DIET
This carnivorous turtle feeds mainly on small fishes. The matamata lies in wait at the murky bottom of its aquatic habitat where fish may be attracted to the fringelike skin on its head. After locating the fish via vibrational cues detected by the skin and enlarged tympana, it uses a gape-and-suck method to violently draw the prey and a large volume of water into the mouth. The water is then expelled and the fish swallowed.

REPRODUCTIVE BIOLOGY
Courtship and mating have not been described; however, nesting generally takes place between October and December. In Venezuela, nests are often constructed in the clay soil of steep riverbanks. Eight to 28 spherical (1.4–1.6 in [3.5–4.0 cm] diameter), brittle-shelled eggs are produced annually. The long incubation (200 or more days) suggests that embryos require a diapause or estivation before hatching occurs.

CONSERVATION STATUS
Not threatened.

SIGNIFICANCE TO HUMANS
The grotesque appearance of this species discourages consumption, even in regions where other turtle species are readily eaten. Matamatas are frequently available in the pet trade where the adults and juveniles may command high prices. ◆

Common snakeneck turtle
Chelodina longicollis

SUBFAMILY
Chelodininae

TAXONOMY
Testudo longicollis Shaw, 1794, Australasia.

OTHER COMMON NAMES
English: Long-neck turtle; French: Chélidés; German: Schlangenhalsschildkröten.

PHYSICAL CHARACTERISTICS
This is a medium-sized (to 10 in [25 cm] carapace length), long-neck species. The oval shell is brown and has a shallow central groove that may be pronounced in some specimens. The wide, cream-colored plastron has a dark pattern that follows the seams of the scutes. In contrast to the wide shell, the neck is relatively thin and the small head is distinctly pointed.

DISTRIBUTION
Eastern Australia from Adelaide, South Australia, to Victoria, New South Wales, and Queensland.

HABITAT
This species prefers slow-moving backwaters, especially weedy lagoons, swamps, and billabongs. It may occasionally be found in the swift currents of streams and rivers.

BEHAVIOR
When the seasonal wetlands dry up, this species is known to make long overland migrations to the nearest water hole or to estivate terrestrially. It may lie dormant for portions of the summer and winter in shallow burrows beneath vegetation. However, in some regions the snakeneck turtle has been observed to hibernate communally in aquatic sites. This species emits a foul-smelling musk from the inguinal and axillary glands which may serve to deter would-be predators.

FEEDING ECOLOGY AND DIET
This species is an opportunistic carnivore, mostly feeding on aquatic invertebrates, fish, tadpoles, and crustaceans. It may also feed upon terrestrial insects and carrion if available.

REPRODUCTIVE BIOLOGY
In the temperate portion of its range, mating has been observed in April and May. Nesting occurs from late September to December. The female deposits eight to 24 brittle-shelled, elongate eggs (up to 1.3 in [34 mm] in length and 0.8 in [20 mm] wide) in nests constructed near the water's edge. A diapause or embryo estivation occurs during development; therefore, incubation may take up to 185 days, although 120 to 150

Elseya dentata

Pseudemydura umbrina

Chelodina longicollis

days are more usual. The observed sex ratio of the resulting hatchlings is independent of incubation temperature.

CONSERVATION STATUS
Not threatened. This species is still quite common.

SIGNIFICANCE TO HUMANS
Although many snakenecks are consumed by Aborigines, this "smelly" snakeneck turtle is generally avoided. ◆

Victoria river snapper
Elseya dentata

SUBFAMILY
Chelodininae

TAXONOMY
Chelymys dentata Gray, 1863, upper Victoria River, Northern Territory, Australia.

OTHER COMMON NAMES
English: Northern Australian snapping turtle, northwest snapping turtle.

PHYSICAL CHARACTERISTICS
This is a medium to large (up to 13 in [34 cm] maximum carapace length) sideneck turtle. The crown of the head has a prominent scale and the thick neck is covered with rough tubercles. The nuchal scute is absent. The posterior rim of the carapace is deeply serrated in juveniles; however, it may become relatively smooth in adults.

DISTRIBUTION
This mostly tropical species is found throughout the Victoria River and nearby drainages of Western Australia and the Northern Territory.

HABITAT
Rivers, fast-moving creeks, and flooded forests of tropical Australia.

BEHAVIOR
Adults, especially the larger females, are strong swimmers. Cloacal respiration is well developed in this species. In fast-moving currents the adults may rarely surface.

FEEDING ECOLOGY AND DIET
These turtles are mostly herbivorous, consuming bark, fruits, pandanus roots, and the seeds and blossoms of flowering plants found along riverbanks.

REPRODUCTIVE BIOLOGY
In the Daly River nesting occurs in February and March. At least one clutch of 10 brittle-shelled eggs is laid each season. The elongate eggs (2 x 1 in [51 x 28 mm]) hatch after 120 days.

CONSERVATION STATUS
Not threatened. Although this species is not on the IUCN Red List, its populations in several river drainages have been negatively affected by pollution.

SIGNIFICANCE TO HUMANS
None known. These highly aquatic turtles live in remote regions of the country and therefore have little direct interaction with humans. ◆

Western swamp turtle
Pseudemydura umbrina

SUBFAMILY
Chelodininae

TAXONOMY
Pseudemydura umbrina Siebenrock, 1901, Australia.

OTHER COMMON NAMES
None known.

PHYSICAL CHARACTERISTICS
This is a very small sideneck turtle (to 6 in [15.5 cm] carapace length). The primitive skull is nearly roofed over with a slight ventral emargination of the temporal region. The intergular scute is large and completely separates the gulars. A dark border is present at the seams of the plastral scutes. The rough skin of the neck is covered with conelike projections.

DISTRIBUTION
Known only from a few wetlands near Perth, Western Australia.

HABITAT
Seasonally inundated swamps and marshes.

BEHAVIOR
The wetlands inhabited by this species experience seasonal periods of intense drought that cause them to dry up completely. Although ambient temperatures may be as high as 104°F (40°C), the turtle survives by burrowing into the sandy soil under decaying plant matter and estivating until the rains return to fill the pools.

FEEDING ECOLOGY AND DIET
This species is carnivorous, feeding on tadpoles, crustaceans, and the adults and larvae of aquatic insects.

REPRODUCTIVE BIOLOGY
Mating has been observed in wild populations during June and August. Nesting occurs in October and November. The female digs a large cavity in a sandy bank with her forelimbs and then enters the chamber to lay her eggs. A single clutch of three to five elongate, brittle-shelled eggs (1.4 x 0.8 in [35 x 20 mm]) is laid annually. Although it is unclear when they hatch, neonates appear after 170–220 days. They may be stimulated to emerge by the onset of winter rains that fill the wetlands and signal the beginning of a period when food is abundant.

CONSERVATION STATUS
Listed as Critically Endangered by the IUCN. The population has stabilized due to an effective captive breeding program; however, fewer than 400 individuals are known to inhabit the protected wetlands.

SIGNIFICANCE TO HUMANS
The rarity, limited distribution, and cryptic coloration and reclusive behavior of this species have limited their interactions with humans. ◆

Resources

Books

Cann, John. *Australian Freshwater Turtles.* Singapore: Beaumont Publishing, 1998.

Pritchard, Peter C. H., and Pedro Trebbau. *The Turtles of Venezuela.* Athens, OH: Society for the Study of Amphibians and Reptiles; Oxford, OH, 1984.

Periodicals

Georges, Arthur, et al. "A Phylogeny for Side-Necked Turtles (Chelonia: Pleurodira) Based on Mitochondrial and Nuclear Gene Sequence Variation." *Biological Journal of the Linnean Society* 67 (1999): 213–246.

McCord, William P., Mehdi Joseph-Ouni, and William W. Lamar. "A Taxonomic Reevaluation of *Phrynops* (Testudines: Chelidae) with the Description of Two New Genera and a New Species of *Batrachemys.*" *Revista de Biologia Tropical* 49 (2001): 715–764.

Patrick J. Baker, MS

Seaturtles
(Cheloniidae)

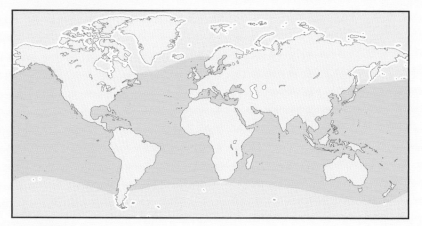

Class Reptilia

Order Testudines

Family Cheloniidae

Thumbnail description
Large marine turtles that have a low, streamlined shell covered with scutes and that have paddle- or flipper-like forelimbs

Size
Up to 84 in (213 cm) carapace length, 1,000 lb (454 kg)

Number of genera, species
5 genera, 6 species

Habitat
Marine ecosystems, circumtropical to temperate regions

Conservation status
Critically Endangered: 2 species; Endangered: 3 species; Data Deficient: 1 species

Distribution
All oceans, Mediterranean Sea

Evolution and systematics

The Cheloniidae are most closely related to the leatherback turtles of the family Dermochelyidae. Fossils are known from as early as the Cretaceous period. Because of uncertainty about the phylogenetic relationships among the five living genera, no subfamilies are currently recognized.

Physical characteristics

Large marine turtles with a low streamlined shell covered with scutes, paddle- or flipper-like forelimbs, a skull with a

solid bony roof, and a head that cannot be retracted within the shell. The plastron is somewhat reduced, lacks a hinge, and is connected to the carapace by ligaments.

Distribution

All tropical and subtropical oceans. Some seaturtles range well into temperate seas, including the Mediterranean Sea.

A green seaturtle (*Chelonia mydas*) digs a body pit before laying eggs. (Photo by Animals Animals ©Adrienne T. Gibson. Reproduced by permission.)

Loggerhead turtle (*Caretta caretta*) with attached remora in the Bahamas. (Photo by Animals Animals ©Herb Segars. Reproduced by permission.)

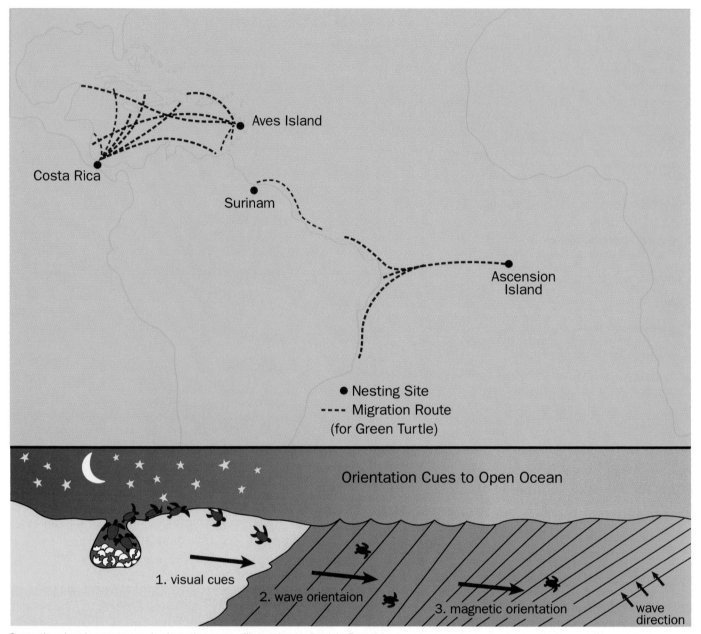

Seaturtle migration routes and orientation cues. (Illustration by Patricia Ferrer)

Habitat

Coastlines on the continental shelves, where feeding and nesting sites are most abundant. Hatchling and small juveniles of most species apparently are mostly pelagic.

Behavior

Seaturtles (males and females) often make extremely long migrations between feeding and nesting grounds (at least 190 mi [300 km] in some cases). Some species congregate off the sandy nesting grounds and then nest en masse in large groups called "arribadas." Most nesting is done at night, but one species is a diurnal nester. Seaturtles enter temperate seas during the

summer but usually either migrate to warmer waters or bury themselves in the mud in shallow coastal areas for the winter.

Feeding ecology and diet

All but one species of seaturtles are primarily carnivorous, feeding on sponges, mollusks, crustaceans, barnacles, sea urchins, or fish. The green seaturtle is primarily herbivorous, feeding mainly on sea grasses.

Reproductive biology

Female seaturtles migrate to nesting beaches in one- (rare) to three-year cycles. Seaturtles nest primarily on tropical dependent sex determination. Warm temperatures produce mostly females, and cool temperatures produce mostly males.

Conservation status

Five species of seaturtles are classified as Threatened, with two listed as Critically Endangered, three as Endangered. The remaining species is listed as Data Deficient.

Significance to humans

Despite international legislation to protect them, seaturtles and their eggs are still eaten locally around the world. The shells of some species, particularly the hawksbill (*Eretmochelys imbricata*), are used for making trinkets. Many more turtles are accidentally killed in fish and shrimp nets. Turtle excluder devices placed on these nets are known to reduce seaturtle drowning by at least an order of magnitude and are legally required by many countries, but universal use is sorely needed. Increasing numbers of seaturtles are being found with fibrous tumors (fibropapilloma) up to 12 in (30 cm) in diameter on the skin, in the mouth, and on the internal organs. More than 70% of the turtles in some areas are infected. The cause of this mysterious disease is not yet fully understand, although it is contagious and certainly linked to human pollution. The long-term effects of these tumors on seaturtle populations are unknown.

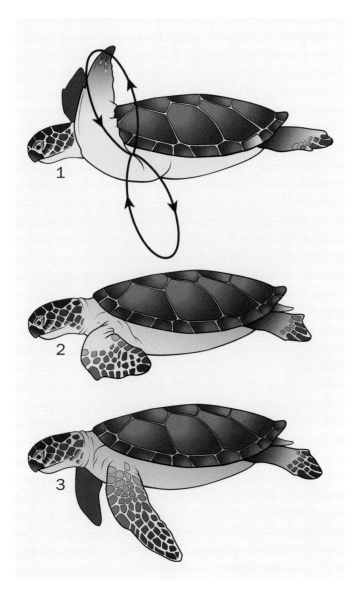

Seaturtle swimming strokes. (Illustration by Patricia Ferrer)

A hawksbill (*Eretmochelys imbricata*) over coral near the Solomon Islands. (Photo by Fred McConnaughey/Photo Researchers, Inc. Reproduced by permission.)

beaches, producing as many as seven or more clutches in a season at intervals of 10–30 days. The eggs are leathery and spherical and measure 1–2 in (30–60 mm) in diameter. Clutches usually contain 90–130 eggs, although maximum clutch size can approach 250. Incubation is generally quite short, only 40–70 days. All species exhibit temperature-

1. Kemp's ridley turtle (*Lepidochelys kempii*); 2. Loggerhead turtle (*Caretta caretta*); 3. Green seaturtle (*Chelonia mydas*). (Illustration by Patricia Ferrer)

Species accounts

Loggerhead turtle

Caretta caretta

TAXONOMY

Testudo caretta Linnaeus, 1758, Insulas Americanas ("American islands"). No subspecies are currently recognized, although the Pacific and Atlantic populations have been considered different races by some authors.

OTHER COMMON NAMES

English: Loggerhead; French: Caouanne; German: Unechte Karettschildkröte; Spanish: Caguama, tortuga boba.

PHYSICAL CHARACTERISTICS

The loggerhead seaturtle is the largest species in this family, reaching 84 in (213 cm) carapace length and weights up to 1,000 lb (454 kg). The head is quite broad posteriorly and short and round in front, hence the common name. Two pairs of prefrontal scales are present on the top of the head forward of the eyes. The heart-shaped carapace is serrate posteriorly and has five or more pairs of pleural scutes, the first pair in contact with the nuchal scute. Eleven to 15 (typically 12 or 13) marginal scutes are present on the rim of the shell. Three inframarginal scutes (all lacking pores) are present on the bridge between the marginal and the plastral scutes.

DISTRIBUTION

All tropical and temperate seas but rare in the eastern and central Pacific Ocean.

HABITAT

Mainly shallow marine waters along tropical continental shores but also around some islands. Loggerhead turtles enter bays, creeks, salt marshes, and the mouths of rivers.

BEHAVIOR

Loggerhead turtles undertake long migrations, often using warm oceanic currents for dispersal. A juvenile released off Okinawa Island was recaptured off San Diego just over two years later, and several adults have been recaptured 1,300–1,700 mi (2,100–2,700 km) from the site of original capture. Females migrate to nesting areas every two or three years. Adults often aggregate off nesting beaches before mi-

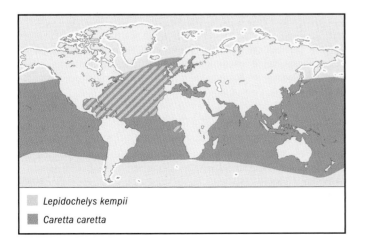

Lepidochelys kempii

Caretta caretta

grating back to feeding habitats. Hatchlings and small juveniles are apparently pelagic and associated with floating plants, animals, and flotsam. This species often ventures into temperate waters and nests farther north than any other seaturtle (solitary nests have been found in New Jersey in the United States). When in open waters, loggerhead seaturtles often float on the surface, presumably sleeping. Mitochondrial DNA studies have shown that turtles from different nesting regions differ genetically. This finding suggests that females return to the nesting beaches on which they hatched.

FEEDING ECOLOGY AND DIET

The loggerhead turtle is primarily carnivorous throughout its life. Hatchlings are known to eat jellyfish, snails, crustaceans, insects, and sargassum (an alga), most obtained while the turtle is floating in sargassum mats. Juveniles and adults feed mostly on the bottom and eat sponges, worms, conch and other snails, clams, squid, octopus, barnacles, horseshoe and other crabs, shrimp, sea urchins, fish, and occasionally hatchling seaturtles, algae, and other aquatic plants.

REPRODUCTIVE BIOLOGY

Loggerhead turtles reach maturity between 10 and 30 years of age. Courtship and mating apparently occur most commonly during the migrations to nesting grounds, several weeks before nesting begins, rather than near the nesting beaches. The male circles the female, bites her neck and shoulders, and mounts her shell from behind. The pair typically floats at the surface during copulation. Mating can occur day or night. Females apparently mate several times. DNA studies have revealed that more than one male may father eggs laid in a single clutch. Nesting usually occurs in spring and summer but with great geographic variation, particularly latitudinal, in timing and duration.

Nests are generally excavated above the high tide line, in front of the first dune, and usually at night. Once the site is chosen, the female first excavates a body pit using all four limbs and then uses only her rear feet to dig the nest chamber in the bottom of the pit. She then deposits 23–198 spherical, leathery eggs (usually 95–130) that measure 1–2 in (2.5–5.1 cm) in diameter and 1–2 oz (26–47 g) in mass. She then covers the nest, first using only her hind legs and eventually using all four limbs to cover and camouflage the entire site. Females may lay up to seven clutches per season at intervals of nine to 28 days, most typically about every two weeks, although four or five clutches per season is more usual. Most females nest only every two or three years. Incubation requires 46–80 days, typically 60–65, depending on the temperature. Hatchlings generally emerge from nests at night to avoid lethal ground temperatures during the day. They then scurry immediately to the surf. This species has temperature-dependent sex determination. Mostly females are produced above 84–86°F (29–30°C), and mostly males are produced below this temperature.

CONSERVATION STATUS

Loggerhead turtles are classified as Endangered on the IUCN Red List.

SIGNIFICANCE TO HUMANS

Although direct consumption of adults and eggs may be declining in many areas, humans are still responsible for much indirect mortality among loggerhead seaturtles through activity on or development of nesting beaches, by contributing to increases in predators such as raccoons and dogs, by drowning the turtles in shrimp or fish nets, and with pollutants. ◆

Green seaturtle
Chelonia mydas

TAXONOMY
Testudo mydas Linnaeus, 1758, Insulas Pelagi: Insulam Adscensionis ("Oceanic Islands: Ascension Island"). No subspecies are currently recognized. Nevertheless, some authors recognize the Pacific green turtle (also called the black turtle) as a distinct species; others consider the Pacific green turtle a subspecies of *Chelonia mydas*. Results of DNA studies do not support recognition of the Pacific turtle as distinct from other green seaturtles.

OTHER COMMON NAMES
English: Green turtle; French: Tortue verte; German: Suppenschildkröte; Spanish: Tortuga verde, Tortuga blanca.

PHYSICAL CHARACTERISTICS
Large, reaching 60 in (152 cm) carapace length and 750 lb (340 kg) body mass. The head is small and rounded anteriorly, and only one pair of elongated prefrontal scales is present on the top of the head forward of the eyes. The heart-shaped carapace is only weakly serrate posteriorly and has four pairs of pleural scutes, the first pair of which does not contact the nuchal scute. Twelve marginal scutes are typically present along each side of the shell. Four inframarginal scutes (all lacking pores) are present on each bridge between the marginal and the plastral scutes. The greenish color of the fat of this turtle is the source of its common name.

DISTRIBUTION
All tropical and temperate seas.

HABITAT
Although green seaturtles venture into temperate seas, adults are found primarily in the tropics. These turtles can be found in the open sea, but they are most commonly seen in areas of shallow water with an abundance of submerged vegetation, especially sea grass flats. Hatchlings are more pelagic and often are found in mats of sargassum.

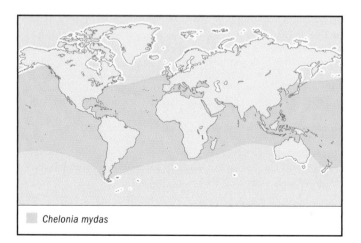

Chelonia mydas

BEHAVIOR
Green seaturtles nest primarily on tropical beaches and may migrate more than 1,900 mi (3,000 km) between feeding and nesting areas. These turtles are known to thermoregulate by basking at the water's surface, but they are the only marine turtle known to leave the water to bask on land. One population in the Gulf of California is known to hibernate under water by partially burying itself in the substrate.

FEEDING ECOLOGY AND DIET
Although it is assumed that hatchlings and juveniles are primarily carnivorous, few data are available. Adults are well known to be almost completely herbivorous, feeding primarily on several genera of sea grasses as well as on algae. Animal matter, such as sponges, jellyfish, mollusks, crustaceans, sea urchins, and sea squirts, is occasionally ingested, some perhaps secondarily while the turtle feeds on sea grasses. Feeding generally occurs during the day.

REPRODUCTIVE BIOLOGY
The age at maturity of green seaturtles is not definitively known but is speculated to be between 20 and 30 years. Courtship and mating take place off the nesting beaches, and females may mate with several males. Courtship involves chasing, nuzzling, rubbing, sniffing, and biting the female. If the female is receptive, the male mounts her shell from behind and swings his much larger tail under hers for intromission. Copulation may last several hours; one report describes a 52-hour mating. Mating may occur at or below the water's surface. Females can store sperm, perhaps for several years, so individual eggs in the same clutch may have different fathers.

There is considerable variation in the timing of the green seaturtle nesting season, both locally and globally. For example, in the western Atlantic, nesting is typically from March to October, with a peak from May to September. In the eastern Pacific, nesting may occur any time between February and January. Females exhibit considerable fidelity to their nesting beaches, and this trait accounts for slight genetic differences between separate nesting colonies. Females nest anywhere from the open sand above the high tide line to fully shaded areas just inland from the beach. Most nesting occurs at night. Nest construction is similar to that of the loggerhead seaturtle, except that the green seaturtle digs the deepest body pit of any cheloniid sea turtle (up to 20 in [50 cm]). Females lay as many as seven clutches in a season, usually at 12–14-day intervals, but laying two to five clutches is most common. Clutch size is generally positively related to the female's size. Clutches range from three to 238 eggs, although clutches of 100–120 eggs are typical. The leathery eggs are nearly spherical and 1–2 in (2.5–5.8 cm), usually 1.4–1.6 in (3.5–4 cm) in diameter and weigh 1–2 oz (28–65 g, usually 35–50 g). Most females nest only every two years, but cycles of one to four years are known. Incubation typically requires 50–70 days but may take between 30 and 90 days, depending on nest temperature. Hatchlings emerge from their nests at night and move immediately to the sea. Green seaturtles exhibit temperature-dependent sex determination. High temperatures produce mostly females, and low temperatures produce mostly males.

CONSERVATION STATUS
Green seaturtles are classified as Endangered on the IUCN Red List. The Mediterranean subpopulation is listed as Critically Endangered.

SIGNIFICANCE TO HUMANS
Despite international protection, green seaturtles and their eggs are still consumed by local peoples in many parts of the

world. Many other human activities also increase mortality. Turtles are killed for sport, drown in shrimp or fish nets, and are wounded by boat propellers. Other turtles are killed by predators whose numbers have increased because of human activities (e.g., raccoons, pigs, and dogs). Nesting grounds are destroyed by hotel and housing developments. ◆

Kemp's ridley turtle

Lepidochelys kempii

TAXONOMY
Thalassochelys (Colpochelys) kempii Graman, 1880, Gulf of Mexico. No subspecies are recognized.

OTHER COMMON NAMES
English: Atlantic ridley, gulf ridley, Mexican ridley; French: Tortue de Kemp; Spanish: Tortuga lora.

PHYSICAL CHARACTERISTICS
The Kemp's ridley is the smallest of the seaturtles, reaching a maximum of only 30 in (76 cm) carapace length, and 108 lb (49 kg) body mass. The head is somewhat pointed anteriorly and has a distinctly hooked upper beak. Two pairs of prefrontal scales are present on the top of the head forward of the eyes. The heart-shaped carapace is serrate posteriorly and has five pairs of pleural scutes; the first pair is in contact with the nuchal scute. Twelve to 14 marginal scutes are present on the rim of the shell. Four inframarginal scutes (each with a posterior pore) are present on each bridge between the marginal and the plastral scutes.

DISTRIBUTION
Gulf of Mexico to north Atlantic Ocean.

HABITAT
Adult ridleys prefer the shallow water of the Gulf of Mexico, although for the first two years of life they drift in floating mats of sargassum or other flotsam in the gulf currents. Subadults venture into temperate waters, such as Chesapeake Bay, to feed.

BEHAVIOR
Little is known about the aquatic behavior of Kemp's ridley turtles. Although mostly confined to the Gulf of Mexico, females still may migrate long distances, often more than 600 mi (1,000 km) to and from the only significant nesting beach, at Rancho Nuevo, Tamaulipas, Mexico.

FEEDING ECOLOGY AND DIET
Kemp's ridley turtles are primarily carnivorous throughout life. Although they feed mainly on crabs, these turtles eat jellyfish, comb jellies, snails, clams, squid eggs, shrimp, insects, barnacles, sea urchins, sea cucumbers, fishes, and diamondback terrapins. They occasionally feed on sargassum weed and other algae or aquatic plants.

REPRODUCTIVE BIOLOGY
Age at maturity of Kemp's ridley turtles is estimated to be 8–12 years. Females then reproduce at one- to three-year intervals, most nesting every one or two years. Courtship and mating occur off the nesting beaches before nesting. These behaviors have not been well described but are similar to those of other seaturtles. Nearly all nesting occurs along a single beach in Tamaulipas, Mexico, from mid April to mid July. Females nest almost exclusively during the day, coming ashore in arribadas, although some come ashore alone. Nests are dug in open sand on the upper beach or in the dunes behind the beach. Nest construction is similar to that by other cheloniid sea turtles, but the Kemp's ridley nests are generally shallower. Females produce up to four clutches, usually one to three, per season at intervals of 10–49 days, although the usual interval is 20–28 days. Clutch size ranges from 51 to 185 eggs, nests of 100–110 eggs being most common. The spherical, leathery eggs measure 1–2 in (2.5–5.1 cm, averaging 3.9 cm) in diameter and weigh 1–1.5 oz (24–41 g, averaging 30 g). Incubation requires 45–70 days depending on temperature, but most eggs hatch in 50–55 days. Kemp's ridley turtles exhibit temperature-dependent sex determination. High temperatures produce mostly females, and low temperatures produce mostly males.

CONSERVATION STATUS
Kemp's ridley turtles are classified as Critically Endangered on the IUCN Red List.

SIGNIFICANCE TO HUMANS
Kemp's ridleys once were the most abundant seaturtle in the Gulf of Mexico, an estimated 42,000 nesting in Mexico in one day. Although these turtles are rigidly protected internationally and the primary nesting beach is a Mexican national reserve, Kemp's ridley turtles remain the most critically endangered marine turtle, perhaps 1,000–2,000 adults remaining. Drowning in shrimp trawl nets is believed to be the most prevalent unnatural cause of death. Recent increases in the nesting population offer hope for recovery. ◆

Resources

Books
Bjorndal, K. J. *Biology and Conservation of Sea Turtles.* Revised edition. Washington, DC: Smithsonian Institution Press, 1995.

Lutz, P. L., and J. A. Musick, eds. *The Biology of Sea Turtles.* Boca Raton, FL: CRC Press, 1997.

Márquez-M., R. *Synopsis of Biological Data on the Kemp's Ridley Turtle,* Lepidochelys kempi *(Garman, 1880).* NOAA Technical Memo NMFS-SEFSC-343. Washington, DC: National Oceanic and Atmospheric Administration, 1994.

Periodicals
Dodd, C. K., Jr. "*Caretta caretta* (Linnaeus) Loggerhead Sea Turtle." *Catalogue of American Amphibians and Reptiles* 483 (1990): 1–7.

Karl, S. A., and B. W. Bowen. "Evolutionary Significant Units versus Geopolitical Taxonomy: Molecular Systematics of an

Resources

Endangered Sea Turtle (Genus *Chelonia*)." *Conservation Biology* 13, no. 5 (1999): 990–9.

Parham, J. F., and D. E. Fastovsky. "The Phylogeny of Cheloniid Sea Turtles Revisited." *Chelonian Conservation Biology* 2, no. 4 (1997): 548–54.

Rostal, D. C., J. S. Grumbles, R. A. Byles, R. Márquez-M., and D. W. Owens. "Nesting Physiology of Kemp's Ridley Sea Turtles, *Lepidochelys kempi*, at Rancho Nuevo, Tamaulipas, Mexico, with Observations on Population Estimates." *Conservation Biology* 2, no. 4 (1997): 538–47.

Van Buskirk, J., and L. B. Crowder. "Life-History Variation in Marine Turtles." *Copeia* (1994): 66–81.

Zug, G. R., and C. H. Ernst. "*Lepidochelys fitzinger* Ridley Sea Turtles." *Catalogue of American Amphibians and Reptiles* 587 (1994): 1–6.

John B. Iverson, PhD

Snapping turtles
(Chelydridae)

Class Reptilia
Order Testudines
Family Chelydridae

Thumbnail description
Large, vicious turtles with a long tail, a hooked beak, and a greatly reduced plastron

Size
7.1–31.5 in (18–80 cm); up to 249 lb (113 kg)

Number of genera, species
2 genera; 4 species

Habitat
Freshwater ponds, lakes, swamps, and rivers

Conservation status
Vulnerable: 1 species

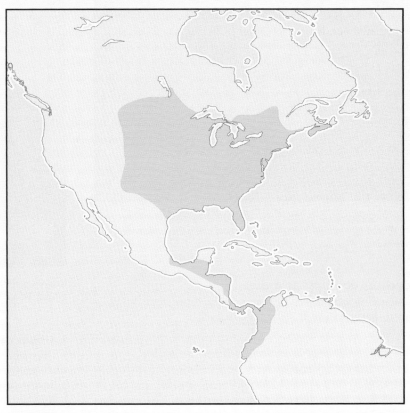

Distribution
Canada to Ecuador

Evolution and systematics

Chelydridae is most closely related to Platysternidae, the family of the big-headed turtles. Some authors have considered the two groups as subfamilies in the same family, although molecular evidence supports their separate family recognition. The fossil record dates from the Paleocene of North America and the Oligocene of Eurasia. The genus *Chelydra* (snapping turtles) is known from as far back as the Pliocene in North America, whereas the genus *Macrochelys* is known from as early the Miocene.

Physical characteristics

Snapping turtles are large aquatic turtles with a long tail with three rows of tubercles; a hooked beak; a keeled, posteriorly serrated carapace; a reduced, cruciform, hingeless plastron; and heavy claws. Males are larger than females. Only 11 marginal scutes are present on each side of the carapace. The abdominal scutes on the plastron are reduced and not in contact medially. The carapace and plastron are connected by a narrow bony bridge. The posterior skull roof is deeply emarginated.

Distribution

Snapping turtles occur in southern Canada across the eastern two-thirds of the United States, and in discontinuous populations from Veracruz, Mexico, to western Ecuador; from sea level to over 6,560 ft (2,000 m) elevation.

Habitat

These turtles are found in almost any kind of freshwater habitat within their range, but also occasionally enter brackish waters.

Behavior

They possess a vicious temperament (hence their common name) and direct their powerful snapping jaws at both their food and their predators. Snapping turtles are highly aquatic, but do leave the water to nest, and one species migrates between bodies of water. They may be active at any time of day or night, but nocturnal activity is rare in northern populations. These turtles hibernate at temperate latitudes, but presumably are active year-round at more tropical sites. They occasionally bask out of water.

Common snapping turtle (*Chelydra serpentina*) swimming with a caught fish. (Photo by Animals Animals ©Zig Leszczynski. Reproduced by permission.)

Snapping turtle (*Chelydra serpentina*) with eggs. (Photo by E. R. Degginger. Bruce Coleman, Inc. Reproduced by permission.)

Feeding ecology and diet

Snapping turtles are primarily carnivorous, but omnivorous to herbivorous in some populations.

Reproductive biology

Male snapping turtles are larger than females, and hence courtship is not elaborate. Females lay up to 109 eggs per clutch during the spring or early summer, with a maximum of one clutch produced per year. Eggs are spherical, hard-shelled (but not brittle), and 0.9–1.6 in (2.3–4 cm) in diameter. Snapping turtles exhibit temperature-dependent sex determination, the sex of hatchlings being determined by the temperature during the middle third of incubation.

Conservation status

Tropical forms are apparently not common; their status is uncertain, but they are apparently not yet endangered. The alligator snapping turtle (*Macrochelys temminckii*) has declined significantly due to overharvesting, and hence is classified as Vulnerable on the IUCN Red List.

Significance to humans

Snapping turtles are harvested primarily for their meat, although some may be removed for the pet trade.

Species accounts

Snapping turtle
Chelydra serpentina

TAXONOMY
Testudo serpentina Linnaeus, 1758, "Calidus regionibus" (warm regions). Two subspecies are recognized.

OTHER COMMON NAMES
English: Common snapping turtle; snapper; French: Chélydre serpentine; German: Schnappschildkröte; Spanish: Tortuga-lagarto común.

PHYSICAL CHARACTERISTICS
The carapace of these large turtles, to 19.3 in (49 cm), bears three low, knobby keels (except in the oldest individuals). The shell is dark, although it may range from brown to olive to black. The head is large, the upper jaw is somewhat hooked anteriorly, and the eyes open dorsolaterally (toward the top and side of the head).

Chelydra serpentina

DISTRIBUTION
Ranges extend from southern Canada across the United States east of the Rocky Mountains to New Mexico, Texas, and

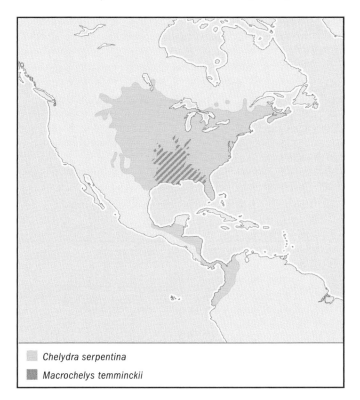

Chelydra serpentina

Macrochelys temminckii

Florida. Tropical snapping turtle populations (until recently considered subspecies of the snapping turtle) range from Veracruz, Mexico, through Central America to western Ecuador in northwestern South America.

HABITAT
Snapping turtles inhabit nearly any body of freshwater within their range, although they also invade brackish water environments in many areas. They seem to prefer warm, shallow, still water, with soft bottoms and abundant aquatic vegetation.

BEHAVIOR
Snapping turtles are highly aquatic, spending most of their time lying camouflaged in the mud in shallow water. Most active during the morning and early evening, these turtles are often active at night in the south, but rarely so in the north. When active, they are primarily bottom-walkers, slowly moving over the bottom in search of food or mates. Most thermoregulation is done by selecting warm, shallow, submerged sites; however, they occasionally bask by floating on the water's surface or even by climbing out of the water onto emergent logs or other objects. They may be active year-round in the south, but hibernate for half the year or more in the north. However, some individuals may be active in the north in midwinter under thick ice cover. In addition to nesting forays by females, snappers frequently travel great distances overland between bodies of water (often almost a mile). They also make equally impressive movements in the water, one observed female moved 2.1 mi (3.4 km) in just 10 days.

FEEDING ECOLOGY AND DIET
Although snapping turtles are probably carnivorous by preference, they can subsist on diets ranging from completely carnivorous to completely herbivorous. Animal foods include sponges, flatworms (planarians), earthworms, leeches, clams, snails, crustaceans, spiders, insects, amphibians and their eggs and larvae, snakes, other turtles, birds, small mammals, and carrion. Snapping turtles also eat algae, duckweed, and other submergent and emergent aquatic plants and their seeds.

REPRODUCTIVE BIOLOGY
Mating is known to occur from April to November, but probably peaks in the spring and fall. Courtship is highly variable, from the male directly mounting the female from behind, to the male trailing along after the female for several minutes before mounting, to face-offs, head-swaying, and/or water gulping and belching before mounting. Eventually the male mounts atop the female's shell and swings his tail under hers to mate. At high latitudes, nesting begins in early June and may extend to early July, whereas at low latitudes it may extend from late April to early June. Females may nest within a few feet of the water's edge or travel overland as far as 9.9 mi (16 km) to nest. The nest is dug (and covered) with the hind feet, and because of its size and depth (to 9.8 in [25 cm]), even a covered nest is quite obvious. Consequently, nests are heavily predated by animals, most within 24 hours of construction.

Females produce at most one clutch per year, with some females apparently skipping some years. They lay spherical, hard-shelled eggs that average 1.1 in (2.8 cm) in diameter and 0.4 oz (11 g) in mass. Egg size does not seem to increase significantly with female size. Clutch size is highly variable, ranging from six to 109, averaging about 32 across the range, and is positively correlated with female body size, latitude, longitude, and elevation (the largest clutches are laid in western Nebraska). Incubation in nature requires 55–125 days (more typically 75–95) depending on nest temperature (development being faster at higher temperatures) and geography (incubation times being longer in the south). Hatchling snapping turtles usually emerge from the nest in the late summer and fall (August to October) and move directly to the water. Hatchlings in northern populations that do not emerge in the fall before the onset of cold weather almost never survive the winter, probably because of their low tolerance of subfreezing body temperatures. High and low incubation temperatures result in the production of all-female offspring, and intermediate temperatures produce all males. Because their clutches are so large, eggs in different parts of the same nest may produce different sex ratios, e.g., all females at the top and all males at the bottom.

CONSERVATION STATUS

This species has such an extensive range and is so prolific that it has so far been able to persist even in habitats significantly altered by humans.

SIGNIFICANCE TO HUMANS

Snapping turtles are exploited by humans primarily for their meat, although some small individuals make their way into the pet trade. ◆

Alligator snapping turtle
Macrochelys temminckii

TAXONOMY

Chelonura temminckii Harlan, 1835, tributary stream of the Mississippi, which enters the river above Memphis, in west Tennessee. No subspecies are recognized.

OTHER COMMON NAMES

English: Alligator snapper; French: Macroclémyde de Temminck; German: Geierschildkröte.

PHYSICAL CHARACTERISTICS

Alligator snapping turtles are the largest freshwater

Macrochelys temminckii

turtles in North America (up to 31.5 in [80 cm]; 250 lb [113 kg]). The carapace bears three prominent knobby keels. The head is very large, has a strongly hooked beak, and eyes that open laterally (toward the side of the head). Three to eight small supramarginal scutes are present laterally between the marginal and pleural scutes. The turtles have a pink, wormlike projection on the tongue, which they can wriggle to lure prey.

DISTRIBUTION

Snapping turtles occur in the Gulf of Mexico drainages in the southeastern United States.

HABITAT

These turtles are primarily found in rivers and large streams, but also in lakes, ponds, swamps, and even brackish water.

BEHAVIOR

Alligator snapping turtles are highly aquatic, almost never leaving the water except to nest. They are capable of making extremely long underwater movements, up to 4.2 mi (6.8 km) in a week, or to 18.7 mi (30 km) or more over several years.

FEEDING ECOLOGY AND DIET

These turtles are primarily carnivorous. In addition to luring fish with their tongue, they also eat crayfish, crabs, clams, snails, salamanders, turtles, snakes, small alligators, birds, and mammals, as well as plant roots, fruits, and seeds (e.g., grapes and acorns).

REPRODUCTIVE BIOLOGY

Mating may occur in the spring or the fall. Courtship is poorly developed and males climb onto the female's shell from behind for mating. Nesting occurs between late April and early June, and most nests are laid within 236 ft (72 m) of the water. The eggs are spherical and 1.3–1.6 in (3.3–4.1 cm) in diameter. Females lay a single clutch each year of nine to 61 eggs (about 35 being typical), larger females producing larger clutches. Laboratory incubation of eggs to hatching requires 79–113 days, depending on temperature. Warm incubation temperatures result in the production of all females, low incubation temperatures produce nearly all females, and intermediate temperatures produce mostly males.

CONSERVATION STATUS

Alligator snapping turtles have declined significantly due to overharvesting, and hence are cited as Vulnerable on the IUCN Red List.

SIGNIFICANCE TO HUMANS

Humans heavily exploit these turtles for their meat, and hatchlings are produced for the pet trade. ◆

Resources

Books

Pritchard, P. C. H. *The Alligator Snapping Turtle, Biology and Conservation*. Milwaukee: Milwaukee Public Museum, 1989.

Periodicals

Congdon, J. D., A. E. Dunham, and R. C. van Loben Sels. "Demographics of Common Snapping Turtles (*Chelydra serpentina*): Implications for Conservation and Management of Long-Lived Organisms." *American Zoologist* 34, no. 3 (1994): 397–408.

Ewert, M. A., and D. R. Jackson. "Nesting Ecology of the Alligator Snapping Turtle (*Macroclemys temminckii*) along the Lower Apalachicola River, Florida." *Florida Game and Fresh Water Fish Commission Non-Game Wildlife Program Final Report* (1994): 1–45.

Resources

Iverson, J. B., H. Higgins, A. G. Sirulnik, and C. Griffiths. "Local and Geographic Variation in the Reproductive Biology of the Snapping Turtle (*Chelydra serpentina*)." *Herpetologica* 53 (1997): 96–117.

Phillips, C. A., W. W. Dimmick, and J. L. Carr. "Conservation Genetics of the Common Snapping Turtles (*Chelydra serpentina*)." *Conservation Biology* 10 (1996): 397–405.

Walker, D., P. E. Moler, K. A. Buhlmann, and J. C. Avise. "Phylogeographic Uniformity in Mitochondrial DNA of the Snapping Turtle (*Chelydra*)." *Animal Conservation* 1 (1998): 55–60.

John B. Iverson, PhD

Central American river turtles
(Dermatemydidae)

Class Reptilia
Order Testudines
Suborder Cryptodira
Family Dermatemydidae

Thumbnail description
This is a large aquatic turtle whose dark shell appears leathery in adults, whose plastron and bridge are both large, and whose head is relatively small with a pointed snout; a row of inframarginal scutes separates the scutes of the carapace from those on the plastron

Size
Up to 26 in (65 cm) and 49 lb (22 kg)

Number of genera, species
1 genus; 1 species

Habitat
Permanent rivers and lakes

Conservation status
Endangered

Distribution
Caribbean Sea and Gulf of Mexico drainages of Mesoamerica

Evolution and systematics

These river turtles (*Dermatemys mawii*) are most closely related to the mud and musk turtles (family Kinosternidae). The fossil record of this family is extensive, with the earliest material being from the Lower Cretaceous in Asia, and abundant remains in North and Central America, Europe, and Africa during the Tertiary. The Central American river turtle is also commonly known as hickety (Belize), jicotea, tortuga aplanada (Mexico), tortuga blanca (Mexico, Guatemala), and tortuga plana (Mexico). No subfamilies are recognized.

The taxonomy of this species is *Dermatemys mawii* Gray, 1847, South America (in error); restricted to Alvarado, Veracruz, Mexico, by Smith and Taylor, 1950.

Physical characteristics

These large turtles (up to 49 lb [22 kg]) have a low, dark, carapace that is unicarinate in juveniles, but smooth and leathery in adults. Females reach larger sizes than males. The posterior margin of the shell is smooth and not serrated. The yellow or cream-colored plastron is large, unhinged, and connected to the carapace by a broad bridge on which four or five inframarginal scutes are located (separating the plastral from the carapacial scutes). The snout is pointed, no chin barbels are present, and the toes are strongly webbed. The top of the head in adult males is yellowish and that in adult females and juveniles is gray.

Central American river turtle (*Dermatemys mawii*). (Illustration by Bruce Worden)

Distribution

Atlantic lowlands from southern Veracruz in Mexico through the southern Yucatán Peninsula, northern Guatemala, and Belize.

Habitat

Permanent water in rivers and large freshwater lakes, occasionally entering brackish water as evidenced by the finding of barnacles on some individuals.

Behavior

This turtle is highly aquatic, is capable of remaining submerged for very long periods, and has a very difficult time moving on land. It is primarily nocturnal, but it can sometimes be seen basking at the surface of the water during the day. It does not bask out of the water.

Feeding ecology and diet

It is almost completely herbivorous, feeding on fruits (e.g., figs), aquatic grasses, and fallen leaves. Insects, fish, and mollusks are occasionally consumed.

Reproductive biology

Courtship and mating apparently occur from March to September, and nesting occurs primarily during the rainy season, from late September to December. However, nesting sometimes occurs from late August to March or April. Females produce one to four clutches per year, with two being usual. Nests are excavated in the soil, usually within 10 ft (3 m) of the water, and are often inundated by rising water. Eggs have been found to be viable after being submerged for up to 28 days. Eggs are brittle-shelled, oblong, 2.1–2.8 in × 1.2–2 in (54–72 mm × 30–50 mm), and weigh 1.2–2.5 oz (34–70 g). Clutch sizes range from two to 20, with eight to 14 being typical. Large females have greater annual reproductive potential than small females. Incubation requires seven to 10 months, and hatching occurs in late May through July, at the beginning of the next rainy season. The sex is determined by nest temperature during the middle third of incubation, with high temperatures producing females and low temperatures producing males.

Conservation status

Listed as Endangered on the IUCN Red List, as Endangered by the U.S. Fish and Wildlife Service, and in Appendix II of the CITES.

Significance to humans

Central American river turtles are intensively exploited for food by indigenous peoples throughout their range, even in areas where the species is legally protected. The flesh is considered a delicacy.

Central American river turtle (*Dermatemys mawii*) in Tabasco, Mexico. (Photo by Jean-Gérard Sidaner/Photo Researchers, Inc. Reproduced by permission.)

Resources

Books

Lee, Julian C. *The Amphibians and Reptiles of the Yucatán Peninsula.* Ithaca, NY: Comstock, 1996.

Polisar, John. "Effects of Exploitation on *Dermatemys mawii* Populations in Northern Belize and Conservation Strategies for Rural Riverside Villages." In *Proceedings: Conservation, Restoration, and Management of Tortoises and Turtles: An International Conference, 11 to 16 July 1993, State University of New York, Purchase, New York,* edited by Jim Van Abbema. Bronx, NY: New York Turtle and Tortoise Society, 1997.

Periodicals

Polisar, John. "Reproductive Biology of a Flood-Season Nesting Freshwater Turtle of the Northern Neotropics: *Dermatemys mawii* in Belize." *Chelonian Conservation and Biology* 2, no. 1 (1996): 13–25.

Polisar, J., and R. H. Horwich. "Conservation of the Large, Economically Important River Turtle *Dermatemys mawii* in Belize." *Conservation Biology* 8, no. 2 (1994): 338–340.

Vogt, R. C., and O. Flores-Villela. "Effects of Incubation Temperature on Sex Determination in a Community of Neotropical Freshwater Turtles in Southern Mexico." *Herpetologica* 48, no. 3 (1992): 265–270.

John B. Iverson, PhD

Leatherback seaturtles

(*Dermochelyidae*)

Class Reptilia
Order Testudines
Suborder Cryptodira
Family Dermochelyidae

Thumbnail description
A huge marine turtle with a smooth, leathery, black, teardrop-shaped carapace, with scattered small white or yellow blotches

Size
Up to 96 in (244 cm) and 1,911 lb (867 kg)

Number of genera, species
1 genus; 1 species

Habitat
Oceans

Conservation status
Critically Endangered

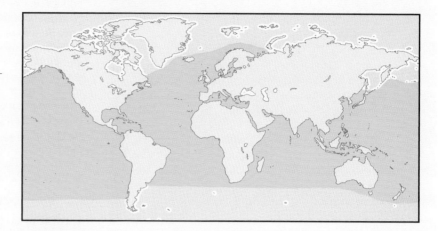

Distribution
Worldwide, oceanic

Evolution and systematics

The leatherback seaturtle (*Dermochelys coriacea*) is most closely related to the other marine turtles of the family Cheloniidae. The family Dermochelyidae is known from the Eocene to the Pliocene in Africa, Europe, Japan, Antarctica, Peru, and North America. The Pacific and Atlantic populations differ genetically and have been recognized as separate subspecies by some authors. This seaturtle is also commonly known as caldon (Trinidad), canal (Caribbean, Latin America), caouana (Carib Indians), cardon (Venezuela), laúd, luth, and trunk turtle. No subfamilies are recognized.

The taxonomy of this species is *Testudo coriacea* Vandelli, 1761, Mediterranean and Adriatic seas.

Physical characteristics

The smooth, streamlined carapace is teardrop shaped (up to 96 in [244 cm] in length), bears seven longitudinal ridges, lacks epidermal scutes, and is covered with leathery skin. Most of the bones of the carapace have been lost, but a mosaic of hundreds of small dermal bones lies just under the leathery skin. The head also lacks scales and has a prominent toothlike cusp on each side of the upper jaw. The shell and skin are generally black with scattered white or yellow blotches or spots. The plastron is predominately white with scattered dark blotches or spots. The forelimbs are paddle-like and lack claws.

Distribution

This pelagic species has perhaps the greatest distribution of any reptile. It is found in all tropical oceans (and in the Mediterranean Sea), and it also ventures into the cold temperate waters around Newfoundland, Iceland, Norway, the Cape of Good Hope, New Zealand, and Alaska. Its nesting is on tropical and subtropical shores.

Habitat

This species is pelagic, and is only rarely found in shallow bodies of water.

Behavior

Leatherback seaturtles make extensive migrations between feeding and nesting areas. Recaptures have been made more than 3,100 mi (5,000 km) away, and individuals have traveled an average of more than 19 mi (30 km) per day for weeks at a time. Feeding takes place in both cold and warm water environments, in large part because leatherbacks can maintain body temperatures above environmental temperatures. Body heat generated by muscle activity is lost very slowly due to the thermal inertia of the large body, excellent insulative properties of the oily skin, and countercurrent circulatory system in the limbs which keeps the heat in the body's core. Leatherbacks sometimes aggregate, probably in association with feeding (e.g., on schools of jellyfish). They also are capable of making dives to depths of more than 3,300 ft (1,000 m) to reach food sources. They are active both day and night, although diving activity is increased at night. As they leave the nest, hatchlings are attracted to open, highly illuminated areas (usually the open sea).

Feeding ecology and diet

Leatherbacks are almost completely carnivorous, preferring oceanic jellyfish. However, they also consume hydrozoans,

Leatherback seaturtle (*Dermochelys coriacea*). (Illustration by Patricia Ferrer)

snails, bivalves, octopi, squid, amphipod crustaceans, crabs, sea urchins, tunicates, and small fish, as well as algae, kelp, and sea grasses. Unfortunately, plastic bags or balloons dumped in the ocean by humans are often mistaken for jellyfish and, if consumed, can block the gastrointestinal tract, causing the turtle's death.

Reproductive biology

Based on growth rings formed annually in the long bones of leatherbacks, sexual maturity is reached at about 13 to 14 years. Courtship and mating have only rarely been observed, and are known to occur in tropical water (e.g., near nesting beaches), but may also occur before or during migration to the nesting beaches. Nesting occurs on a great number of tropical beaches, but the number of females using a given beach is often small, and the numbers at even the best sites are declining. Atlantic leatherbacks nest from April to November; those from the Pacific nest at various times of the year depending on the location. Nesting is usually at night, and takes place just above the high tide line, on open sandy beaches free of rocks, etc., that would abrade the soft plastron. The female digs a huge body pit with both forelimbs and hind limbs, and then excavates the actual nest cavity with her hind feet. Once the eggs are laid, the opening to the nest cavity is covered with sand by the hind limbs, and eventually the forelimbs are also used to fill the body pit.

Eggs are spherical, with pliable shells, and are 1.9–2.6 in (49–65 mm) in diameter and weigh 2.5–3.2 oz (70–90 g); however, from one to 103 undersized (0.6–1.8 in [15–45 mm]) and yolkless eggs may be produced along with the normal egg complement, which numbers from 46 to 160. Clutch and egg size both tend to increase with female size. Clutch sizes are also generally larger on Atlantic beaches than on eastern Pacific beaches, and clutches laid during the middle of the nesting season are usually larger than those in early or late nests. Females oviposit up to 11 times during a nesting season, although half that many clutches per season is more usual. Females nest only every two or three years. Incubation requires 50–78 days, but most undisturbed nests apparently hatch in 60–68 days. Most hatchlings leave the nest at night. Sex is determined by temperature during the middle third of incubation, with only females produced at high temperatures and only males at low ones.

Conservation status

This species is listed as Critically Endangered on the IUCN Red List, as Endangered by the U.S. Fish and Wildlife Service, and in Appendix I of the CITES.

Significance to humans

Population estimates suggest that this species has declined by more than 70% in less than one generation. Egg poaching has been the primary reason for the decline in this species, although adults are also harvested in many areas for their flesh and/or the oil in their shells.

Leatherback seaturtle hatchling (*Dermochelys coriacea*) in Michoacan, Mexico. (Photo by Animals Animals ©George H. H. Huey. Reproduced by permission.)

Resources

Books

Pritchard, Peter C. H., and Pedro Trebbau. *The Turtles of Venezuela*. Athens, OH: Society for the Study of Amphibians and Reptiles; Oxford, OH, 1984.

Periodicals

Spotila, James R., et al. "Worldwide Population Decline of *Dermochelys coriacea:* Are Leatherback Turtles Going Extinct?" *Chelonian Conservation and Biology* 2, no. 2 (1996): 209–222.

Steyermark, Anthony C., et al. "Nesting Leatherback Turtles at Las Baulas National Park, Costa Rica." *Chelonian Conservation and Biology* 2, no. 2 (1996): 173–183.

Tucker, A. D., and N. B. Frazer. "Reproductive Variation in Leatherback Turtles, *Dermochelys coriacea*, at Culebra National Wildlife Refuge, Puerto Rico." *Herpetologica* 47 (1991): 115–124.

Wood, Roger C., Jonnie Johnson-Gove, Eugene S. Gaffney, and Kevin F. Maley. "Evolution and Phylogeny of the Leatherback Turtles (Dermochelyidae), with Descriptions of New Fossil Taxa." *Chelonian Conservation and Biology* 2, no. 2 (1996): 266–286.

Zug, George R., and James F. Parham. "Age and Growth in Leatherback Turtles, *Dermochelys coriacea* (Testudines: Dermochelyidae): A Skeletochronological Analysis." *Chelonian Conservation and Biology* 2, no. 2 (1996): 244–249.

John B. Iverson, PhD

New World pond turtles

(Emydidae)

Class Reptilia
Order Testudines
Suborder Cryptodira
Family Emydidae

Thumbnail description
Small- to medium-sized turtles; carapace may be depressed, domed, or strongly keeled; plastron may or may not be hinged; double articulation found between the fifth and sixth cervical vertebrae

Size
10–24 in (25–60 cm)

Number of genera, species
12 genera; 35 species

Habitat
Freshwater rivers, streams, lakes, and ponds; other species are semiaquatic to fully terrestrial; still other species frequent estuaries and coastal waters

Conservation status
Endangered: 6 species; Vulnerable: 7 species; Lower Risk/Near Threatened: 14 species

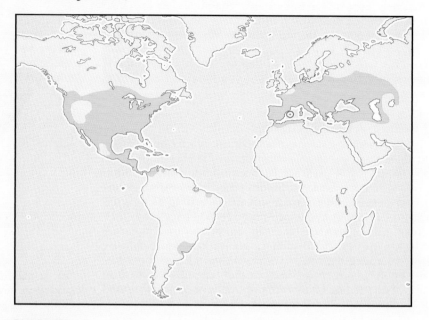

Distribution
Temperate and tropical regions of North and South America, Europe, western Asia, and northern Africa

Evolution and systematics

The oldest fossils are known from the Upper Cretaceous and Paleocene of North America. These modern cryptodires are most closely related to the Geoemydidae of South America, Europe, Africa, and Asia and the Testudinidae, which also are found in North America.

Morphological and molecular evidence suggests a close relationship among western pond turtles (*Actinemys marmorata*), European pond turtles (*Emys orbicularis*), and Blanding's turtles (*Emydoidea blandingii*). Some researchers consider them to be members of a single genus (*Emys*), while others recognize them individually as the sole representatives of monotypic genera. Two subfamilies are recognized: the Emydinae (palatine excluded from the triturating surface) and the Deirochelyinae (humeropectoral sulcus excluded from the entoplastron).

Physical characteristics

There are typically eight pleurals, five vertebrals, and 24 marginals on the carapace and 12 scutes on the plastron. The seam between the posterior marginal scutes and the last vertebral overlap the pygal bone. A double articulation is found between the fifth and sixth cervical vertebrae. Most species have at least some webbing between the toes, and some species have a hinged plastron.

Bog turtles (*Glyptemys muhlenbergii*) attain a maximum size of 5 in (12 cm), whereas adult Gray's sliders (*Trachemys*

venusta grayi) may reach 24 in (60 cm) or more. Males are generally smaller than females in the aquatic emydids; however, among semiaquatic and terrestrial species this may be reversed.

Distribution

These turtles are found in the lowland temperate regions of North America, North Africa, southern Turkey, the Middle East, and throughout Europe to southern Russia. They were formerly more widespread in Europe, but the Scandinavian populations were extirpated during the Pleistocene.

Habitat

This extremely diverse family is found in many habitats. They occur in abundance in most permanent freshwater rivers, streams, lakes, and ponds. One species is only found in estuaries and coastal waters, and a few species are semiaquatic to fully terrestrial.

Behavior

Whether fully aquatic or terrestrial, most emydids have a well-developed basking habit. Some species are active year-round; others are seasonally inactive (dry season or winter). Males of many species exhibit elaborate courtship displays.

Among the temperate northern species, hibernacula are generally located in well-oxygenated areas; however, painted turtles (*Chrysemys picta*) and Blanding's turtles are tolerant of extremely hypoxic, or low oxygen, conditions. At least two aquatic species, the chicken turtle (*Deirochelys reticularia*) and the western pond turtle, are known to hibernate terrestrially. The eastern box turtle (*Terrapene carolina*) burrows beneath leaf litter and hibernates in shallow soil where it may experience subfreezing temperatures.

Feeding ecology and diet

This family includes species that are strictly herbivorous to those that are strictly carnivorous. Hatchlings of many species are highly carnivorous, but switch to a more omnivorous diet as they mature. Some species have diverse, generalized diets and others are highly specialized. In map turtles (genus *Graptemys*), the females may develop huge heads with broad palates that enable them to crush large mollusks. Chicken turtles and Blanding's turtles have independently evolved a long neck with a well-developed hyoid apparatus, an elaborate bony structure that rapidly expands the throat to suck in prey items. This feeding adaptation is frequently found in piscivorous (fish-eating) turtle species.

Reproductive biology

In sexually dimorphic aquatic species, the female is larger than the male. The size difference is most extreme among species of the genus *Graptemys*. Mating generally occurs in the spring; however, some species may store sperm from an earlier mating for several years. The male is brightly colored and may possess long thin claws on the forelimbs that are vigorously waved before the female during courtship. A unique pattern of head bobs also may be exchanged before

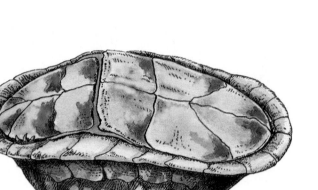

Hinged plastron of a box turtle (*Terrapene* sp.)—the turtle is closed up inside its shell. (Illustration by Gillian Harris)

the female allows the male to mount. This elaborate courtship suggests that females choose their mate. The elongate eggs, which may be flexible or brittle shelled, are generally laid in nests dug in the soil away from the water (sometimes more than 0.6 mi [1 km] away). Most species that have been investigated exhibit temperature-related sex determination.

Conservation status

Seven species are listed as Vulnerable and six as Endangered on the IUCN Red List; 14 others are listed as Lower Risk/Near Threatened. Human activities (e.g., pollution, habitat destruction, road mortality, and collecting for the pet

A red-bellied turtle (*Pseudemys nelsoni*) climbs aboard an alligator (*Alligator mississippiensis*) in Everglades National Park, Florida, USA. (Photo by J. H. Robinson/Photo Researchers, Inc. Reproduced by permission.)

The European pond turtle (*Emys orbicularis*) is native to Spain, France, northern Italy, southern Germany, Poland, Turkey, Iran, Russia, Morocco, and Algeria. (Photo by Henri Janssen. Reproduced by permission.)

A river cooter hatchling (*Pseudemys concinna*). (Photo by Animals Animals ©Mella Panzella. Reprodced by permission.)

trade) are responsible for declines in most species. No species demonstrates the destructive effect that human exploitation may have on a turtle population better than the diamondback terrapin (*Malaclemys terrapin*), which once faced extinction throughout its range due to overcollection for human consumption. This turtle recovered as it fell out of favor with the wealthy.

Significance to humans

Many species are prominent in the international pet trade; the red-eared slider (*Trachemys scripta elegans*) has been the world's pet turtle for several decades. The hatchlings are exported by the tens of thousands from ranching operations in Louisiana. This turtle has established breeding populations throughout the world and is considered an invasive pest because it may harm native species. A few species are consumed by humans locally.

1. Painted turtle (*Chrysemys picta*); 2. European pond turtle (*Emys orbicularis*); 3. Pond slider (*Trachemys scripta*); 4. Female diamondback terrapin (*Malaclemys terrapin*); 5. Eastern box turtle (*Terrapene carolina*); 6. Female spotted turtle (*Clemmys guttata*). (Illustration by Gillian Harris)

Species accounts

Painted turtle
Chrysemys picta

SUBFAMILY
Deirochelyinae

TAXONOMY
Testudo picta Schneider, 1783, location unknown, although said to be England (in error). Four subspecies are recognized.

OTHER COMMON NAMES
French: Chrysémydes peint; German: Zierschildkröte; Spanish: Tortuga pinta.

PHYSICAL CHARACTERISTICS
A small- to medium-sized (maximum carapace length 10 in [25 cm]) turtle with a dark olive to black carapace. The upper and lower surfaces of the marginals are adorned with a pattern of red markings. The plastron may be plain yellow, yellow with a central pattern, or with complex designs of red and yellow. The dark head has a pattern of thin yellow lines and a distinctive yellow spot behind the eye in most subspecies. The males are smaller than the females and have long claws on the forelimbs.

DISTRIBUTION
This widespread species is found from southwestern British Columbia to Nova Scotia and throughout the central and southern regions of temperate North America. Disjunct populations occur in the U.S. Southwest.

HABITAT
Ponds, streams, slow-flowing portions of rivers and estuaries.

BEHAVIOR
By absorbing solar radiation with their dark carapaces, painted turtles thermoregulate by basking on almost any exposed surface. They bask early in the morning to elevate their body temperature, forage for food, and then return to basking sites to facilitate digestion. In the northern populations, the juveniles and adults spend a majority of the winter trapped below thick ice. This species does not readily absorb oxygen from the water; therefore, it must tolerate long periods of hypoxia or anoxia. The mineralized shell buffers the accumulation of lactic acid formed under anaerobic conditions to maintain a stable blood pH through the winter. The hatchlings remain within the shallow nest chamber over the winter and may be exposed to temperatures of 10°F (–12°C) or lower. Although they tolerate freezing at high subzero temperatures (e.g., to 25°F [–4°C]), they must remain supercooled (i.e., without the tissues freezing) in order to survive colder temperatures.

FEEDING ECOLOGY AND DIET
Omnivorous, feeding upon aquatic vegetation, insects, tadpoles, small fish, and carrion.

REPRODUCTIVE BIOLOGY
Although individual females may not reproduce every year, nesting is annual and seasonal. Courtship and mating occur in the autumn and spring, but nesting usually occurs in the spring and early summer. Females can store sperm in their oviducts for years and may not need to mate annually. The size of the elongate, flexible eggs (1.1–1.4 in [28–35 mm] long and 0.6–0.9 in [16–23 mm] wide) decreases with increasing latitude and clutch size. As many as five, but typically one or two, clutches of one to 20 eggs are deposited in nests constructed in sand or loamy soil. The eggs hatch after 72 to 80 days of incubation. This species has temperature-related sex determination, where males are produced below 82°F (28°C) and mostly females are produced at higher temperatures. Paternity analysis using DNA has shown that eggs within the same clutch are sometimes fertilized by more than one male.

CONSERVATION STATUS
Not threatened. This species remains common, in part because it tolerates disturbance due to human activity and its reproductive output is exceptional.

SIGNIFICANCE TO HUMANS
The colorful painted turtle hatchlings and adults are often available in the international pet trade. As a result of their small size and their considerable overlap with the larger and presumably more palatable common snapping turtle (*Chelydra serpentina*), they are rarely eaten by humans. ◆

Diamondback terrapin
Malaclemys terrapin

SUBFAMILY
Deirochelyinae

TAXONOMY
Testudo terrapin Schoepff, 1793, Philadelphia, Pennsylvania, and Long Island, New York, later restricted by Schmidt (1953, 95) to the coastal waters of Long Island. Seven subspecies are recognized.

OTHER COMMON NAMES
French Malaclémyde terrapin; German: Diamantschildkröte.

PHYSICAL CHARACTERISTICS
A small- to medium-sized turtle (maximum carapace length 9 in [24 cm]) with a rough, slightly keeled carapace and smooth speckled skin. The light brown, gray, or black carapace has raised concentric rings because the scutes are not shed each year, and new larger scutes form below. The rigid plastron varies in color from yellow to black and may have a distinctive pattern of blotches. The color of the soft skin is also quite variable, ranging from black to a light gray with black flecks. The females are much larger than the males and have larger heads with broad crushing plates. There are specialized salt glands near the eyes that excrete excess salt.

DISTRIBUTION
East Coast of temperate North America from southern Texas along the Gulf of Mexico and around Florida to Cape Cod, Massachusetts.

HABITAT
Salt marshes in brackish coastal waters and estuaries.

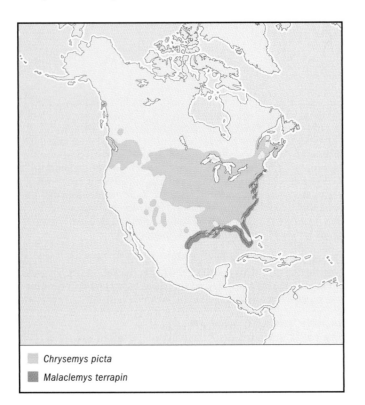

Chrysemys picta

Malaclemys terrapin

BEHAVIOR

Normally active during the day, this species may haul out on rocks or the banks of tidal creeks to bask. Terrapins range widely while foraging for food; however, they are often found in the same small area over consecutive years. Adults hibernate communally on the muddy bottom of creek beds. In southern climates the terrapins may become active on warm winter days, but in northern populations they remain in dormancy.

FEEDING ECOLOGY AND DIET

This species is highly specialized to feed upon mollusks. The females develop especially large heads with broad jaws for crushing the shells of marine gastropods. The diet of the smaller males differs from that of the female in that they consume different size classes of gastropods and they may supplement with a variety of aquatic insects.

REPRODUCTIVE BIOLOGY

The female is considerably larger than the male and may therefore choose her mate. In temperate regions, nesting occurs from mid-May to late July, but southern populations may nest as late as September. The females can store sperm in their oviducts for at least four years; however, fertility rates drop precipitously after the second year. The elongate eggs (1.0–1.7 in [26–42 mm] long and 0.6–1.1 in [16–27 mm] wide) have flexible shells. Two or more clutches of up to 20 eggs are deposited annually in the sandy dunes above the winter high tide mark. The eggs hatch after 61 to 104 days of incubation. Sex is dependent upon the incubation temperature; mostly males are produced from 77 to 84°F (25 to 29°C), however at 86°F (30°C) the hatchlings are all female. Most hatchlings emerge in the autumn and are presumed to hibernate aquatically, but some hatchlings may overwinter in the nest.

CONSERVATION STATUS

This species is listed as Lower Risk/Near Threatened on the IUCN Red List. Once considered a delicacy among aristocrats,

it was decimated during the early twentieth century by overcollection for human consumption. The terrapin has largely recovered; however, populations continue to be threatened by the destruction and degradation of the tidal marshes they inhabit, and many thousands are needlessly drowned in crab traps that could be made safe by a simple modification.

SIGNIFICANCE TO HUMANS

This species was once prized for its delicate flesh; however, it fortunately fell out of favor with the wealthy. It is still consumed locally and is often sold in the Asian markets of large North American cities. ◆

Pond slider

Trachemys scripta

SUBFAMILY
Deirochelyinae

TAXONOMY
Testudo scripta Schoepff, 1792, location unknown, later designated as Charleston, South Carolina, by Schmidt (1953, 102). Three subspecies are recognized.

OTHER COMMON NAMES
English: Red-eared slider, yellow-bellied slider.

PHYSICAL CHARACTERISTICS
This is a medium-sized (maximum carapace length 12 in [30 cm]) turtle with a green carapace, usually marked with yellow. The upper and lower surfaces of the marginals are adorned with a pattern of yellow lines. The plastron may be plain yellow or yellow with a dark pattern, or black in melanistic males. The green head has a pattern of thin yellow lines. There is a distinctive red stripe behind the eye in one subspecies (*Trachemys scripta elegans*),

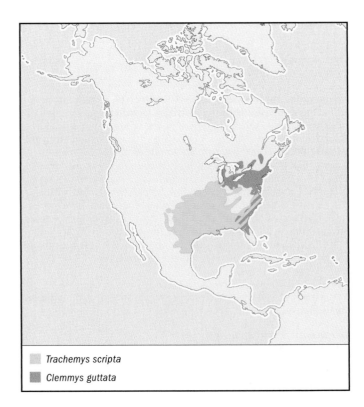

Trachemys scripta

Clemmys guttata

and a yellow blotch in another (*Trachemys scripta scripta*). The males are smaller than the females and have long claws on the forelimbs.

DISTRIBUTION
This widespread species is found from southern Michigan throughout the central, southern, and southeastern regions of temperate North America.

HABITAT
Ponds, streams, slow-flowing portions of rivers and estuaries.

BEHAVIOR
Their daily activity consists of alternately basking to elevate body temperatures and foraging. Higher body temperatures facilitate digestion. The home range may be extensive and include aquatic habitats that can only be reached by overland migration. In northern populations the slider turtle hibernates at the bottom of the aquatic habitat beneath sheets of ice. Adults and juveniles are often frozen to death when drawdowns or extreme temperatures cause the pond to freeze solid.

FEEDING ECOLOGY AND DIET
Omnivorous, feeding upon aquatic vegetation, insects, tadpoles, small fish, and carrion.

REPRODUCTIVE BIOLOGY
The smaller males become darker, or melanistic, as they mature. They possess long, thin claws on their forelimbs which they wave in front of the female during courtship. Several males may court a female simultaneously. Courtship and mating occur in the autumn and spring, but nesting usually occurs in the spring and early summer. Females can store sperm in their oviducts for years and do not need to mate annually. The size of the elongate, flexible eggs (1.2–1.7 in [31–43 mm] long and 0.7–1.0 in [19–26 mm] wide) is related to the size of the female. Up to five clutches of two to 23 eggs are deposited annually in dry sand, clay, or loamy soil. The female moistens the substrate with water carried in an accessory bladder as she digs. The eggs hatch after 60 to 80 days of incubation under field conditions. Although other environmental factors are known to influence the hatchling sex ratio, it is largely related to the incubation temperature; mostly males are produced from 72.5 to 80.6°F (22.5 to 27.0°C), however at 86°F (30°C) the hatchlings are all female. Even in the most northern portions of their range, hatchlings generally overwinter in the nest, where they are often exposed to subzero temperatures.

CONSERVATION STATUS
Listed as Lower Risk/Near Threatened by the IUCN. Sliders are still quite common. This resilient species is abundant even in polluted and greatly disturbed habitats.

SIGNIFICANCE TO HUMANS
This species is widespread in the pet trade. Many thousands of red-eared slider hatchlings are produced by ranching operations in Louisiana each summer. Introduced populations occur in nearly every temperate or tropical country to which they are exported, and they may represent a threat to native species. ◆

Spotted turtle
Clemmys guttata

SUBFAMILY
Emydinae

TAXONOMY
Testudo guttata Schneider, 1792, type locality not stated origi-

nally. No subspecies are recognized.

OTHER COMMON NAMES
French: Clemmyde à gouttelettes;, Tortue ponctuée; German: Tropfenschildkröte.

PHYSICAL CHARACTERISTICS
A small (maximum carapace length to 5 in [12.5 cm]) turtle with a smooth, dark carapace that is punctuated with yellow or orange spots. The upper surfaces of the skin are similarly dark and spotted. The digits are weakly webbed.

DISTRIBUTION
Great Lakes region from northern Illinois to southern Ontario, and the eastern United States from Maine to central Florida.

HABITAT
Bogs, wet meadows, and woodland streams.

BEHAVIOR
Equally at home on the land or in the water, the spotted turtle migrates between habitat types throughout the year. It basks on mats of vegetation floating at the surface of the water or on tussocks of marsh grass. This species has one of the shortest activity seasons of any temperate species. Although it is active at low temperatures in early spring, by mid- to late June the spotted turtle seeks out muskrat burrows or buries itself beneath the ground to estivate until the summer heat has subsided. It is briefly active during the autumn before hibernation.

FEEDING ECOLOGY AND DIET
The spotted turtle is omnivorous, consuming a variety of animal matter (aquatic and terrestrial insects, worms, slugs, crustaceans, and tadpoles) as well as aquatic grasses and filamentous algae.

REPRODUCTIVE BIOLOGY
Courtship behavior has been observed at low temperatures during March and April and continues through June. Several males may aggressively pursue a single female across terrestrial and aquatic portions of the habitat. After the male bites the female repeatedly on the hind limbs and tail, copulation proceeds either on land or water. Nesting generally occurs in the early morning or late evening from May to July. The size of the elongate, flexible eggs (1.0–1.3 in [25–34 mm] long and 0.6–0.7 in [16–18.5 mm] wide) is related to the carapace length of the female. Up to two clutches of one to eight eggs are deposited annually in the grass tussocks, hummocks, and sphagnum moss of the wetland or in upland nests constructed in loamy soil. The eggs hatch after 70 to 83 days of incubation. Sex is dependent upon the incubation temperature; mostly males are produced from 72.5 to 80.6°F (22.5 to 27.0°C), however at 86°F (30°C) the hatchlings are all female. Hatchlings generally emerge from the nest in the autumn, but may occasionally overwinter in the nest.

CONSERVATION STATUS
This species is listed as Vulnerable on the IUCN Red List. It is protected by state and local laws throughout its range, but the regulations are not rigorously enforced and entire populations are often collected for the pet trade.

SIGNIFICANCE TO HUMANS
This species is popular in the international pet trade because of its small size, attractive pattern, and docile nature. ◆

European pond turtle

Emys orbicularis

SUBFAMILY
Emydinae

TAXONOMY
Testudo orbicularis Linnaeus, 1758, southern Europe. About 14 subspecies are variably recognized.

OTHER COMMON NAMES
None known.

PHYSICAL CHARACTERISTICS
A small- to medium-sized (maximum carapace length to 12 in [30 cm]) turtle with a smooth, dark carapace that is punctuated with yellow spots or streaks. The upper surfaces of the skin are similarly dark and spotted. The digits are webbed and the tail is relatively long.

DISTRIBUTION
Northwestern Africa (Tunisia to Morocco), Europe (Portugal to Greece to Lithuania) to northern Iran, and the Aral Sea region in southern Russia.

HABITAT
These turtles are found in most aquatic habitats with soft bottoms and abundant vegetation, including rivers, streams, drainage canals, ponds, and marshes.

BEHAVIOR
This species basks to elevate body temperature, but quickly retreats to the bottom when disturbed. It is active at low temperatures in early spring; by mid- to late June in southern portions of the range it may estivate until the summer heat has subsided. It is briefly active during the autumn before hibernation, which may last six months at the northern limits of the range.

FEEDING ECOLOGY AND DIET
The European pond turtle is predominantly carnivorous, consuming aquatic and terrestrial insects, worms, crustaceans, fish, frogs, salamanders, and tadpoles.

Emys orbicularis

REPRODUCTIVE BIOLOGY
Courtship behavior has been observed at low temperatures during March to May. The males are aggressive breeders; after biting the female repeatedly on the hind limbs and tail and bumping her with his shell, copulation proceeds either on land or water. Nesting generally occurs in the early morning or late evening from May and June. The size of the elongate, flexible eggs (1.2–1.5 in [30–39 mm] long and 0.7–0.9 in [18–22 mm] wide) is related to the carapace length of the female. Up to two clutches of three to 16 eggs are deposited annually in loamy soil. The incubation temperature determines the sex of hatchlings; mostly males are produced from 75 to 82°F (24 to 28°C), but at 86°F (30°C) nearly all the hatchlings are female. In northern habitats the eggs hatch in late summer or early autumn after an extended incubation. Hatchlings generally emerge from the nest in the autumn, but may occasionally overwinter in the nest.

CONSERVATION STATUS
This species is listed as Lower Risk/Near Threatened on the IUCN Red List. Although they are not considered to be threatened, many European pond turtle populations must contend with the red-eared slider, an aggressive competitor that has been introduced through the pet trade.

SIGNIFICANCE TO HUMANS
This species is popular in the pet trade because it is relatively small, attractive, and docile. ◆

Eastern box turtle

Terrapene carolina

SUBFAMILY
Emydinae

TAXONOMY
Testudo carolina Linnaeus, 1758, Carolina (exact location not cited). Six subspecies are recognized.

OTHER COMMON NAMES
French: Tortue tabatière; Spanish: Tortuga de Carolina.

PHYSICAL CHARACTERISTICS
This is a small- to medium-sized (up to 9 in [22 cm] maximum carapace length) turtle. The high domed shell is adorned with a variable pattern of yellow striations on a black background. The plastron has two hinges that allow it to draw both lobes against the carapace, forming a tight seal that protects the limbs from predators. The smaller males have red eyes, a concave plastron, and a long thick tail.

DISTRIBUTION
Southern, central, and eastern United States, northeastern Mexico, and the Yucatán Peninsula.

HABITAT
Woodlands, pastures, and wet meadows.

BEHAVIOR
Eastern box turtles are active throughout the day, but retreat to short, temporary burrows when temperatures reach extremes of hot and cold. They bask early in the morning, forage, and then bask to improve digestion. Their activity increases after periods of rain. Although they are dependent upon available

Terrapene carolina

FEEDING ECOLOGY AND DIET
Omnivorous, feeding on grasses, flowers, berries, insects, and earthworms.

REPRODUCTIVE BIOLOGY
Beginning early in May, the male displays a stereotypical courtship pattern. He first circles and bites at the female's shell, occasionally bumping her with his carapace. Males may raise their head to pulsate their colorful throat. Olfactory cues and behavioral postures of the female precipitate mounting and intromission. The male may continue to bite at the head and shell of the female throughout copulation. Nesting occurs from mid-May to late July. The size of the elongate, flexible eggs (1.0–1.6 in [25–40 mm] long and 0.7–1.0 in [17–25 mm] wide) is related to the carapace length of the female. As many as five, but usually two, clutches of 1 to 11 eggs are deposited annually in flask-shaped nests constructed in sandy or loamy soil. Although incubation may be completed in as little as 57 days, the eggs generally hatch after 70 to 80 days. Sex is dependent upon the incubation temperature; mostly males are produced from 72.5 to 80.6°F (22.5 to 27.0°C), however at 83.3°F (28.5°C) the hatchlings are all female.

CONSERVATION STATUS
This species is listed as Lower Risk/Near Threatened on the IUCN Red List. Suburban sprawl has led to localized habitat destruction; small populations are extirpated as wet meadows are filled and stands of trees are cleared for housing developments. Box turtles that migrate from declining habitats are often killed on the roadways that bisect their range.

SIGNIFICANCE TO HUMANS
The small adult size, attractive colors and patterns, and gentle disposition of this species make it a popular choice among turtle fanciers. ◆

microhabitats, their home range is generally less than 3.7 acres (1.5 ha). This species digs a shallow burrow in the soil, but in regions where hard soil prevents deep penetration and snow is transient, they may be exposed to subzero temperatures. The adults tolerate brief freezing episodes where the heart stops beating and the majority of extracellular water is ice.

Resources

Books

Dodd, C. Kenneth, Jr. *North American Box Turtles: A Natural History.* Norman: University of Oklahoma Press, 2001.

Gibbons, J. Whitfield. *Life History and Ecology of the Slider Turtle.* Washington, DC: Smithsonian Institution Press, 1990.

Schmidt, Karl P. *A Check List of North American Amphibians and Reptiles.* 6th edition. Chicago: University of Chicago Press, 1953.

Periodicals

Buhlmann, Kurt A., and J. Whitfield Gibbons. "Terrestrial Habitat Use by Aquatic Turtles from a Seasonally Fluctuating Wetland: Implications for Wetland Conservation Boundaries." *Chelonian Conservation and Biology* 4 (2001): 115–127.

Costanzo, J. P., J. D. Litzgus, J. B. Iverson, and R. E. Lee. "Cold-Hardiness and Evaporative Water Loss in Hatchling Turtles." *Physiological and Biochemical Zoology* 74 (2001): 510–519.

Holman, J. Alan, and Uwe Fritz. "A New Emydine Species from the Middle Miocene (Barstovian) of Nebraska, USA, with a New Generic Arrangement for the Species of *Clemmys sensu* McDowell (1964)." *Zoologische Abhandlungen (Dresden)* 51 (2001): 331–353.

Pearse, D. E., F.J. Janzen, and J. C. Avise. "Multiple Paternity, Sperm Storage, and Reproductive Success of Female and Male Painted Turtles (*Chrysemys picta*) in Nature." *Behavioral Ecology and Sociobiology* 51 (2002): 164–171.

St. Clair, R. C. "Patterns of Growth and Sexual Size Dimorphism in Two Species of Box Turtles with Environmental Sex Determination." *Oecologia* 115 (1998): 501–507.

Patrick J. Baker, MS

Eurasian pond and river turtles, and Neotropical wood turtles

(Geoemydidae)

Class Reptilia

Order Testudines

Suborder Cryptodira

Family Geoemydidae

Thumbnail description
A diverse family of hard-shelled turtles with a single articulation between the fifth and sixth cervical vertebrae, webbing between the toes, and posterior marginal scutes that extend up on to the suprapygal bone

Size
Up to 32 in (80 cm) carapace length and 110 lb (50 kg)

Number of genera, species
27 genera; 62 species

Habitat
Freshwater to coastal marine systems to primary and secondary forests

Conservation status
Extinct: 1 species; Critically Endangered: 13 species; Endangered: 18 species; Vulnerable: 11 species

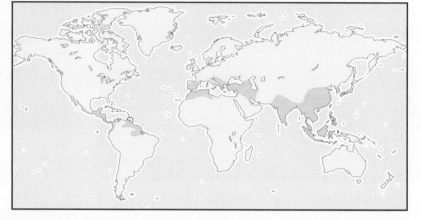

Distribution
Eurasia and North Africa and the tropical Americas

Evolution and systematics

Formerly known as the Bataguridae, this family is most closely related to the tortoises of the family Testudinidae. Together these two families are next most closely related to the pond turtles of the family Emydidae. Fossils are known from as long ago as the Eocene. Geoemydid turtles previously were divided into two subfamilies based on the upper jaw width (i.e., wide alveolar crushing surfaces versus narrow ones). However, molecular studies have demonstrated that similarities in jaw width do not precisely reflect phylogenetic relationships. Hence, no subfamilies currently are recognized.

Physical characteristics

The well-developed shell includes 24 marginal and 12 plastral scutes. The pectoral and abdominal scutes contact the marginal scutes. In addition, the posterior marginal scutes extend up on to the suprapygal bone. A single articulation is found between the fifth and sixth cervical vertebrae. Most included species have at least some webbing between the toes and some species have a hinged plastron. Sexual dimorphism is extraordinary in some species; for example, males of the Indian tent turtle (*Kachuga tentoria*) are usually less than 4 in (10 cm) in shell length, whereas females approach 12 in (30 cm).

Distribution

Europe and North Africa to southern China and the East Indies; the Americas from northern Mexico to Brazil and Ecuador.

Habitat

From almost any freshwater ecosystem to coastal marine systems to fully terrestrial in primary and secondary forests; mainly tropical and subtropical.

A Malayan box turtle (*Cuora amboinensis*) hatchling. (Photo by Henri Janssen. Reproduced by permission.)

A female black-breasted leaf turtle (*Geoemyda spengleri*). (Photo by Henri Janssen. Reproduced by permission.)

The yellow-headed temple turtle (*Hieremys annandalii*) is one of the species that is ritually released into ponds near Buddhist temples. (Photo by Henri Janssen. Reproduced by permission.)

Behavior

These turtles are fully aquatic to terrestrial. Some species are active year-round, while others are seasonally inactive (dry season or winter). The adult males are aggressive breeders, often intimidating unresponsive females by bumping or biting during courtship.

Feeding ecology and diet

This family includes species that are strictly herbivorous to those that are strictly carnivorous, with many omnivorous species as well. Some species have diverse, generalized diets and others have highly specialized diets.

Reproductive biology

The reproductive biology of most included species is very poorly known. Several species exhibit seasonal changes in the color of the head and soft parts that are apparently related to breeding. The production of multiple clutches per year is typical (at least five in some cases). In general, clutch size is positively related to body size. Many smaller species lay only one or two large eggs per clutch, but some of the larger species may lay up to 35 eggs in a clutch. Incubation times are highly variable, ranging from 60 to 272 days. Those with the longer times experience embryonic diapause, in which development arrests for various lengths of time early during incubation. Two species are known to have dimorphic sex chromosomes, and therefore genetically determined sex. However, five species also are known to have temperature-dependent sex determination, with warm temperatures producing females and cooler temperatures resulting in males. In one of these five species, still cooler temperatures again produce females. These turtles hybridize readily in captivity, even between distant genera, and wild populations of hybrids may exist.

Conservation status

On the IUCN Red List, 11 species are listed as Vulnerable, 18 as Endangered, 13 as Critically Endangered, and one as Extinct. These turtles are in precipitous decline in Asia, where they are ruthlessly exploited for food, traditional medicines, and the pet trade. Habitat loss further compounds this crisis situation. Beyond the species presumed to be extinct, several species have not recently been seen in the wild. The extinction in nature of perhaps one-third of the included species is virtually a certainty unless this exploitation and the forces behind it can be curbed. Captive breeding programs for many of these threatened species are underway and may be the only hope for their long-term existence.

Significance to humans

Most species are collected by local humans for food or medicine whenever they are encountered.

1. Yellow-margined box turtle (*Cistoclemmys flavomarginata*); 2. Painted terrapin (*Callagur borneoensis*); 3. Chinese stripe-necked turtle (*Ocadia sinensis*). (Illustration by Wendy Baker)

Species accounts

Painted terrapin
Callagur borneoensis

TAXONOMY
Emys borneoensis Schlegel and Müller, 1844, Borneo. No subspecies are recognized.

OTHER COMMON NAMES
English: Painted batagur, saw-jawed turtle, three-stripe batagur, Sungei tuntong.

PHYSICAL CHARACTERISTICS
This is a large geoemydid turtle (up to 30 in [76 cm] carapace length) with a rigid plastron and bridge (i.e., no plastral hinge); a fourth vertebral scute that is wider than long; crushing surfaces of the upper jaw that are broad along their entire length and bear a single, well-developed medial ridge; five claws on the forefeet; and no neck stripes.

DISTRIBUTION
Southern Thailand, Malay Peninsula, Sumatra, and Borneo.

HABITAT
Tidal sections of rivers and estuaries.

BEHAVIOR
The females migrate considerable distances to nest (e.g., 9–31 mi [15–50 km] in the Malay Peninsula).

FEEDING ECOLOGY AND DIET
This species is primarily herbivorous, feeding on leaves and fruits, but also apparently eating clams and other shellfish on occasion.

REPRODUCTIVE BIOLOGY
There are color differences between the males and females; those colors intensify incredibly during the breeding season. For example, nonbreeding males have a gray head, whereas during the breeding season the male's head becomes white with a black-edged red stripe between the eyes. Courtship and mating have not been described, but apparently occur in at least January and February on the east coast of the Malay Peninsula. Females migrate great distances to nest, either upstream on river sandbanks above tidal influence or on coastal beaches. Nesting occurs from late May to August, is nocturnal, and apparently peaks during the night at low tides. During the nesting season, one to three clutches of six to 25 (usually 10 to 15) large, elongate, hard-shelled eggs (measuring 2.4–3.1 in × 1.4–1.8 in [61–79 mm × 36–45 mm], and 2–3 oz [56–83 g]) are laid. The female digs no body pit, but rather uses her hind legs to dig a nest chamber to a depth of 9–13 in (24–34 cm). The internesting interval averages 26 days. Incubation requires 85–98 days at 88°F (31°C). The means by which hatchlings produced on sea beaches migrate to freshwater rivers is unknown, as is the effect of incubation temperature on sex.

CONSERVATION STATUS
This species is listed as Critically Endangered on the IUCN Red List. A government-sponsored hatching program in the Malay Peninsula seeks to counteract local exploitation in the wild.

SIGNIFICANCE TO HUMANS
The eggs and meat of these turtles are relished by local people. In addition, since it is believed by some that this turtle brings good luck to its owner, a local pet trade has developed. ◆

Yellow-margined box turtle
Cistoclemmys flavomarginata

TAXONOMY
Cistoclemmys flavomarginata Gray, 1863, Mainland China and Taiwan. Two or three subspecies are recognized.

OTHER COMMON NAMES
Japanese: Hakogame.

PHYSICAL CHARACTERISTICS
A small turtle (up to 7 in [17 cm] carapace length) with a high-domed carapace, unserrated posteriorly, with a distinct yellow vertebral stripe. The plastron is large and unnotched posteriorly, with a movable hinge between the pectoral and abdominal scutes, and the plastral lobes are capable of closing the shell opening completely. A single lemon-yellow stripe passes posteriorly from the eye onto the neck.

DISTRIBUTION
Southern China, Taiwan, and the Ryukyu Islands.

HABITAT
This species may occasionally inhabit freshwater ponds and streams, as well as rice paddies, but spends most of its time terrestrially in primary and secondary forests.

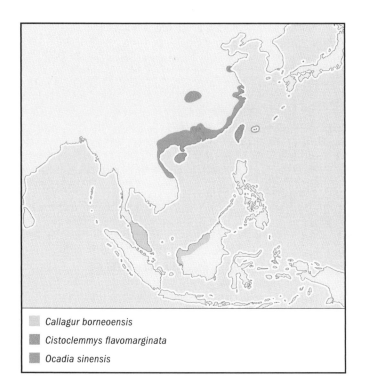

☐ *Callagur borneoensis*
▨ *Cistoclemmys flavomarginata*
▨ *Ocadia sinensis*

BEHAVIOR

These turtles are often considered to be semiaquatic, but some populations may be almost exclusively terrestrial in their habits. Their ability to close up the plastron like a box is an adaptation for terrestriality, both for predator protection and desiccation resistance. On Taiwan terrestrial home ranges average 1.2 acres (0.5 ha) for females and 8.6 acres (3.5 ha) for males. Adults overwinter terrestrially by burying themselves under leaf litter or fallen logs or in the abandoned burrows of other animals.

FEEDING ECOLOGY AND DIET

These turtles are apparently omnivorous. They are attracted to traps baited with bananas in the field in Taiwan, but they eat both plant and animal matter in captivity. They are reported to consume worms, insects, snails, and fruit.

REPRODUCTIVE BIOLOGY

During courtship the male rams the front of the female's shell to subdue her, and then moves to mount her shell from behind for copulation. On Taiwan this species is estimated to mature at 12 or 13 years. Females apparently nest from May through at least July on Taiwan, May to perhaps September on mainland China, and June to perhaps September in the Ryukyu Islands. On Taiwan, females produce one to three clutches per season; however, some females apparently do not reproduce every year. The shallow nests (2–3 in [5–8 cm]) are constructed at well-drained, open sites at the edges of forests. The clutch size ranges from one to four eggs. Eggs are elongate, brittle-shelled, and measure 1.6–2.1 in (40–54 mm) in length, 0.9–1.1 in (23–28 mm) in diameter, and 0.4–1.0 oz (12–27 g) in mass. Incubation apparently takes about two months, but the effect of incubation temperature on sex is unknown. This species apparently hybridizes with *Geoemyda japonica*.

CONSERVATION STATUS

This species is listed as Endangered on the IUCN Red List. It is removed by humans for consumption for food and traditional medicine and for pets, and is also affected by habitat destruction (i.e., forest cutting).

SIGNIFICANCE TO HUMANS

These turtles are eaten by local people and also are exploited for the pet trade. ◆

Chinese stripe-necked turtle
Ocadia sinensis

TAXONOMY

Emys sinensis Gray, 1834, China. No subspecies are recognized.

OTHER COMMON NAMES

None known.

PHYSICAL CHARACTERISTICS

This is a medium-sized geoemydid turtle (up to 9 in [24 cm] carapace length) with the plastron rigidly attached to the carapace and lacking a hinge; a fourth vertebral scute that is wider than long; the crushing surfaces of the upper jaw are broad along their entire length, with a single, well-developed medial ridge present on each surface; five claws on the forefoot; and numerous dark-bordered, narrow yellow stripes on the head and neck.

DISTRIBUTION

Taiwan, southern China (including Hainan Island), and northern Vietnam.

HABITAT

This species inhabits slow-moving lowland freshwater habitats, from ponds and rivers to marshes and human-made canals.

BEHAVIOR

This is an aquatic turtle that often climbs out of the water to bask. Its behavior is in great need of study.

FEEDING ECOLOGY AND DIET

The only detailed study of this turtle's diet is from Taiwan; this study suggests that significant dietary differences exist between the sexes. Juveniles of both sexes tend to be carnivorous, eating primarily insects along with plant roots, shoots, and leaves. The males remain primarily carnivorous into adulthood, consuming mainly dipteran (mosquito) larvae and other insects, as well as some plant leaves, seeds, and roots. The females become increasingly herbivorous as they mature, feeding primarily on the leaves of terrestrial plants that grow along riverbanks, but also occasionally eating insects (especially dipteran larvae).

REPRODUCTIVE BIOLOGY

Males apparently mature after three to four years and females after five to eight years. On Taiwan the females apparently nest from April to early June, although they are reported to nest from April to August on mainland China. On Taiwan the relatively deep nests (6–9 in [15–22 cm]) are constructed on open sandbars along the river or in open areas away from the water. The clutch size ranges from seven to 17 eggs on Taiwan, whereas three to 14 eggs per clutch are apparently laid on mainland China. The eggs are elongate, brittle-shelled, and on Taiwan measure 1.2–1.5 in (30–39 mm) in length, 0.7–0.9 in (18–22 mm) in diameter, and 0.2–0.4 oz (6–10 g) in mass. Three eggs laid in captivity by a female from Hainan Island were 1.6 × 1.0 in (40 × 25 mm), suggesting some geographic variation in reproductive parameters. The effect of temperature on sex is unknown. This species apparently hybridizes with at least the Chinese three-striped box turtle (*Cuora trifasciata*) and the Annam leaf turtle (*Mauremys annamensis*).

CONSERVATION STATUS

This species is listed as Endangered on the IUCN Red List. It is surprisingly tolerant of polluted aquatic systems.

SIGNIFICANCE TO HUMANS

These turtles are frequently consumed by humans and small numbers still enter the pet trade. They may be ranched in impoundments on Taiwan and Hainan Island for commercial purposes. Studies have shown that plastrons of this species are frequently sold fraudulently as "tortoise shell" in medicine shops. ◆

Resources

Books

Liat, Lim Boo, and Indraneil Das. *Turtles of Borneo and Peninsular Malaysia.* Kota Kinabalu, Malaysia: Natural History Publications (Borneo), 1999.

Zhou, J., and T. Zhou. *Chinese Chelonians Illustrated.* Nanjing, China: Jiangsu Science and Technology Publishing House, 1992.

Periodicals

Chen, Tien-Hsi, and Kuang-Yang Lue. "Ecology of the Chinese Stripe-Necked Turtle, *Ocadia sinensis* (Testudines: Emydidae), in the Keelung River, Northern Taiwan." *Copeia* 1998, no. 4 (1998): 944–952.

———. "Population Characteristics and Egg Production of the Yellow-Margined Box Turtle, *Cuora flavomarginata flavomarginata,* in Northern Taiwan." *Herpetologica* 55, no. 4 (1999): 487–498.

Iverson, J. B., W. P. McCord, P. Spinks, and H. B. Shaffer. "A New Genus of Geoemydid Turtle from Asia (Testudines)." *Hamadryad* 25, no. 2 (2000): 86–90.

Yasukawa, Yuichirou, Ren Hirayama, and Tsutomu Hikida. "Phylogenetic Relationships of Geoemydine Turtles (Reptilia: Bataguridae)." *Current Herpetology* 20, no. 2 (2001): 105–133.

John B. Iverson, PhD

▲
American mud and musk turtles
(*Kinosternidae*)

Class Reptilia
Order Testudines
Suborder Cryptodira
Family Kinosternidae

Thumbnail description
Small- to medium-sized turtles with a reduced or
hinged plastron that are capable of producing a
foul-smelling musk

Size
4–15 in (10–38 cm)

Number of genera, species
4 genera; 25 species

Habitat
Freshwater ponds, rivers, or marshes

Conservation status
Vulnerable: 4 species

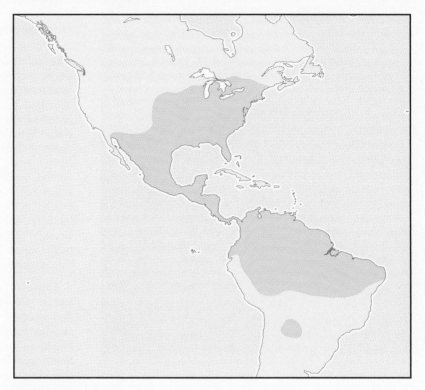

Distribution
North and South America

Evolution and systematics

The family is most closely related to the Dermatemydidae. Two monophyletic subfamilies are recognized: the Kinosterninae (including the living genera *Kinosternon* and *Sternotherus*) and the Staurotypinae (including the living genera *Staurotypus* and *Claudius*). Fossils of each subfamily are known from as early as the Eocene.

Physical characteristics

These turtles are generally small (usually less than 8 in [20 cm]) with oblong, moderately domed shells. The Mexican giant musk turtle (*Staurotypus triporcatus*) is the largest species, reaching a shell length of 15 in (38 cm). The plastron has one (*Staurotypus, Sternotherus*, some *Kinosternon*), two (most *Kinosternon*), or no hinges (*Claudius*), and is generally reduced (with 11 or fewer epidermal scutes), although some *Kinosternon* have a plastron extensive enough to close the shell opening completely. The kinosternine plastron lacks the entoplastral bone found in staurotypines. All members produce a pungent musk from glands located in front of and behind the bridge area between the plastron and the carapace. The head of some species may be greatly enlarged (an advantage for mollusk feeding), and all have sensory barbels on the chin.

Distribution

Eastern and southern North America to Argentina, from sea level to 8,500 ft (2,600 m).

Habitat

Within their ranges, these turtles can be found in nearly any freshwater aquatic system. Because of their relatively poor swimming ability, they prefer slow-moving or still waters (e.g., permanent ponds with lush vegetation or backwaters of lotic systems). However, some species inhabit highly seasonal ephemeral ponds which may only contain water for a few months of each year.

Behavior

In seasonal environments (high latitude or deserts), these turtles have short annual activity periods (three months or less in some cases) and spend the rest of the year hibernating or estivating underground. In wetter, more tropical habitats, they are active year-round. Most are highly aquatic, rarely leaving the water except to nest, although a few species may spend considerable time foraging terrestrially (especially during the wet seasons). Some species are active primarily

Striped mud turtles (*Kinosternon baurii*) range from the Florida Keys up the Atlantic coast into Virginia. (Photo by Joe McDonald. Bruce Coleman, Inc. Reproduced by permission.)

during the day, others primarily at night, and still others may be found active at any time.

Feeding ecology and diet

These turtles are all primarily carnivorous, feeding mainly on mollusks (snails and clams), crustaceans, insects, annelids, and even fish (usually as fresh carrion). Some are highly specialized mollusk feeders and eat little else. However, some

species, especially those with extensive terrestrial habits and those in highly seasonal ponds, include plant material (primarily seeds) in their diets.

Reproductive biology

In most species the males are larger than the females, have a long muscular tail with an epidermal "nail" at the end, and have rough patches of scales (clasping organs) on the back of their hind legs. As a result, male courtship is forceful and not very elaborate; females may select mates based on the male's ability to subdue or restrain her movements. All mating occurs in the water. Females lay one to 12 eggs (usually three to six) in each clutch and the clutch size tends to be greater in larger turtles. Many species lay multiple clutches (up to six per year). Eggs are oblong, brittle shelled, and range in size from 0.9 × 0.6 in (2.3 × 1.4 cm) in the stinkpot (*Sternotherus odoratus*) to 1.7 × 1.0 in (4.4 × 2.6 cm) in the Mexican giant musk turtle. In some species the nests are poorly constructed and the eggs are simply laid among leaf litter; however, others dig more typical flask-shaped nests, and still others dig a body pit or bury themselves completely below the ground before digging a nest chamber.

Incubation is generally quite long (e.g., 75 days to a year) and the embryos of some species exhibit diapause during early development or estivation later in development, presumably as adaptations to avoid inhospitable times of the year for a hatchling turtle. *Staurotypus* is unique in this family in having heteromorphic sex chromosomes (as in humans). All others that have been studied exhibit temperature-dependent sex determination, with warm temperatures producing females, intermediate temperatures producing mostly males, and still cooler temperatures again producing mostly females.

Conservation status

Most kinosternid species are common, reaching amazing population densities (as high as 1,200 per 2.5 acres [1 ha]).

Eastern mud turtle hatchlings (*Kinosternon* sp.). (Photo by Animals Animals ©Zig Leszczynski. Reproduced by permission.)

A newly hatched juvenile common musk turtle or stinkpot (*Sternotherus odoratus*). (Photo by Henri Janssen. Reproduced by permission.)

However, two tropical, one subtropical, and one temperate species are listed as Vulnerable on the IUCN Red List. The two tropical species (*K. dunni* and *K. angustipons*) are lowland forms with very restricted ranges and hence are probably affected most negatively by habitat destruction. The subtropical species (*K. sonoriense*) lives primarily in permanent water systems in the deserts of the U.S. Southwest; human competition for water resources has eliminated most of the habitat for this species. The temperate species (*Sternotherus depressus*) also has a restricted distribution in the permanent streams of north-central Alabama; habitat destruction associated with coal mining and forest clear-cutting seems to have caused the declines in this species.

Significance to humans

Most species of mud and musk turtles have no significance to humans, except as aquarium pets. A few tropical species may be eaten locally by indigenous people, but not at levels high enough to eliminate populations.

1. White-lipped mud turtle (*Kinosternon leucostomum*); 2. Stinkpot (*Sternotherus odoratus*); 3. Yellow mud turtle (*Kinosternon flavescens*). (Illustration by Jacqueline Mahannah)

Species accounts

Yellow mud turtle
Kinosternon flavescens

SUBFAMILY
Kinosterninae

TAXONOMY
Platythyra flavescens Agassiz, 1857, Rio Blanco, near San Antonio, Texas.

OTHER COMMON NAMES
English: Illinois mud turtle; Spanish: Tortuga-pecho quebrado amarilla.

PHYSICAL CHARACTERISTICS
This is a small kinosternid turtle (maximum shell length 6 in [16 cm]) with a yellowish chin and throat and a low carapace, with the ninth marginal scute raised distinctly higher than the eighth but equal to the height of the tenth. The plastron has two hinges and is not greatly reduced and the pectoral scutes are triangular in shape and are only narrowly in contact.

DISTRIBUTION
Ranges nearly continuously from southern Nebraska to southern New Mexico and to northeastern Mexico, with relict populations known in northwestern Nebraska, Iowa, Illinois, Missouri, Kansas, and east Texas.

HABITAT
An inhabitant of grassland habitats, preferring still water, this turtle also is found in permanent to very temporary pools, even those created by humans (e.g., ditches and cattle ponds). It is only rarely found in streams or rivers, and then only in backwaters or cutoffs.

BEHAVIOR
This turtle spends the majority of each year inactive, buried underground (hibernating or estivating). Activity is stimulated by warm rains (and filling ponds); these turtles will migrate considerable distances from emergence sites to bodies of water, and hence are most often seen moving terrestrially.

The activity season for some populations is as short as for any known turtle—as little as two months in very dry years. As ponds dry up these turtles again bury terrestrially and estivate and/or hibernate until the next warm rainy season. Yellow mud turtles are also frequently seen basking at the edge of the water, even in subtropical locations.

FEEDING ECOLOGY AND DIET
This species is decidedly carnivorous, and plant matter found in their stomach was probably only accidentally ingested when foraging for small animals such as snails, clams, insects, crustaceans, earthworms, tadpoles, and fish. They often forage on carrion.

REPRODUCTIVE BIOLOGY
Yellow mud turtles mate rapaciously when they emerge from dormancy and reach the water. Warm rainfall escalates courtship behavior; in captivity males can be induced to court females simply by changing the water in their containers. Females leave the water and may migrate considerable distances to nest in May and June. In Nebraska, gravid females bury themselves completely underground 6–10 in (15–25 cm) and deposit their eggs while buried at these depths. Females remain buried for variable periods once the eggs are laid. Some females may dig out and return to the water the next day after nesting, but some apparently remain in estivation with the eggs through the summer and then dig deeper in the autumn to hibernate for the winter. The eggs are small and elliptical (0.9–1.2 × 0.6–0.7 in [23–31 × 14–18 mm]) with white, brittle eggshells. The clutch size ranges from one to nine (typically four to six), with larger females producing larger clutches. A maximum of a single clutch is produced per year in most populations, and some females do not nest every year. However, two clutches per year may be produced in southern populations. Eggs hatch after 90 to 118 days; at least in Nebraska, hatchlings then dig straight down as much as 3 ft (1 m) below the nest to avoid freezing temperatures during the winter. They dig back out and head for the water during the following spring.

CONSERVATION STATUS
Not threatened. This species has a wide distribution and exhibits high densities in many populations, and so is not in need of protection over most of its range. However, most relict populations in northwestern Nebraska, southeastern Iowa, western Illinois, and northeastern Missouri are small, vulnerable to extirpation, and hence in great need of protection.

SIGNIFICANCE TO HUMANS
Exploited only minimally by humans for the pet trade. ◆

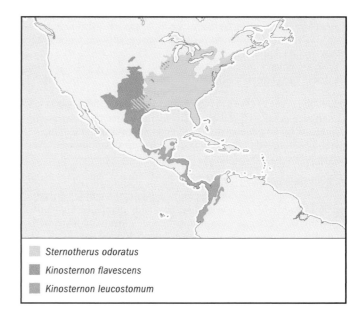

Sternotherus odoratus
Kinosternon flavescens
Kinosternon leucostomum

White-lipped mud turtle
Kinosternon leucostomum

SUBFAMILY
Kinosterninae

TAXONOMY
Cinosternum leucostomum Duméril and Bibron, 1851, Rio Usumacinta, El Peten, Guatamela.

OTHER COMMON NAMES
Spanish: Chachanya, pochitoque.

PHYSICAL CHARACTERISTICS
This is a medium-sized kinosternid turtle (maximum shell length 8 in [20 cm]), with a smooth carapace (sometimes with a weak medial keel), a large plastron with two hinges capable of fully closing the carapacial opening, a raised eleventh marginal scute, the axillary and inguinal scutes on the bridge not in contact, and usually with a broad yellowish lateral head stripe extending from the orbit to the neck.

DISTRIBUTION
Ranges from central Veracruz, Mexico, southward in Atlantic drainages to Nicaragua, and then in both Atlantic and Pacific drainages southward to Colombia, Ecuador, and extreme northwestern Peru.

HABITAT
It inhabits nearly any freshwater aquatic habitat except fast-flowing rivers and streams. It prefers still waters, but also wanders extensively on land in some populations.

BEHAVIOR
This turtle may be active year-round and is primarily nocturnal. If water levels recede, it often leaves the water and estivates terrestrially under vegetation for up to 80 days.

FEEDING ECOLOGY AND DIET
This mud turtle is omnivorous, eating mollusks, insects, worms, and carrion, but also seeds, fruits, leaves, and stems of plants. It is not known whether they feed out of the water.

REPRODUCTIVE BIOLOGY
Courtship and mating have not been described. Although nesting may occur in any month of the year, it appears to be concentrated in August to September and February to March in Mexico. Females produce multiple clutches during the year and lay their eggs (usually at night) in shallow nests or under leaf litter. The eggs are relatively large, elongate (1.3–1.5 × 0.6–0.7 in [34–37 × 16–19 mm]), and have brittle shells. The clutch size ranges from one to five, with larger clutches being produced by larger females. The eggs hatch after 90 to 265 days, and embryos exhibit diapause early in development or estivation late in development.

CONSERVATION STATUS
Not threatened. Very little is known about the status of this turtle in the field. However, because of its extensive distribution across huge tracts of undisturbed or minimally disturbed habitats, it is not currently in need of protection.

SIGNIFICANCE TO HUMANS
This turtle is occasionally eaten by humans and can sometimes be seen for sale in food markets within its range. Small numbers are also exported to the pet trade. ◆

Stinkpot
Sternotherus odoratus

SUBFAMILY
Kinosterninae

TAXONOMY
Testudo odorata Latrielle, 1801, Charleston, South Carolina.

OTHER COMMON NAMES
English: Common musk turtle; French: Sternothère odorant; German: Gewöhnliche Moschusschildkröte.

PHYSICAL CHARACTERISTICS
This is a small kinosternid turtle (maximum shell length 5.4 in [13.7 cm]) with two light stripes on each side of the head (sometimes observed in old individuals) and with barbels on the chin and the throat. The plastron is reduced, has 11 epidermal scutes, a single weakly movable anterior hinge, and a pectoral scute with a quadrangular shape.

DISTRIBUTION
Ranges in North America from New England and southern Ontario to Wisconsin and south through Texas and Florida.

HABITAT
This turtle inhabits nearly any permanent body of fresh water, but prefers ponds or lakes with muddy bottoms and extensive submergent vegetation.

BEHAVIOR
It is most commonly observed foraging along the bottom, but is sometimes seen basking as high as 7 ft (2 m) above the water on the boles of trees in wooded (and hence shaded) aquatic habitats. It rarely leaves the water except to bask or nest. It hibernates at high latitudes, but is active year-round in the south.

FEEDING ECOLOGY AND DIET
These turtles prefer animal food (e.g., earthworms, snails, clams, crayfish, crabs, insects, tadpoles, fish, and fish eggs) and even scavenge on dead animals. However, they also often feed on algae and aquatic plants (particularly seeds). They have perhaps the most generalized diet of all kinosternids.

REPRODUCTIVE BIOLOGY
Courtship and mating apparently can occur any time the turtles are active, with peaks in the spring and autumn. Nesting occurs in the spring, earlier and longer in the south (February to July) than in the north (May to July). Some females simply drop their eggs in leaf litter, whereas others dig well-formed nests up to 4 in (10 cm) deep. The eggs are very small and elliptical (0.9–1.2 × 0.5–0.7 in [22–31 × 13–17 mm]) with white, porcelain-like eggshells. The clutch size ranges from one to nine (typically two to five) and tends to increase with the female's body size. As many as four clutches may be laid each year in the south, but one or two is the norm farther north. Hatchlings emerge in the summer or autumn after about 65 to 85 days of incubation and move directly to the water.

CONSERVATION STATUS
Not threatened. This species is so widespread and reaches such high densities that human impact has mostly been via habitat loss (e.g., draining swamps or ponds).

SIGNIFICANCE TO HUMANS
Exploited by humans only minimally for the pet trade. ◆

Resources

Periodicals

Edmonds, Jonathan H., and Ronald J. Brooks. "Demography, Sex Ratio, and Sexual Size Dimorphism in a Northern Population of Common Musk Turtles (*Sternotherus odoratus*)." *Canadian Journal of Zoology* 74, no. 5 (1996): 918–925.

Iverson, John B. "Life History and Demography of the Yellow Mud Turtle, *Kinosternon flavescens.*" *Herpetologica* 47, no. 4 (1991): 373–395.

———. "Molecules, Morphology, and Mud Turtle Phylogenetics (Family Kinosternidae)." *Chelonian Conservation and Biology* 3, no. 1 (1998): 113–117.

Morales-Verdeja, S. A., and R. C. Vogt. "Terrestrial Movements in Relation to Aestivation and the Annual Reproductive Cycle of *Kinosternon leucostomum.*" *Copeia* (1997): 123–130.

Van Loben Sels, Richard C., Justin D. Congdon, and Josiah T. Austin. "Life History and Ecology of the Sonoran Mud Turtle (*Kinosternon sonoriense*) in Southeastern Arizona: A Preliminary Report." *Chelonian Conservation and Biology* 2, no. 3 (1997): 338–344.

Vogt, R. C., and O. Flores-Villela. "Effects of Incubation Temperature on Sex Determination in a Community of Neotropical Freshwater Turtles in Southern Mexico." *Herpetologica* 48, no. 3 (1992): 265–270.

John B. Iverson, PhD

African sideneck turtles

(Pelomedusidae)

Class Reptilia
Order Testudines
Family Pelomedusidae

Thumbnail description
Medium-sized, sideneck turtles with five claws on the hind feet, four to eight neural bones present, the pleural bones almost always meeting at the midline behind the neurals, mesoplastral bones present, and the pelvis fused to the plastron

Size
Up to 21.6 in (55 cm) carapace length

Number of genera, species
2 genera; 18 species

Habitat
Freshwater habitats, from permanent rivers and lakes to ephemeral ponds

Conservation status
Vulnerable: 2 species

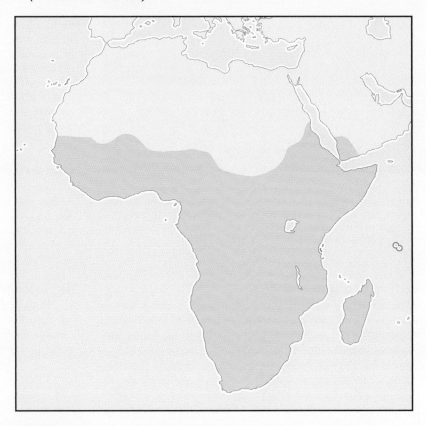

Distribution
Africa, Madagascar, and the Seychelles Islands

Evolution and systematics

Pelomedusidae is most closely related to Podocnemidae, but diverged from that family at least by the Cretaceous (at least 110 million years ago). Fossils are known from the Miocene to the Recent, but all belong to the two living genera. No subfamilies are recognized.

Physical characteristics

These turtles are generally small to medium in size, usually less than 12 in (30 cm), with a relatively extensive plastron that may (in *Pelusios*) or may not (in *Pelomedusa*) have a hinge present between the pectoral and abdominal scutes. The neural series is highly variable (with four to eight present), and the pleural bones almost always meet at the midline posterior to the neurals. A pair of mesoplastral bones are present between the hyo- and hypoplasta, and may (in *Pelusios*) or may not (in *Pelomedusa*) be in contact. Five claws are present on the hind feet.

Distribution

Africa up to at least 10,200 ft (3,100 m) elevation, Madagascar, and the Seychelles Islands.

Habitat

As a group, these turtles occupy nearly any freshwater aquatic system, from permanent lakes or rivers to ephemeral pools which may contain water only a few weeks at a time.

Behavior

Surprisingly little is known about the behavior of these turtles. They frequently bask at the water's edge, and several species make extensive overland movements during the wet season. During the dry season, many species estivate underground. In the temperate climates of southern Africa, they will hibernate terrestrially or aquatically. When captured, they produce a pungent musk from glands located near the bridges (bony structures that connect the plastron and carapace).

An East African black mud turtle (*Pelusios subniger*) in Nigeria. Sideneck turtles fold their necks to the side under their shells for concealment, rather than draw their heads straight back into the shell, as do straight necked turtles. (Photo by P. Ward. Bruce Coleman. Reproduced by permission.)

Feeding ecology and diet

These turtles are all primarily carnivorous, feeding mainly on annelids, mollusks, crustaceans, insects, fish, amphibians, and carrion. Some are at least partly herbivorous, eating water lettuce, aquatic grasses, or fruits.

Reproductive biology

No species in this family has been well studied; most of what is known is based on anecdotal reports. Most species apparently nest in late spring or summer, from October to January. The eggs are elongate with leathery shells, and clutch sizes range from six to 48. Multiple annual clutches have not been confirmed for any species, but seem likely considering the length of the potential nesting season. All species that have been studied exhibit temperature-dependent sex determination, with warm temperatures producing females, intermediate temperatures producing mostly males, and still cooler temperatures again producing females.

Conservation status

The actual field status of most of the species in this family has not been adequately assessed. The only two species listed as Vulnerable on the IUCN Red List are the two with the most restricted distributions: *Pelusios broadleyi* in Lake Turkana in Kenya, and *P. seychellensis* in the Seychelles.

Significance to humans

These turtles are occasionally eaten by indigenous people, but their foul-smelling musk secretions probably serve to discourage more regular consumption. They are also in low demand for the pet trade.

1. Helmeted turtle (*Pelomedusa subrufa*); 2. East African serrated mud turtle (*Pelusios sinuatus*); 3. East African black mud turtle (*Pelusios subniger*). (Illustration by Barbara Duperron)

Species accounts

Helmeted turtle
Pelomedusa subrufa

TAXONOMY
Testudo subrufa Lacepède, 1788, "Indes" [in error; restricted to Cape of Good Hope]. Three subspecies are recognized.

OTHER COMMON NAMES
English: Cape terrapin, helmeted terrapin; French: Roussâtre; German: Starrbrust-Pelomeduse; Afrikaans: Gewone water-skilpad.

PHYSICAL CHARACTERISTICS
Small to medium turtles, with a maximum shell length of 13 in (33 cm), and a broad, flattened, brown to olive carapace. The plastron is rigid (i.e., unhinged), and firmly attached to the carapace. A pair of small triangular mesoplastral bones are present between the hyo- and hypoplastra, but they are widely separated.

DISTRIBUTION
Africa from Senegal and Ethiopia to South Africa, Madagascar, southern Saudi Arabia, and Yemen.

HABITAT
Helmeted turtles are semiaquatic, inhabiting ponds, marshes, and streams, as well as temporary rain pools.

BEHAVIOR
These turtles commonly migrate overland between bodies of water, and therefore most frequently are seen on land or basking at the water's edge. When their habitat dries up, they estivate in

the mud until the next rains (which may be longer than a year). They hibernate terrestrially in the ground or under leaves during the winter in southern Africa. Adults feed mostly at dawn or during the night, but hatchlings forage day and night. Helmeted turtles are occasionally very aggressive in captivity.

FEEDING ECOLOGY AND DIET
These turtles are primarily carnivorous, feeding on earthworms, snails, clams, crustaceans, insects, fishes, frogs and tadpoles, small reptiles, birds, mammals, and carrion of any kind. They have even been observed to attack larger prey, such as birds, as a group, as well as to feed on the ticks on the hide of rhinoceroses when the latter enter waterholes. They also occasionally feed on various parts of aquatic plants and the fruits of terrestrial species.

REPRODUCTIVE BIOLOGY
In subtropical environments, courtship and mating occur in the spring. The male chases the female from behind, touching or biting her posterior extremities with his head, and eventually mounting the top of her shell. He extends his head in front of hers, sways it from side to side, and expels water through his nostrils. His longer tail then swings under hers for intromission. The well-formed, flask-shaped nest is constructed in the late spring to early summer, and apparently only one clutch is laid per year. From 13 to 30 oblong, leathery eggs, averaging 1.5 by 0.9 in (38 by 22 mm) are laid in a clutch. Hatching requires 75–110 days, with hatchlings emerging in February to June. This species is known to exhibit temperature-dependent sex determination, with warm temperatures producing females, intermediate temperatures producing mostly males, and still cooler temperatures again producing females.

CONSERVATION STATUS
This species is so widespread across Africa, occupies such a variety of aquatic habitats, and reaches such high densities that human impact has apparently not been extensive. In fact, the construction of ponds and waterholes for livestock and wildlife has benefited this species by providing new habitats.

SIGNIFICANCE TO HUMANS
This species is eaten by some indigenous people, often by roasting the whole animal under hot coals. Some groups believe the blood to have medicinal properties. The turtles are exploited only minimally for the pet trade. ◆

East African serrated mud turtle
Pelusios sinuatus

TAXONOMY
Sternotherus sinuatus Smith, 1838, "rivers to the north of 25° south latitude" [South Africa]. No subspecies are recognized.

OTHER COMMON NAMES
English: Serrated hinged terrapin, serrated turtle; Afrikaans: Groot waterskilpad.

PHYSICAL CHARACTERISTICS
These turtles are medium to large size, with a maximum shell length of 2.2 in (5.5 cm), and an elongate, oval, variably

Pelomedusa subrufa

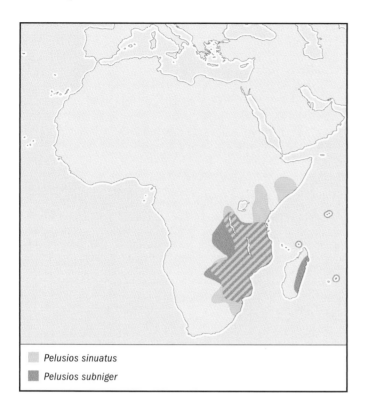

Pelusios sinuatus

Pelusios subniger

keeled, posteriorly serrated carapace. The large, posteriorly notched plastron has a well-developed hinge between the pectoral and abdominal scutes, and the pair of meosplastral bones between the hyo- and hypoplastra are in contact on the midline. The anterior plastral lobe is relatively short, being less than twice as long as the interabdominal seam. A small axillary scute is present on each bridge.

DISTRIBUTION
Eastern and southeastern Africa (southern Somalia to northeastern South Africa).

HABITAT
These turtles inhabit permanent rivers and lakes.

BEHAVIOR
These turtles are most commonly seen basking on rocks, logs, or the shoreline. They inhabit permanent water, thus they apparently do not estivate. They are frequently eaten by the Nile crocodile.

FEEDING ECOLOGY AND DIET
This species is primarily carnivorous, feeding on earthworms, snails, insects, ticks (from the hides of wallowing ungulates), fish, frogs, and carrion. They also occasionally feed on aquatic plants, as well as fruits that fall into the water.

REPRODUCTIVE BIOLOGY
Very little is known about the natural history of this species. Females apparently nest during the summer, from at least October through January, and perhaps until April. Nests are dug by the female as far as 1,640 ft (500 m) from the water, and seven to 30 eggs, averaging 1.7 by 1 in (43 by 25 mm) and 0.7 oz (20 g), are laid in a clutch. Incubation takes only 48 days at 32–91.4°F (33°C). Hatchlings are most commonly seen after emergence in March and April.

CONSERVATION STATUS
This species is not listed internationally for protection; but the actual status has not been formally surveyed.

SIGNIFICANCE TO HUMANS
These turtles are occasionally eaten by humans. ◆

East African black mud turtle
Pelusios subniger

TAXONOMY
Testudo subnigra Lacepède, 1788, no type locality [restricted to Tamatave, Madagascar]. Two subspecies are recognized.

OTHER COMMON NAMES
English: Pan hinged terrapin, pan terrapin; Afrikaans: Panwaterskilpad.

PHYSICAL CHARACTERISTICS
East African black mud turtles are small turtles, with a maximum shell length of 7.9 in (20 cm), and an elongate, oval, unkeeled, unserrated carapace. The medium-sized plastron is posteriorly notched, has a well-developed hinge between the pectoral and abdominal scutes, and the pair of meosplastral bones between the hyo- and hypoplastra are in contact on the midline. The anterior plastral lobe is much broader than the posterior lobe, and it is only slightly longer than the interabdominal seam. The plastron is strongly constricted at the level of the abdominal-femoral seam. Axillary scutes are not present on the bridges.

DISTRIBUTION
Eastern and southeastern Africa (Tanzania to South Africa), Madagascar, and the Seychelles Islands.

HABITAT
These turtles inhabit nearly any freshwater aquatic habitat, from permanent lakes and rivers to streams, marshes, swamps, and even temporary pools.

BEHAVIOR
These turtles are reported to be nocturnal, but they are known to bask at the water's edge and to migrate overland between bodies of water. They are also known to estivate underground until favorable conditions return.

FEEDING ECOLOGY AND DIET
This species is primarily carnivorous, eating worms, mollusks, insects, crustaceans, fish, amphibians, and carrion. They also occasionally consume aquatic plants, as well as ripe fruit that has fallen into the water.

REPRODUCTIVE BIOLOGY
Little is known about the biology of this species. Courtship has not been described. One captive female laid eggs in February and March, but nesting in nature probably occurs during the summer (December and January). Clutch size ranges from eight to 12 eggs, which are elliptical, leathery, and average 1.4 by 0.8 in (36 by 21 mm). Incubation in the laboratory at 86°F (30°C) lasted 58 days.

CONSERVATION STATUS
Not threatened, but the actual status in nature has not been formally surveyed.

SIGNIFICANCE TO HUMANS
These turtles are occasionally eaten by humans. ◆

Resources

Books

Boycott, R. C., and O. Bourquin. *The Southern African Tortoise Book: A Guide to Southern African Tortoises, Terrapins and Turtles.* Privately published, KwaZulu-Natal, South Africa, 2000.

Branch, B. *Field Guide to the Snakes and Other Reptiles of Southern Africa.* Cape Town: Struik Publishers, 1998.

Spawls, S., K. Howell, R. Drewes, and J. Ashe. *A Field Guide to the Reptiles of East Africa.* San Diego: Academic Press, 2002.

Periodicals

Anderson, N. B. "Pelomedusidae; *Pelusios sinuatus*; Serrate Hinged Terrapin; Reproduction." *African Herp News* 23 (1995): 46.

Rödel, M. O. "Predation on Tadpoles by Hatchlings of the Freshwater Turtle *Pelomedusa subrufa.*" *Amphibia-Reptilia* 20 (1999): 173–183.

John B. Iverson, PhD

Big-headed turtles
(Platysternidae)

Class Reptilia
Order Testudines
Suborder Cryptodira
Family Platysternidae

Thumbnail description
A turtle of moderate size with a huge head that cannot be drawn inside its shell, a long scaly tail, and a large plastron

Size
Up to 8 in (20 cm)

Number of genera, species
1 genus; 1 species

Habitat
Mountain streams

Conservation status
Endangered

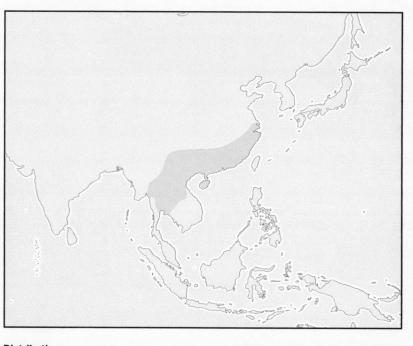

Distribution
Southeast Asia

Evolution and systematics

Although big-headed turtles (*Platysternon megacephalum*) were once considered closely related to pond turtles (family Emydidae), most recent studies suggest that they are the sister group to snapping turtles (Chelydridae). Some authorities even consider big-headed turtles to be a subfamily within the Chelydridae. However, differences in skeletons, chromosomes, and DNA sequences argue for separate family recognition. One genus of fossil turtles from the Cretaceous of Southeast Asia is tentatively assigned to this family. No subfamilies are recognized.

The taxonomy of this species is *Platysternon megacephalum* Gray, 1831, China. Three subspecies are recognized.

Physical characteristics

The most distinctive characteristic of this turtle is its huge head (half as wide as the shell) which cannot be withdrawn into its shell. The top of the skull is completely roofed over in bone and covered with a large epidermal scale. The upper jaw has a strongly hooked beak. The carapace is very low and sometimes bears a single medial keel. The plastron is relatively large, but it is only narrowly connected to the carapace by ligaments. The tail is nearly as long as the shell and is covered by large epidermal scales. Well-developed musk glands are present in front of and behind the bridge.

Distribution

Southern China south to Thailand and southern Myanmar.

Habitat

Known only from small, cool, rocky streams in mountainous areas up to 6,600 ft (2,000 m).

Behavior

Although big-headed turtles have not been well studied in the wild, they appear to be primarily nocturnal and crepuscular (active at twilight), spending the day under rocks or logs. When disturbed, they bite viciously. Their climbing ability is incredible, no doubt an advantage when clambering over rocks in fast-flowing streams. In captivity they have even been known to climb over wire fences and up window curtains to the ceiling. When molested in captivity these turtles are known to produce squeal-like noises. They apparently hibernate in the northern portion of their range, and there is speculation that they might hibernate terrestrially.

Feeding ecology and diet

This turtle is probably a strict carnivore in the wild, although its feeding habits are known only from captive

Big-headed turtle (*Platysternon megacephalum*). (Illustration by Michelle Meneghini)

studies. It will eat fish, frogs, and invertebrates. Although it surely feeds primarily on stream bottoms, it may also forage terrestrially along stream banks.

Reproductive biology

Big-headed turtles lay one to four large elongate eggs (1.5–1.7 by 0.9 in [37–44 by 22 mm]) per clutch, although one- or two-egg clutches are most common. Nesting is speculated to occur from May to August; a single captive egg hatched in September. Whether the turtles produce more than one clutch per year is unknown.

Conservation status

This species is listed as Endangered on the IUCN Red List.

Significance to humans

Big-headed turtles are considered to be a delicacy in many parts of their range; the flesh is thought to have the same aphrodisiac properties as rhinoceros horns. These turtles are also exploited for the pet trade.

Resources

Periodicals

Ernst, C. H., and A. F. Laemmerzahl. "Geographic Variation in the Asian Big-Headed Turtle, *Platysternon megacephalum* (Reptilia: Testudines: Platysternidae)." *Proceedings of the Biological Society of Washington* 115 (2002): 18–24.

Haiduk, M. W., and J. W. Bickham. "Chromosome Homologies and Evolution of Testudinoid Turtles with Emphasis on the Systematic Placement of *Platysternon*." *Copeia* (1982): 60–66.

Schleich, H. H., and U. Gruber. "Eine neue Großkopfschildkröte, *Platysternon megacephalum tristernalis* nov. ssp., aus Yünnan, China." *Spixiana* 7, no. 1 (1984): 67–73.

Walsh, T., and M. Russell. "The Natural History and Captive Husbandry of the Big-Headed Turtle (*Platysternon megacephalum*)." *Reptiles* 6, no. 2 (1998): 66–75.

Whetstone, K. N. "*Platysternon* and the Evolution of Chelydrid Turtles." *Herpetological Review* 8 Suppl. (1997): 20.

John B. Iverson, PhD

Afro-American river turtles

(Podocnemididae)

Class Reptilia

Order Testudines

Family Podocnemididae

Thumbnail description
Large sideneck turtles with four claws on the hind feet, a complete series of seven neural bones posterior to which the pleurals meet at the midline, mesoplastral bones present, and the pelvis is fused to the plastron

Size
Up to 42.1 in (107 cm) carapace length

Number of genera, species
3 genera; 8 species

Habitat
Freshwater rivers, streams, flooded forests, lakes, and ponds

Conservation status
Endangered: 2 species; Vulnerable: 4 species; Lower Risk/Conservation Dependent: 1 species

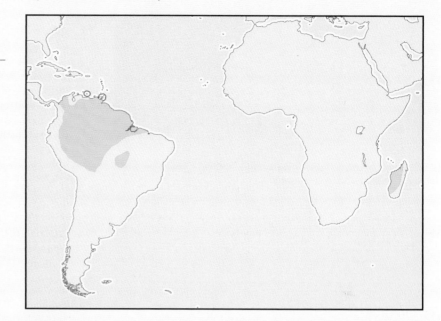

Distribution
Madagascar and northern South America, although fossils have been discovered in freshwater, marine, and terrestrial habitats nearly worldwide

Evolution and systematics

Podocnemididae is most closely related to Pelomedusidae, but diverged from that family at least by the Cretaceous (at least 110 million years ago). Although the living forms occur in fresh water in South America and Madagascar, this family, including freshwater and marine forms, has an extensive fossil record in North and South America, Europe, Asia, and Africa. The largest turtle that ever lived, *Stupendemys geographicus*, at up to 7.5 ft (2.3 m) shell length, belongs to this family.

Physical characteristics

These large sideneck turtles have a complete series of seven neural bones, posterior to which the pleurals meet at the midline; mesoplastral bones are present; the pelvis is fused to the plastron; there is no cervical scute; an intergular scute touches the dorsal rim of the plastron; and there are four claws on the hind feet.

Distribution

Madagascar and Atlantic drainages of northern South America, although fossils are known from freshwater, marine, and terrestrial habitats nearly worldwide.

Habitat

Most Podocnemididae are riverine (living on the banks of a river) or large lake species, although one species inhabits ponds and streams.

Behavior

The behavior of these turtles is highly variable. Some species are almost totally aquatic, migratory, and riverine, leaving the water only to nest colonially on seasonally inundated midstream sandbars. Others inhabit smaller tributaries or flooded forest pools, nesting in groups on sandy river banks; and one species (*Podocnemis vogli*) inhabits streams and ponds, migrates considerable distances overland to nest, and even estivates underground when the aquatic habitat dries up. Some species (and mainly females) are known to bask.

Feeding ecology and diet

These turtles are primarily herbivorous, consuming mainly the fruits of riparian trees, but also leaves, stems, and grasses. They occasionally eat freshwater sponges, mollusks, crustaceans, insects, and fish.

Reproductive biology

The courtship and mating of the Podocnemididae have not been described. The more aquatic species may migrate considerable distances to nesting beaches. The females nest predominantly on sandbars and river banks, often in large numbers. The leathery eggs range from nearly spherical in the largest species, to oblong in the others. Clutch size ranges from five to 20 in the smallest species, to as many as 156 in the largest. Incubation is completed in 40–149 days. All studied species exhibit temperature-dependent sex determination,

A yellow-spotted river turtle (*Podocnemis unifilis*). (Photo by Henri Janssen. Reproduced by permission.)

with warm temperatures producing females, and intermediate temperatures producing males. In some species still cooler temperatures may again produce mostly females.

Conservation status

All species are exploited by humans for their meat or their eggs. At one time nearly 124,000 of these turtles nested on a single beach during a six-week season. However, an estimated 33 million eggs were once harvested annually from just three beaches, and an average of over 6,000 adult females were slaughtered annually from a single beach. Because of this harvesting intensity, all species have been reduced in numbers, four to the point of being listed at Vulnerable to extinction on the IUCN Red List, and two more considered Endangered. Efforts are being made to protect the remaining colonial nesting beaches.

Significance to humans

These turtles are exploited primarily for their eggs and their meat.

1. South American river turtle (*Podocnemis expansa*); 2. Yellow-spotted river turtle (*Podocnemis unifilis*); 3. Madagascan big-headed turtle (*Erymnochelys madagascariensis*). (Illustration by Jonathan Higgins)

Species accounts

South American river turtle
Podocnemis expansa

TAXONOMY
Emys expansa Schweigger, 1812, "America meridionali" (South America). No subspecies are recognized.

OTHER COMMON NAMES
English: Arrau, giant South American river turtle; French: Podocnémide élargie; German: Arrauschildkröte; Spanish: Arrau.

PHYSICAL CHARACTERISTICS
South American river turtles are large sideneck turtles, up to 42.1 in (107 cm) in shell length, with a broad, flat carapace that is wider posteriorly than anteriorly, usually having two barbels on the chin, a broad skull, and with the front of the upper jaw squared off (as opposed to notched or rounded). Juveniles have large lemon-yellow spots on the head within which are one or two black spots.

DISTRIBUTION
Orinoco and Amazon River basins of northern South America.

HABITAT
These turtles primarily inhabit large rivers, but will venture into flooded areas adjacent to rivers during high water.

BEHAVIOR
This species migrates considerable distances up or down rivers to localized, colonial nesting beaches. Two to three weeks prior to nesting, females bask on sand beaches in the morning and the late afternoon.

FEEDING ECOLOGY AND DIET
This primarily herbivorous species apparently feeds predominantly on the fruits of riparian trees. Leaves and stems are also taken, as are freshwater sponges and occasional insects.

REPRODUCTIVE BIOLOGY
Courtship has not been described, but DNA studies reveal that eggs in the same clutch are often fertilized by different males. As water levels subside after the peak of the rainy season, females migrate to the main river channels and upstream or downstream to sandbars to nest. The nesting season is short, lasting only 10–60 days (typically 25–45) in February and March on the Orinoco, and September to October (or even December), at various Amazon River branches. Most nesting occurs at night, after midnight. Females use all four limbs to excavate a body pit about 3.3 ft (1 m) in diameter and 1.6 ft (0.5 m) deep. They then excavate the actual nest chamber in the bottom of this pit, using only the hind limbs. Following egg laying, the female covers the chamber and usually fills the body pit, in either case using her hind limbs. This species is unique in its family in producing eggs that are nearly spherical, averaging 1.6 in (4 cm) in diameter. However, females occasionally produce one or two "giant" eggs, which can be as large as 3.2 in (8 cm) in longest diameter. Clutch size ranges from 48 to 156 eggs, although around 80 is typical. Larger females lay larger clutches of larger eggs, which are buried deeper than those produced by smaller females. Only a single

Podocnemis expansa

Podocnemis unifilis

clutch is laid per year. Incubation is rapid in the very warm sand of the nesting beaches, requiring only about 45 days. Over the following two to three days, the hatchlings dig out as a group, generally emerging at night or in the early morning to avoid lethally high ground temperatures. This species exhibits temperature-dependent sex determination, with temperatures above 90.7°F (32.6°C) producing females, and temperatures below that producing mostly males; however, females may be produced again at still cooler temperatures. The pivotal temperature of 90.7° (32.6°C) is the highest known for any turtle.

CONSERVATION STATUS
This species is listed as Lower Risk/Conservation Dependent by the IUCN. Exploitation by humans has virtually eliminated this turtle from most of the upper Amazon River basin, and populations across the range are much reduced, in spite of national legal protection and concerted efforts to protect the remaining nesting beaches from disturbance, harvesting, predation, and flooding.

SIGNIFICANCE TO HUMANS
Adults and even hatchlings are harvested for their flesh, and eggs are collected for the oil that can be extracted from them. This harvesting is technically illegal and at the local subsistence level, but it remains to be seen whether this species can ever recover from the last four centuries of overexploitation. ◆

Yellow-spotted river turtle
Podocnemis unifilis

TAXONOMY
Podocnemis unifilis Troschel, 1848, "im Rupununi und Takutu" (rivers in Guyana). No subspecies are recognized.

OTHER COMMON NAMES
English: Terecay, yellow-headed sideneck, yellow-spotted Amazon turtle; Spanish: Tracaja.

PHYSICAL CHARACTERISTICS
Yellow-spotted river turtles are large sideneck turtles, with a maximum shell length of 26.8 in (68 cm), an oval carapace bearing a low keel on the second and third vertebral scutes, and a slight medial indentation anteriorly. Only a single barbel is usually present under the chin. Juveniles have yellow-orange spots on the head.

DISTRIBUTION
These turtles occur in the tropical lowlands of northern South America, including the Orinoco and Amazon River basins.

HABITAT
Yellow-spotted river turtles inhabit freshwater rivers, streams, lakes, ponds, and flooded forests.

BEHAVIOR
These turtles frequent rivers and large lakes during times of low water, but during the rainy season they migrate far into flooded forest areas. Females migrate to sand beaches along the main river courses to nest, and bask on the shoreline prior to nesting season.

FEEDING ECOLOGY AND DIET
Primarily herbivorous, this species feeds on the fruits of riparian trees, water hyacinths, and grasses. They also occasionally eat clams and dead fish.

REPRODUCTIVE BIOLOGY
Nesting season is timed with the period of low water and hence varies geographically, occurring in late January to early March in Venezuela, July to December in Colombia, October to February in Peru, and June to July, September to October, or during December at Brazilian sites. Females typically emerge to nest on sandbars just after dark, and excavate and cover their shallow nests only with their hind limbs. The eggs are elongate and average about 1.8 by 1.2 in (4.5 by 3.0 cm), and 0.9 oz (25 g). Clutch size ranges from four to 49 eggs, 20–30 being usual, with larger females producing larger clutches. At least two clutches may be produced each season. This species exhibits temperature-dependent sex determination, with warm temperatures (greater than 89.6°F [32°C]) producing females, and cool temperatures producing males.

CONSERVATION STATUS
Although listed as Vulnerable, this species is still exploited by local people. Females are captured on nesting beaches, fished for with hooks, and speared from the water, and eggs are harvested from nests.

SIGNIFICANCE TO HUMANS
Adults and eggs are harvested for human consumption. ◆

Madagascan big-headed turtle
Erymnochelys madagascariensis

TAXONOMY
Dumerilia madagascariensis Grandidier, 1867, "Mouroundava Tsidsibouque flumina in occidentali insulae Madagascar littore" (Morondava and Tsidibou rivers on the western coast of Madagascar). No subspecies are recognized.

OTHER COMMON NAMES
None known.

PHYSICAL CHARACTERISTICS
Madagascan big-headed turtles are medium-sized, sideneck turtles, with a maximum shell length of 19.7 in (50 cm), and a flattened oval carapace that lacks keels. The connection of the pelvis to the carapace contacts the suprapygals.

DISTRIBUTION
These turtles occur only in the western drainages of Madagascar.

HABITAT
This species inhabits slow-moving rivers, streams, ponds, and swamps.

BEHAVIOR
Very little is known about the behavior of Madagascan big-headed turtles. They apparently often spend the dry season buried in the mud.

FEEDING ECOLOGY AND DIET
These turtles are generally omnivorous, with adult females feeding primarily on the shoots of *Phragmites*, and the adult males feeding mainly on snails and fish. Juveniles eat both plant and animal material, including insects and crustaceans.

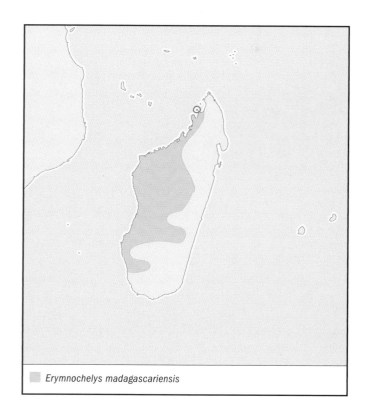

☐ *Erymnochelys madagascariensis*

REPRODUCTIVE BIOLOGY
Nesting occurs primarily during October and November, but may extend beyond that month. Females lay one to three clutches in a season, but apparently reproduce only every other year. Clutch size ranges from 10 to 30, and eggs average 1.5 by 0.9 in (38 by 24 mm), and 0.4 oz (10.2 g). The effects of temperature on sex determination have not been studied.

CONSERVATION STATUS
This species is cited as Endangered on the IUCN Red List.

SIGNIFICANCE TO HUMANS
These turtles are captured by humans for food in baited hoop nets or traps, on hook and line, with spears or harpoons, by diving for them, and on nesting beaches. The eggs are also taken from the nesting beaches. ◆

Resources

Periodicals

Escalona, T., and J. E. Fa. "Survival of Nests of the Terecay Turtle (*Podocnemis unifilis*) in the Nichare-Tiwadu Rivers, Venezuela." *Journal of Zoology* 244 (1998): 303–312.

Kuchling, G. "Possible Biennial Ovarian Cycle of the Freshwater Turtle *Erymnochelys madagascariensis*." *Journal of Herpetology* 27, no. 4 (1993): 470–472.

——. "Biologie und Lebensraum von *Erymnochelys madagascariensis* (Grandidier, 1867) und Vergleich mit den andern Wasserschildkröten Madagaskars." *Salamandra* 28, nos. 3 and 4 (1993): 231–250.

Thorbjarnarson, J. B., N. Perez, and T. Escalona. "Nesting of *Podocnemis unifilis* in the Capanaparo River, Venezuela." *Journal of Herpetology* 27, no. 3 (1993): 344–347.

Valenzuela, N. "Multiple Paternity in Side-Necked Turtles *Podocnemis expansa*: Evidence from Microsatellite DNA Data." *Molecular Ecology* 9 (2000): 99–105.

——. "Constant, Shift, and Natural Temperature Effects on Sex Determination in *Podocnemis expansa* Turtles." *Ecology* 82, no. 11 (2001): 3,010–3,024.

——. "Maternal Effects on Life-History Traits in the Amazonian Giant River Turtle *Podocnemis expansa*." *Journal of Herpetology* 35, no. 3 (2001): 368–378.

John B. Iverson, PhD

Tortoises
(Testudinidae)

Class Reptilia
Order Testudines
Suborder Cryptodira
Family Testudinidae

Thumbnail description
Terrestrial turtles with elephantine hind legs, flattened forelegs, and unwebbed toes

Size
Up to 55 in (140 cm) carapace length and 562 lb (255 kg)

Number of genera, species
12 genera; ca. 47 species

Habitat
Terrestrial ecosystems

Conservation status
Critically Endangered: 1 species; Endangered: 7 species; Vulnerable: 16 species

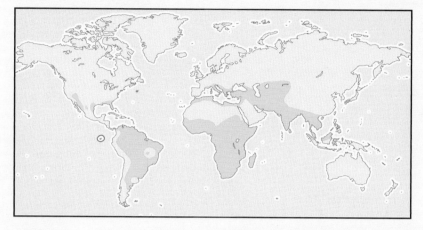

Distribution
All major land masses except Australia and Antarctica

Evolution and systematics

This family is most closely related to the pond, river, and wood turtles of the family Geoemydidae. It is well represented in the fossil record, with material dating back to the Eocene. No subfamilies are recognized.

Physical characteristics

These are small to very large terrestrial turtles with a high-domed shell in all but one species; columnar hind limbs, ele-

phantine in appearance; forelimbs somewhat flattened and armored with large scales; and short and unwebbed toes, each with two or fewer phalanges.

Distribution

Mainly tropical and subtropical in North and South America, Europe, Asia, and Africa, as well as numerous oceanic islands.

Habitat

Terrestrial, from deserts and grasslands to shrublands to the floors of primary forests.

Behavior

Tortoises often engage in male-to-male combat, usually involving shell ramming, and sometimes even biting of the extremities. Temperate species spend the winter underground either buried in the soil or in burrows they have constructed.

Feeding ecology and diet

Most species are herbivorous, eating grasses, fruits, flowers, seeds, or foliage, but a few species also eat animal matter (even carrion) opportunistically.

Reproductive biology

Courtship usually includes the male chasing the female, often head bobbing, biting at her extremities, and/or ramming

A Greek tortoise (*Testudo graeca*) hatchling in Morocco. (Photo by Animals Animals ©M. Linley. Reproduced by permission.)

Giant tortoises (*Geochelone nigra vandenburghi*) gathering in a seasonal rain pond on the caldera floor of Alcedo Volcano, Isabela Island, Galápagos. (Photo by Tui De Roy. Bruce Coleman, Inc. Reproduced by permission.)

her shell with his. Eggs are brittle-shelled, spherical to elongate, typically measuring 1–2 in (3–6 cm) in greatest diameter. Clutch sizes are generally small, ranging from one to 51 eggs per clutch, and are generally related to female size. Multiple clutches in one season are produced by many species, although some females apparently do not reproduce every year. Incubation is typically 100–160 days and is reported to last as long as 18 months in one species. Although most species have not been studied, those that have exhibit temperature-dependent sex determination, with females produced at high incubation temperatures and males at low temperatures.

Conservation status

Sixteen species are listed as Vulnerable, seven as Endangered, and one as Critically Endangered on the IUCN Red List. Because of their terrestrial habits, tortoises are especially vulnerable to human exploitation for food, traditional medicine, and the pet trade. Island populations have been especially prone to extirpation. Habitat destruction also takes its toll, especially for species inhabiting forested environments. Most species are legally protected by local countries, but illegal harvesting continues in most locales.

Significance to humans

Eaten by local people on every continent within its range.

1. Desert tortoise (*Gopherus agassizii*); 2. Hermann's tortoise (*Testudo hermanni*); 3. Galápagos tortoise (*Geochelone nigra*); 4. Pancake tortoise (*Malacochersus tornieri*); 5. South American yellow-footed tortoise (*Geochelone denticulata*). (Illustration by Joseph E. Trumpey)

Species accounts

South American yellow-footed tortoise
Geochelone denticulata

TAXONOMY
Testudo denticulata Linnaeus, 1766, Virginia. No subspecies are recognized.

OTHER COMMON NAMES
English: Morrocoy amarillo, yellow-footed tortoise.

PHYSICAL CHARACTERISTICS
This is a large tortoise (up to 10 in [26 cm] carapace length) with an elongate, high-domed shell with the twelfth marginal scutes fused to form a supracaudal scute, the fifth and sixth marginal scutes touching the second pleural scutes, no nuchal scute, no carapacial or plastral hinge, a divided gular scute that does not strongly project anteriorly and does not reach the entoplastron, a humeropectoral seam that does not cross the entoplastron, the external narial opening basically rounded, the premaxilla lacking a medial ridge, but the maxilla bearing one, an unflattened tail that lacks an enlarged terminal scale, five claws on the forefeet, the forelegs with large yellow or orange scales, and the carapace with the older scutes areas yellow or orange.

DISTRIBUTION
Atlantic versant of northern South America, including the Amazon River basin; southern populations may be disjunct.

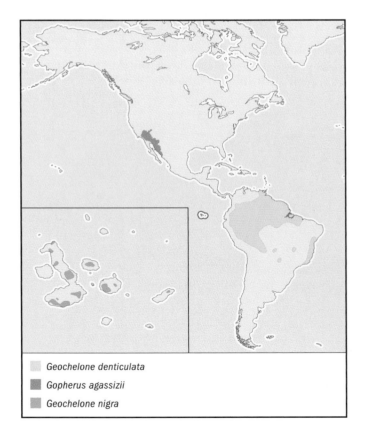

Geochelone denticulata

Gopherus agassizii

Geochelone nigra

HABITAT
Tropical evergreen and deciduous forests.

BEHAVIOR
Males use stereotyped head movements to identify other males. If a turtle does not respond with head movements, the male sniffs its cloacal region, presumably to confirm the sex and species. A returned head response elicits combat, in which males ram one another even to the point of one overturning the other.

FEEDING ECOLOGY AND DIET
This species is mostly herbivorous, feeding on fallen fruit, succulent plants, grasses, and mushrooms, but also eating termites and carrion.

REPRODUCTIVE BIOLOGY
This tortoise may mate throughout the year. During courtship the male pushes or rams the female's shell and bites at her limbs to immobilize her. He mounts the female's shell from the rear for copulation, during which time his head and neck are fully extended forward and downward, his mouth is open, and he may make clucklike vocalizations. The nesting season is very extended and it has been suggested that it might occur year-round. Multiple clutches are produced at intervals of about one to two months. Captive females have produced up to four clutches per year. The eggs are brittle-shelled but their shape varies from spherical to elongate and from 1.6 to 2.4 in (40 to 60 mm) in greatest length by 1.4 to 2.2 in (35 to 56 mm) in width. Egg mass may range from 1.4 to 4.0 oz (41 to 112 g), averaging about 2.5 oz (72 g). Clutch size ranges from one to 12 eggs, although clutches of four to six are most common. Incubation requires 128–152 days (mean 136). The effect of temperature on sex during development is not known.

CONSERVATION STATUS
This species is listed as Vulnerable on the IUCN Red List. Forest cutting is increasingly affecting this species.

SIGNIFICANCE TO HUMANS
These tortoises are eaten regularly by local peoples and also are collected in low numbers for the pet trade. ◆

Galápagos tortoise
Geochelone nigra

TAXONOMY
Testudo nigra Quoy and Gaimard, 1824, Hawaiian Islands (in error). Twelve subspecies are variably recognized, and the taxonomy of the various island populations is controversial.

OTHER COMMON NAMES
Spanish: Tortuga galápago.

PHYSICAL CHARACTERISTICS
This is a large tortoise (up to 51 in [130 cm] carapace length) with a uniform black, dark brown, or gray carapace that varies in shape from domed and rounded to saddle shaped. The twelfth marginal scutes are fused to form a supracaudal scute, the fifth and sixth marginal scutes touch the second pleural scutes, no nuchal scute is present, no carapacial or plastral hinge is present,

the divided gular scute does not strongly project anteriorly, the humeropectoral seam does not cross the entoplastron, the external narial opening is basically rounded, the premaxilla lacks a medial ridge, but the maxilla bears one, the unflattened tail lacks an enlarged terminal scale, and five claws are present on the forefeet.

DISTRIBUTION
Occurs only on the Galápagos Islands off Ecuador.

HABITAT
Volcanic islands, from semiarid lowlands to moist uplands.

BEHAVIOR
These tortoises are active by day, but sleep under vegetation or overhanging rocks at night. The females in some populations migrate from feeding areas to nesting areas with appropriate soil and sunlight. Combat and courtship behaviors are similar to those in the yellow-footed tortoise, although the males are even more vocal, generating deep basal roars. When approached by Darwin's finches, these tortoises stand erect and allow the finches to glean skin, ticks, and other ectoparasites from their bodies.

FEEDING ECOLOGY AND DIET
This species is almost totally herbivorous, feeding on grasses, forbs, cacti and other succulents, sedges, fruits, and even the leaves of bushes.

REPRODUCTIVE BIOLOGY
Courtship and mating are known from December to August, but vary by island. Nesting has been observed from late June to December, although the season varies in length by island. Possibly as many as four clutches are laid each year. The female excavates a flask-shaped nest, 7–12 in (18–30 cm) deep, with her hind feet. The eggs are almost spherical, brittle-shelled, and measure 2.2–2.6 in (56–65 mm) in greatest diameter. Clutch size ranges from two to 19 eggs depending on the female's size and island, with clutches of six to 10 being most typical. Incubation may take 85 to more than 200 days.

CONSERVATION STATUS
This species is listed as Vulnerable on the IUCN Red List, but one subspecies is listed as Extinct, two as Extinct in the Wild, four as Vulnerable, four as Endangered, and one as Critically Endangered. Predation by cats, rats, dogs, and pigs and competition from goats, donkeys, and pigs all remain as problems.

SIGNIFICANCE TO HUMANS
This species is probably only rarely eaten by locals today, and perhaps their primary current significance is as an attraction for ecotourists. ◆

Desert tortoise
Gopherus agassizii

TAXONOMY
Xerobates agassizii Cooper, 1863, southern California mountains, near Fort Mojave. No subspecies are recognized.

OTHER COMMON NAMES
Spanish: Tortuga del desierto, tortuga de los cerros, tortuga del monte.

PHYSICAL CHARACTERISTICS
This is a medium-sized tortoise (up to 19 in [49 cm] carapace length) with no plastral or carapacial hinge, a domed shell, the twelfth marginal scutes fused into a single supracaudal scute,

the fifth vertebral scute the broadest in the series, the carapacial scutes with light centers, a single axillary scute on each bridge, a divided gular scute that does not strongly project anteriorly, the premaxilla and the paired maxilla each bearing a medial ridge, the forelimbs flattened and shovel-like, the width of the forefeet and hind feet approximately equal, and the tail not flattened.

DISTRIBUTION
Southwestern United States and northwestern Mexico.

HABITAT
Thorn scrub to cactus deserts.

BEHAVIOR
This species constructs a burrow for use as a retreat from predators and the weather. Burrows may be barely long enough to accommodate the tortoise or as long as 33 ft (10 m) in northern sites. These tortoises remain inactive during the hottest parts of the day in summer and migrate to deeper burrows to hibernate for the winter. Desert tortoises are surprisingly social and head-bob at one another whenever they meet. Male-to-male encounters begin with head-bobbing, but escalate into ramming contests using the gular projections on the anterior plastron. This combat is occasionally intense enough to result in the overturning of one of the males.

FEEDING ECOLOGY AND DIET
Desert tortoises are mainly herbivorous, feeding on various grasses, cacti and other succulents, and flowers.

REPRODUCTIVE BIOLOGY
Maturity in desert tortoises is reached after 15 to 20 years. Courtship and mating are continuous from the spring to the autumn. The male follows the female, bobbing his head as he overtakes and circles her. He then bites at her head, forelimbs, or shell, and may even ram her with the gular projections on his plastron. Eventually if she is receptive and remains stationary, he moves behind her and mounts her shell for copulation. He produces hissing and grunting sounds while mounted. Nesting occurs mainly from May though July, with one to three clutches of two to 15 eggs (usually five or six) being laid each year. However, some females apparently skip reproduction in some years. The eggs are brittle-shelled, elliptical to nearly spherical, and average 1.6–1.8 in (40–45 mm) by 1.3–1.5 in (34–38 mm). Incubation in the field requires 90–120 days.

CONSERVATION STATUS
This species is listed as Vulnerable on the IUCN Red List. Habitat destruction (for road or housing construction, agriculture, livestock grazing, and by off-road vehicle use) is increasingly affecting natural populations. In addition, tortoises in many populations are afflicted with upper respiratory tract disease, caused by a mycoplasmic bacterium.

SIGNIFICANCE TO HUMANS
These tortoises increasingly are collected illegally for the Asian food markets in larger western U.S. cities. ◆

Pancake tortoise
Malacochersus tornieri

TAXONOMY
Testudo tornieri Siebenrock, 1903, Busisi, Tanzania. No subspecies are recognized.

Testudo hermanni

Malacochersus tornieri

OTHER COMMON NAMES
English: African pancake tortoise.

PHYSICAL CHARACTERISTICS
This is a small tortoise (up to 7 in [18 cm] carapace length) with no plastral or carapacial hinge, a very low, flexible shell with great reduction in bones (unique among all tortoises), and a nuchal scute present.

DISTRIBUTION
Kenya and Tanzania in East Africa.

HABITAT
Rock outcrops in savannas and scrublands.

BEHAVIOR
These tortoises are well-adapted to their rocky habitat. They are amazingly adept climbers and, when disturbed, quickly enter rock crevices and wedge themselves in place using their claws and sturdy limbs. They also estivate under rocks during hot, dry weather. Pancake tortoises also are known to bask.

FEEDING ECOLOGY AND DIET
This species is herbivorous, feeding on grasses, succulents, and leaves.

REPRODUCTIVE BIOLOGY
Courtship has been observed in captivity in January and February and consists of the male trailing the female, biting at her limbs, climbing on her back, and biting at her head. Mating has been recorded only in December in the field. Nesting in the wild apparently occurs in July or August. A single brittle-shelled, elongate egg (1.7–1.9 in [44–48 mm] by 1–1.1 in [26–28 mm]) is generally laid per clutch, although two are produced on occasion. Captives produce up to six clutches per year, but nesting in the wild is unknown. Incubation requires 113–221 days.

CONSERVATION STATUS
This species is listed as Vulnerable on the IUCN Red List. Removal for the pet trade and habitat alteration for agriculture pose the greatest threats to this tortoise.

SIGNIFICANCE TO HUMANS
These tortoises have been heavily exploited for the pet trade. ◆

Hermann's tortoise
Testudo hermanni

TAXONOMY
Testudo hermanni Gmelin, 1789, no type locality given. Two subspecies are recognized.

OTHER COMMON NAMES
None known.

PHYSICAL CHARACTERISTICS
This is a small tortoise (up to 10 in [26 cm] carapace length) with no plastral or carapacial hinge, a domed shell with the twelfth marginal scutes separate (i.e., not fused into a supracaudal scute), a long narrow nuchal scute present, no enlarged tubercles on the thigh, a large horny scale on the end of the tail, five claws (usually) on the forefoot, and five to 10 longitudinal rows of small scales on the anterior surface of the foreleg.

DISTRIBUTION
Southern Europe from southeastern France to eastern Turkey and Romania.

HABITAT
Dry habitats with dense vegetation, from scrublands to woodlands.

BEHAVIOR
The males engage in combat during the breeding season, by ramming other males and/or by biting at their heads and legs. The species hibernates during the winter by burrowing underground.

FEEDING ECOLOGY AND DIET
These tortoises are primarily herbivorous, feeding on legumes, buttercups, grasses, and fruits from trees. They also occasionally eat earthworms, snails, slugs, insects, and carrion.

REPRODUCTIVE BIOLOGY
Courtship and mating are apparently concentrated in the spring. Courting males chase the female, ramming her shell, biting at her head and legs, and finally mounting her shell from behind. While mounted, the male may make high-pitched squeaks. Flask-shaped nests 3 in (7–8 cm) deep are constructed in April to June. The clutch size ranges from two to 12 eggs and is related to female size, with small females in western populations having smaller clutches than the larger eastern females. Up to three clutches may be laid per season. The brittle-shelled eggs are usually elongate, measuring 1.1–1.6 in (28–40 mm) by 0.8–1.2 in (21–30 mm). Incubation takes 55 to 90 days. This species exhibits temperature-dependent sex determination, with females produced at high incubation temperatures and males at lower temperatures.

CONSERVATION STATUS
This species is noted as Lower Risk/Near Threatened on the IUCN Red List, although the subspecies *Testudo hermanni her-*

manni is categorized as Endangered. The numbers of this tortoise are declining due to habitat destruction, wildfires, predation by feral animals, and removal for the pet trade.

SIGNIFICANCE TO HUMANS
These tortoises were once heavily exploited for the pet trade, but this occurs much less frequently today. ◆

Resources

Books

Ballasina, D., ed. *Red Data Book on Mediterranean Chelonians.* Bologna, Italy: Edagricole, 1995.

Pritchard, Peter C. H., and Pedro Trebbau. *The Turtles of Venezuela.* Athens, OH: Society for the Study of Amphibians and Reptiles; Oxford, OH, 1984.

Periodicals

Moll, D., and M. W. Klemens. "Ecological Characteristics of the Pancake Tortoise, *Malacochersus tornieri,* in Tanzania." *Chelonian Conservation and Biology* 2, no. 1 (1996): 26–35.

Pritchard, Peter C. H. "The Galápagos Tortoises: Nomenclatural and Survival Status." *Chelonian Research Monographs* 1 (1996): 1–85.

Wallis, I. Z., B. T. Henen, and K. A. Nagy. "Egg Size and Annual Egg Production by Female Desert Tortoises (*Gopherus agassizii*): The Importance of Food Abundance, Body Size, and Date of Egg Shelling." *Journal of Herpetology* 33, no. 3 (1999): 394–408.

John B. Iverson, PhD

Softshell turtles
(Trionychidae)

Class Reptilia
Order Testudines
Suborder Cryptodira
Family Trionychidae

Thumbnail description
Medium to very large aquatic turtles with a rounded, flattened carapace covered with leathery skin; fleshy skin on the head that covers the jaws; and streamlined forelimbs with three pronounced claws

Size
4–47 in (10–120 cm)

Number of genera, species
13 genera; 25 species

Habitat
Rivers, streams, lakes, marshes, and temporary ponds

Conservation status
Critically Endangered: 5 species; Endangered: 5 species; Vulnerable: 6 species

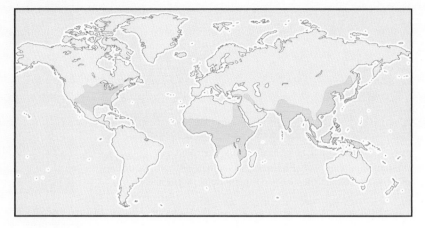

Distribution
North America, Africa, Asia, Indo-Australian archipelago

Evolution and systematics

These primitive cryptodires are most closely related to the Carettochelyidae of New Guinea and Australia. The oldest softshell fossil is from the late Jurassic. The recent revision of the phylogenetic arrangement proposed by Meylan has yet to gain universal acceptance and use, but it is generally supported by both morphological and molecular evidence. Two subfamilies are recognized: Trionychinae (without flexible flaps on plastron below hind limbs) and Cyclanorbinae (with flexible flaps on plastron below hind limbs).

Physical characteristics

Softshell turtles gain their name from the leathery layer of epidermis that covers the bony portion of the carapace and extends, in most species, to form a flexible disk overhanging the feet and tail. Flap-shelled species have a limited leathery carapace; however, they have developed retractable, flaplike hinges on the plastron that protect the limbs from below. When considering the entire leathery shell, the diversity of sizes found among softshell species covers an order of magnitude in carapace length. The smallest species attains a maximum size of 5 in (12 cm), whereas adults of the largest species may reach 47 in (120 cm) or more.

Most species possess a long retractile neck; this is especially well developed in the giant ambush-feeding species. However, the head of the Malayan softshell (*Dogania subplana*) is too large to be drawn completely within the shell. The long, tubelike proboscis found in most species allows them to breathe air from the water's surface without moving from the bottom. A thick layer of fleshy skin covers an incredibly sharp horny beak. The digits are strongly webbed with three claws present on each of the forelimbs. Most species are fairly uniform in coloration, allowing them to blend in with the substrate; however, several species in India and Southeast Asia have unique patterns on the carapace that may consist of broad stripes, vibrant spots, or elaborate designs. The plastron may exhibit two to nine callosities—thickened areas of epidermis overlying the plastral bones that develop as the turtle grows. Plastral callosities vary in number and shape among species and, when comparing adults, may be useful in taxonomic identification.

Distribution

Softshell turtles are found in lowland temperate regions east of the Rocky Mountains of North America, in sub-Saharan Africa and along the Nile River into Egypt, and across the Mediterranean Sea to southern Turkey and the Middle East. They also are present from eastern Pakistan, throughout India and southern Asia along the coast of China into southeastern Russia, and throughout the Indo-Australian archipelago to the southern shores of Irian Jaya, but there are no living species in Australia. The Malayan softshell (*Dogania subplana*) is the only species known to inhabit streams at high elevations. Trionychids formerly were distributed throughout Europe, South America, and Australia, as well as the three continents where they occur today. The Chinese softshells (*Pelodiscus sinensis, Palea steindachneri*) have been established by accidental or intentional introductions outside the known range. Extralimital populations of both species, farmed throughout Southeast Asia, have been documented in the Hawaiian Islands.

The eastern spiny softshell turtle (*Trionyx spiniferus*) has a leathery shell rather than the typical hard scutes of other turtles. (Photo by Robert P. Carr. Bruce Coleman, Inc. Reproduced by permission.)

A narrow-headed softshell turtle (*Chitra indica*), locally called "chitra chitra," breaks out of its shell. This rare turtle is claimed to be the world's biggest softshell turtle at its full-grown weight of 440 lb (200 kg). (Photograph. AP Wide World/Fishery Department. Reproduced by permission.)

Habitat

These highly aquatic turtles may be found in all freshwater habitats. Most species prefer clean, well-oxygenated waterways with soft sand or mud bottoms. They commonly inhabit large rivers, streams, lakes, and ponds, but they also may occur in swamps, marshes, temporary ponds, and drainage ditches. A few species tolerate brackish water, but only the Asian giant softshell (*Pelochelys cantorii*) is a permanent resident of coastal waters.

Behavior

Softshell turtles may remain buried at the bottom of their aquatic habitat for most of the day; however, some species will bask on sandbars or steep riverbanks. Prolonged submergence is possible because they can absorb oxygen across the skin and lining of the throat. This is especially important for temperate species that hibernate for months beneath the ice. Conversely, the flap-shelled turtles may be adapted to long periods of drought. The protective shell enables these turtles to burrow deep into the mud and may prevent desiccation while estivating.

Feeding ecology and diet

Most species are primarily carnivorous; however, plant material may occasionally be consumed. The majority of softshell species are generalists; all forms of animal matter (live or as carrion) are taken opportunistically. A few species are highly specialized for ambushing fish and other free-swimming aquatic animals. Adaptations for this mode of prey acquisition are most highly developed in the Asian narrow-headed softshell (*Chitra indica*); these turtles have small, forward-set eyes and exceptionally long necks that can be thrust from the sandy bottom with explosive speed and deadly accuracy.

Reproductive biology

Mating generally occurs in the spring; however, some species may store sperm for several years. The spherical eggs are generally laid in sandy banks above the flood line along a river or lake. Clutch size ranges from as few as 3 eggs in some species to more than 100 in others. Some species lay several clutches in a single year. Most species that have been investigated exhibit genetic sex determination; however, heteromorphic sex chromosomes are not present.

Conservation status

Six species are listed as Vulnerable, five as Endangered, and five as Critically Endangered on the IUCN Red List. As with most turtle species endemic to Asia, the softshells of China, Southeast Asia, and India are heavily exploited for food and medicine. Most softshells are dependent upon high levels of dissolved oxygen in their aquatic environment; therefore, they are particularly vulnerable to the effects of habitat destruction and degradation. They are extremely sensitive to rotenone, a chemical that is intentionally added to streams to kill fish, and pesticides (e.g., DDT) in farmland runoff. Human consumption has taken a heavy toll on the populations of many Asian species. Many species are legally protected by local countries, but exploitation continues. Some species are used to meet the demands of food markets.

Significance to humans

Softshells are used for food throughout their range; however, *Pelodiscus sinensis* is intensively farmed throughout Asia to supply consumer demand in China.

1. Female Chinese softshell turtle (*Pelodiscus sinensis*); 2. Spiny softshell (*Apalone spinifera*); 3. Indian flapshell turtle (*Lissemys punctata*). (Illustration by Gillian Harris)

Species accounts

Indian flapshell turtle
Lissemys punctata

SUBFAMILY
Cyclanorbinae

TAXONOMY
Testudo punctata Lacépède, 1788, India. Two subspecies are recognized.

OTHER COMMON NAMES
None known.

PHYSICAL CHARACTERISTICS
This is a small turtle (maximum carapace length 11 in [27.5 cm]) with a relatively deep oval shell. This species is unique among softshells because the posterior margin of the bony carapace is ringed by peripheral bones. The evolutionary origins (primitive or derived) of this feature are unresolved; however, it provides additional protection for the hind limbs, which may be completely retracted when the plastral flaps are pulled tightly against the bony rim of the carapace. Seven plastral callosities develop on the hyoplastra and hypoplastra, xiphiplastra, epiplastra, and entoplastron.

DISTRIBUTION
The Indus and Ganges river drainages of Pakistan, India, Sri Lanka, southeastern Nepal, Bangladesh.

HABITAT
Shallow backwaters of rivers, ponds, and marshes with a soft muddy bottom, temporary ponds.

BEHAVIOR
The retractable flaps over the limbs and the thick callosities on the plastron may be adapted to long periods of drought. The protective shell enables these turtles to burrow deep into the mud and may prevent desiccation. They have been observed in India to survive for 160 days while buried during estivation.

FEEDING ECOLOGY AND DIET
Like many flap-shelled species, these turtles tend to be omnivorous. Small aquatic animals and carrion make up a large portion of their diet; however, aquatic vegetation is also a common food item.

REPRODUCTIVE BIOLOGY
Courtship and mating have been observed in April and continue through the summer. The smaller male initiates mating by stroking the female's carapace with his chin. While facing the male, a receptive female extends her neck and they both bob their heads in a stereotypical pattern before settling to the bottom for copulation. The elaborate courtship observed for this species may be rare among softshells, but it is similar to that of many sexually dimorphic species where the larger female chooses her mate. The spherical, brittle-shelled eggs range from 0.9 to 1.3 in (24 to 33 mm) in diameter. Multiple clutches of two to 14 eggs are produced annually.

CONSERVATION STATUS
Not threatened.

SIGNIFICANCE TO HUMANS
These turtles constitute the largest volume of any species in Indian food markets. They are consumed locally in Bangladesh, but also may be exported to Chinese markets. ◆

Spiny softshell
Apalone spinifera

SUBFAMILY
Trionychinae

TAXONOMY
Trionyx spiniferus LeSueuer, 1827, New Harmony, Posey County, Indiana, on the Wabash River. Six subspecies are recognized.

OTHER COMMON NAMES
English: Goose-neck turtle, leatherback turtle; French: Tortue luth, tortue-molle à épines; German: Lederschildkröte; Spanish: Tortuga-casco suave espinosa.

PHYSICAL CHARACTERISTICS
This is a medium-sized softshell with spiny protuberances on the anterior rim of a drab olive carapace. A pattern of dark circles, which is especially distinct among the males, fades in the adult females. At least two pairs of callosities (hyo-hypoplastral and xiphiplastral) are present in adults.

DISTRIBUTION
Northern Mexico across most of the southern, central, and eastern United States to the Great Lakes region of southern Canada.

HABITAT
Slow-moving rivers, shallow streams, and large ponds and lakes.

BEHAVIOR
This species may remain buried at the bottom of its aquatic habitat for most of the day; however, it occasionally basks on sandbars or steep riverbanks. The spiny softshell remains submerged for long periods by absorbing oxygen through the skin

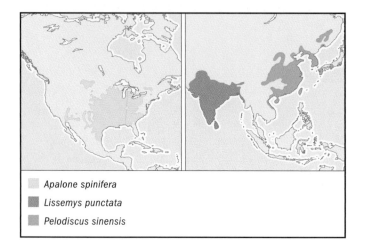

Apalone spinifera
Lissemys punctata
Pelodiscus sinensis

and lining of the throat while carbon dioxide diffuses across the skin. In northern regions, this species hibernates beneath the ice for several months each winter.

FEEDING ECOLOGY AND DIET
These turtles are predominantly carnivorous, ingesting all available aquatic animals; however, plant material including acorns and leaves is also consumed. In Iowa, crustaceans, fish, and insects were the most important food items, but plant matter was found in 61% of all turtles sampled.

REPRODUCTIVE BIOLOGY
Mating occurs in the early spring and eggs are generally laid in June and July. The brittle, spherical eggs are approximately 1.1 in (28 mm) in diameter. At least one clutch of four to 32 eggs is produced annually. The eggs hatch in late summer and hatchlings emerge in the autumn to hibernate underwater.

CONSERVATION STATUS
Not threatened. The widespread distribution and relative abundance of this species in its preferred habitats suggest that it is not currently at risk. Because these turtles are dependent upon high levels of dissolved oxygen in their aquatic environment, water pollution may be the most significant threat to softshell populations.

SIGNIFICANCE TO HUMANS
This species may be consumed locally; however, it is collected in large numbers for sale in Asian food markets in large North American cities and for export. Hatchlings are often available in the pet trade. ◆

Chinese softshell turtle
Pelodiscus sinensis

SUBFAMILY
Trionychinae

TAXONOMY
Trionyx (Aspidonectes) sinensis Wiegmann, 1835, small island in the Tiger River near Macao. No subspecies are recognized.

OTHER COMMON NAMES
Japanese: Suppon.

PHYSICAL CHARACTERISTICS
This is a small to medium turtle (maximum leathery shell length 10 in [25 cm]) with an oval carapace. The shell is rela-

tively smooth; however, the anterior portion of the carapace is studded with blunt knobs. Up to seven plastral callosities may develop on the hyoplastra and hypoplastra, xiphiplastra, and epiplastra. The tubelike proboscis has a horizontal ridge that projects from either side of the septum.

DISTRIBUTION
Widespread in the lowland areas of southern China, northern Vietnam, Hainan Island, Taiwan, and Japan.

HABITAT
Rivers, streams, lakes, marshes, and rice fields.

BEHAVIOR
Although this species occasionally basks on the banks, it remains buried at the bottom of its aquatic habitat for most of the day. The Chinese softshell remains submerged for long periods by absorbing oxygen through the skin (33%) and lining of the throat (67%), while most carbon dioxide passively diffuses out across the skin. In Japan this species hibernates from October to April.

FEEDING ECOLOGY AND DIET
This species is primarily carnivorous, ingesting all available aquatic animals, especially fish and crustaceans; however, the seeds of marsh plants are also consumed.

REPRODUCTIVE BIOLOGY
Mating has been observed from May to July in Japan. The male uses the claws of the forelimbs to clasp the front rim of the female's shell during copulation and may bite at her neck and limbs. Females begin nesting early in the spring and continue through the late summer. A shallow, boxlike nest is excavated and a clutch of 15–28 brittle, spherical eggs (average 0.8 in [20 mm] in diameter) is deposited. As many as four clutches are produced in a single year. Although incubation is generally completed in 60 days, hatching may occur after 40 to 80 days.

CONSERVATION STATUS
Chinese softshell turtles are listed as Vulnerable on the IUCN Red List. Despite the heavy consumer pressure on this species, it is still relatively abundant in eastern China and northern Vietnam.

SIGNIFICANCE TO HUMANS
This species is consumed locally throughout its range; however, it is intensively farmed in Southeast Asia to supply the food and medicine market in China. Hatchlings are reared to subadulthood in two to three years before being shipped for processing into prepared soups, or more commonly to restaurants and markets where they are bought fresh and butchered alive. ◆

Resources

Books
Das, Indraneil. *Die Schildkröten des Indischen Subkontinents.* Frankfurt, Germany: Chimaira, 2001.

Periodicals
Greenbaum, E., and J. L. Carr. "Sexual Differentiation in the Spiny Softshell Turtle (*Apalone spinifera*), a Turtle with Genetic Sex Determination." *Journal of Experimental Zoology* 290 (2001): 190–200.

Meylan, Peter A. "Phylogenetic Relationships of Soft-Shelled Turtles (Family: Trionychidae)." *Bulletin of the American Museum of Natural History* 186 (1987): 1–101.

Pritchard, Peter C. H. "Observations on Body Size, Sympatry, and Niche Divergence in Softshell Turtles (Trionychidae)." *Chelonian Conservation and Biology* 4 (2001): 5–27.

Patrick J. Baker, MS

Crocodilians
(Crocodiles, alligators, caimans, and gharials)

Class Reptilia
Order Crocodylia
Number of families 3
Number of genera, species 8 genera; 23 species

Photo: A false gharial (*Tomistoma schlegelii*). (Photo by Animals Animals ©C. C. Lockwood. Reproduced by permission.)

Evolution and systematics

There are 23 widely recognized species of crocodiles, alligators, caimans, and gharials, all members of the order Crocodylia. Superficially they resemble reptiles, yet their closest cousins are birds and extinct Dinosauria, a group known as archosaurs ("ruling reptiles"). Modern Crocodylia are the latest iteration of the Crocodylomorpha, a major group whose evolutionary heritage spans almost 240 million years. Crocodylia are often described as "living fossils," unchanged in millions of years, but this description is inaccurate. The Crocodylomorpha were a diverse and successful group occupying terrestrial, freshwater, and marine ecosystems, and their modern counterparts are barely less fantastic. This group is informally referred to as "crocodilians," although the term "crocodylians" technically refers to members of the order.

Every successful group has a beginning. The earliest crocodyliforms were terrestrial hunters that shared an ankle with modern Crocodylia, but little else, yet they were dominant predators whose legacy diversified throughout the Jurassic and Cretaceous periods. Their success is evident in excavations, as more crocodyliform material than dinosaur bones often turns up. Discoveries have been remarkable, such as curious peglike teeth and spiked protective plates from *Desmatosuchus* that indicate a defensive, vegetarian lifestyle. Despite dabbling in herbivory, it was in carnivory that crocodyliforms excelled. The awe-inspiring skulls of *Sarcosuchus* and *Deinosuchus* paint a picture of massive killers over 35 ft (10.7 m) in length. From terrestrial beginnings, crocodyliforms branched out into freshwater and marine habitats. In extreme thalattosuchians (marine crocodiles), limbs were replaced with paddles and a fluked tail to compete with sharks and ichthyosaurs for dominance of the sea. However, these marine forms were an evolutionary dead end, and freshwater species with limited marine ability proliferated from the Juras-

sic onward. Modern Crocodylia first appeared over 100 million years ago, and despite experiments with various and occasionally bizarre forms, the semiaquatic predator has become their signature role.

Scientists disagree about crocodyliform classification and evolutionary relationships. In 2003, 23 species of Crocodylia are widely recognized, divided into three families: Alligatoridae (alligators and caimans; eight species), Crocodylidae (crocodiles; 14 species), and Gavialidae (gharial; one species). The Crocodylidae are further divided into two subfamilies, Crocodylinae and Tomistominae. Some taxonomists also divide Alligatoridae into two subfamilies: Alligatorinae and Caimaninae.

Physical characteristics

A crocodile may be thought of as an elegant solution to the problem of catching prey, surviving unpredictable environments, conserving limited energy, and reproducing successfully. In appearance, crocodiles superficially resemble lizards, having scales, a long tail, and four limbs. But appearances can be deceptive, and a closer look reveals that crocodilians are unique.

All 23 species are broadly similar in appearance, varying mainly in size, scale patterns, color, and skull morphology. The smallest species is Cuvier's dwarf caiman (*Paleosuchus palpebrosus*); adult males rarely exceed 5 ft (1.6 m) in length and females 4 ft (1.2 m). Within the same family, the black caiman (*Melanosuchus niger*) and American alligator (*Alligator mississippiensis*) vie for largest size, yet rarely exceed 14 ft (4.3 m). Crocodylidae range from the diminutive dwarf crocodile (*Osteolaemus tetraspis*) at 6 ft (1.8 m) to the massive estuarine crocodile (*Crocodylus porosus*) that can exceed 16 ft (5 m). The

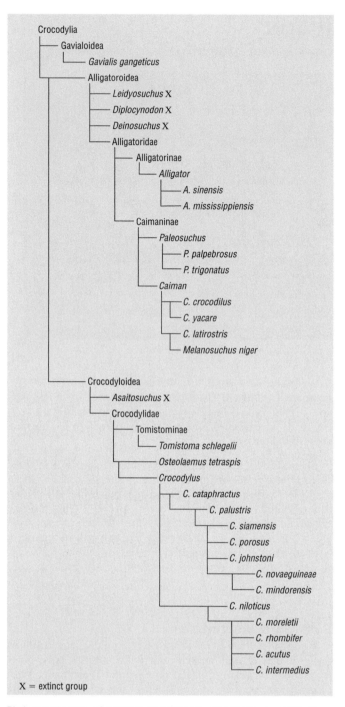

Phylogenetic tree of modern crocodilians. (Illustration by GGS. Courtesy of Gale.)

arine crocodiles and Indian gharials, are capable of attaining such sizes, evidence of these giants is scarce. The largest crocodile reliably measured and published in the literature, an estuarine crocodile from Papua New Guinea, was 20.7 ft (6.3 m) long. While unlikely to be the maximum possible size for this species, stories of even larger animals are difficult to verify. One fact is certain—crocodiles over 20 ft (6 m) are exceptionally rare.

Crocodilians undergo a dramatic increase in size from hatchling to adult. Over its lifetime, an estuarine crocodile may grow from a 12-in, 2.8-oz (30-cm, 80-g) hatchling to a 20-ft, 2,650-lb (600-cm, 1,200-kg) adult. A 20-fold increase in length and 15,000-fold increase in weight is quite a feat in the animal kingdom. Imagine, then, how this compares with the extinct *Sarcosuchus*, which reached 35 ft (10.7 m) and over 19,800 lb (9,000 kg)! Growth is most rapid when young, yet scientists are unsure whether adults reach a maximum size or continue to grow slowly until they die. The enormous sizes attained by extinct species such as *Sarcosuchus* and *Deinosuchus* may have been possible by maintaining those fast juvenile growth rates throughout a greater percentage of their lives.

Crocodilians are covered in a thick, leathery skin broken into various sizes and shapes of scales in particular areas. Scales on the back are large and rectangular, lying in parallel rows from shoulders to pelvis and continuing onto the tail. These dorsal "scutes" each contain a bony plate called an osteoderm ("skin bone") just below the surface. A tough covering of beta-keratin helps minimize water loss, although the more flexible alpha-keratin is found between the scales. Osteoderms not only offer protection, they are infused with blood vessels and function as solar panels, transporting heat from the surface to the body core during basking. Adjacent osteoderms are closely integrated like the beams of a bridge, providing support for the spinal column. Large nuchal plates protect the nape. Scales on the flanks and limbs are generally smaller, rounder, and softer to allow bending. Those on the belly are even, rectangular, and smooth to reduce friction sliding over the ground. Small osteoderms are found in the belly scales of most species. Thick, rectangular scales are present on the tail, with sharp, upward-pointing scutes providing extra surface area as the tail sweeps through water. Scales on the head are small, irregular in shape and thin, housing blood vessels and sensory nerves. Each species has a unique pattern of scales and osteoderms.

Deceptively, a layer of mud and dust often covers dry, basking adults, suggesting a bland coloration. However, most species exhibit distinctive color patterns, which enhance camouflage and aid communication. Dorsal color is typically tanned yellow to dark brown, overlaid with characteristic dark bands, spots, or speckles. Juveniles of all species are more vivid, their bright colors fading in adulthood. Ventral scales are creamy white with varying degrees of black pigmentation, except for the almost black bellies of dwarf caimans and dwarf crocodiles. Color mutations where pigment is usually absent are rare, genetic anomalies. Leucistic and albino crocodiles are as tempting to predators as their "white chocolate" appearance suggests, but they are popular tourist attractions in captivity, where they must be shielded from excess sunlight. Both short- and long-term changes in skin color have been

sole gavialid, the Indian gharial (*Gavialis gangeticus*) can also reach 16 ft (4.9 m). Females are always the smaller sex, and this is most apparent in estuarine crocodiles, where females of 10 ft (3 m) are considered to be very large.

As the largest living reptiles, males over 16 ft (4.9 m) may tip the scales at 1,100 lb (500 kg), but these are lightweights compared to rare individuals that exceed 18 ft (5.5 m) and 2,200 lb (1,000 kg). Although several species, including estu-

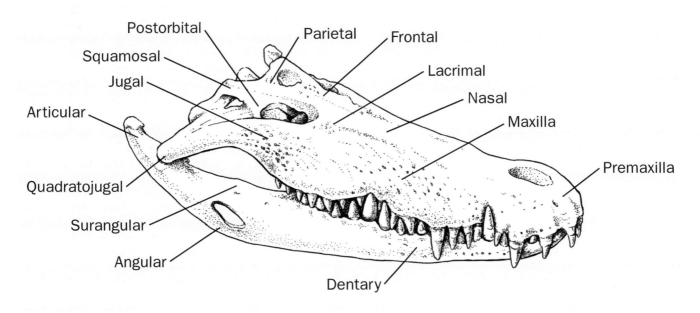

A crocodilian skull. (Illustration by Brian Cressman)

recorded in several species. Changes in mood, such as those caused by stress, and environmental temperature can dull skin color. Long-term change can be effected by the environment, with individuals from shaded areas becoming darker as black pigment (melanin) accumulates in the skin.

The crocodilian head always draws attention. The skull, although massive and sturdy, is infiltrated with air spaces. These spaces reduce weight without compromising strength, and provide extensive areas for muscle attachment and expansion. Two pairs of openings on either side of the cranium classify the skull as diapsid. There is considerable variation in skull and jaw morphology across all 23 species, and this has an ecological significance: broad jaws are reinforced by bony ridges to resist strong bite forces for crushing prey, while slender jaws slice with little resistance through water to seize slippery prey.

The head houses all the major sense organs, vital for navigation, communication, and hunting. Senses are concentrated on the dorsal surface, so they remain exposed even when the head is partially submerged. Remarkably, a crocodile can hide its entire body below the water while maintaining maximal sensory input from its surroundings. As masters of stealth and ambush, crocodiles have no equal.

The eyes of the crocodile are placed high on each side of the head, turrets that provide 270° of widescreen coverage plus 25° of binocular overlap directly ahead to accurately judge distance. The pupil, round and dilated at night to permit maximum light entry, is compressed to a thin vertical slit during daylight to protect the sensitive retina. Inside the eye, cone cells on the retina provide color acuity by day, and high densities of rod cells give excellent low-light sensitivity at night. These rods can change shape to further alter sensitivity. A layer behind the retina, the tapetum lucidum, is impregnated with guanine crystals to reflect light back across the visual cells. This effectively doubles visual sensitivity at night, and shining a beam of light directly into a crocodile's eye rewards the observer with a fiery red eyeshine. Visual cells are most concentrated in a horizontal band across the back of the retina, a fovea providing highest visual acuity where crocodiles need it most—along the same plane as water. To focus (accommodate), crocodilians change the shape of their lens using the ciliary body.

Three eyelids cover each eye. The upper lid contains bony ossification to protect the eye, large bony palpebrals in caimans lending "eyebrows" to their appearance. The lower lid lacks ossification and is responsible for closing the eye. The third eyelid, the nictitating membrane, sweeps laterally over the cornea to clean the eye and protect it from abrasion underwater. Although the nictitating membrane is transparent except for the ossified leading edge, crocodilians still see poorly through it. Lachrymal (tear) glands lubricate its passage via ducts connected to the nasal cavities. Fluids may even accumulate when the crocodile remains out of water—real "crocodile tears," yet an unlikely source for the popular myth.

The ears are located immediately behind the eyes, the eardrum protected by an elongated flap of skin. Hearing sensitivity can be altered by opening a slit in front of the flap, or lifting the flap upward. When submerged, the ears normally close, as hearing becomes secondary to the ability to feel vibrations through the water. Detectable frequencies range from below 10 Hz to over 10 kHz, and sound pressure levels below -60 dB can be detected within certain bandwidths. In other words, crocodilians have excellent hearing, on a par with birds and mammals. Peak sensitivities range from 100 Hz to 3 kHz depending on the species, which coincides with the bandwidth of calls produced by juveniles. Vocalization is well developed in crocodilians, with over 20 different call types from both juveniles and adults recognized.

Crocodilians can breathe when submerged by exposing the dorsal margin of their head and hence their raised nostrils. Inhaled air passes through sinuses separated from the mouth by a bony secondary palate, where any chemicals in the air are detected by sensory epithelial cells. The presence and direction to food is easily discerned, and smell plays an important role in chemical communication. In early crocodyliforms, the internal nostrils (choanae) opened in the front of the mouth, but over millions of years they moved back to the throat, a phenomenon termed post-nasal drift. The palatal valve, a fleshy extension of the tongue, completely seals the throat from the mouth, hence crocodilians breathe easily near the surface even if the mouth is flooded with water. The glottis, an opening to the trachea and lungs, is located directly beneath the choanae. By varying tension in muscles lining the opening, exhaled air is forced through a constriction capable of relatively complex vocal sounds. Amplification is provided by expanding the throat using the hyoid apparatus, a curved cartilage beneath the glottis. A curious bend in the trachea of several species may further amplify the sound, similar to the long, curved necks of cranes.

The tongue lies between each mandibular bone of the lower jaw, behind the mandibular symphysis (fusion). Hence in slender-snouted species with extended sympheses, the tongue is greatly reduced. Although relatively immobile, the tongue can be pushed against the roof of the mouth to manipulate objects or pulled down to create a pouch for hatchlings. Typically bright yellow or orange, the tongue's color may provide a social or warning signal when the jaws gape. Pores cover the surface of the tongue, through which "salt glands" produce a saline fluid in brackish or sea water in Crocodylidae and Gavialidae. Alligatorid pores play no role in salt secretion, supporting theories of a more recent marine dispersal phase for the Crocodylidae. Chemoreceptors lining the tongue detect chemicals in water, yet little is known of their sensitivity. Their importance is implied in their ability to detect food underwater, and in the role of pheromones secreted from chin and paracloacal musk glands.

High densities of dome pressure receptors (DPRs) cover scales on the head, particularly around the jaws. Disturbances of the water surface create pressure waves easily detected by DPRs, rapidly alerting the crocodile to potential prey near the head. Crocodilians also react rapidly to movement underwater (such as fishes) even when vision is unavailable. Similar pressure receptors, Integumentary Sense Organs (ISOs), are located on the caudal margin of body scales in Crocodylidae and Gavialidae, but not Alligatoridae. Their function is not fully understood, nor is the reason why alligatorids entirely lack them. However, evidence of DPRs exists in extinct crocodyliforms, suggesting their sensory role in water has long been part of their repertoire.

Betraying their terrestrial origins, crocodilians are surprisingly mobile predators on land. Although lacking the stamina for pursuit, their explosive force catches most prey unaware. Like all archosaurs, the hind limbs are significantly larger and stronger than the forelimbs, suited to the croco-

dilian propensity for launching the body forward at speed. Five toes are present on the front feet and four on the back, although residual bones from the fifth still exist. The inner three toes terminate with strong, blunt nails that provide traction; the outer one (back feet) or two (front feet) lack claws and bend backward during walking. There is extensive webbing between the toes on the back feet, but webbing is minimal or absent from the front feet.

The limbs are used to crawl, walk, and gallop. The crawl employs the limbs alternately to slide the body across mud, sand, or grass. When sufficiently motivated, the limbs can propel the body forward in a slithering manner at much greater speed, up to 6.2 mph (10 kph). In the uniquely crocodilian high walk, the feet rotate inward toward the body and support it from below. Lifting the head and belly clear of the ground enables the crocodile to traverse obstacles or rough terrain. In a few species, the front and hind limbs move in tandem to gallop. This springlike gait accelerates the crocodile up to 10.6 mph (17 kph) for several seconds until the safety of water can be reached. Cuban crocodiles (*C. rhombifer*) remind us of the frightening aggression of their terrestrial ancestors when they gallop toward a threat.

Water is clearly the crocodilian's preferred domain, a home for prey that live in the water and a magnet for prey that live on land. Mobility is possible through the powerful tail, which makes up half the body's total length. Flattened dorsoventrally to provide extensive surface area for propulsion, the tail is undulated laterally by powerful muscles. Limbs are swept back during rapid swimming, although when moving slowly they help the crocodile steer, brake, reverse, or walk across the bottom. So powerful is the tail that it can drive hundreds of pounds (or kilograms) of crocodile vertically out of the water to capture prey several feet (or meters) overhead.

Internally, the pleural cavity contains the lungs, and the visceral cavity houses major organs associated with digestion and reproduction. These cavities are separated by the bilobed liver and diaphragmaticus, a sheet of muscle analogous to the diaphragm in mammals. Inhalation is achieved by contracting the diaphragmaticus, which pulls the liver backward and expands the pleural cavity. Thoracic (intercostal) muscles also expand the chest, and reduced pressure in the lungs draws air in through open airways. To exhale, the diaphragmaticus and thoracic muscles relax, compressing the pleural cavity and forcing air out of the airways. Crocodilians control their buoyancy primarily through the volume of air in their lungs. By moving the liver, hind legs, and tail, subtle postural changes are also possible. When diving, air is forced out of the lungs and the crocodilian, which is considerably heavier than water, sinks rapidly. Swimming may facilitate this sinking, and by sweeping the hind legs forward the crocodilian can reverse and submerge simultaneously. Swimming or pushing off the bottom returns the crocodilian to the surface, and positive buoyancy is achieved by filling the lungs with air. Stones called gastroliths in the stomach typically comprise 1–5% of the crocodile's total weight, and their presence may provide additional ballast.

Situated between the lungs is the most complex heart in the animal kingdom, apparently the result of adaptation to

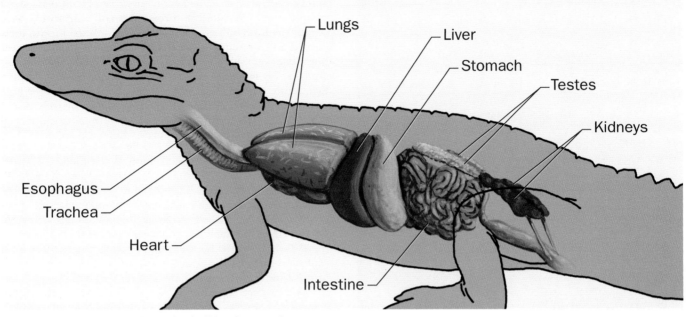

Crocodilian internal organs. (Illustration by Brian Cressman)

the demands of the crocodiles' semiaquatic lifestyle and their size. Unlike other reptiles, the crocodilian heart is fully divided into four chambers, as are the hearts of birds and mammals. Uniquely, valves under nervous and hormonal control can alter blood flow. These ensure that vital oxygenated blood circulates between essential areas during oxygen stress, such as while diving, while deoxygenated blood is sent to nonessential areas. During rest and normal exercise, blood in the right ventricle passes via a coglike valve to the pulmonary arteries and lungs to acquire oxygen. During diving, this valve constricts, and deoxygenated blood is diverted to the left aortic arch that leads to the nonessential visceral organs—a pulmonary-to-systemic shunt. Only a small volume is used to collect residual oxygen in the lungs. A second valve, the foramen of Panizza, connects the base of left and right aortic arches. The right aorta directs blood to the head, limbs and tail, and these vital areas require oxygenated blood during oxygen stress. The foramen of Panizza allows oxygenated blood to pass from right to left aortas (to visceral organs) only during rest and normal exercise, cutting them off when not needed.

Biochemical adaptations complement the action of the heart. Crocodilian blood contains complex hemoglobin molecules capable of carrying more oxygen molecules than those of any other vertebrate. Crocodilians also endure much higher levels of lactic acid (produced when oxygen is scarce) in their blood than any other vertebrate. Blood pH has been measured below 6.1 without serious consequences, a level that would kill any other vertebrate. The result? A submergence time of nearly two hours when quiescent, even longer under cool conditions. American alligators have remained trapped under ice for eight hours and survived. Heavy activity substantially reduces submergence time, but crocodilians need only to outlast their prey. The blood also houses complex an-

tibacterial proteins capable of fighting off infection. Living in bacteria-filled waters, where injuries from fights are common, a strong immune system is essential. Crocodile blood has even been shown to kill "superbugs" for which scientists have no known cure.

Distribution

Crocodilians are found in over 90 countries and islands, generally in tropical and subtropical regions warm enough for successful reproduction. American and Chinese alligators, found in the highest latitudes of any species, do have a limited ability to survive seasonal freezing where deep water or shelter is available. Large adults can even endure being trapped in ice, as long as their internal organs do not freeze and their nostrils project above the surface.

Alligatoridae are restricted to North, Central, and South America, except for the Chinese alligator (*A. sinensis*), which occurs in eastern China. The stronghold of the Crocodylidae is Africa, India, and Asia, although a handful are found in the Americas. The single member of Gavialidae is found in India and adjacent countries. Estuarine crocodiles have the widest distribution (India to Vanuatu), although Nile crocodiles cover the greatest area (most of Africa and Madagascar). Spectacled caimans (*Caiman crocodilus*) number in the millions and are the most numerous throughout Central and South America. Introduced populations of certain species, particularly caimans, are established outside their natural ranges.

Crocodilians favor freshwater habitats, although several members of Crocodylidae tolerate higher salinity. Estuarine (also known as saltwater) crocodiles live in freshwater tidal rivers, hypersaline creeks, and along coastlines, and can travel at sea. Their ability to excrete excess salt through lingual salt

American crocodiles (*Crocodylus acutus*) along the river bank in the Everglades National Park. (Photo by Guido Cozzi. Bruce Coleman, Inc. Reproduced by pemission.)

glands and produce concentrated urine make this possible. All species frequent freshwater and low-salinity areas where available, including tidal rivers, freshwater marshes, and both natural and artificial lakes and pools. Distribution is influenced by the density and diversity of prey, available nesting habitat, shelter for juveniles and adults, thermal conditions, seasonal changes, and competition between species. Temporary range extensions have occurred when competition from one species over another has been reduced due to hunting, etc.

Feeding ecology and diet

Crocodilians are renowned for their ability to acquire food, often violently. All species are carnivorous, and mostly generalist. A wide range of mammals, birds, reptiles, amphibians, crustaceans, mollusks, fishes, and insects are taken readily by adults of most species. There are various restrictions, however. Young juveniles are limited to small prey that enter or approach water, primarily insects, spiders, crustaceans, fishes, small reptiles, and amphibians. Juveniles eat regularly, each day if possible. As they grow, the size and range of available prey increases. Species with specialized foraging strategies as adults (such as gharials) begin to exhibit characteristic preferences.

Although anything that moves within striking range is often considered fair game for adult crocodilians, most species display some selection criteria. These may include prey avail-

ability, but also species-specific preferences influenced by morphology and ecology. Broad-snouted alligatorids with strong bites and blunt teeth include hard-shelled prey in their diet; slender-snouted species such as gharials have weaker bites, but their sharp, undifferentiated teeth and slender jaws are ideal for sweeping quickly through water to seize slippery fish. Many Crocodylidae possess jaws between these two extremes, reflecting a generalist diet influenced by prey distribution and seasonal availability. Species with more specialized jaws will, however, take other prey where available.

Crocodilians display several hunting techniques. Surprisingly, most prey are small and taken as they approach the head, even in very large adults. The kill zone is an arc traced by the head and neck, although some species literally dive onto prey just below the surface. Terrestrial prey are ambushed at the water's edge, the hunter is either submerged or exposes only the eyes and nostrils. Once within striking range, there is an explosion of teeth and water as the crocodilian propels itself forward using its tail and limbs. Larger prey are dragged into the water where they drown. Although not pack hunters, the presence of several crocodilians in the water speeds a prey's demise. However, capture is not always successful. Misses are common, and large prey bitten on the head, limbs, or body may escape, only to die later from their injuries.

Although there are stories of crocodiles using their tails to sweep prey off their feet, hard evidence is sorely lacking, though the tail is important in hunting. Larger crocodiles are often seen using the tail to herd small fish into shallow water to be scooped up with a sweep of open jaws. The tail can be used to push the body vertically out of the water, ideal for catching prey flying over the water or hanging in low branches. This behavior has been witnessed in many species. Several species are reported to form a living dam with their bodies, preventing migrating fish from escaping. By cooperating, many individuals increase their chances of success. Nile crocodiles frequently cooperate after large prey is captured, taking turns to hold the carcass while others spins their bodies to rip off mouthfuls of flesh. Crocodilians learn rapidly to associate events with outcomes, often attending predictable events such as prey migrations.

Once captured, small prey is deftly manipulated by the jaws for immediate swallowing; head raised, the prey is flicked into the throat under gravity. Larger prey is first crushed several times by the back teeth, perforating skin and shell to assist digestion. Prey too large to be swallowed is typically held firmly in the jaws, then the head is whipped violently to one side. This tears prey apart, and each piece is swallowed once retrieved. Very large prey is first dismembered by holding onto a piece with the jaws, then spinning the body axis several times to twist it off. The carcass is anchored by its own weight, or by other feeding crocodilians. Live prey are easily incapacitated by rolling, as defense against this maneuver is impossible. Once inside the stomach, food is subjected to highly acidic gastric juices. The action of the muscular stomach plus gastrolith stones slowly macerates flesh and renders bone and shell into smaller pieces. Only keratin (found in hair, nails, and turtle shell) is immune. Compacted balls of indigestible material are regularly coughed up.

Frequently classed as man-eaters, only a handful of species are considered a significant threat to humans. Most fatalities are reported from American alligators, estuarine crocodiles, and Nile crocodiles, the latter responsible for the greatest number of crocodile-related deaths each year with several hundred people estimated killed or seriously injured. Threats can be reduced significantly by educating people to the danger posed by crocodilians, and providing alternative means of accessing water.

Reproductive biology

The basic crocodilian breeding system is polygynous (one male, multiple females), although short-term monogamous pair bonding has been described in Nile crocodiles and possibly other species. Multiple paternity (several males, same female) has also been recorded in some social situations. Once sexual maturity is reached, dictated by size rather than age, reproductive activity follows an annual cycle. Environmental triggers such as changing temperature, rainfall, humidity, and day length trigger hormonal changes in each sex. In males, testosterone levels rise, testes increase in weight, and sperm production increases. In females, estradiol levels rise, triggering the liver to produce vitellogenin for yolk production in the ovaries, and follicle size increases prior to ovulation. Double clutches have been reported in mugger crocodiles (*C. palustris*) and Nile crocodiles, influenced by extended environmental conditions favorable to breeding.

As habitat and climate change geographically, the timing and duration of reproductive activities varies between species and even within a species' distribution. Courtship, mating, and nesting may occur over a period of several months (as in Nile and estuarine crocodiles), or may be concentrated into several weeks (as in American alligators). Johnstone's crocodiles, or Australian freshwater crocodiles (*C. johnstonii*), nest within a two to three week period, a phenomenon known as "pulse nesting."

Crocodilians exhibit two nesting strategies: hole nesting and mound nesting. In the former the female selects a soft substrate, such as sand or mulch, and excavates a chamber several inches deep using her hind legs. After laying her eggs, the chamber is concealed again. In mound-nesting species, the female first scrapes material, such as vegetation, soil, or mud, into a mound using her front and hind legs. She also rips up fresh vegetation with her jaws. The resulting mound can be over 3.3 ft (1 m) high and 6.6–9.8 ft (2–3 m) in diameter. Once the mound is complete, the female behaves like a hole nester: she excavates a chamber into the top of the mound with her hind legs, lays her eggs, conceals them, and compacts the nest using her hind legs and body. Only Crocodylidae and Gavialidae excavate hole nests, in those species that nest during the dry season or when little vegetative matter is available. Mound nests are built by Alligatoridae and some Crocodylidae, typically nesting during the wet season or in areas that inundate easily, as the additional height reduces the risk of eggs being drowned by floods. American and Cuban crocodiles have been reported to choose either hole or mound nests, depending on climate and habitat.

Nesting location may be determined by available materials, proximity of water, temperature, and even social factors.

Females of more territorial species choose solitary nesting sites, isolated visually or by distance from those of other females. Favorite sites may be reused each year, although not necessarily by the same female. More gregarious species may use communal sites. Although under the vigilant watch of several females, there are disadvantages to communal sites. Late nesters may inadvertently dig up older eggs, and predators have an easier time finding eggs where nests are concentrated. Frequency of nesting is also under social pressure. In the wild, between 10% and 80% of females of a given species may nest each year; the percentage determined by the amount of nesting habitat available, the territorial nature of the females, and species-specific differences.

As with porcupines, people are curious how crocodilians manage to mate without causing each other grief! In reality, it is a gentle affair (once the competition has been dispensed with, that is). To copulate successfully, males must court females to gain their consent. Males of some species, such as estuarine crocodiles, establish territories that contain a number of females, others, such as American alligators, display competitively to attract females. Courtship may be elaborate or subtle, involving mutual signaling on a visual, olfactory, auditory, and tactile level. Alligators combine rumbling bellows with infrasonic vibrations, the water dancing across their backs proving a potent aphrodisiac for females. Gular musk glands are rubbed across the head and neck in mutual appeasement, and several minutes of head- and tail-raising postures are necessary for consent. Once she consents, the female allows the male to press her underwater. To align his vent with hers, the male rolls the female's body in one direction while rotating his tail in the other, using limbs for purchase. Male crocodilians have a single penis, unlike the hemipenes of most other reptiles. Copulation can last several minutes, and may be repeated over a period of days, although most males attempt copulation with as many females as possible. In captivity, males have been reported copulating with almost 20 females. Copulation may take place in shallow or deep water.

After fertilization, eggs are retained in the oviduct for two to four weeks, although periods of six months have been reported in some spectacled caimans. Embryos grow to around 20 somites (cells) before laying, with development believed to resume once eggs are exposed to air. Freshly laid eggs are covered in a fraction of an inch (several millimeters) of mucus, which cushions their impact as they slide into position. Mucus may also prevent gas exchange and hence further development until laying is complete and the mucus dissolves. Clutch size varies greatly: larger species lay up to 70 eggs, smaller ones as few as 10. It can take between 20 and 90 minutes to lay the entire clutch, during which time the female becomes curiously docile. Each egg contains the yolk (nutrient storage), albumen (water supply), leathery inner shell membrane, and hard, calcified outer shell membrane (protection, control of water and gas exchange). The embryo, lying atop the yolk, attaches to the inner shell membrane 24 hours after laying. As development continues, an opaque white band spreads around the egg's axis before eventually encompassing the entire egg.

Several important variables influence embryo development. Nest temperatures may fluctuate between 84.2°F (29°C) and 93.2°F (34°C), but small changes have significant effects. On average, incubation time lasts 70–90 days, yet higher temperatures reduce this and lower ones increase it. Temperature also determines sex. The phenomenon of temperature-dependent sex determination (TSD) is found in crocodilians, marine turtles, and some lizards. Unlike genetic sex determination (GSD), the sex of the embryo is determined not by sex chromosomes at fertilization, but by a critical temperature-sensitive period during incubation (the middle third). In all crocodilians, the greatest percentage of males is produced around 87.8–89.6°F (31–32°C), with more females produced above and below this temperature. Temperatures above 93.2°F (34°C) and below 84.2°F (29°C) produce almost 100% females, although females produced at higher temperatures suffer higher mortality and genetic deformities. Above 95°F (35°C) and below 80.6°F (27°C) embryos rarely survive. The size of hatchlings, their growth rate, and even preferred basking temperatures are influenced by incubation temperature.

All species nest during warm seasonal climates, providing appropriate ambient temperatures for incubation. Solar radiation provides additional heat, although many nests are built in the shade to reduce overheating, and nesting substrate buffers the eggs from extreme temperature fluctuations each day. Mound nests generate heat from the breakdown of plant materials, and even the developing embryos provide some metabolic heat during the later stages of incubation. The smooth-fronted caiman (*Paleosuchus trigonatus*) typically nests in closed-canopy forests, where ambient temperatures may be insufficient for optimal egg development. To address this, caimans build nests within or adjacent to termite mounds—metabolic heat produced by the termites helps to warm the nest. The female's presence is often required to break open the nest and free the hatchlings. Rainfall helps to cool nests, and some females have been observed urinating on the nest. As a suggested cooling mechanism the volume involved may be insufficient, but it may play a role in chemical marking as hatchlings recognize chemicals in contact with their eggs.

Pores permeating inner and outer shell membranes allow gas exchange, and both high humidity and oxygen are necessary for development. Embryo demands increase further as development continues. Development seems highly susceptible to perturbation, yet crocodilian nests provide an effective environment for incubation. But not always. Exposed nests can overheat, and those built in areas prone to flooding are easily submerged. More than 12 hours underwater, particularly later in development when oxygen demands are higher, can spell disaster. Another threat faces developing eggs—nest predators such as monitor lizards, wild pigs, raccoons, even ants. Females of most species defend the nest, often fasting for over two months to remain vigilant, but predators can still catch her off guard.

Eventually the demands of the embryo exceed the capability of the egg and hatching occurs. Shortly before emergence, the fully developed embryos may vocalize. Calls propagate from one egg to the next, producing a chorus audible to the adult from 165 ft (50 m) away. The female is typically much closer to the nest, often observed resting her throat directly above the nest chamber close to hatching. The female scrapes back sand, mud, or vegetation from above the eggs with her front legs, and the vibration stimulates the eggs to hatch. Some hatchlings head for water, but others remain and vocalize, often with head raised to encourage the female to carry them in her jaws. By lowering the tongue, a gular pouch is created to hold hatchlings. Some species transport hatchlings one at a time, others, such as Nile crocodiles, transport several. The female carries them to water, opens her mouth, and washes the hatchlings out with a sweeping motion of the head before returning to the nest. Hatchlings form small pods or crèches, hiding amongst shoreline vegetation. Not all species perform hatchling transport, and puncture marks caused inadvertently by sharp-toothed females may indicate why not. Infertile or dead eggs are normally eaten by the female, which led to early speculation that females ate their young! In captivity, territorial males of certain species have been observed assisting the female to open nests, hatch eggs, and transport hatchlings. Wild Nile crocodile males occasionally guard juveniles after hatching, but male parental care is atypical.

Female protection of juveniles after hatching may last for days (as in Johnstone's crocodiles), weeks (as in Nile crocodiles), or even up to two years (as in American alligators). Hatchlings remain in close proximity to the female, often using her as a convenient basking platform. Vocalization is well developed in crocodilians, and is an important component of juvenile life. Contact calls maintain group cohesion and alert siblings to the presence of food. Distress calls scatter individuals and bring the adult female aggressively to bear. The level of parental care in crocodilians is fascinating. In spectacled caimans, pods from different females may combine into larger crèches tended by resident adults. Even more remarkable, adult females of three species (broad-snouted caimans, Orinoco, and Siamese crocodiles) have been observed feeding juveniles. The female macerates the carcass and juveniles tear off small pieces. This level of parental care is unprecedented for reptiles, and perhaps reminds us of the closer taxonomic affinities between crocodilians and birds.

Conservation status

Four of the 23 crocodilian species (Chinese alligator, Orinoco crocodile, Philippine crocodile, Siamese crocodile) are considered Critically Endangered, with a further three (Cuban crocodile, false gharial, Indian gharial) listed as Endangered and considered at risk of extinction. In what is considered the most dramatic recovery of any large vertebrae group, 16 of the 23 species went from Endangered to abundant or not threatened in the last 30 years of the twentieth century. These species' previous decline was attributed primarily to overhunting (for skins) and habitat loss. Recovery was due to a combination of species protection, habitat protection, suppression of illegal trade, and enlightened management programs promoting sustainable use of wild populations as an incentive for their conservation. Efforts to improve the status of the 7 remaining endangered species continue as of 2003.

Resources

Books

Alderton, D. *Crocodiles and Alligators of the World.* New York: Facts on File, 1991.

Behler, J. L., and D. A. Behler. *Alligators and Crocodiles.* Stillwater, MN: Voyager Press, 1998.

Campbell, G., and A. L. Winterbotham. *Jaws Too: The Natural History of Crocodilians with Emphasis on Sanibel Island's Alligators.* Ft. Myers, FL: Sutherland Publishing, 1985.

Grigg, G., F. Seebacher, and C. E. Franklin. *Crocodilian Biology and Evolution.* Sydney: Surrey Beatty and Sons, 2001.

Guggisberg, C. A. W. *Crocodiles: Their Natural History, Folklore, and Conservation.* Harrisburg, PA: Stackpole Books, 1972.

McIlhenny, E. A. *The Alligator's Life History.* Boston: Christopher Publishing, 1935.

Minton, S. A., Jr., and M. R. Minton. *Giant Reptiles.* New York: Scribner, 1973.

Neill, W. T. *The Last of the Ruling Reptiles.* New York: Columbia University Press, 1971.

Richardson, K., G. J. W. Webb, and S. C. Manolis. *Crocodiles: Inside Out. A Guide to the Functional Morphology of Crocodilians.* Sydney: Surrey Beatty and Sons, 2002.

Ross, C. *Crocodiles and Alligators.* New York: Facts on File, 1989.

Webb, G. J. W., and S. C. Manolis. *Crocodiles of Australia.* Sydney: Surrey Beatty and Sons, 1989.

Organizations

Crocodile Specialist Group, Florida Museum of Natural History. Box 117800, Gainesville, FL 32611-7800 USA. Phone: (352) 392-1721. Fax: (352) 392-9367. E-mail: prosscsg@flmnh.ufl.edu Web site: <http://www.flmnh.ufl.edu/natsci/herpetology/crocs.htm>

Other

Bibliography of Crocodilian Biology. January 18, 1996 [cited January 2003]. <http://utweb.ut.edu/faculty/mmeers/bcb/index.html>

Crocodilians: Natural History and Conservation. December 2002 [cited January 2003]. <http://www.crocodilian.com>

Adam R. C. Britton, PhD

Gharials
(Gavialidae)

Class Reptilia
Order Crocodylia
Family Gavialidae

Thumbnail description
A large crocodilian with extremely narrow jaws; adult males have a distinctive knob over the nostrils

Size
13.1–16.4 ft (4–5 m)

Number of genera, species
1 genus; 1 species

Habitat
Rivers

Conservation status
Endangered

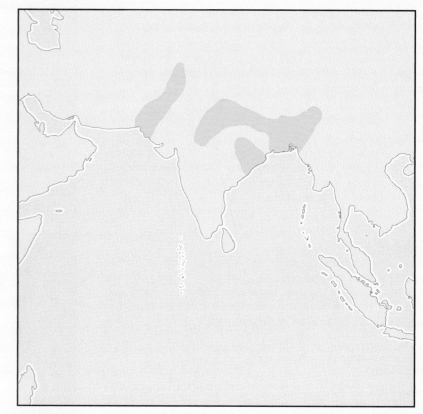

Distribution
Northern part of the Indian subcontinent

Evolution and systematics

In the geologic past the family Gavialidae was widespread. Fossils of some 12 species, all from the Cenozoic era (65 million years ago till the present), have been recovered from India, South America, Africa, and Europe. Paleontologists have argued that the Indian gharial separated from the rest of the crocodilians in the Mesozoic era (251–65 million years ago). Recent molecular studies, however, indicate a divergence between *Gavialis gangeticus* and its nearest relative, the Malayan gharial (*Tomistoma schlegelii*) as recently as 20 million years ago. Buccal cavity (mouth) morphology of the gharial suggests that the species is able to excrete salt, hence transoceanic dispersal might have aided speciation.

The taxonomy for this species is Indian gharial *Gavialis gangeticus* Gmelin, 1768, Ganges River.

Other common names include: French: Gavial du Gange; German: Schnabelkrokodil; Spanish: Gavial del Ganges.

Physical characteristics

Gharials have the longest snout of all crocodilians and are among the largest. Males average 13–15 ft (4–4.5 m) with a

maximum recorded length of over 19.7 ft (6 m). Females average 11.5–13 ft (3.5–4 m). Adults are dark or brownish olive; hatchlings are grayish brown with five irregular transverse bands on the body and nine on the tail. They are white or yel-

Gharial (*Gavialis gangeticus*). (Illustration by Brian Cressman)

lowish white on the underside. The body form is sleek, the scales smooth, and the head is well differentiated from the body. *Gavialis* is unique among crocodilians in that it is the only species to display visible sexual dimorphism: mature males develop a cartilaginous knob, or narial excrescence, on the tip of the snout, females do not. This knob creates a buzzing sound during exhalation, a social signal during courtship. It has no other known function. The most aquatic of all crocodilians, *Gavialis* do not migrate long distances overland, and adults are unable to lift their bodies off the ground.

Distribution

Current known distribution of the gharial is noncontinuous, up to 1,640 ft (500 m) above sea level in India, Nepal, and Pakistan. A few isolated individuals have been reported from Bangladesh and Bhutan.

Habitat

Gharials are found in deep, fast-flowing rivers, and prefer to occupy junctions and bends in these rivers where pools are deep and current is reduced. Exposed midriver banks are used as basking sites, particularly in the cooler winter months, and steep sand banks are used as nesting sites.

Behavior

Gharials are social crocodilians and groups of adults congregate at suitable sites. They rarely venture far from the safety of the aquatic environment into which they can dive when alarmed. Males become territorial in the courtship/mating season, and spar for territory in shallow water. Lying parallel to each other, males raise heads well out of the water, and attempt to push each other off balance with their snouts. Violent "snout bashing" and biting has been observed. Female *Gavialis* become territorial in the nesting season, and will also defend nests from predators.

Feeding ecology and diet

Gharials are the only known crocodilians to subsist almost entirely on a diet of fish. In the juvenile stage, the species may also consume tadpoles, shrimp, and water insects. Prey is sought by either the "sit and wait" technique, or by actively foraging, and fish are captured with lightning fast lateral swipes of the jaw. Fish are impaled on the sharp teeth and the gharial throws them to the back of the throat by jerking the head upward. The extremely narrow snout of *Gavialis* reduces friction in the water as compared to broad-snouted crocodilians, another factor contributing to its efficiency as a fish catcher.

Reproductive biology

Gavialis males mature at about 11.5 ft (3.5 m) at the age of 15, when the narial excrescence becomes large and rounded. Females mature at 10 ft (3 m), as young as eight years old. The breeding season is markedly seasonal throughout the

The gharial (*Gavialis gangeticus*) is one of the largest of all crocodilian species. (Photo by E&P Bauer. Bruce Coleman, Inc. Reproduced by permission.)

range of the species; courtship and mating occur in December to January and nesting from March to May. The duration of the nesting season is shorter at higher latitudes (22 days on the Narayani River in Nepal) and longer at southerly latitudes (57 days at the Madras Crocodile Bank). Evidence from captive-breeding studies indicates that larger females nest first, inferring that these females are mated with first. Females select nest sites at least 5 ft (1.5 m) above water level, and may dig trial nests before nesting. Clutch size closely correlates with the size of the female, ranging from 12 to nearly 100 eggs in very large females. The first clutches laid by females typically contain a high proportion of infertile eggs. Eggs measure 2.2 by 3.4 in (55 by 86 mm), and weigh 5.5 oz (156 g). Embryonic development is rapid, particularly in the later stages of development. Incubation temperature determines the incubation period and sex of hatchlings, as well as hatchling size. Incubation period is shorter at higher temperatures (an average of 53 days at 91°F/33°C) and longer at lower temperatures (93 days at 82°F/28°C). Females have been observed to guard nests in the wild, dig out hatchlings from nests, and display prolonged protection of their crèche of hatchlings. Males may also protect the young. Eggs are

A pair of gharials (*Gavialis gangeticus*) enjoy the waters of India. (Photo by Michael Fogden. Bruce Coleman, Inc. Reproduced by permission.)

preyed upon by jackals, pigs, hyenas, mongooses, and monitor lizards. Hatchling mortality is high, largely due to floods in the monsoon period (when the eggs hatch), and predation by large wading birds and soft-shelled turtles.

Conservation status

In Nepal, *Gavialis* populations are in decline, and fewer than 1% of all gharials hatched in nature reach a length of 6.6 ft (2 m). Egg loss by animal and human predators is high, and flooding causes significant loss of nests. A total of 55 wild gharial and 50 released gharials were observed in 1999 at Babai, Kali, Karnali, Koshi, Mahakali, Narayani, and Rapti River systems. The low number of males in the wild population (sex ratio one male to 10 females) may be adversely affecting the population. Reasons for decline are habitat destruction due to increasing human pressure on the environment from extensive agriculture, firewood collection, and cattle grazing.

In Bangladesh, *Gavialis* was believed to be extinct in the 1970s, but continued survival of the species was confirmed in the Padma (Ganges) and Jamuna (Brahmaputra) Rivers in 1981. A survey in 1985 recorded 18 individuals in the Padma River at Changhat, Charkhidipur, and Godagari localities. However, known nesting areas that produced up to 12 nests as recently as 1985 have seen no nests since 1990. Factors affecting declining gharial populations are fishing activities and habitat degradation.

In Bhutan, the species is near extinction in the wild, with isolated individuals reported from rivers near Bhutan's southern border with India.

In Pakistan, gharials have been extirpated from most of the country, with the main population of about 20 individuals occuring on the Nara Canal, part of the Indus River Dolphin Sanctuary. Small populations are believed to survive around Gudder Barrage (Sind) and Taunsa Barrage (Punjab). The Pakistan government is currently planning a restocking effort, perhaps using surplus captive stock from India.

In the Orissa state in India, *Gavialis* is now restricted to the Mahanadi River. Over 700 juvenile gharial have been released in the Satkoshia Gorge Sanctuary, but survival has been only 5% due to heavy human usage of the river for bamboo rafting and fishing. In the tristate National Chambal Sanctuary, encompassing the states of Rajasthan, Uttar Pradesh, and Madhya Pradesh, over 1,200 wild gharial were counted in recent surveys. Over 3,500 captive-reared gharial from wild eggs have been released in protected areas of these states, and several other small but vital breeding populations survive on the Girwa, Ken, and Son Rivers.

The gharial is cited as Endangered on the IUCN Red List, and is on Appendix I of CITES. There are an estimated 1,500–2,500 gharials in the wild.

Significance to humans

Gharial eggs are collected for food by indigenous people in some parts of their range, and the Hindu river goddess Ma Ganga is sometimes depicted riding on a gharial.

Resources

Books

Daniel, J. C. *The Book of Indian Reptiles and Amphibians.* New Delhi, India: Oxford University Press, 2002.

Ross, C. A., ed. *Crocodiles and Alligators.* New York: Facts on File Inc., 1989.

Periodicals

Negi, B., ed. "Indian Crocodilians." *Envis* 2, no 1 (June 1999).

Whitaker, N., and H. V. Andrews. "Reproductive Biology, Growth, and Incubation Temperature and Its Influence on Sex and Size of the Gharial (*Gavialis gangeticus*)." *Gmelin at the Madras Crocodile Bank/Center for Herpetology* (2002).

Whitaker, R., and D. Basu. "The Gharial (*Gavialis gangeticus*): A Review." *Journal of the Bombay Natural History Society* 79, no. 3 (1983).

Organizations

Crocodile Specialist Group, Florida Museum of Natural History. Box 117800, Gainesville, FL 32611-7800 USA. Phone: (352) 392-1721. Fax: (352) 392-9367. E-mail: prosscsg@flmnh.ufl.edu Web site: <http://www.flmnh.ufl.edu/natsci/herpetology/crocs.htm>

Madras Crocodile Bank/Center for Herpetology. P. O. Box 4, Tamil Nadu, 603 104 South India. E-mail: mcbtindia@vsnl.net

Other

Crocodiles: Status Survey and Conservation Action Plan, 1998 (Online). 1998 [cited January 2003]. <http://www.flmnh.ufl.edu/natsci/herpetology/act-plan/plan1998a.htm>

Ross, J. P., ed. *Crocodiles. Status Survey and Conservation Action Plan,* 2nd ed. Gland, Switzerland, and Cambridge, UK: IUCN/SSC Crocodile Specialist Group, 1998.

Florida Museum of Natural History: Herpetology. [cited January 2003]. <http://www.flmnh.ufl.edu/natsci/herpetology/herpetology.htm>

Romulus Earl Whitaker III, BSc
Nikhil Whitaker

Alligators and caimans
(Alligatoridae)

Class Reptilia
Order Crocodylia
Suborder Eusuchia
Family Alligatoridae

Thumbnail description
Powerful animals with a long and muscular tail, four short limbs straddling a scaly body, and strong jaws lined with obvious teeth

Size
4–20 ft (1.2–6 m) in total length

Number of genera, species
4 genera; 8 species

Habitat
Calm or slow-moving freshwater areas

Conservation status
Critically Endangered: 1 species; Lower Risk/Conservation Dependent: 1 species

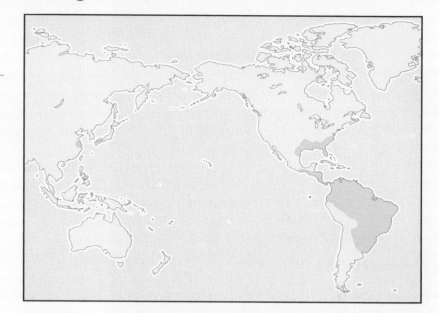

Distribution
Central to northern South America, parts of southern and western Central America and Mexico, the southeastern United States, and a small area of eastern China

Evolution and systematics

Although the order Crocodylia dates back at least 200 million years to the Age of Reptiles, its living members, including those of the family Alligatoridae, can hardly be described as primitive. Instead, they survived the mass extinction 65 million years ago that ended the dinosaurs' reign and evolved over the centuries into animals well suited to their current place in the natural world. Like other members of the order, the family Alligatoridae are the descendants of the Archosauria, or "ruling lizards," which included the dinosaurs. A defining characteristic of these animals is a diapsid skull, which has two temporal openings. Turtles, by comparison, have anapsid skulls with no temporal openings.

Within the crocodilians, the family Alligatoridae can be followed as far back as the Paleocene (57–65 million years ago), when caiman ancestors are thought to have roamed the earth. Ancestors of other species, including the American alligator (*Alligator mississippiensis*) and Chinese alligator (*Alligator sinensis*), date back to the Miocene and Pleistocene, respectively. The alligatorids are separated into two major groups: the alligators (subfamily Alligatorinae) and the caimans (subfamily Caimaninae). The former group has two living representatives in the *Alligator* genus. The other six species of alligatorids fall under three genera within the caimans. (Some systematists list only five caimans, with the Yacaré as a subspecies of the common caiman.) In all, the eight species are as follows:

- American alligator, *Alligator mississippiensis*
- Chinese alligator, *A. sinensis*
- common caiman, *Caiman crocodilus*
- broad-snouted caiman, *C. latirostris*
- Yacaré, *C. yacare*
- black caiman, *Melanosuchus niger*
- Cuvier's dwarf caiman, *Paleosuchus palpebrosus*
- smooth-fronted caiman, *P. trigonatus*

Physical characteristics

In general appearance, alligators are similar to crocodiles, with stout bodies and powerful tails that are at least as long as their bodies. They have long snouts and noticeably toothed upper and lower jaws. Alligatorids are distinguished most notably from the crocodiles by their mandibular teeth, all of which slide inside the upper jaw and out of view when the mouth is closed. In contrast, the fourth mandibular teeth in crocodiles are visible outside the closed jaw.

Alligatorids are grayish, sometimes tending toward green, brown, yellow, or black, depending on the species. The young are often banded. Adult size ranges from about 4 ft (1.2 m) in Cuvier's dwarf caiman (*P. palpebrosus*) to 13 ft (4 m) in the American alligator (*A. mississippiensis*).

Like the crocodile's body, the alligatorid body is armored with tough osteoderms and, frequently, large scales that do not overlap. The osteoderms in some species do not extend onto the belly, making this smooth part of the skin highly

American alligators (*Alligator mississippiensis*) in the Everglades in Florida. (Photo by Wendell Metzen/Photo Researchers, Inc. Reproduced by permission.)

The Yacaré caiman (*Caiman yacare*) in Brazil. (Photo by E. & P. Bauer. Bruce Coleman, Inc. Reproduced by permission.)

desirable as leather for human uses. Alligatorids have short legs tipped with claws. The forelimbs are smaller than the hind limbs and have five, rather than four, partially webbed toes. Their body form allows them to glide in a sinuous pattern through the water, normally with just the side-to-side motion of the tail providing the locomotive force. On land the strength of their legs makes them quick and formidable predators.

Distribution

Primarily a New World group, all but the two *Alligator* species occur in southern Mexico, parts of Central America, or northern to central South America. *Alligator mississippiensis* is the only member of this family to be found in the United States, where it exists in southeastern states, from the Carolinas to Texas. *A. sinensis* makes its home in eastern China, far distant from its New World relatives.

Habitat

Alligatorids are restricted to freshwater areas and frequently are found in lakes, slow-moving streams and rivers,

swamps, marshes, and other wetland habitats. Some species even make use of roadside ditches. They prefer sites with slow-moving or still waters. They often inhabit vegetated areas, sometimes with muddy or murky water.

Behavior

Alligatorids are ectotherms ("cold-blooded" animals) and most often are seen basking on the shoreline to raise their body temperature. Sometimes they are seen sliding along the shoreline on their bellies, using their feet to push them through the mud and muck to the water. They also do the "high walk," which is somewhat similar to a lizard's walk; alligatorids, however, hold their legs more upright than straddled. Although they may look sedate much of the time, their short legs can give them quick acceleration for grasping a passing mammal.

Careful observers also see them floating at the surface of the water, where only their most dorsal surface and occa-

An American alligator (*Alligator mississippiensis*) in southern Florida, USA, with its eyes just above the water. (Photo by Joe McDonald. Bruce Coleman, Inc. Reproduced by permission.)

sionally just the nose and the tip of the head are exposed. Often, the animal actually is maintaining its internal temperature through this activity, either lying in the sun-heated upper layers of the water column to warm up or moving to shady, chillier waters to cool off. Their presence is made known when they begin to sweep their tails slowly and propel themselves gracefully forward. While they usually are motionless or swim slowly, they can make quick movements in the water. One noticeable trait is their ability to jettison almost vertically out of the water. This maneuver typically is accompanied by a quick chomp of the jaws around a startled bird or other prey item.

Alligatorids, including those in more temperate climates, do not hibernate. While temperatures in the southeastern United States and China can approach freezing in the winter, American and Chinese alligators remain active all year, though they are more subdued as temperatures dip and may even become dormant. To beat the cold, they move to shallow water and lie motionless, with just the nose poking into the cold air. Young alligators, on the other hand, may retreat to the mother's den to survive cold snaps. Juveniles and adults make use of burrows during winter months.

Alligatorids often live in groups and form dominance hierarchies, at least during the breeding season and sometimes all year. The highest-ranking individuals exert their dominance through various ritualized behaviors, which may include slaps of the head against the surface of the water, loud vocalizations, and open-mouthed charges.

Feeding ecology and diet

A typical juvenile diet includes snails and other invertebrates, whereas adults of various species commonly eat fish, small mammals, other reptiles (including smaller alligatorids), or birds. Opportunistic feeders, alligatorids continue to eat clams, snails, and invertebrates into adulthood. The larger species, including the black caiman, are known to take large prey, such as small deer and cattle.

Predation of alligatorids occurs primarily among eggs and hatchlings. Raccoons, coati, skunks, foxes, and other mammals, as well as snakes and various raptors, are known to raid nests or snatch up young alligatorids. Once an alligatorid reaches about 3 ft (1 m) in length, the risk of predation decreases. Nonetheless, anacondas in South America occasionally kill adult caimans, and alligatorids have been reported to have cannibalistic tendencies. Cannibalism is rare, however, and alligatorids frequently live peacefully in large groups.

Reproductive biology

Alligatorids begin the reproductive season in the spring. Following courtship, which may include loud bellows, tactile types of behavior, and underwater vibrations of the male's trunk, alligatorids use vegetation to construct nest mounds, where they lay from one dozen to five dozen eggs, depend-

Hatchling American alligators (*Alligator mississippiensis*). (Photo by Animals Animals © Lynn M. Stone. Reproduced by permission.)

ing on the species. Egg laying generally takes place once a year, in midsummer, with hatching one to two months later. Female alligatorids typically respond to sounds emanating from the neonate, dig up the nest, and assist in their hatching. Temperature-dependent sex determination has been associated with several species, including the American alligator and common caimans. Low nest temperatures (below 88°F, or 31°C) produce female hatchlings, and high temperatures (above 90°F, or 32°C) produce males.

Conservation status

The Chinese alligator is listed by the IUCN Red Book as Critically Endangered. This status stems primarily from loss of habitat due to human encroachment. According to the IUCN Crocodile Specialist Group, "The Chinese alligator is the most critically endangered crocodilian in the world. Thousands are bred in captivity, but fewer than 150 remain in the wild." The group is working with the Chinese government to protect the species and has launched the Chinese Alligator Fund to assist in these efforts. In addition, many conservation efforts over the past three decades have been implemented to prevent overharvesting of other alligatorids. The only other species listed by the IUCN is the black caiman, which is listed as Lower Risk/Conservation Dependent.

Significance to humans

Although the benefit is difficult to quantify, some species of alligatorids play an important role in the tourism industry. American alligators in the southeastern United States, for example, have become tourist attractions, drawing visitors to the Everglades of southern Florida and the bayous of Louisiana. Several members of this family are hunted, especially for their skin, which is used as leather for shoes, bags, and various accessories. Humans also hunt these animals for meat and, recently, for their gonads, which are used to make perfume.

1. American alligator (*Alligator mississippiensis*); 2. Common caiman (*Caiman crocodilus*); 3. Chinese alligator (*Alligator sinensis*); 4. Smooth-fronted caiman (*Paleosuchus trigonatus*). (Illustration by Brian Cressman)

Species accounts

American alligator
Alligator mississippiensis

SUBFAMILY
Alligatorinae

TAXONOMY
Alligator mississippiensis Daudin, 1801, "les bords du Mississipi," United States. No subspecies are recognized.

OTHER COMMON NAMES
English: Gator, pike-headed alligator, Florida alligator, Mississippi alligator, Louisiana alligator; French: Alligator de Amérika; German: Hechtalligator, Mississippi-Alligator; Spanish: Aligator de Mississippi.

PHYSICAL CHARACTERISTICS
With a broad snout and heavy, armorlike, dorsal scales, American alligators are dark grayish green to black, with pale whitish bellies. The young commonly have conspicuous yellow markings on the back and tail. In their geographic range, the only other crocodilian is the American crocodile (*Crocodylus acutus*), which has a noticeably narrower and tapering snout. Adults generally reach lengths of about 8–13 ft (2.4–4 m), though some individuals may top 19 ft (5.8 m).

DISTRIBUTION
The American alligator is found throughout the coastal plains of the southeastern United States from the Carolinas south to Florida and west to Texas.

HABITAT
This species inhabits freshwater, especially marshes, swamps, lakes, and slow-moving rivers.

BEHAVIOR
Alligators often form extended families of sorts, with several generations living in the same vicinity for many years. When the breeding season arrives, the courtship ritual includes a se-
ries of tactile behaviors, including gentle bumping and rubbing between the male and female. Both males and females bellow, with the male's roars a bit louder than the female's and more plentiful during mating season. Females often utter low grunts when calling the young. Males and females of all ages hiss when threatened. This species may become dormant during the winter, but it does not hibernate.

FEEDING ECOLOGY AND DIET
The largest reptiles in North America, adult American alligators are at the top of the food chain in their habitat. They are carnivorous and eat almost anything that is in or near the water, including turtles; fish; small mammals, such as otters; and even young alligators. If possible, the alligator swallows its prey whole. If the prey item is large, however, it first drowns the victim, then tears off bite-sized chunks. Younger alligators eat primarily fish and small invertebrates.

REPRODUCTIVE BIOLOGY
Males and females mature at 10 years or older. Mating occurs each spring. Each nest contains about three dozen to four dozen eggs, of which two-thirds or more typically survive to hatching. Egg gestation is about two months. Females provide parental care by guarding the nest and young, by opening the buried nest to assist in hatching, and by transporting hatchlings to water. Young remain near their mother in a "pod" for at least two to three months and often as long as two to three years. Their life span can run 50 or more years.

CONSERVATION STATUS
Not listed by the IUCN.

SIGNIFICANCE TO HUMANS
The species is a source of meat and hides for such uses as shoes, belts, and purses. In some places, they also have become a boon to the tourist industry. ◆

Chinese alligator
Alligator sinensis

SUBFAMILY
Alligatorinae

TAXONOMY
Alligator sinensis Fauvel, 1879, "Chinkiang" (= Zhenjiang/ Chinkjang/Chenchiang), Kiangsu Province, People's Republic of China. No subspecies are recognized.

OTHER COMMON NAMES
English: Yangtze alligator, Tou lung, Yow lung, T'o, China alligator; French: Alligator de Chine; German: China-Alligator; Spanish: Alligator de China.

PHYSICAL CHARACTERISTICS
A yellowish gray alligator with osteoderms on the belly as well as on the back and a heavy snout that tapers toward its vaguely upturned end. This is a small alligator that has an average total length of about 5 ft (1.5 m) in males and 4.5 ft (1.4 m) in females. Maximum lengths have been recorded at

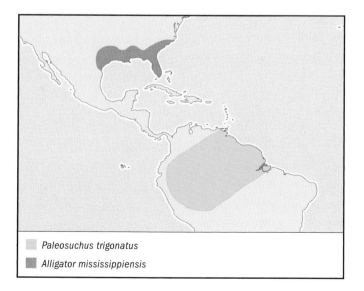

Paleosuchus trigonatus
Alligator mississippiensis

Alligator sinensis

about 6.6 ft (2 m) in the male and 5.7 ft (1.7 m) in the female. Young are similar to adults in appearance but have noticeable yellow banding.

DISTRIBUTION
This species occupies a small area in the Yangtze River basin along China's central Atlantic coastline.

HABITAT
The Chinese alligator inhabits the subtropical temperate ecotone in marshy areas, ponds, lakes, and languid rivers.

BEHAVIOR
This species is dormant during the late fall, winter, and early spring and relatively inactive during much of the rest of the year owing to the cool temperatures in its geographic area. Each year in April, the alligators emerge from their winter burrows, which line still waters, and find sunny spots in which to bask. As summer begins, they switch to more nocturnal habits and begin their annual mating rituals. During courtship, bellowing from the males and females becomes pronounced, though they bellow at other times of the year as well.

FEEDING ECOLOGY AND DIET
The young prefer insects and other small invertebrates, whereas larger individuals also take fish, clams, and the occasional small mammal or waterfowl. They have blunt teeth adapted well to crushing shelled animals.

REPRODUCTIVE BIOLOGY
Females mature at about five to seven years. Mating occurs in early summer, with the females building nests about two to three weeks later and laying up to four dozen eggs; fewer than two dozen is common. Hatchlings generally emerge in September. Females provide parental care, including assisting in the hatching process and carrying the newly hatched young from the nests, which are located on land, to the water. The life span of these alligators in captivity nears 70 years, and they can reproduce into their 50s.

CONSERVATION STATUS
The species is listed as Critically Endangered by the IUCN. Their decline is associated primarily with habitat destruction.

SIGNIFICANCE TO HUMANS
None known. ◆

Common caiman
Caiman crocodilus

SUBFAMILY
Caimaninae

TAXONOMY
Caiman crocodilus Linnaeus, 1758, type locality not specified. Four subspecies are recognized.

OTHER COMMON NAMES
English: Spectacled caiman, brown caiman; French: Caiman è lunettes; German: Brillenkaiman, Krokodilkaiman; Spanish: Caimán común.

PHYSICAL CHARACTERISTICS
Dark cross-banding, tough dorsal armor, and a bony facial ridge are the most distinguishing features of this greenish gray to brownish gray crocodilian. Adults can reach 4–6 ft (1.2–1.8 m), with rare individuals growing up to 10 ft (3 m).

DISTRIBUTION
The species is found primarily in the Amazon and Orinoco River basins, but it extends from southern Mexico to northern Argentina and to the islands of Trinidad and Tobago north of eastern Venezuela. It also inhabits southern Florida and has been introduced in Cuba and Puerto Rico.

HABITAT
The common caiman inhabits freshwater areas, particularly swamps but also lakes, rivers, and even water-filled roadside ditches.

BEHAVIOR
During the breeding season, the males bellow to help establish their territories. More than one female may mate with a single

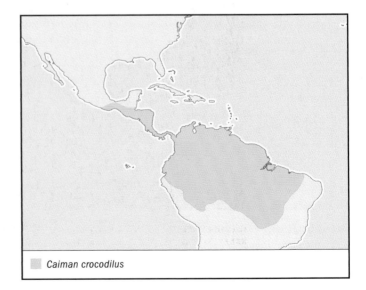

Caiman crocodilus

male and build a nest in his territory. During the remainder of the year, which usually is much drier than the wet breeding season, caimans congregate in whatever freshwater pools are available.

FEEDING ECOLOGY AND DIET

The caiman diet varies from land invertebrates among the youngest individuals to snails among juveniles and mainly fish among adults.

REPRODUCTIVE BIOLOGY

Caimans begin breeding during the annual wet season. Females lay about one dozen to three dozen eggs in a terrestrial nest made of vegetation. The mother provides parental care by helping in the hatching process and by carrying the neonates, which are typically about 8 in (20 cm) long, to nearby water. The male guards the nest. After hatching, the young generally remain near the parents until they are almost a year old. Common caimans have been known to live into their 60s, though this is rare.

CONSERVATION STATUS

Not listed by the IUCN. One of the four subspecies (*C. c. apaporiensis*) of the common caiman, however, is under threat by range overlap and cross-breeding with another subspecies, *C. c. crocodilus*.

SIGNIFICANCE TO HUMANS

The common caiman is harvested, sometimes illegally, as food or for its skin. It also has been seen in the pet trade. ◆

Smooth-fronted caiman
Paleosuchus trigonatus

SUBFAMILY
Caimaninae

TAXONOMY
Paleosuchus trigonatus Schneider, 1801, type locality not specified. No subspecies are recognized.

OTHER COMMON NAMES

English: Schneider's smooth-fronted caiman; French: Caiman è front lisse de Schneider; German: Keilkopfkaiman; Spanish: Jacaré coroa cachirre.

PHYSICAL CHARACTERISTICS

These caimans are called smooth-fronted because they lack the bony ridge typically seen between the eyes of other caimans. They are dark gray to black, with highly ridged dorsal scales and a tail that is barely as long as the body. Males reach 4.9–5.5 ft (1.5–1.7 m), and females attain a length of 3.9–4.6 ft (1.2–1.4 m).

DISTRIBUTION

The species is distributed throughout northern and north-central South America.

HABITAT

The smooth-fronted caiman usually is found in very shallow streams of heavily vegetated rainforests.

BEHAVIOR

Unusual for alligatorids, this species spends much of its time in hiding spots on land (such as under logs), rather than basking along the shoreline. It does, however, follow the typical alligator pattern of building nest mounds of vegetation, where it lays its eggs.

FEEDING ECOLOGY AND DIET

Hatchlings have an insectivorous diet, but juveniles and adults share a diet of reptiles and birds. Adults also commonly take mammals, including porcupines.

REPRODUCTIVE BIOLOGY

The timing of this species' breeding period coincides with the annual fluctuation of dry and wet seasons. Courtship, nest building, and egg laying take place toward the end of the dry season, and hatching ensues shortly after the rains begin to fall. A typical nest contains about a dozen eggs.

CONSERVATION STATUS

Not listed by the IUCN.

SIGNIFICANCE TO HUMANS

None known. ◆

Resources

Books

Alderton, D. *Crocodiles and Alligators of the World.* New York: Facts on File, 1991.

Ashton, Ray E., and Patricia Sawyer Ashton. *Handbook of Reptiles and Amphibians of Florida.* Part 2: *Lizards, Turtles and Crocodilians.* 2nd edition. Miami: Windward Publishing Co., 1991.

Behler, J. L., and D. A. Behler. *Alligators and Crocodiles.* Stillwater, MN: Voyager Press, 1998.

Campbell, G., and A. L. Winterbotham. *Jaws Too: The Natural History of Crocodilians with Emphasis on Sanibel Island's Alligators.* Ft. Myers, FL: Sutherland Publishing, 1985.

Guggisberg, C. A. W. *Crocodiles: Their Natural History, Folklore, and Conservation.* Harrisburg, PA: Stackpole Books, 1972.

Hirschhorn, Howard H. *Crocodilians of Florida and the Tropical Americas.* Miami: Phoenix Publishing Co., 1986.

King, F. Wayne, and Russell L. Burke, eds. *Crocodilian, Tuatara and Turtle Species of the World: A Taxonomic and Geographic Reference.* Washington, DC: Association of Systematics Collections, 1989.

McIlhenny, E. A. *The Alligator's Life History.* Boston: Christopher Publishing, 1935.

Minton, S. A., Jr., and M. R. Minton. *Giant Reptiles.* New York: Scribner, 1973.

Neill, W. T. *The Last of the Ruling Reptiles.* New York: Columbia University Press, 1971.

Ross, Charles A., ed. *Crocodiles and Alligators.* New York: Facts of File, Inc., 1989.

Zug, George R, Laurie J. Vitt, and Janalee P. Caldwell. *Herpetology: An Introductory Biology of Amphibians and Reptiles.* 2nd edition. San Diego: Academic Press, Inc., 2001.

Resources

Periodicals

Brazaitis, P., M. Watanabe, and G. Amato. "The Caiman Trade." *Scientific American* 278 (March 1998): 70–76.

Stewart, D. "Visiting the Heart of Alligator Country." *National Wildlife* 38, no. 4 (June/July 2000): 20–27.

Thorbjarnarson, J. "The Hunt for Black Caiman." *International Wildlife* 29 no. 4 (July/August 1999): 12–19.

Thorbjarnarson, J., X. Wang, and L. He. "Reproductive Ecology of the Chinese Alligator (*Alligator sinensis*) and Implications for Conservation." *Journal of Herpetology* 35, no. 4 (December 2001): 553–558.

Zimmer, C. "Prepared for the Past." *Natural History* 110, no. 3 (April 2001): 28–29.

Organizations

Crocodile Specialist Group, Florida Museum of Natural History. Box 117800, Gainesville, FL 32611-7800 USA. Phone: (352) 392-1721. Fax: (352) 392-9367. E-mail: prosscsg@flmnh.ufl.edu Web site: <http://www.flmnh.ufl.edu/natsci/herpetology/crocs.htm>

Other

Animal Diversity Web. [cited December 2002]. <http://animaldiversity.ummz.umich.edu>

"Crocodilian Species List." 2002 [cited December 2002]. <http://www.flmnh.ufl.edu/cnhc/csl.html>

"Alligatoridae." *The Reptipage.* August 2002 [cited December 2002]. <http://reptilis.net/crocodylia/gators/alligatoridae.html>

"Alligatoridae: Alligators, Caimans, and Their Prehistoric Relatives." December 1999 [cited December 2002]. <http://www.kheper.auz.com/gaia/biosphere/vertebrates/crocodylia/Alligatoridae.html>

"Crocodilian Biology Database." August 2002 [cited December 2002]. <http://www.flmnh.ufl.edu/cnhc/cbd.html>

Leslie Ann Mertz, PhD

Crocodiles and false gharials
(*Crocodylidae*)

Class Reptilia

Order Crocodylia

Family Crocodylidae

Thumbnail description
Medium-sized to very large, egg-laying, aquatic reptiles characterized by well-toothed jaws and dorsal armor

Size
5–20 ft (1.5–6.1 m) total length

Number of genera, species
3 genera; 14 species

Habitat
Marshes, rivers, streams, lakes, and creeks

Conservation status
Critically Endangered: 3 species; Endangered: 2 species; Vulnerable: 3 species; Lower Risk/ Conservation Dependent: 1 species; Data Deficient: 1 species

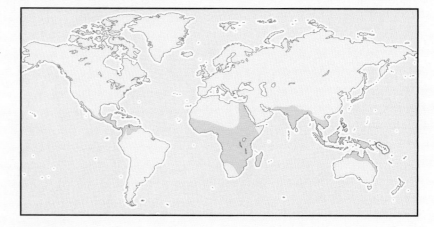

Distribution
North and South America, Africa, Asia, and Australia

Evolution and systematics

Centers of abundance for the Crocodylidae are South America (four species), and the Indo-Pacific region (six species), although the family is postulated to have radiated from Africa in the Miocene epoch. Studies on the physiology of crocodiles have revealed that all species are able to concentrate and excrete salt, and this feature may well have aided in further speciation via transoceanic migrations. Some adaptations for continued survival of the Crocodylidae are the position of the eyes and nostrils, parental care, and their cost-effective metabolisms, but they are capable of immense speed on land or in water when necessary. Morphological and molecular data support a close relationship between *Crocodylus* and *Osteolaemus*. In addition, the Indo-Pacific crocodilians (*Crocodylus palustris, C. porosus, C. siamensis, C. novaeguineae, C. mindorensis,* and *C. johnstonii*) are closely related. In the past, subspecies status has been proposed for both *Crocodylus palustris* and *Crocodylus niloticus,* but these were not recognized as of 2002. The addition of a new species to the family Crocodylidae, the Borneo crocodile (*Crocodylus raninus,*), has also not been recognized scientifically, pending further investigation. The New World species (*Crocodylus acutus, C. intermedius, C. moreletii,* and *C. rhombifer*) are a monophylectic assemblage, that is, they share a common ancestor. Two subfamilies are recognized: Crocodylinae, containing two genera and 12 species; and Tomistominae, containing one genus and two species.

Physical characteristics

Crocodiles range from the small (6.2 ft [1.9 m] total length) West African dwarf crocodile (*Osteolaemus tetraspis*) to the huge (20 ft [6.1 m] total length) saltwater crocodile (*Crocodylus porosus*). Crocodiles are characterized by smaller, weak front limbs and strong hind limbs with webbed toes. Dorsal scales backed by osteoderms form heavy armor plating on neck and back. All crocodiles are elongate with snout-vent length almost equal to tail length. The tail is strongly muscled and flattened for swimming. Other adaptations for aquatic life include nostril and ear valves, a nictitating membrane to cover the eye, and a glottal valve in the throat. Crocodiles have effective visual, auditory, and olfactory systems. Crocodile teeth are replaced throughout life. Crocodiles can live for 70–80 years.

Distribution

Crocodiles are found in the tropical regions of the world, although two species, the American crocodile (*Crocodylus acutus*) and the mugger crocodile (*Crocodylus palustris*) also occur in the subtropics. There are four species in the Americas, three in Africa, and seven in the Asia and Pacific region.

Habitat

Although some species of crocodiles, such as the mugger crocodile and Nile crocodile (*Crocodylus niloticus,*) are highly adaptable and can live well in a variety of aquatic habitats, others such as the Malayan gharial (*Tomistoma schlegelii*) and the West African dwarf crocodile are restricted to forested, swampy areas. Deep water is a prime requisite for safety and security from drought, but some species live in places where the water dries up annually. In these habitats, crocodiles excavate deep tunnels to wait out inhospitable weather conditions. Crocodiles have found niches in habitats including hill

A Nile crocodile (*Crocodylus niloticus*) feeds on a Grant's gazelle (*Gazella granti*) in Kenya. (Photo by Fritz Polkina. Bruce Coleman, Inc. Reproduced by permission.)

streams, large rivers, marshes, ponds, lakes, canals, reservoirs, as well as saline habitats such as mangrove creeks and saltpans.

Behavior

Crocodiles are mainly nocturnal hunters and scavengers. They raise their body temperatures by hauling themselves out of the water and basking in the sun, usually both morning and evening. Crocodiles often appear torpid while they are in fact in a state of readiness to either flee or attack. Crocodiles are variably vocal. Adult males may bellow to establish dominance. All size classes will growl and hiss when intimidated. Hatchlings grunt to each other and will squawk loudly if caught. There is evidence that they also communicate via ultrasound.

Feeding ecology and diet

Crocodiles are opportunistic feeders and their diet includes a huge range of both invertebrates and vertebrates. The young are agile and will jump to catch dragonflies and flying termites. The bulk of their diet is insects and spiders. As the young grow, they prey on crabs, fish, frogs, reptiles, birds, and mammals. Crocodiles are active hunters that stalk prey both above and below water. Most species eat carrion, even

leaving the safety of the water to feed on a carcass. All species have teeth and jaws designed for seizing, tearing, and crushing rather than chewing. Some species, like the Malayan gharial, Johnstone's crocodile (*Crocodylus johnstonii*), and African slender-snouted crocodile (*Crocodylus cataphractus*) have narrow jaws and very sharp teeth specially adapted to catch fish. Sensory pores in and around the mouth help crocodiles find prey under water. Several species of crocodiles have been seen to herd fish to shore with their bodies, sometimes communally. Crocodiles play a vital role at the apex (top) of the aquatic food chain. They control predators on commercially important fishes and keep the habitat clean as part-time scavengers.

Reproductive biology

Crocodiles are territorial. Males are larger than females and will defend territories and compete for mates. Males mate with as many females as possible during the fixed breeding seasons. Crocodiles use a variety of social signals, especially at mating time. These include jawslaps, roaring, dominance, and subordination displays. Females initiate courtship in some species.

All crocodiles lay eggs. Females lay their eggs 40–70 days after mating. The incubation period depends on nest tem-

Crocodilians may care for their young for over two years. From left to right, top to bottom: 1. Female lays eggs; 2. Female uncovers nest and helps hatchlings out of their eggs; 3. Hatchlings break out of their shells and make their way to the water's edge; 4. Female carries the newly hatched young in her mouth to the water; 5. An adult provides food for the young, and; 6. Protects the young crocodilians. (Illustration by Wendy Baker)

perature and is typically 60–90 days. The sex of the developing embryo is determined by temperature; in general, higher temperatures produce males and lower temperatures result in females.

Five species of crocodiles dig holes in the sand, earth, or gravel embankments above the high-water line to lay their eggs. The hole is L-shaped and as deep as the length of the female's hind limbs, which are heavily clawed for digging. The eggs emerge covered with a lubricating mucous, which cushions the brittle shells from cracking as they fall into the nest hole. These species are dry-season nesters whose young hatch with the coming of the rains when many small prey items also emerge.

Nine species of crocodiles are mound nesters. The female gathers together a heap of vegetation, soil, and compost using her legs and jaws. She compacts the mound by crawling over it and then excavates a hole on the top in which to lay her eggs. Mound nesters lay their eggs at the start of the wet season, and the young hatch when the water is highest, a time when there is plenty of small prey.

Female crocodiles (and sometimes males) generally stay near their nest during incubation, repeatedly visiting the site,

especially when hatching time approaches. When ready to hatch, the young call with quacking grunts and the parent releases the young from the nest and carries them to the water. Crèche formation is essential for hatchling survival and sometimes both parents guard the young for several weeks or more until they disperse. Adult crocodiles are conditioned to respond to distress calls of the young. Despite parental care, mortality in hatchling crocodiles is generally over 90% due to predators like fishes, crabs, snakes, monitor lizards, raptors, large wading birds, mongooses, foxes, and jackals.

Conservation status

Most of the 14 species of crocodiles have been greatly reduced in numbers due to overhunting and habitat loss. The fact that some of the larger species such as the saltwater crocodile and Nile crocodile sometimes kill domestic animals and humans has contributed to the unpopularity of these predators in many parts of the world. The hunting of crocodiles for skins was a major reason for the decline of most species. Now, however, sustainable-use programs are responsible for the recovery and continued survival of several species, notably the Nile crocodile, saltwater crocodile, and New Guinea crocodile (*C. novaeguineae*). In the 1980s the world trade in crocodile skins was close to 500,000 per year, most from the wild. In 1999 the annual trade was about 398,000, the majority from managed wild populations and farms such as the Samut Prakan Crocodile Farm in Thailand and Mainland Holdings in Papua New Guinea, which together have over 80,000 crocodiles. In general, countries that harvest crocodiles from the wild and on farms on a sustainable basis are successfully protecting their species.

The 2002 IUCN Red List includes 10 species of crocodiles: 3 are Critically Endangered, 2 are Endangered, 3 are Vulnerable, 1 is Lower Risk/Conservation Dependent, and 1 is Data Deficient.

Significance to humans

Crocodiles have long figured in religious beliefs, myths, and folktales around the world. It is not surprising that such huge, ancient looking, and dangerous reptiles should inspire fear, respect, and awe from the human race. Nile crocodiles were mummified and buried with the pharaohs of Egypt. The Hindu river goddess, Ma Ganga, sometimes sits astride a mugger crocodile, and more than one New Guinean indigenous creation myth features the saltwater crocodile. At the same time, indigenous people all over the world have considered crocodiles a wonderful source of tasty eggs and meat. The real challenge to the survival of this ancient reptile's lineage started in the later nineteenth century when the slaughter for skin began. Crocodile leather demand reached its peak

A Nile crocodile (*Crocodylus niloticus*) carrys young in its mouth. (Photo by Animals Animals ©Roger De La Harpe. Reproduced by permission.)

in the mid-twentieth century, when over a million skins were being traded per year.

Conservation efforts with sustainable use at the forefront started in the early 1970s and quickly reversed the fate of several endangered crocodiles. Today, up to half a million crocodile skins are traded internationally each year, more than half from farms and ranches.

1. Mugger crocodile (*Crocodylus palustris*); 2. Nile crocodile (*Crocodylus niloticus*); 3. Saltwater crocodile (*Crocodylus porosus*); 4. American crocodile (*Crocodylus acutus*); 5. Johnstone's crocodile (*Crocodylus johnstonii*). (Illustration by Brian Cressman)

Species accounts

American crocodile
Crocodylus acutus

TAXONOMY
Crocodylus acutus Cuvier, 1807 Antilles.

OTHER COMMON NAMES
English: American saltwater crocodile; French: Crocodile amèricain; German: Mittelamerikanisches Krokodil; Spanish: Cocodrilo americano.

PHYSICAL CHARACTERISTICS
The American crocodile has a somewhat narrow snout and slender build. Adults average 10–11 ft (3–3.5 m) in length, but males can grow to 20 ft (6 m). Females average 8–10 ft (2.5–3 m). A distinctive hump on the skull of adults just in front of the eyes is diagnostic of the species. Yellowish to gray with dark cross-marks when young; older American crocodiles often lose the bands and are a uniform sandy color or dark brown. The underside is white.

DISTRIBUTION
United States, Mexico, Central America, northern South America, islands of the Caribbean Sea.

HABITAT
Coasts along mangroves, estuaries, large rivers, and sometimes inland lakes.

BEHAVIOR
This is a social animal, coexisting in large groups. The American crocodile has a high tolerance to salt water and excretes excess salt from lingual glands on the tongue.

Crocodylus acutus

FEEDING ECOLOGY AND DIET
This species eats large fish, frogs, turtles, birds, and small mammals. In some places they take livestock, and there are occasional reports of very large individuals attacking humans. Hatchlings feed on crabs, insects, small frogs, tadpoles, and fish.

REPRODUCTIVE BIOLOGY
Although usually a hole nester, the female American crocodile may make a mound nest of sand, vegetation, and compost. The breeding season is late April to early May in Florida. In South America the breeding season is March to May. They lay 30–60 eggs, which hatch in 80–90 days, depending on incubation temperature. The female crocodile guards her nest, assists the hatchlings, and thereafter protects them from predators. Juveniles appear to disperse from the vicinity of the nest site within a few days.

CONSERVATION STATUS
This species is fairly widespread, but only small populations of the American crocodile are left throughout its range. The species is protected in most countries where it occurs, but enforcement is often inadequate. Threats are habitat fragmentation and direct human disturbance. Estimated wild population is 10,000 to 20,000. It is listed as Vulnerable on the IUCN Red List, and appears under Appendix I of CITES.

SIGNIFICANCE TO HUMANS
Eggs are used as a food source by indigenous people in parts of the range. ◆

Johnstone's crocodile
Crocodylus johnstonii

TAXONOMY
Crocodylus johnsoni Krefft, 1873, Herbert River, Queensland.

OTHER COMMON NAMES
English: Freshie, Australian freshwater crocodile; French: Crocodile de Johnstone; German: Australien Krokodi; Spanish: Cocodrilo de Johnstone.

PHYSICAL CHARACTERISTICS
Johnstone's crocodile has a very narrow snout. Adults average 6.5 ft (2 m) in length. Males grow to over 10 ft (3 m); females grow to 6.5 ft (2 m) and average 5 ft (1.5 m). Dark or light brown in color with black bands or spots on the body and tail. The underside is white. The dorsal scales are quite smooth, arranged in six neat rows.

DISTRIBUTION
Tropical regions of northern Australia.

HABITAT
This species inhabits freshwater rivers, streams, and pools, but occasionally estuarine habitats.

BEHAVIOR
Social animals, Johnstone's crocodiles may form dense aggregations during the annual dry season. Large males and females assert territoriality and dominance by chasing and biting the tails

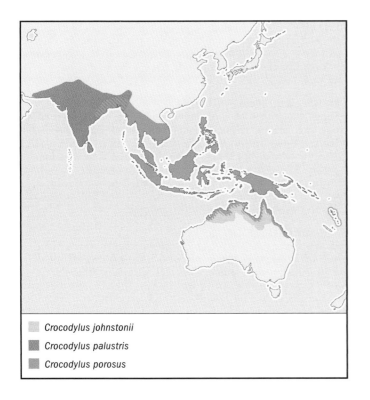

Crocodylus johnstonii

Crocodylus palustris

Crocodylus porosus

of smaller ones. These crocodiles migrate considerable distances overland in the dry season. Where there is no water they seek refuge under embankments and in piles of leaves and dense vegetation. Johnstone's crocodiles are quite vocal and growl in response to loud noise or a human presence in "their" pond.

FEEDING ECOLOGY AND DIET
Crustaceans, insects, spiders, fish, frogs, lizards, snakes, birds, and mammals. Insects are the most common food item, followed by fish, which are caught by using a "sit-and-wait" strategy. Johnstone's crocodiles are also known to stalk larger prey; large crocodiles catch and eat wallabies and water birds.

REPRODUCTIVE BIOLOGY
These crocodiles court and mate at the beginning of the dry season (around May). Females dig nest holes in sand banks, sometimes communally, and lay an average of 13 eggs. Average incubation time is 74 days. Females rarely guard the nest and only become attentive when the young signal they are ready to hatch and emerge. The female opens the nest and carries the young to the water, sometimes inflicting puncture wounds on the hatchlings with her sharp teeth. The female stays with her crèche of hatchlings for a month or more. Juveniles that survive to maturity are known to return to the same breeding and nesting areas from which they were hatched.

CONSERVATION STATUS
Seriously depleted by the skin trade but recuperating. Threats to the species are habitat destruction and the introduction of the poisonous cane toad (*Bufo marinus*), which has led to mortality in otherwise healthy populations. Estimated wild population is at least 100,000. This species no longer appears on the IUCN Red List, but it is on Appendix II of CITES.

SIGNIFICANCE TO HUMANS
Johnstone's crocodiles and their eggs remain a traditional source of food for several indigenous groups and as part of their folklore. ◆

Mugger crocodile
Crocodylus palustris

TAXONOMY
Crocodylus palustris Lesson, 1834, Ganges River, India.

OTHER COMMON NAMES
English: Marsh crocodile; French: Crocodile à front large; German: Sumpfkrokodil; Spanish: Cocodrilo marismeño.

PHYSICAL CHARACTERISTICS
Relatively stocky reptiles, adult males average 10 ft (3 m). Adult females average 7.4 ft (2.25 m), and may reach 10 ft (3 m). The maximum size on record is 18 ft (5.5 m). Gray, olive, or brownish above with dark markings that become less distinct as the animal gets older. The underside is white or yellowish. The dorsal scutes are prominent and often quite irregular in arrangement, giving the reptile an untidy appearance. Presence of large postoccipital scutes differentiate this species from the sympatric (in India and Sri Lanka) saltwater crocodile.

DISTRIBUTION
Mainly India and Sri Lanka. Small populations still occur in southeastern Iran, and in parts of Pakistan, Nepal, Bhutan, and Bangladesh.

HABITAT
A very adaptable crocodile, muggers are found in clear hill streams up to 1,975 ft (600 m) above sea level, as well as in sewage-treatment ponds, large rivers, lakes, swamps, and even saline habitats.

BEHAVIOR
These are social animals. Unless under hunting pressure, muggers are conspicuous, basking in the open to thermoregulate. Younger muggers are more solitary and cryptic. Muggers can migrate overland for 6 mi (10 km) or more to find water during the dry season. They also dig horizontal tunnels 33 ft (10 m) or deeper in lake and river banks to allow them to survive dry periods and even severe drought.

FEEDING ECOLOGY AND DIET
Muggers are opportunistic feeders. The diet of the young consists of insects, crabs, frogs, fish, carrion, as well as occasional small reptiles, birds, and mammals. Mature muggers over 6.5 ft (2 m) long catch larger prey such as mammals (from monkeys to deer), water birds, and large catfishes. Near human habitation, muggers will sometimes take farm animals and birds. Muggers have rarely been known to prey on humans.

REPRODUCTIVE BIOLOGY
The annual mating season is December/January on mainland south Asia, variable in Sri Lanka. After about 40 days of gestation, the female mugger digs a hole nest and lays 20–40 eggs between March and May, the number increasing with her size. The female typically remains close to the nest for the 60–70 days of incubation to release and care for the young.

CONSERVATION STATUS
This species has vanished from most of its range. In 2002 there were over 8,000 crocodiles at various rearing stations, including India's largest breeding center, the Madras Crocodile Bank, with no protected areas to release them. The mugger crocodile is strictly protected in the seven countries in which it occurs. The species is listed as Vulnerable on the IUCN Red List and is on CITES Appendix I. Estimated wild population is 5,000–10,000.

SIGNIFICANCE TO HUMANS

When muggers were still plentiful, their eggs and meat were important food resources for many of south Asia's indigenous people. Attacks on humans are rare, but muggers are treated with caution, respect, and sometimes animosity by local people. A well-planned and managed industry based on muggers as a sustainable resource would benefit both humans and crocodiles in south Asia. ◆

Nile crocodile
Crocodylus niloticus

TAXONOMY

Crocodylus niloticus Laurenti, 1768, India and Egypt.

OTHER COMMON NAMES

French: Croco afrique; German: Nordöstliches Nilkrokodil; Spanish: Cocodrilo del Nilo; Swahili: Mamba.

PHYSICAL CHARACTERISTICS

This is a heavyset species with prominent dorsal scales or scutes arranged in even rows. Average adult lengths are 11.5 ft (3.5 m) for males and 8 ft (2.5 m) for females. Males grow to a maximum length of 18 ft (5.5 m); females to 11.5 ft (3.5 m). Young Nile crocodiles are brown or olive with strong darker markings. In adults the markings are vague and older animals are uniformly dark brown or gray. The belly is yellowish, white, or gray.

DISTRIBUTION

South of the Sahara desert throughout much of tropical and southern Africa and Madagascar.

HABITAT

This species inhabits wetlands, rivers, and lakes, inclusive of coastal areas.

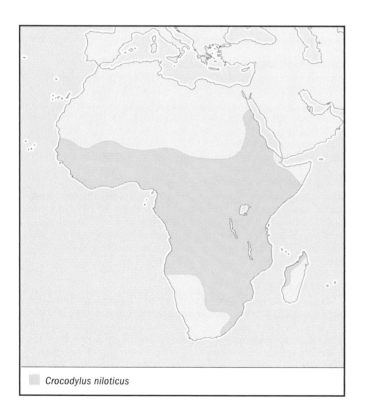

Crocodylus niloticus

BEHAVIOR

Although a social animal, adult males square off for territory and mates during the breeding season. It is common for large groups of adults of both sexes to bask and feed in one area.

FEEDING ECOLOGY AND DIET

Young Nile crocodiles feed largely on insects and spiders with frogs, fish, snakes, and other small vertebrates. Subadults and adult crocodiles mainly eat fish, although very large adults kill and eat antelopes, warthogs, domestic animals, and occasionally humans. These crocodiles use various techniques to catch their prey, including the "hide and wait" method at water holes and river crossings, waiting underwater for fish, and following scent trails to carrion, even when far from water.

REPRODUCTIVE BIOLOGY

The nesting season across Africa occurs from August to January. The female digs a nest hole above the high water line to lay 50–80 eggs. The female stays close to the nest for the 80–90 days of incubation. When the young hatch, she digs them out of the underground nest and carries them to the water. Nests are raided by a variety of predators, namely monitor lizards, hyenas, and humans. The crèche of young stay together for a month or more under the watchful eye of the parent.

CONSERVATION STATUS

The Nile crocodile forms the basis of a successful sustainable-use program for skins and meat in several African countries. In other parts of its range, notably Central and West Africa, the crocodile is in decline and there is very little status information. It is afforded some level of local protection throughout its range, and is no longer included on the IUCN Red List. As of 2002, the Nile crocodile was on Appendix I of CITES, except for ranched animals in Botswana, Ethiopia, Kenya, Malawi, Mozambique, South Africa, Tanzania, Zambia, and Zimbabwe. Nile crocodiles are on Appendix II in Madagascar and Uganda, where there is an annual hunting quota. The estimated population in the wild is 250,000–500,000.

SIGNIFICANCE TO HUMANS

There is a long history of reverence for the Nile crocodile, dating back to the time of the pharaohs when hundreds of crocodiles were mummified along with dead kings. Today there is less reverence and more fear and intolerance toward these crocodiles, which sometimes prey on livestock and humans. ◆

Saltwater crocodile
Crocodylus porosus

TAXONOMY

Crocodylus porosus Schneider, 1801, Ceylon and western India.

OTHER COMMON NAMES

English: Estuarine crocodile, Indo-Pacific crocodile; French: Crocodile marin; German: Leistenkrokodil; Spanish: Cocodrilo poroso Bahasa.

PHYSICAL CHARACTERISTICS

This is the world's largest crocodilian, growing to over 20 ft (6.1 m) in length and weighing up to 2,200 lb (1 metric ton). Adult males average 13–16.5 ft (4–5 m) in length, and females 10–11.5 ft (3–3.5 m). The saltwater crocodile may be black, dark brown, or yellowish dorsally, with lighter flanks spotted with black. The underside is white or yellowish. Unlike all other crocodilians, the enlarged postoccipital scales are usually absent. Instead, the postoccipital region on the back of the neck is covered with small beadlike scales. The hind feet are strongly webbed.

DISTRIBUTION
Scattered populations from the east coast of India to Australia. This is the most widely distributed crocodilian in the world.

HABITAT
Although most at home in brackish tidal mangrove waterways, saltwater crocodiles can be found in freshwater habitats hundreds of miles (or kilometers) inland.

BEHAVIOR
Large, adult males are territorial and solitary. Females typically have small home ranges of a few square miles (or kilometers) while adult males have been known to patrol territories of 100 mi^2 (260 km^2).

FEEDING ECOLOGY AND DIET
Young saltwater crocodiles feed mainly on small crabs and fish. Adult saltwater crocodiles may lie in wait for large mammals, both wild and domestic, at water holes. Humans are sometimes taken.

REPRODUCTIVE BIOLOGY
The breeding season occurs from April to May in India, and January to February in Australia. The female lays 40–70 eggs in a mound of vegetation. A nesting female will typically make herself a wallow near the nest from where she will guard it for the 70–80 days incubation. She releases the hatchlings and then guards them, but predation on hatchlings is still high.

CONSERVATION STATUS
Although widely distributed, saltwater crocodiles have been killed out of fear and for their valuable skins throughout their range. Only in Australia and Papua New Guinea are there effective management programs in place, which allow sustainable use of saltwater crocodiles for skin and meat. Other Asian countries in the range of the saltwater crocodile need to enforce existing protective legislation while educating people on how to live with crocodiles. The saltwater crocodile is afforded protection on paper throughout its range. It is no longer listed on the IUCN Red List, but it appears on CITES Appendix I. Estimated wild population is 200,000–300,000.

SIGNIFICANCE TO HUMANS
Large saltwater crocodiles consider humans and their livestock as prey items. This has led to animosity toward this species; yet the saltwater crocodile forms the basis for a skin industry for a large number of indigenous people in the Asia/Pacific region. ◆

Resources

Books
Brochu, C. A. "Congruence between Physiology, Phylogenetics and the Fossil Record on Crocodylian Historical Biogeography." In *Crocodilian Biology and Evolution*, edited by G. C. Grigg, F. Seebacher, and C. E. Franklin. New South Wales, Australia: Surrey Beatty & Sons, 2001.

Brochu, C. A., and L. D. Densmore III. "Crocodile Phylogenetics: A Summary of Current Progress." In *Crocodilian Biology and Evolution*, edited by G. C. Grigg, F. Seebacher, and C. E. Franklin. New South Wales, Australia: Surrey Beatty & Sons, 2001.

Daniel, J. C. *The Book of Indian Reptiles and Amphibians.* New Delhi, India: Oxford University Press, 2002.

Ross, C. A., ed. *Crocodiles and Alligators.* New York: Facts on File, Inc., 1989.

Schmidt, K. P. *A Check List of North American Amphibians and Reptiles.* Chicago: University of Chicago Press, 1953.

Smith, H. M., and E. H. Taylor. *An Annotated Checklist and Key to the Reptiles of Mexico Exclusive of Snakes.* Washington, DC: Bulletin of the U.S. National Museum, 1950.

Webb, G. J. W., and S. C. Manolis. *Crocodiles of Australia.* New South Wales, Australia: Reed Books Pty, Ltd., 1989.

Periodicals
Fuchs, K., R. Mertens, and H. Wermuth. "Zum Status vom *Crocodylus cataphractus* und *Osteolaemus tetraspis.*" *Stuttgarter Beiträge zur Naturkunde* Serie A. (Biologie) 266 (1974): 1–8.

Kushlan, J., and F. J. Mazzotti. "Historic and Present Distribution of the American Crocodile in Florida." *Journal of Herpetology* 23, no. 1 (1989): 1–7.

Lesson, R. P. "Reptiles." In I. G. S. Bélanger, ed., "Voyage aux Indes-Orientales, par le Nord de l'Europe, les Provinces du Caucase, la Géorgie, l'Armenie, et la Perse, . . . pendant les anné es 1825–29." *Zoologie.* (1834). A. Bertrand, Paris: 291–336, pls. 1–7.

Mertens, R. "Zur Systematik und Nomenclatur der Ceylon-Krokodile." *Senckenbergiana, Biologica* 41, nos. 5–6 (1960): 267–272.

Negi, B., ed. "Indian Crocodilians." *Envis* 2, no 1. (June 1999).

Patnaik, R., and H. H. Schleich. "Fossil Crocodile Remains from the Upper Siwaliks of India." *Mitteilungen der Bayerischen Staatssammlung für Paläontologie und Historische Geologie* 33 (1993): 91–117.

Webb, G. J. W., R. Buckworth, and S. C. Manolis. "*Crocodylus johnstoni* in the McKinlay River, N.T. VI. Nesting Biology." *Australian Wildlife Research* 10 (1983): 607–637.

Wermuth, H. "Systematik der Rezenten Krokodile." *Mitteilungen aus dem Zoologischen Museum im Berlin* 29, no. 2 (1953): 375–514, figs. 1–66.

Organizations
Crocodile Specialist Group, Florida Museum of Natural History. Box 117800, Gainesville, FL 32611-7800 USA. Phone: (352) 392-1721. Fax: (352) 392-9367. E-mail: prosscsg@flmnh.ufl.edu Web site: <http://www .flmnh.ufl.edu/natsci/herpetology/crocs.htm>

Madras Crocodile Bank/Center for Herpetology. P. O. Box 4, Tamil Nadu, 603 104 South India. E-mail: mcbtindia@ vsnl.net

Resources

Other

Crocodiles: Status Survey and Conservation Action Plan, 1998. (Online) 1998 [cited January 6, 2003]. <http://www.flmnh .ufl.edu/natsci/herpetology/act-plan/plan1998a.htm>

Florida Museum of Natural History: Herpetology [cited January 5, 2003]. <http://www.flmnh.ufl.edu/natsci/herpetology/ herpetology.htm>

Ross, J. P., ed. *Crocodiles. Status Survey and Conservation Action Plan.* 2nd ed. IUCN, Gland, Switzerland and Cambridge, U.K.: IUCN/SSC Crocodile Specialist Group, 1998.

Romulus Earl Whitaker III, BSc
Nikhil Whitaker

▲
Sphenodontia
Tuatara
(Sphenodontidae)

Class Reptilia
Order Sphenodontia
Family Sphenodontidae
Number of families 1

Thumbnail description
Neither sex has ear holes, males lack a copulatory organ, both sexes have rear-pointing extensions on ribs, teeth tightly fused to the surface of the jawbone, and a double row of teeth on the upper jaw

Size
Males are about 24 in (60 cm) in length and 2 lb (1 kg) in weight; females are smaller, measuring less than 16 in (40 cm) in length and rarely exceeding 1 lb (454 g) in weight

Number of genera, species
1 genus; 2 species

Habitat
Islands with low-growing, salt-tolerant trees with a relatively open forest floor

Conservation status
Vulnerable: 1 species

Distribution
Islands off the North and South Islands of New Zealand

Evolution and systematics

Tuatara are confusing animals. They certainly confused J. E. Gray of the British Museum, when, in 1831, he described the skull of a tuatara as a species of Agamidae. Then, in 1840, when a whole tuatara came to hand, he placed it in a separate genus from the one assigned to the previously described skull. It was Gray's successor at the British Museum, Albert Günther, who realized, in 1867, that tuatara belonged to an array of fossils in an order of its own, today referred to as Sphenodontia.

The sphenodontians have a fossil record of about 225 million years and, as a group, were most diverse in the late Triassic and Jurassic 180–220 million years ago, when they inhabited Europe, Africa, and North America. Sphenodontians were already in decline during the age of the dinosaurs, and almost all of them became extinct by the early Cretaceous. A single lineage in the family Sphenodontidae survived on a landmass that separated from the southern continent of Gondwana 60–80 million years ago. This lineage gave rise to tuatara, and the Gondwanian fragment now forms the islands of New Zealand. Whether the tuatara have changed significantly from

their ancient ancestors is unclear, however. No early sphenodontids have been found in the New Zealand fossil record.

The name tuatara was bestowed by the Maori people when they arrived in New Zealand about 1,000 years ago. In the Maori language, tuatara is both a singular and plural noun. Two living species are recognized: *Sphenodon guntheri* (Buller, 1877) and *Sphenodon punctatus* (Gray, 1842). Within *S. punctatus* there are two distinguishable genetic forms.

Physical characteristics

In general appearance, tuatara resemble some agamid and iguanid lizards, but many morphologic features distinguish tuatara from most or all lizards. For example, tuatara have no ear holes, and the males lack a copulatory organ. Furthermore, tuatara have uncinate processes (rear-pointing extensions) on the ribs and their teeth are acrodont (tightly fused to the surface of the jawbone), and the young have a horny caruncle, or egg-breaker, to assist with hatching. The most

Tuatara means "spiny back" in Maori. (Photo by Hans Reinhard/Photo Researchers, Inc. Reproduced by permission.)

unusual feature, however, is a double row of teeth on the upper jaw into which fit those of the lower jaw.

Tuatara males are distinguishable from females. The males are larger than females and are often twice as heavy. Both sexes have a crest on the midline of the head and back, merging into toothlike projections down the tail. The crest of the males is larger than in females and can be inflated during aggressive and mating behavior. The Maori name, tuatara, refers to these crests on the back.

The largest tuatara are some populations of the northern form, males of which can reach more than 24 in (60 cm) in length and 2 lb (1 kg) in weight. Females are usually less than 16 in (40 cm) in length and rarely exceed 1 lb (454 g) in weight. In the northern form of *S. punctatus*, size varies broadly with latitude; the largest animals are also the most northern. Nonetheless, differences in mean body size can be found within the same island groups. The cause of this variation is unclear.

Tuatara vary in color from olive green to reddish to gray to almost black. The base color usually is overlaid with irregular darker markings and speckled with paler spots. Compared with *S. punctatus*, *S. guntheri* is more often olive green and tends to be more heavily speckled with paler spots. The crest is typically white in both species. Juveniles are often pale gray-brown, with paler V-shaped markings on the upper surface and distinctive darker markings radiating from the eyes. When newly hatched, they sometimes have a large triangular pale patch on top of the head from the snout to the eyes.

Distribution

Subfossil remains of tuatara, not identifiable by species, are present in caves, sand dunes, and Maori midden (rubbish) sites throughout the North and South Islands of New Zealand and on some offshore islands. Only a single wild population of *S. guntheri* is now known, on North Brother Island in Cook Strait off northern South Island. Populations of the northern form of *S. punctatus* are confined to about 26 islands off northeastern North Island. The Cook Strait form of tuatara is confined to four islands off northern South Island.

Habitat

Islands inhabited by viable populations of tuatara have four features in common: low-growing, salt-tolerant trees that form a complete canopy over a relatively open forest floor; large invertebrates that include giant flightless orthopterans, carnivorous snails, giant centipedes, and large numbers of flightless tenebrionid beetles; high densities and wide diversities of lizards; and high densities and wide diversities of burrowing seabirds. The seabirds, dominated by petrels, prions, and shearwaters, are regarded as "ecosystem engineers" that soften and aerate the soils, enrich the soils with guano, and provide burrows used by tuatara.

Behavior

In their habitat, the tuatara are active at night. On such islands, tuatara are the top terrestrial predator, reaching densities of up to 810 per acre (2,000 per ha). Encounters between males may be limited to visual displays of animals taking parallel positions in opposite directions, then inflating the lungs and throat, and gaping the mouth and snapping it shut. If one male does not retreat, the two animals resort to combat that includes lunge-and-bite sequences.

Feeding ecology and diet

Tuatara are often sit-and-wait predators but will also actively forage. On average, 75% of the diet consists of invertebrates, especially large, flightless insects, such as orthopterans and tenebrionid beetles. Less than 5% of the diet includes seabirds or lizards. These proportions vary with season.

Reproductive biology

Each tuatara uses a burrow, or series of burrows; on islands densely inhabited by seabirds, these burrows may be less than 3 ft (0.9 m) apart. Males display before combat and

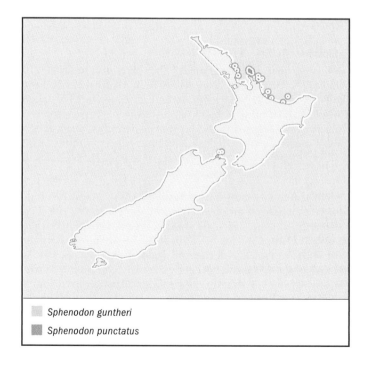

▢ Sphenodon guntheri
▢ Sphenodon punctatus

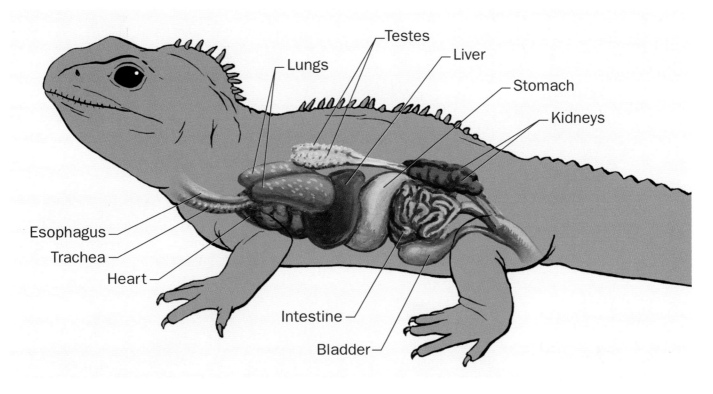

Tuatara internal organs. (Illustration by Brian Cressman)

mating. Combat is particularly common in summer and autumn, when males establish territories to include several females. Displays during combat and courtship include stiffening and inflation of the skin and spines down the neck and back.

Mating is in late summer or early autumn every two to five years, depending on form and location. Receptive females are attracted by an exaggerated walk, called "proud walk" or *stolzer Gang*, during which males with crest erect and throat inflated stiffly and slowly circle females. After about 20 minutes, the females either leave or remain to mate. During mating, a male mounts a female, raising the base of her tail using his hind legs, to align his vent with hers. As in birds, sperm transfer is between the vents of each sex.

The females lay their eggs the spring after mating. Cook Strait populations of *S. punctatus* lay eight to 15 eggs, whereas the northern populations and *S. guntheri* lay four to 13 eggs. The eggs are laid in rookeries, where females may gather from distances of at least 200 yards (180 m). The rookeries include unforested areas or gaps in forest canopy exposed to the sun. Each nest is a shallow depression or short burrow up to 20 in (50 cm) deep, with the eggs covered by loose soil. Incubation takes 12–15 months. Temperature in the nest is critical for egg survival and influences the sex of the hatchlings. Laboratory studies show that the highest hatching success is achieved when soil temperatures vary between 64°F and 72°F (18–22°C). Nests at temperatures of about 68°F (20°C) produce a strong female bias among juvenile *S. punctatus*. Similar studies of *S. guntheri* found that eggs that hatched at lower temperatures of around 64°F (18°C) produced more females, but those in temperatures that fluctuated between 64°F and 74°F (18–23°C) produced more males. Tuatara take up to 13 years to reach sexual maturity. Unlike the females, males have an annual reproductive cycle. The average life span of adults is at least 60 years.

A tuatara skull. (Illustration by Brian Cressman)

1. *Sphenodon guntheri*; 2. *Sphenodon punctatus*. (Illustration by Brian Cressman)

Conservation status

Currently, *S. guntheri* is ranked as Vulnerable by the IUCN Red Data Book, whereas *S. punctatus* has been delisted. The decline of tuatara on the main North and South Islands of New Zealand and on many offshore islands has been attributed to predation by introduced mammals, especially rats. Innovative and aggressive campaigns against introduced mammals by the New Zealand Department of Conservation, combined with research on behavior and reproductive biology by universities and partnerships with zoos in New Zealand and the United States, have led to the beginnings of a resurgence by both species. Within the range of the northern form, introduced rats have been removed from seven islands where tuatara populations were dangerously low or were showing recruitment failure. An additional population has been established on an island within the form's historic range. Attempts are under way to establish two new populations of *S. guntheri* on islands from which rats have been removed, and additional restoration plans are being formed that will benefit the Cook Strait *S. punctatus*.

Significance to humans

Tuatara are culturally important to New Zealand indigenous people (Maori), who regard them as living treasures. They are also of particular scientific importance because of their unique biological features. This interest was such in the nineteenth century that tuatara were collected heavily for export to scientific institutions, leading the government to give tuatara full legal protection in 1895.

Resources

Books

Benton, M. *Vertebrate Palaeontology.* London: Chapman and Hall, 1997.

Gaze, Peter. *Tuatara Recovery Plan 2001–2011*, Vol. 47. Wellington, New Zealand: Department of Conservation Threatened Species Recovery Plan, 2001.

Newman, Don. *Tuatara.* Dunedin, New Zealand: John McIndoe and Department of Conservation, 1987.

Robb, Joan. *New Zealand Amphibians and Reptiles.* Auckland, New Zealand: William Collins, 1986.

Periodicals

Cree, Alison, and Charles Daugherty. "Tuatara Sheds Its Fossil Image." *New Scientist* (October 20, 1990): 22–26.

Cree, Alison, Michael Thompson, and Charles Daugherty. "Tuatara Sex Determination." *Nature* 375 (June 15, 1995): 543.

Daugherty, Charles, Alison Cree, J. M. Hay, and Michael B. Thompson. "Neglected Taxonomy and Continuing Extinctions of Tuatara (*Sphenodon*)." *Nature* 347 (September 13, 1990): 177–179.

Gillingham, James C., Cristopher Carmichael, and Tracy Miller. "Social Behavior of the Tuatara, *Sphenodon punctatus.*" *Herpetological Monographs* 9 (1995): 5–16.

Towns, David, Charles H. Daugherty, and Alison Cree. "Raising the Prospects for a Forgotten Fauna: A Review of 10 Years of Conservation Effort for New Zealand Reptiles." *Biological Conservation* 99, no. 1 (2001): 3–16.

Tyrrell, Claudine, Alison Cree, and David R. Towns. "Variation in Reproduction and Condition of Northern Tuatara (*Sphenodon punctatus punctatus*) in the Presence and Absence of Kiore." *Science for Conservation* 153 (August 2000): 1–42.

Organizations

The Tuatara Recovery Group, c/o Department of Conservation. P.O. Box 10 420, Wellington, New Zealand. Phone: 64 (4) 471-0726. Web site: <http://www.doc.govt.nz>

David Robert Towns, PhD

Squamata
(Lizards and snakes)

Class Reptilia

Order Squamata

Number of families About 42

Number of genera, species About 1,880 genera; 7,200 species

Photo: A Mozambique spitting cobra (*Naja mossambica*) in South Africa. (Photo by Animals Animals © Anthony Bannister. Reproduced by permission.)

Evolution and systematics

Squamates are the most diverse living clade of reptiles, including about 1,440 genera and 4,450 species of lizards plus 440 genera and 2,750 species of snakes. Although snakes are commonly considered to constitute their own group, they clearly have descended from lizards and are merely limbless lizards. Squamates exhibit more than 70 shared derived traits, which indicate that they are descendants of a common ancestor, forming a large natural monophyletic group. (Snakes and lizards once were classified as different suborders, but since snakes are embedded within lizards, this classification is no longer tenable under the monophyletic standard of modern phylogenetic systematics.)

Sister group

The sister group to squamates is Rhynchocephalia, represented today by only two species of tuatara (*Sphenodon*) from New Zealand. Superficially, tuatara resemble lizards, in that they have a dorsal crest of scales. They possess two temporal arches (the "diapsid" condition), however, making their skulls quite rigid. They do not have copulatory organs. In tuatara the lower jaw articulates directly with the upper skull, resulting in a narrow gape and slow jaw movements. Tuatara are visual ambush predators that eat large prey, which they pick up with their sticky and fleshy tongues ("lingual prehension"). Because these traits are shared with basal lizards, they are probably ancestral states.

Streptostyly and Jacobson's organ

In all squamates, the lower temporal arch has been lost, and the lower jaw hinges on the quadrate bone, which hangs down from the cranium, a situation known as "streptostyly." This hanging jaw setup provides a mechanical advantage that results in faster jaw opening and closure, a wider gape, and a stronger bite, presumably greatly enhancing prey capture and

feeding. All lizards and snakes also have a pair of eversible copulatory organs (hemipenes) at the base of their tails. Another novel and important feature of squamates is a vomeronasal olfactory organ (Jacobson's organ) in the palate, separate from the nasal capsule. Thus, squamates possess three chemosensory systems: (1) taste buds on their tongues, (2) nasal smelling of volatile airborne scents, and (3) vomeronasal analysis of heavy non-airborne chemicals picked up by the tongue and transferred into the Jacobson's organ in the roof of the mouth, where the signal is processed.

Vomerolfaction ability differs greatly among squamates, remaining relatively undeveloped in some basal groups but becoming acutely sensitive in more derived groups, such as varanids and snakes. Ancestral squamates were ambush predators that detected prey by movement using visual cues and had relatively low activity levels and poorly developed chemosensory systems. The large clade Iguania retained these ancestral features, whereas in the more advanced Scleroglossa a suite of derived characteristics developed. Ancestral scleroglossans captured and manipulated prey with their jaws (jaw prehension), thus freeing the tongue to evolve along other lines and facilitating the evolution of sensitive vomerolfaction. Active foraging and higher activity levels were further consequences. They also possessed the combined abilities to flex the skull (mesokinesis), and discriminate prey based on chemical cues, ultimately producing more than 20 remaining families of other lizards plus all snakes (13–18 families), making up 80% of all squamates. Most snakes are entirely legless, and no living snake has functional legs (although a fossil did), eardrums, or movable eyelids. Many other lizards, especially burrowers, also have become limbless and lost their eardrums and eyelids.

Fossil record and biogeography

Squamate history dates back at least to the early Jurrasic, if not the late Triassic. Squamata is the sister taxon to

Rhynchocephalia, together making up Lepidosauria. About 200 million years ago (mya), before true lizards existed, one lineage of lepidosaurs gave rise to the ancestor of squamates. No fossils of this common ancestor ("stem group") of all squamates are known. It must have existed during the Upper Triassic to the Lower Jurassic, about 200–250 million mya, but a 50-million-year gap in the fossil record precludes exact dating. Most squamate fossils are considerably more recent. Much of early squamate evolution occurred during the Jurassic and Cretaceous on the supercontinent of Pangaea. Diversification of arthropods during the Jurassic provided a literal banquet for terrestrial vertebrates that could find and capture them. Five late Jurassic (about 150 mya) fossil lizards represent ancient extinct lineages. They are scattered across the squamate phylogenetic tree, indicating that early diversification into major clades (Iguania, Gekkota, Scincomorpha, and Anguimorpha) had taken place by the end of the Jurassic. These early lineages gave rise to the currently recognized 23–24 lizard families and 13–18 snake families.

Squamate evolution is intertwined intimately with continental drift. Some squamate families appear to be of Gondwanan origin, but others arose in the Northern Hemisphere and subsequently dispersed to southern continents. Australia's snakelike pygopodids arose from diplodactylid geckos within the island continent. Four lizard groups reached Australia and underwent extensive adaptive radiations. Some groups, such as skinks and varanids, became more speciose there than they are today in their probable ancestral source areas.

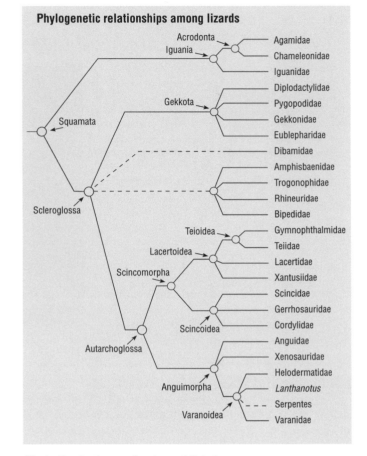

Phylogenetic relationships among lizards

(Illustration by Argosy. Courtesy of Gale.)

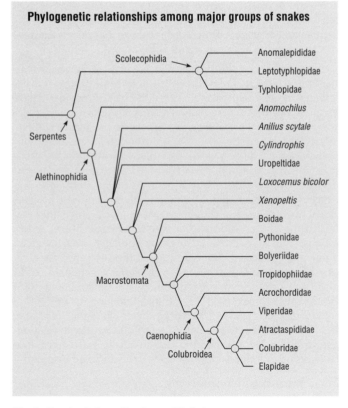

Phylogenetic relationships among major groups of snakes

(Illustration by Argosy. Courtesy of Gale.)

When the northern Laurasian plate separated from the southern Gondwanan plate in the mid-Jurassic, two isolated landmasses were formed. Gondwana presumably held primitive iguanians and gekkotans, whereas Laurasia must have contained ancestral eublepharid geckos, scincomorphans, and anguimorphans. When Gondwana broke apart, its iguanians and gekkotans became isolated on the three large southern landmasses, Africa (agamids, chameleons, and gekkonids), South America (iguanids and sphaerodactyline geckos), and the Australian region (agamids and diplodactylid geckos).

Gekkonids and skinks dispersed widely and became virtually cosmopolitan. Both crossed oceans by rafting and moving across land bridges. Other groups either remained confined to the landmass of origin (cordylids, corytophanines, crotaphytines, diplodactylids, gymnophthalmids, helodermatids, hoplocercines, lanthanotids, leiocephalines, leiosaurines, liolaemines, oplurines, phrynosomatines, pygopodids, sphaerodactylines, tropidurines, and xantusiids) or exhibited a more limited dispersal (agamids, anguids, chamaeleonids, iguanids, lacertids, teiids, and varanids). Exactly when and how snakes diversified and colonized the continents remains poorly understood, but within snakes, scleroglossan evolution produced groups as diverse as fossorial (adapted to digging) burrowers that live in social-insect colonies (scolecophidians) and sea snakes that inhabit the world's warm oceans (Hy-

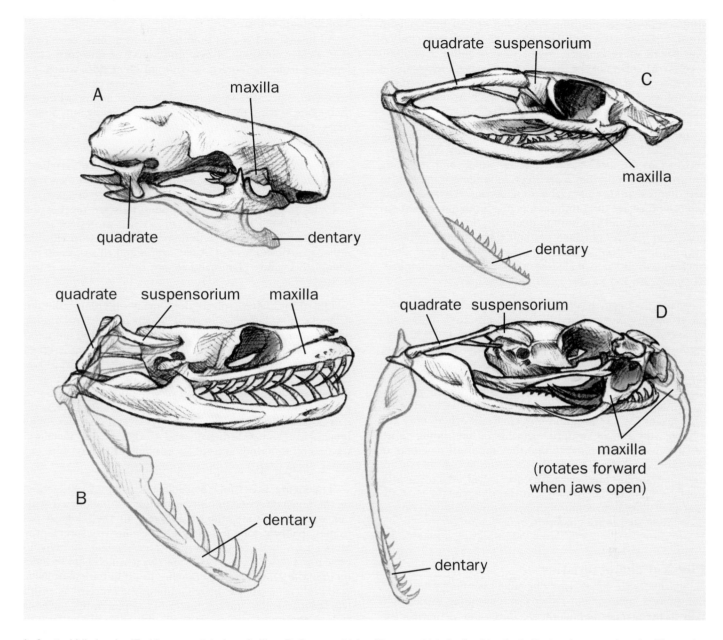

A. Spotted blindsnake (*Typhlops punctatus*)—primitive; B. Boa constrictor (*Boa constrictor*)—fixed teeth; C. Eastern hog-nosed snake (*Heterodon platirhinos*)—moveable rear teeth; D. Western diamondback rattlesnake (*Crotalus atrox*)—front-fanged. Primitive skulls have no suspensorium bone, no elongated quadrate bone, and a solid, compact skull. Front-fanged skulls have a suspensorium that is angled up, elongated quadrate that is angled back (allows jaws to open wider), and a less solid skull with more moveable bones. Snake skulls show variations of these traits on a continuum, with primitive and front-fanged at the extremes, and rear-fanged somewhere in the middle. (Illustration by Bruce Worden)

drophiinae). Snakes are arboreal, terrestrial, and aquatic and are top predators in almost all natural communities.

Phylogenetic relationships and number of families

A basal split in squamate phylogeny produced Iguania (99 genera and approximately 1,230 species), which retained ancestral traits (visual ambush predators with lingual prehension and poorly developed vomerolfaction), and Scleroglossa (almost 6,000 species), which adopted innovative new methods of finding and eating prey as well as acutely sensitive vomerolfaction and hydrostatic (operated by liquid pressures) forked tongues. Scleroglossa includes dibamids, amphisabaenians, and snakes, but their exact affinities within Scleroglossa remain uncertain. Remaining scleroglossans, in turn, bifurcated into two large clades, Gekkota and Autarchoglossa. Gekkota (about 1,000 species) evolved elliptical pupils and the ability to operate at low temperatures, allowing nocturnal activity. They use their tongues to clean their lips and eye spectacles. Geckos took to the night, where they found a cornucopia of nocturnal arthropods. (Some geckos have reverted to a diurnal way of life.) The largest and most advanced clade, Autarchoglossa (about 4,800 species), is composed of two smaller sister clades, Scincomorpha (seven families of lizards, with about 1,800 species) and Anguimorpha (five lizard families plus 15–18 snake families, with a total of more than 3,000 species). Members of the three clades, Iguania, Gekkota, and Autarchoglossa, differ considerably in morphologic features, physiological characteristics, behavior, life histories, and ecological niches, especially in foraging mode.

Iguanians are sit-and-wait ambush foragers that catch mobile prey as they move past their hunting stations. Most autarchoglossans are more active, foraging widely and searching for prey; as a result, they have access to sedentary and hidden prey items that are unavailable to iguanians. Active foraging is more expensive than ambush foraging, both in terms of energy expended and exposure to predators, but the returns are greater in calories obtained per unit time. Autarchoglossans have evolved flexible joints in their muzzles and skulls (mesokinesis and cranial kinesis), further improving their ability to capture and subdue large and agile prey.

Snake phylogeny has not yet been resolved, but three major groups are recognized: blindsnakes, primitive snakes, and advanced snakes. Blindsnakes (Scolecophidia) include three families (Anomalepididae, Leptotyphlopidae, and Typhlopidae). These are specialized burrowers and are considered sister to all other snakes (Alethinophidia), which are organized into four superfamilies (Anilioidea, Booidea, Acrochordoidea, and Colubroidea). Anilioids and booids are considered primitive snakes. As an indication of their lizard origin, boas and pythons still possess vestigial remnants of hind limbs, called "anal claws," which indicate that they are basal members of the ophidian clade. Colubroids, the most diverse snakes, are more advanced. Higher snakes are called Macrostomata, which includes boids, pythons, acrochordids, and the most advanced of all snakes, colubrids, viperids, and elapids. The vast majority of snakes are colubrids.

Physical characteristics

Squamates are ectotherms, obtaining their bodily heat from the external environment rather than generating it metabolically in the manner of endotherms (birds and mammals). Ectothermy is sometimes seen as a disadvantage, and lay people often erroneously refer to snakes and lizards as "cold-blooded." Snakes and lizards regulate their body temperatures behaviorally. When it is cold, they actively bask and seek out warm microclimates. When it is too hot, they look for shade and cooler places. When they are active, many squamates have body temperatures just as high as those of birds and mammals. In fact, ectothermy has real advantages over endothermy, especially in warm, dry, unproductive environments, such as deserts and semiarid regions. Ectotherms enjoy a low-energy lifestyle that endotherms can only envy. The food consumed by a 0.17-oz (5 g) bird in a single day will last a 0.17-oz squamate a month. An unproductive habitat, where endotherms cannot exist, will support viable populations of ectotherms. Snakes can go for many months, or even an entire year, without feeding, simply by allowing the body temperature to drop. The economic lifestyle offered by ectothermy enables squamates to thrive and persist during droughts and periods of resource shortages, which a high-energy endotherm simply could not tolerate.

Morphologic features

All lizards and snakes shed their skins at least once each year. Before shedding, a new inner layer of skin forms and separates from the outermost layer of older skin, which appears pale. Eye spectacles of snakes become cloudy. Snakes hook their lip scales on an immobile object and crawl out of their skins, leaving them behind intact and inside out. Sometimes, a snake can be identified to a species from its shed skin. Most lizards shed their skins in patches.

Most lizards have fracture planes in the caudal vertebrae, which facilitate tail loss. Some can even autotomize their own tails voluntarily. Tail loss often allows a lizard to escape from a would-be predator. In such species, tails are regenerated rapidly, although the regrown tail is never as perfect as the original tail. Regenerated tails do not have bony vertebrae but rather a cartilaginous rod. A few lizards and all snakes do not have such fracture planes and cannot regenerate a tail if it is broken off.

Lizard toes have evolved along many different pathways. Adhesive toe lamellae evolved independently in anoles, geckos, and skinks; improved climbing ability; and led to an arboreal lifestyle. Gecko foot hairs, apparently overdesigned by orders of magnitude, exploit intermolecular Van der Waals forces (the subatomic analogue of gravitational forces) to provide powerful purchase even on very smooth surfaces. (Some geckos can run up a pane of glass.) Similarly, independent acquisition of fringed toe scales in several families (e.g., Iguanidae, Teiidae, Scincidae, and Gerhosauridae) increased traction on soft sands and enhanced climbing ability, even enabling some lizards to run across water.

The highly successful Chamaeleonidae (about 130 species) set their own direction in lizard evolution, taking lingual feeding and sit-and-wait ambush foraging to their logical extremes. Ballistic tongues allow chameleons to capture prey more than a body length away without moving, and having their toes bound together (zygodactly) permits them to hang on to and balance on thin branches to exploit arboreal habitats. Prehensile tails facilitate climbing and are used as a fifth leg. Independently moving, turret-like eyes allow chameleons to look all around without even moving their heads. By capturing their prey without moving, staying completely concealed in vegetation, chameleons have eliminated the riskiest aspect of sit-and-wait ambush foraging—pursuit movements.

Jaw prehension of prey freed the tongue from its role in prey capture, permitting it to evolve to become a specialized chemical sampler, carrying non-airborne chemical particles into the mouth to be received and deciphered by the vomeronasal system. The vomeronasal system was present in squamate ancestors, but it remained relatively rudimentary in most iguanians. In scleroglossans, the foretongue became specialized for protrusion and for picking up and transporting chemical signals, and the skull became less robust and more flexible than in iguanians. The temporal region is less broad, owing to reduction or loss of the upper temporal bar. Additional points of potential flexibility arose in scleroglossan skulls, a condition known as "cranial kinesis," or mesokinesis. This allows the muzzle and upper jaw to flex upward and downward, making the jaws more efficient in capture and

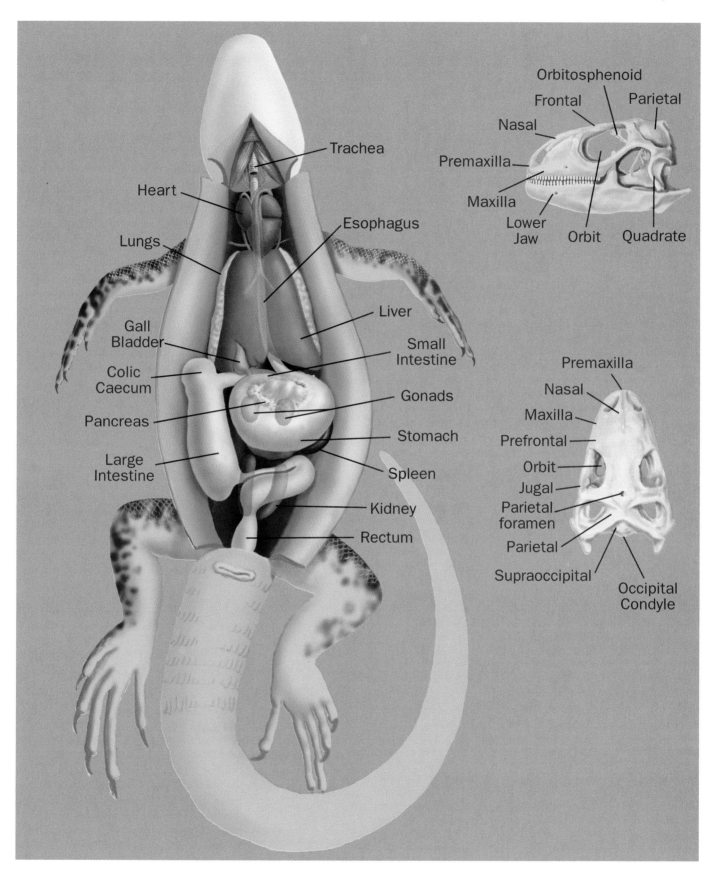

Lizard skull, internal organs, and digestive system. (Illustration by John Megahan)

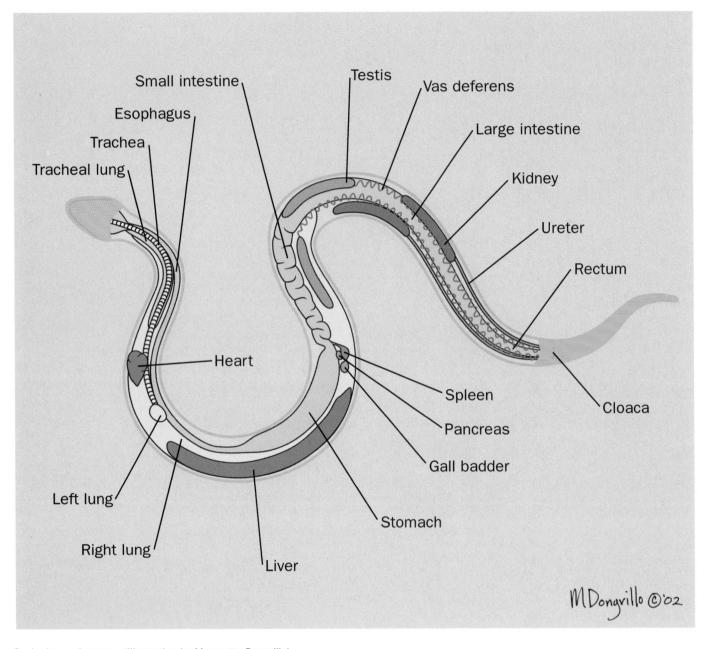

Snake internal organs. (Illustration by Marguette Dongvillo)

manipulation of agile prey. Such jaws bend and better conform to prey, enhancing feeding success.

This combination of skull and chemosensory modifications gave scleroglossans access to microhabitats and prey previously unavailable to iguanians and predisposed them to higher activity levels. For example, an ability to detect and discriminate prey chemically gave scleroglossans access to prey that could not be detected visually. No longer limited to prey moving across their field of vision, squamates could now find highly cryptic invertebrates and vertebrates living in crevices, in the ground, or in water. Remaining in one place for long periods of time has a low-energy payoff compared with searching actively through the habitat for hidden and sedentary prey. Active or wide foraging provided these lizards with a competitive advantage and selected for higher levels of activity. Moving about searching for prey is energetically costly and also increases the risk of exposure to potential predators. Alert behavior and rapid response to predators evolved to enable increased activity levels. Widely foraging scleroglossans find and consume more prey calories per unit time than do iguanians. Gekkota evade both competition and predation by being nocturnal, whereas Autarchoglossa evade potential diurnal predators by being exceedingly alert and agile. Elongation of the body and increased jaw flexibility permitted varanoid lizards to swallow large prey and set the stage for the evolution of snakes.

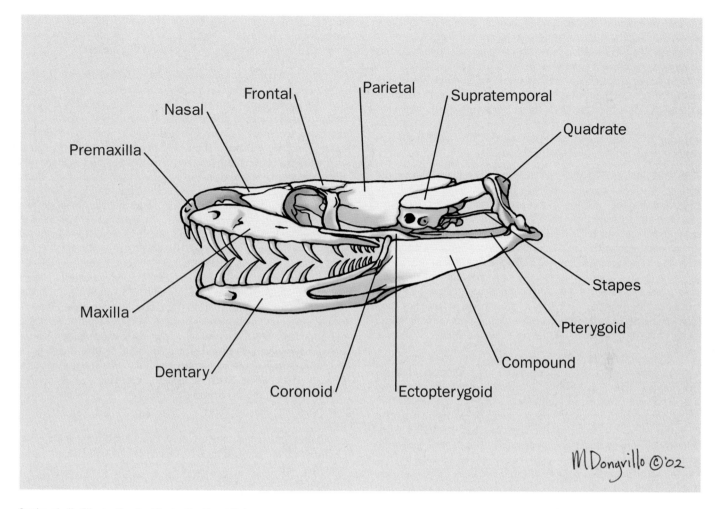

Snake skull. (Illustration by Marguette Dongvillo)

Snakes carried cranial kinesis to an even higher level than did their lizard ancestors, evolving numerous flexible joints in their skulls. Liberation of the mandibular symphysis (the tendons connecting the two lower jaws) set off snake evolution. Unlike lizards, most snakes also have independent movement of bones on the left and right sides of their skulls. Coupled with streptostyly, these adaptations allow snakes to swallow exceedingly large prey. Snake skulls have diversified widely. Snakes lost both temporal arches and apertures, permitting greater independent movements of the head bones. Many snakes have highly flexible jaws and snouts with many joints and considerable cranial kinesis. The musculature of a snake's head is quite complex, allowing for independent movements of cranial bones. When swallowing large prey, snakes "walk" their way down a prey item, first opening one side of their jaws, extending the jawbones forward, biting down, and then repeating the process on the other side.

Owing to lack of limbs, snake diversity is restricted by morphologic features. Nevertheless, snakes have accomplished some rather spectacular things. Some snakes (*Dipsas*) pull snails out of their shells. Snail-eating lizards (*Dracaena*) crush snail shells with molariform teeth. Many lizards are termite specialists. Some, such as certain geckos, catch termites at night when they are active above ground. Others, such as lacertids and teiids, break into termite tunnels during the day. Still others, such as some fossorial skinks, find termites in tunnels and termitaria below ground. All termite-specialized snakes find termites below ground, and many actually spend most of their lives inside termitaria.

Skulls of burrowing snakes are secondarily compacted. Two major sister clades of snakes are Scolecophidia (blindsnakes in the families Leptotyphlopidae, Typhlopidae, and Anomalepididae) and Alethinophidia (all others). Blindsnakes have solid, blunt, and nearly toothless skulls. Considerable variation exists in scolecophidian skulls. Leptotyphlopids manipulate and transport prey with their mandibles (lower jaw), whereas typhlopids and, presumably, anomalepidids rake prey into their mouths with teeth in the upper jaw by rapidly protracting and retracting their maxillae. Mouths of other snakes (Alethinophidia) are filled with dozens of sharp recurved teeth arrayed along several different bones. Snake maxillae vary widely and are movable: hollow hypodermic fangs attached to these bones in viperids swing through almost a full 90° from the folded back, closed-mouth position to the fully erect, stabbing position.

To understand the origin of snakes, one must examine snakelike lizards. Burrowing lizards have small appendages or no limbs at all. They also have no external ear openings, and their eyes often are capped over with a clear spectacle. Ancestral snakes probably were fossorial. Snake eyes have been rebuilt after degenerating during an extensive subterranean existence. All other tetrapods focus by changing the lens curvature using muscles within the eye, but snakes have no such muscles and focus instead by moving the lens back and forth with another set of muscles in the iris.

A rare autarchoglossan lizard from Borneo known as the earless monitor (*Lanthanotus*) has been identified as a likely candidate for the position of sister group to snakes. *Lanthanotus* are cylindrical, long-tailed lizards with long necks and short legs. Like snakes, they have a hinge in the lower jaw and no external ear opening. They have forked tongues and tails that do not regenerate, and they shed their skins in one piece, just like snakes. *Lanthanotus* is the only anguimorphan lizard with a clear brille in the lower eyelid, which could be a precursor to the spectacle of snakes. Other snakelike traits of *Lanthanotus* include a solidly encased brain, loss of the upper temporal arch, and teeth on the palatine and pterygoid bones.

If snake ancestors were subterranean, ancestors of snakes were the most successful among many scleroglossans that experimented with fossoriality. Considering the many times limblessness has arisen in autarchoglossans, why did evolution of limblessness in varanoids set off such an extensive adaptive radiation as that seen in snakes? Varanoid lizards share a combination of characteristics that opened up a unique opportunity for them, compared with other subterranean lizards. Possession of a forked tongue allowed for keen chemosensory discrimination of prey and detection of airborne chemical signals as well as the ability to follow chemical trails by using the deeply forked tongue as an edge detector. Because fossorial lizards tend to be relatively small, a fossorial varanoid (ancestral snake) would probably be small as well. A fossorial varanoid encountering termites could determine what they were and feed on them, and it could trace their chemical trail back to the colony. Other fossorial autarchoglossans might be able to identify termites, but the lack of forked tongues would inhibit their ability to trace prey to the nest.

Evolution of a body small enough to allow movement through termite passageways, along with a correspondingly small head, would permit access to a rich food resource base. Extreme elongation of the trunk is restricted primarily to subterranean autarchoglossans, but none has taken it to the extremes that snakes did. Limbless or nearly limbless terrestrial scleroglossans (*Ophisaurus* and pygopodids) have relatively truncated bodies compared with most snakes. Locomotion through existing passageways would favor a concertina-like movement, which in turn would select for longer bodies (as opposed to longer tails) in these snake ancestors. This set of traits describes fairly accurately the three primitive snake families Typhlopidae, Leptotyphlopidae, and Anomalepididae. Elongation of the body most likely preadapted these reptiles for a return to the surface, where a banquet of large vertebrate prey (amphibians, lizards, birds, and mammals) had diversified. Once on the surface, these snake ancestors underwent selection for increased ability to ingest large prey and evolved larger body sizes. They also evolved a loose mandibular symphysis, which allows the two lower jawbones to spread apart, facilitating ingestion of large prey.

Two evolutionary innovations contributed to the success of snakes above ground, efficient locomotion and their highly derived feeding mechanism. An elongated and very flexible body provides much more trunk control over locomotion. Limbed tetrapods expend considerable energy working against gravity to move their own body mass up and down with each step. Across lizard species, the net cost of locomotion (per gram) decreases linearly with increased body mass. The energetic cost of snake locomotion is much more variable. Snakes using concertina locomotion expend more energy than similarly sized lizards, whereas others, who use sidewinding locomotion, expend much less. Snakes can move quite rapidly, and, using an S-shaped loop in the neck, they can strike quickly to capture prey. After returning to the surface, not only could snakes eat large prey relative to their body and head diameter, they also could move their highly flexible bodies around in a manner that few elongated lizards could. Any crevice, hole, or passageway into which they could get their heads was accessible. Increased numbers of vertebrae and associated musculature facilitated swimming, climbing, and other types of locomotion that were either poorly developed or nonexistent in lizards: rectilinear, concertina, sidewinding, and lateral undulation.

Size

Snakes and lizards vary widely in size, from diminutive to gigantic. The smallest lizards, such as the Australian skink *Menetia*, are among the smallest of terrestrial vertebrates. Neonates have a snout-vent length of only 0.4 in (10 mm) and weigh less than 0.0035 oz (0.1 g), and adults have a snout-vent length of 1 in (25 mm) and a weight of 0.01 oz (0.3 g). Contrast these tiny skinks with Komodo dragons (*Varanus komodoensis*), at 5 ft (1.5 m) in snout-vent length with a weight of up to 154 lb (70 kg). The largest living squamate is the South American green anaconda (*Eunectes murinus*), with a snout-vent length of over 30 ft (10 m) and a weight of more than 330 lb (150 kg). Reticulated pythons are almost as large but not as massive. Both constrictors kill and swallow extremely large prey.

General body form

The ancestral condition was that of a tetrapod with four limbs, each with five toes. Reduced limbs and leglessness have arisen repeatedly among squamates, especially in skinks. Except for some pythons and boas, which possess rudimentary vestigial remnants of hind limbs, all snakes are completely limbless. Elongation of the body or tail generally accompanies limb reduction, as it facilitates locomotion without limbs.

Fossoriality has arisen independently many times among scleroglossans. Chemosensory abilities and narrowing of the skull through loss of the temporal arches preadapted scleroglossan clades to burrowing. Chemoreception allowed them to find and pursue prey underground and also to eliminate potentially noxious prey from their diets, opening up yet another adaptive zone. Ultimately, species in nine scleroglossan families (Pygopodidae, Scincidae, Dibamidae, Amphisbaenidae, Trogonophidae, Rhineuridae, Bipedidae, Gymnop-

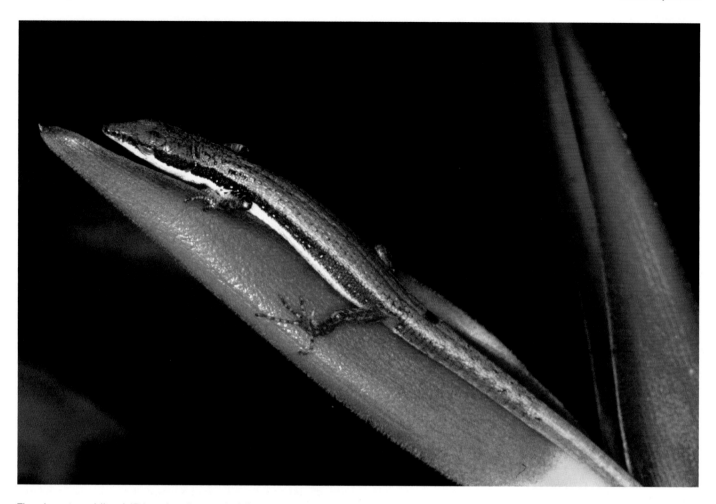

The elegant-eyed lizard (*Prionodactylus argulus*) lives on low vegetation and vines in undisturbed Amazon rainforest where it feeds on small spiders and insects. (Photo by Laurie J. Vitt. Reproduced by permission.)

thalmidae, and Anguidae), and snakes in several families, took maximum advantage of a new underground world. Scleroglossans with strikingly similar body plans swim through sand, burrow in tropical soils, and haunt the nests of social insects.

Coloration

Lizards come in a wide variety of colors, including red, orange, yellow, green, blue, indigo, and violet. Most match the color of substrates on which they live, offering camouflage, which confers some degree of protection from predators. Snakes are equally colorful, with some, such as coral snakes (*Micrurus*), being warningly colored with bright bands of red, yellow, and black.

Distribution

Snakes and lizards are found everywhere in the world, except at very high latitudes, on cold mountaintops, and in the Arctic and Antarctic. At high latitudes and elevations, temperature becomes a limiting factor for animals that rely on external heat sources (ectotherms). Nevertheless, many lizards and snakes have evolved adaptations, such as viviparity (bearing live young), that facilitate living in cold environments.

Biogeography

Throughout the world, most lizard assemblages contain mixtures of iguanians, gekkotans, and autarchoglossans. More diverse squamate faunas tend to have proportionally greater numbers of species of autarchoglossans, whereas less diverse faunas have relatively more iguanians. In assemblages with a substantial number of autarchoglossans, most of the iguanian and gekkotan fauna is arboreal, saxicolous (lives among rocks), nocturnal, or active in the shade. In contrast, where squamate assemblages lack or have few autarchoglossans, such as in North American deserts and high-elevation habitats in South America, iguanians occupy many microhabitats held by autarchoglossans in mixed assemblages elsewhere. Iguanians and gekkotans probably have been displaced by autarchoglossans throughout their evolutionary history, explaining much of their current ecological and geographical distribution. At the same time, the set of traits that provides autarchoglossans with a competitive advantage throughout the world may constrain their ability to persist in the environments and microhabitats dominated by iguanians and gekkotans.

Gekkotan and autarchoglossan lizards are more species rich in the Old World (30% and 51–52%, respectively) than in the New World (16–19% and 31–33%, respectively). Iguanians display the opposite pattern, with considerably fewer species in the Old World (18–19%) than in the New World (49–51%). In the New World, Amazonia and Venezuela have high percentages of autarchoglossans (mostly teiids and gymnophthalmids) and a low percentage of iguanians. Iguanians outnumber scleroglossans in the Caribbean and Central America, and they dominate in Argentina, where there is little scleroglossan diversity, probably because warm seasons are too short to maintain rich autarchoglossan faunas. In the Old World, autarchoglossans are somewhat impoverished in Madagascar, where iguanians and gekkotans are relatively diverse. In South Africa cordylids have reverted to sit-and-wait foraging, possibly owing to a lack of other diurnal sit-and-wait ambush foragers (e.g., agamids) and competition with other actively foraging lizards (scincids, gerrhosaurids, and varanids). Regional trends are even more pronounced at a local level when lizards from particular study sites are considered. Iguania constitute 74% of the saurofauna at twelve New World desert study sites in the Great Basin, Mojave, and Sonoran deserts, but only 8% and 18% at Old World desert study sites in Africa and Australia, respectively.

Habitat

Most natural habitats, including tropical and subtropical islands, support a diversity of squamates. Squamates are terrestrial, arboreal, fossorial, saxicolous, aquatic (both freshwater and marine), diurnal, crepuscular (active at twilight), and nocturnal. They live in deserts, grasslands, chaparral, thornscrub, savannas, forests, and rainforests. Their body plans and lifestyles predispose them to being especially diverse in open, warm, semiarid areas.

Behavior

The shift in feeding behavior from catching prey with the tongue to jaw prehension had numerous ramifications, ultimately leading to the scleroglossan suite of innovations: enhanced chemosensory ability, active foraging, high active body temperatures, selection of high-payoff food, an enhanced ability to find hidden and sedentary prey, and an improved capacity to capture agile prey. As sit-and-wait ambush foragers, iguanians find mobile prey visually. High numbers of ants, insect larvae, grasshoppers, spiders, beetles, and other hymenopterans in their diets suggest that they sample somewhat randomly among arthropods in their immediate microenvironments. Dietary specialization on ants has occurred several times among iguanians, and, in a few cases, entire clades were generated. For example, all species of horned lizards in the North American iguanian genus *Phrynosoma* are ant specialists, suggesting that ant specialization evolved early in the evolutionary history of this clade and was carried through to all present day descendants.

Many insect larvae, eusocial termites, and other nonmobile arthropods escaped detection by iguanians but could not evade scleroglossans. Access to these resources allowed explosive diversification within the Scleroglossa. In contrast to iguanians, scleroglossans are active foragers (with a few exceptions) with keen chemosensory systems that can add nonmobile prey to

their diets. Use of chemical cues by scleroglossans to discriminate prey also facilitates avoidance of noxious prey items. Dietary differences between Iguania and Scleroglossa are subtle, but some abundant prey (ants, other hymenopterans, and beetles) eaten by iguanians are underrepresented in scleroglossan diets. These prey often contain noxious chemicals (particularly alkaloids) and may be discriminated against based on chemical signals detectable by scleroglossans but not by most iguanians. Because alkaloids are metabolic toxins, avoidance of them may have opened up new metabolic opportunities for scleroglossans, allowing for higher activity levels as well as prolonged activity at high body temperatures.

Just as iguanians took lingual feeding to its logical end point, autarchoglossans took jaw prehension and chemoreception to their logical extremes in varanid lizards and snakes. Having evolved superior chemosensory abilities, autarchoglossans became fierce competitors and awesome predators of their more primitive relatives, iguanians and gekkotans. Many gekkotans escaped from autarchoglossans and other diurnal predators by becoming nocturnal. Switching to a nighttime existence, geckos found an unexploited virtual cornucopia of nocturnal insects, such as crickets, moths, and spiders. To avoid autarchoglossans, iguanians became arboreal, shifted to shady microhabitats, or moved up into colder habitats at higher elevations. Some became herbivorous and evolved large body size (iguanines and leiolepidines).

Herbivory evolved several times within Iguania, producing the subfamilies Iguaninae and Liolaeminae within Iguanidae and the subfamily Leiolepidinae within Agamidae. Most of these herbivores are larger in body size compared with their carnivorous relatives. Herbivory either released these iguanians from body size constraints associated with reliance on arthropod prey or drove the evolution of large body size, perhaps as an antipredator tactic—these are the largest iguanians. These herbivorous lizards shifted their foraging behavior, becoming grazers, and enhanced their chemosensory abilities, using the tongue-vomeronasal system to detect chemical signals. Numerous other correlates of herbivory developed, including an enlarged fermentation chamber in the gut and use of microorganisms for digestion of cellulose.

Other iguanians diversified, maintaining rudimentary vomerolfaction, relatively small size, crypsis, sit-and-wait foraging, and relatively low activity levels while subsisting on a wide variety of arthropods. Most insectivorous iguanians eat some ants, and ant specialization has occurred several times. Herbivory also has evolved several times within Scleroglossa, with similar results. Avoidance of plants containing noxious chemicals could have been a driving force behind evolution of chemosensory food discrimination in these lizards. Remaining scleroglossans were dominated by carnivorous, actively foraging clades, although numerous evolutionary reversals took place. Cordylids and some snakes, for example, armed with the scleroglossan arsenal of innovations, reverted to ambush foraging. Evolutionary reversals in diet and foraging modes occurred in cordylids and xenosaurids, along with the associated loss of ability to discriminate prey chemically. Such reversion back to sit-and-wait ambush foraging demonstrates the attractiveness of low-energy requirements and camouflage offered by the iguanian lifestyle.

A bearded dragon with young (*Pogona vitticeps*) in Australia. (Photo by Animals Animals ©Zig Leszczynski. Reproduced by permission.)

Like their ancestors, snakes rely heavily on chemosensory cues to locate prey. Not all snakes are active foragers, however; boas, pythons, and vipers have reverted to the iguanian sit-and-wait mode of ambush foraging but armed with a keen chemosensory ability. Two snake subfamilies have evolved infrared receptors ("pits") wired to the optical receptor region of their brains, allowing them virtually to "see" endothermic prey in the dark. Similarly to other snakes, these predatory snakes exploit their sophisticated vomeronasal chemosensory system to locate scent trails and find ideal sites for ambush attacks.

Many snakes are larger than most lizards, and many are dietary specialists. Most eat various vertebrate prey, including fishes, amphibians, lizards, birds, mammals, and even other snakes. Like most lizards, a few snakes consume arthropods, including ants, termites, spiders, centipedes, and scorpions. Some snakes have specialized in other invertebrates, such as earthworms, slugs, and snails. Snake skull morphologic characteristics and dentitions have evolved along many different pathways, each presumably adapting its bearer to efficient exploitation of its own particular prey. Diets of various species of snakes are restricted to amphibian and reptilian eggs, avian eggs, snails, frogs, toads, lizards, other snakes, birds, and mammals. Many snakes will not eat anything outside their own particular prey category. As examples, hog-nosed snakes (*Heterodon*) eat only toads, mussuranas (*Clelia*) eat mostly other snakes, and several snake species (*Liophidium*, *Scaphiodontophis*, and *Sibynophis*) feed almost exclusively on scincid lizards.

Skinks have bony plates called osteoderms embedded within their scales, which overlap in the manner of shingles on a roof, providing a sort of armor. Most have smooth scales and are difficult to grasp and hold on to, especially when they are squirming. Nevertheless, some species of snakes have specialized in skinks as prey items. Several of these skink specialists have evolved hinged teeth that fold back when they encounter an osteoderm but ratchet upright between scales, offering a firm purchase. One clade of gekkotan lizards, Pygopodidae, has converged on the limbless snake body plan. Pygopodids are known as flap-footed lizards because they have no forelimbs and greatly reduced hind limbs. Two species in one genus of pygopodids, *Lialis*, feed largely on skinks and have independently evolved hinged teeth.

Many snakes kill their prey by constriction, which requires short vertebrae; heavy, supple bodies; and slow movements. Very fast snakes, such as cobras and racers, have elongated vertebrae with musculature extending considerable distances between vertebrae; such snakes are slender and not as supple and seldom can constrict their prey. Another potent solution

A young green python (*Morelia viridis*) feeding on a wild mouse. While adults are bright green, young may be yellow or red. (Photo by Karl H. Switak/Photo Researchers, Inc. Reproduced by permission.)

to prey capture, used by about 20% of snakes (the inspiration for the hypodermic needle), is envenomation, which has evolved repeatedly among snakes. Some snake venoms are actually powerful protein enzymes, which begin digesting a prey item even before the snake swallows it. Injecting venom into a large and potentially dangerous prey and then releasing it to run away and die elsewhere protects a snake from being injured by its prey. Using their keen vomeronasal sensory systems, snakes can follow the trail left by the departing envenomated prey with considerable accuracy to find the dead and partially digested food item. Snakes, monitor lizards, and large teiids use their hydrostatic, long, forked tongues as edge detectors to follow scent trails.

Rear-fanged snakes (opistoglyphs) are thought to have a primitive condition—their fangs are too far back in their mouths for efficient delivery of venom. Such snakes have to chew to inject venom. The family Elapidae, which includes coral snakes and cobras, has permanently erect short fangs (proteroglyph) in the front of the mouth; they also must chew to inject venom. Vipers and pitvipers have by far the most efficient means of injecting venom deep into their prey. They

have long, hollow front fangs attached to the maxillary bone, which hinges backward when a snake closes its mouth but swings forward as the mouth is opened. Fangs in such solenoglyph snakes swing through an arc of 90° from the resting position to the fully erect stabbing position. With use, fangs fall out and are ingested while embedded in prey items, but they are replaced quickly. (A venomous solenoglyph snake has a set of replacement fangs in the roof of its mouth.)

Reproductive biology

The ancestral condition was egg laying, but live bearing has arisen repeatedly among squamates in at least nine different families of lizards as well as among many snakes—many times within some genera. Nearly 20% of all lizard species are viviparous, representing at least 56 independent origins. At least 35 additional origins exist among snakes. In a few species, some populations are oviparous, and others are viviparous. Viviparity arises via egg retention. A female that can "hold" her eggs can bask, warming them and enhancing development as well as protecting them from nest predators. Live bearing and egg retention have allowed squamates to invade cooler regions and to live at higher elevations and higher latitudes. Live bearing allowed transcontinental migrations of some squamates across cold, high-latitude land bridges. Examples include New World natricine snakes and Boinae, among others. Several oviparous squamates (some anguids, skinks, and a few snakes) "guard" or attend nests, protecting developing eggs; a few, including skinks in the genus *Eumeces* and snakes in the subfamily Pythoninae, enhance development by providing water or heat to developing eggs.

Within Iguania, viviparity clearly arose in connection with invasion of cold habitats. Iguanian eggs simply are held in the oviducts until they hatch and neonates are laid or extruded (ovoviviparity). The most complex known squamate placenta is found in the South American skink, *Mabuya heathi*. Within scleroglossans, viviparity has not arisen in teiids, varanids, or helodermatids, and it occurs in only one species of lacertid. An ability to seek out good nest sites, thereby increasing juvenile survivorship, could be an alternative to viviparity. Carrying offspring for long periods of time, as is necessary in viviparous species, probably increases predation risk to females, because gravid females burdened with embryos cannot run as fast or escape as well. Nearly all cordylids are viviparous, as are many anguids and skinks. Cordylids as a group are ecologically more similar to iguanians than to other autarchoglossans, and evolution of viviparity may reflect a long history of high levels of predation or predictable predation on nests of their ancestors, likely from other autarchoglossans. Complex placentas, as seen in some skinks and all xantusiids, do not occur within Iguania.

Because iguanians rely on crypsis to evade predators, filling their body cavity with eggs does not negatively affect detectability; hence, many iguanians, such as ctenosaurs and horned lizards, produce massive clutches and are extremely fecund. At the low end of relative clutch mass are those of anoles, unusual iguanians that lay a single egg at one time but produce many clutches. The single evolutionary event that resulted in the small and genetically fixed clutch size of the more than 300 *Anolis* species may have been related to arboreality. Alternatively, high numbers of deaths in the nest could have driven small clutch sizes, providing an advantage to individuals that distributed their eggs in time and space (bet hedging). Nevertheless, production of a single egg does not interfere with crypsis provided by the elongated, twiglike morphologic features of *Anolis*. For unknown reasons, probably historical, all members of the clade Gekkota have a fixed clutch size of one or two. This, too, may reflect evolution of smaller clutch size associated with mass-related maneuverability on vertical surfaces in a gekkotan ancestor. As a consequence of their more active lifestyle, autarchoglossans typically are streamlined; there are no horned lizard counterparts, although some cordylids that have reverted to ambush foraging approach a spiny tanklike body form. As a consequence, relative clutch mass is constrained and usually smaller, on average, than in non-*Anolis* iguanians.

Reduced clutch volume in scleroglossans and its attendant reproductive consequence (lower investment per reproductive episode) might at first glance appear to be costly. Iguanians and autarchoglossans use space in fundamentally different ways. Iguanians are territorial and live most of their lives in a relatively small area. Clutches usually are deposited within their territories or nearby; thus, nest site selection is limited by iguanian behavior. Moreover, chemical cues play a minor role in nest selection, because most iguanian species have poorly developed vomerolfaction systems. In contrast, most autarchoglossans are not constrained in their use of space by terroriality, and they have well-developed chemosensory systems. Using their superior vomerolfactory abilities, they can seek out and choose the best possible nest sites. Many teiid nests in exposed sandy areas along streams and skink (*Eumeces*) nests in rotting logs are examples. Both *Tupinambis* and *Varanus* deposit clutches in termite nests, which provide heat, humidity, and protection. A survey of squamates using social-insect nests for egg-deposition sites found no iguanians; however, scleroglossans in five lizard families and three snake families use them regularly. Among iguanians, communal nesting occurs in a few iguanines, an herbivorous clade with well-developed chemosensory abilities, and a few species in which limited nest sites appear to be the foci of territories. Thus, scleroglossans could have an advantage in nest site selection that offsets any cost resulting from reduced clutch volume.

Because of dramatic differences in the use of space between iguanians (territorial) and scleroglossans, particularly autarchoglossans (nonterritorial and free ranging), social systems are drastically different. Iguanians typically have polygynous mating systems centered on defendable resources in which social signals are visual. Autarchoglossans typically have polygynous mating systems centered on sequential female defense in which a combination of visual and chemical cues predominate as social signals. A few have monogamous mating systems. Among cordylids that have reverted to ambush foraging and territoriality, the social system has also switched to resource defense polygyny.

Parthenogenesis (reproduction without fertilization) has evolved many times among lizards and in at least one snake. In such all-female unisexual "species," a female lays eggs that develop into exact genetic clones of herself. Many such parthenoforms have arisen via hybridization of sexual parental species. Because no energy is invested in making males, parthenogenetic forms have a much faster rate of increase than do sexual species. Without sexual reproduction, however, they cannot evolve. If heterozygosity itself confers fitness, asexual reproduction maintains existing heterozygosity (acquired in hybridization), whereas in sexual reproduction recombination will disrupt heterozygosity.

Conservation status

Squamate reptiles have suffered substantial habitat loss due to the extensive encroachment of humans into almost all natural biomes. Some species of lizards and snakes have been negatively affected much more severely than others. Many species that live on islands, such as those in the Caribbean, Southeast Asia, and Madagascar, are now endangered, often because humans have introduced competitors such as goats or predators that include rats, cats, dogs, and mongoose. Some species of endemic Caribbean land iguanas (genus *Cyclura*) are perilously close to extinction. Other species, such as those that dwell in extensive desert regions, have been more fortunate because humans have not yet been able to turn arid areas into green fields of crops. However, flat-tailed horned lizards (*Phrynosoma mcalli*) and fringe-toed lizards (*Uma inornata*) in the U.S. Southwest are threatened by habitat fragmentation and loss due to development. Biodiversity in tropical regions is high, but humans are rapidly destroying rainforests and other tropical habitats. Some species of Mexican cloud forest anguid lizards (*Abronia*) may well have gone extinct before they were ever collected or described.

Resources

Books

Arnold, E. N. "Cranial Kinesis in Lizards: Variations, Uses, and Origins." Vol. 30, *Evolutionary Biology*, edited by Max K. Hecht, Ross J. MacIntyre, and Michael T. Clegg. New York: Plenum Press, 1998.

Estes, R. "The Fossil Record and Early Distribution of Lizards." In *Advances in Herpetology and Evolutionary Biology*, edited by G. J. Rhodin and K. Miyata. Cambridge: Museum of Comparative Zoology, Harvard University, 1983.

Estes, R., K. de Queiroz, and J. Gauthier. "Phylogenetic Relationships Within Squamata." In *Phylogenetic Relationships of the Lizard Families*, edited by R. Estes and G. Pregill. Stanford, CA: Stanford University Press, 1988.

Greene, Harry W. *Snakes: The Evolution of Mystery in Nature.* Berkeley: University of California Press, 1997.

Pianka, E. R. *Ecology and Natural History of Desert Lizards: Analyses of the Ecological Niche and Community Structure.* Princeton, NJ: Princeton University Press, 1986.

Resources

Pianka, E. R., and L. J. Vitt. *Lizards: Windows to the Evolution of Diversity.* Berkeley: University of California Press, 2003.

Schwenk, K. "Feeding in Lepidosaurs. In *Feeding: Form, Function, and Evolution in Tetrapod Vertebrates,* edited by K. Schwenk. San Diego: Academic Press, 2000.

Zug, George R., Laurie J. Vitt, and Janalee P. Caldwell. *Herpetology: An Introductory Biology of Amphibians and Reptiles.* 2nd edition. San Diego: Academic Press, 2001.

Periodicals

Autumn, K., Y. A. Liang, S. T. Hsieh, W. Zesch, W. P. Chan, T. W. Kenny, R. Fearing, and R. J. Full. "Adhesive Force of a Single Gecko Foot-Hair." *Nature* 405 (2000): 681–685.

Gans, C. "The Feeding Mechanism of Snakes and Its Possible Evolution." *American Zoologist* 1 (1961): 217–227.

Huey, R. B. "Egg Retention in Some High Altitude Anolis Lizards." *Copeia* 1977 (1977): 373–375.

Huey, R. B., and E. R. Pianka. "Ecological Consequences of Foraging Mode." *Ecology* 62 (1981): 991–999.

Huey, R. B., and M. Slatkin. "Cost and Benefits of Lizard Thermoregulation." *Quarterly Review of Biology* 51 (1976): 363–384.

McDowell, S., and C. Bogert. "The Systematic Position of *Lanthanotus* and the Affinities of the Anguimorphan Lizards." *Bulletin of the American Museum of Natural History* 105, no. 1 (1954): 1–142.

Patchell, F. C., and R. Shine. "Feeding Mechanisms in Pygopodid Lizards: How Can *Lialis* Swallow Such Large Prey?" *Journal of Herpetology* 20 (1986): 59–64.

Savitsky, A. H. "Hinged Teeth in Snakes: An Adaptation for Swallowing Hard-Bodied Prey." *Science* 212 (1981): 346–349.

Schwenk, K. "Why Snakes Have Forked Tongues." *Science* 263 (1994): 1573–1577.

Smith, K. K. "Mechanical Significance of Streptostyly in Lizards." *Nature* 283 (1980): 778–779.

Eric R. Pianka, PhD

Angleheads, calotes, dragon lizards, and relatives
(*Agamidae*)

Class Reptilia
Order Squamata
Suborder Sauria
Family Agamidae

Thumbnail description
Small to large lizards

Size
1.6–13.8 in (40–350 mm) in snout-vent length

Number of genera, species
52 genera; about 420 species

Habitat
Versatile

Conservation status
Endangered: 2 species; Vulnerable: 1 species;
Data Deficient: 2 species

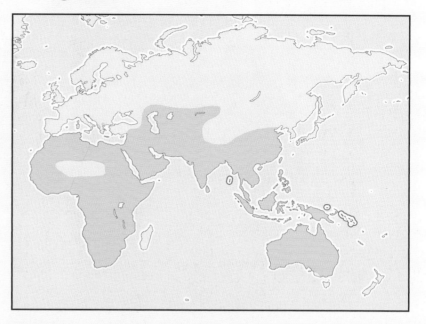

Distribution
Europe, Africa, Asia, and Australia

Evolution and systematics

Agamids are derived descendents of ancestors of New World Iguanidae. They are Old World ecological counterparts of iguanids, with numerous highly convergent ecological equivalents, such as *Phrynosoma* and *Moloch*, *Hydrosaurus* and *Basiliscus*, *Ctenosaura* and *Uromastyx*, *Pogona* and *Agama*, and *Corytophanes* and *Acanthosaura*. A unique shared derived feature that ties Agamidae to Chamaeleonidae (chameleons are derived from within agamids) is acrodont dentition, in which teeth are fused to the top of the jawbones and are not replaced after they are formed. As a lizard grows, new teeth are added posteriorly. Agamids also have caniform (sometimes fanglike) pleurodont teeth set in sockets anteriorly, which are replaced continuously. Two subfamilies are recognized:

Agaminae

These are small to large terrestrial and arboreal lizards distributed in Africa, Asia, and Australia. There are 50 genera with more than 400 species.

Leiolepidinae

These are medium to large terrestrial lizards found in northern Africa east to Southeast Asia. There are two genera with 21 species.

Physical characteristics

Agamids range from commonplace to bizarre to spectacular. They are small, medium, and large diurnal terrestrial

The flying lizard (*Draco volans*) has folds of skin that enable it to glide through the air. (Photo by Stephen Dalton/Photo Researchers, Inc. Reproduced by permission.)

Frilled lizard display. (Illustration by Wendy Baker)

lizards. (Some are arboreal or saxicolous.) None is fossorial. All agamids have movable eyelids, and all have legs. Scales are irregularly shaped and are rough, spiny, or keeled in many but not all species. Enlarged scales are scattered across the dorsal surface in many species, and some have dorsal crests of enlarged scales along their spines. Tails do not regenerate. Many agamids possess intrascalar pores. Head scales are usually smooth and are seldom enlarged. The external ear opening and tympanic membrane is conspicuous in most agamids (with the exception of some Australian *Tympanocryptis* and Indian and Sri Lankan *Otocryptis*). Many agamids exhibit striking sexual dimorphism in body size as well as dichromatism (sexual dimorphism in color patterns).

Distribution

Agamids are found in Europe, Africa, Asia, throughout Southeast Asia, including Indonesia and the Philippines, New Guinea, the Solomon Islands, and Australia. On Madagascar, they are replaced by oplurines.

Habitat

Agamids thrive in both sandy and rocky deserts. They also occur in savannas and in thornscrub and dry forest habitats.

Flying lizards (*Draco* spp.) glide from tree to tree using patagia—wing-like structures formed as they extend their ribs outward. (Illustration by Bruce Worden)

Behavior

The low-energy, sedentary lifestyle of agamids allows them to get by on small amounts of food. Agamids rely on camouflage to escape the attention of potential predators.

Feeding ecology and diet

Agamids are visual, sit-and-wait, ambush predators, and, as a consequence, they encounter only relatively mobile prey. Most are insectivorous, but a few eat plants. Except for the few herbivorous species, agamids take prey items into their mouths using their sticky tongues.

Reproductive biology

Males of most species are territorial, defending a space that includes the home ranges or territories of several females. Under such "resource defense polygyny," a male mates with most females residing within his territory. Males typically engage in territorial displays with other males, either head bobbing or using an extendable dewlap.

All agamids lay eggs, except for a few species of high-latitude northern Eurasian *Phrynocephalus*, which are live-bearers. (*Cophotis* may also be a live-bearer, but this has not been confirmed.) Small species have small clutch sizes, but some larger species lay more than a dozen eggs. Some agamids lay several clutches of eggs during a growing season.

Conservation status

Most agamids are not threatened. The IUCN lists only five species as being in trouble: two are Endangered, one is Vulnerable, and two are listed as Data Deficient. Both the endangered species (*Calotes liocephalus* and *Ceratophora tennentii*) are endemic to Sri Lanka. Several other Sri Lankan endemic agamid species have suffered habitat loss.

Significance to humans

Agamids are major insectivores in many different habitats. Some species are kept as pets in terraria, and others are eaten by humans.

1. Long-snouted dragon (*Lophognathus longirostris*); 2. Bearded dragon (*Pogona minor*); 3. Thorny devil (*Moloch horridus*); 4. Frilled lizard (*Chlamydosaurus kingi*); 5. Military dragon (*Ctenophorus isolepis*); 6. Central netted dragon (*Ctenophorus inermis*). (Illustration by Amanda Humphrey)

1. Toad-headed agama (*Phrynocephalus mystaceus*); 2. Brown garden lizard (*Calotes versicolor*); 3. Spiny-tailed agama (*Uromastyx acanthinurus*); 4. Earless dragon (*Tympanocryptis cephalus*); 5. Spiny agama (*Agama hispida*); 6. Flying lizard (*Draco volans*); 7. Sailfin lizard (*Hydrosaurus amboinensis*); 8. Butterfly agama (*Leiolepis belliana*); 9. Leaf-horned agama (*Ceratophora tennentii*). (Illustration by Amanda Humphrey)

Species accounts

Spiny agama
Agama hispida

SUBFAMILY
Agaminae

TAXONOMY
Agama hispida Bocage, 1896, Angola. Two subspecies are recognized.

OTHER COMMON NAMES
English: Desert agama, ground agama.

PHYSICAL CHARACTERISTICS
Spiny agamas are medium-size, usually terrestrial lizards; however, they do climb up perches, such as fence posts. Males have a blue head, a bright red nape, and yellow shoulders. Females are drab, with orange, brown, and cream splotches. These lizards possess two fanglike canine teeth large enough to draw blood, which probably are used to pierce the hard elytra of beetle prey.

DISTRIBUTION
The species occurs in western Cape Horn and adjacent regions of southern Africa, including most of the Namib and Kalahari Deserts in Namibia, South Africa, and southern Botswana.

HABITAT
They inhabit interdunal streets in arid semidesert, open sandy veld, salt pans, and coastal sand dunes.

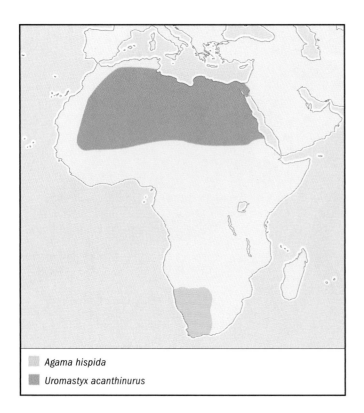

▨ *Agama hispida*
▪ *Uromastyx acanthinurus*

BEHAVIOR
These lizards dig short tunnels at the base of bushes. They do not live in colonies but are solitary.

FEEDING ECOLOGY AND DIET
Spiny agamas are sit-and-wait ambush predators. Their diet consists mostly of ants, beetles, and termites. Small amounts of plant foods also are eaten

REPRODUCTIVE BIOLOGY
During the breeding season males are colorful, with blue heads, bright red napes, and yellow shoulders. Males defend territories and mate with several females that reside inside the territory. Females lay large numbers of fairly small eggs. The average size of 45 clutches was reported as 13.4 eggs.

CONSERVATION STATUS
Not threatened.

SIGNIFICANCE TO HUMANS
Local myth has it that spiny agamas climb trees or posts to scan the horizon for rain. Locals also warn that these lizards are venomous, saying that agamas do not make their own poison but rather obtain venom by "milking" cobras. When questioned further, Afrikaaners explain that the lizards are often seen with their heads inside a cobra's mouth, extracting cobra venom. Of course, a more likely and simpler alternative explanation for such an observation is simply that cobras eat these lizards! ◆

Brown garden lizard
Calotes versicolor

SUBFAMILY
Agaminae

TAXONOMY
Calotes versicolor Daudin, 1802, Pondicherry, India.

OTHER COMMON NAMES
English: Beauty lizard, bloodsucker, Indian garden lizard.

PHYSICAL CHARACTERISTICS
The brown garden lizard is a medium-size, brownish lizard with a laterally flattened body and a crest of scales on the neck and partway down the back.

DISTRIBUTION
The species occurs from eastern Iran to southern China and south to the Maldives and Sumatra.

HABITAT
They live in open habitats, such as light, sun-drenched forests.

BEHAVIOR
Brown garden lizards are agile climbers and are adept at hiding behind branches. They follow humans and thrive in open parks and gardens and on date palms.

FEEDING ECOLOGY AND DIET
These sit-and-wait ambush predators feed largely on insects, but they also eat other, smaller lizards.

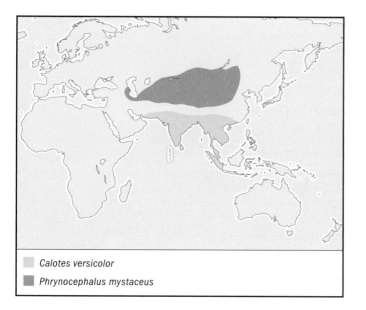

Calotes versicolor
Phrynocephalus mystaceus

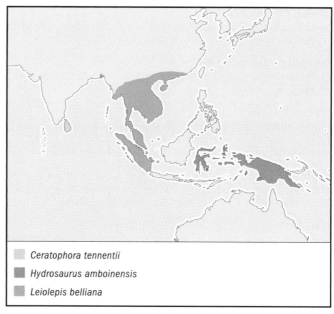

Ceratophora tennentii
Hydrosaurus amboinensis
Leiolepis belliana

REPRODUCTIVE BIOLOGY

Brown garden lizards sometimes are called "bloodsuckers," because males have bright red heads during the breeding season, just before the rainy season. They staunchly defend their territories against other males. They nod their heads and extend the gular pouch in a threat posture toward other males. Two males first watch each other from a distance and then suddenly walk straight toward each other. Fighting males stand erect on their hind legs and tails, grasping each other with their front legs and trying to bite their opponents. If one male does not back away, a serious biting fight ensues. Males court females using pushups and head-bobbing displays. Females lay from one to 25 eggs in the middle of the rainy season. Hatchlings mature at nine to 12 months.

CONSERVATION STATUS

Not threatened.

SIGNIFICANCE TO HUMANS

These lizards have lived in close association with humans for centuries and are kept in captivity in terraria. ◆

Leaf-horned agama

Ceratophora tennentii

SUBFAMILY

Agaminae

TAXONOMY

Ceratophora tennentii Gunther, 1861, Sri Lanka.

OTHER COMMON NAMES

None known.

PHYSICAL CHARACTERISTICS

These small, tropical, arboreal agamids have evolved unusual appendages on their snouts. (*Ceratophora* means "horn carrying.") Horns are larger in males than in females (in some other species, horns are entirely missing among females). The rostral appendage of the leaf-horned agama is covered with scales and shaped like a leaf, flattened from side to side and coming to a point.

DISTRIBUTION

This lizard inhabits the Knuckles Mountains, Sri Lanka.

HABITAT

The leaf-horned agama occurs in cloud forests between 2,950 ft (900 m) and 3,940 ft (1,200 m) in elevation.

BEHAVIOR

These arboreal lizards live in forests, where they frequent the lower branches of trees. These slow-moving chameleon-like lizards rely on camouflage to evade enemies.

FEEDING ECOLOGY AND DIET

These lizards are sit-and-wait ambush predators, eating insects that move past their perches. Occasionally, they jump down to the ground to catch an insect.

REPRODUCTIVE BIOLOGY

Males expand an erectile sail-like crest on the back of the neck and back when displaying. The snout appendage probably also is used in sexual displays.

CONSERVATION STATUS

The leaf-horned agama has lost much of its habitat to human encroachment and logging. They have a very small geographic range and now are officially listed as Endangered by the IUCN.

SIGNIFICANCE TO HUMANS

None known. ◆

Frilled lizard

Chlamydosaurus kingii

SUBFAMILY

Agaminae

TAXONOMY

Chlamydosaurus kingii Gray, 1825, Port Nelson, northwestern coast of Australia.

OTHER COMMON NAMES
English: Frill-necked lizard, cloaked lizard; Australian aboriginal dialect: Bemmung.

PHYSICAL CHARACTERISTICS
These are large, pale or dark gray lizards. Juveniles sometimes are reddish.

DISTRIBUTION
The species occurs in all of tropical northern Australia, down the east coast as far as Brisbane, Queensland. Cogger (1992) says that they are "extra-limital" in southern New Guinea.

HABITAT
The frilled lizard inhabits semihumid grassy woodlands. These arboreal lizards are seldom found very far from trees.

BEHAVIOR
The species spends most of its time on tree trunks, descending to the ground after rain. When threatened, these lizards erect a large, reddish, fanlike frill around their necks. Two modified elongated hyoid bones form rods used to expand the frill. Like many long-legged lizards, frilled lizards can run bipedally.

FEEDING ECOLOGY AND DIET
This lizard feeds mostly on invertebrates but also on some small vertebrates.

REPRODUCTIVE BIOLOGY
A clutch of 13 eggs has been recorded.

CONSERVATION STATUS
The species is not threatened at present, but it may be devastated soon by introduced toxic cane toads.

SIGNIFICANCE TO HUMANS
Fire prevention signs along roads in the Northern Territory announce, "We like our lizards frilled, not grilled." ◆

Central netted dragon
Ctenophorus inermis (sometimes called *C. nuchalis*)

SUBFAMILY
Agaminae

TAXONOMY
Ctenophorus inermis De Vis, 1884, Delta Station, Bogantungan, Queensland, Australia.

OTHER COMMON NAMES
None known.

PHYSICAL CHARACTERISTICS
These are medium-size yellowish brown terrestrial desert lizards with relatively short legs, a low crest along the top of the neck, a narrow vertebral stripe, and a blunt snout. Breeding males have orange to reddish heads and throats.

DISTRIBUTION
The species occurs in desert regions of central Australia.

HABITAT
They inhabit red sandy deserts with spinifex grass vegetation.

BEHAVIOR
These lizards dig several shallow dead-end burrows near favored basking sites, to which they retreat when threatened. If disturbed in their burrows, they dash off and hide in another nearby burrow. Body temperatures correlate with ambient air temperatures, averaging about 96.8°F (36°C). When ambient air temperatures are high, the lizards climb up as high as 3.3 ft (1 m) above ground and face directly into the sun. Varanid monitor lizards prey upon central netted dragons.

FEEDING ECOLOGY AND DIET
These lizards are omnivorous. They eat such insects as ants, grasshoppers, beetles, and termites and are partially herbivorous, taking about 25% of their diet by volume in plant food.

REPRODUCTIVE BIOLOGY
Males defend territories during the breeding season in the spring months. Clutch sizes range from two to six eggs, depending on the size of the female; the average is about four eggs.

CONSERVATION STATUS
Not threatened.

SIGNIFICANCE TO HUMANS
None known. ◆

Military dragon
Ctenophorus isolepis

SUBFAMILY
Agaminae

TAXONOMY
Ctenophorus isolepis Fisher, 1881, Nickol Bay, Western Australia. Three subspecies are recognized.

OTHER COMMON NAMES
None known.

PHYSICAL CHARACTERISTICS
These are small reddish or reddish brown terrestrial lizards with long hind legs. Males are more colorful than females,

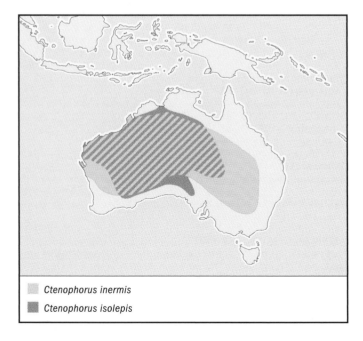

☐ *Ctenophorus inermis*
■ *Ctenophorus isolepis*

with dark black patches on their sides and bellies. Males of the desert subspecies also have yellowish stripes on the sides of their heads and shoulders.

DISTRIBUTION
Military dragons occur in desert regions of central Australia.

HABITAT
The species inhabits red sand plain deserts with spinifex grass vegetation.

BEHAVIOR
In the early morning, these lizards bask in the open sun, but as temperatures climb during midday, they position themselves in the dense shade offered by spinifex grass tussocks. Their active body temperature is about 100°F (37.8°C), but it is lower during winter months and higher in the summer. Lizards are active during midday in winter, but they are most active early and late in the day in the summer. When pursued, they make long, zigzag runs through the open between grass tussocks. If chased until they experience oxygen debt, they seek cover by trying to climb into spinifex tussocks, but their long legs impede their ability to move inside such grasses.

FEEDING ECOLOGY AND DIET
These sit-and-wait ambush predators feed mainly on ants, but they also eat grasshoppers, termites, beetles, and other insects.

REPRODUCTIVE BIOLOGY
The military dragon has two clutches of three to four eggs per year, laid during spring and early summer. Hatchlings mature by the next year.

CONSERVATION STATUS
Not threatened.

SIGNIFICANCE TO HUMANS
None known. ◆

Flying lizard
Draco volans

SUBFAMILY
Agaminae

TAXONOMY
Draco volans Linnaeus, 1758, Java, Indonesia.

OTHER COMMON NAMES
None known.

PHYSICAL CHARACTERISTICS
These are slender, long-legged, small lizards with folding ribs that expand to form a winglike structure. At rest, these dermal sails are folded along the body, giving the lizards a slim appearance.

DISTRIBUTION
The species inhabits the Indonesian islands, including Borneo, Java, Sumatra, Sulawesi, and Timor. They also occur in Thailand, western Malaysia, and the Philippines.

HABITAT
Their habitat is open forests and dense rainforests of both lowlands and highlands.

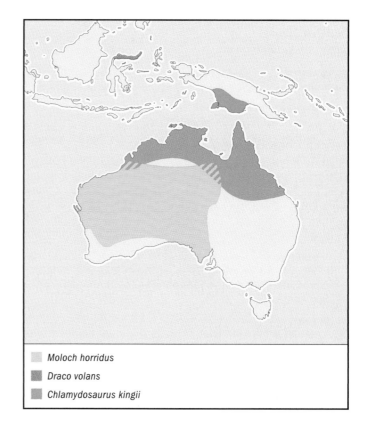

◻ *Moloch horridus*
◼ *Draco volans*
◼ *Chlamydosaurus kingii*

BEHAVIOR
With their "wings" extended, these long-tailed, lightly built agamids glide gracefully between trees, losing altitude along the way. When gliding, these delicate, slender lizards use their tails to steer and sometimes can travel as far as 55 yd (50 m). Expert hang gliders, they rise up and stall at exactly the right moment to make a gentle landing. Upon landing on an adjacent tree, with the head up, they scamper up the tree, gaining elevation in preparation for their next flight. When on the ground, flying lizards are clumsy and vulnerable to predators.

FEEDING ECOLOGY AND DIET
The species feeds almost exclusively on ants and termites.

REPRODUCTIVE BIOLOGY
Wings of males and females are of different colors, which allows these lizards to identify the sex of another at a distance. Males defend territories, courting females by extending their brightly colored throat dewlap appendages, much like anoles do in the New World. Females lay one to four eggs. Their unusual eggs are elongated and spindle-shaped, with dense calcium carbonate "caps" at each end. The function of the caps has not been studied, but they could be deposits for developing embryos.

CONSERVATION STATUS
Not threatened.

SIGNIFICANCE TO HUMANS
None known. ◆

Sailfin lizard
Hydrosaurus amboinensis

SUBFAMILY
Agaminae

TAXONOMY
Hydrosaurus amboinensis Schlosser, 1768.

OTHER COMMON NAMES
English: Soa soa.

PHYSICAL CHARACTERISTICS
These are large, semiaquatic lizards with a pronounced crest on the neck and an enlarged sailfin down the back to the base of the tail. Both sexes have a black and dark green reticulated pattern.

DISTRIBUTION
The sailfin lizard occurs in Southeast Asian islands (Celebes, Moluccas, and New Guinea). Another population occurs in the Philippines.

HABITAT
This species is found in trees in the vicinity of water.

BEHAVIOR
These lizards are expert swimmers and take refuge in the water. Using their fringed toes, juveniles can run across the surface of water. They spend most of their time in branches overhanging water.

FEEDING ECOLOGY AND DIET
These herbivorous lizards harbor intestinal endosymbiotic microbes that produce cellulases that aid in digestion of plant foods.

REPRODUCTIVE BIOLOGY
Crests and sailfins of males are larger than those of females and may be used in courtship displays or in fighting with other males. Females lay three to nine eggs. Hatchlings are about 8 in (20 cm) long, about two-thirds of which is tail.

CONSERVATION STATUS
Not listed by the IUCN. Owing to habitat loss and hunting, however, these lizards are now uncommon and could be threatened.

SIGNIFICANCE TO HUMANS
These large lizards are considered a delicacy. ◆

Long-snouted dragon
Lophognathus longirostris

SUBFAMILY
Agaminae

TAXONOMY
Lophognathus longirostris Boulenger, 1883, Champion Bay, Western Australia.

OTHER COMMON NAMES
English: Long-snouted lashtail.

PHYSICAL CHARACTERISTICS
This species is a medium-size grayish lizard with a long snout and a very long tail, up to three times its snout-vent length,

☐ *Tympanocryptis cephalus*
■ *Pogona minor*
▨ *Lophognathus longirostris*

which is used as a counterbalance in climbing. It has whitish stripes along the side.

DISTRIBUTION
The species occurs in central arid regions of Australia.

HABITAT
This lizard inhabits open savanna woodlands, riparian habitats, and red sandy deserts, usually associated with sand ridges in desert habitats.

BEHAVIOR
These uncommon, large, agile, and fast agamids are never found very far from trees. They can attain speeds of 15 mph (24 km/h) running bipedally on their hind legs.

FEEDING ECOLOGY AND DIET
The species eats wasps, beetles, grasshoppers, mantids, hemipterans, and various insect larvae. Occasionally, they also eat some plant food.

REPRODUCTIVE BIOLOGY
Seven clutches averaged 3.9 eggs. Little else is known about reproduction in this species.

CONSERVATION STATUS
Not threatened.

SIGNIFICANCE TO HUMANS
None known. ◆

Thorny devil
Moloch horridus

SUBFAMILY
Agaminae

TAXONOMY
Moloch horridus Gray, 1841, Western Australia.

OTHER COMMON NAMES
English: Mountain devil.

PHYSICAL CHARACTERISTICS
The thorny devil is a very spiny, moderately sized, reddish and yellowish lizard with a round body and a short tail, about 4–6 in (10–15 cm) long. Adult females are larger and stouter than adult males; they range from 3.1 to 4.3 in (80–110 mm) in snout-vent length and weigh 1.2–3.1 oz (33–88.7 g). Adult males are all less than 3.8 in (96 mm) in snout-vent length and never weigh more than 1.7 oz (49 g).

DISTRIBUTION
The species occurs in the southern section of the Northern Territory, the northern section of South Australia, and in Western Australia.

HABITAT
Thorny devils are found through most of arid inland Australia, particularly on sandy soils, but they seldom occur on stony soils. They prefer two quite different habitats: spinifex sandy plain and the sand ridge deserts of the interior and the mallee belt of southern South Australia and southwestern Western Australia. The geographic distribution of the species corresponds more closely to the distribution of sandy and sandy loam soils than to any climatological field.

BEHAVIOR
Thorny devils display a bimodal seasonal pattern of activity. These lizards move little during the coldest winter months (June and July) or the hottest summer months (January and February). They are active for a three-month Austral autumnal period (March, April, and May) and a five-month period that spans late winter, spring, and early summer (August through December), during which mating and egg deposition take place. During hot summer days, thorny devils are inactive, retreating into shallow underground burrows that they dig for themselves.

These lizards posses a curious knoblike spiny appendage on the backs of their necks, which has been likened to a false head. When threatened, they tuck their real heads down between their forelegs, leaving this false head in the position of the real head. This makes them difficult for most predators to swallow. When disturbed, thorny devils also inflate themselves with air, puffing up like little puffer fish. They can also change color rapidly; when warm and active, they are usually a pale yellow and red. When they are alarmed or when they are cold, however, they turn dark olive drab.

FEEDING ECOLOGY AND DIET
Thorny devils are obligate ant specialists, eating virtually nothing else. They consume several species of ants but are especially partial to very small *Iridomyrmex* ants, especially *Iridomyrmex flavipes*. Large numbers of these tiny ants are eaten per meal by an individual thorny devil (estimates range from 675 to 2,500).

REPRODUCTIVE BIOLOGY
Mating has been observed in the autumn, which suggests that thorny devils may have a mechanism of sperm storage. In contrast to the relatively sedentary summer to autumn existence, thorny devils move over much greater distances during August and September, when most mating takes place. Female thorny devils excavate nest chambers and lay clutches of eggs in September, October, and November. Eggs are laid from mid-September through late December. Only a single clutch is laid per year, and clutch size varies from three to 10, with a mode of eight eggs per clutch. Seven clutches had reported incubation times of 90–132 days.

Hatchlings emerge in January and February, weighing an average of 0.06 oz (1.8 g) and measuring 2.5–2.6 in (63–65 mm) in total length (snout to tail tip). Hatchlings may eat their own egg cases to obtain calcium and other materials to support early growth.

CONSERVATION STATUS
Not threatened.

SIGNIFICANCE TO HUMANS
These lizards are interesting to biologists because they are independently evolved ecological equivalents of North American horned lizards and one of the best examples of convergent evolution. ◆

Bearded dragon
Pogona minor

SUBFAMILY
Agaminae

TAXONOMY
Pogona minor Sternfeld, 1919, Hermannsburg, Northern Territory, Australia.

OTHER COMMON NAMES
None known.

PHYSICAL CHARACTERISTICS
The bearded dragon is a medium-size, grayish to dark gray lizard with splotches of cream.

DISTRIBUTION
It is distributed through the central deserts of Australia.

HABITAT
These lizards occur in a variety of habitats, including red sandy deserts and shrub acacia woodlands.

BEHAVIOR
Bearded dragons are semiarboreal, often using perches as basking sites and vantage points. They have a distinctive throat pouch, which is extended in defensive displays, but they do not appear to defend territories.

FEEDING ECOLOGY AND DIET
These sit-and-wait ambush foragers prey on grasshoppers, beetles, termites, and insect larvae.

REPRODUCTIVE BIOLOGY
The average clutch size of 73 females was reported as 7.6 eggs.

CONSERVATION STATUS
Not threatened.

SIGNIFICANCE TO HUMANS
Bearded dragons are popular in the pet trade and often are kept and bred in captivity. ◆

Earless dragon
Tympanocryptis cephalus

SUBFAMILY
Agaminae

TAXONOMY
Tympanocryptis cephalus Gunther, 1867, Nickol Bay, Western Australia.

OTHER COMMON NAMES
None known.

PHYSICAL CHARACTERISTICS
The earless dragon is a small, roundish lizard that mimics rocks. It may be gray or reddish. There is no external ear opening. Its head is rounded and bulbous, and its tail usually is obscurely banded.

DISTRIBUTION
The species is distributed in rocky, arid regions of the central portion of Western Australia and across the southern Northern Territory, with isolated populations in western Queensland.

HABITAT
Earless dragons are found in rocky habitats, such as the Gibber Desert.

BEHAVIOR
This lizard mimics rocks, relying on camouflage to escape detection.

FEEDING ECOLOGY AND DIET
The species is an insectivorous sit-and-wait ambush predator.

REPRODUCTIVE BIOLOGY
Nothing is known about its reproductive habits.

CONSERVATION STATUS
Not threatened.

SIGNIFICANCE TO HUMANS
None known. ◆

Butterfly agama
Leiolepis belliana

SUBFAMILY
Leiolepidinae

TAXONOMY
Leiolepis belliana Gray, 1827, Penang. Two subspecies are recognized.

OTHER COMMON NAMES
None known.

PHYSICAL CHARACTERISTICS
This is a dorsoventrally flattened, medium-size lizard with a round tail that is about twice as long as the snout-vent length. There are three yellow, dark-bordered dorsal stripes with yellowish ocelli scattered across the back and orange-red and black splotches along the sides. The long hind legs have light yellow speckling on the upper surface. The scientific name *Leiolepis* means "smooth scaled." The bodies of these lizards are covered with small, smooth body scales, although the scales on their tails are keeled.

DISTRIBUTION
The species occurs on the Malaysian peninsula and in Southeast Asia (Laos, Thailand, Vietnam, Burma [Myanmar], Sumatra, and Bangka Island) east to southern China (Hainan Island).

HABITAT
Butterfly agamas inhabit open coastal tropical areas with sandy soils.

BEHAVIOR
These handsome, agile, active lizards dig deep burrows. They are reputed to live in colonies. They are active during midday at high temperatures. Their elongated ribs allow them to flatten their bodies; they are even reputed to be able to parachute and make gliding jumps, although this behavior requires confirmation. It may be related to the origin of their common name, butterfly agama.

FEEDING ECOLOGY AND DIET
Butterfly agamas feed on insects and small crabs in nature. They also eat some plant foods.

REPRODUCTIVE BIOLOGY
Little is known about their reproductive biology, but they do lay eggs.

CONSERVATION STATUS
Not threatened.

SIGNIFICANCE TO HUMANS
These lizards thrive in captivity in large terraria. In Bangkok, Thailand, the species is sold regularly as food for humans. ◆

Toad-headed agama
Phrynocephalus mystaceus

SUBFAMILY
Leiolepidinae

TAXONOMY
Phrynocephalus mystaceus Pallas, 1776, "Arenosis Naryn" and "deserti Comani." Two subspecies are recognized.

OTHER COMMON NAMES
English: Bearded toad head.

PHYSICAL CHARACTERISTICS
These small, brownish terrestrial lizards have large heads and peculiar pouches, or "beards," at the corners of their mouths that are expanded in threat displays.

DISTRIBUTION
The species occurs in extreme western Asia and the northern Caucasus along the Caspian Sea and across northern Iran and northern Afghanistan to central Asia.

HABITAT
It inhabits sand dunes and semideserts with sparse vegetation and also is found on hard soils covered with small pebbles.

BEHAVIOR
These lizards dig burrows in sandy areas between dunes. Fleeing individuals run long distances, up to 66 yd (60 m); pause briefly with the tail rolled up; and then, vibrating their bodies, rapidly sink into loose sand. If threatened, they adopt a defensive stance, raising up on their hind legs, opening their mouths wide, and

hissing and sometimes lunging, with folds of skin spread out at both sides of their mouth. These red folds make the mouth look much wider than it actually is. They rock back and forth on their legs, whirling their tail in a spiral, and jump toward their antagonist. The entire performance is very menacing.

FEEDING ECOLOGY AND DIET
These are sit-and-wait insectivores.

REPRODUCTIVE BIOLOGY
The species reaches sexual maturity at two years. Breeding takes place from late April to early July. Females are slightly larger than males and lay two clutches of two to three eggs. Young hatch in about 70 days.

CONSERVATION STATUS
Not threatened.

SIGNIFICANCE TO HUMANS
The species can be maintained in terraria, where some herpetoculturists have succeeded in keeping them alive for seven to eight years. ◆

Spiny-tailed agama
Uromastyx acanthinurus

SUBFAMILY
Leiolepidinae

TAXONOMY
Uromastyx acanthinurus Merrem, 1820, Egypt. Six subspecies are recognized.

OTHER COMMON NAMES
English: Bell's dab lizard, dob lizard; French: Fouette-queue; German: Veranderlicher Dornschwanz.

PHYSICAL CHARACTERISTICS
These large, flattened lizards have very spiny tails. They are gray, yellowish, or reddish.

DISTRIBUTION
The species is found in the Sahara Desert of northern Africa.

HABITAT
These lizards inhabit rocky slopes in small mountain valleys with rich vegetation.

BEHAVIOR
These large lizards dig extensive burrows, into which they retreat (rock crevices are sometimes used as well) when threatened, blocking off the entrance with their very spiny, sharp, and muscular tails. If pursued, they thrash their tails vigorously from side to side; a few blows from this sharp, spiny club discourages most small enemies. These lizards sometimes are found in small colonies of about 20 individuals.

FEEDING ECOLOGY AND DIET
This lizard forages over large areas, walking a zigzag path. They are vegetarians, eating flower buds of a wide variety of plants. They are especially fond of yellow flowers. In droughts they even subsist by eating wood. The species can survive a fast of an entire year. They have powerful jaws that can shear and crush even hard grains. They also eat insects such as beetles, when they are available.

REPRODUCTIVE BIOLOGY
Courtship takes place from April to June. A complete courtship sequence may take from two hours to an entire day. During courtship the male bites the female's neck and back and curls the base of his tail underneath hers. Females lay two clutches of about 23 eggs in a lateral chamber dug off the main tunnel at a depth of about 23.6 in (60 cm). Hatchlings appear in September or October and have a total length of about 3 in (75 mm). The young eat both insects and flowers. They grow rapidly, reaching 8 in (20 cm) in their first year of life.

CONSERVATION STATUS
Not threatened.

SIGNIFICANCE TO HUMANS
This species is kept in captivity in terraria. ◆

Resources

Books

Cogger, H. G. *Reptiles and Amphibians of Australia*. Ithaca, NY: Cornell University Press, 1992.

Daniel, J. C. *The Book of Indian Reptiles*. Bombay, India: Bombay Natural History Society, Oxford University Press, 1983.

Estes, R. "The Fossil Record and Early Distribution of Lizards." In *Advances in Herpetology and Evolutionary Biology: Essays in Honor of Ernest E. Williams*, edited by A. G. J. Rhodin and K. Miyata. Cambridge, MA: Museum of Comparative Zoology (Harvard University), 1983.

Pianka, E. R. *Ecology and Natural History of Desert Lizards: Analyses of the Ecological Niche and Community Structure*. Princeton, NJ: Princeton University Press, 1986.

Pianka, E. R., and L. J. Vitt. *Lizards: Windows to the Evolution of Diversity*. Berkeley: University of California Press, 2003.

Saville-Kent, W. *The Naturalist in Australia*. London, 1897.

Schleich, H. Hermann, Werner Kästle, and Klaus Kabisch. *Amphibians and Reptiles of North Africa*. Koenigstein, Germany: Koeltz Scientific Books, 1996.

Schmidt, K. P., and R. F. Inger. *Living Reptiles of the World*. Garden City, NY: Hanover House, 1957.

Witten, G. J. "Family Agamidae." In *Fauna of Australia*. Vol. 2A, *Amphibia and Reptilia*, edited by C. J. Gasby, C. J. Ross, and P. L. Beesly. Canberra: Australian Biological and Environmental Survey, 1993.

Zug, George R., Laurie J. Vitt, and Janalee P. Caldwell. *Herpetology: An Introductory Biology of Amphibians and Reptiles*. 2nd edition. San Diego: Academic Press, 2001.

Periodicals

Bentley, P. J., and W. F. Blumer. "Uptake of Water by the Lizard *Moloch horridus*." *Nature* 194 (1962): 699–700.

Bursey, C. R., S. R. Goldberg, and D. N. Woolery. "*Oochoristica piankai* sp. n. (Cestoda: Linstowiidae) and

Resources

Other Helminths of *Moloch horridus* (Sauria: Agamidae) from Australia." *Journal of the Helminthological Society of Washington* 63, no. 2 (1966): 215–221.

Huey, R. B., and E. R. Pianka. "Seasonal Variation in Thermoregulatory Behavior and Body Temperature of Diurnal Kalahari Lizards." *Ecology* 58 (1977): 1066–1075. (With an appendix by J. A. Hoffman.)

Pianka, E. R. "Australia's Thorny Devil." *Reptiles* 5, no. 11 (1997): 14–23.

———. "Comparative Ecology of Two Lizards." *Copeia* (1971): 129–138.

———. "Ecology of the Agamid Lizard *Amphibolurus isolepis* in Western Australia." *Copeia* (1971): 527–536.

Pianka, E. R., and H. D. Pianka. "The Ecology of *Moloch horridus* (Lacertilia: Agamidae) in Western Australia." *Copeia* (1970): 90–103.

Pianka, G. A., E. R. Pianka, and G. G. Thompson. "Egg Laying by Thorny Devils (*Moloch horridus*) Under Natural Conditions in the Great Victoria Desert." *Journal of the Royal Society of Western Australia* 79 (1998): 195–197.

Sporn, C. C. "Additional Observations on the Life History of the Mountain Devil, *Moloch horridus*, in Captivity." *Western Australian Naturalist* 9 (1965): 157–159.

———. "The Breeding of the Mountain Devil in Captivity." *Western Australian Naturalist* 5 (1955): 1–5.

———. "Further Observations on the Mountain Devil in Captivity." *Western Australian Naturalist* 6 (1958): 136–137.

Eric R. Pianka, PhD

Chameleons
(Chamaeleonidae)

Class Reptilia
Order Squamata
Suborder Sauria
Family Chamaeleonidae

Thumbnail description
Very small to large, arboreal, semiterrestrial, and terrestrial reptiles with pincer-like feet, prehensile tails, and long projectile tongues

Size
0.75–28 in (19–711.2 mm)

Number of genera, species
6 genera; 180 species and subspecies

Habitat
Forest, savanna, and desert

Conservation status
Vulnerable: 4 species

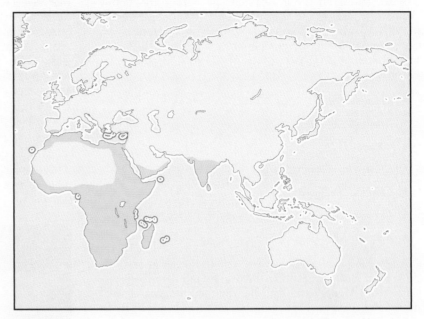

Distribution
Southern Portugal and Spain, Sicily, Malta, southern Greece (southern Peloponnese) Samos, Chios, Crete, southern Turkey, Cyprus, Syria, Lebanon, Jordan, Israel, Sinai Peninsula, Yemen, Saudi Arabia, Iraq, Iran, Pakistan, India, Sri Lanka, Africa and neighboring islands, Fernando Póo, Canary Islands, Socotra, Pemba, Zanzibar, Mafia, Seychelles, Comoro Islands; Madagascar and adjacent islets, Nosy Be, Nosy Boraha (Île Sainte-Marie-de-Madagascar), Nosy Faly, Nosy Ambariovato, Nosy Mangabe, Nosy Tanikely, Nosy Alanana (Île aux Prunes), and Nosy Sakatia. Introduced to Hawaii, Réunion, and possibly Mauritius and several other islands near Africa, Madagascar, and Greece.

Evolution and systematics

The fossil record is sparse, but the history of chameleons may be more than 60 million years old. Although chameleons are believed to have originated in Africa or Madagascar, the oldest known fossil, 26 million years old, is *Chamaeleo caroliquarti* from western Bohemia. Based on the known fossil record, chameleons were distributed in Africa but also in parts of the world where they are not found today, such as China, Bavaria, and western Bohemia. Chameleons eventually disappeared from the latter three regions, perhaps as a consequence of changing climatic conditions that favored cooler temperatures and lower humidity.

New areas were inhabited as chameleons radiated to more hospitable climates and evolved into new forms. Mountains, forests, and savannas isolated some species, and their morphologic characteristics evolved to include rows of scales called crests that were high, wavy, or spiky on the back (dorsal crest), throat (gular crest), or belly (ventral crest). A number of chameleons developed one, two, three, four, or six bony horns of different shapes and sizes, flexible extensions on the snout, movable flaps of skin on the side of the head, and other differentiating characteristics, such as patterns, coloration,

and body shape and size. All chameleons retained certain prominent features, however, that in combination distinguish them from all other lizards, including projectile tongues used to capture prey, large protruding eyes encased in an eyelid with a tiny aperture referred to as eye turrets, toes fused in bundles of two and three to form grasping pincers, and a prehensile tail.

The classification of this diverse group of lizards has undergone many revisions in genera, families, subfamilies, species, and subspecies throughout the nineteenth and twentieth centuries. The naturalist John E. Gray introduced 16 genera in 1865, and Franz Werner later reduced the number to three genera but also created three new families in 1902. The numbers of species discovered and described increased over the years, but some of them later were eliminated as synonyms of taxa already described and named by previous authors. Werner recognized 70 species in 1902 and raised the number to 88 by 1911; by 1981 Vincent A. Wager recognized 113. By 1986 the number of recognized species had risen to 128, according to Charles Klaver and Wolfgang Böhme, who revised the entire phylogeny of the family based on the morphologic features of the male sexual organs and the lung morphology, bone structure, and chromosome characteristics. While elements of this classification system are not finalized and may be subject to change, it has been accepted worldwide as best representing the relationships within the family Chamaeleonidae. At the last published revision of the system in 1997, Klaver and Böhme recognized no subfamilies, six genera, and two subgenera (*Chamaeleo* and *Trioceros*). Within these groupings they cited a total of 171 forms (species plus subspecies):

- *Bradypodion*: 27
- *Brookesia*: 24
- *Calumma*: 25
- *Chamaeleo* (*Chamaeleo*): 24
- *Chamaeleo* (*Trioceros*): 37
- *Furcifer*: 20
- *Rhampholeon*: 14

A few subspecies were elevated to full species after 1997, and several new species were discovered and described, primarily from Madagascar. By 2002 the total number of valid species and subspecies was 180, but this number is likely to change in the future.

Physical characteristics

Chameleons are best known for their ability to change colors. The palette of any species is limited to only certain colors, however. In the case of members of the genera *Brookesia* and *Rhampholeon*, the palette consists mainly of shades of tan, brown, and black. The coloration of juvenile chameleons is usually more cryptic than adults of the same species, which may help conceal them from predators. The most dramatic and varying coloration probably belongs to the panther chameleon, *Furcifer pardalis*, from Madagascar. Within their wide geographic range in the northern third of the island, the coloration of adult males varies significantly from locale to locale, with numerous distinctive color palettes, such as pink and blue, green and red, aqua blue and green, red-orange and white, and turquoise and navy blue.

Chameleons display variations of their color palettes in response to psychological or physiological stimuli and to communicate, not to match their background as was once believed. Although chameleons lack vocal chords, some species are capable of vibration that produces an audible sound or can expel air forcibly from the lungs to generate a hissing or squeaking noise. Chameleons also are known to make sounds in a frequency inaudible to the human ear, but they cannot hear very well, because they lack eardrums and external ear openings.

A male communicating his intentions to a mate often sports the most vivid colors at his disposal. A female likewise will display coloration to communicate her willingness to mate. Calm, subdued colors may indicate receptivity, where dark, intense coloration warns her suitor to stay away. In a few species, such as *Calumma boettgeri* and *C. nasuta*, females show striking purplish blue spots, called "threat spots," on the head

to deter males. Competing males exhibit bright and intense colors, but the loser usually changes to drab coloration and slinks away to indicate that the contest is over. Chameleons that have a range of colors in their palette may manifest them on different parts of the body, such as the legs, throat, or head. Some are capable of showing stripes and patterns that recede when the chameleon is not in an excited or stressed state. Calm chameleons typically display the least vivid colors. An ill chameleon may become dark or pale in coloration, and sleeping chameleons are often very pale. Color also plays a part in thermoregulation; dark colors absorb the sun's rays when chameleons are cold, and paler colors deflect sunlight.

The epidermis does not grow, and the chameleon sheds completely from time to time as it outgrows this layer of skin. Two cell layers beneath the transparent outer layer can contain red and yellow cells containing pigment granules called

A veiled chameleon (*Chamaeleo calyptratus calyptratus*) uses its ballistic tongue for prey capture. The tongue pad is withdrawn to form a pouch (invaginated) immediately before prey contact. The pouch engulfs the prey and wet adhesion and interlocking muscles maintain grip while suction created by the pouch retractor muscle transfers the prey deeper into the pouch. The tongue retracts until it returns to its resting position on the hyoid bone and the prey is in the chameleon's mouth. (Photo by John H. Tashjian, courtesy of Reptile Haven, Escondido. Reproduced by permission.)

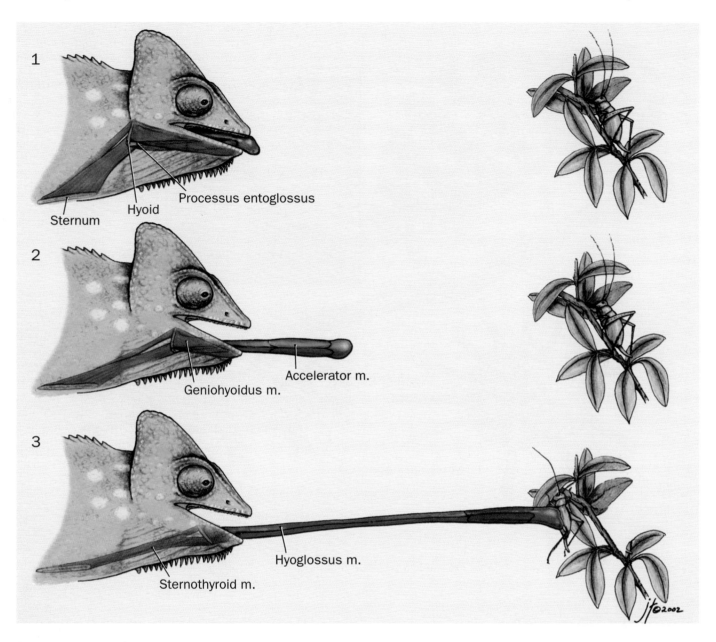

The tongue anatomy of a chameleon hunting. Top: tongue is in resting position; middle: tongue moves out of mouth; bottom: tongue is fully extended and stuck onto insect. (Illustration by Joseph E. Trumpey)

chromatophores, and, beneath them, there are cell layers that reflect blue and white light. Below these layers are black or brown pigment called melanin. As these layers of cells expand, contract, and overlap due to result of stimulation (or the lack of it), the chameleon can rapidly change color.

Sex determination in adult chameleons is not a difficult matter for most chameleon species, because they most often are sexually dimorphic, meaning males and females are different in form or size. For example, in the majority of species where males have horns, females lack horns. Males are usually larger than females, except in the genera *Brookesia* and *Rhampholeon*. Species that are not sexually dimorphic may be different in coloration, or sexually dichromatic, such as *Furcifer*

pardalis. Females of this species are typically a reddish orange or tan marked with brown or black, regardless of geographic locale. Determining the sex of species in the genera *Brookesia* and *Rhampholeon* depends primarily on the presence of a bulge at the base of the tail created by paired sexual organs called hemipenes. It is much more difficult to ascertain the sex of juveniles of most species from birth to about six months of age, or whenever the first indication of adult coloration, horns, crests, or a hemipenial bulge becomes apparent.

The most important physical feature of a chameleon is its large and protruding eyes. A chameleon can move its eyes independently and is able to process two images at once. This ability is the chameleon's best defense against predators,

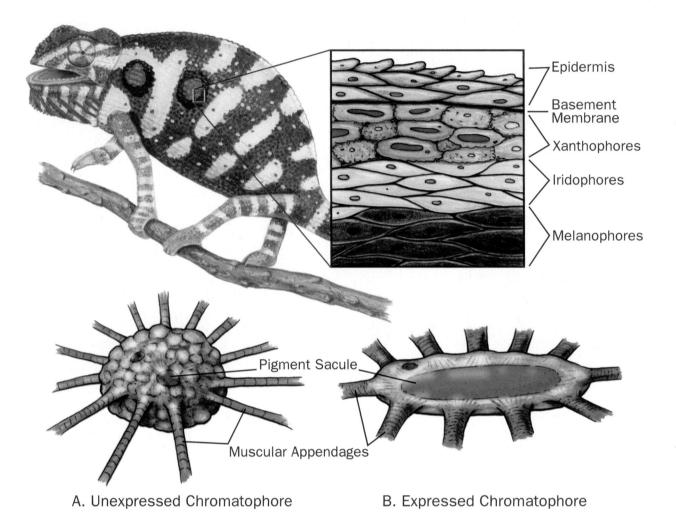

Epidermis

Basement Membrane

Xanthophores

Iridophores

Melanophores

Pigment Sacule

Muscular Appendages

A. Unexpressed Chromatophore

B. Expressed Chromatophore

Chromatophores are found in chameleons' skin. (Illustration by Joseph E. Trumpey)

because its hearing is very poor. It scans the surrounding environment with telescopic vision that enables it to plan and execute a defense (usually concealment or flight) well in advance of the predator's approach. Phenomenal eyesight also facilitates locating prey from a great distance. As chameleons target prey, two separate images merge into one to gauge distance. Then chameleons engage their most fascinating feature—the tongue.

In 2000 a group of researchers published the results of a study on the mechanics of prey prehension in chameleons that unraveled the mysteries of how a chameleon's tongue really works. The hyoid bone is a piece of cartilage that extends into the mouth from the throat bones (called the hyolingual apparatus) and is attached to a chameleon's long tongue. This is where the tongue rests when it is not in use. The tongue is launched from the hyoid bone with the use of ringed muscles in the tongue. This highly complex structure, composed of cartilage, muscles, nerves, glands, and tissue, is used this way in prey capture:

- The central cylindrical accelerator muscle is responsible primarily for projecting the tongue for prey capture.

- The tip of the tongue, or tongue pad, sits atop the accelerator muscle, connected by several pairs of muscles. As the chameleon launches the tongue pad at prey, it turns inside out (evaginates) and actively reverses (invaginates) to form a pouch immediately before prey contact.

- The pouch engulfs the prey, and wet adhesion and interlocking maintain grip while suction created by the largest paired muscles, the pouch retractor, transfers prey deeper into the pouch.

- The tongue retractors attached to the accelerator muscle return the tongue to the resting position on the hyoid bone, and the prey is in the chameleon's mouth to be crushed and swallowed.

Two Nosy Island panther chameleons (*Furcifer pardalis*) display defensive colors. (Photo by Animals Animals ©Zig Leszczynski. Reproduced by permission.)

Before this study, the capture of prey often was attributed only to adhesion to the tongue pad. The withdrawal of the tongue pad to form a pouch not only creates suction forces on the prey but also increases the adhesive properties of the tongue. Suction accounts for more than two-thirds of the total force generated by a chameleon's tongue. This permits capture of larger prey, such as lizards and birds, than is possible using just adhesion. Chameleons also employ the tongue pad to lap drinking water from leaves or other surfaces.

The skeletal structure of chameleons is remarkable for it's flexibility. They can compress their bodies to bask in sunlight or inflate their lungs and expand the rib cage to bluff potential predators. Chameleon feet are designed to grasp, with five toes on each foot fused in bundles of two and three toes to form a pincer. On the front feet the bundle of three toes is on the inside of the foot, and the bundle of two toes is on the outside. This is reversed on the rear foot, giving them a secure and strong grasp and allowing them to navigate horizontally or vertically on a wide variety of vegetation or structures. Sharp claws on each toe help them climb and grip surfaces that they cannot grasp tightly, such as tree trunks, and are used by females to excavate tunnels to lay eggs.

In the genera *Bradypodion*, *Calumma*, *Chamaeleo*, and *Furcifer*, tail length is roughly equal to or slightly longer than body length. These species can use their prehensile tail as a fifth limb and to anchor themselves while launching their long tongue at prey. Some species, such as Parson's chameleon (*C. parsonii parsonii*), use their tails to communicate. Males engaged in ritualistic threat displays repeatedly coil the tips of their long tails tightly, curl them up and over the back, and whip them forward. Sleeping chameleons often roll their long tails into a perfect coil like a watch spring, and chameleons may intertwine tails during copulation. Members of the genera *Brookesia* and *Rhampholeon* have much shorter, less prehensile tails that nevertheless can be used as a grasping hook in some species.

Distribution

Chameleons occur naturally only in the Old World. Africa (including offshore islands) has the highest concentration of species and subspecies, with all 27 members of the genus *Bradypodion*, all 14 forms in the genus *Rhampholeon*, and 59 forms in the genus *Chamaeleo*, totaling 99 species and subspecies. Forty percent of the world's species inhabit Madagascar and offshore islands, including 19 of the 21 members of the genus *Furcifer*, all 27 forms in the genus *Brookesia*, and 28 forms in the genus *Calumma*, for a total of 73 species and subspecies. The eight remaining forms are from Yemen (one),

A veiled chameleon (*Chamaeleo calyptratus*) catches a cricket. (Photo by Animals Animals ©Stephen Dalton. Reproduced by permission.)

Saudi Arabia (two), India and Sri Lanka (one), the Comoro Islands (two), Socotra (one), and the Seychelles (one).

One species, *Chamaeleo chamaeleon*, or the common chameleon, is found in Europe, the Middle East, Greece, northern Africa, southwestern Saudi Arabia, and Yemen. This makes up the widest range of distribution of any chameleon species. Jackson's chameleon (*Chamaeleo jacksonii*) has the dubious distinction of being the first chameleon species to be introduced and become well established in the New World. A few dozen specimens imported for the pet trade in the 1970s escaped into the wilderness in Hawaii, creating a large feral population that has continued to thrive more than 30 years later.

Habitat

Chameleon habitat is as varied as the species in this diverse family of reptiles. One species, *Chamaeleo namaquensis*, lives in one of the most inhospitable regions on earth, the Namib Desert in Africa. This sturdy and aggressive chameleon tolerates extremely high temperatures by day and near freezing temperatures at night and lives a terrestrial existence near the sparse vegetation of sand dunes. Other species are far less tolerant of such extreme temperatures and require high humidity, particularly species that are montane or rainforest specialists. These species may not survive the loss of their complex environmental niches in the future, whether from deforestation, modification, or climatic change. Some chameleon species have adapted to the degraded vegetation that invariably accompanies the burgeoning human population in underdeveloped countries. Unprotected natural forest is frequently burned or cut for grazing, agriculture, fuel, and housing. Sometimes agriculture, such as fruit or coffee trees, provides alternate habitat for arboreal creatures like chameleons, but such crops as rice do not.

There are chameleon species that utilize vegetation in or near virtually every forest type, including lowland evergreen broadleaf rainforest, semi-evergreen moist broadleaf forest, deciduous or semi-deciduous broadleaf forest, thorn forest, upper and lower montane forest, cloud forest, disturbed natural forest, and exotic or native species plantations. Other species live in grassland, scrub, or semidesert conditions and can be found from sea level all the way up to elevations of nearly 15,000 ft (4,500 m).

Behavior

Chameleons are diurnal, and many species begin the day at dawn by seeking a spot to bask in the sun to increase body temperature and metabolism. Chameleons are ectothermic (cold-blooded) and must regulate their temperature by ex-

Common chameleons (*Chamaeleo chamaeleon*) on a thorny plant in Tunisia. (Photo by J.C. Carton/Carto. Bruce Coleman, Inc. Reproduced by permission.)

posing their body to sun or shade. Once they reach a comfortable temperature, they may begin seeking prey or lap dew or rain to quench thirst. A typical day is spent resting or seeking prey while keeping a watchful eye for predators as they move about in the environment. All moving objects must be analyzed as a potential threat. If the object advances in their direction, chameleons may move off into foliage or attempt to conceal themselves by swiveling behind their perches,

known as "squirreling." The main predators of chameleons are birds and snakes.

At dusk chameleons seek a place to roost and sleep and often return to the same location every night. Many species roost at the ends of branches to sleep. Small species may wrap their tails around the stems and drape their bodies on top of large leaves, with their heads pointed down. If a predator

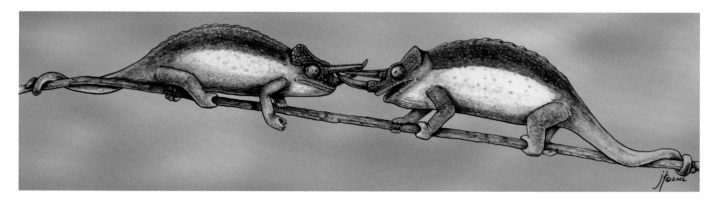

Pair of male Jackson's chameleons (*Chamaeleo jacksonii*) fighting. (Illustration by Joseph E. Trumpey)

Panther chameleon (*Furcifer pardalis*) male and female copulating. Chamaeleonidae males have paired hemipenes (reproductive organs) that allow them to approach a female from either side and use the hemipene closest to the female for copulation. Fused toes that form grasping feet allow them to achieve mating even on a precarious perch high in a tree. (Photo by John H. Tashjian, courtesy of Texas Christian University. Reproduced by permission.)

touches the branch, leaf, or plant, they release their grip and slide or drop to the ground and play dead or scurry into the underbrush.

Suitable chameleon habitat, whether disturbed or undisturbed, must include enough space to support a viable population of these often asocial and territorial reptiles. Males are usually intolerant of other males of the same species within visual proximity, especially during mating season. Females normally avoid males when they are gravid or unreceptive to mating. Habitat partitioning by height and life stage may relieve some of these pressures in dense populations within a restricted area of habitat, but conflicts can result in injuries or even death.

Feeding ecology and diet

Chameleons consume a wide variety of flying and crawling insects, butterflies, moths, larvae, snails, and spiders in nature, and larger species consume some vertebrates as well. Chameleons prey on smaller chameleons, lizards, birds, and even snakes. Captive chameleons will accept young mice, but it is unlikely that this is a natural prey. Chameleons also ingest vegetation, including leaves, flowers, and fruits. Other organic matter, such as bark, twigs, moss, and soil are sought out and consumed by chameleons, but the nutritional or medicinal value of some of these items is unknown. Chameleons are sit-and-wait ambush predators, but many species are quite mobile and travel long distances seeking prey along the way, while others are much more sedentary and utilize a smaller range. There is anecdotal evidence that chameleons congregate in areas where insects appear only at certain times of the year, such as when insects are attracted to coffee blooming or when cicadas hatch. When these food supplies are no longer present, the chameleons disperse.

Reproductive biology

Reproduction in chameleons typically begins with ritualistic courtship displays by males. In many species this entails the display of bright colors and a series of jerking or bobbing head movements while advancing on a female. Some males advance slowly with a halting or jerky gait, but others move very quickly and can be aggressive toward females. Females that are unreceptive or gravid may flee or may face the suitor with gaping mouth while hissing, rearing up on the hind legs, and rocking to discourage the male's advances. Females are known to approach males and grasp their forelegs or horns

A flap-necked chameleon (*Chamaeleo dilepis*) seen in KwaZulu-Natal, South Africa. Chameleons can move their two eyes independently. (Photo by Nigel J. Dennis/Photo Researchers, Inc. Reproduced by permission.)

to stop their pursuit. In some cases, unreceptive or gravid females attack males and inflict bite wounds that can be fatal.

If the female remains passive to the courtship of the male, he will mount the female by grasping her flanks and position himself on the right or left side of her body. He then everts the nearest of his two sexual organs while inserting it in her cloaca, the common chamber into which the intestinal, urinary, and generative canals discharge in reptiles, and copulation ensues. Some species copulate for a few minutes and others for as long as several hours, after which they usually go their separate ways. A few species form pair bonds for a period of time during the mating season.

The majority of chameleon species are oviparous, meaning they produce offspring by laying eggs in tunnels or pits in the ground or under rocks or leaves after a gestation period that can last a few weeks or several months, depending on the species. Females excavate tunnels and pits by digging with their

front feet and then back into them to deposit eggs. When they are finished, they bury the eggs, fill in the tunnel or pit, and stomp the soil down to conceal the location of the nest. Some females drag leaves and twigs over the site. This is the final act of motherhood for a chameleon, and her young will be independent at birth. Incubation times vary according to species and according to the stage of embryonic development at the time the eggs are deposited. The shortest known incubation period is around one month, and the longest is 18 months. The young emerge by slitting a star-shaped opening in the end of the eggshell with the egg tooth, a sharp, calcified protrusion on the tip of the upper jawbone that later falls off.

Ovoviviparous species (those with eggs that hatch within the mother's body, or immediately after being laid) are found primarily in climates that experience greater extremes of cold and likely represent a reproductive strategy to increase the survival rate of neonates where eggs deposited in the ground might not hatch. Basking gravid females often position themselves so that sunlight is directed on their swollen abdomens to warm the developing babies. A female paces nervously while giving birth to her young, which emerge encased in thin, transparent membranes. The neonates wriggle free of the membrane and begin moving and climbing about immediately, usually seeking food within 24 hours. They instinctively disperse, perhaps to avoid predation by the mother; this rarely occurs, however, in the confines of a cage in captive births. While live-bearing females are in contact with their offspring at birth (unlike egg-laying females), they do not nurture them in any way.

Conservation status

In 1996 three chameleon species, *Furcifer campani*, *F. labordi*, and *F. minor*, were classified as Vulnerable by the IUCN, based on a 20% population decline in 10 years, or three generations. A fourth species, *Brookesia perarmata*, was classified as Vulnerable for this reason and also because they occupy an area of less than 39 sq mi (100 sq km) and fewer than five locations. All chameleons in the genera *Bradypodion*, *Calumma*, *Chamaeleo*, and *Furcifer* are listed on CITES Appendix II, indicating that they are threatened with extinction unless commercial trade is tightly controlled. A moratorium on importation for commercial trade of all but four species of chameleons (*F. pardalis*, *F. lateralis*, *F. oustaleti*, and *F. verrucosus*) from Madagascar was imposed by CITES in 1995, owing to escalating levels of trade and concerns that extinction might result. This moratorium remained in effect in 2002. Although *Brookesia perarmata* is included on the IUCN Red List as Vulnerable, no members of the genera *Brookesia* or *Rhampholeon* received formal protection from CITES to prevent unsustainable commercial trade as of 2002.

The main threats to chameleons are ongoing loss, modification, and fragmentation of acceptable habitat and collection for the legal and illegal commercial pet trade. When the extent of occurrence is small, the number of known sites is few, the distribution structure is fragmented, and the species is a specialist within a declining habitat, the risk of extinction is high. This paradigm is applicable to numerous species of chameleons worldwide. The majority of chameleon species have not survived or reproduced in captivity and should not be considered candidates for captive breeding projects aimed at preservation of the species. Habitat preservation and conservation management in the wild for these vulnerable species are critical to preventing future extinction.

Significance to humans

There are relatively few traditional uses for chameleons by local people within their range of distribution, but these uses generally involve burning or killing chameleons for folk medicine or to ward off evil spirits. Some cultural traditions dictate that chameleons must not be harmed. Chameleons are not used very often as food. The major consumption of wild chameleons is for the international commercial live pet trade that reached its apex in the 1990s, when more than 260,000 chameleons were exported from Madagascar and 345,000 from Africa and Yemen. The major consumers are the United States, western Europe, and Asia. The commercial trade in reptiles, particularly those captured in the wild, has been criticized by conservation, scientific, and animal rights organizations as inhumane and because it is detrimental to the survival of wild populations. It has been estimated that less than 1% of chameleons taken in the wild live longer than a few months in captivity. This is primarily the result of captivity-related stress, injury, diseases, parasites, and failure or inability to meet the highly specialized environmental and nutritional requirements necessary for survival in a captive setting. For humane and conservation reasons, chameleons should not be considered appropriate as pets.

1. Male Parson's chameleon (*Calumma parsonii*); 2. Common chameleon (*Chamaeleo chamaeleon*); 3. Male veiled chameleon (*Chamaeleo calyptratus*); 4. Male Jackson's chameleon (*Chamaeleo jacksonii*); 5. Male panther chameleon (*Furcifer pardalis*). (Illustration by Joseph E. Trumpey)

1. Male KwaZulu-Natal Midlands dwarf chameleon (*Bradypodion thamnobates*); 2. Armored chameleon (*Brookesia perarmata*); 3. Female minor chameleon (*Furcifer minor*); 4. Short-tailed chameleon (*Rhampholeon brevicaudatus*) hatchling; 5. Female jeweled chameleon (*Furcifer campani*). (Illustration by Joseph E. Trumpey)

Grzimek's Animal Life Encyclopedia

Species accounts

KwaZulu-Natal Midlands dwarf chameleon
Bradypodion thamnobates

TAXONOMY
Bradypodion thamnobates Raw, 1976, Nottingham Road, Natal, South Africa.

OTHER COMMON NAMES
Afrikaans: Natalse Middelveld.

PHYSICAL CHARACTERISTICS
This species is 6–7.5 in (152–191 mm) in length. Females are marginally smaller than males. Male coloration is dark green or blue-green with white markings on the head and a prominent gular crest composed of lobed scales. There is also a spiky dorsal crest. Large conical or rounded scales, which are white, blue, or reddish in color, cover the body and legs. Female coloration is olive green, and the markings, scalation, and crests are less pronounced. Juveniles are brownish, yellowish, or light green.

DISTRIBUTION
The species occurs in South Africa in the southeastern province of KwaZulu-Natal in the midlands between the Mooi River, Bulwer, Howick, and Dargle at an elevation of 3,000–4,200 ft (915–1,281 m).

HABITAT
These chameleons live among shrubs, bushes, and anthropogenic vegetation, including gardens and hedges in residential areas.

BEHAVIOR
The species is calm and docile, but males are aggressive toward other males.

FEEDING ECOLOGY AND DIET
The dwarf chameleon preys upon small crawling and flying insects.

REPRODUCTIVE BIOLOGY
Males court females by head bobbing while approaching a female. Unreceptive females display defensively or flee. Receptive females remain passive and allow males to copulate. Gestation is from five to eight months, depending on ambient climate, and females give live birth to as many as 18 neonates.

CONSERVATION STATUS
Listed as Near Threatened by the IUCN. Limited numbers have been exported to zoos and a few private people as of 2002. The range of distribution of this rare species is poorly protected, continues to erode, and is under serious threat, which may affect its abundance in the future.

SIGNIFICANCE TO HUMANS
These chameleons are used occasionally in traditional medicine by indigenous people in Africa. ◆

Armored chameleon
Brookesia perarmata

TAXONOMY
Brookesia perarmata Angel, 1933, Antsingy at an elevation of 984 ft (300 m), Province du Ménabé, Madagascar.

OTHER COMMON NAMES
English: Antsingy leaf chameleon.

PHYSICAL CHARACTERISTICS
At 4–6 in (102–152 mm), this is the largest and most easily identifiable member of the genus *Brookesia*. It is reddish brown, brown, and tan in coloration. The most remarkable physical feature is a row of pointed scales projecting outward from the spine that continue onto the tail, diminishing in size. The remainder of the body and tail bear numerous thorny scales, giving it the armored appearance for which it is named.

DISTRIBUTION
The only known locale is l'Antsingy d'Antsalova in the Tsingy de Bemaraha Nature Reserve, Madagascar.

HABITAT
The armored chameleon inhabits bushes, shrubs, and leaf litter in or near dense, dry, deciduous forest, where it is primarily terrestrial. According to E. R. Brygoo, three specimens were collected in 1952 "on the trail from Antsalova to Tsiandro, at the level of the Ambodiriana clearing, in the rocks to the north of the road, among the vegetation between the stones."

BEHAVIOR
The species is docile, secretive, and often sedentary.

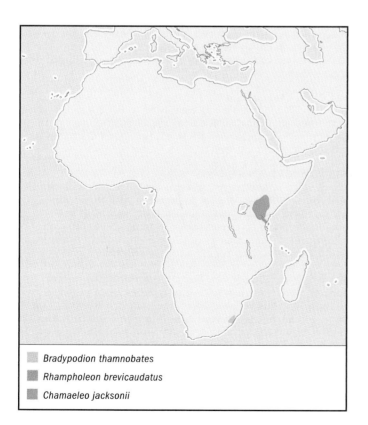

Bradypodion thamnobates
Rhampholeon brevicaudatus
Chamaeleo jacksonii

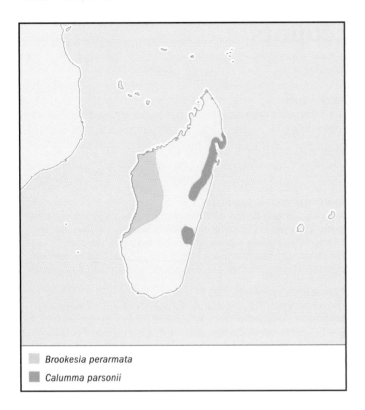

Brookesia perarmata

Calumma parsonii

FEEDING ECOLOGY AND DIET

The armored chameleon feeds on invertebrates within its prey size range.

REPRODUCTIVE BIOLOGY

Little is known of the reproductive biology of this species.

CONSERVATION STATUS

This species is classified as Vulnerable by the IUCN based on a 20% population decline in 10 years, or three generations, owing to levels of exploitation and because the population is characterized by an acute restriction in area of occupancy and number of sites. Despite this listing, it receives no formal protection from CITES to prevent unsustainable commercial trade. It has been exported regularly to the pet trade beginning around 1993, but quantities of harvested specimens were unrecorded as of 2002. Protection from commercial trade and preservation of habitat are critical to preventing extinction.

SIGNIFICANCE TO HUMANS

The only member of the genus *Brookesia*, the armored chameleon is in high demand for the commercial pet trade. Longevity and reproduction in captivity are poor. ◆

Parson's chameleon

Calumma parsonii

TAXONOMY

Calumma parsonii Cuvier, 1832, Madagascar. Three geographically discrete forms are recognized, based on size, eye turret color, and/or color of the scales outlining the mouth. One subspecies is recognized.

OTHER COMMON NAMES

French: Caméléon de Parson; German: Parsons Chamäleon; Spanish: Camaleón Parson; Malagasy: Ranotoetrabe, ranotohitra, taintrotro, tampamaly.

PHYSICAL CHARACTERISTICS

This species grows to 12–28 in (305–711 mm) in length. The nominate form is one of the three largest and the heaviest of all species of chameleons, at weights up to 24.5 oz (700 g). Males are tan, turquoise, or blue-green, with white, tan, or yellow scales outlining the edge of the mouth, orange or blue-green eye turrets, and two robust bony horns protruding from the snout. Females are red-brown, green, or blue-green. Three or more prominent dark stripes curve down from the dorsum toward the anterior of the body. There is no dorsal crest. Juveniles are orange, blue-green, or tan. The subspecies, *Calumma parsonii cristifer*, is 10–18 in (254–457 mm). Males are deep blue-green with a rust blotch on the flank and a complete dorsal crest. Females are tan, orange, or green.

DISTRIBUTION

C. p. parsonii occurs in Ifanadiana and Ikongo, southeastern Madagascar, and Vavatenina, Nosy Boraha, and Maroantsetra, northeastern Madagascar. *C. p. cristifer* is found in Andasibe.

HABITAT

C. p. parsonii has been described as originating from primary rainforest, but as of 2002 little intact forest remained within the known range of distribution. Observations after the 1960s have been limited to fragments of disturbed forest and shady mature tree plantations, such as coffee, lychee, and mango. This species preferentially selects dense vegetation and avoids direct sunlight. The *C. p. cristifer* habitat is largely within the boundaries of a well-protected national park retaining undisturbed moist montane rainforest as of 2002.

BEHAVIOR

Parson's chameleon is sedentary and shy. Males engage in dramatic territorial displays but are otherwise passive. Females are easily stressed and display with yellow blotches distributed over the head and body.

FEEDING ECOLOGY AND DIET

The species consumes a wide variety of vertebrate and invertebrate prey, including small lizards and birds. Malagasy residents report consumption of flowers, fruits, and dry leaves as a component of the total diet.

REPRODUCTIVE BIOLOGY

No detailed long-term research has been performed on the reproductive biology of *C. p. parsonii*, but data from captive management indicate that sexual maturity and adult size are attained between three and five years of age, substantially later than any other species of Chamaeleonidae. Courtship and copulation begin from two to four months after winter brumation (dormancy) ends, and the mating season lasts approximately eight weeks. Gestation is from three to five months, and incubation has varied from 13 to 24 months in captivity. Clutch size is 20–60 eggs, and the interval between clutches is one year. Less is known of the reproductive biology for *C. p. cristifer*. In 1997 one clutch of 37 eggs produced 35 live hatchlings in captivity after 13 months of incubation.

CONSERVATION STATUS

Not listed by the IUCN, but listed on Appendix II of CITES; proposals to uplist it to Appendix I were submitted in 1998 and 2001. The 1995 CITES moratorium on importation failed to halt sporadic commercial trade, at least through 2002. The continuing levels and patterns of exploitation place small populations at risk of extirpation and decreased reproductive potential. As of 2002 *C. p. parsonii* had not been documented in a protected area for more than 30 years. This species is at risk of extinction because it is known from relatively few sites, the distribution structure is highly fragmented, and it is a specialist of a rapidly declining habitat. *C. p. cristifer* is vulnerable to extinction because population densities are low, and it is known only from a few sites. Conservation is dependent on preservation of suitable habitat and prevention of commercial trade.

SIGNIFICANCE TO HUMANS

From 1986 to 1995, 19,000 wild-taken specimens of *C. parsonii* were exported legally for the commercial pet trade, and sporadic illegal trade continued subsequent to the 1995 CITES moratorium on importation, owing to high international demand and escalating retail selling prices. Despite long life expectancy, fewer than 1% of specimens imported before 1995 remained alive by 2002, and fewer than 300 neonates had hatched alive in captivity. ◆

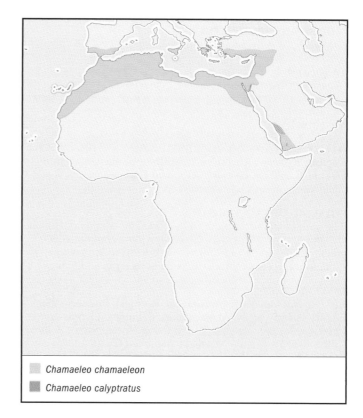

☐ *Chamaeleo chamaeleon*
☐ *Chamaeleo calyptratus*

Veiled chameleon
Chamaeleo calyptratus

TAXONOMY

Chamaeleo (Chamaeleo) calyptratus Duméril & Bibron, 1851, Yemen and southwestern Saudi Arabia. One subspecies is recognized.

OTHER COMMON NAMES

English: Cone-head chameleon, Yemen or Yemeni chameleon; German: Jemen-Chamäleon.

PHYSICAL CHARACTERISTICS

The species grows to 10–24 in (254–610 mm) in length. The most prominent feature is a high, prominent casque that is much larger in males than females and the tallest of any chameleon species. Male coloration is shades of green, turquoise, yellow, orange, white, and black with bold stripes and spots. Nongravid females are green with pale patterning, but gravid females display vivid yellow, blue, and green spots and patterns against a dark background. Juveniles are green at birth, and their sex is distinguished easily by the presence of a short, fleshy projection called a tarsal spur on the hind feet of males. It produces audible sounds.

Although the subspecies *Chamaeleo calyptratus calcarifer* (Peters, 1871) was still considered valid in 1997, there is anecdotal evidence that it is a hybrid of *Ch. calyptratus* and *Ch. arabicus* (Matschie, 1893); this requires further investigation.

DISTRIBUTION

Ch. c. calyptratus occurs in Yemen, centered around Ta'izz and Ibb. *Ch. c. calcarifer* inhabits Saudi Arabia and possibly Yemen.

HABITAT

The veiled chameleon primarily utilizes *Acacia* species in areas of heavy cultivation and exotic plantations, such as *Catha edulis*. The climate is arid, but the habitat is the greenest part of the Arabian Peninsula, and these chameleons usually are found near a water source.

BEHAVIOR

The veiled chameleon is considered very aggressive and defensive toward members of its species (conspecifics) and humans. Males can inflict serious injuries or death in territorial or courtship disputes. An adult at the Dallas Zoo attacked a mirror and regularly coiled and uncoiled his tail while threatening his image.

FEEDING ECOLOGY AND DIET

The species feeds on insects, some vertebrates, and significant amounts of vegetation, particularly acacia.

REPRODUCTIVE BIOLOGY

Courtship consists of head bobbing, intensified coloring, and approaching the female with a jerky gait. Females unreceptive to mating may gape, swing from side to side, hiss, or even attack the males. Receptive females slowly move off but allow the male to mount and copulate. Males may become aggressive and are known to head butt or bite, sometimes causing grave injury to females.

CONSERVATION STATUS

Not listed by the IUCN, but listed on CITES Appendix II. In 2001 it was not considered rare within its range and is adapted to highly disturbed habitat in close proximity to humans. Reports of increasing use of agricultural pesticides may have a detrimental impact on future abundance.

SIGNIFICANCE TO HUMANS

Fewer than 10,000 wild-taken specimens were exported for the commercial pet trade from 1985 to 1999. The species reproduced consistently in captivity and is well established in herpetoculture. ◆

Common chameleon
Chamaeleo (Chamaeleo) chamaeleon

TAXONOMY
Chamaeleo (Chamaeleo) chamaeleon Linnaeus, 1758, Europe, Middle East, Greece, northern Africa, Egypt (Sinai Peninsula), southwestern Saudi Arabia, and Yemen. Two subspecies are recognized.

OTHER COMMON NAMES
English: European chameleon, Mediterranean chameleon; French: Caméléon commun; German: Europäisches Chamäleon, Gemeines Chamäleon.

PHYSICAL CHARACTERISTICS
The species attains a length of 8–15 in (200–400 mm). Females are often larger than males. Coloration varies but includes green, yellow, gray, and brown with numerous stripes and spots forming consistent patterns on each form.

DISTRIBUTION
Ch. c. chamaeleon occurs in southern Portugal, southern Spain, Canary Islands, Sicily, Malta, southern Turkey, Cyprus, Syria, Lebanon, Jordan, Israel, southern Peloponnese, Samos, Chios, Crete, western Sahara, Morocco, Algeria, Tunisia, Libya, and Egypt. *Ch. c. musae* inhabits the Sinai Peninsula and Egypt. *Ch. c. orientalis* is found in southwestern Saudi Arabia and Yemen.

HABITAT
Owing to a wide distribution, the species is found in numerous habitat types at elevations up to 8,500 ft (2,600 m), including forests, plantations, and scrub in semidesert or coastal regions, usually near a source of water.

BEHAVIOR
This chameleon is fairly aggressive toward conspecifics.

FEEDING ECOLOGY AND DIET
The common chameleon feeds on a wide range of invertebrate and vertebrate prey, including young birds and reptiles, and some vegetation.

REPRODUCTIVE BIOLOGY
Specimens exposed to seasonal temperatures near freezing experience a dormant period (brumation) until temperatures become warmer. As they return to normal activity, males seek females for mating. Unreceptive females may gape, inflate with air, and butt males. Receptive females remain passive and permit copulation. Several copulations may occur per day until the female adopts dark coloration with orange markings and rejects the male's advances. Females produce up to 60 eggs after a gestation period of two months. Young emerge six to 11 months later, depending on climatic conditions and length of diapause, a dormant period in embryonic development.

CONSERVATION STATUS
The species is listed on Appendix II of CITES. In Greece it is strictly protected from collection, killing, abuse, wounding, captivity, and export by a presidential decree, and it is listed in the Greek *Red Data Book of Threatened Vertebrates*. Populations on Chios and Crete are believed to have been extirpated as of the late 1980s. It is included in the "European Community Habitats Directive" in Appendix IV as a priority species, requiring establishment of protected areas. The species has been proposed for an IUCN classification of Lower Risk: Conservation Dependent in Andalusia, Spain, and the European Union, with an additional classification of Vulnerable in the provinces

of Huelva and Cádiz, Spain, based on a study conducted from 1993 to 1999. Major threats are habitat destruction and modification, road mortality, and translocation. The conservation status is unknown in northern Africa, the Middle East, and the Saudi Arabian Peninsula.

SIGNIFICANCE TO HUMANS
The common chameleon was exported from Algeria, Egypt, Jordan, and Morocco in limited numbers from 1986 to 1999 for the commercial pet trade. Life span and reproduction in captivity are considered very poor. Large numbers of specimens are collected and sold at markets in northern Africa for purposes of traditional folk medicine and to ward off bad luck. A common practice in Morocco is to throw live chameleons into a fire. Hundreds of dried chameleons are strung in groups of 30–50 using stout twine and hung in the food markets in Morocco. ◆

Jackson's chameleon
Chamaeleo (Trioceros) jacksonii

TAXONOMY
Chamaeleo (Trioceros) jacksonii Boulenger, 1896, Uganda, later amended to Kikuyu, near Nairobi, Kenya. Two subspecies are recognized.

OTHER COMMON NAMES
English: Mt. Meru chameleon, three-horned chameleon; French: Caméléon de Jackson; German: Ostafrikanisches Dreihornchamäleon.

PHYSICAL CHARACTERISTICS
This species grows to 6–14 in (152–356 mm) in length. *Ch. j. xantholophus* is the largest, followed by *Ch. j. jacksonii* and then *Ch. j. merumontanus*. Males have three annulated (composed of rings) horns, two preorbital and one nasal in all three forms. Females may have no horns, a single nasal horn, or three well-developed horns that are slightly smaller than those of the males, depending on the subspecies. The body coloration is shades of green or brown with a dark red, yellow, or blue wash on the head, flanks, or tail. Juveniles are brown, black, and off-white and show an infusion of adult coloration at about the age of six months.

DISTRIBUTION
Ch. j. jacksonii occupies the highland of central Kenya, except the eastern slopes of Mount Kenya. *Ch. j. xantholophus* lives on the southern, eastern and northeastern faces of Mount Kenya and in Kenya and has been introduced to Hawaii. *Ch. j. merumontanus* inhabits Mount Meru in Tanzania at an elevation of 7,500–9,000 ft (2,288–2,745 m).

HABITAT
This is a montane, arboreal species that utilizes various habitat types, including moist forest, exotic plantations, and dense bushes.

BEHAVIOR
Jackson's chameleon is docile and gentle. Males settle territorial and mating disputes by engaging in pushing contests with their horns, but they rarely resort to biting or inflict injury. Females generally are very calm and tolerant of conspecifics.

FEEDING ECOLOGY AND DIET
Jackson's chameleon preys on a wide variety of invertebrates.

REPRODUCTIVE BIOLOGY

Males engage in ritualistic head bobbing to court females. Nonreceptive females become dark, rock, gape, and may grasp the forelegs or horns of an advancing male. If the female is receptive, she lifts her tail to facilitate copulation. Gestation varies from five to nine months, depending on ambient temperatures. The species is ovoviviparous (live-bearing), with births of three to 50 young per clutch, depending on the subspecies.

CONSERVATION STATUS

Jackson's chameleon is listed on CITES Appendix II. Kenya ceased exporting specimens of *Ch. j. jacksonii* and *Ch. j. xantholophus* in the early 1980s. Hawaii considers feral Jackson's chameleons detrimental to the ecosystem. The species lives mainly in highly degraded habitat and exotic plantations, where it is believed to have adapted successfully to living in close proximity to humans.

SIGNIFICANCE TO HUMANS

The species was exported from Kenya in very high numbers for the commercial pet trade beginning in the 1960s, which was accompanied by very high levels of morbidity and mortality through the 1980s, when Kenya stopped exportation. From 1993 to 1999 more than 12,000 *Ch. j. merumontanus* individuals were exported from Tanzania, but it is unknown as of 2002 whether this has had a detrimental effect on population densities. This subspecies did not adapt well to captivity, and reproduction is poor. Several generations of *Ch. j. xantholophus* originating from Hawaii were bred in captivity in the 1990s but only by a very few people worldwide. ◆

Jeweled chameleon
Furcifer campani

TAXONOMY

Furcifer campani Grandidier, 1872, Massif de l'Ankaratra, Madagascar.

OTHER COMMON NAMES

English: Campan's chameleon; Malagasy: Kamora, soamarandrana.

PHYSICAL CHARACTERISTICS

This species attains a length of 4–5.5 in (107–133 mm). There are three longitudinal white or yellowish stripes on each flank, and rows of colored dots cover the body. Red markings outline the ridge above the eyes. The body color in calm females is green; they are black when excited or stressed. Males are brown.

DISTRIBUTION

The species inhabits the high plateaus south of Antananarivo, Madagascar in Ambatolampy, Ambohimitombo, Andringitra, Ankaratra, Antobeba, Ibity, and Manjakatompo.

HABITAT

The jeweled chameleon inhabits grasslands, shrubs, and savanna and is semiterrestrial.

BEHAVIOR

This is a small and shy species.

FEEDING ECOLOGY AND DIET

The jeweled chameleon preys upon small invertebrates.

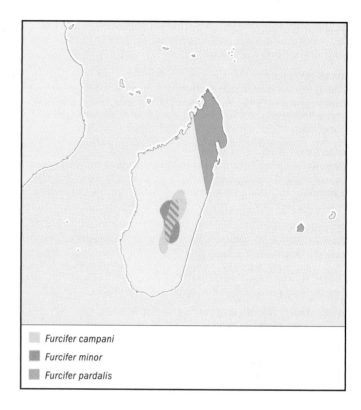

Furcifer campani

Furcifer minor

Furcifer pardalis

REPRODUCTIVE BIOLOGY

The reproductive biology is unknown.

CONSERVATION STATUS

The species is classified as Vulnerable by the IUCN based on a 20% population decline in 10 years, or three generations. It is listed on CITES Appendix II and included in the 1995 CITES moratorium on importation, but specimens are known to have been imported for the commercial pet trade at least through 2002.

SIGNIFICANCE TO HUMANS

More than 15,000 wild-taken specimens were exported legally to the commercial pet trade from 1986 to 1996. There was no documented long-term survivability or reproduction in captivity. ◆

Minor chameleon
Furcifer minor

TAXONOMY

Furcifer minor Gönther, 1879, Fianarantsoa, Betsileo, southeastern Madagascar.

OTHER COMMON NAMES

English: Lesser chameleon; Malagasy: Sakosotoha.

PHYSICAL CHARACTERISTICS

The species is 6–10 in (140–254 mm) in length, half of which is tail. Males have paired rostral horns, and females are hornless. Coloration in nongravid females is green with faint yellow and pink markings. Gravid females are spectacularly colored with vivid yellow bands and spots against a dark green or black background, with two purple spots adorning the anterior flank and a reddish-pink hue on the top of the head. Males are tan or blue with rust or dark brown stripes and blue eye turrets; they exhibit an orange patch bordered in black at the shoulder. Juveniles are green at birth.

DISTRIBUTION
The species is distributed across south-central Madagascar at elevations up to 5,414 ft (1,650 m) in western Ambositra, Ambatofinandrahana, Ambatomenaloha, Itremo, and northern Vinanitelo.

HABITAT
Extensive deforestation within the known range of distribution has reduced the habitat of this species to sparse anthropogenic vegetation, including fruit trees and bushes in close proximity to human habitation.

BEHAVIOR
These chameleons may be aggressive toward conspecifics, but they are otherwise docile.

FEEDING ECOLOGY AND DIET
The minor chameleon preys upon invertebrates.

REPRODUCTIVE BIOLOGY
Females lay 10–16 eggs as many as three times annually; eggs incubate approximately nine months before the young emerge.

CONSERVATION STATUS
The species is classified as Vulnerable by the IUCN based on a 20% population decline in 10 years, or three generations. It is listed on CITES Appendix II and included in the 1995 CITES moratorium on importation, but specimens are known to have been imported for the commercial pet trade through 2002.

SIGNIFICANCE TO HUMANS
Before 1993, net exports of wild-taken specimens equaled five, but 2,400 of the species had been harvested for the pet trade before the 1995 CITES moratorium on importation went into effect. Limited captive reproduction occurred with a second generation of offspring bred in captivity, but by 1998 no legal specimens remained alive in captivity. ◆

Panther chameleon
Furcifer pardalis

TAXONOMY
Furcifer pardalis Cuvier, 1829, Madagascar.

OTHER COMMON NAMES
French: Caméléon panthère; German: Pantherchamäleon; Malagasy: Amboalava, tana.

PHYSICAL CHARACTERISTICS
This species grows to 8–17 in (305–381 mm) in length. The coloration of adult males can vary significantly between geographic locales, but usually it includes two or three predominant colors from the following palette: pink, red, maroon, orange, yellow, green, turquoise, cobalt blue, brown, tan, gray, black, and white. Excited coloration may produce additional colors on various parts of the body that are not obvious in a calm male. There is a broken white, bluish, or cream lateral stripe midflank. Regardless of geographic locale, females are shades of reddish orange or tan marked with brown or black. Juveniles are reddish brown or gray and black at birth. These color variants are not considered valid subspecies as of 2002, but there is anecdotal evidence that interbreeding specimens

from distant locales results in sterile offspring. The snout terminates in a short bony process that is more prominent in males than females, and the chameleons possess complete dorsal, ventral, and gular crests.

DISTRIBUTION
The species occupies roughly the northern third of Madagascar, from Toamasina on the central-eastern coast to Ambanizana in the Masoala Peninsula, Sambava on the northeastern coast, Antsiranana in the north, and Ankaramy on the northwestern coast. They also inhabit islets adjacent to Madagascar: Nosy Be, Nosy Komba, Nosy Tanikely, Nosy Faly, Nosy Mangabe, and Nosy Boraha. They have been introduced to the island of Réunion.

HABITAT
The species is arboreal and primarily inhabits degraded or anthropogenic vegetation, including forest and scrub as well as the undisturbed edges of rainforest on Nosy Be, Nosy Mangabe, and the Masoala Peninsula.

BEHAVIOR
Males are active and highly mobile and frequently are seen crossing roads. Females are more sedentary and secretive. The species is not particularly aggressive toward conspecifics or humans.

FEEDING ECOLOGY AND DIET
The panther chameleon feeds on invertebrate and vertebrate prey typical for medium-size to large chameleons as well as some vegetation.

REPRODUCTIVE BIOLOGY
Panther chameleons lay 10–46 eggs about 45 days after copulation and can produce more than four clutches of eggs annually. The young hatch four to nine months later, depending on climatic conditions. Growth is rapid, and sexual maturity occurs between the ages of six and nine months.

CONSERVATION STATUS
This species is listed on Appendix II of CITES. Before 1999 commercial trade was unrestricted, and more than 34,000 wild-taken specimens were exported from Madagascar in 1998 alone, raising concerns that local populations of distinctive color variants would be extirpated. An annual export quota of 2,000 specimens was established by CITES in 1999, to reduce trade volume. Conservation is not dependent on preserving natural habitat, owing to the mostly successful adaptation of this species to heavily degraded habitat.

SIGNIFICANCE TO HUMANS
The panther chameleon is captured extensively for the international commercial pet trade. More than 100,000 specimens were exported between 1986 and 1999. Captive breeding programs experienced limited success by producing several generations, but there were no established commercial closed-cycle operations as of 2002 where only captive-born adult animals lay eggs to provide the stock that is subsequently sold. ◆

Short-tailed chameleon
Rhampholeon brevicaudatus

TAXONOMY
Rhampholeon brevicaudatus Matschie, 1892, Derema, Usambara Mountains, Tanganyika (Tanzania).

OTHER COMMON NAMES
English: Bearded pygmy chameleon.

PHYSICAL CHARACTERISTICS
The species is 2–3 in (51–76 mm) in length, most of which is the snout-to-vent length. Females are brownish-orange, tan, or yellow, and males are cream or brown with darker brown stripes. The interior of the mouth is blue-gray. A fleshy spike protrudes from the chin, and the tail is very short in females and slightly longer, with a prominent hemipenial bulge, in males.

DISTRIBUTION
This species is found in Derema in the Usambara Mountains of East Tanzania, Africa, at elevations up to 4,200 ft (1,300 m).

HABITAT
The short-tailed chameleon lives in grasses, bushes, and leaf litter in humid forests. It is semiterrestrial.

BEHAVIOR
This is one of a number of species that vibrate rapidly and audibly when they are disturbed or stressed, which may act as a deterrent to predators.

FEEDING ECOLOGY AND DIET
The short-tailed chameleon preys upon very small invertebrates.

REPRODUCTIVE BIOLOGY
Males have been observed courting females several days before mating. Male courtship display may include pursuing the receptive female, rearing up on the hind legs, and approaching the female with a rhythmic gate. Copulation lasts only a few minutes, but it can be repeated several times over subsequent days. Clutch size is two to five eggs, which are laid in a shallow pit after a gestation period of three to four weeks. The young emerge about three months later.

CONSERVATION STATUS
This species receives no formal protection from CITES to prevent unsustainable trade. Habitat status and population densities were unknown as of 2002.

SIGNIFICANCE TO HUMANS
The short-tailed chameleon has been captured for the commercial pet trade since at least 1990, but recording export quantities has not been required, and the total number exported is unknown as of 2002. ◆

Resources

Books

Brady, L. D., and R. A. Griffiths. *Status Assessment of Chameleons in Madagascar.* Cambridge: International Union for the Conservation of Nature and Natural Resources, 1999.

Brygoo, E. R. *Faune de Madagascar.* Vol. 33, *Reptiles Sauriens Chamaeleonidae: Genre Chamaeleo.* Paris: ORSTOM et CNRS, 1971.

———. *Faune de Madagascar.* Vol. 47, *Reptiles Sauriens Chamaeleonidae: Genre Brookesia et Complement pour le Genre Chamaeleo.* Paris: ORSTOM et CNRS, 1978.

Franke, Joseph, and T. M. Telecky. *Reptiles as Pets: An Examination of the Trade in Live Reptiles in the United States.* Washington DC: Humane Society, 2001.

Glaw, F., and M. Vences. *A Fieldguide to the Amphibians and Reptiles of Madagascar.* 2nd ed. Cologne: M. Vences & F. Glaw, 1994.

Klaver, Charles J., and Wolfgang Böhme. *A Compilation and Characterization of the Recent Animal Groups: Chamaeleonidae.* Part 112. Berlin, New York: Das Tierreich, 1997.

Mellado, J., L. Gimenez, J. J. Gomez, et al. *El Camaleón en Andalucía: Distribución actual y amenazas para su supervivencia.* Rota: Fundacion Alcalde Zoilo Ruiz-Mateos, 2001.

Nečas, P. *Chameleons: Nature's Hidden Jewels.* Malabar, FL: Krieger Publishing Co., 1999.

Periodicals

Abate, Ardith. "Assessing the Health of Wild-Caught Chameleons." *Chameleon Information Network* 31 (Spring 1999): 9–17.

———. "Reports from the Field: Parson's Chameleon." *Chameleon Information Network* 29 (Fall 1998): 17–25.

———. "The Exportation of Chameleons from Madagascar: Past and Present." *Chameleon Information Network* 32 (Summer 1999): 9–17.

———. "The Fate of Wild-Caught Chameleons Exported for the Pet Trade." *Chameleon Information Network* 42 (Winter 2001): 15–18.

Abate, Ardith, and K. Kalisch "Chameleon Profile: *Chamaeleo (T.) jacksonii.*" *Chameleon Information Network* 14 (Winter 1994): 19–28.

Deas, J. "In Search of the Veiled or Yemeni Chameleon, *Chamaeleo calyptratus calyptratus.*" *Chameleon Information Network* 43 (Spring 2002): 10–20.

Dimaki, M. "Reports from the Field: The European Chameleons." *Chameleon Information Network* 41 (Fall 2001): 11–13.

Green, G. M., and R. W. Sussman. "Deforestation History of Eastern Rain Forests of Madagascar from Satellite Images." *Science* 248 (April 1990): 212–215.

Herrel, A., J. J. Meyers, P. Aerts, et al. "The Mechanics of Prey Prehension in Chameleons." *Journal of Experimental Biology* 203 (2000): 3255–3263.

Kalisch, K. "The Captive Care and Breeding of *Chamaeleo [Calumma] parsonii cristifer.*" *Chameleon Information Network* 27 (Spring 1998): 15–17.

Raxworthy, Christopher J., and Ronald A. Nussbaum. "Extinction and Extinction Vulnerability of Amphibians and Reptiles in Madagascar." *Amphibian and Reptile Conservation* 2, no. 1 (2000): 15–23.

Risley, T. "Chameleon Profile: *Brookesia* and *Rhampholeon.*" *Chameleon Information Network* 31 (Spring 1999): 21–30.

———. "Field Observations on the Panther Chameleon, *Chamaeleo [Furcifer] pardalis* Cuvier, 1829." *Chameleon Information Network* 24 (Summer 1997): 17–29.

Resources

———. "Preliminary Observations on a Small Population of *Chamaeleo [Calumma] parsonii parsonii* Cuvier, 1824 in Eastern Madagascar." *Chameleon Information Network* 25 (Fall 1997): 25–28.

Organizations

The Chameleon Information Network. 13419 Appalachian Way, San Diego, CA 92129 USA. Phone: (858) 484-2669. Fax: (858) 484-4757. E-mail: chamnet1@aol.com Web site: <http://www.animalarkshelter.org/cin>

World Conservation Union (IUCN). Rue Mauverney 28, Gland, 1196 Switzerland. Phone: 41 (22) 999-0000. Fax: 41 (22) 999-0002. E-mail: mail@hq.iucn.org Web site: <http://www.iucn.org>

Other

Convention on International Trade in Endangered Species (CITES). "Notification to the Parties No. 833.: Significant Trade in Animal Species Included in Appendix II. Recommendations of the Standing Committee." [cited June 8, 2002]. <http://www.cites.org>

United Nations Environmental Program—World Conservation Monitoring Centre (UNEP-WCMC). "Significant Trade in Animals, Net Trade Outputs 1994–1999." Report directed to the CITES Animals Committee regarding the implementation of Resolution Conf. 8.9. [cited June 8, 2002]. <http://www.cites.org/eng/cttee/animals/17/sigtrade2001.shtml>

UNEP/WCMC/UNESCO. "World Heritage Sites. Tsingy de Bemaraha Strict Nature Reserve." ([cited June 8, 2002]. <http://www.wcmc.org.uk/protected_areas/data/wh/bemaraha.html>

World Wildlife Fund (WWF). "Maputaland-Pondoland Bushland and Thickets (AT1012)." [cited June 8, 2002]. <http://www.worldwildlife.org/wildworld/profiles/terrestrial/at/at1012_full.html>

Ardith L. Abate

Anoles, iguanas, and relatives
(Iguanidae)

Class Reptilia

Order Squamata

Suborder Sauria

Family Iguanidae

Thumbnail description
Small to large lizards, the majority of which are terrestrial and oviparous, that are marked by pleurodont teeth, which lie in inner-jaw grooves rather than in sockets

Size
1.6–30 in (30–750 cm) in snout-vent length (svl), and some have tails reaching twice the svl

Number of genera, species
69 genera; approximately 900 species

Habitat
Diverse, with most species occurring either in arid locales or wooded habitats

Conservation status
Extinct: 2 species; Critically Endangered: 6 species; Endangered: 4 species; Vulnerable: 12 species; Lower Risk/Near Threatened: 1 species; Data Deficient: 17 species

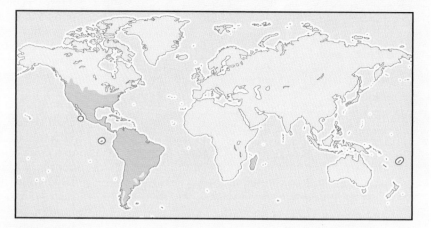

Distribution
Mainly a New World family, extending into much of North America, throughout Central America, and into South America; some species exist in other areas, including Madagascar and the islands of Polynesia

Evolution and systematics

The Iguanidae is a large family of some 860–900 species and nearly 70 genera. The iguanids comprise eight subfamilies:

- Corytophaninae, the casquehead lizards of Central America

- Crotaphytinae, the leopard and collared lizards

- Hoplocercinae, a small assemblage of South American tropical lizards

- Iguaninae, the true iguanas and spinytails

- Oplurinae, the Madagascar iguanas

- Phrynosomatinae, the tree, side-blotched, sand, spiny, and horned lizards

- Polychrotinae, the anoles

- Tropidurinae, the ground lizards of Neotropical South America and the Antilles Islands of the West Indies

The Iguanidae family is most closely related to the families Agamidae of Africa, Asia, and Australia, and the family Chamaeleonidae, mainly of Africa and Madagascar. These three families, collectively grouped as Iguania, diverged within the snake and lizard lineage very early in their evolution. Some taxonomists now classify each Iguanidae subfamily as a separate family, in which case the endings of the subfamily names listed above end in "-idae" rather than "-inae."

Physical characteristics

This large family contains species with varied appearances. Members of the Iguanidae family range from squat, toadlike horned lizards (*Phrynosoma* spp.) that fit in the palm of a hand, to iguanas (*Iguana* spp.) that are as long as a man is tall, and long-tailed and sleek anoles often seen climbing on a window screen. However, the family does have one characteristic that sets it apart from the other Iguania families. That trait is pleurodont teeth, which lie in inner-jaw grooves rather than in sockets.

The subfamilies within Iguanidae can be described in more detail. Members of the Iguaninae include the green iguana (*Iguana iguana*) and marine iguana (*Amblyrhynchus cristatus*). The typical iguanine is an imposing beast with a strong, muscular tail and (in males) a row of noticeable spines down the center of the back.

The corytophanines and polychrotines are slender-bodied lizards with long tails and long limbs. Corytophanines are generally the larger of the two, ranging in size from about 12 to 25.5 in (30.5 to 64.8 cm), compared to the 3–16 in (7.6–40.6 cm) range in the anole subfamily. The most obvious difference is the casque, or helmet, seen in the Corytophaninae subfamily. The casque is actually a tall fin on the top of the head. Corytophanines also generally have exaggerated dorsal and tail fins.

A pair of green anoles (*Anolis carolinensis*); male has extended dewlap. (Photo by Animals Animals ©James H. Robinson. Reproduced by permission.)

The crotaphytines and phrynosomatines are the typical lizards associated with southwestern United States. The former, which include the collared and leopard lizards, are quick, long tailed, and fairly large headed. The leopards are noted, and named, for their abundant dorsal spots, and the collared lizards usually display incomplete neck bands. The phrynosomatines are more diverse, and include more hefty-bodied species. Perhaps the most well-known phrynosomatines are the horned lizards (*Phrynosoma* sp.), which are often mistaken for toads because of their wide bodies and very short limbs.

The final three groups of Iguanidae are the hoplocercines, oplurines, and tropidurines. They are all small to moderately sized lizards. The first group has spiny scales; the second and third groups are distinguished more by geographic distribution (Madagascar for the second, and extreme South American areas for the third) than their appearance.

Distribution

Iguanids are largely lizards of the Western Hemisphere. They range from far southwestern Canada, into much of the United States, throughout Central America and the West Indies, and in almost all of South America except the southwestern edge. Some, like the iguanines, have wide ranges, with representatives from the southwestern United States, Central America, the northern half of South America, and the Galápagos Islands. Others, like the hoplocercines, are more restricted, with species occurring only in tropical South America.

The exceptions to the Western Hemisphere rule are the oplurines, which occur in Madagascar off the southeastern coast of Africa, and a few other species, such as *Brachylophus* species, that live on the Fiji Islands.

Habitat

Iguanids are almost all land-dwelling lizards. Many species, including most of the iguanines, crotaphytines, and phrynosomatines, prefer arid areas. These desert dwellers often seek sites with at least some vegetation, rocks, or other cover to provide escape routes from predators. Other iguanids seek a wooded habitat, with some, like many corytophanines, living in rainforests. Within these varied habitats, iguanids may be either terrestrial or arboreal, with some switching between the two depending on time of day and outdoor temperature. The most unusual habitat among the iguanids is that of the well-known marine iguana of the Galápagos Islands. This lizard actually frequents salt water, where it forages marine algae to fill its diet.

Behavior

Iguanids are frequently seen stretched out on a rock or otherwise basking in the morning sun. This behavior falls under the heading of thermoregulation. Because lizards are poikilothermic, or "cold blooded," they engage in basking to raise their body temperature to prepare for the day's activities of feeding, perhaps breeding, and evading ever-present predators. Many iguanids spend the night in burrows and emerge each morning to find a sunny spot and raise their internal temperature. As the day gets hotter, ground-dwellers will move into shadier spots so they do not overheat. The more arboreal species may move from a basking spot on an outer tree limb or the sunward wall of a house to the inner branches of a tree or shrub, or closer to the ground where the temperature is lower.

The typical iguanid spends the day in a high state of alert, as predators abound, particularly for the small- and medium-

The Barrington iguana (*Conolophus pallidus*) is endemic to Barrington Island, Galápagos. (Photo by R. A. Mittermeier. Bruce Coleman, Inc. Reproduced by permission.)

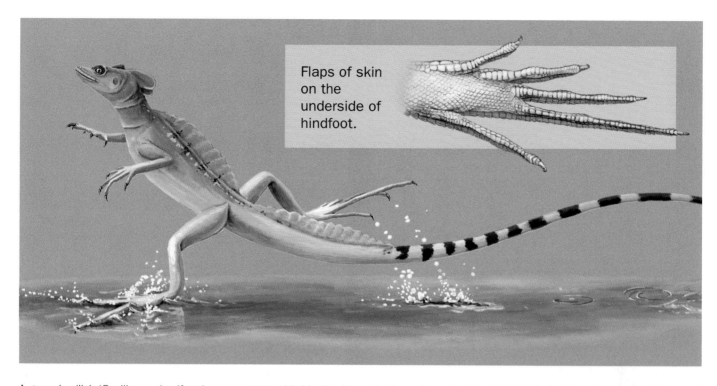

Flaps of skin on the underside of hindfoot.

A green basilisk (*Basiliscus plumifrons*) runs on water aided by the flaps of skin on the underside of its hindfoot. (Illustration by Emily Damstra)

sized species. Many iguanids are cryptically colored and patterned, and are best served by remaining still when a predator passes. Others are quick runners and dash off almost immediately after they spot an animal that is even remotely threatening. Favorite hiding places for these runners are crevices between or under rocks, tangles of vegetation, or anything else that provides adequate cover. Some, like the zebra-tailed lizard (*Callisaurus draconoides*), start and stop while running or abruptly change direction when momentarily out of view to confound their pursuer. Various lizards, including *Uma* species, conceal themselves by squirming under the sand. A few inflate their bodies to avert capture. Common chuckwallas (*Sauromalus obesus*) are well known for this practice, and inflate their bodies when hiding in a narrow crevice. The enlarged size makes them nearly impossible to extract. The horned lizards also puff up their bodies, but the result is an erection of their numerous spines, an excellent deterrent against biting predators. Several species of these small lizards will go a step further if the attacker is persistent, and squirt up to one-third of their blood supply out of their eye pores. Foxes and coyotes, who are the most common recipients of this behavior, seem to find the blood distasteful and release the lizards.

Another unusual predator-avoidance strategy is employed by the so-called Jesus Christ lizards (*Basiliscus* sp.) that appear to "walk on water" while making their escape. Normally terrestrial, the basilisks drop into the water when threatened and race on their hind legs across the water surface. Studies have shown that fringes on the hind toes trap a bubble of air beneath their feet and keep them from sinking if they are running quickly enough. Basilisks have been known to race across

water as far as 100 ft (30.5 m). Numerous other iguanids engage in bipedal running, and some are very quick sprinters, but none run across the water like the basilisks.

Behaviors between members of the same species also vary among the iguanids. Territoriality is prevalent, and many species exhibit stereotyped behaviors, such as doing repeated "push-ups" or marking their turf with secretions from femoral pores, to defend specific areas for feeding or breeding. In many species, territoriality becomes more pronounced when food resources are low. In some cases, dominance hierarchies may develop. Research on green iguanas (*Iguana iguana*) has shown that dominant males typically have higher levels of testosterone, as well as particularly strong jaws and enlarged femoral pores. These males were able to attain and defend prime territories against other males, and also to attract more females.

Among many species, including *Cyclura carinata* and *Sauromalus obesus*, territoriality becomes more pronounced during the breeding season. The males of some species captivate potential mates by certain actions. One of the most noticeable mating displays is the flaring of the throat fan, or dewlap, which is often seen in anoles (*Anolis* sp.). Anoles also flare the dewlap during territorial displays, which may include head bobbing and posturing.

Feeding ecology and diet

Most iguanids are either insectivorous or herbivorous, with the larger lizards tending toward a completely vegetarian diet. The iguanines, for instance, are primarily herbivorous as

Green iguanas (*Iguana iguana*) hatching in Central America. (Photo by Animals Animals ©Kevin & Suzette Hanley. Reproduced by permission.)

adults, but the young also eat arthropods. Some members of this family, such as *Gambelia* species, will eat other lizards, and a few species are cannibalistic.

The insectivorous iguanids typically either stalk their prey or wait in ambush, particularly if the lizard is cryptically colored. After striking out at an insect or other arthropod, most lizards swallow the prey quickly. The common side-blotched lizard (*Uta stansburiana*) engages in the unusual practice of banging the prey on the ground several times before consuming it. Horned lizards are distinguished by their diet, which consists almost wholly of ants. These small lizards must eat considerable numbers of ants to gain adequate nutrition. Since ants produce formic acid to deter predators, the lizards may smell of formic acid as a result.

Reproductive biology

Most iguanids lay eggs, but a few give birth to live young. These include some *Phrynosoma*, *Sceloporus*, and *Leiolamus* species, as well as *Corytophanes percarinatus*. The number of eggs ranges widely. Many iguanids, such as the green anole (*A. carolinensis*), lay only one or two eggs at a time, but others lay dozens. The horned lizard (*P. cornutum*) commonly lays 24 eggs, but the number sometimes reaches more than

36 per clutch. A green iguana female may lay more than 60 eggs at a time.

In many species, optimal environmental conditions can spawn additional clutches. For example, some *Dipsosaurus*, *Crotaphytus*, and *Gambelia* species that live in cooler climates have only one clutch each year, while warmer-climate populations may have two, three, four, or more clutches.

Parental care among most iguanids is either nonexistent, or is restricted to simply burying the eggs or engaging in short-term nest guarding. The iguanines are most known for nest guarding. The rhinoceros iguana (*Cyclura cornuta cornuta*), for example, lays from two to nearly three dozen eggs, then tenaciously guards her nest for several days. The Jamaican iguana (*Cyclura collei*) guards her nest for up to two weeks. Among the iguanines, nest-guarding behaviors may range from displays to physical attacks.

Conservation status

The 2002 IUCN Red List includes 2 iguanids as Extinct, 6 as Critically Endangered, 4 as Endangered, 12 as Vulnerable, 1 as Lower Risk/Near Threatened, and 17 as Data Deficient. The extinct species are both tropidurines: *Leiocephalus eremitus* and *L. herminieri*. In addition, other iguanid species that are not listed by the IUCN are declining in either number or range due to overcollecting or habitat disruption and destruction. Introduced species also may play a role in these declines. For example, numbers of horned lizards in southern California have dropped in association with the proliferation of Argentine ants, which are replacing the native ants that make up the bulk of the lizards' diet.

A number of conservation activities are under way to protect threatened iguanids. For example, the Jamaican iguana (*Cyclura collei*), was believed extinct until a dead lizard was found in 1970 and a second individual, this time a living male, was discovered 20 years later. Since that time, the Jamaican Iguana Research and Conservation Group formed to survey the lizard's present habitat, which is now limited to a small peninsula west of Kingston. The population, estimated at about 100 individuals, is faced with several threats in the wild, including predation of the eggs (and likely the young) by the Indian mongoose (*Herpestes javanicus* [=*auropunctatus*]). Researchers now hope to boost the population by gathering eggs from the wild and rearing the hatchlings at the Hope Zoo in Jamaica until the lizards are large enough to avoid predation. The Natural Resources Conservation Authority is also taking measures, including promoting the designation of the area as a national park, to protect the lizard's current habitat from deforestation and development. Other proposals include introducing the lizards to areas such as Great Goat Island, which

supported a population of Jamaican iguanas through the middle of the twentieth century.

Another at-risk iguanid is the Turks and Caicos iguana (*Cyclura carinata carinata*), which is native to these islands north of Haiti and the Dominican Republic. The lizards were once widespread, but the population is now estimated to number about 30,000 adults, and these are restricted to the smaller islands. Introduced mammals are the largest threat. Cats and dogs prey on the lizards, and now-wild livestock competes with the mainly herbivorous lizards for vegetation. Habitat destruction is also taking a toll. In an attempt to prevent the continued decline of the lizard population, the National Trust for the Turks and Caicos Islands worked with the government to draft legislation protecting the lizards from additional introduced mammals. The trust has also trapped cats in vulnerable areas. Other efforts include recently installed viewing platforms to keep tourists away from the lizards, the introduction of tourist fees to support further conservation efforts, and new educational programs to spread the word about the lizards and their plight.

Significance to humans

Many iguanid lizards are prized as sources of food meat, particularly in the southwestern United States and Mexico. They are also a mainstay in the pet lizard trade. Perhaps most importantly, they are valued by nature lovers, who can spend hours watching the antics of the lizards.

1. Male common chuckwalla (*Sauromalus obesus*); 2. Desert iguana (*Dipsosaurus dorsalis*); 3. Male zebra-tailed lizard (*Callisaurus draconoides*); 4. Coachella Valley fringe-toed lizard (*Uma inornata*); 5. Green anole (*Anolis carolinensis*); 6. Male common lesser earless lizard (*Holbrookia maculata*); 7. Male cape spinytail iguana (*Ctenosaura hemilopha*). (Illustration by Bruce Worden)

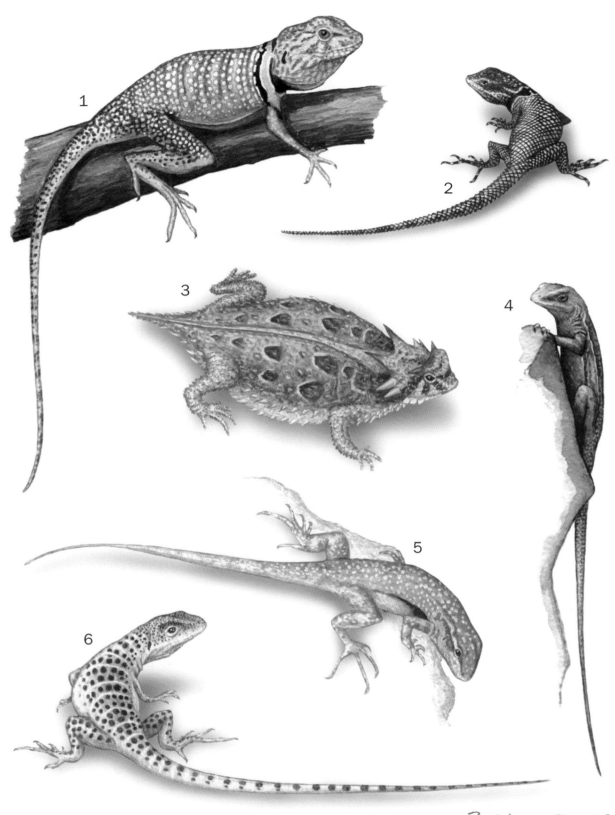

1. Male common collared lizard (*Crotaphytus collaris*); 2. Mountain spiny lizard (*Sceloporus jarrovii*); 3. Texas horned lizard (*Phrynosoma cornutum*); 4. Female common sagebrush lizard (*Urosaurus graciosus*); 5. Male common side-blotched lizard (*Uta stansburiana*); 6. Male long-nosed leopard lizard (*Gambelia wislizenii*). (Illustration by Bruce Worden)

Species accounts

Common collared lizard
Crotaphytus collaris

SUBFAMILY
Crotaphytinae

TAXONOMY
Crotaphytus collaris Say, 1823, Verdigris River, Oklahoma. Six subspecies are recognized.

OTHER COMMON NAMES
English: Mountain boomer, yellowhead collared lizard, western collared lizard, eastern collared lizard, Chihuahuan collared lizard, Sonoran collared lizard; French: Lézard à collier; German: Halsbandleguan; Spanish: Lagartija de collar.

PHYSICAL CHARACTERISTICS
Collared lizards are sturdy-looking lizards with a large head, long, round tail, and two dark, often-broken rings around the neck. The body ranges from yellowish green to blue, olive, or grayish brown, is dotted with small whitish spots, and is sometimes tinged with orange. Males are typically more brightly colored than females. Adults may reach 14 in (35.5 cm), with the tail comprising two-thirds of that length.

DISTRIBUTION
Collared lizards are found in the central and west-central United States and into northeastern Mexico.

HABITAT
These lizards prefer hilly, rocky glades and prairies with little shade.

BEHAVIOR
These are exceptionally cautious lizards that need little provocation to run (often on two feet at high speed) to hiding places under and between rocks. They are aggressive when cornered, and will either bite, or show their intent to bite by gaping at the attacker. Despite the common name of mountain boomer, these lizards are silent.

FEEDING ECOLOGY AND DIET
Collared lizards stalk or ambush various insects and spiders, along with occasional small lizards. They are known to wave their long tails before snapping out at their prey.

REPRODUCTIVE BIOLOGY
The breeding season of the collared lizard is in May and June. In July, females lay about a half dozen eggs per clutch, either in a burrow beneath rocks or buried several inches deep in the sand. Some females, particularly those in warmer climates, have two clutches per year. The eggs hatch two to three months later.

CONSERVATION STATUS
Not listed by the IUCN.

SIGNIFICANCE TO HUMANS
None known. ◆

Long-nosed leopard lizard
Gambelia wislizenii

SUBFAMILY
Crotaphytinae

TAXONOMY
Gambelia wislizenii Baird and Girard, 1852, Sante Fe, New Mexico. Four subspecies are recognized.

OTHER COMMON NAMES
English: Leopard lizard, small-spotted lizard, Cope's lizard, Lahontan Basin lizard, pale leopard lizard; French: Lézard léopard; German: Leopardleguan; Spanish: Cachora leopardo.

PHYSICAL CHARACTERISTICS
Long-nosed leopard lizards have dark-gray to black spots covering a light-gray to olive-gray back and tail. The coloration varies between individuals, and depending on temperature, even in the same lizard. These lizards frequently sport a series of white bands on the back and tail, and females often develop orange splotches along the sides during breeding season. Adults can reach 17 in (43 cm), including a long tail that makes up slightly more than two-thirds of the total length.

DISTRIBUTION
Long-nosed leopard lizards occur from the west-central United States to north-central Mexico.

HABITAT
These lizards live in flat desert areas with little vegetation.

BEHAVIOR
To escape predators, long-nose leopard lizards will either race (often on two feet) into burrows or dash to the cover of low-

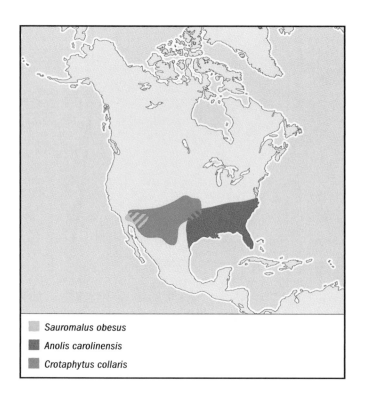

Sauromalus obesus
Anolis carolinensis
Crotaphytus collaris

Sceloporus jarrovii
Gambelia wislizenii

Phrynosoma cornutum
Ctenosaura hemilopha

lying vegetation. Cornered or captured lizards become aggressive and will bite their attacker.

FEEDING ECOLOGY AND DIET
The long-nosed leopard lizard is a diurnal species that stalks its meals of insects and spiders, frequently lunging airborne after a tasty arthropod. It will also eat other lizards, including those of its own species.

REPRODUCTIVE BIOLOGY
The breeding season of the long-nosed leopard lizard runs through May. In July, a female will lay about two to four eggs, although clutches of nearly a dozen are possible. Individuals in warmer climates may lay a second clutch. The eggs hatch within one to two months.

CONSERVATION STATUS
Not listed by the IUCN.

SIGNIFICANCE TO HUMANS
None known. ◆

Cape spinytail iguana
Ctenosaura hemilopha

SUBFAMILY
Iguaninae

TAXONOMY
Ctenosaura hemilopha Cope, 1863, Cape St. Lucas, Baja California. Five subspecies are recognized (some taxonomists list six).

OTHER COMMON NAMES
English: Northern false iguana, spiny-tailed iguana; French: Iguane communà queue espineuse; German: Schwarze Leguane; Spanish: Iguana del Cabo.

PHYSICAL CHARACTERISTICS
Cape spinytail iguanas are large, stocky, wrinkly-looking lizards, with a strongly ridged tail and crest of scales running along the top of the back. The crest is more pronounced in males than females. Adults can reach 3 ft (0.9 m) in total length, including their long tails, which can extend about one and a half times the length of the body.

DISTRIBUTION
These iguanas occur in northwestern Mexico, including the state of Sonora, and on islands in the Gulf of California.

HABITAT
Cape spinytail iguanas live in areas with numerous rocky crevices and frequently with some trees.

BEHAVIOR
Cape spinytail iguanas are territorial, and the males form dominance hierarchies when habitat is limited. Although they will fight back aggressively with their strong legs, jaws, and tail if cornered or handled, they usually opt to run into a rock crevice for cover when they feel threatened.

FEEDING ECOLOGY AND DIET
These lizards are diurnal, and herbivorous, dining on flowers, fruits, and leaves of native vegetation. They occasionally eat invertebrates.

REPRODUCTIVE BIOLOGY
These colonial lizards typically form groups with one lead male, several subordinate males, and a harem of females who are dominant to the subordinate males. They are oviparous, laying two dozen or more eggs in one clutch per year. The eggs hatch in about three months.

CONSERVATION STATUS
Not listed by the IUCN.

SIGNIFICANCE TO HUMANS
These lizards are sold as pets and are a minor source of food for humans. ◆

Desert iguana
Dipsosaurus dorsalis

SUBFAMILY
Iguaninae

TAXONOMY
Dipsosaurus dorsalis Baird and Girard, 1852, Colorado Desert, California. Four subspecies are recognized.

OTHER COMMON NAMES
English: Northern crested lizard, crested lizard, desert lizard; French: Iguane du désert; German: Wüstenleguan; Spanish: Iguana del desierto, cachoron guero.

PHYSICAL CHARACTERISTICS
Desert iguanas are robust lizards with a crest of raised, enlarged scales along the top of the back. They have whitish bellies, slate-colored backs that are spotted with white, and striped tails. Males have reddish, posterior markings. Adults can reach 15 in (38 cm) long, including a tail that is almost twice as long as the body.

DISTRIBUTION
Desert iguanas occur in the southwestern United States and northwestern Mexico, including islands in the Gulf of California.

HABITAT
These lizards live in open desert, often amidst scrubby bushes, where they occupy burrows and rock crevices.

Dipsosaurus dorsalis
Holbrookia maculata

BEHAVIOR
Like the chuckwallas, desert iguanas will scurry into a crevice when threatened and puff up their bodies to avert being extracted by a predator.

FEEDING ECOLOGY AND DIET
Desert iguanas are diurnal, primarily herbivorous lizards that feed on the foliage and fruit of bushes and other desert plants, but will also eat invertebrates. They are territorial over feeding areas.

REPRODUCTIVE BIOLOGY
The breeding season of the desert iguana runs from spring to midsummer. Females typically lay three to eight eggs from early to late summer, and under optimum conditions may have a second clutch.

CONSERVATION STATUS
Not listed by the IUCN.

SIGNIFICANCE TO HUMANS
None known. ◆

Common chuckwalla
Sauromalus obesus

SUBFAMILY
Iguaninae

TAXONOMY
Sauromalus obesus Baird, 1858, Fort Yuma, Arizona. Four subspecies are recognized. Discussions are underway over the placement of this species under the name *Sauromalus ater.*

OTHER COMMON NAMES
English: Glen canyon chuckwalla, western chuckwalla, chuck, chuckawalla, iguana; French: Chuckwalla; German: Chuckwalla; Spanish: Chacahuala.

PHYSICAL CHARACTERISTICS
Chuckwallas are robust lizards, with many loose folds of skin on the body, and a tail that stretches as long as the body. Their body color varies geographically, from the common grayish brown hue to brownish red and black. They are also known to change colors in response to environmental conditions. Males typically have dark heads and forelimbs. Females and juveniles are commonly distinguished by some banding, but adult males in some areas are also banded. Adults can reach 16–18 in (41.6–45.7 cm) in total length.

DISTRIBUTION
These lizards occur in the southwestern United States to northwestern Mexico, including parts of Arizona, Nevada, Utah, the Mojave and Sonoran Deserts, and in Baja California.

HABITAT
Chuckwallas prefer a rocky desert setting with plenty of crevices and other small hiding places.

BEHAVIOR
These are quick lizards with a proficient predator-avoidance strategy. When pursued or otherwise threatened, a chuckwalla will race into a crevice, take a deep breath, and inflate its body so that extraction from the point of entry is nearly impossible. They spend the winter deep within rock crevices, but individu-

als will occasionally poke a nose outside on a particularly warm day. Chuckwallas are sometimes territorial.

FEEDING ECOLOGY AND DIET
Chuckwallas are herbivorous and feed diurnally, foraging for vegetation after warming up by basking in the desert sun. Limited food resources can negatively affect their growth rate.

REPRODUCTIVE BIOLOGY
Mating among the chuckwallas begins in the spring. About two months later, each female lays five to 16 eggs (usually fewer than 10) in depressions or tunnels under stones. The young typically hatch in early fall.

CONSERVATION STATUS
Not listed by the IUCN.

SIGNIFICANCE TO HUMANS
These lizards are used in the pet trade and are also a minor source of human food. ◆

Zebra-tailed lizard
Callisaurus draconoides

SUBFAMILY
Phrynosomatinae

TAXONOMY
Callisaurus draconoides Blainville, 1835, northern Baja California. Nine subspecies are recognized.

OTHER COMMON NAMES
English: Gridiron-tailed lizard, northern zebra-tailed lizard, western zebra-tailed lizard, eastern zebra-tailed lizard, Nevada zebratail, Mojave zebratail; German: Zebraschwanzleguan; Spanish: Lagartija cachora, perrita.

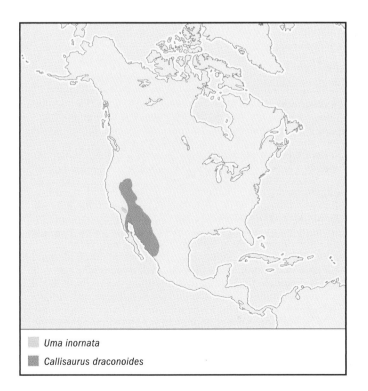

▨ *Uma inornata*
▨ *Callisaurus draconoides*

PHYSICAL CHARACTERISTICS
Zebra-tailed lizards are speckled, light gray to brown lizards with particularly long forelimbs and a flat, broadly banded tail. The dark tail bands continue on to the light-colored ventral tail, where they are particularly noticeable. Males typically are distinctly blue on the belly, and display a pink throat fan during breeding season. Adults reach about 10 in (25.4 cm) in total length, with the tail about one and a half times as long as the body.

DISTRIBUTION
These lizards occur in the southwestern United States, northwestern to west-central Mexico.

HABITAT
The zebra-tailed lizard prefers a sand or gravel substrate in desert canyon washes, scrubby plains, and other areas that are dry for much of the year.

BEHAVIOR
This wary species is noted for its swiftness. When threatened by a predator, the zebra-tailed lizard will arch its tail over its back and wave the particularly conspicuous ventral banding at the pursuer. As the predator's attention is drawn to that location, the lizard speeds off to cover in a dizzying pattern of stops, starts, and sharp turns.

FEEDING ECOLOGY AND DIET
This mainly diurnal lizard will consume tender spring vegetation, but mostly feeds on insects and spiders. A sit-and-wait hunter, it may suddenly leap at prey, sometimes jumping a foot or more to secure its meal.

REPRODUCTIVE BIOLOGY
Male zebra-tailed lizards flare the throat fan during the breeding season. During the summer, females typically lay two to six eggs per clutch, and a female may have up to five clutches in one year.

CONSERVATION STATUS
Not listed by the IUCN.

SIGNIFICANCE TO HUMANS
None known. ◆

Common lesser earless lizard
Holbrookia maculata

SUBFAMILY
Phrynosomatinae

TAXONOMY
Holbrookia maculata Girard, 1851, opposite Grand Island, Platte River, Nebraska. Nine subspecies are recognized.

OTHER COMMON NAMES
English: Mountain earless lizard, speckled earless lizard, band-tailed earless lizard, bleached earless lizard, Bunker's earless lizard, Huachuca earless lizard, eastern earless lizard, northern earless lizard, western earless lizard, spotted lizard; Spanish: Lagartija.

PHYSICAL CHARACTERISTICS
Common lesser earless lizards are gray, with rather regular dark blotches down the back and onto the sides and tail. They lack

the ear openings typical of other lizards. Males have orange to red throat patches. Adults average about 5 in (12.7 cm) in total length, including a tail that is about as long as the body in males and slightly shorter in females. Subspecies have slightly different scalation patterns.

DISTRIBUTION
These lizards occur in the central United States as far north as Nebraska and Wyoming, and also in northern Mexico.

HABITAT
Lesser earless lizards live in sandy, open areas with sparse vegetation that provides shady retreats.

BEHAVIOR
These are quick lizards that race between burrows or the shade of a bush during the day. The males are often territorial. As the breeding season progresses, the females become more vividly colored, eventually developing a bright yellow head and reddish-orange sides and hind legs. The males use a series of rapid head bobs to court females who are just beginning to show the breeding coloration. Later-season, high-intensity coloration appears to signal the end of a female's receptivity to the male's approach.

FEEDING ECOLOGY AND DIET
The diet of the lesser earless lizard consists of insects, especially grasshoppers and ants, and spiders. Occasionally it will eat other, smaller lizards.

REPRODUCTIVE BIOLOGY
Breeding in the lesser earless lizard occurs in early summer. About two months later, the females lay between three and 11 eggs in an underground burrow. Younger females have one clutch per year; older females may have a second clutch.

CONSERVATION STATUS
Not listed by the IUCN.

SIGNIFICANCE TO HUMANS
None known. ◆

Texas horned lizard
Phrynosoma cornutum

SUBFAMILY
Phrynosomatinae

TAXONOMY
Phrynosoma cornutum Harlan, 1825, Great Plains, east of the Rocky Mountains. No subspecies are recognized.

OTHER COMMON NAMES
English: Horned toad, horny toad, San Diego horned lizard, California horned lizard; French: Lézard crapaud ou cornu; German: Texas-Krötenechse; Spanish: Torito de la Virgen.

PHYSICAL CHARACTERISTICS
Texas horned lizards are squat and spiny, with a short tail and two large occipital "horns" on the head. They also have a large blotch on either side of the neck, and large, dark spots on the back. The general body color is brown, sometimes with a yellowish or reddish tint, but color varies with external temperature. Adults are typically about 3 in long (7.6 cm), but some may reach 5 in (12.7 cm).

DISTRIBUTION
Texas horned lizards occur in the west-central United States, and in northern Mexico.

HABITAT
These lizards are found nearly anywhere throughout their range where it is dry, flat, and sparsely vegetated.

BEHAVIOR
Texas horned lizards are diurnal. They peculiarly squirt blood from their eyes. This behavior may be observed while the lizard is shedding, or when it is seriously threatened. Other defensive tactics include hissing, inflating the body, and erecting the horns at an attacker.

FEEDING ECOLOGY AND DIET
Ants are the primary component of the diet of the Texas horned lizard, although the lizards will also eat other arthropods. They drink by gathering raindrops on their back, then channeling the drops through their scales and into the mouth.

REPRODUCTIVE BIOLOGY
Texas horned lizards mate in spring, and females lay an unusually large number of eggs (two to three dozen) which they bury about 6 in (15.2 cm) deep. The young typically hatch in July and early August.

CONSERVATION STATUS
Not listed by the IUCN, but the species has disappeared from about one-half of its historical range. Their disappearance coincides with a loss of habitat, as well as a proliferation of fire ants. Fire ants have displaced harvester ants, a mainstay of the lizard's diet.

SIGNIFICANCE TO HUMANS
An unusual and well-known reptile, it is the state reptile of Texas. ◆

Mountain spiny lizard
Sceloporus jarrovii

SUBFAMILY
Phrynosomatinae

TAXONOMY
Sceloporus jarrovii Cope, 1875, Southern Arizona. Eight subspecies are recognized.

OTHER COMMON NAMES
English: Yarrow's spiny lizard; French: Lézard épineux, lézard de palisades; German: Jarrov Zaunleguan; Spanish: Lagartija espinosa de Yarrow.

PHYSICAL CHARACTERISTICS
Mountain spiny lizards are robust lizards with finely patterned skin. The dorsal body is dark gray to black and has large scales, each of which contains a central, white dot. The male's belly and sides range from blue to black to gray; the female's belly is usually whitish. The overall coloration of these lizards can vary depending on environmental conditions. Adults can reach about 9 in (23 cm) long, with the tail making up one-half to three-fifths of that length.

DISTRIBUTION
These lizards occur in the western United States to northwest and west-central Mexico.

HABITAT
Mountain spiny lizards are found in mountainous areas at elevations between 5,000 and 10,000 ft (1,520 and 3,050 m). They prefer rocky, well-forested areas.

BEHAVIOR
These lizards engage in extended morning basking periods, and are less wary than most other lizards. They are territorial and will defend larger areas when food is limited. During lean times, they may also alter their diet and become less active.

FEEDING ECOLOGY AND DIET
Mountain spiny lizards eat insects and spiders, as well as lizards, including other mountain spiny lizards.

REPRODUCTIVE BIOLOGY
Most lizards lay eggs, but mountain spiny lizards bear live young, which are born in the spring. The gestation period is about five months, but since females can store sperm, breeding can occur several months earlier. A typical litter ranges from five to eight, but a few more or less is not unusual.

CONSERVATION STATUS
Not listed by the IUCN.

SIGNIFICANCE TO HUMANS
None known. ◆

Coachella Valley fringe-toed lizard
Uma inornata

SUBFAMILY
Phrynosomatinae

TAXONOMY
Uma inornata Cope, 1895, Colorado Desert, Riverside County, California. No subspecies recognized.

OTHER COMMON NAMES
English: Coachella uma, Coachella sand-lizard; German: Fransenzehenleguan.

PHYSICAL CHARACTERISTICS
Coachella Valley fringe-toed lizards are light gray with delicate black speckles on the back and dorsal tail. These speckles may appear as black-outlined, white circles with small, black, central dots. The speckles typically run together to form discontinuous, lateral stripes that extend from the head to about the middle of the body. The Coachella Valley fringe-toed lizard is also distinguished by a fringed fourth toe. Adults can reach about 13 in (33 cm) in total length, of which one-half to two-thirds may be tail.

DISTRIBUTION
These lizards occur only in the Coachella Valley in southern California's Riverside County.

HABITAT
Coachella Valley fringe-toed lizards live in dunes and other desert areas with loose, sandy soil and some low vegetation.

BEHAVIOR
When not scurrying about the dunes, this lizard is either hiding in a bush or a burrow, or buried in the sand. It accomplishes the latter avoidance tactic by maneuvering its hind limbs and body (so-called sand swimming) to wiggle itself down into the substrate.

FEEDING ECOLOGY AND DIET
Primarily insectivorous, these diurnal lizards occasionally eat vegetation.

REPRODUCTIVE BIOLOGY
Female Coachella Valley fringe-toed lizards lay one to four eggs from late spring to early fall.

CONSERVATION STATUS
Listed as Endangered by the IUCN.

SIGNIFICANCE TO HUMANS
None known. ◆

Common sagebrush lizard
Urosaurus graciosus

SUBFAMILY
Phrynosomatinae

TAXONOMY
Urosaurus graciosus Baird and Girard, 1852, Southern California. Two subspecies are recognized.

OTHER COMMON NAMES
English: Long-tailed uta, western sagebrush lizard, northern sagebrush lizard, southern sagebrush lizard, brush lizard, Arizona brush lizard; Spanish: Lagartija de Matorral, cachora coluda.

PHYSICAL CHARACTERISTICS
Sagebrush lizards are light grayish brown lizards with a very long tail. They have a series of thin, dark lines branching across the head and running down the back and sides. The lizards are generally darker in color in the early morning and become lighter as the temperature rises. Blue blotches are

▨ *Urosaurus graciosus*
■ *Uta stansburiana*

frequently noticeable on the belly. Adults can reach 10.5 in (26.7 cm), including a tail that makes up about 70% of the total length.

DISTRIBUTION
Sagebrush lizards occur in the southwestern United States and in Baja California, Mexico.

HABITAT
These lizards live among sagebrush and other shrubs, as well as in mesquite and creosote trees, in otherwise open desert areas.

BEHAVIOR
This lizard is known as a climber and is frequently seen stretched out perfectly still on branches, tree trunks, and other vertical surfaces. When threatened, they scamper into thicker vegetation, or abandon their arboreal habits altogether and run into burrows hidden beneath vegetation.

FEEDING ECOLOGY AND DIET
Sagebrush lizards mostly eat insects. They do much of their foraging in vegetation, but will venture onto the desert ground when necessary.

REPRODUCTIVE BIOLOGY
Female sagebrush lizards commonly lay clutches of two to five eggs from late spring to late summer, although clutch sizes of a dozen eggs have been reported. In cooler climates, one clutch is common. Females in warmer climates may lay up to six clutches a year.

CONSERVATION STATUS
Not listed by the IUCN.

SIGNIFICANCE TO HUMANS
None known. ◆

Common side-blotched lizard
Uta stansburiana

SUBFAMILY
Phrynosomatinae

TAXONOMY
Uta stansburiana Baird and Girard, 1852, Salt Lake Valley. Six subspecies are recognized.

OTHER COMMON NAMES
English: Northern ground uta, western ground uta, western side-blotched lizard, Nevada side-blotched lizard, northern side-blotched lizard, eastern side-blotched lizard, plateau side-blotched lizard, uta, Stansbury's swift, northern brown-shouldered lizard; German: Seitenfleckleguan; Spanish: Cachora del suelo, Lagartija de manchas lateralis.

PHYSICAL CHARACTERISTICS
Side-blotched lizards are mostly brown, with five, light-colored, dorsal stripes running from the head at least to the beginning of the tail. These stripes are sometimes indistinct. During breeding season, the males often take on a distinctly bluish hue on the head, tail, top of the back, and sides.

DISTRIBUTION
Side-blotched lizards occur in the western United States, from Washington State south to Baja California and northern Mexico.

HABITAT
The side-blotched lizard is a wide-ranging species that lives in the flat desert, as well as rocky outcrops, hills, and even mountainous regions. It usually prefers sites with some vegetation.

BEHAVIOR
Side-blotched lizards have the unusual habit of grasping their prey in their mouths, then beating it on the ground before consuming it. They are less skittish than many other lizards, but if approached too closely, they will run for a burrow, bush, or other cover.

FEEDING ECOLOGY AND DIET
These lizards are primarily insectivorous, but they are also known to eat other arthropods, including scorpions.

REPRODUCTIVE BIOLOGY
Breeding in side-blotched lizards begins in early spring, and females commonly lay one to two clutches of three or four eggs each from spring to midsummer. Females in warmer climates sometimes lay up to seven clutches in a single year. Most eggs hatch from June to August. They have unusually speedy development, and the hatchlings reach maturity before the year is out.

CONSERVATION STATUS
Not listed by the IUCN.

SIGNIFICANCE TO HUMANS
These lizards help control the population of insects that are agricultural pests. ◆

Green anole
Anolis carolinensis

SUBFAMILY
Polychrotinae

TAXONOMY
Anolis carolinensis Voigt, 1832, Carolina. Two subspecies are recognized.

OTHER COMMON NAMES
English: American chameleon, Carolina anole; French: Anolis de Caroline; German: Rotkehlanolis; Spanish: Anolis verde.

PHYSICAL CHARACTERISTICS
Slim lizards, green anoles have pointed heads and long tails that can be more than twice the length of the body. They have the ability to switch from their usual green coloration to brown or gray. Males display a conspicuous usually reddish-pink throat fan, known as a dewlap. Adults typically range from 5 to 8 in (12.7 to 20.3 cm) in total length, and about 2.5 in (6.3 cm) from snout to vent. Males are usually longer than females of similar age.

DISTRIBUTION
Green anoles occur in the southeastern United States as far west as Texas and north to Virginia. They are also found in Cuba and on islands in the Caribbean.

HABITAT
Green anoles prefer moist, vegetated habitats with climbing surfaces, including trees, vines, and bushes, as well as fences and house walls. They are commonly found near the ground, but climb higher to bask.

BEHAVIOR

Green anoles are territorial, and males exert their dominance by flaring their dewlaps, bobbing their heads, erecting the dorsal skin into a crest, and engaging in stereotypical posturing to enlarge their body image. During the breeding season, males also use their dewlaps to entice females, which appear to choose a mate based on the coloration and possibly the ultraviolet reflectance of the dewlap.

FEEDING ECOLOGY AND DIET

The diet of the green anole includes small insects and spiders. Diurnal feeders, these lizards typically stalk their prey.

REPRODUCTIVE BIOLOGY

Breeding occurs from spring to fall. Females typically lay a single egg, and bury it in leaf litter and loose soil. The young hatch five to seven weeks later. Females can continue to lay eggs throughout the long reproductive season, as often as once every two weeks.

CONSERVATION STATUS

Not threatened.

SIGNIFICANCE TO HUMANS

These lizards are common in the pet trade. ◆

Resources

Books

Burghardt, G., and A. Rand, eds. *Iguanas of the World: Their Behavior, Ecology, and Conservation.* Park Ridge, NJ: Noyes Publications, 1982.

Campbell, J. A. *Amphibians and Reptiles of Northern Guatemala, the Yucatan and Belize.* Animal Natural History Series, Vol. 4. Norman: University of Oklahoma Press, 1998.

Cei, J. M. *New Species of* Tropidurus *(Sauria, Iguanidae) from the Arid Chacoan and Western Regions of Argentina.* Lawrence: University of Kansas, Museum of Natural History, 1982.

Cogger, H., and R. Zweifel, eds. *Encyclopedia of Reptiles and Amphibians.* San Diego: Academic Press, 1998.

Conant, R., J. T. Collins, I. H. Conant, T. R. Johnson, and S. L. Collins. *A Field Guide to Reptiles & Amphibians of Eastern & Central North America.* Peterson Field Guide Series. Boston: Houghton Mifflin Co., 1998.

Crother, B. I., ed. *Caribbean Amphibians and Reptiles.* San Diego: Academic Press, 1999.

Frank, N., and E. Ramus. *A Complete Guide to Scientific and Common Names of Reptiles and Amphibians of the World.* Pottsville, PA: NG Publishing Inc., 1996.

Grismer, L. L., and H. W. Greene. *Amphibians and Reptiles of Baja California, Including Its Pacific Islands and the Islands in the Sea of Cortés (Organisms and Environments).* Berkeley: University of California Press, 2002.

Lee, J. C. *A Field Guide to the Amphibians and Reptiles of the Maya World: The Lowlands of Mexico, Northern Guatemala, and Belize.* Ithaca, NY: Cornell University Press, 2000.

Milstead, W., ed. *Lizard Ecology: A Symposium.* Columbia: University of Missouri Press, 1967.

Pianka, E. *Ecology and Natural History of Desert Lizards: Analyses of the Ecological Niche and Community Structure.* Princeton, NJ: Princeton University Press, 1986.

Pianka, E. R., and L. J. Vitt, eds. *Lizard Ecology.* Berkeley: University of California Press, 1994.

———. *Lizards: Windows to the Evolution of Diversity.* Berkeley: University of California Press, 2003.

Savage, J. M., M. Fogden, and P. Fogden. *The Amphibians and Reptiles of Costa Rica: A Herpetofauna Between Two Continents, Between Two Seas.* Chicago: University of Chicago Press, 2002.

Smith, H. M. *Handbook of Lizards: Lizards of the United States and Canada.* Ithaca, NY: Comstock Publishing Co., 1946.

Stebbins, R. *A Field Guide to Western Reptiles and Amphibians.* Boston: Houghton Mifflin Co., 1985.

Periodicals

Bealor, M. T., and C. Krekorian. "Chemosensory Identification of Lizard-Eating Snakes in the Desert Iguana, *Dipsosaurus dorsalis* (Squamata: Iguanidae)." *Journal of Herpetology* 36, no.1 (2002): 9–15.

Carothers, J. H. "Dominance and Competition in an Herbivorous Lizard." *Behavioral Ecology and Sociobiology* 8 (1981): 261–266.

Frost, D. R., and R. Etheridge. "A Phylogenetic Analysis and Taxonomy of Iguanian Lizards (Reptilia: Squamata)." *University of Kansas, Museum of Natural History Miscellaneous Publications* 81 (1989): 1–65.

Hager, S. B. "The Role of Nuptial Coloration in Female *Holbrookia maculata*: Evidence for a Dual Signaling System." *Journal of Herpetology* 35, no. 4 (2001): 624–632.

Hazen-Hammond, S. " 'Horny Toads' Enjoy a Special Place in Western Hearts." *Smithsonian* 25 (1994): 82–86.

Ramírez-Bautista, A.; O. Ramos Flores, J. W. Sites, Jr. "Reproductive Cycle of the Spiny Lizard *Sceloporus jarrovii* (Sauria: Phrynosomatidae) from North-Central Mexico." *Journal of Herpetology* 36, no. 2 (June 2002): 225–233.

Organizations

Iguana Specialist Group of the World Conservation Union (IUCN). Web site: <http://www.iucn-isg.org/index.php>

The International Iguana Society. 133 Steele Rd., West Hartford, CT 06119 USA. Web site: <http://www.iguana society.com>

Other

Pianka, E. R., and W. L. Hodges. *Horned Lizards.* (cited November 4, 2002). <http://uts.cc.utexas.edu/varanus/phryno.html>

Proliferation of Argentine Ants in California Linked to Decline in Coastal Horned Lizards. February 26, 2002 (cited November 4, 2002). <http://ucsdnews.ucsd.edu/newsrel/science/mclizard.htm>

Leslie Ann Mertz, PhD

Geckos and pygopods
(Gekkonidae)

Class Reptilia

Order Squamata

Suborder Lacertilia

Family Gekkonidae

Thumbnail description
Small to medium-size lizards, often drab in color
and with stocky, flattened bodies and large
heads (geckos) or elongate with no forelimbs
and with hind limbs reduced to tiny flaps
(pygopods)

Size
0.6–10.0 in (16–255 mm) in snout-vent length

Number of genera, species
116 genera; 1,109 species

Habitat
Desert, forest, savanna, and grassland

Conservation status
Extinct: 3 species; Extinct in the Wild: 1
species; Critically Endangered: 1 species;
Endangered: 3 species; Vulnerable: 15 species;
Lower Risk/Near Threatened: 8 species

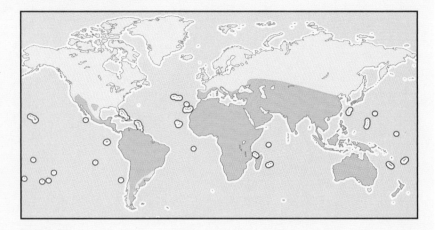

Distribution
Worldwide in tropical, subtropical, and some warm temperate areas

Evolution and systematics

One of the earliest lizards to show some of the features of
geckos is the late Jurassic *Eichstaettisaurus* from Germany. Although it was similar to modern geckos in general appearance, it lacked the derived features that characterize the living
forms, and its true affinities remain the subject of controversy.
The oldest definitive gecko represented in the fossil record is
Hoburogecko suchanovi, which lived in Mongolia about 100 million years ago. Tertiary gecko fossils as well as geckos imbed-
ded in amber have been recorded from numerous localities
around the world, and many of them belong to living genera.

Geckos constitute the bulk of the Gekkota, the sister
group of the Autarchoglossa and one of the three major lineages of lizards. Some researchers regard the xantusiids,
dibamids, and amphisbaenids as allied to the geckos, but the
evidence is equivocal. Geckos (including pygopods) are par-
titioned into 116 genera in four subfamilies: the Eublepharinae (22 species), Diplodactylinae (121 species), Gekkoninae
(930 species), and Pygopodinae (36 species), each of which is
sometimes treated as a separate family. Morphologic features
and at least some DNA sequence evidence support the Eublepharinae as the sister group of the remaining geckos. The
diplodactylines, which are restricted to Australia, New Caledonia, and New Zealand, are allied most closely to the py-
gopodines, which occur chiefly in Australia, with two species
reaching New Guinea. Of the four subfamilies, Gekkoninae
contains the greatest number of species and has, by far, the
widest distribution, occurring throughout the tropics and subtropics worldwide. The origin of the major clades within the

Gekkonidae may be linked to the breakup of the supercontin-
nent of Pangea in the late Jurassic and the subsequent fragmentation of the southern continent of Gondwana during the
Cretaceous and early Tertiary.

Physical characteristics

Most geckos are relatively small (1.2–3.5 in, or 30–90 mm)
with short, somewhat flattened bodies; large heads; large eyes;
and well-developed limbs. Eublepharines have movable eyelids, but they have been replaced by a transparent spectacle
(eye cover) in other members of the subfamily. The smallest
geckos are the sphaeros of the West Indies; the Jaragua
sphaero (*Sphaerodactylus ariasae*) averages only 0.63 in (16 mm)
in snout-vent length. At the other end of the spectrum is the
New Caledonian giant gecko (*Rhacodactylus leachianus*), the
largest living species at more than 10 in (250 mm). The recently extinct Delcourt's giant gecko (*Hoplodactylus delcourti*)
of New Zealand, however, was much larger still at 14.6 in
(370 mm) in snout-vent length.

The feet are one of the most striking and varying aspects
of the morphologic characteristics of the gecko. Eublepharines and some gekkonines and diplodactylines have slender
digits with well-developed claws, but many species have expanded pads on the base and or tips of the toes that permit
adhesion to smooth surfaces. These pads may be distal (at the
tips of the toes) and fan- (*Ptyodactylus*) or leaf-shaped (*Phyllodactylus*), or they may be basal (at the base of the toes) and
arranged in single (*Gekko*) or divided (*Hemidactylus*) rows. The

A white-striped gecko (*Gekko vittatus*) demonstrates the toes and climbing ability of geckos. A. Underside of each foot has specialized toe pads; B. Overlapping scales on each toe; C. Each scale is made up of thousands of hair-like setae; D. Each seta is divided into microscopic filaments. (Illustration by Patricia Ferrer)

first digit of the hands and feet may be reduced in size, but it is never lost entirely.

Pygopods have features that are an exception to the typical gecko body plan. In this lineage the forelimbs have been lost entirely, whereas the hind limbs are reduced to small flaps lying lateral to the vent. Pygopods also have smaller heads and much more elongate bodies (and especially tails) than other members of the family. Pygopods are covered with smooth, imbricating (overlapping) scales. Such scaling is extremely rare among geckos, which usually are covered with small granules, with or without larger keeled tubercles (enlarged scales with a raised ridge) intermixed.

Most geckos are drab in color, in keeping with their nocturnal habits. Browns and grays are the most common colors, and diffuse patterns of chevrons or crossbars characterize many species; in diurnal forms, such as the Malagasy day gecko (*Phelsuma*), bright greens, yellows, reds, and blues may be encountered. The males of many species of geckos possess a series of precloacal glands or femoral pores, or both, on the ventral surface of the groin and thighs.

Distribution

Geckos are chiefly tropical and subtropical in their distribution, but species range as far north as the southwestern United States, southern Europe, and southern Siberia. To the south, geckos reach Stewart Island in New Zealand and approach the southern tip of South America. Although they are most common at lower elevations, geckos are found up to 12,000 ft (3,700 m) in the Himalayas. Geckos also have reached most tropical and subtropical islands and, along with skinks, are often the only land reptiles on remote oceanic islands. Some species, such as the house gecko (*Hemidactylus frenatus*) and the mourning gecko (*Lepidodactylus lugubris*) have wide distributions, but most species are restricted to small geographic ranges. Many geckos are substrate-limited and prefer only particular types of rocks, trees, or soils. Gecko diversity is especially high in arid and semiarid habitats in Africa and Australia and in forested parts of Southeast Asia and Madagascar. Few species live in North America, Europe, or temperate Asia. Some geckos live commensally with humans and are transported easily. The Mediterranean gecko (*Hemidactylus turcicus*) has been introduced into many places in North America, and a colony survives as far north as Baltimore, Maryland.

Habitat

Geckos require egg-laying sites, adequate supplies of arthropod prey, and retreats that protect against temperature extremes and predators, all of which can be found in a diversity of habitats. In arid zones, geckos often occupy narrow rock crevices or else they burrow, creating shallow tunnels or occupying those of other animals in sandy soils. A few species, such as the web-footed gecko (*Palmatogecko rangei*), are specialists of dune faces. Some arid-zone geckos and pygopods live and forage in grass hummocks. Humid tropical forest habitats also are used widely by geckos, which may live on the trunks or branches or in the canopy of trees, under rotting logs, or on rocks along streams and rivers. In savannas and grasslands geckos are less numerous and often patchily distributed, using trees, rocks, or termite nests as shelter. A small, but conspicuous minority of geckos favor the walls of buildings or other manmade structures, where artificial lighting attracts insect prey.

Behavior

Most geckos are nocturnal and emerge from hiding in the early evening to forage and seek mates. Because they gain most of their heat via conduction from warm surfaces, their body temperatures drop as the night progresses, and activity may be limited to just a few hours at cooler times of the year. Diurnal geckos may have one or two peaks of activity during the day, often in the late morning and again in the mid- to late afternoon. Tropical species are active year-round, but in the north and south of the family's distribution, geckos remain inactive, deep in burrows or crevices, during cold periods. They rarely cease all activity, however, and can emerge to take advantage of warmer nights.

Many geckos are relatively solitary, though Bibron's gecko (*Pachydactylus bibronii*) and some other species can reach very high densities and may share retreat sites. These geckos have

A Tokay gecko (*Gekko gecko*) in defensive posture. (Photo by Kim Taylor. Bruce Coleman, Inc. Reproduced by permission.)

reduced levels of aggression toward one another, but there is little evidence of a complex social structure. In the Indian golden gecko (*Calodactylodes aureus*), only the largest male in an area is brightly colored. If he is removed, the next largest male assumes this color and associated dominant status.

Many geckos, especially males, actively defend important resources, such as retreat sites and feeding areas. These geckos stave off rivals of their own and other species by vocalizing, using complex patterns of clicks and chirps. Geckos also use vocalizations, in combination with bites and defecation, to deter predators. Many geckos have cryptic coloration or outline-concealing skin folds and flaps to avoid detection, and a few, such as the Namib day gecko (*Rhoptropus afer*), can outrun most predators. Tail autotomy, the ability to shed or drop the tail, is common among geckos. Threatened geckos lure predators to attack the tail, which continues to wriggle after it is shed, distracting the predator and allowing the animal to escape. The loss of the tail usually carries with it a significant energetic cost, but in the marbled gecko (*Christinus marmoratus*), tail loss has the immediate benefit of increased running speed. Certain geckos, mostly island species, also can shed large portions of their body skin if they are grasped by predators, and they can regrow skin over the large wounds. Members of two genera of diplodactyline geckos (*Strophurus* and *Eurydactylodes*) can ooze or squirt a sticky fluid from the tail that can entangle the mouth parts of such predators as spiders.

Feeding ecology and diet

Nearly all geckos survive on insectivorous diets. Most small species eat only arthropods, but some larger species take small vertebrate prey. Tokay geckos (*Gekko gecko*), for example, can

A parachute or flying gecko (*Ptychozoon* sp.) in the East Indies. (Photo by Animals Animals ©Stephen Dalton. Reproduced by permission.)

The knob-tailed gecko (*Nephrurus stellatus*) eats spiders, insects, scorpions and smaller geckos. (Photo by G. E. Schmida. Bruce Coleman, Inc. Reproduced by permission.)

overpower and eat small snakes, lizards, and mammals as well as nestling birds. One species of New Caledonian gecko (*Rhacodactylus auriculatus*) has specialized fanglike teeth for piercing the bodies of other lizards. Burton's snake lizard (*Lialis burtonis*) has hinged teeth that permit it to feed on hardbodied prey, such as small skinks, whose skins are reinforced by osteoderms (bony plates embedded in the skin). In New Zealand and on other islands, both diurnal and nocturnal geckos often supplement their diet with the fruits, nectar, or pollen of plants. In some cases, these lizards may play important roles as both pollinators and seed dispersers.

Geckos hunt using a combination of visual and chemical cues. Eublepharines and probably some other geckos forage widely and use chemical cues to locate prey. Most other species, however, are ambush predators, moving little and relying on vision to identify arthropod prey that come within striking range.

Reproductive biology

Males of some species attract females by calling. This reaches an extreme in the bell geckos (*Ptenopus*) of southern Africa, in which males participate in choruses. Individual males try to attract mates by calling from their burrow entrances, which serve as resonating chambers to amplify the sound. Less vocal geckos, such as leopard geckos (*Eublepharis macularius*), can identify members of the opposite sex by chemical cues, and many others identify mates visually at close range. Male geckos rub or lick females before mating and restrain them during copulation by biting them on the nape of the neck or the back. In the mourning gecko (*Lepidodactylus*

lugubris) and a few other species, there are no males. Such unisexual species have arisen from the hybridization of two bisexual parental species and, once established, reproduce clonally by parthenogenesis.

Most geckos and all pygopods lay eggs. In gekkonine geckos the eggs are hard-shelled, but in the remaining subfamilies they are leathery. Females lay eggs in protected sites that often provide a high-humidity microclimate, such as on the axils of leaves, under bark, or in shallow nests in the soil. Desert geckos lay eggs in burrows or rock crevices or lay flattened, adherent eggs on vertical or overhanging rock surfaces. All geckos have fixed clutch sizes. Most species produce two young in a clutch, but a few groups of mostly smaller species produce a single egg at a time. Tropical species may produce several clutches a year, sometimes only during wetter periods, but those in cooler climates often have only a single clutch in a year.

Geckos typically abandon their eggs, and development takes one to six months, depending on temperature. In eublepharines and some gekkonines, the sex of the offspring is temperature-dependent. The average temperature experienced by developing embryos during the second trimester of development determines what sex the geckos will become, with higher temperatures yielding males and lower temperatures yielding females. Hatchling geckos slit their eggshells with paired egg teeth that are shed shortly after eclosion (hatching). The geckos of New Zealand and one species in neighboring New Caledonia are unique in being viviparous (live-bearing) and possessing a simple placenta. These species always produce twins, which may gestate for four to 14 months.

Conservation status

Population estimates exist for very few geckos, and the conservation status of most species is unknown. Many geckos live in desert areas that are affected little by humans or, like some tropical species, actually exploit human habitations for their own use. Many island-dwelling geckos with restricted distributions, however, are imperiled by habitat destruction, particularly deforestation, and by the introduction of rats, cats, and other predatory mammals. Among the only geckos believed to have become extinct in historical times are the giant gecko of Round Island in the Mascarenes (*Phelsuma edwardnewtoni*) and the largest gecko that ever lived, Del-

A gecko (*Gekko vittatus*) in the Solomon Islands. (Photo by W. Cheng. Bruce Coleman, Inc. Reproduced by permission.)

court's giant gecko from New Zealand. In each case, introduced predators probably were to blame. Geckos of the genus *Phelsuma*, which are especially brightly colored and attractive, are all internationally protected as CITES Appendix II species because they are popularly sold in the pet trade.

Significance to humans

Large geckos, such as the voracious gecko (*Gehyra vorax*) of Fiji, once were hunted for food, but most modern human consumption of geckos is for medicinal purposes. Tokay geckos and other species are sold dried or pickled in wine or spirits to increase vitality and cure kidney ailments in China and parts of Southeast Asia. In much of the tropics house geckos are welcome as predators on insect pests, and in Europe and North America geckos are favorites of herpetoculturalists. All geckos are harmless, but their mysterious nocturnal habits, large eyes, and climbing abilities have been interpreted as signs of evil; in some cultures they are regarded incorrectly as venomous to the touch.

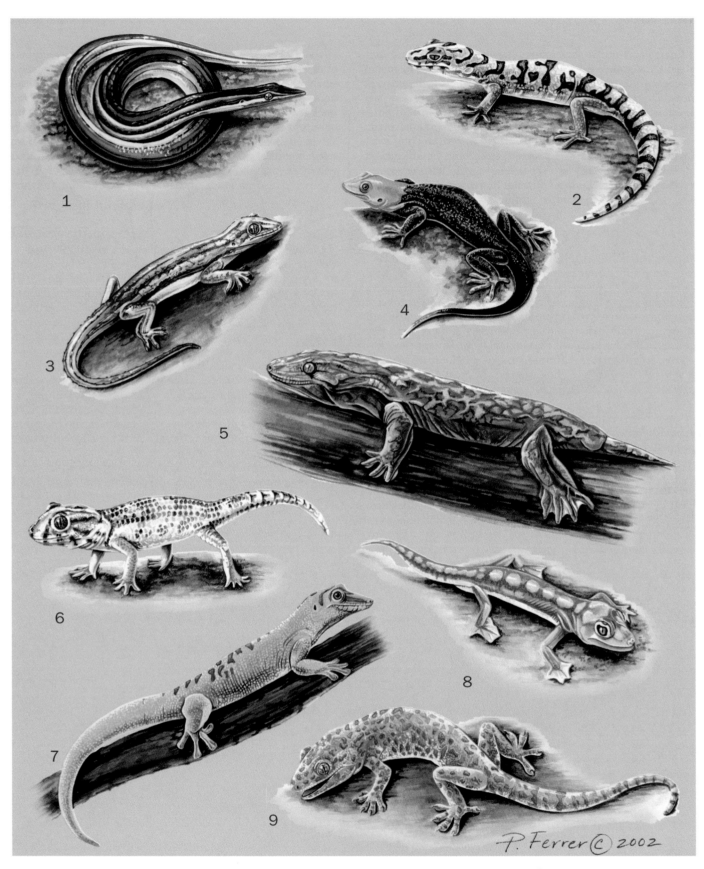

1. Burton's snake lizard (*Lialis burtonis*); 2. Western banded gecko (*Coleonyx variegatus*); 3. House gecko (*Hemidactylus frenatus*); 4. Yellow-headed gecko (*Gonatodes albogularis*); 5. New Caledonian giant gecko (*Rhacodactylus leachianus*); 6. Common plate-tailed gecko (*Teratoscincus scincus*); 7. Madagascar day gecko (*Phelsuma madagascariensis*); 8. Web-footed gecko (*Palmatogecko rangei*); 9. Tokay gecko (*Gekko gecko*). (Illustration by Patricia Ferrer)

Grzimek's Animal Life Encyclopedia

Species accounts

Western banded gecko
Coleonyx variegatus

SUBFAMILY
Eublepharinae

TAXONOMY
Stenodactylus variegatus Baird, 1859, Rio Grande and Gila Valleys, Arizona, United States. Five subspecies are recognized.

OTHER COMMON NAMES
French: Coleonyx varié; German: Gebänderter Krallengecko; Spanish: Salamanquesa de franjas.

PHYSICAL CHARACTERISTICS
This species reaches 3 in (75 mm) in snout-vent length. The body is covered in fine granules; the digits are slender, without pads; and the tail is constricted at the base. These geckos have movable eyelids and vertical pupils. They are pink to pale yellow, with brown bands, blotches, or irregular markings.

DISTRIBUTION
The subspecies *Coleonyx variegatus variegatus* lives in southeastern California, southwestern Nevada, and western Arizona in the United States and the northern Gulf of California, Mexico. *C. v. abbotti* occurs in southwestern California in the United States and northern Baja California, Mexico. *C. v. bogerti* lives in southeastern Arizona and southwestern New Mexico in the United States and northern Sonora, Mexico. *C. v. sonoriensis* inhabits western Sonora and southern Baja California, Mexico. *C. v. utahensis* occurs in Utah, southern Nevada, and northwestern Arizona.

HABITAT
The species inhabits rocky desert and semidesert.

BEHAVIOR
These geckos are nocturnal; several individuals may aggregate in a single burrow.

FEEDING ECOLOGY AND DIET
The species is a generalist arthropod feeder. Energy is stored as fat in the tail.

REPRODUCTIVE BIOLOGY
The female lays two leathery-shelled eggs per clutch, usually between May and September.

CONSERVATION STATUS
Although not listed by the IUCN, the San Diego banded gecko (*C. v. abbotti*) is of special concern in California.

SIGNIFICANCE TO HUMANS
None known. ◆

New Caledonian giant gecko
Rhacodactylus leachianus

SUBFAMILY
Diplodactylinae

TAXONOMY
Ascalabotes leachianus Cuvier, 1829, type locality unknown.

OTHER COMMON NAMES
English: Leach's giant gecko; French: gecko géant de Leach, caméléon géant; German: Neukaledonischer Riesengecko.

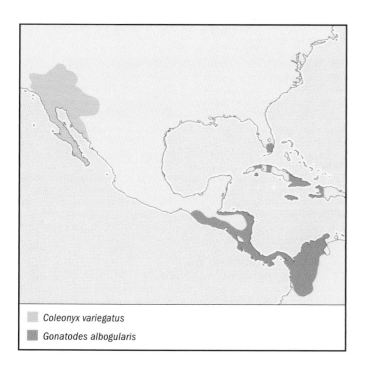

Coleonyx variegatus

Gonatodes albogularis

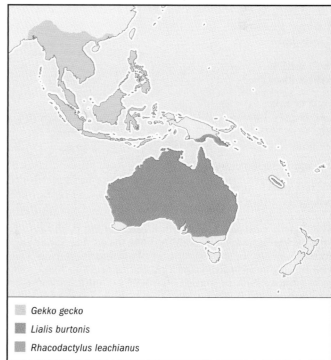

Gekko gecko

Lialis burtonis

Rhacodactylus leachianus

PHYSICAL CHARACTERISTICS
The species grows to 10 in (255 mm) in snout-vent length and is considered the largest living gecko. It is heavy-bodied, with extensive skin folds on the flanks and legs and partially webbed digits. The head is elongate; the feet are large, with broad, undivided pads; and the tail is very short. These geckos are brownish, greenish, or gray with darker punctuations (speckles) or reticulations (net-like patterns) or with white or pinkish bars on the flanks.

DISTRIBUTION
The species is native to New Caledonia.

HABITAT
These geckos inhabit humid forest.

BEHAVIOR
The geckos are nocturnal, spending daylight hours motionless on tree trunks or branches. They make a variety of croaks, growls, and whistles.

FEEDING ECOLOGY AND DIET
The species is insectivorous (insect-eating) and frugivorous (fruit-eating). It may eat fruits exclusively at certain times of year and also prey on nestling birds and lizards.

REPRODUCTIVE BIOLOGY
Clutches of two large (up to 1.6 in, or 40 mm) leathery-shelled eggs are laid in shallow nests in the ground. Sex determination may be temperature-dependent.

CONSERVATION STATUS
Not threatened. The species is widespread on the mainland and offshore islands, but deforestation and illegal collecting for the pet trade are causes for concern.

SIGNIFICANCE TO HUMANS
This is a highly desirable species in the herpetocultural trade. ◆

Tokay gecko
Gekko gecko

SUBFAMILY
Gekkoninae

TAXONOMY
Lacerta gecko Linnaeus, 1758, Java, Indonesia. Two subspecies are recognized, but some populations currently included in *Gekko gecko gecko* probably are distinct.

OTHER COMMON NAMES
French: Tokay; German: Tokeh.

PHYSICAL CHARACTERISTICS
The species grows to 7.1 in (180 mm) in snout-vent length. The body is flattened, the head is broad, and the pupils are vertical. Precloacal glands are present in males. The scales are granular, and there are several rows of enlarged tubercles. The toes are dilated broadly, with large, undivided pads. The species has a bluish or grayish body with both orange or red markings as well as white ones.

DISTRIBUTION
G. g. gecko occurs in tropical Asia from northeastern India to eastern Indonesia. It was introduced into southern Florida in the United States. *G. g. azhari* lives in Bangladesh.

HABITAT
These geckos inhabit trees in tropical forests and disturbed areas (any areas affected by humans); they also live on buildings.

BEHAVIOR
The species is nocturnal and highly aggressive. Individuals gape, lunge, and bite both other geckos and predators. They are highly vocal in both defensive and courtship situations.

FEEDING ECOLOGY AND DIET
The diet includes insects and a variety of small vertebrates.

REPRODUCTIVE BIOLOGY
Several clutches of two hard-shelled eggs are attached firmly to substrate and, at least in captivity, defended.

CONSERVATION STATUS
Not threatened.

SIGNIFICANCE TO HUMANS
The species is used in traditional Asian medicine and traded commercially as pets. ◆

Yellow-headed gecko
Gonatodes albogularis

SUBFAMILY
Gekkoninae

TAXONOMY
Gymnodactylus albogularis Duméril and Bibron, 1836, Martinique and Cuba. Four subspecies are recognized.

OTHER COMMON NAMES
French: Gonatode à gorge blanche; German: Weisskehlgecko; Spanish: Geco cabeza-amarilla

PHYSICAL CHARACTERISTICS
The species reaches 1.3–1.8 in (32–45 mm) in snout-vent length. The dorsal scales are granular, the pupils are round, and the digits are narrow, without enlarged toe pads. Males are gray to black with a yellow head, and females are gray with dark brown mottling.

DISTRIBUTION
Gonatodes albogularis albogularis exists in Aruba, Curaçao, eastern Colombia, and northwestern Venezuela. *G. a. bodinii* lives in the Archipelago de Los Monjes, Venezuela. *G. a. fuscus* inhabits Cuba and Central America to northwestern Colombia; it was introduced into southern Florida. *G. a. notatus* occurs in Hispaniola, Jamaica.

HABITAT
The geckos live under debris and on trees in a variety of habitats. They also occur on buildings and in piles of lumber or trash.

BEHAVIOR
The species is diurnal and basks in partly exposed positions on houses or trees.

FEEDING ECOLOGY AND DIET
The diet consists of small arthropods.

REPRODUCTIVE BIOLOGY
The female lays several clutches of single hard-shelled eggs a year.

CONSERVATION STATUS
Not threatened.

SIGNIFICANCE TO HUMANS
None known. ◆

House gecko
Hemidactylus frenatus

SUBFAMILY
Gekkoninae

TAXONOMY
Hemidactylus frenatus Duméril and Bibron, 1836, Java, Indonesia.

OTHER COMMON NAMES
English: Half-toed geckos; French: Margouillat; German: Asiatischer Halbzehengecko.

PHYSICAL CHARACTERISTICS
The species grows to 2.6 in (65 mm) in snout-vent length. The body is flattened. The toe pads are divided, and the first digit is much smaller than the others. The tail has enlarged ventral plates and a denticulate margin. These geckos are grayish, pinkish, or pale brown with darker flecks.

DISTRIBUTION
The species exists in Southeast Asia and the Indo-Australian archipelago, the Philippines, Taiwan, and much of Micronesia, Melanesia, and Polynesia. It was introduced into tropical Australia, eastern Africa, Mexico, the United States and elsewhere.

HABITAT
These geckos live among many types of vegetation but often are found around human habitations.

BEHAVIOR
Males may be aggressive, especially in areas of high density near abundant food sources. They have complex multiple click vocalizations.

FEEDING ECOLOGY AND DIET
The species is a generalist insectivore, occasionally eating smaller lizards.

REPRODUCTIVE BIOLOGY
Females lay clutches of hard-shelled eggs throughout the year. There is an incubation period is 45–71 days. The female is capable of sperm storage.

CONSERVATION STATUS
Not threatened. This species may cause declines in native geckos where it is introduced.

SIGNIFICANCE TO HUMANS
This commensal species is found frequently in and around houses and other manmade structures. ◆

Web-footed gecko
Palmatogecko rangei

SUBFAMILY
Gekkoninae

TAXONOMY
Palmatogecko rangei Andersson, 1908, Lüderitz, Namibia

OTHER COMMON NAMES
French: Gecko du désert; German: Schwimmfußgecko.

PHYSICAL CHARACTERISTICS
This species usually is 2.4–2.6 in (60–65 mm) in snout-vent length, with a maximum of 3.2 in (78 mm). The body is elongate and cylindrical, with slender legs terminating in broad webbed feet. The skin is translucent and pinkish with brown markings on top and bright white below.

DISTRIBUTION
The species is distributed through western Namibia and adjacent northwestern South Africa and western Angola.

HABITAT
These geckos inhabit the Namib Desert dunes.

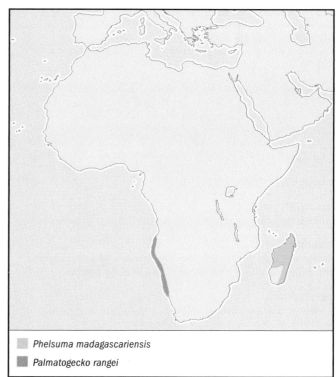

Phelsuma madagascariensis
Palmatogecko rangei

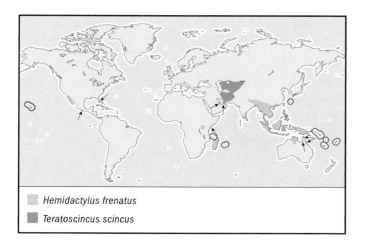

Hemidactylus frenatus
Teratoscincus scincus

BEHAVIOR
This species is nocturnal and retreats by day into burrows up to 20 in (50 cm) long, excavated in windward dune faces using webbed feet. It assumes a stiff-legged posture when it is alarmed. Males attack and bite to defend the area around burrows.

FEEDING ECOLOGY AND DIET
This gecko preys on dune-dwelling spiders and beetles. All water is obtained from condensed fog or from prey.

REPRODUCTIVE BIOLOGY
The species breeds in spring. Females lay two calcareous (hard-shelled) eggs, 0.8 × 0.4 in (21 × 10 mm), in burrows in the sand in summer (November to March). Incubation lasts about 90 days.

CONSERVATION STATUS
The species is widespread and common in most of its range in Namibia and probably in Angola. A small South African population is threatened by diamond-mining operations.

SIGNIFICANCE TO HUMANS
None known. ◆

Madagascar day gecko
Phelsuma madagascariensis

SUBFAMILY
Gekkoninae

TAXONOMY
Gecko madagascariensis Gray, 1831, Madagascar. The nominate form (the first subspecies to be named) and three additional subspecies—*Phelsuma madagascariensis boehmei*, *P. m. grandis*, and *P. m. kochi*—are all from Madagascar.

OTHER COMMON NAMES
French: Phelsume de Madagascar; German: Madagassischer Taggecko.

PHYSICAL CHARACTERISTICS
This species grows to 3.9–5.1 in (100–130 mm) in snout-vent length. The body is heavy and the tail thick. The toes have pads that are broadened distally, with a greatly reduced first digit on each foot. The scales are small and granular, and the pupils are circular. Males have precloacal glands. These geckos are bright green with red markings on the snout, head, and back.

DISTRIBUTION
The species ranges across northern and eastern Madagascar.

HABITAT
These arboreal geckos are found on trees in disturbed areas and in primary forest; they also are found on houses.

BEHAVIOR
These geckos are diurnal and may congregate in high-density populations. Males maintain territories.

FEEDING ECOLOGY AND DIET
This is a generalist insectivore, but it also eats fruit or nectar, if available.

REPRODUCTIVE BIOLOGY
Several clutches of one or two hard-shelled eggs are laid every four to six weeks during the summer or early fall (November to May). The incubation period is about two months.

CONSERVATION STATUS
Not threatened. CITES regulates international trade.

SIGNIFICANCE TO HUMANS
This species is common in the herpetocultural trade. It breeds very well in captivity and can occur around human dwellings. ◆

Common plate-tailed gecko
Teratoscincus scincus

SUBFAMILY
Gekkoninae

TAXONOMY
Stenodactylus scincus Schlegel, 1858, I-li River, Kazakhstan.

OTHER COMMON NAMES
English: Frog-eyed gecko, wonder gecko; German: Wundergecko.

PHYSICAL CHARACTERISTICS
This species reaches 4.6 in (116 mm) in snout-vent length. The head is large, with prominent eyes. The digits are straight, without pads. The body is covered with large cycloid, imbricate scales (which is uncommon for geckos) extending along the dorsum of the tail. There are no precloacal glands. The color and pattern of these geckos vary, but they usually have brown, orange, or bluish stripes or bands on a whitish, yellow, or gray background.

DISTRIBUTION
Teratoscincus scincus scincus lives in southern Kazakhstan, Uzbekistan, Turkmenistan, Tadzhikistan, Kyrgyzistan, and northern Afghanistan. *T. s. keyserlingii* occurs in eastern Iran, southern Afghanistan, western Pakistan, the United Arab Emirates, and Qatar. *T. s. rustamovi* exists in the Fergana Valley of Uzbekistan and Tadzhikistan.

HABITAT
The habitat is sandy arid and semiarid areas.

BEHAVIOR
This species is nocturnal and burrowing. The enlarged scales of the tail can be rubbed together to create a buzzing noise that may deter predators. The fragile skin is damaged easily and can be shed to escape predators.

FEEDING ECOLOGY AND DIET
This species is a generalist insectivore.

REPRODUCTIVE BIOLOGY
Females lay clutches of two hard-shelled eggs.

CONSERVATION STATUS
Not threatened.

SIGNIFICANCE TO HUMANS
These geckos are traded commercially as pets. ◆

Burton's snake lizard
Lialis burtonis

SUBFAMILY
Pygopodinae

TAXONOMY
Lialis burtonis Gray, 1834, Round Hill Fauna Reserve, New South Wales, Australia.

OTHER COMMON NAMES
English: Burton's legless lizard; French: Lialis de Burton; German: Spitzkopf-Flossenfuß.

PHYSICAL CHARACTERISTICS
The species reaches 10 in (250 mm) in snout-vent length. It is elongate, with no forelimbs and with hind limbs reduced to small flaps. The tail is longer than the body. The head is long, with a very elongate snout, and the pupils are vertical. It is brown or gray, with or without a pattern of regular spots or stripes.

DISTRIBUTION
The species ranges across Australia and southeastern New Guinea.

HABITAT
The geckos live in terrestrial habitats, from deserts to humid forest.

BEHAVIOR
They are active day or night.

FEEDING ECOLOGY AND DIET
The species feeds on lizards, chiefly small skinks.

REPRODUCTIVE BIOLOGY
Mating occurs in the spring, and females lay two elongate, leathery-shelled eggs in summer.

CONSERVATION STATUS
The species is widespread and common throughout most of its range.

SIGNIFICANCE TO HUMANS
None known. ◆

Resources

Books

Greer, A. E. *The Biology and Evolution of Australian Lizards.* Chipping Norton, Australia: Surrey Beatty and Sons, 1989.

Rösler, H. *Geckos der Welt, alle Gattungen.* Leipzig: Urania-Verlag, 1995.

Szczerbak, Nicolai N., and Michael L. Golubev. *Gecko Fauna of the USSR and Contiguous Regions.* Ithaca, NY: Society for the Study of Amphibians and Reptiles, 1996.

Periodicals

Bauer, A. M., and I. Das. "A Review of the Gekkonid Genus *Calodactylodes* (Reptilia: Squamata) from India and Sri Lanka." *Journal of South Asian Natural History* 5, no. 1 (2000): 25–35.

Bauer, A. M., and A. P. Russell. "*Hoplodactylus delcourti* n. sp. (Reptilia: Gekkonidae), the Largest Known Gecko." *New Zealand Journal of Zoology* 13, no. 1 (1986): 141–148.

———. "The Evolutionary Significance of Regional Integumentary Loss in Island Geckos: A Complement to Caudal Autotomy." *Ethology, Ecology, and Evolution* 4, no. 4 (1992): 343–358.

Cooper, W. E., Jr. "Prey Chemical Discrimination and Foraging Mode in Gekkonoid Lizards." *Herpetological Monographs* 9 (1995): 120–129.

Daniels, C. B. "Running: An Escape Strategy Enhanced by Autotomy." *Herpetologica* 39, no. 2 (1983): 162–165.

Frankenberg, E., and D. L. Marcellini. "Comparative Analysis of the Male Multiple Click Calls of Colonizing House Geckos *Hemidactylus turcicus* from the Southern U.S.A. and Israel and *Hemidactylus frenatus.*" *Israel Journal of Zoology* 37, no. 2 (1990): 107–118.

Girling, J. E., A. Cree, and L. J. Guillette Jr. "Oviductal Structure in a Viviparous New Zealand Gecko, *Hoplodactylus maculatus.*" *Journal of Morphology* 234, no. 1 (1997): 51–68.

Haacke, W. D. "The Burrowing Geckos of Southern Africa, 1. (Reptilia: Gekkonidae)." *Annals of the Transvaal Museum* 29, no. 12 (1975): 197–243.

Hedges, S. B., and R. Thomas. "At the Lower Size Limit in Amniotes: A New Diminutive Lizard from the West Indies." *Caribbean Journal of Science* 37, no. 3–4 (2001): 168–173.

Ineich, I. "La Parthénogenèse chez les Gekkonidae (Reptilia, Lacertilia): Origine et Evolution." *Bull. Soc. Zool. France* 117, no. 3 (1992): 253–266.

Kluge, A. G. "Cladistic Relationships in the Gekkonoidea (Squamata, Sauria)." *Miscellaneous Publications, Museum of Zoology, University of Michigan* 173 (1987): i–iv, 1–54.

———. "Gekkotan Lizard Taxonomy." *Hamadryad* 26 (2001): 1–209.

Patchell, F. C., and R. Shine. "Food Habits and Reproductive Biology of the Australian Legless Lizards (Pygopodidae)." *Copeia* 1986, no. 1 (1986): 30–39.

Petren, K., D. T. Bolger, and T. J. Case. "Mechanisms in the Competitive Success of an Invading Sexual Gecko over an Asexual Native." *Science* 259, no. 5093 (1993): 354–358.

Russell, A. P. "A Contribution to the Functional Analysis of the Foot of the Tokay, *Gekko gecko* (Reptilia: Gekkonidae)." *Journal of Zoology London* 176, no. 3 (1975): 437–476.

Werner, Y. L. "Observations on Eggs of Eublepharid Lizards, with Comments on the Evolution of the Gekkonoidea." *Zool. Meded.* 47, no. 17 (1972): 211–224, pl. I.

Whitaker, A. H. "The Roles of Lizards in New Zealand Plant Reproductive Strategies." *New Zealand Journal of Botany* 25, no. 2 (1987): 315–328.

Organizations

Global Gecko Association. 4920 Chester Street, Spencer, Oklahoma 73084-2560 USA. Web site: <http://www.gekkota.com>

Aaron Matthew Bauer, PhD

Blindskinks

(Dibamidae)

Class Reptilia

Order Squamata

Family Dibamidae

Thumbnail description
Small, secretive brown fossorial lizards with blunt heads and tails

Size
Small to medium-sized lizards, about 2–10 in (50–250 mm) in snout-vent length

Number of genera, species
2 genera; 11 species

Habitat
Forests, rainforests, semi-arid deciduous brush, semi-arid open scrubland, dense bush, deciduous scrub, dense forest, and pine-oak forest

Conservation status
Not threatened

Distribution
Disjunct: northeastern Mexico and Southeast Asia

Evolution and systematics

Exact affinities of these small, wormlike, burrowing scleroglossans remain unknown. They could be related to other fossorial scincomorphans, amphisbaenians, or even to snakes. Many of the traits they share with other squamate groups could have evolved convergently, however, in response to their subterranean habits. Although the geographic distributions of the two genera of dibamids are widely separated, their relationship to each other is supported strongly by numerous shared derived characteristics, some of which also are shared with members of various other lizard families. Dibamids have diverged greatly from other squamates and could represent remnants of an extremely ancient lineage. No subfamilies are recognized.

Physical characteristics

Blindskinks are small to medium-sized lizards, about 2–10 in (50–250 mm) in snout-vent length. They are brown, with short, blunt tails, which break at fracture planes in tail vertebra. Forelimbs and bones of the pectoral girdle are entirely absent. Males have small flaplike hind limbs somewhat reminiscent of those of some pygopodids, perhaps used in courtship or copulation. Females are entirely limbless. Remnants of the pelvic girdle are present. *Dibamus* are the only squamates with pores on their lower legs. As in many burrowers, the skulls consist of massively fused bones, parts of

which have lost the ability to move with respect to one another. The heads are blunt, and eyes are vestigial and lie beneath an immovable head scale. Paired frontal and nasal bones and scales are present. The parietal bone is fused, without a foramen. Teeth are pleurodont, set in sockets, and small, and they curve backward to a single point. There are no teeth on the palatal or pterygoid bones. Tongues are covered dorsally with filamentous papillae, without lingual scales. The foretongue is nonretractable. The nostrils are located at the tip of the snout on an enlarged rostral scale. They have no external ear openings. Their bodies are covered with shiny, smooth, overlapping scales without osteoderms.

A blindskink, *Dibamus bourreti,* from Vietnam. (Photo by Robert W. Murphy. Reproduced by permission.)

Blindskink, *Dibamus* sp. (Illustration by Michelle Meneghini)

Distribution

Commonly known as blindskinks, only two genera of dibamids exist. The single species of *Anelytropsis* lives in a tiny area in northeastern Mexico. (Only 19 specimens had been collected by 1985.) The 10 species of *Dibamus* are scattered around Southeast Asia, from the Nicobar Islands across the islands of the Indo-Australian archipelago, including the Sunda Islands. They also occur in Vietnam, southern Thailand, Malaysia, Sumatra, most of Indonesia, Borneo, the Philippines, and westernmost New Guinea (Irian Jaya).

Habitat

Dibamus requires humus in moist condition, burrowing deep beneath rocks or logs during dry seasons. *Anelytropsis* is adapted to a wider variety of drier habitats, including semi-arid deciduous brush, semi-arid open scrubland, dense bush, deciduous scrub, dense forest, and pine-oak forest. *Anelytropsis* has been found beneath rocks, in or under rotting logs, in the decayed bases of yuccas, and in loose litter.

Behavior

Dibamids are secretive, living on the forest floor, often underneath stones but sometimes underneath leaf litter, or moving about underground, sometimes inhabiting burrows and crevices or the catacombs of social insect nests. They exploit existing burrows and openings in the soil but can burrow in loose, friable soils or decomposing leaf litter.

Feeding ecology and diet

Like most lizards, they are thought to eat insects.

Reproductive biology

Dibamus lays a clutch of just a single egg but may deposit several clutches during a season. *Dibamus* eggs are soft and elongated when laid, but they become more spherical and calcify and harden after deposition. Nothing is known about reproduction in *Anelytropsis*.

Conservation status

Not threatened; although many species are susceptible to habitat loss. There are currently no conservation efforts underway to protect dibamids.

Significance to humans

None known

Resources

Books

Estes, R., K. de Queiroz, and J. Gauthier. "Phylogenetic Relationships Within Squamata." In *Phylogenetic Relationships of the Lizard Families*, edited by R. Estes and G. Pregill. Stanford, CA: Stanford University Press, 1988.

Pianka, E. R., and L. J. Vitt. *Lizards: Windows to the Evolution of Diversity*. Berkeley: University of California Press, 2003.

Pough, F. Harvey, Robin M. Andrews, John E. Cadle, Martha L. Crump, Alan H. Savitzky, and Kentwood D. Wells. *Herpetology*. Upper Saddle River, NJ: Prentice Hall, 1998.

Zug, George R., Laurie J. Vitt, and Janalee P. Caldwell. *Herpetology: An Introductory Biology of Amphibians and Reptiles*. 2nd edition. San Diego: Academic Press, 2001.

Periodicals

Greer, A. E. "The Relationships of the Lizard Genera *Anelytropsis* and *Dibamus*." *Journal of Herpetology* 19, no 1 (1985): 116–156.

Eric R. Pianka, PhD

Wormlizards
(*Amphisbaenidae*)

Class Reptilia
Order Squamata
Suborder Amphisbaenia
Family Amphisbaenidae

Thumbnail description
Elongate, slender, limbless, fossorial reptiles with scales arranged in annular rings; eyes greatly reduced or absent; no external ear openings; head either rounded, spatulate, or laterally compressed; and short tails

Size
Amphisbaenids exhibit a huge size range, from tiny worm-like species of 4 in (100 mm) maximum length to some of the largest known amphisbaenians of more than 32 in (800 mm) maximum length in some species; most species fall within the range of 10–16 in (250–400 mm)

Number of genera, species
18 genera; 160 species

Habitat
Amphisbaenids are found in a variety of habitats ranging from lowland rainforests to deciduous forests to arid deserts

Conservation status
No species listed by the IUCN

Distribution
Amphisbaenids occur in Africa, South and Central America, Europe, and the West Indies

Evolution and systematics

Most recent phylogenetic analyses have placed amphisbaenians as one of three suborders of Squamates (the clade that includes snakes, lizards, and amphisbaenians), but their exact placement within that clade is not well understood. The interrelationships among the four amphisbaenian families are also poorly understood. The Amphisbaenidae has no significant fossil record; a few isolated remains occur in Europe and in Africa. Amphisbaenids are possibly most closely related to trogonophids based on a recent phylogenetic analysis of the group. This is the largest and most heterogeneous amphisbaenian family. No subfamilies are recognized.

Physical characteristics

Certain features of amphisbaenids are common to most or all amphisbaenians. These include: a unique modification of the middle ear in which an elongated structure, the extracolumella, attaches to the stapedial bone of the middle ear extending forward to attach to tissue along the sides of the face and allowing the reception and transmission of vibrations to the inner ear; reduction or absence of the right lung; an en-

larged, medial, premaxillary tooth; the periodic shedding of the skin in a single piece; a heavily ossified and robust skull; the absence of eyelids; a forked tongue; and the absence of external ear openings.

The family Amphisbaenidae contains some of the smallest and largest amphisbaenians and even some of the most highly miniaturized reptiles known. One exceptionally tiny amphisbaenid is the African *Chirindia*, with some species attaining an adult body length of only 4 in (100 mm) and a body diameter of 0.125 in (3 mm). The largest amphisbaenian is probably the South American *Amphisbaena alba*, which attains an adult body length of over 30 in (800 mm). In general, amphisbaenians show little or no sexual dimorphism in body size. All amphisbaenids are completely limbless, but some retain internal vestiges of the pectoral and pelvic girdles. In some species, a small eye is visible under a translucent head scale, but in others the eye is not visible at all. Amphisbaenids exhibit a variety of cranial shapes (keel-headed, round-headed, shovel-headed) and a generally cylindrical body shape. In some species the snout is conical and in others it is blunt. The tail is always short but may exhibit

A wormlizard (*Amphisbaena* sp.) in Brazil. (Photo by Animals Animals ©Fabio Colombini. Reproduced by permission.)

a variety of shapes, including rounded and blunt-tipped, pointed, or dorsoventrally flattened. Most amphisbaenid species are capable of autotomizing the tail, but none can regenerate a new tail. All amphisbaenids have pleurodont dentition, with varying numbers of teeth occurring on the tooth-bearing elements of the skull.

Many species are pallid in appearance, presumably related to their subterranean existence. However, some species exhibit striking patterns of coloration, including dark checkerboard patterns on light backgrounds. Still others are a solid brown, yellow, or grey dorsally, with a paler underside.

Beyond those easily recognizable features, amphisbaenids are also characterized by a number of internal conditions, including the presence of pelvic vestiges in all species, the absence of a sternum, and a heavily ossified skull.

Distribution

The family Amphisbaenidae is both the largest and most widespread family of amphisbaenians. Wormlizards occur in tropical and subtropical areas of Central and South America, Africa, Europe, Asia, and the West Indies. *Blanus cinereus* is the only amphisbaenian occurring in Europe.

Habitat

Wormlizards are known to occur in a variety of habitats, including deserts, tropical rainforests, and woodlands. They are usually found by digging into the soil but may also be found hidden under surface litter or debris, under rocks and logs, or even within ant and termite nests. Wormlizards are sometimes found after heavy rains when the animals are driven from their burrow systems.

Behavior

Because amphisbaenians are fossorial, relatively little is known of their behavior or ecology compared to other rep-

tiles. However, a few insights into their behavior are notable. The family Amphisbaenidae contains the greatest diversity of burrowing specializations, which can be loosely categorized in concert with head shapes. Round-headed species burrow by driving or ramming the head forward. Those species with shovel-shaped heads burrow by driving the head downward, lifting the head up, compressing the soil onto the roof of the tunnel with the flattened snout, and smoothing the tunnel walls with the pectoral region. Keel-headed forms with laterally compressed heads burrow by driving the head forward alternately to the left and right of center and then smoothing the tunnel walls by pressing the sides of the head and trunk against them.

Feeding ecology and diet

Natural diets of amphisbaenians are poorly studied. Some laboratory studies of captive amphisbaenians have led researchers to suggest that amphisbaenians are specialized for feeding on large vertebrate prey, which they apparently handle and eat effectively in captivity. Interestingly, however, all direct studies performed to date on wild-caught animals indicate that amphisbaenians rarely exploit large prey in the wild but feed primarily on small invertebrates instead. The few direct diet studies completed for amphisbaenids indicate a remarkable degree of consistency in diet; all species studied so far appear to feed primarily on such prey items as termites, ants, adult and larval beetles, cockroaches, and lepidopteran larvae. A few irregular occurrences are notable. One study found six specimens of an amphisbaenid species with stomachs filled with fungi, and another study found one specimen with a lizard limb in its stomach and a second specimen containing an entire burrrowing snake.

For amphisbaenids, chemical and auditory cues are the most important means used in locating prey. The uniquely adapted middle ear system allows prey movements to be detected, while the forked tongue and the Jacobson's organ allow the detection of chemical odors. Airborne sounds are picked up and transmitted to the inner ear along the specialized extracolumellar apparatus, which may also amplify the vibrations as well. This unique anatomy is consistent with behavioral studies conducted in laboratory experiments, which suggest that amphisbaenians can hear prey movements through the soil.

Reproductive biology

Information on the reproductive biology of amphisbaenids is extremely limited. From what little we know, amphisbaenids are typically egg-layers, although some species of *Loveridgea* and *Monopeltis* bear live young. Clutch sizes average one to four elongate eggs. Wormlizard eggs have been discovered in ant and termite colonies, implying that amphisbaenians may utilize these underground colonies not only as food sources but perhaps as egg-laying chambers as well.

Conservation status

No species of Amphisbaenidae is listed by the IUCN.

Significance to humans

Wormlizards are of no economic significance to humans, but they may benefit humans ecologically by feeding on populations of ants and termites and potentially helping to keep these populations in check.

Species accounts

White-bellied wormlizard
Amphisbaena alba

TAXONOMY
Amphisbaena alba Linné, 1758, "America."

OTHER COMMON NAMES
French: Amphisbène blanche; German: Rote Doppelschleiche.

PHYSICAL CHARACTERISTICS
Amphisbaena alba is the largest South American amphisbaenian, with a total length of 15.7–33.4 in (400–850 mm) and a midbody diameter ranging from 1–2 in (20–25 mm). The tail is approximately 6% of total length. It has smooth, square scales arranged in annular rings that are separated by grooves. The species is uniformly pale-

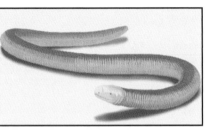

Amphisbaena alba

colored, although some specimens are white ventrally. The head is more or less rounded in shape. Caudal autotomy is absent in this species. Dentition is pleurodont, with conical, sharply pointed, and slightly recurved teeth. It has a large, median premaxillary tooth flanked by six smaller premaxillary teeth. Five teeth occur on the maxilla; seven teeth on the lower jaw.

DISTRIBUTION
Throughout South America east of the Andes, Panama, and the West Indies.

HABITAT
The white-bellied wormlizard inhabits forested lowlands, soil, and leaf litter. *Amphisbaena alba* is rarely seen on the surface.

BEHAVIOR
Amphisbaena alba is well known to herpetologists for several unusual and interesting behavioral traits, one of which involves a unique ecological relationship with leaf-cutter ants (genus *Atta*). *A. alba* apparently follows the pheromone-marked trails of these ants to their nests, where it then takes up residence in one of the nest chambers, feeding primarily on beetle larvae and other arthropods that are also inhabitants of *Atta* nests.

The defensive behavior of *Amphisbaena alba* when threatened is also of scientific interest to behavioral biologists. When disturbed, *A. alba* raises its tail off the ground and waves it, while simultaneously raising its head with an open, gaping mouth. The head and tail are close together, the body held almost in a circle, which almost gives the impression of a two-headed animal. This stereotyped behavior has led to the common name "two-headed snake" applied by people who live near these animals.

FEEDING ECOLOGY AND DIET
Only a few studies examining the natural diet of *Amphisbaena alba* exist. These have documented that *A. alba* feeds primarily on small arthropods such as beetles, ants, termites, spiders, crickets, and the larvae of various insects. On the other hand, laboratory-based behavioral studies indicate that, when offered larger vertebrate prey items such as rodents, *A. alba* will prey on them with enthusiasm, even biting and tearing pieces out of them.

REPRODUCTIVE BIOLOGY
Amphisbaena alba is oviparous, laying eggs in clutches of eight to 16. This is the largest known clutch size for any amphisbaenian and may possibly be related to the large body size of this species. According to some studies, reproduction is seasonal and primarily restricted to the dry season.

CONSERVATION STATUS
Not listed by the IUCN.

SIGNIFICANCE TO HUMANS
None known. ◆

Amphisbaena alba

Resources

Books

Gans, C. *Biomechanics: An Approach to Vertebrate Biology.* Philadelphia: J. B. Lippincott Company, 1974.

Linné, C. V. *Systema naturae per regna tria naturae.* Editio decima, reformata. Vol. 1. Stockholm, 1758.

Schwenk, K. "Feeding in Lepidosaurs." In *Feeding: Form, Function, and Evolution in Tetrapod Vertebrates.* San Diego: Academic Press, 2000: 175–291.

Vanzolini, P. E. *Evolution, Adaptation and Distribution of the Amphisbaenid Lizards (Sauria: Amphisbaenidae).* Ph.D. diss. Harvard University, 1951.

Periodicals

Colli, G. R., and D. S. Zamboni. "Ecology of the Worm-lizard *Amphisbaena alba* in the Cerrado of Central Brazil." *Copeia* 1999: 733–742.

Gans, C. "The Characteristics and Affinities of the Amphisbaenia." *Transactions of the Zoological Society of London* 34 (1978): 347–416.

Gans, C., and E. Wever. "The Amphisbaenian Ear: *Blanus cinereus* and *Diplometopon zarudnyi.*" *Proceedings of the National Academy of Sciences* 72 (1975): 1487–1490.

———. "The Ear and Hearing in Amphisbaenia (Reptilia)." *Journal of Experimental Zoology* 179 (1972): 17–34.

Kearney, M. "The Appendicular Skeleton in Amphisbaenians." *Copeia* 2002, no. 3 (2002): 719–738.

Montero, R., and C. Gans. "The Head Skeleton of *Amphisbaena alba.*" *Annals of the Carnegie Museum* 68 (1999): 15–80.

Renous, S., J. P. Gasc, and A. Raynaud. "Comments on the Pelvic Appendicular Vestiges in an Amphisbaenian: *Blanus cinereus* (Reptilia, Squamata)." *Journal of Morphology* 209 (1991): 23–28.

Riley, J., J. M. Winch, A. F. Stimson, and R. D. Pope. "The Association of *Amphisbaena alba* (Reptilia: Amphisbaenia) with the Leaf-cutting Ant *Atta cephalotes* in Trinidad." *Journal of Natural History* 20 (1986): 459–470.

Zangerl, R. "Contributions to the Osteology of the Postcranial Skeleton of the Amphisbaenidae." *American Midland Naturalist* 33 (1945): 764–780.

———. "Contributions to the Osteology of the Skull of the Amphisbaenidae." *American Midland Naturalist* 31 (1944): 417–454.

Maureen Kearney, PhD

Mole-limbed wormlizards
(Bipedidae)

Class Reptilia
Order Squamata
Suborder Amphisbaenia
Family Bipedidae

Thumbnail description
Elongate, slender, fossorial reptiles with scales arranged in annular rings, short robust forelimbs, hindlimbs absent, no external ear openings, a rounded head, blunt snout, and short tail

Size
4.7–9.4 in (120–240 mm)

Number of genera, species
1 genus; 3 species

Habitat
Arid scrub lands or desert, arroyos, alluvial sediments around river basins

Conservation status
Not listed by the IUCN

Distribution
Bipedids have a restricted distribution in coastal southwestern Mexico and Baja California

Evolution and systematics

Most recent phylogenetic analyses have placed amphisbaenians as one of three suborders of Squamates (the clade that includes snakes, lizards, and amphisbaenians), but their exact placement within that clade is not well understood. The interrelationships among the four amphisbaenian families are also poorly understood. On the one hand, the retention of forelimbs in bipedids, in contrast to the completely limbless condition in all other amphisbaenians, suggests that those forms may be the most primitive amphisbaenians. On the other hand, the only amphisbaenian family with a good fossil record is the Rhineuridae, and some of those fossil forms, although limbless, retain different primitive features, such as fully formed, enclosed orbits, suggestive of a potentially basal position for this group within amphisbaenians. Unfortunately, bipedids have no fossil record at all. No subfamilies are recognized.

Physical characteristics

Certain features of bipedids are common to most or all amphisbaenians. These include: reduction or absence of the right lung; an enlarged, medial, premaxillary tooth; the peri-odic shedding of the skin in a single piece; a heavily ossified and robust skull; absence of eyelids; a forked tongue; and the absence of external ear openings.

The family Bipedidae includes three species that attain a body length of 4.5–9.4 in (115–240 mm) and a body width of 0.27–0.39 in (7–10 mm). The most striking feature of these species is the presence of short, robust forelimbs positioned close to the back of the head, a unique condition among amphisbaenians. Bipedids retain all typical elements of the forelimb, including the humerus, radius, ulna, carpals, metacarpals, and phalanges. In addition, *Bipes* exhibits a unique condition termed hyperphalangy, which refers to an extra element occurring in the first digit of the hand, one more than is typically found in reptiles. This condition is presumably related to digging functions. The tail of bipedids is very short, as in most amphisbaenians, representing between 10–20% of the entire length of the animal. The eyes are reduced and sometimes covered by a head scale. The head is rounded and blunt. The teeth are conical, slightly recurved, and attached to the jaw in pleurodont fashion. There is no pigment or pattern to the skin, making the external appearance pale pink or flesh-colored. Caudal autotomy occurs at a single autotomy

A tropical wormlizard (*Bipes canaliculatus*) in Mexico. (Photo by Joseph T. Collins/Photo Researchers, Inc. Reproduced by permission.)

constriction site, but bipedids do not regenerate tails. Bipedids are also recognized by a number of unique internal anatomical characteristics, which are evident through x rays or dissection. For example, despite the absence of external rearlimbs, bipedids retain elements of the pelvic girdle and a small vestige of the leg bone (the femur) internally. Also, the skull of bipedids is characterized by fusion of the frontal and parietal bones.

Distribution

The genus *Bipes* is endemic to Mexico, with a relatively narrow distribution in western Mexico. One species, *Bipes biporus*, occurs throughout the cape region of Baja California, Mexico; *B. canaliculatus* occurs in Guerrero and Michoacán; and *B. tridactylus* occurs in coastal Guerrero.

Habitat

Mole-limbed wormlizards are known to occur in loose, sandy, or loamy soils. They are usually found in a narrow range of microhabitats, including those associated with the root systems of shrubs and small trees. Traditionally, *Bipes* has been thought to be restricted to sandy soils, but recent reports suggest the additional occurrence of *Bipes* in different microhabitats, including under rocks and rotting logs along gravel slopes and in coarser, rockier soils near river basins. Also, mole-limbed wormlizards have been collected outside of their tunnel systems and are sometimes found on the surface of the soil, especially nocturnally when they emerge from their burrow systems.

Behavior

All bipedid species are fossorial, living in self-constructed tunnel systems that range from 0.78–7.8 in (2–20 cm) below the surface. *Bipes* uses its digging front limbs to create an initial opening in the soil; once below the surface, it folds the limbs back against the body and uses the head (as all other amphisbaenians do) to burrow through the soil. Rectilinear movements of the body are also used for locomotion within the tunnel.

Often, burrows lead to the soil surface under rocks. When the rocks are lifted, *Bipes* will retreat down into the burrow system quickly.

Feeding ecology and diet

Direct examination and field studies of feeding in amphisbaenians are almost nonexistent, since these animals are so secretive. Therefore, we must rely mainly on indirect studies and on studies of captive animals. Laboratory studies based on dissections of wild-caught animals and examination of stomach contents indicate that bipedids feed primarily on small invertebrates such as termites, beetle larvae, and ants. *Bipes biporus* fits the pattern of a generalist predator that exploits prey items found both under the soil and on the soil surface covered by objects such as fallen bark or debris.

For bipedids, chemical cues are the most important means used in locating prey. The forked tongue and the Jacobson's organ allow the detection of chemical odors.

Reproductive biology

All species within this family are oviparous, laying 1–4 eggs per clutch. Eggs are laid in January and hatching occurs in April for *Bipes canaliculatus* and *B. tridactylus*. In *B. biporus* eggs are laid in July and hatchlings appear in September.

Conservation status

No species of Bipedidae is listed by the IUCN.

Significance to humans

Mole-limbed wormlizards are of no economic significance to humans, but they may benefit humans ecologically by feeding on populations of ants and termites and potentially helping to keep these populations in check.

Species accounts

Two-legged wormlizard

Bipes biporus

TAXONOMY
Bipes biporus Cope, 1894, Cape San Lucas, Baja California, Mexico.

OTHER COMMON NAMES
English: Mole lizard; French: Bipédidés; German: Zweifuss-Doppelschleichen; Spanish: Ajolote.

PHYSICAL CHARACTERISTICS
The two-legged wormlizard has an average body length of 7.5–8.3 in (190–210 mm). Its tail is approximately 10% of total length. It has a midbody diameter of 0.23–0.27 in (6–7 mm). It is typically pink or

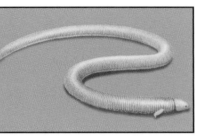

Bipes biporus

flesh-colored uniformly, but some specimens are white ventrally. It has five claw-bearing digits on each limb and two preanal pores. It has a caudal autotomy restriction site at a single point

on the tail less than 10 annuli posterior to the cloacal opening. The tail is not regenerated subsequent to autotomization.

DISTRIBUTION
Western half of the Baja California peninsula.

HABITAT
The two-legged wormlizard inhabits sandy soils. These reptiles can be found by digging to depths up to 6 in (15 cm) below the soil surface. *Bipes biporus* is often found in association with the roots of mesquite shrubs. It can sometimes be found on the soil surface under debris or rocks, especially at night.

BEHAVIOR
Bipes biporus are active throughout the year, in any season. They inhabit shallow burrow systems, with an average depth of 0.98 in (2.5 cm) below the soil surface. *B. biporus* exhibit diurnal vertical movements; studies indicate that they may be found closer to the surface in early morning hours, which is suggestive of thermoregulatory behavior. It is possible that *B. biporus* also moves horizontally with respect to shade around surface objects such as trees or shrubs in order to thermoregulate. Some data suggest that they are more likely to be found in sunny locations and/or nearer to the surface during early parts of the day.

Bipes biporus leaves it burrow systems at times and travels on the surface of the soil, especially at night. These animals move clumsily on the surface, using a combination of rectilinear locomotion of the body and overhand strokes of the short front limbs.

FEEDING ECOLOGY AND DIET
Bipes biporus feeds mainly on termites, insect larvae, and ants found below the soil surface. Evidence also exists that *Bipes* occasionally feeds on spiders and insects, which are surface-active prey, indicating that at least some feeding occurs outside of the tunnel system. Field studies indicate that *B. biporus* are more likely to be found outside of their tunnel systems during the night.

REPRODUCTIVE BIOLOGY
Female *Bipes biporus* become sexually mature once a body size of approximately 7.3 in (185 mm) is achieved, usually at approximately 45 months of age. Females lay eggs in clutches of one to four during the dry season. Hatchlings appear in late September, which is just before the rainy season, ensuring food availability for growing young. Field observations indicate that only about half the adult females are gravid in any given year, which has led to the suggestion that *B. biporus* are reproductively active every other year.

CONSERVATION STATUS
Not listed by the IUCN.

SIGNIFICANCE TO HUMANS
None known. ◆

Bipes biporus

Resources

Books

Gans, C. *Biomechanics: An Approach to Vertebrate Biology.* Philadelphia: J. B. Lippincott Company, 1974.

Schwenk, K. "Feeding in Lepidosaurs." In *Feeding: Form, Function, and Evolution in Tetrapod Vertebrates.* San Diego: Academic Press, 2000: 175–291.

Vanzolini, P. E. *Evolution, Adaptation and Distribution of the Amphisbaenid Lizards (Sauria: Amphisbaenidae).* Ph.D. diss. Harvard University, 1951.

Periodicals

Castañeda, M. R., and T. Alvarez. "Contribucion al conocimiento de la osteologia apendicular de *Bipes* (Reptilia: Amphisbaenia)." *Anales de la Escuela Nacional de Ciencias Biologicas Mexico* 17 (1968): 189–206.

Cope, E. D. "On the Genera and Species of Euchirotidae." *American Naturalist* 28 (1894): 436–437.

Gans, C. "Amphisbaenians—Reptiles Specialized for a Burrowing Existence." *Endeavour* 28 (1969): 146–151.

———. "The Characteristics and Affinities of the Amphisbaenia." *Transactions of the Zoological Society of London* 34 (1978): 347–416.

Gans, C., and E. Wever. "The Ear and Hearing in Amphisbaenia (Reptilia)." *Journal of Experimental Zoology* 179 (1972): 17–34.

———. "The Amphisbaenian Ear: *Blanus cinereus* and *Diplometopon zarudnyi.*" *Proceedings of the National Academy of Sciences* 72 (1975): 1487–1490.

Hodges, W. L., and E. Perez-Ramos. "New Localities and Natural History Notes on *Bipes canaliculatus* in Guerrero, México." *Herpetological Review* 32, no. 3 (2001): 153–156.

Kearney, M. "The Appendicular Skeleton in Amphisbaenians." *Copeia* 2002, no. 3 (2002): 719–738.

———. "Diet in the Amphisbaenian *Bipes biporus.*" *Journal of Herpetology.* In press.

Papenfuss, T. "The Ecology and Systematics of the Amphisbaenian Genus *Bipes.*" *Occasional Papers of the California Academy of Sciences* 136 (1982): 1–42.

Renous, S., J. P. Gasc, and A. Raynaud. "Comments on the Pelvic Appendicular Vestiges in an Amphisbaenian: *Blanus cinereus* (Reptilia, Squamata)." *Journal of Morphology* 209 (1991): 23–28.

Taylor, E. H. "Does the Amphisbaenid Genus *Bipes* Occur in the United States?" *Copeia* 1938 (1938): 202.

Zangerl, R. "Contributions to the Osteology of the Postcranial Skeleton of the Amphisbaenidae." *American Midland Naturalist* 33 (1945): 764–780.

———. "Contributions to the Osteology of the Skull of the Amphisbaenidae." *American Midland Naturalist* 31 (1944): 417–454.

Maureen Kearney, PhD

Florida wormlizards
(Rhineuridae)

Class Reptilia
Order Squamata
Suborder Amphisbaenia
Family Rhineuridae

Thumbnail description
Elongate, slender, limbless, fossorial reptiles with scales arranged in annular rings, eyes greatly reduced or absent, no external ear openings, a spatulate head covered with a keratinized shield, and a dorsoventrally flattened, short tail

Size
9.5–11 in (240–280 mm)

Number of genera, species
1 genus; 1 species

Habitat
Sandy soils, xerix scrub forest, mesic hammock forest

Conservation status
Not listed by the IUCN

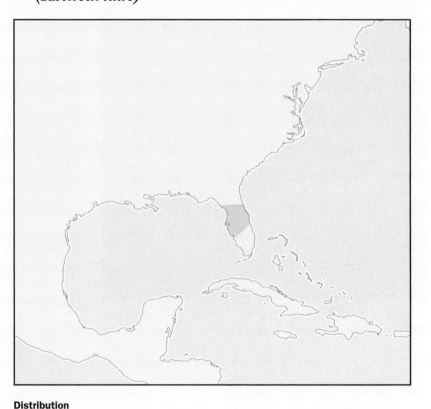

Distribution
Rhineurids have a restricted distribution in northeastern and north-central Florida

Evolution and systematics

Most recent phylogenetic analyses have placed amphisbaenians as one of three suborders of Squamates (the clade that includes snakes, lizards, and amphisbaenians), but their exact placement within that clade is not well understood. The interrelationships among the four amphisbaenian families are also poorly understood. Past studies disagree on whether rhineurids are the most primitive amphisbaenians or are derived within the group. A recent phylogenetic analysis suggests that rhineurids are derived amphisbaenians.

The family Rhineuridae is the only amphisbaenian family with a significant fossil record. This fossil record is entirely from the western and central United States, indicating that rhineurids were once widely distributed across the United States, whereas today a single relict species occurs in Florida. Rhineurids are found in the fossil record extending as far back as the Paleocene (about 60 million years ago). These fossils are already quite derived, exhibiting a shovel-headed cranial shape similar to that occurring in the extant species from Florida, but they also exhibit some intriguing primitive cranial features that are absent in most amphisbaenian species living today. For example, some fossil rhineurids have a complete orbit, enclosed posteriorly by the jugal bone, an element of the skull that is absent in

all extant forms. The presence or absence of limbs in these fossil forms is not definitively known due to vagaries of fossil preservation, but no limb elements have been found so far associated with these fossils. If rhineurid fossils are ever found with preserved limb elements, this would significantly change our understanding of relationships within Amphisbaenia.

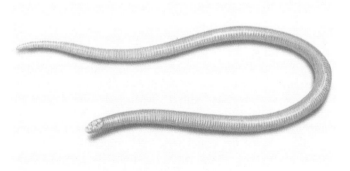

Florida wormlizard (*Rhineura floridana*). (Illustration by John Megahan)

A Florida wormlizard (*Rhineura floridana*). (Photo by Dave Norris/Photo Researchers, Inc. Reproduced by permission.)

The taxonomy of this species is *Rhineura floridana* Baird, 1859, Micanopy, Florida.

Physical characteristics

Certain features of rhineurids are common to most or all amphisbaenians. These include: a unique modification of the middle ear in which an elongated structure, the extracolumella, attaches to the stapedial bone of the middle ear, extending forward to attach to tissue along the sides of the face and allowing the reception and transmission of vibrations to the inner ear; reduction or absence of the right lung; an enlarged, medial, premaxillary tooth; the periodic shedding of the skin in a single piece; a heavily ossified and robust skull; absence of eyelids and external ear openings; and a forked tongue.

Rhineurids exhibit a cylindrical body shape, similar to most amphisbaenians. The snout is dorsoventrally depressed, and the head bears a strong craniofacial angulation. The snout region may also exhibit a zone of keratinization over fused head scales, thus resembling a spatulate digging shield. The tail is short, flattened, and bears conical dorsal tubercles. Caudal autotomy is absent. The ventral throat scales are elongated. Rhineurids are usually pallid or pink in appearance. The dentition in rhineurids is pleurodont. The lower jaw is "countersunk" beneath the upper jaw, and the nostrils occur in a ventral position. A single, median, premaxillary tooth occurs on the premaxilla, sometimes flanked by two smaller teeth. There are four or five teeth on the maxillary and six teeth on the lower jaw.

Beyond those easily recognizable features, rhineurids are also characterized by a number of unique internal conditions, including the complete absence of any pectoral girdle elements and a uniquely modified pelvic girdle.

The Florida wormlizard has a body length ranging from 9.5–11 in (240–280 mm). The tail is less than 10% of the total length; it is flattened and covered dorsally by conical tubercles. The Florida wormlizard has a midbody diameter of approximately 0.5 in (12 mm).

Distribution

The single extant species, *Rhineura floridana*, ranges throughout northeastern and northcentral Florida. The recent discovery of the Florida wormlizard in southern Georgia means that this species is no longer considered a Florida endemic.

Habitat

This species lives in sandy hammocks; high pine; and dry, sandy, easily burrowed soils.

Behavior

Rhineurids are completely fossorial, using the spatulate head to tunnel into the soil. The shovel-shaped snout is an effective aid in penetrating the soil. Additionally, the snout bears a hard covering that forms a keratinous shield, thus hardening the penetrating edge even more. The keratin usually wears off with use and is continually replaced. The irregular, raised scales on the dorsal surface of the tail are used to plug the mouth of the tunnel, with soil particles blending into these raised tubercles and camouflaging both the tail and the tunnel entrance. When threatened, the Florida wormlizard retreats quickly into the soil and uses its tail to block the tunnel entrance.

Feeding ecology and diet

The natural history of rhineurids is very poorly known, and we have little information on diet for this species. Most amphisbaenians feed primarily on termites, beetle larvae, and ants, and presumably this is also true of rhineurids.

For rhineurids, chemical and auditory cues are the most important means used in locating prey. The uniquely adapted middle ear system allows prey movements to be detected, while the forked tongue and Jacobson's organ allow the detection of chemical odors. Airborne sounds are picked up and transmitted to the inner ear along the specialized extracolumellar apparatus, which may amplify the vibrations as well. This unique anatomy is consistent with behavioral studies conducted in laboratory experiments that suggest that amphisbaenians can hear prey movements through the soil.

Reproductive biology

Due to their secretive habits, little is known of the reproductive biology of rhineurids. Rhineurids are oviparous, laying a clutch that is usually comprised of two eggs in subterranean burrows. Detailed data are not available.

Conservation status

Rhineura floridana is not listed by the IUCN.

Significance to humans

The Florida wormlizard is of no economic significance to humans, but it may benefit humans ecologically by feeding on populations of ants and termites and potentially helping to keep these populations in check.

Resources

Books

Gans, C. *Biomechanics: An Approach to Vertebrate Biology.* Philadelphia: J. B. Lippincott Company, 1974.

Schwenk, K. "Feeding in Lepidosaurs." In *Feeding: Form, Function, and Evolution in Tetrapod Vertebrates.* San Diego: Academic Press, 2000: 175–291.

Vanzolini, P. E. *Evolution, Adaptation and Distribution of the Amphisbaenid Lizards (Sauria: Amphisbaenidae).* Ph.D. diss. Harvard University, 1951.

Periodicals

Baird, S. F. "Description of New Genera and Species of North American Lizards in the Museum of the Smithsonian Institution." *Proceedings of the Academy of Natural Sciences of Philadelphia* 1858 (1859): 253–256.

Berman, D. "A New Amphisbaenian (Reptilia: Amphisbaenia) from the Oligocene-Miocene John Day Formation, Oregon." *Journal of Paleontology* 50 (1976): 165–174.

———. "*Hyporhina tertia*, New Species (Reptilia: Amphisbaenia) from the Early Oligocene (Chadronian) White River Formation of Wyoming." *Annals of the Carnegie Museum* 44 (1972): 1–10.

———. "*Spathorhynchus natronicus*, a Middle Eocene Amphisbaenian (Reptilia) from Wyoming." *Copeia* 1973 (1973): 704–721.

Carr, A. "Notes on Eggs and Young of the Lizard *Rhineura floridana*." *Copeia* 1949 (1949): 77.

Cope, E. D. "Remarks on Reptiles." *Proceedings of the Academy of Natural Sciences, Philadelphia* 13 (1861): 75.

Gans, C. "The Characteristics and Affinities of the Amphisbaenia." *Transactions of the Zoological Society of London* 34 (1978): 347–416.

Gans, C., and E. Wever. "The Ear and Hearing in Amphisbaenia (Reptilia)." *Journal of Experimental Zoology* 179 (1972): 17–34.

Gilmore, C. W. "Fossil Lizards of North America." *Memoirs of the National Academy of Science* 22 (1928): 1–201.

Kearney, M. "The Appendicular Skeleton in Amphisbaenians." *Copeia* 2002, no. 3 (2002): 719–738.

Taylor, E. H. "Concerning Oligocene Amphisbaenid Reptiles." *University of Kansas Science Bulletin* 34 (1951): 521–579.

Zangerl, R. "Contributions to the Osteology of the Postcranial Skeleton of the Amphisbaenidae." *American Midland Naturalist* 33 (1945): 764–780.

———. "Contributions to the Osteology of the Skull of the Amphisbaenidae." *American Midland Naturalist* 31 (1944): 417–454.

Maureen Kearney, PhD

Spade-headed wormlizards

(*Trogonophidae*)

Class Reptilia
Order Squamata
Suborder Amphisbaenia
Family Trogonophidae

Thumbnail description
Elongate, limbless, fossorial reptiles with scales arranged in annular rings, eyes greatly reduced or absent, no external ear openings, a spade-shaped head with sharp cutting edges, and an extremely short, pointed tail

Size
3.1–9.4 in (80–240 mm)

Number of genera, species
4 genera; 8 species

Habitat
Loose, sandy, or loamy soils

Conservation status
No species listed by the IUCN

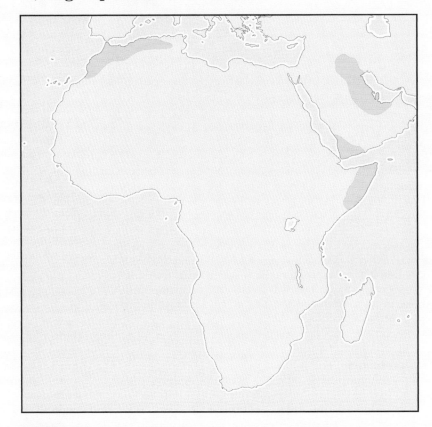

Distribution
Trogonophids occur in northern Africa, the island of Socotra, and a disjunct distribution in the Middle East from western Iran to eastern Somalia

Evolution and systematics

Most recent phylogenetic analyses have placed amphisbaenians as one of three suborders of Squamates (the clade that includes snakes, lizards, and amphisbaenians), but their exact placement within that clade is not well understood. The interrelationships among the four amphisbaenian families are also poorly understood. A recent phylogenetic analysis indicates that trogonophids may be most closely related to the amphisbaenids among amphisbaenians. No significant fossil record exists for this family; a single fossil specimen from Africa is known. *Trogonophis* is probably the most primitive member of the Trogonophidae, with the highly specialized *Agamodon* and *Diplometopon* being the most derived members of the group. No subfamilies are recognized.

Physical characteristics

Certain features of trogonophids are common to most or all amphisbaenians. These include: a unique modification of the middle ear in which an elongated structure, the extracolumella, attaches to the stapedial bone of the middle ear ex-

tending forward to attach to tissue along the sides of the face and allowing the reception and transmission of vibrations to the inner ear; reduction or absence of the right lung; an enlarged, medial, premaxillary tooth; the periodic shedding of the skin in a single piece; a heavily ossified and robust skull; absence of eyelids and external ear openings; and a forked tongue.

Trogonophids are all limbless but retain both pectoral and pelvic girdle vestiges. The head is spade-shaped, with sharp lateral edges on the spade. The body shape of trogonophids is unique among amphisbaenians, being higher than wide, and resembling an upside down "U" in cross-section rather than the more common circular shape. The ventral surface of the body appears excavated, or concave. This shape is partially due to the elongation of the ribs, the ends of which dig into the ground, helping to balance the animal against the forces created during the oscillating motion that trogonophids use in burrowing. The dentition of trogonophids is acrodont, another unique feature among amphisbaenians. The tail is extremely short, pointed, and downward curving, and may be keeled in some species. Caudal autotomy is always absent in

A spade-headed (or shorthead) wormlizard (*Trogonophis wiegmanni*). (Photo by Axel Kwet. Reproduced by permission.)

these species. Striking pigmentation with spotting and checkerboard patterns occurs in some species. Beyond those easily recognizable features, trogonophids are also characterized by a number of unique internal conditions, including an enlarged, pectoral sternal plate and a greatly enlarged premaxilla in the facial portion of the skull.

Distribution

Trogonophids occur in northern Africa, Socotra Island, and the eastern Arabian peninsula.

Habitat

Spade-headed wormlizards are known to occur mainly in loose, sandy soils. *Trogonophis wiegmanni*, the least specialized member of this group, may be the most likely member of this family to be found above ground. *Agamodon* tends to be found in very fine, sandy soils and at greater depths below the surface.

Behavior

Trogonophids are unique among amphisbaenians in using mainly oscillating, rather than rectilinear, movements while tunneling. Oscillatory movements of the head consist of rotating the head alternatively left and right, which effectively shaves soil off the end of the tunnel via the sharp edges along the sides of the face, and simultaneously compacts the loose shavings onto the walls of the tunnel with the sides of the

head. This behavior explains some of the specialized and unique anatomy of these forms. For example, the noncircular shape of the trunk in cross-section is useful in force resistance during oscillatory movements of the head, preventing the entire body of the animal from spinning. Trogonophids also use their exceptionally short tails as anchors to apply force when burrowing with their heads. Some trogonophids display interesting defensive behaviors when threatened, such as rolling over on their backs and freezing.

Feeding ecology and diet

Because amphisbaenians are fossorial, little is known of their behavior or ecology. Most species feed primarily on small arthropods, such as termites and beetle larvae. However, laboratory studies suggest that captive trogonophids are capable of preying on much bigger animals by biting off pieces from the prey animal, and their skull anatomy and dentition seem to support this ability. However, direct examination of stomach contents and field studies are rare for trogonophids, and what they actually eat in the wild is not known.

For trogonophids, chemical and auditory cues are the most important means used in locating prey. The uniquely adapted middle ear system allows prey movements to be detected, while the forked tongue and the Jacobson's organ allow the detection of chemical odors. Airborne sounds are picked up and transmitted to the inner ear along the specialized extracolumellar apparatus, which may also amplify the vibrations as well. This unique anatomy is consistent with behavioral studies conducted in laboratory experiments, which suggest that amphisbaenians can hear prey movements through the soil.

Reproductive biology

The reproductive biology of trogonophids is very poorly known. All trogonophids are believed to be oviparous except for some species of *Trogonophis*, which bear live young (typically five neonates per litter).

Conservation status

No species of Trogonophidae are listed by the IUCN.

Significance to humans

Spade-headed wormlizards are of no economic significance to humans, but they may benefit humans ecologically by feeding on populations of ants and termites and potentially helping to keep these populations in check.

Species accounts

No common name
Agamodon anguliceps

TAXONOMY
Agamodon anguliceps Peters, 1882, "Barava (African orientalis)" Brava, Somali Republic.

OTHER COMMON NAMES
None known.

PHYSICAL CHARACTERISTICS
This wormlizard has a steep, wedge-shaped head with sharp, raised edges. Its average body length is 4–8 in (100–180 mm). The tail is approximately 8% of total length. The species is pink ventrally, and it has dark blotches on a yellow background dorsally. Its dentition is acrodont, with a semifused row of teeth on both upper and lower jaws. The median premaxillary tooth is usually flanked by two other, smaller premaxillary teeth. It has three to four teeth on the maxillary, and six to eight teeth on the lower jaw.

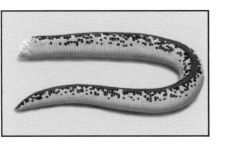

Agamodon anguliceps

DISTRIBUTION
Southcentral coast of Somali Republic, eastern Ethiopia.

HABITAT
This species inhabits loose sandy soils in sandy scrub forests and deserts.

BEHAVIOR
Agamodon anguliceps is a highly derived trogonophid, exhibiting the most specialized anatomy and behavior within this group. Much of this specialization is related to the oscillatory locomotion that is used in the relatively uncompressible, sandy soils in which it lives. *Agamodon anguliceps* exhibits a fright reaction when disturbed, consisting of rolling onto its back, lying still, and exhibiting the pink underside of the body. Some evidence indicates that *A. anguliceps* exhibits vertical migratory movements through the soil corresponding to times of day. For instance, movements in the soil peak at dawn and dusk and individuals are generally closer to the surface around dusk, while they are found much deeper in the soil after dawn.

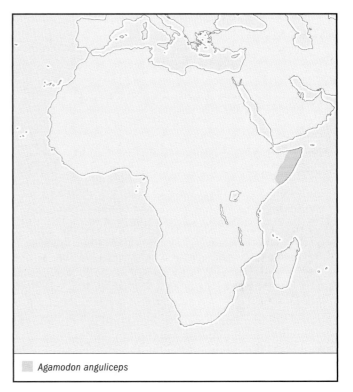

Agamodon anguliceps

FEEDING ECOLOGY AND DIET
No studies examining the natural diet of *Agamodon anguliceps* exist. If we presume that the diet of *A. anguliceps* is similar to that of other amphisbaenians, then it would consist of small invertebrates such as termites, beetles, and beetle larvae. On the other hand, laboratory-based behavioral studies indicate that, when offered larger prey items, *A. anguliceps* is capable of attacking, killing, biting, and efficiently eating various vertebrates. This is not surprising considering the heavy jaws and chewing muscles exhibited in this species.

REPRODUCTIVE BIOLOGY
No specific data exist on the reproductive behavior of *Agamodon anguliceps*. This species is believed to be oviparous.

CONSERVATION STATUS
Not listed by the IUCN.

SIGNIFICANCE TO HUMANS
None known. ◆

Resources

Books

Gans, C. *Biomechanics: An Approach to Vertebrate Biology.* Philadelphia: J. B. Lippincott Company, 1974.

Schwenk, K. "Feeding in Lepidosaurs." In *Feeding: Form, Function, and Evolution in Tetrapod Vertebrates.* San Diego: Academic Press, 2000: 175–291.

Vanzolini, P. E. *Evolution, Adaptation and Distribution of the Amphisbaenid Lizards (Sauria: Amphisbaenidae).* Ph.D. diss. Harvard University, 1951.

Periodicals

Gans, C. "The Characteristics and Affinities of the Amphisbaenia." *Transactions of the Zoological Society of London* 34 (1978): 347–416.

———. "Notes on a Herpetological Collection from the Somali Republic. I. Introduction and Itinerary." *Mus. Roy. Afrique Centrale, Ann.* 8, no. 134 (1965): 1–14.

———. "Studies on Amphisbaenids (Amphisbaenia: Reptilia). I. A Taxonomic Revision of the Trogonophidae and a Functional Interpretation of the Amphisbaenid Adaptive Pattern." *Bulletin of the American Museum of Natural History* 119 (1960): 129–204.

Gans, C., and E. Wever. "The Amphisbaenian Ear: *Blanus cinereus* and *Diplometopon zarudnyi.*" *Proceedings of the National Academy of Sciences* 72 (1975): 1487–1490.

———. "The Ear and Hearing in Amphisbaenia (Reptilia)." *Journal of Experimental Zoology* 179 (1972): 17–34.

Kearney, M. "The Appendicular Skeleton in Amphisbaenians." *Copeia* 2002, no. 3 (2002): 719–738.

Peters, W. "Über eine neue Art und Gattung der Amphisbaenoiden, *Agamodon anguliceps*, mit eingewachsenen Zähnen, aus Barava (Ostafrica) und über die zu den trogonophides gehörigen Gattungen." *Math. Nat. Mitteil. Sitzber. K. Preussische Akad. Wiss. Berlin*, no. 3 (1882): 321–326; Sitzber.

Zangerl, R. "Contributions to the Osteology of the Postcranial Skeleton of the Amphisbaenidae." *American Midland Naturalist* 33 (1945): 764–780.

———. "Contributions to the Osteology of the Skull of the Amphisbaenidae." *American Midland Naturalist* 31 (1944): 417–454.

Maureen Kearney, PhD

Night lizards
(Xantusiidae)

Class Reptilia
Order Squamata
Suborder Scincomorpha
Family Xantusiidae

Thumbnail description
Small to medium-sized lizards lacking eyelids and having enlarged plates on the dorsal surface of the head and transverse rows of enlarged scales on the belly

Size
1.5–5 in (3.7–12.7 cm) snout-vent length

Number of genera, species
3 genera, 23 species

Habitat
Deserts to rainforests

Conservation status
Vulnerable: 1 species

Distribution
North America, Central America, Cuba

Evolution and systematics

Xantusiidae represents an ancient reptile lineage of uncertain relationships. The question what group of lizards constitutes the nearest relatives of Xantusiidae has long puzzled herpetologists. Night lizards resemble geckos in some features, but in other characteristics they appear more allied with teiids, lacertids, and skinks. Although the question continues to be debated by systematists, current evidence seems to indicate that xantusiids are probably most closely related to one of the latter three families, but which one remains a mystery. The fossil record extends back to the Paleocene, but it has shed little light on the origins of the family.

Also controversial is the question of relationships among the three living genera: *Cricosaura* (1 species), *Lepidophyma* (17 species), and *Xantusia* (5 species). Evidence from DNA sequences and from chromosomes indicates that the Cuban night lizard (*Cricosaura typica*) is the oldest separate lineage in the family, whereas characteristics based on internal and external morphology have been interpreted as evidence that *Xantusia* is the earliest branch. Subfamilies have been recognized within Xantusiidae but are not used in this account because of the small number of genera and the ambiguity of the evidence of relationships among them.

Physical characteristics

The most distinctive external feature of Xantusiidae is the lack of eyelids. The eyes are covered by a large, transparent scale ("spectacle" or *brille*) similar to that in snakes and many geckos. The family often is erroneously said to be characterized by having vertically elliptical pupils, but this condition is present only in *Xantusia*. *Cricosaura* and *Lepidophyma* have round pupils. The belly is covered by a series of transverse rows of large, rectangular scales, and there are enlarged plates on the dorsal surface of the head. Body size differs tremendously among the species, from a maximum snout-vent length of only 1.5 in (3.7 cm) for the desert night lizard (*Xantusia vigilis*) to 5 in (12.7 cm) for the yellow-spotted night lizard (*Lepidophyma flavimaculatum*).

Distribution

Xantusiidae is a strictly New World family. Members of *Xantusia* are found in the southwestern United States and northwestern Mexico. The Cuban night lizard occurs only in a small area on Cuba. The species of *Lepidophyma* are found from Nuevo Leon, Mexico, south to Panama. Xantusiids live from sea level to an elevation of more than 9,000 ft (2,700 m).

The granite night lizard (*Xantusia henshawi*) is active in rock crevices where it lives during the day, but is out on the rock surfaces at night. (Photo by Animals Animals © Zig Leszczynski. Reproduced by permission.)

Habitat

Night lizards are habitat specialists. Many of the species are restricted to particular microhabitats. Some species are found primarily in decaying yuccas and agaves. Others live only in rock crevices, caves, or rainforest litter.

Behavior

Because of their secretive and reclusive nature, night lizards are seldom found active outside cover, and little is known about their behavior. Despite the name, night lizards are not strictly nocturnal and may be active day or night, depending on temperature and other conditions. The English name appears to be based on the presence of elliptical pupils in members of the genus *Xantusia*. Night lizards have a low preferred body temperature and low metabolic rate.

Feeding ecology and diet

Some species of Xantusiidae are strictly insectivorous. Others have a diet that includes plant material. Whether any are exclusively herbivorous is unclear. These lizards search for food primarily within the confines of their cover, such as decaying logs.

Reproductive biology

All species for which reproductive information is available are live-bearing, with one notable exception. Recent observations indicate that the Cuban night lizard is egg-laying.

Conservation status

Only one species, the island night lizard (*Xantusia riversiana*), is officially listed as Vulnerable. The listing of this species is based on its limited distribution (three California islands) and threats to the populations posed by introduced animals, particularly pigs and goats. Most of the species in the genus *Lepidophyma* have restricted distribution and have been severely affected by the decimation of forests in tropical America.

Significance to humans

Because of their secretive habits, night lizards are seldom seen or appreciated by humans. In some regions they are feared because of the mistaken notion that they are venomous.

1. Cuban night lizard (*Cricosaura typica*); 2. Yellow-spotted night lizard (*Lepidophyma flavimaculatum*); 3. Desert night lizard (*Xantusia vigilis*). (Illustration by John Megahan)

Species accounts

Desert night lizard
Xantusia vigilis

TAXONOMY
Xantusia vigilis Baird, 1858, Fort Tejon, California, United States.

OTHER COMMON NAMES
English: Yucca night lizard; French: Xantusie du désert; German: Yucca-Nachtechse; Spanish: Lagartija nocturna del desierto.

PHYSICAL CHARACTERISTICS
The desert night lizard is small (maximum snout-vent length, 1.5 in [3.7 cm]) and has vertically elliptical pupils lacking eyelids. The lizard is covered with small, granular dorsal scales and 12 longitudinal rows of ventral scales. The body usually is brown with small dark spots.

DISTRIBUTION
North America, spotty in southwestern United States and northwestern Mexico.

HABITAT
The desert night lizard inhabits desert and chaparral. It is common in decaying yucca logs or dead agaves. The species was considered rare until the discovery of its close association with these plants.

BEHAVIOR
The desert night lizard is rarely found outside cover.

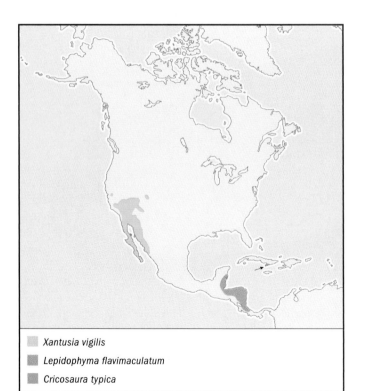

☐ *Xantusia vigilis*
☐ *Lepidophyma flavimaculatum*
☐ *Cricosaura typica*

FEEDING ECOLOGY AND DIET
The desert night lizard is insectivorous, feeding primarily on ants and beetles within the confines of yucca logs and agaves.

REPRODUCTIVE BIOLOGY
In the Mohave Desert, copulation takes place in May and early June. Gestation is approximately 90 days. One to three young (average, 1.9) are born in September and early October. During dry years there may be no reproduction.

CONSERVATION STATUS
Not threatened. Despite the low reproductive rate, populations are quite dense (approximately 12,000 per square mile [4,000 per square kilometer]) in favorable habitat in the Mohave Desert, but extensive areas in the western part of the range have been cleared for housing.

SIGNIFICANCE TO HUMANS
None known. ◆

Yellow-spotted night lizard
Lepidophyma flavimaculatum

TAXONOMY
Lepidophyma flavimaculatum Duméril, 1851.

OTHER COMMON NAMES
Spanish: Escorpión nocturno puntos amarillos, lepidofima.

PHYSICAL CHARACTERISTICS
The yellow-spotted night lizard is medium-sized (maximum snout-vent length, 5 in [13 cm]) with round pupils lacking eyelids. The sides of the body are covered with large, tubercular scales. Ten longitudinal rows of scales cover the ventral surface. The lizard is dark brown to black with yellow spots.

DISTRIBUTION
Southern Mexico to Panama.

HABITAT
The yellow-spotted night lizard inhabits wet tropical forests (rainforests, cloud forests) in decaying logs, tree stumps, leaf litter, rock crevices, caves, and ruins.

BEHAVIOR
The yellow-spotted night lizard seldom is found outside cover.

FEEDING ECOLOGY AND DIET
The yellow-spotted night lizard eats insects, spiders, scorpions, and other arthropods found in decaying logs.

REPRODUCTIVE BIOLOGY
The yellow-spotted night lizard bears five to eight live young. All-female populations occur in Costa Rica and Panama.

CONSERVATION STATUS
The yellow-spotted night lizard is not officially listed as threatened but is severely affected by the clearing of forests.

SIGNIFICANCE TO HUMANS
None known. ◆

Cuban night lizard

Cricosaura typica

TAXONOMY

Cricosaura typica Gundlach and Peters, 1863, Cabo Cruz, Cuba.

OTHER COMMON NAMES

None known.

PHYSICAL CHARACTERISTICS

The Cuban night lizard is small (maximum snout-vent length, 1.6 in [3.9 cm]). It has round pupils and lacks eyelids. The lizard is covered with small, flattened dorsal scales and eight longitudinal rows of rectangular ventral scales. It is dark brown with a vague dorsal-lateral stripe.

DISTRIBUTION

Cabo Cruz and Sierra Maestra, Cuba.

HABITAT

The Cuban night lizard lives under rocks in dry, subtropical forest.

BEHAVIOR

In captivity the Cuban night lizard is active in the evening and remains under cover or buried in soil during the day.

FEEDING ECOLOGY AND DIET

The Cuban night lizard eats insects, primarily ants, and spiders.

REPRODUCTIVE BIOLOGY

The Cuban night lizard is the only known egg-laying xantusiid. The lizard lays a single egg in a 2.5–5 in (1–2 cm) hole. The egg hatches two months after being laid.

CONSERVATION STATUS

Although not threatened, the Cuban night lizard has a limited distribution and is affected by forest degradation.

SIGNIFICANCE TO HUMANS

None known. ◆

Resources

Books

Alvarez del Toro, M. *Los reptiles de Chiapas.* 3rd edition. Chiapas: Instituto de Historia Natural, Tuxtla Gutierrez, 1982.

Campbell, J. A. *Amphibians and Reptiles of Northern Guatemala, the Yucatan, and Belize.* Norman: University of Oklahoma Press, 1998.

Estes, R. *Sauria terrestria, Amphisbaenia.* Vol. 10A, *Handbuch der Palaeoherpetologie.* Stuttgart: Gustav Fisher Verlag, 1983.

Mautz, W. J. "Ecology and Energetics of the Island Night Lizard, *Xantusia riversiana*, on San Clemente Island." In *Third California Islands Symposium: Recent Advances in Research on the California Islands,* edited by F. G. Hochberg. Santa Barbara: Santa Barbara Museum of Natural History, 1993.

Periodicals

Ansel, Fong G., M. Rolando Viña, and B. Angel Arias. "Aspectos de la Historia Natural de *Cricosaura typica* (Sauria: Xantusiidae) de Cuba." *Caribbean Journal of Science* 35 (1999): 148–150.

Bezy, R. L. "*Xantusia vigilis.*" *Catalogue of American Amphibians and Reptiles* 302 (1982): 1–4.

——. "The Natural History of the Night Lizards, Family Xantusiidae." In *Proceedings of the Conference on California Herpetology: Southwestern Herpetologists Society Special Publication* no. 4 (1988): 1–12.

——. "Night Lizards: The Evolution of Habitat Specialists." *Terra* 28 (1989): 29–34.

Bezy, R. L., and J. L. Camarillo R. "Systematics of Xantusiid Lizards of the Genus *Lepidophyma.*" *Contributions in Science* 493 (2002): 1–41.

Brattstrom, B. H. "The Number of Young of *Xantusia.*" *Herpetologica* 7 (1951): 143–144.

——. "The Food of the Night Lizards, Genus *Xantusia.*" *Copeia* (1952): 168–172.

Crother, B. I. "*Cricosaura, C. typica.*" *Catalogue of American Amphibians and Reptiles* 443 (1988): 1–3.

Crother, B. I., M. M. Miyamoto, and W. F. Presch. "Phylogeny and Biogeography of the Lizard Family Xantusiidae." *Systematic Zoology* 35 (1986): 37–45.

de Armas, L. F., A. Rams, and A. Torres. "Primeras observaciones sobre la alimentacion de *Cricosaura typica* (Sauria: Xantusiidae) en condiciones naturales." *Miscellaneous Zoology* 32 (1987): 1–2.

Estrada, A. R., and L. F. de Armas. "Apuntos ecologicos sobre *Cricosaura typica* (Sauria: Xantusiidae) de Cuba." *Caribbean Journal of Science* 34 (1998): 160–162.

Fellers, G. M., and C. A. Drost. "*Xantusia riversiana.*" *Catalogue of American Amphibians and Reptiles* 518 (1991): 1–4.

——. "Ecology of the Island Night Lizard, *Xantusia riversiana*, on Santa Barbara Island, California." *Herpetological Monographs* 5 (1991): 28–78.

Hass, C. A., and S. B. Hedges. "Karyotype of the Cuban Lizard *Cricosaura typica* and Its Implication for Xantusiid Phylogeny." *Copeia* (1992): 563–565.

Hedges, S. B., R. L. Bezy, and L. R. Maxson. "Phylogenetic Relationships and Biogeography of Xantusiid Lizards, Inferred from Mitochondrial DNA Sequences." *Molecular Biology and Evolution* 8 (1991): 767–780.

Lee, J. C. "The Diel Activity Cycle of the Lizard, *Xantusia henshawi.*" *Copeia* (1974): 934–940.

——. "The Autecology of *Xantusia henshawi* (Sauria: Xantusiidae)." *Transactions of the San Diego Society of Natural History* 17 (1975): 259–278.

——. "*Xantusia henshawi.*" *Catalogue of American Amphibians and Reptiles* 189 (1976): 1–2.

Mautz, W. J. "The Metabolism of Reclusive Lizards, the Xantusiidae." *Copeia* (1979): 577–584.

——. "Thermal Biology and Microhabitats of Xantusiid Lizards: Herpetology of the North American Deserts,

Resources

Proceedings of a Symposium. *Southwest Herpetologists Society Special Publication* 5 (1994): 227–238.

Mautz, W. J., and T. J. Case. "A Diurnal Activity Cycle in the Granite Night Lizard, *Xantusia henshawi.*" *Copeia* (1974): 243–251.

Petzold, H.-G. "*Cricosaura typica* Gundlach & Peters, eine herpetologische Kostbarkeit aus Kuba." *Die Aquarien und Terrarien Zeitschrift* 22 (1969): 82–85.

Telford, S. R., and H. W. Campbell. "Ecological Observations on an All Female Population of the Lizard *Lepidophyma flavimaculatum* (Xantusiidae) in Panama." *Copeia* (1970): 379–381.

Zweifel, R. G., and C. H. Lowe. "The Ecology of a Population of *Xantusia vigilis*, the Desert Night Lizard." *American Museum Novitates* 2247 (1966): 1–57.

Robert L. Bezy, PhD
L. Lee Grismer, PhD

Wall lizards, rock lizards, and relatives
(Lacertidae)

Class Reptilia
Order Squamata
Suborder Sauria
Family Lacertidae

Thumbnail description
Small- to medium-sized diurnal, heliothermic, terrestrial lizards, all with well-developed legs and a tail usually much longer than the body

Size
Usually less than 8 in (20 cm), maximum around 20 in (50 cm)

Number of genera, species
28 genera; 225+ species

Habitat
Varied; most abundant in scrublands and desert, but also entering forest and montane grassland

Conservation status
Critically Endangered: 1 species; Endangered: 1 species; Vulnerable: 5 species

Distribution
Old World, throughout Africa, sub-Arctic Eurasia, and the northern East Indies

Evolution and systematics

Lacertid lizards are the sister groups of teioids (teiids and relatives of the New World). Together they form the Lacertiformes. Molecular studies have allowed the recent recognition of three subfamilies, although the relationships of a number of genera (e.g., *Takydromus*) remain unresolved. The most primitive subfamily is the Gallotiinae, which comprises the giant lizards of the Canary Islands and the smaller, but closely related *Psammodromas* from Iberia and adjacent North Africa. They share a suite of unusual, but non-unique features, as well as the strange ability to squeak. The Eremiainae includes many genera endemic to the Ethiopian region. The basal genera occur in tropical forests (e.g., *Holaspis*, *Gastropholis*, and *Poromera*) or montane habitats (e.g., *Adolfus* and *Tropidosaura*), with a progression of more specialized genera showing increasing adaptation to arid habitats (e.g., *Ichnotropis*, *Heliobolus*, *Meroles*, *Pedioplanis*, *Ophisops*, *Mesalina*, and *Acanthodactylus*). The remaining subfamily, Lacertinae, includes the typical lacertids of Eurasia. The relationships and composition of many of the genera in the subfamily remain in question. The large genus *Lacerta* has long been paraphyletic but has been increasingly subdivided in recent years, with the creation of a number of genera and subgenera, e.g., *Omanosaura*, *Timon*, *Darevskia*, *Caucasilacerta*, *Parvalacerta*, and so on.

Lacertids are believed to have arisen in Eurasia and invaded Africa and the Ethiopian region, where they evolved more xeric forms (adapted to arid conditions). One or more derivatives then re-entered the dry areas of Eurasia. The family has a long but fragmentary fossil history, extending back at least as far as the Palaeocene. Fragments of a recently described 120 million-year-old fossil lizard in amber from the Lower Cretaceous of Lebanon shows that common external features of lacertids and other autarchoglossan lizards had already evolved.

Physical characteristics

Lacertids are conservative in morphology, and no species have lost limbs, ears, or eyes. They are small- to medium-sized with slender bodies, well-developed limbs, and a long tail that in oriental grass lizards (*Takydromus*) may be nearly five times as long as the head and body. The dorsal scales are usually small, smooth, and granular (although in some genera they are large, rough, and overlapping). The belly scales are always large and quadrangular and almost always arranged in distinct longitudinal and transverse rows. The head is covered with large, symmetrical scales that have osteoderms, and most species have a pineal "eye" on the top of the head. The tongue is quite deeply notched. A skin-fold "collar" of enlarged scales on the neck is also usually present. In the skull the upper temporal arches are completely ossified, and the pterygoid and palatine bones are paired and lie close together anteriorly. The pterygoids often bear a group of palatal teeth. The dentition of the jaws is pleurodont (the teeth attached to the sides of the jawbone). Femoral pores are usually present in both sexes. The tail has whorls of keeled scales, which may be spiny; the tail is easily shed but can be regenerated. A complex armature supports the hemipenes in the Eremiainae.

A wall lizard (*Podarcis lilfordi*) from southern Europe. (Photo by Animals Animals © E. R. Degginger. Reproduced by permission.)

Distribution

This family occurs throughout Africa and most of Eurasia. A few species occur on some off-shore islands, including the British Isles, the Canaries, Madeira, many Mediterranean islands, Socotra, Sri Lanka, and many islands of the Sunda Shelf. They are absent from Madagascar, however. They are the only lizards to enter the Arctic Circle.

Habitat

Members of this family are found from tundra and high montane grasslands through heath, Mediterranean vegetation, tropical forests, semi-arid scrublands, and true desert.

Behavior

Most species are active terrestrial or rock-living species, while a few others are arboreal. All are diurnal heliotherms. The exceptionally long tails of Asian grass lizards are prehensile to allow them to climb in vegetation. Of the many desert-living species, the most superbly adapted is the shovel-snouted lizard (*Meroles anchietae*) of the Namib Desert. It has large hindlimbs with fringed toes to help it run quickly over loose sand, and an aptly named snout that allows it to dive under the sand to escape predators and to sleep in cooler, deeper sand. Hatchlings of the Kalahari sand lizard (*Heliobolus lugubris*) are boldly marked in black and white and walk with a stiff-legged gait. They mimic the noxious oogpister ("eye squirter") beetle (*Anthia*), but when adult they become colored in cryptic tans and brown.

When attacked, lacertids can discard their tail. From the sixth vertebrae backwards, every tail bone has a special plane of weakness through its body. There are corresponding weak points in the surrounding connective tissue and musculature. If the tail is held, circular muscles at that position contract strongly, the tissues break, and the tip of the tail falls off. The discarded fragment continues to twist violently to attract the attention of the attacker while the tailless lizard escapes. The shortened tail can regrow, but it lacks bone and is supported only by a central rod of cartilage.

Feeding ecology and diet

Lacertids are active, diurnal lizards that hunt insects in open situations. A few are sit-and-wait ambush predators, usually capturing their prey with a quick dash from shaded cover. All are primarily insectivores, although some are also partial herbivores, eating dried seeds or fruit. The giant lizards (*Gallotia*) of the Canary Islands are almost entirely vegetarian, at least as adults.

Reproductive biology

The males of many species are brightly colored, with dominant males developing bright breeding colors. When defending its territory, the lizard displays by turning broadside, lifting its head, and expanding its throat. The flanks also flatten so as to present the most massive, threatening appearance. If the threat display does not deter a rival, an attack follows. The combatants bite each other and shake, but wounds are seldom serious because the head is well-armored.

With the exception of a few European species, all lacertids are oviparous, which is surprising as many live in temperate and montane environments where viviparity could be expected to evolve. Most lay small clutches (usually fewer than 10) of soft-shelled eggs in a small chamber dug in moist soil, often beneath a sun-warmed stone. As usual in lizards, the larger species lay more eggs, and in the eyed lizard (*Lacerta lepida*) this may be up to 20. There is no brood-care behavior. The viviparous lizard of northern Europe (*Lacerta vivipara*) is one of the few lacertids to give birth to live babies. A small brood of four to 11 young is born from late July to early October after a three to four month gestation. Seven Caucasian lacertids of the genus *Darevskia* are all-female species. They reproduce by parthenogenesis and result from interspecific hybridization.

Over much of the savannas of the northern part of southern Africa, the Cape rough-scaled lizard (*Ichnotropis capensis*) and common rough-scaled lizard (*I. squamulosa*) live together. It is a simple rule in ecology that two species do not inhabit the same niche, as they would compete for the same resources. However, adult and juvenile lizards do not usually compete for the same food, and even if they eat the same species they take different size classes. The two rough-scaled lizards grow to maturity, mate, and lay their eggs within a year, and after reproducing die. Incubation of the eggs takes about two months. During any sixth-month period, one species is represented by adults that mate and lay eggs. At this time the other species' eggs are hatching, and the adults have died after reproducing. There is therefore no competition for food between the adults of one species and the hatchlings of the other species. As these hatchlings grow to maturity, the adults of the first species die, while within the ground their eggs develop. The two "annual" rough-scaled lizards have therefore evolved an elegant rotating solution to living together that minimizes competition.

Conservation status

Simony's giant lizard (*Gallotia simonyi*) of the Canary Islands was for many years considered extinct, until a small population was discovered in 1975 on a few inaccessible cliffs on El Hierro. It occurs in low numbers and is categorized as Critically Endangered by the IUCN. The Gomeran giant lizard (*G. gomerana*) was only known from sub-fossil material until a very small population of live lizards was discovered in 2001, again on inaccessible sea cliffs. It appears to be even rarer than Simony's giant lizard and, while not yet listed by the IUCN, is one of the most endangered reptiles in the world. Both species are threatened by the introduction of predators such as cats and rats to the islands. Clark's lacerta (*Darevskia clarko-*

rum), which is listed as Endangered, is restricted to the Black Sea region. Five other lacertids are considered Vulnerable in the IUCN Red List 2000, while over 30 lacertids are included on the Bern Convention on Conservation of European Wildlife.

Significance to humans

Few lacertids grow big enough to eat, although the giant lizards of the Canary Islands were eaten by the early settlers.

1. Sand lizard (*Lacerta agilis*); 2. Six-lined grass lizard (*Takydromus sexlineatus*); 3. Western sandveld lizard (*Nucras tessellata*). (Illustration by Gillian Harris)

Species accounts

Western sandveld lizard
Nucras tessellata

SUBFAMILY
Eremiainae

TAXONOMY
Nucras tessellata A. Smith, 1838, eastern parts of the Cape Colony.

OTHER COMMON NAMES
English: Striped sand lizard.

PHYSICAL CHARACTERISTICS
A slender, brilliantly colored lizard with a tail almost twice as long as the body. The lizard has a black body with four thin cream stripes on the back and brilliant white bars on the sides of the head and body. The rear of the body, hind limbs, and long tail are rich red-brown.

DISTRIBUTION
This species occurs in the western arid region of South Africa, just extending into adjacent Namibia and Botswana.

HABITAT
This lizard inhabits succulent scrubland in a winter-rainfall area.

BEHAVIOR
This is an unusual lacertid that spends long periods underground, sheltering in its burrow. It forages on rocky hillsides among the stones and succulent vegetation.

FEEDING ECOLOGY AND DIET
Despite its relatively small size, this lizard specializes in feeding on scorpions, although grasshoppers and beetles are also eaten. It searches actively for the insects' burrows and then digs them out. As it spends a lot of time locating and digging for food, it is vulnerable to predators such as birds and mongoose. Its

bright red tail helps to deflect sudden attacks away from the vulnerable head towards the expendable tail. Safety in this lizard is therefore not based on camouflage or mimicry but on distraction.

REPRODUCTIVE BIOLOGY
A small clutch of three to four eggs are laid under a sun-warmed rock and take two to three months to develop.

CONSERVATION STATUS
Its habitat is sparsely populated and it is not threatened.

SIGNIFICANCE TO HUMANS
None known. ◆

Sand lizard
Lacerta agilis

SUBFAMILY
Lacertinae

TAXONOMY
Lacerta agilis Linnaeus, 1766, southern Sweden (restricted by Mertens and Muller, 1928).

OTHER COMMON NAMES
French: Lézard agile; German: Zauneidechse.

PHYSICAL CHARACTERISTICS
This is a heavy-bodied lacertid, reaching almost 12 in (30 cm) in eastern populations but smaller in the west. The head is blunt and short. Color is variable; in western races the female is brown with darker spots, while males develop bright green flanks; in eastern forms the whole back is almost completely green.

DISTRIBUTION
It has one of the largest ranges of all lacertids, and occurs from the British Isles in the west to northwest China in the east, and from southern Sweden and Karelia in the north (62°N) to the Pyrenees, Balkans, and central Greece (39°N) in the south.

HABITAT
As can be expected from its common name, this lizard favors sandy soils in the western part of its range, inhabiting heathlands and coastal dunes. In the east, however, it prefers clay soils.

BEHAVIOR
The sand lizard is very inconspicuous, although it often clambers and basks in thick shrub. At night or when threatened it retreats into a hole in the ground or cavities amid the roots of bushy shrubs. Males are territorial and will drive off rival males.

FEEDING ECOLOGY AND DIET
This species actively searches for prey among cover and eats a wide variety of insects, and it is occasionally cannibalistic.

REPRODUCTIVE BIOLOGY
Clutches of from three to 14 eggs are buried in a hole excavated in sun-warmed sand and take 40–60 days to develop.

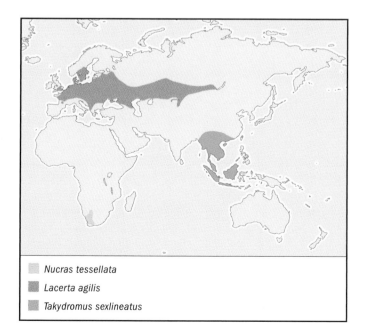

Nucras tessellata

Lacerta agilis

Takydromus sexlineatus

CONSERVATION STATUS
Not threatened; although locally threatened in many Western European countries (e.g., Britain, Holland, and Germany) by scrub encroachment, deforestation, urban development, and recreational activities.

SIGNIFICANCE TO HUMANS
This is a "flagship" species for protecting threatened heathlands in Western Europe. ◆

Six-lined grass lizard
Takydromus sexlineatus

TAXONOMY
Takydromus sexlineatus Daudin, 1802.

OTHER COMMON NAMES
English: Oriental six-lined runner.

PHYSICAL CHARACTERISTICS
This is a long, slender lizard reaching nearly 14 in (36 cm) in length, of which less than a quarter is head and body. Except for the head shields, all the body and limb scales as well as the tail are roughly keeled. The body is olive green to reddish brown above with light dorsolateral stripes and sometimes spots; the legs and tail are often reddish.

DISTRIBUTION
This species is widespread, extending from southern China through the Malay peninsula to Borneo, Sumatra, and Java.

HABITAT
Grasslands.

BEHAVIOR
This is a very active and acrobatic lizard that climbs easily in vegetation, using its very long and prehensile tail for balance and support.

FEEDING ECOLOGY AND DIET
Insect prey is actively hunted in thick vegetation. It may stand upright, balanced on its tail base, and lunge at flying insects.

REPRODUCTIVE BIOLOGY
Some populations breed throughout the year, producing up to six small clutches of up to 10 eggs. Fewer clutches and fewer eggs are laid in cooler northern parts of its range.

CONSERVATION STATUS
Not threatened.

SIGNIFICANCE TO HUMANS
This species is frequently offered for sale as bird food in pet shops in Hong Kong. ◆

Resources

Books

Böhme, W., ed. *Handbuch der Reptilien und Amphibien Europas.* 2 vols. Wiesbaden, Germany: AULA Verlag, 1984–1986.

Branch, B. *Field Guide to the Snakes and Other Reptiles of Southern Africa.* Struik Publishers, 1998.

Valakos, E.D., W. Böhme, V. Perez-Mellado, and P. Maragou, eds. *Lacertids of the Mediterranean Region: A Biological Approach.* Athens, Greece: Hellenic Zoological Society, 1993.

Periodicals

Arnold, E. N. "Towards a Phylogeny and Biogeography of the Lacertidae: Relationships Within an Old-World Family of Lizards Derived from Morphology." *Bulletin of the British Museum of Natural History, Zoology* 55 (1989): 209–257.

————. "Interrelationships and Evolution of the East Asian Grass Lizards, *Takydromus* (Squamata: Lacertidae)." *Zoological Journal of the Linnean Society* 119 (1997): 267–296.

Harris, D. J., E. N. Arnold, and R. H. Thomas. "Relationships of Lacertid Lizards (Reptilia: Lacertidae) Estimated from Mitochondrial DNA Sequences and Morphology." *Proceedings of the Royal Society, London* 265 (1998): 1930–1948.

Bill Branch, PhD

Microteiids
(Gymnophthalmidae)

Class Reptilia
Order Squamata
Suborder Lacertiformes
Family Gymnopthalmidae

Thumbnail description
Small, terrestrial, egg-laying lizards, usually fully
limbed but some with reduced limbs, most of
which are secretive inhabitants of the tropical
forest floor

Size
Adults are less than 2.3 in (60 mm) in snout-
vent length

Number of genera, species
36 genera; 175+ species

Habitat
Microteiids are usually found in leaf litter or
under logs in tropical forests; several species
live in wet areas and escape into the water

Conservation status
Not listed by IUCN

Distribution
Neotropical, ranging from southern Mexico to north-central Argentina east of the
Andes, including Caribbean islands

Evolution and systematics

The Gymnopthalmidae were considered part of the family Teiidae throughout most of the twentieth century but are
now considered distinct by most practicing herpetologists.
The gymnophthalmids are sister group to the teiids, and these
two families comprise a group called Teioidea, whose closest
relatives are lizards in the family Lacertidae, which is restricted to the Old World.

There are 36 genera (or more depending on the source)
in the Gymnopthalmidae and more than 175 species. The
family is not well known. New forms are discovered and described regularly, and taxonomic rearrangements undertaken
by experts will result in changes in the names and number of
genera and species. The most speciose genera are *Proctoporus*
(28 species), *Bachia* (18 species), *Anadia* (15 species), *Ptychoglossus* (15 species), *Leposoma* (12 species), and *Neusticurus* (11
species). These account for about half of all the gymnophthalmids described by 2002. Twelve genera are monotypic,
containing only one described species. No subfamilies are recognized.

Physical characteristics

The gymnophthalmids are small terrestrial lizards. Most
species have fully developed limbs, with the exception of two
genera, *Bachia* and *Calyptommatus*, that have limbs reduced or
absent, respectively. Dorsal scalation is variable; many species
have small dorsal scales. Strongly keeled dorsal scales characterize the genus *Arthrosaura*. Gymnophthalmids have ventral
scales larger than dorsal scales that can be either smooth or
keeled. There are no osteoderms dorsally or ventrally. Tail
length varies, and each tail vertebra has a fracture plane that
allows the tail to be easily broken. Gymnophthalmids have
good visual and olfactory systems. They have well-formed
eyes and eyelids.

Distribution

Gymnophthalmids are strictly Neotropical lizards that occur from southern Mexico to north-central Argentina east of
the Andes. Several genera and species occur on Caribbean islands, for example *Gymnophthalmus* spp., *Tretioscincus bifasciatus*, and *Proctoporus* spp.

Vanzosaura rubricauda. (Photo by Laurie J. Vitt. Reproduced by permission.)

Habitat

Gymnophthalmids are inhabitants of the forest floor or wet areas associated with tropical forests. They are denizens of leaf litter and detritus and can be found under logs, rocks, or other debris. In an analysis of an ecological community of Amazonian lizards, Vitt and Zani reported in 1991 that three species of gymnophthalmids were restricted to leaf litter. One species was found in relatively sunny spots whereas another was more commonly found in shade. A third, *Leposoma per-carinatum*, was found along edges of swamps with fluctuating water levels.

Behavior

Gymnophthalmids are actively foraging lizards but secretive and hard to observe. These small lizards forage for small arthropod prey on the forest floor in leaf litter. Species of *Alo-poglossus* and *Neusticurus* escape predators by diving into the water.

Feeding ecology and diet

All gymnophthalmids are insectivorous, foraging for arthropod prey found in the microhabitats where they live.

Reproductive biology

Reproductive biology is described for only a few species. All known species lay eggs, and clutch size is probably two for most species. Species of *Gymnophthalmus* and *Leposoma* are parthenogenetic, consisting only of females that produce fertile eggs.

Conservation status

No species are presently listed by the IUCN, but gymnophthalmids are susceptible to habitat alteration.

Significance to humans

The significance of the gymnophthalmids to people is not readily apparent. The ecological role these lizards play in tropical systems may never be completely understood but may be important nonetheless. Gymnophthalmids are prey to myriad predators, and themselves consume a wide variety of invertebrate prey.

Alopoglossus angulatus. (Photo by Laurie J. Vitt. Reproduced by permission.)

1. *Gymnophthalmus underwoodi*; 2. *Bachia bresslaui*; 3. *Neusticurus ecpleopus*. (Illustration by Barbara Duperron)

Species accounts

No common name
Bachia bresslaui

TAXONOMY
Apatelus bresslaui Amaral, 1935, Estado de Sao Paolo, Brazil.

OTHER COMMON NAMES
None known.

PHYSICAL CHARACTERISTICS
This gymnophthalmids has an elongated body and tail, reduced limbs, and no external ear opening. Colli and others reported information on this rare species in 1998. The largest individual in their sample had a snout-vent length of 4.2 in (10.6 cm). The tail is more than 1.5 times the body length. This species belongs to an ancestral group of *Bachia*, as evidenced by the presence of head shields that are lost in derived taxa.

DISTRIBUTION
Cerrado ecosystem of central Brazil and northeastern Paraguay, South America.

HABITAT
The species is known to live in sandy soils. Distributional information is lacking and it is possible the species also uses other substrate types.

BEHAVIOR
Colli and others reported in 1998 that based on the presence of scorpions, ants, beetles, and other prey in the diet, the species probably forages above ground.

FEEDING ECOLOGY AND DIET
The diet of five individuals that were examined consisted of ants, beetles, beetle larvae, scorpions, and wolf spiders.

REPRODUCTIVE BIOLOGY
The species is presumed to lay eggs, but the average clutch size is unknown.

CONSERVATION STATUS
Not listed by the IUCN. This species may be more common and more widely distributed throughout the Cerrado ecosystem than previously thought. Threats include habitat destruction.

SIGNIFICANCE TO HUMANS
None known. ◆

No common name
Gymnophthalmus underwoodi

TAXONOMY
Gymnophthalmus underwoodi Grant, 1958, Barbados. *Gymnophthalmus underwoodi* is actually a complex of species, or independent evolutionary units. Some populations of *Gymnophthalmus underwoodi* are bisexual, and others are unisexual, all-female species (parthenogens). The species was described from a series of female specimens from Barbados; no males were found. It was later proved that the Caribbean populations and some South American populations are parthenogenetic, reproducing without males or sperm.

Charles J. Cole and others (1983, 1990) and Laurence M. Hardy and others (1989) showed that this species is a product of hybridization. They suggested that *G. underwoodi* evolved from *G. speciosus* and a yet-undescribed *Gymnophthalmus*. Cole not only predicted what the undiscovered parent species would look like, he also predicted the specific arrangement of its 22 pairs of chromosomes and the nature of 33 of the unknown species' proteins!

Later, Cole and others (1993) reported the missing ancestor of *G. underwoodi* to be *G. cryptus*, a bisexual species described in 1992 from the Orinoco River drainage in Venezuela. After it arose in the upper Orinoco River watershed, *G. underwoodi* dispersed throughout the Guiana region and reached islands in the West Indies. Because multiple hybridization events could have occurred, there may be other clonal lineages of *G. underwoodi* that exist in a complex of species. Experts presume that more cryptic species exist that have not been described by 2002.

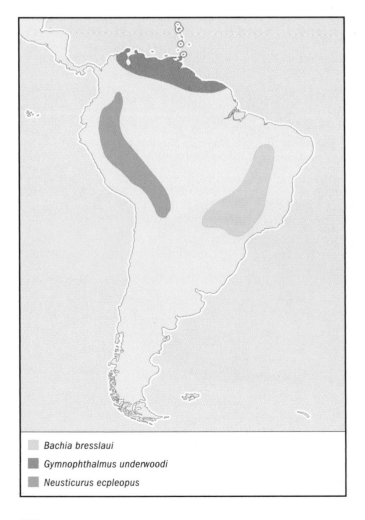

■ *Bachia bresslaui*

■ *Gymnophthalmus underwoodi*

■ *Neusticurus ecpleopus*

OTHER COMMON NAMES
None known.

PHYSICAL CHARACTERISTICS
These are small microteiids with a snout-vent length of 1.4–1.7 in (3.6–4.3 cm). The tail is about 1.5 times body length. The body is cylindrical, and the dorsal and ventral scales are smooth. The limbs are fully developed but small, with four fingers and five toes. These lizards usually are shiny bronze or olive on back, darker on the flanks, with a light dorsolateral stripe. The tail may be the same color as the body or range from bluish to orange or red.

DISTRIBUTION
The *Gymnophthalmus underwoodi* species complex occurs in the Guianan region of South America and in Trinidad and other islands of the West Indies.

HABITAT
These lizards are found in open types of tropical forest in leaf litter or grass. They are often found in microhabitat patches exposed to direct sunlight.

BEHAVIOR
Gymnophthalmus underwoodi, like almost all gymnophthalmids, is a secretive denizen of leaf litter, and its behavior has never been studied. Because their diet consists of small surface-dwelling arthropods, it is likely these lizards actively search for their prey among leaf litter and under logs and other objects. They do not obviously bask, but they do occur mostly in leaf litter receiving direct sunlight. These lizards are part of a large group of lizards that are heliothermic, the Teoidea, almost all of which are strictly diurnal and not territorial.

FEEDING ECOLOGY AND DIET
In a study of an Amazonian lizard community, Vitt and Zani (1991) documented several orders of small insects in the diet of this species, specifically dermapterans, collembolans, and dipterans.

REPRODUCTIVE BIOLOGY
Like other microteiids, this species lays eggs. Clutch size may range from one to four with an average clutch size of two.

CONSERVATION STATUS
Not listed by the IUCN. Lack of knowledge about the ecology of this species complex impedes fully informed decisions about conservation needs. Threats include habitat destruction, and it will be difficult to assess the importance of habitat loss to cryptic species in the *G. underwoodi* complex until the distributions of different forms are understood.

SIGNIFICANCE TO HUMANS
None known. ◆

No common name
Neusticurus ecpleopus

TAXONOMY
Neusticurus ecpleopus Cope, 1876, Peru.

OTHER COMMON NAMES
None known.

PHYSICAL CHARACTERISTICS
The body is cylindrical and the limbs are fully formed with five fingers and five toes. The maximum snout-vent length in males is 3.3 in (8.4 cm) and the tail is 1.4–1.8 times snout-vent length. There are six longitudinal rows of tubercles along the back and tubercles along the flanks. The tail is moderately compressed with a double crest formed by tubercles. The snout is blunt and the gular region is enlarged in adult males. These lizards are brown with black and lighter spots on the dorsum and flanks. The belly is orangish to reddish. Avila-Pires (1995) reported a relatively high degree of geographical variation.

DISTRIBUTION
Western Amazon along the slopes of the Andes from southern Colombia south to Bolivia.

HABITAT
Neusticurus ecpleopus is found along the banks of forest streams living in leaf litter and using muddy stream banks.

BEHAVIOR
In their study of the ecology of *Neusticurus ecpleopus*, Vitt and Avila-Pires in 1998 found that the species was active throughout the day and inactive at night. They do not bask in the sun. These lizards frequently enter water and swim to the bottom to escape predators. The flattened tail is thought to facilitate swimming.

FEEDING ECOLOGY AND DIET
Although a variety of small insect prey are consumed, the diet information provided by Vitt and Avila-Pires (1998) showed the lizards mostly consumed fly larvae, crickets, and ants. It is likely the diet reflects the availability of small insect prey where the lizards happen to be living.

REPRODUCTIVE BIOLOGY
Like other gymnophthalmids, *Neusticurus ecpleopus* lays multiple clutches in a reproductive season, most likely in clutches of two, but the exact number and frequency is unknown.

CONSERVATION STATUS
Not listed by the IUCN. The species is not uncommon along primary forest streams within its range. Threats include habitat destruction, particularly since *Neusticurus ecpleopus* lives in primary forest.

SIGNIFICANCE TO HUMANS
None known. ◆

Resources

Books

Avila-Pires, T. C. S. *Lizards of Brazilian Amazonia (Reptilia: Squamata)*. Leiden, Germany: Zoologische Verhandelingen, 1995.

Cogger, H. G., and R. G. Zweifel, eds. *Encyclopedia of Reptiles and Amphibians*. 2nd ed. San Diego: Academic Press, 1998.

Dixon, J. R., and P. Soini. *The Reptiles of the Upper Amazon River Basin, Iquitos Region, Peru*. 2nd rev. ed. Milwaukee: Milwaukee Public Museum, 1986.

Murphy, J. C. *Amphibians and Reptiles of Trinidad and Tobago*. Malabar, FL: Krieger, 1997.

Resources

Pough, F. H., R. M. Andrews, J. E. Cadle, M. L. Crump, A. H. Savitzky, and K. D. Wells. *Herpetology.* 2nd ed. Upper Saddle River, NJ: Prentice Hall, 2001.

Pianka, E. R., and L. J. Vitt. *Lizards: Windows to the Evolution of Diversity.* Berkeley: University of California Press, 2003.

Powell, R., and R. W. Henderson, eds. *Contributions to West Indian Herpetology: A Tribute to Albert Schwartz.* Contributions to Herpetology, Volume 12. Ithaca, NY: Society for the Study of Amphibians and Reptiles, 1996.

Schwartz, A., and R. W. Henderson. *Amphibians and Reptiles of the West Indies: Descriptions, Distributions, and Natural History.* Gainesville, FL: University of Florida Press, 1991.

Vitt, L. J., and S. de la Torre. *A Research Guide to the Lizards of Cuyabeno.* Museo de Zoologia (QCAZ) Centro de Biodiversidad y Ambiente Pontificia Universidad Catolica del Ecuador, 1996.

Vitt, L. J., and E. R. Pianka. *Lizards: Windows to the Evolution of Diversity.* Berkeley: University of California Press, 2003.

Zug, G. R., L. J. Vitt, and J. L. Caldwell. *Herpetology: An Introductory Biology of Amphibians and Reptiles.* 2nd ed. San Diego: Academic Press, 2001.

Periodicals

Cole, C. J. "A Lizard Foretold." *Natural History* 5 (1989): 12, 14–17.

———. "A Lizard Found." *Natural History* (August 1994): 66–69.

Cole, C. J., H. C. Dessauer, and A. L. Markezich. "Missing Link Found: The Second Ancestor of *Gymnophthalmus underwoodi* (Squamata: Teiidae), a South American Unisexual Lizard of Hybrid Origin." *Natural History* 3055 (1993): 13.

Colli, Guarino R., Mariana G. Zatz, and Helio J. daCunha. "Notes on the Ecology and Geographical Distribution of the Rare Gymnophthalmid Lizard, *Bachia bresslaui.*" *Herpetologica* 54, no. 2 (1998): 169–174.

Dixon, J. R. "A Systematic Review of the Teiid Lizards, Genus *Bachia*, with Remarks on *Heterodactylus* and *Anotosaura.*" *Miscellaneous Publication of the University of Kansas Museum of Natural History* 57 (1973): 1–47.

Hardy, L. M., C. J. Cole, and C. R. Townsend. "Parthenogenetic Reproduction in the Neotropical Unisexual Lizard, *Gymnophthalmus underwoodi* (Reptilia: Teiidae)." *Journal of Morphology* 201 (1989): 215–234.

Hoogmoed, M. S. "Notes on the Herpetofauna of Suriname IV. The Lizards and Amphibians of Suriname." *Biogeographica* 4 (1973): 1–419.

Hoogmoed, M. S., C. J. Cole, and J. Ayarzaguena. "A New Cryptic Species of Lizard (Sauria: Teiidae: *Gymnophthalmus*) from Venezuela." *Zoologische Mededelingen (Leiden)* 66 (1992): 1–8.

Kizirian, D. A. "A Review of Ecuadorian *Proctoporus* (Squamata: Gymnophthalmidae) with Descriptions of Nine New Species." *Herpetological Monographs* 10 (1996): 85–155.

Vitt, L. J., and P. A. Zani. "Ecological Relationships Among Sympatric Lizards in a Transitional Forest in the Northern Amazon of Brazil." *Journal of Tropical Ecology* 14 (1998): 63–86.

Vitt, L. J., and T. C. S. Avila-Pires. "Ecology of Two Sympatric Species of *Neusticurus* (Sauria: Gymnophthalmidae) in the Western Amazon of Brazil." *Copeia* no. 3 (1998): 570–582.

Lee A. Fitzgerald, PhD

Whiptail lizards, tegus, and relatives
(Teiidae)

Class Reptilia
Order Squamata
Suborder Lacertiformes
Family Teiidae

Thumbnail description
Medium to large-sized diurnal, actively foraging, egg-laying, terrestrial lizards with well-developed limbs, long tails, and rectangular scales on the belly

Size
2–24 in (55–600 mm) snout-to-vent length; 5–51 in (120–1,300 mm) total length

Number of genera, species
9 genera; 18 species

Habitat
Forest, savanna, desert, and grassland

Conservation status
Extinct: 2 species; Critically Endangered: 1 species; Vulnerable: 1 species; Data Deficient: 2 species

Distribution
Middle North America (45° north) to southern South America (40° south), including Caribbean islands

Evolution and systematics

The whiptails, tegu lizards, and their allies make up the Teiidae, sister family to the Gymnophthalmidae. Throughout most of the twentieth century these families were classified together. Although many herpetologists still use the vernacular names, macroteiids and microteiids, these groups are now considered distinct families by practically all herpetologists. Teiidae and Gymnophthalmidae together form a lineage, Teioidea, which is sister to the Old World family Lacertidae (wall lizards, rock lizards, and their allies). Teiids and lacertids are so similar in appearance and ecology that it can be difficult to identify specimens to family without knowing their geographic origin. Because of their common ancestry, Teioidea (teiids plus gymnophthalmids) and Lacertidae belong to the suborder Lacertiformes.

An extinct subfamily of teiids, the polyglyphanodontines, existed in both North America and Mongolia in the Cretaceous period, an indication that ancient teiids once were widespread in the northern hemisphere. New and Old World polyglyphanodontines were distinguishable by the mid to late Cretaceous, but there is no consensus among experts whether teiids evolved in the New or the Old World. It is agreed, however, that teiids were extinct in North America by the end of the Cretaceous, whereas the family diversified in tropical America. North American *Cnemidophorus* species must have recolonized North America from South American ancestors.

Two subfamilies of Teiidae, Teiinae and Tupinambinae, are recognized on the basis of several characteristics of the skull bones and mandible. Members of Teiinae are distributed in North, Central, and South America, whereas Tupinambinae occur only in South America. Teiinae includes the genera *Ameiva*, *Teius*, *Cnemidophorus*, *Dicrodon*, and *Kentropyx*. With 56 known species in 2001, and others being described from South America, *Cnemidophorus* accounts for more than half of all the teiids. *Ameiva* and *Kentropyx* are well represented, whereas *Dicrodon* and *Teius* each contain three named species. The subfamily Tupinambinae contains the genera *Tupinambis* (six species), *Dracaena* (two species), *Callopistes* (two species), and *Crocodilurus* (one species).

The California whiptail lizard (*Cnemidophorus tigris mundus*). (Photo by Animals Animals ©Zig Leszczynski. Reproduced by permission.)

Physical characteristics

Teiids range from small (2.1 in [55 mm] snout-to-vent length, 4.7 in [120 mm] total length), such as in *Cnemidophorus inornatus*, to large (23.6 in [600 mm] snout-to-vent, 59 in [1,500] mm total length), such as in *Tupinambis rufescens*. They are fully limbed, terrestrial lizards that are diurnal, active foragers. All teiids lay eggs. Teiids are distinguished from Old World lacertids by having head scales not fused to the skull bones (fused in Lacertidae) and teeth that are solid on the base (hollow in Lacertidae). Teiid teeth are held to the jaws with cementum, a characteristic so distinctive that fossil teiid jawbones can be identified through the presence of this feature alone. Teiids are characterized as having small granular scales on the dorsum and rectangular plate-like scales on the belly. In one genus, *Kentropyx*, the belly scales are modified into pointed and keeled scales hypothesized to be an adaptation for climbing in bushy vegetation. Despite interesting morphological differences among genera, all teiids are relatively long bodied and long limbed with relatively narrow heads. All teiids have long tails, often more than 1.5 times body length. Fracture planes in each tail vertebra allow their tails to be easily broken. Teiids have good visual and olfactory systems. They have well-formed eyes and eyelids and long, forked tongues.

Distribution

Teiids are strictly New World lizards, distributed from the northern United States and through Mexico, Central America, and South America, except the extreme southern cone beyond approximately 45° latitude. Teiids, especially *Ameiva* and *Cnemidophorus* occur on many Caribbean islands. Teiids are widespread east of the Andes in South America and occur in the interandean valleys and coastal areas of Peru and southern Ecuador (e.g., *Dicrodon*, *Callopistes flavipunctatus*, and *Ameiva*). *Callopistes maculatus* occurs from northern to central Chile. *Dracaena guianensis* is Amazonian, whereas *Dracaena paraguayensis* inhabits the Pantanal ecoregion in Brazil, Bolivia, and Paraguay.

Habitat

As their continuous geographic range indicates, teiids are found in a variety of habitats, including wet and dry forests, primary and secondary forests, savannas, grasslands, deserts, and beaches. Regardless of habitat type, teiids need warm microhabitats where they can bask in direct sun to raise their body temperature for activity. For this reason, teiids usually are found using relatively open areas. In tropical forests, for example, they are frequently observed around treefall, along roads, and in clearings. The open habitats of deserts and beaches are well

A black and white tegu (*Tupinambis teguixin*) eating an egg in South America. (Photo by Animals Animals ©Zig Leszczynski. Reproduced by permission.)

suited to the teiid lifestyle, and teiids occupy almost all such areas in North and South America within their latitudinal range. Habitat use by teiids is clearly tied to their thermal biology. In Costa Rica, one researcher demonstrated the effect of body size on the ability of three species of *Ameiva* to use different habitats. The smallest species could heat and cool rapidly and used the hottest, most open habitat. The largest species used the most shaded forest, where it would not be susceptible to overheating, and the medium-sized species used forest that was intermediate in shadiness. Juveniles of the large species shared microhabitat with adults of the small species, a finding that added support to the idea that thermoregulatory needs are coupled to habitat use among the sun-loving teiids.

Two genera of teiids are semiaquatic, *Crocodilurus* and *Dracaena*.

Behavior

Teiids are classic examples of actively foraging lizards. Teiids generally live in burrows they excavate themselves or that are made by other animals. A typical macroteiid day begins with the lizard basking in direct sun to raise its body temperature. Whiptails, especially *Cnemidophorus* and *Ameiva*,

prefer relatively high body temperatures for activity, commonly measured in the field at 98.6°F–140°F (37°C–40°C). Once activity temperatures are achieved, a macroteiid embarks on long foraging or mate-seeking expeditions within its home range. Teiids maintain high body temperature while active by shuttling between sun and shade.

Feeding ecology and diet

Teiids are opportunistic feeders, and they are very good at finding and taking advantage of concentrated patches of prey such as insect larvae, ants, and termites that they dig out of the leaf litter or other cover. An individual may find several food sources during a foraging bout, and their meals are made up of different kinds of prey. Across the family, prey size correlates with body size. The large species of *Ameiva*, *Teius*, and the tegus (*Tupinambis*) include large amounts of fallen fruit in their diets. Tegus are omnivorous, consuming vertebrate prey and carrion as they encounter it. Tegus also are known to be important egg predators and have been reported to be the most important predator of caiman nests in the Venezuelan Llanos. Tegus have heterodont dentition as adults with pointed teeth in the front of their mouths for seizing prey and molariform teeth in the back of their jaws for crushing hard prey. Exceptions to the rule of opportunism among teiids are the caiman lizards (*Dracaena* spp.). These very large (more than 12 in [300 mm] snout-to-vent length), spectacular teiids are aquatic specialists that live around streams and swamps and feed primarily on snails. They have laterally compressed tails for swimming and foraging in water. Caiman lizards are named for the enlarged dorsal scales that look like crocodilian skin. Caiman lizards have a blunt head and molariform teeth for crushing their molluscan prey.

Reproductive biology

Teiids are not territorial, and several individuals' home ranges may overlap. Males are larger than females and compete for mates. Males follow receptive females and guard them against competing males. All teiids are egg layers. The number of eggs laid by females correlates with body size both among and within species. The largest species, *Tupinambis merianae* and *T. rufescens*, may lay approximately a dozen eggs when they reach sexual maturity, but by the time a female reaches maximum size, she may lay 30 or more eggs. The nesting ecology of most teiids is simple. Females deposit their clutches in the ground or within logs or debris. Tegus in southern South America, *T. merianae*, *T. rufescens*, and *T. duseni*, build elaborate nests of vegetation in their underground burrows into which they deposit their eggs. Females attend the nests throughout the incubation period. *T. teguixin* in northern South America lays its eggs in active termite mounds in trees.

The reproductive biology of whiptails and their allies is particularly noteworthy because of the existence of unisexual species. Unisexual species have no males, and individual lizards have no sperm. Mothers lay fertile eggs that develop into identical daughters, that is, clones. This mode of asexual reproduction is called parthenogenesis, and biologists sometimes refer to parthenogenetic species as parthenoforms. Parthenogenetic teiids arise when two sexual species hybridize. Parthenogenesis has been an important mode of speciation for whiptails of the genus *Cnemidophorus*, which contains at least 12 unisexual species. It is known, for example, that the unisexual desert grassland whiptail (*C. uniparens*) originated from hybridization events between the Texas spotted whiptail (*C. gularis*) and the little striped whiptail (*C. inornatus*). Hybridization events could easily have happened multiple times, hence parthenogenetic species exist in clonal complexes, as in the Laredo striped whiptail (*C. laredoensis*) complex. Some parthenoforms have the typical condition of two sets of chromosomes (diploid), whereas others have three sets (triploid). The advantage of parthenogenesis is that when a mother produces identical daughters, each of her genes doubles in frequency in each descendant generation. Because all the individuals are reproducing females, teiid populations grow more rapidly than do populations of sexual species. Within the Teiidae, there is a parthenogenetic species of *Kentropyx* and one of *Teius*; parthenogenesis also is known in seven other families of squamates.

Conservation status

The Teiidae, for the most part, are common lizards that do well in a variety of habitats and in most circumstances ap-

pear to endure human influences. The *Ameiva* and *Cnemidophorus* endemic to Caribbean islands and the tupinambines subject to the pet trade and exotic leather trade are two major exceptions. Two species of *Ameiva* are categorized as Extinct by the IUCN, and another, the St. Croix ground lizard (*Ameiva polops*), is Critically Endangered. The St. Lucia whiptail (*Cnemidophorus vanzoi*) is listed as Vulnerable. There are not enough data to determine the conservation status of *Callopistes* and other macroteiids that have been exploited for the pet trade. Island-dwelling lizards are clearly sensitive to human impact, and conservationists need to be aware of threats to teiids on islands or in otherwise restricted geographic ranges.

Several species of tegu are commercially exploited in very large numbers as pets or for skins. There is a long history of commercial trade in two species of tegu lizards (*Tupinambis merianae* and *T. rufescens*) from Argentina and Paraguay. In the 1980s, an average of 1.9 million tegus were traded yearly for the exotic skin trade, making tegus among the most exploited reptiles in the world. Tegu skins are prized for the pattern of tile-like belly scales, and they are used for cowboy boots, shoes, belts, and other exotic leather accessories in North America, Europe, and southeast Asia. Tegu lizards are listed in CITES Appendix II, and the trade is legal and monitored internationally. Harvest quotas are 1,000,000 for Argentina and 300,000 for Paraguay, and both countries have established management programs for the lizards that depend on trade controls and harvest monitoring. The caiman lizards (*Dracaena* spp.) have been exploited for their skins but not as extensively as the tegus. Tegus, *Callopistes*, and several species of *Ameiva* have appeared in the pet trade in large numbers. Mainland macroteiids appear to have a life history that enables their populations to withstand harvest by humans, but prudent conservation will require careful monitoring and management programs to ensure the take is sustainable over the long term.

Significance to humans

People have used tegu lizards for as long as there are historical records. South American Indians hunt and eat tegu, and tegu lizards are exploited commercially for their skins. In the areas where skins are traded, hunters sell the skins and consume the meat. Tegu fat is prized throughout Argentina and Paraguay for medicinal purposes. The trade in tegu lizards is economically important to local people and to the tanning industry. Thousands of hunters contribute to the total harvest of one million skins annually, and the export value of tegu skins is several millions of dollars.

The significance of the smaller teiids to people is less apparent, but these animals may still be important. Teiids are prey to myriad predators and themselves consume a variety of invertebrate prey and disperse seeds of the fruit they eat. Whiptails and their allies can occur at relatively high population densities and probably play an ecological role in their habitats. *Cnemidophorus tigris* has been studied in the Mojave desert, for example, and its populations have been shown to track its key habitat resources remarkably closely.

1. Giant ameiva (*Ameiva ameiva*); 2. Six-lined racerunner (*Cnemidophorus sexlineatus*); 3. Desert grassland whiptail (*Cnemidophorus uniparens*); 4. Crocodile tegu (*Crocodilurus lacertinus*); 5. Caiman lizard (*Dracaena paraguayensis*). (Illustration by Jacqueline Mahannah)

Species accounts

Giant ameiva
Ameiva ameiva

SUBFAMILY
Teiinae

TAXONOMY
Ameiva ameiva Linnaeus, 1758, America, restricted by Hoogmoed, 1973, to the confluence of the Cottica River and the Perica Creek, Suriname.

OTHER COMMON NAMES
None known.

PHYSICAL CHARACTERISTICS
The largest males reach snout-vent lengths of nearly 7.87 in (200 mm), with the tail about twice body length. Males are larger than females with enlarged jaw musculature. The head is pyramid-shaped with a blunt snout. The tympana are well developed. The eyes are fully developed with functional eyelids. The limbs are fully developed with five fingers and toes with claws. As in all teiids, the teeth are pleurodont, being attached to the inner side of the jaw. The tongue is forked and covered with scale-like papillae except for the smooth tips. Well-developed femoral pores that produce a waxy secretion are present on the underside of the hind limbs. Head scales are large and smooth. Dorsal scales are small and granular. The ventral scales are plate-like, smooth, and rectangular. Scales around the tail are rectangular and mostly keeled except near where the tail joins the body. Color pattern varies across the geographic range of this species and also with age. Juveniles are mostly brown, sometimes with the head and anterior part of the back green, with a dark strip on each side running from the eye down the flanks to the hind limbs. Larger individuals may be completely green dorsally, and the dark strip is less prevalent in larger individuals. Adults have a brown reticulated pattern anteriorly, with green back, hind limbs, and tail. The underside of the head, chest, and forelimbs are white. The belly can be pale turquoise and the underside of the tail bright turquoise.

DISTRIBUTION
Central and South America.

HABITAT
Wet and dry forests, primary and secondary forests.

BEHAVIOR
These are diurnal lizards with high operating temperatures around 98.6°F (37°C). Giant ameivas live in burrows they excavate themselves. Males are larger than females and compete for mates. These active lizards bask in the morning to reach preferred operating temperature and embark on long foraging expeditions seeking patchily distributed prey that they find by using a combination of visual and olfactory cues.

FEEDING ECOLOGY AND DIET
Giant ameivas are active, widely foraging lizards that are opportunistic feeders on insects, fallen fruits, and small vertebrates.

REPRODUCTIVE BIOLOGY
This species is oviparous, laying eggs in soft soil, leaf litter, or rotting logs.

CONSERVATION STATUS
The giant ameiva is common throughout its range in open areas in wet or dry forests. Threats include habitat destruction and alteration.

SIGNIFICANCE TO HUMANS
None known. ◆

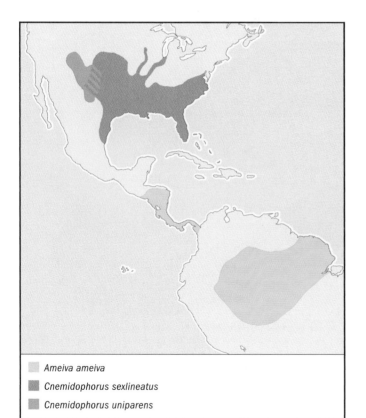

Ameiva ameiva
Cnemidophorus sexlineatus
Cnemidophorus uniparens

Six-lined racerunner
Cnemidophorus sexlineatus

SUBFAMILY
Teiinae

TAXONOMY
Cnemidophorus sexlineatus Linneaus, 1766.

OTHER COMMON NAMES
French: Cnémidophore à six raies; German: Sechsstreifen-Rennechsen.

PHYSICAL CHARACTERISTICS
Adults average 2.1–2.9 in (55–75 mm) in snout-to-vent length, with a maximum size of 3.3 in (85 mm). Females are slightly

larger than males. Individuals are striped without spots, with seven longitudinal light stripes on the greenish brown to black ground color. The head, neck, and anterior part of the body are bright yellowish green. The tail is bright blue in hatchlings and fades to brown in adults.

DISTRIBUTION
Eastern North America in the United States from Chesapeake Bay south to Key West, west to southern South Dakota, south to eastern New Mexico.

HABITAT
This species occurs in xeric habitats that are relatively open with patchy vegetation and well-drained soil.

BEHAVIOR
These teiids overwinter in burrows that they excavate themselves or that were made by other animals. They emerge in April over most of their range, and activity peaks in midsummer. Palmer and Braswell (1995) reported that in North Carolina they are the "last lizards to become active in the spring and the first to enter hibernation in the fall." Hatchlings appear in late summer, by which time adults are not nearly as active. Six-lined racerunners, like other teiids, are heliothermic lizards that prefer relatively high body temperatures. They are active on hot sunny days. Mark Paulissen (1988) studied foraging ecology, activity, and temperature selection by these lizards and reported a mean body temperature of 98.2–98.8°F (36.8–37.1°C). While active, six-lined racerunnners thermoregulate by shuttling between sun and shade. They hide under rocks, logs, trash piles, or any suitable object that gives them safe refuge. Six-lined racerunners use speed as their defense.

FEEDING ECOLOGY AND DIET
These are active foraging lizards that feed opportunistically on insects and other arthropods, often digging up hidden prey. Dietary studies document feeding on grasshoppers, spiders, butterflies, moths, land snails, beetles, beetle larvae, and ants.

REPRODUCTIVE BIOLOGY
Females reach maturity in their second season and lay one to six eggs in a clutch depending on their body size and reproductive frequency. Nesting takes place in spring and summer, and hatchlings appear by mid July. Hatchlings are 1.2–1.8 in (31–45 mm) snout-to-vent length.

CONSERVATION STATUS
Not threatened.

SIGNIFICANCE TO HUMANS
None known. ◆

Desert grassland whiptail
Cnemidophorus uniparens

SUBFAMILY
Teiiinae

TAXONOMY
Cnemidophorus uniparens Wright and Lowe, 1965.

OTHER COMMON NAMES
Spanish: Huico de pastizal-desértico.

PHYSICAL CHARACTERISTICS
This is a small, all-female whiptail with a maximum snout-to-vent length of 3.4 in (86 mm). It is striped without spots, containing six cream to white stripes on the olive-brown to black background. The venter is white. The tail is bright blue in hatchlings and blue-green to olive-green in adults.

DISTRIBUTION
Chihuahua, Mexico, north in Chihuahua desert to central New Mexico, United States, west to Sonora, Mexico, and southeastern Arizona, United States.

HABITAT
Desert grassland.

BEHAVIOR
This whiptail is mostly active during morning, with a smaller peak of activity in late afternoon. As do those of most teiids, the home ranges of individuals overlap.

FEEDING ECOLOGY AND DIET
As do other whiptails, these active foraging lizards feed opportunistically on insects and other arthropods.

REPRODUCTIVE BIOLOGY
The desert grassland whiptail is an all-female species that reproduces parthenogenetically. Reproductive individuals may express both male-like behavior and initiate pseudocopulation with other females. This behavior stimulates reproduction in captivity, but its significance in natural populations is unknown. Females attain reproductive maturity at 2.4 in (60 mm) snout-to-vent length and lay one to four eggs depending on size.

CONSERVATION STATUS
Not listed by IUCN.

SIGNIFICANCE TO HUMANS
None known. ◆

Crocodile tegu
Crocodilurus lacertinus

SUBFAMILY
Tupinambinae

TAXONOMY
Crocodilurus lacertinus Daudin, 1802. Islands adjacent to tropical South America.

OTHER COMMON NAMES
None known.

PHYSICAL CHARACTERISTICS
The crocodile tegu is the only species in its genus. It has a typical marcoteiid body form anteriorly, but the tail is at least twice body length and laterally compressed. Dorsal scales along the tail converge into one row, as in a crocodile's tail. Adults are brown or greenish above with orange mottling on limbs. The venter is yellow or whitish.

DISTRIBUTION
Amazon basin and upper Orinoco River drainage, South America.

Crocodilurus lacertinus

Dracaena paraguayensis

Paraguayan caiman lizard
Dracaena paraguayensis

SUBFAMILY
Tupinambinae

TAXONOMY
Dracaena paraguayensis Amaral, 1950. São Lourenço, Mato Grosso, Brazil.

OTHER COMMON NAMES
French: Dracène de la Guyane; German: Krokodilteju.

PHYSICAL CHARACTERISTICS
These are large teiids that may reach more than 39 in (1 m) total length and more than 17.7 in (450 mm) snout-to-vent length. Individuals have large, blunt heads and molariform crushing teeth for dealing with hard-shelled molluscan prey. The dorsum has enlarged scales that look like a crocodile's scutes. The tail is laterally compressed with two rows of crest-like scales.

DISTRIBUTION
Pantanal region of southwestern Brazil and northern Paraguay.

HABITAT
Seasonally flooded freshwater marshes and savannas in the Pantanal region.

BEHAVIOR
This semiaquatic lizard basks on tree limbs and on the banks of pools and water courses. It forages in the water for snails. Excellent swimmers, caiman lizards use the laterally compressed tail to propel themselves through the water. They forage in water and escape predators by diving into the water.

FEEDING ECOLOGY AND DIET
Specialists on snails as prey, caiman lizards capture large freshwater snails and crush them in powerful jaws. They have specialized molariform teeth for crushing hard prey.

REPRODUCTIVE BIOLOGY
This species is oviparous, but few details of its natural history are known.

CONSERVATION STATUS
Not threatened, and common in the Pantanal ecoregion. Caiman lizards are valued by the exotic leather trade for their skins, which are used to make leather for boots and other fashion accessories. Threats include destruction of wetlands in the Pantanal and direct overexploitation of the species for the skin trade. Trade levels in the 1990s were not high, but no management plans for the species are in place.

SIGNIFICANCE TO HUMANS
Local people in parts of the species' range often believe the myth that caiman lizards are venomous and dangerous. Their skins are valued by the exotic leather trade. ◆

HABITAT
Forested stream sides.

BEHAVIOR
This species is semiaquatic, foraging along water's edge or in the water. Crocodile tegus are excellent swimmers, using the laterally compressed tail to propel them through the water. These lizards may forage in water and escape predators by diving into the water.

FEEDING ECOLOGY AND DIET
These are opportunistic feeders of insects and other arthropods.

REPRODUCTIVE BIOLOGY
This species is oviparous, but details of its natural history are not well studied.

CONSERVATION STATUS
Not threatened, and widespread in most of its range. Threats include habitat destruction.

SIGNIFICANCE TO HUMANS
None known. ◆

Resources

Books
Cogger, H. G., and R. G. Zweifel, editors. *Reptiles and Amphibians.* New York: Smithmark, 1992.

Degenhardt, W. G., C. W. Painter, and A. H. Price.

Amphibians and Reptiles of New Mexico. Albuquerque: University of New Mexico Press, 1996.

Fitzgerald, L. A., J. M. Chani, and O. E. Donadio. "*Tupinambis* Lizards in Argentina: Implementing Management of a

Resources

Traditionally Exploited Resource." In *Neotropical Wildlife Use and Conservation*, edited by J. G. Robinson and K. H. Redford. Chicago: University of Chicago Press, 1996.

Palmer, W. M., and A. L. Braswell. *Reptiles of North Carolina*. Chapel Hill: University of North Carolina Press, 1995.

Pianka, E. R., and L. J. Vitt. *Lizards: Windows to the Evolution of Diversity*. Berkeley: University of California Press, 2003.

Pough, F. H., R. M. Andrews, J. E. Cadle, M. L. Crump, A. H. Savitzky, and K. D. Wells. *Herpetology*. 2nd edition. Upper Saddle River, NJ: Prentice Hall, 2001.

Vitt, L. J, and E. R. Pianka. *Lizards: Windows to the Evolution of Diversity*. Berkeley: University of California Press, 2003.

Wright, J. W., and L. J. Vitt. *Biology of Whiptail Lizards, Genus Cnemidophorus*. Norman: Oklahoma Museum of Natural History, 1993.

Zug, G. R., L. J. Vitt, and J. L. Caldwell. *Herpetology: An Introductory Biology of Amphibians and Reptiles*. 2nd edition. San Diego: Academic Press, 2001.

Periodicals

Beebe, W. "Field Notes on the Lizards of Kartabo, British Guiana, and Caripito, Venezuela: Part 3, Teiidae, Amphisbaenidae and Scincidae." *Zoologica* 30 (1945): 7–32.

Carpenter, Charles C. "Aggressive Behaviour and Social Dominance in the Six-lined Racerunner (*Cnemidophorus sexlineatus*)." *Animal Behavior* 8 (1960): 61–66.

Duellman, William E., and Richard G. Zweifel. "A Synopsis of the Lizards of the Sexlineatus Group (genus *Cnemidophorus*).

"*Bulletin of the American Museum of Natural History* 123, no. 3 (1962): 155–210.

Echternacht, Arthur C. "Middle America Lizards of the Genus *Ameiva* (Teiidae) with Emphasis on Geographic Variation." *Miscellaneous Publications of the Museum of Natural History of the University of Kansas*, no. 55 (1971): 1–86.

Fitch, Henry S. "Natural History of the Six-lined Racerunner (*Cnemidophorus sexlineatus*)." *University of Kansas Publication Museum of Natural History* No. 11 (1958): 11–62.

Paulissen, M. A. "Diet of Adult and Juvenile Six-lined Racerunners, *Cnemidophorus sexlineatus* (Sauria: Teiidae)." *Southwestern Naturalist* 32, no. 3 (1987): 395–397.

———. "Ontogenetic and Seasonal Shifts in Microhabitat Use by the Lizard *Cnemidophorus sexlineatus. Copeia* 1988, no. 4 (1988): 1021–1029.

———. "Ontogenetic Comparison of Body Temperature Selection and Thermal Tolerance of *Cnemidophorus sexlineatus.*" *Journal of Herpetology* 22, no. 4 (1988): 473–476.

———. "Optimal Foraging and Intraspecific Diet Differences in the Lizard *Cnemidophorus sexlineatus.*" *Oecologia* 71, no. 3 (1987): 439–446.

Peters, James A., Roberto Donoso-Barros, and Braulio Orejas-Miranda. "Catalogue of the Neotropical Squamata: Part II, Lizards and Amphisbaenians." *United States National Museum Bulletin* 297 (1970): viii, 293.

Lee A. Fitzgerald, PhD

Girdled and plated lizards

(Cordylidae)

Class Reptilia
Order Squamata
Suborder Sauria
Family Cordylidae

Thumbnail description
Small to large diurnal, heliothermic, terrestrial lizards, mostly with well-developed legs, although some groups are snakelike and have vestigial limbs

Size
Girdled and flat lizards, 5–13 in (13–33 cm); grass lizards, 22 in (56 cm); plated lizards, 6–28 in (15–71 cm)

Number of genera, species
7 genera; 88 species

Habitat
Forest, savanna, scrubland, desert, and grassland

Conservation status
Extinct: 1 species; Vulnerable: 5 species; Lower Risk/Near Threatened: 5 species

Distribution
Southern and tropical Africa and Madagascar

Evolution and systematics

The Cordylidae is the only lizard family restricted to Africa, with one subfamily also occurring on Madagascar. The fossil history is very poor, although some fossils (*Pseudolacerta* and *Palaeocordylus*) from the early Eocene to the early Miocene of Europe are provisionally assigned to the family. Relationships to other lizards also remain problematic, but they appear to be close to the Scinicidae and less confidently with the lacertiforms (Teiidae, Gymnophthalmidae, and Lacertidae). The family is relatively ancient and evolved before the separation of the southern supercontinent Gondwana and the separation of Madagascar from Africa in the middle Cretaceous epoch (80–100 million years ago).

Two well-defined subfamilies are recognized, and have often been treated as separate families within a superfamily—the Cordyliformes.

The first subfamily is Cordylinae, or girdled lizards. The head has four parietal scales and the nostril is enclosed in a single, or between only two, scales. Cordylines are restricted to southern and tropical Africa. Within cordylines, the flat lizards (*Platysaurus*, 16 species) are the basal stock and retain

oviparity. Crag lizards (*Pseudocordylus*) and grass lizards (*Chamaesaura*) evolve from within girdled lizards (*Cordylus*), and to reflect this evolutionary relationship these genera have recently been transferred to *Cordylus*. This large genus now contains more than 40 species.

The second subfamily is Gerrhosaurinae, or plated lizards. The head has two parietal scales and the nostril is surrounded by three to four scales. Two genera are restricted to Madagascar and another three genera inhabit savanna and semiarid regions of Africa south of the Sahel. The desert plated lizard, previously included in a separate genus (*Angolosaurus*), is now transferred to *Gerrhosaurus*.

Physical characteristics

Girdled lizards are the most typical cordylines, with well-developed limbs and stout bodies covered with overlapping, spiny scales. The bodies of flat lizards, however, are covered with small, granular scales, although the legs may still have a few scattered, spiny scales. Scales on the belly and back are strengthened with bony elements (osteoderms). A longitudinal, expandable fold of granular scales runs along the flanks

A common grass lizard (*Chamaesaura anguina*) eating a grasshopper and showing elongate form, Port Elizabeth, South Africa. (Photo by Bill Branch. Reproduced by permission.)

to give flexibility (it is reduced in size in some genera). The head is usually triangular with large, symmetrical, bony head shields. The tongue has a simple notch and is covered in papillae. Femoral pores are conspicuous in males, and sometimes also in females. The tail can be shed and regenerated. In girdled lizards it is ringed with spiny scales and is not much longer than the head-body length. Plated lizards have much longer tails, ringed with elongate, rectangular scales.

Among cordylines, flat lizards are unmistakable. The body is very flat and covered with granular scales, while the legs often have scattered spiny scales. These depressed dandies are clothed in Jacobean splendor, the colors varying from species to species. They are most vivid on the belly, where their intensity is hidden from predators. Females and juveniles have black backs, usually attractively marked with three pale, longitudinal, dorsal stripes. Males grow slightly larger than females.

Even though they are not closely related, grass and snake lizards have evolved elongate bodies and long tails that allow them to move freely in long grass. They move with the speed and agility of snakes. The vestigial limbs are often reduced to minute spikes, and although they appear useless they give stability when the lizard is at rest. Their rustic colors camouflage them in dried grass.

Plated lizards (*Gerrhosaurus*) have stout bodies and well-developed limbs. The desert plated lizard (*G. skoogi*) is adapted for desert life. The spadelike snout, relatively short, conical tail, and cylindrical body allow it to dive into the loose sand of slip faces, while the long limbs and fringed toes allow it to run at lightning speed across wind-compacted sand. The dwarf plated lizard (*Cordylosaurus subtessellatus*) is a very small (to 6 in [15 cm]), brightly colored species restricted to the arid western regions of southern Africa. Plated snake lizards or seps (*Tetradactylus*, six species) show a progressive evolutionary transition from normal-limbed species to others that have become snakelike with vestigial limbs. Madagascan plated lizards (*Tracheloptychus*, two species, and *Zonosaurus*, 13 species) are more gracile and skinklike than their African cousins and have reduced body armor.

Distribution

The greatest diversity of girdled lizards, and their probable center of origin, is south of the Zambezi River, although a number of species extend into central and East Africa. Flat lizards have speciated explosively in Zimbabwe and adjacent areas, where 14 species (some with local races) occur. Two isolated species occur in the Northern Cape province of South Africa, one extending into southern Namibia. Grass and snake lizards are restricted mainly to southern Africa, although both have representatives in the savannas of central Africa. Plated lizards extend through the savannas of southern and central Africa, reaching Ethiopia in the north and Ghana in the west. Two genera also occur on Madagascar and have been introduced to a number of other Indian Ocean islands (e.g., Aldabra).

Habitat

Flat lizards live in rock outcrops in either savanna or rocky desert. Most girdled lizards occupy similar habitats, although a few live under dead tree bark or in tunnels in grassland. Grass lizards prefer short heath or grassland. Most plated lizards inhabit savanna or scrubland. However, one species is adapted to desert dunes, some Madagascan species inhabit forest floors, and another is semiaquatic and favors riverbanks. The snakelike seps forage in grassland or mountain heathland. The dwarf plated lizards live among succulents and rocks from the Karoo of South Africa to Angola.

Behavior

Girdled lizards' thick scales with bony plates protect them from abrasion against rough rock. To evade predators, many species jam themselves into rock cracks by inflating the body and shortening and thickening the skull, which has an unusual hinged structure. Armadillo lizards (*Cordylus cataphractus*) are very social and form large groups (up to 43) that inhabit the

An armadillo lizard (*Cordylus cataphractus*), showing its typical defensive posture from which it gets its common name, in Namaqualand, South Africa. (Photo by Bill Branch. Reproduced by permission.)

An armadillo lizard (*Cordylus cataphractus*) defends itself from predators by rolling up in a tight ball, holding its tail in its mouth, and exposing the rings of spiny armor that surround its body. The predator shown here is a black-backed jackal (*Canis mesomelas*). (Illustration by Bruce Worden)

same rock crack. They are very wary and retreat at the first sign of danger. If caught in the open, like their namesakes, they bite their tail and roll into a tight ball. It makes it difficult for predators to swallow them.

The shape of flat lizards permits them to squeeze under thin rock flakes where they are safe from predators. Up to 12 individuals may squeeze into the same crack, although it is unusual to find adult males together during the breeding season. They are restricted to certain types of rock (e.g., granite, gneiss, and some sandstone), and are therefore found in isolated populations. Sociable, they form dense colonies. Prime territory on a rock face is defended by a dominant male during the breeding season. In confrontations, males circle each other and expose their brightly colored bellies by tilting sideways.

Snake and grass lizards hunt grasshoppers and other insects in grassland. They are diurnal, retreating at night into a grass tussock or beneath a stone. The vestigial limbs are minute, and are used for support when stationary and to as-

sist small movements in long vegetation. The very long tail (up to three times the body length) is used for propulsion, rendering them almost impossible to catch when they "swim" through grass. When startled they also "spring" by flexing stiff coils against the ground. If grabbed, the tail is readily shed. However, regeneration is very rapid; it must be, as they are helpless without their tails.

Except for a few small Madagascan species that clamber on tree trunks, all plated lizards are terrestrial. They dig holes in loose sand around bushes or excavate leaf litter from large rock cracks or under boulders. When foraging they move slowly, often sliding down slopes on their smooth bellies. When basking they rest on the belly with the limbs flexed upward off the ground. The thick tail base is used for fat storage. Rarely common, they are shy and usually solitary, although the giant and desert plated lizards often form loose colonies. When disturbed they dash into bush clumps or to their retreats. In danger the desert plated lizards dive into loose sand, disappearing with a swimming motion. They can remain buried for up to 24 hours, sheltering from both danger and temperature extremes.

Feeding ecology and diet

Flat and girdled lizards are ambush predators that make short dashes to capture small invertebrates (flies, beetles, caterpillars, etc.). They also eat berries when seasonally available. The slow-moving plated lizards eat large insects, snails, etc. The larger species also eat leaves and berries, and will even eat small lizards and snakes if they can be caught. They are active foragers, and scrape away loose soil or leaf litter looking for hidden prey. Desert plated lizards forage at the base of dunes feeding on beetles, dry plant debris, grass seeds, and stems of the succulent Nara plant.

Reproductive biology

Girdled lizards are viviparous, giving birth to a few (one to six) large babies each year. Some live in diffuse colonies,

A Cape flat lizard (*Platysaurus capensis*) breeding male in stunning color and showing flattened habitus, Richtersveld National Park, South Africa. (Photo by Bill Branch. Reproduced by permission.)

in which the males are territorial during the breeding season. Although they usually have drab coloration, adult males do have active femoral and glandular pores and appear to use chemical clues to signal status and territorial boundaries. Sexual maturity is reached in two to four years and they are long-lived (up to 25 years is known in captivity). In courtship male flat lizards present to females, raising their head and forebody to reveal the bright colors of the throat and chest. Unlike other cordylines, they are oviparous and lay only two eggs, usually in November to December. The relatively large, elongate eggs are 0.3–0.4 in (7–10 mm) wide, 0.7–0.9 in (17–22 mm) long, soft-shelled, and laid in deep cracks, usually in damp leaf mold. Numerous females may nest in the same crack. Grass lizards are also viviparous. Surprisingly for such thin, active lizards, they have relatively large litters (up to 12 young, each 6 in [15 cm] long). Birth may take two to three days, and the young often escape by wriggling from their mother's body.

All gerrhosaurines are oviparous, and lay a few soft-shelled eggs in moist sites. Little is known of the reproduction of seps or plated snake lizards. One species lays a small clutch of two to five eggs in midsummer in a live *Anochetus faurei* ant nest. Several females may use the same site. The eggs develop safely in the warm, protected ant nest and hatch between February and April; hatchlings measure 4.5–5.5 in (114–140 mm).

Conservation status

The IUCN lists five species as Vulnerable and another five as Lower Risk/Near Threatened. Because they live on isolated rocky outcrops, many girdled and flat lizards have very restricted distributions. Due to their bright colors and hardy nature, they are also popular in the international reptile pet trade. As a consequence many populations are threatened, and trade in girdled lizards (*Cordylus*) is therefore protected under CITES, Appendix II. Eastwood's seps (*Tetradactylus eastwoodae*) is one of the few extinct African reptile species. It lived in montane grassland along the eastern escarpment of South Africa, but this is now all covered by pine plantations.

Significance to humans

The large colonies of colorful flat lizards are important tourist attractions at scenic sights in southern Africa (e.g., at World's View in the Matobo National Park, Zimbabwe, and Augrabies Falls in South Africa).

1. Cape flat lizard (*Platysaurus capensis*); 2. Madagascan plated lizard (*Zonosaurus madagascariensis*); 3. Giant girdled lizard (*Cordylus giganteus*). (Illustration by Wendy Baker)

Species accounts

Giant girdled lizard
Cordylus giganteus

SUBFAMILY
Cordylinae

TAXONOMY
Cordylus giganteus A. Smith, 1844, interior districts of southern Africa.

OTHER COMMON NAMES
English: Sungazer; French: Cordyle géant; German: Riesengürtelschweif.

PHYSICAL CHARACTERISTICS
Giant girdled lizards have very spiny scales, as well as four very large occipital spikes that adorn the back of their triangular head. The tail has whorls of very large spines. The body is yellow to dark brown, clouded with dark brown. Juveniles are more intensely marked, with irregular crossbars of red-brown on the back.

DISTRIBUTION
Restricted to the Highveld region of central South Africa.

HABITAT
They live in flat or gently sloping grassland, with deep soil for their burrows.

BEHAVIOR
Found in diffuse colonies, they dig long burrows (up to 6 ft [1.8 m]) in deep soil. Usually only a single adult lives in a burrow, but it may be shared with juveniles. If a predator enters the burrow when the lizard is inside, the lizard backs along the tunnel toward the entrance, lashing its spiny tail from side to side in the face of the intruder. They often bask at the entrance to their burrow or on a nearby termite mound, staring at the sun—hence their common name. They are dormant during winter and rarely seen above ground from May to mid-August.

FEEDING ECOLOGY AND DIET
Sit-and-wait ambushers, they feed mainly on invertebrates (beetles, grasshoppers, millipedes, termites, and spiders), although they will take small vertebrates if the opportunity arises.

REPRODUCTIVE BIOLOGY
One or two young, measuring up to 6 in (15 cm), are born from January to April, possibly only every two to three years.

CONSERVATION STATUS
This species is listed as Vulnerable by the IUCN. Their numbers are declining due to habitat destruction by agriculture and, to a limited extent, because of illegal collecting for the pet trade.

SIGNIFICANCE TO HUMANS
Because it was familiar and conspicuous to the early settlers of South Africa, the giant girdled lizard was chosen as the national lizard, and its images have adorned postage stamps and conservation posters. ◆

Cape flat lizard
Platysaurus capensis

SUBFAMILY
Cordylinae

TAXONOMY
Platysaurus capensis A. Smith, 1844, Great Namaqualand, Namibia.

OTHER COMMON NAMES
None known.

PHYSICAL CHARACTERISTICS
The cape flat lizard is a very flattened lizard. Females and juveniles have a dark brown back with three broad, cream stripes, a straw-colored tail, and a white belly with a central black patch. Sexually mature males are beautifully colored. The head and most of the body are Prussian blue, sometimes with numerous pale spots and the juvenile stripes may faintly persist. The rear of the body, hind limbs, and tail are red-brown. Beneath the throat is light blue, the chest dark blue, and the belly black-centered.

DISTRIBUTION
Namaqualand as far south as Garies, Northern Cape, South Africa, and extending into southern Namibia along the Fish River canyon.

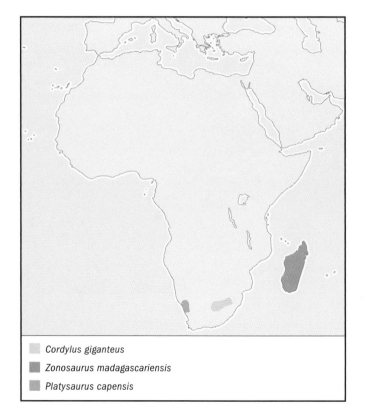

■ *Cordylus giganteus*
■ *Zonosaurus madagascariensis*
■ *Platysaurus capensis*

HABITAT
Large granite outcrops in rocky desert.

BEHAVIOR
These beautiful lizards are common on granite, but rarely form dense colonies. The males are shy and difficult to approach.

FEEDING ECOLOGY AND DIET
It is an ambush predator that sits motionless in the shade of a rock crack and makes a quick dash to collect small insects. Flowers and berries are also eaten in season.

REPRODUCTIVE BIOLOGY
A small clutch of only two large, oval eggs are laid in moist soil beneath a sunny rock crack in November to December; a second clutch may be laid later in the summer.

CONSERVATION STATUS
Not threatened. Some small colonies may be threatened by mining activities, but much of this lizard's rocky desert home is uninhabited.

SIGNIFICANCE TO HUMANS
None known. ◆

Madagascan plated lizard
Zonosaurus madagascariensis

SUBFAMILY
Gerrhosaurinae

TAXONOMY
Zonosaurus madagascariensis Gray, 1831, Madagascar.

OTHER COMMON NAMES
None known.

PHYSICAL CHARACTERISTICS
A medium-sized plated lizard (up to 14 in [36 cm]) with an elongate, slightly flattened body with a prominent lateral fold running the length of the flank. The long tail is almost twice as long as the body. The ground color is brown, with two yellow lateral stripes that run from the eye to the base of the tail. The color of the flanks is variable, but is usually mottled with white and dark scales. The belly is grayish white, but may be reddish on the throat in some areas.

DISTRIBUTION
More common in eastern Madagascar, but also known in the west and south and on some offshore islands. It has also been introduced to Aldabra.

HABITAT
Normally found in dry, open landscapes, they adapt well to field edges and secondary thickets.

BEHAVIOR
A solitary but common, terrestrial species that forages in sun-warmed clearings. It shelters in a small tunnel at the base of a bush, and may climb onto the base of trees.

FEEDING ECOLOGY AND DIET
They are active foragers that eat a wide variety of insects, as well as fruits and berries.

REPRODUCTIVE BIOLOGY
A small clutch (up to five) of large, elongate eggs are laid in moist soil.

CONSERVATION STATUS
Not threatened. Widespread and tolerant of agriculture, this species remains common throughout most of Madagascar.

SIGNIFICANCE TO HUMANS
None known. ◆

Resources

Books
Branch, Bill. *Field Guide to Snakes and Other Reptiles of Southern Africa.* South Africa: Struik Publishers, 1998.

Glaw, Frank, and Miguel Vences. *Field Guide to the Amphibians and Reptiles of Madagascar.* 2nd edition. Privately printed, 1994.

Periodicals
Frost, Darrel, et al. "A Molecular Perspective on the Phylogeny of the Girdled Lizards (Cordylidae, Squamata)." *American Museum Novitates* 3310 (2001): 1–10.

Organizations
Herpetological Association of Africa. P.O. Box 20142, Durban North, 4016 South Africa. Web site: <http://www.wits.ac.za/haa>

Bill Branch, PhD

Skinks

(Scincidae)

Class Reptilia
Order Squamata
Suborder Sauria
Family Scincidae

Thumbnail description
Tiny to moderately large lizards, both limbed and limbless, usually with smooth scales

Size
0.9–19.3 in (23–490 mm) in snout-vent length

Number of genera, species
126+ genera; about 1,400 species

Habitat
Versatile

Conservation status
Extinct: 3 species; Critically Endangered: 2 species; Endangered: 3 species; Vulnerable: 21 species; Lower Risk/Near Threatened: 5 species; Data Deficient: 7 species

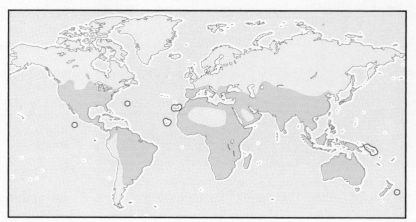

Distribution
Cosmopolitan, except at high elevations and latitudes

Evolution and systematics

With well over 100 genera and more than 1,400 species, skinks are by far the largest family of lizards, an exceedingly diverse group. Their diversity is evident in all aspects of their biology. Terrestrial, arboreal, fossorial, and even semiaquatic skinks exist. Skinks have radiated to fill niches in all types of environments, including arid deserts, savannas, lowland rainforests, temperate forests, and cool montane habitats. Some skinks are diminutive, but others are large. These lizards vary in morphologic characteristics from short and robust with strong, well-developed limbs to elongated and fragile with tiny or no vestigial limbs. In some arid regions (e.g., Australia), they dominate the lizard fauna; in others, such as the Sonoran and Great Basin Deserts, they are essentially absent. Skinks have dispersed widely, many having rafted across oceans to colonize other continents and even remote islands in the Pacific. Four subfamilies are recognized; one (Scincinae) is probably not monophyletic but rather a paraphyletic group:

Acontinae

This is a subfamily of moderately large, limbless, fossorial skinks found in South Africa. There are three genera and 17 species. They are derived from unknown scincine ancestors.

Feylininae

This subfamily comprises moderate to large limbless skinks that inhabit tropical western and central Africa. There is one genus with six species. They are derived from unknown scincine ancestors.

Lygosominae

These skinks are very diverse, ranging from small to large, advanced skinks. They are found worldwide, with more than 82 genera and about 900 species.

Scincinae

This diverse subfamily occurs in North America, Africa, southwest Asia, southern Asia, eastern Asia, and the Philippines. There are more than 30 genera, with more than 300 species.

Scincinae is not a natural group but is based on shared ancestral characters and probably does not contain all descendents of a common ancestor (i.e., it is paraphyletic). Scincines are primitive skinks with smooth cylindrical bodies and small legs. The scincine genus *Eumeces* displays several important ancestral character states, which could resemble the common ancestral state for all skinks. Other candidates for basal scincids include certain African and Asian scincines (*Brachymeles* and *Chalcides*).

Both acontines and feylinines appear to be derived from scincine ancestors. Acontines (*Acontias*, *Acontophiops*, and *Typhlosaurus*), found in Africa, are specialized legless burrowing skinks found only in leaf litter and loose, sandy soil or underneath logs. The small subfamily Feylininae contains a single genus, *Feylinia*, consisting of six species, also burrowers, from central tropical Africa.

By far the largest subfamily is the Lygosominae, with more than 80 genera and more than 900 species. Lygosomines are derived, highly advanced skinks, with five distinct and presumably monophyletic lineages: the *Egernia* group, the *Lygosoma* group, the *Mabuya* group, the *Sphenomorphus* group, and the *Eugongylus* group. Most of these groups occur in Australia (a major center for skink diversity). Some members of each group also occur outside Australia, especially in New Guinea and New Caledonia. Exactly where *Lygosoma* and *Mabuya* (a paraphyletic genus) and several other Old World lygosomines

A female broad-headed skink (*Eumeces laticeps*) guards her eggs. (Photo by Larry L. Miller/Photo Researchers, Inc. Reproduced by permission.)

should be placed within Lygosominae remains uncertain. New skink genera are still being described.

Physical characteristics

All skinks presumably derived from a single common ancestor (i.e., they are monophyletic). They have large, symmetric, shieldlike scales on the head. Most skinks have smooth, glossy, circular scales, but a few have sharp or keeled scales. Bony plates, known as osteoderms, underlie skink scales. One distinctive feature of skinks is the bony secondary palate in the roof of the mouth, which separates the respiratory and digestive passages. (These are confluent in most other lizards.) Other lizards pant when they are thermally stressed (which cools the roof of the mouth, into which large blood sinuses can dissipate heat), but skinks do not pant, perhaps because the secondary palate impedes heat exchange.

Skinks have repeatedly evolved reduced appendages; numerous different evolutionary groups have produced completely limbless forms (*Acontias, Anomalopus, Barkudia, Brachymeles, Coeranoscincus, Feylinia, Lerista, Melanoseps, Ophiomorus, Ophioscincus, Scelotes, Scolecoseps, Sepsophis, Typhlacontias,* and *Typhlosaurus*). These 15 genera are found in Africa, Asia, and Australia. A single North American genus, *Neoseps*, approaches limblessness, with both front and hind limbs reduced to tiny appendages. The same is true of many skinks in Southeast Asia (e.g., species of *Larutia* and *Riopa*). Legless skinks, as well as those with reduced limbs, are typically burrowers and usually have small eyes and consolidated head shields. All degrees of limb reduction can occur within a single genus.

In most lizards, including some skinks, inguinal fat bodies protrude into the abdominal cavity from the pelvic area. (These store valuable energy reserves used in reproduction.) Members of the large *Sphenomorphus* group of skinks, however, have lost these fat bodies and rely on their tails to store fat. Tail loss thus can be costly. In many skinks, the tails of juveniles are markedly brighter than the tails of adults. Red, blue, and yellow tails are thought to lure the attention of predators away from the body.

Skinks have a wide variety of eye types. As in other lizards, only the lower eyelid moves; it is lowered to open the eye and raised to close it. The ancestral condition is a freely movable, scaly, opaque eyelid. Several other derived states exist among skinks: some species have a freely movable eyelid with a clear, disclike central scale, or window, through which the lizard can see even when the eye is closed. In other species of skinks the eyelid is fused immovably in the raised position, with an expanded clear area through which they can see. Still other skinks have a clear eyelid fused all around, forming a spectacle similar to those of geckos and snakes. Larger skinks tend to display the ancestral condition, with movable, opaque eyelids, but most smaller skink species have more derived eyelid conditions. Permanently capped eyes in small skinks limit evaporative water loss and protect the eye.

Distribution

Skinks are cosmopolitan, occurring on all continents except Antarctica as well as on many oceanic islands. They have colonized the New World only a handful of times. As a result, only a few genera are represented (e.g., *Eumeces, Mabuya, Neoseps, Scincella,* and *Sphenomorphus*). Nevertheless, two genera have been particularly successful, the scincine *Eumeces* in North America and the lygosomine *Mabuya* in South America. About 30 species of *Eumeces* and 15 species of *Mabuya* are found in the New World. Both genera are also represented in the Old World.

Blue-tongued skink (*Tiliqua scincoides*) threat display. (Illustration by Amanda Humphrey)

A bobtail skink (*Tiliqua rugosa*) in defensive posture. (Photo by Animals Animals ©Patti Murray. Reproduced by permission.)

Habitat

Skinks may be terrestrial, fossorial, or arboreal. Some skinks (*Amphiglossus astrolabi*, *Eulamprus quoyi*, and *Tropidophorus grayi*) are actually semiaquatic.

Behavior

Skinks have adopted a wide variety of habits. Most are diurnal, but others, such as some species of *Egernia* and *Eremiascincus*, are nocturnal. Many North American skinks, especially *Eumeces*, are secretive, spending much of their time under fallen logs or rocks; thus, they are not very conspicuous. Central and South American skinks (*Mabuya* and *Sphenomorphus*) are active during the day and frequently are observed basking or foraging. In some parts of Africa and Australia, skinks are very conspicuous and diverse, being active during the heat of midday, often in very arid areas.

Almost all skinks exploit tail autotomy as a means of escape from predators, although a few skinks, such as some species of *Egernia*, *Corucia*, and *Tiliqua*, do not shed their tails. Some *Egernia* species have flattened, spiny tails, which are used to block off entrances to their crevice retreats. Australian *Ctenotus* and North American *Scincella* sometimes return to the site where their tails were lost and swallow the remains of them. Few, if any, other vertebrates display auto-amputation and self-cannibalism.

Visual cues are used in individual discrimination, particularly the choice of mate, as well as in prey discrimination. Skinks have exceedingly well developed olfactory abilities and can recognize and determine species, sex, and sexual receptivity of other individuals by scent. They also can detect predators and discriminate prey on the basis of chemical cues.

All skinks possess bony plates within their scales, known as osteoderms, which are composed of compound plates of several interconnected bones underneath each scale. Scales overlap in the fashion of shingles on a roof. Such body armor doubtless confers protection from predators. One pygopodid lizard and several species of snakes have evolved hinged teeth to facilitate obtaining a firm grip on their skink prey. (Hinged teeth fold when they hit an osteoderm.) Indeed, if a skink struggles backward during ingestion, the teeth lock into place. Skinks are swallowed rapidly, suggesting that they might actually facilitate their own ingestion, crawling away from the ratchet-like teeth down a predator's gullet. Some skinks have very loose skin and scales that tear away when they are attacked by predators, allowing for escape.

Feeding ecology and diet

Skinks are active, widely foraging predators. Most skinks are insectivorous, but a few large Australian species (*Tiliqua*) are omnivorous. Some burrowing species, such as *Lerista* and *Typhlosaurus*, feed largely on termites.

Reproductive biology

Live bearing (viviparity) has arisen many times among skinks, although many species retain the ancestral condition and lay eggs (oviparity). In two Australian species of skinks (*Lerista bougainvillii* and *Saiphos equalis*), both egg laying and live bearing occur among different populations within each species. Brood sizes vary greatly among skinks, from one to two in some species (such as *Lobulia* and *Prasinohaema*, which appear to have a fixed clutch size, as in anoles and geckos) to 53 or even 67 in others (such as the Australian *Tiliqua gerrardii*). Some skinks lay their eggs in communal nests, which could be an indication that suitable nest sites are in short supply. This behavior also could represent repeated use by one or more females of nest sites with a history of high hatching success.

Some viviparous skinks give birth to a single, extremely large neonate (*Corucia zebrata*, *Tiliqua rugosa*, and *Typhlosaurus gariepensis*). Within viviparous skinks, the entire range of fetal nutritional types occurs. Many species ovulate large eggs containing all the nutrients necessary for development; thus, developing offspring feed on their own yolk (lecithotrophy, ovoviviparity) before being born alive. Various degrees of placental development take place, in which nutrients are passed from mother to offspring (matrotrophy) during development. For example, *Chalcides* have complex placental connections between mother and progeny through which pass substantial amounts of nutrients required for development. Still other species, such as *Mabuya heathi* in Brazil, have an advanced placental arrangement, passing more than 99% of nutrients necessary for neonatal development from the mother through a placenta.

Conservation status

Most species of skinks are not threatened, but a few, especially island forms, are threatened. The IUCN lists 41 species, 3 as Extinct, 2 as Critically Endangered, 3 as Endangered, 21 as Vulnerable, 5 as Lower Risk/Near Threatened, and 7 as Data Deficient.

Significance to humans

Skinks are important insectivores in many natural habitats. Some large species, such as *Corucia*, are commonly kept as pets. Others were once considered to have medicinal value.

1. Prickly forest skink (*Gnypetoscincus queenslandiae*); 2. Striped blind legless skink (*Typhlosaurus lineatus*); 3. Sandfish (*Scincus scincus*); 4. *Menetia greyii*; 5. Broad-headed skink (*Eumeces laticeps*); 6. Striped skink (*Mabuya striata*). (Illustration by Barbara Duperron)

1. Broad-banded sand swimmer (*Eremiascincus richardsonii*); 2. Bobtail (*Tiliqua rugosa*); 3. *Cyclodomorphus branchialis*; 4. Night skink (*Egernia striata*); 5. Prehensile-tailed skink (*Corucia zebrata*); 6. Fourteen-lined comb eared skink (*Ctenotus quattuordecimlineatus*). (Illustration by Barbara Duperron)

Species accounts

Striped blind legless skink
Typhlosaurus lineatus

SUBFAMILY
Acontinae

TAXONOMY
Typhlosaurus lineatus Boulenger, 1887, Cape of Good Hope, Africa. Four subspecies are recognized.

OTHER COMMON NAMES
None known.

PHYSICAL CHARACTERISTICS
This is a small, legless, blind skink with a shovel-nosed snout and countersunk lower jaw. The eyes are vestigial. The hard, smooth body is yellowish, reddish, or black, with varying numbers of dark longitudinal stripes and a short blunt tail.

DISTRIBUTION
The species occurs in the Kalahari Desert and adjoining areas.

HABITAT
It inhabits sandveld semidesert.

BEHAVIOR
These subterranean fossorial skinks burrow just beneath the surface, often under logs and fallen debris.

FEEDING ECOLOGY AND DIET
These skinks are termite specialists.

REPRODUCTIVE BIOLOGY
Females give birth to two to three living young in mid-January through early March.

CONSERVATION STATUS
Not threatened.

SIGNIFICANCE TO HUMANS
None known. ◆

Prehensile-tailed skink
Corucia zebrata

SUBFAMILY
Lygosominae

TAXONOMY
Corucia zebrata Gray, 1855, Makira Island (San Cristobal), Solomon Islands.

OTHER COMMON NAMES
None known.

PHYSICAL CHARACTERISTICS
This big skink has a large head, well-developed and strongly clawed limbs, a robust body, and a prehensile tail. The lower eyelid is scaly. It has no supranasals; the prefrontals are narrowly separated or in contact, and the parietals are widely separated. The dorsal ground color varies, ranging from khaki to

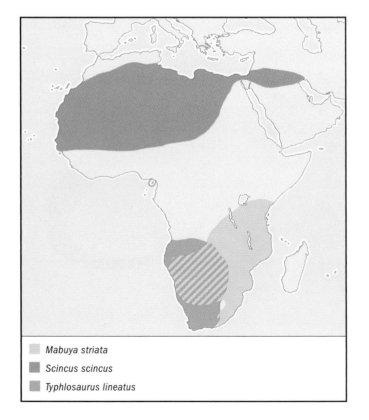

Mabuya striata
Scincus scincus
Typhlosaurus lineatus

Corucia zebrata
Tiliqua rugosa
Ctenotus quattuordecimlineatus

gray-green to pale olive green, with lighter and darker flecks dorsally. Rostral and nasal scales, often the frontonasal scales, are cream or light yellow. The tail is olive green or brown, without markings. The ventral color is yellow-green to light gray-green. The iris is golden yellow to lime green.

DISTRIBUTION
The species is endemic to the Solomon Islands: Bougainville, Shortland Islands, Choiseul, Vella Lavella, New Georgia, Isabel, Guadalcanal, Ngela, Malaita, Makira, Ugi, and Santa Ana.

HABITAT
This large skink is fairly common, but because it is nocturnal, sheltering during the day in hollows or among dense foliage in the larger forest trees, it is seldom seen. Its preferred habitat is the strangler fig tree (*Ficus* sp.). These lizards are almost completely arboreal, though they occasionally are encountered on the ground, moving between trees at night.

BEHAVIOR
These skinks are nocturnal. They possess strongly prehensile tails and are excellent climbers. They move slowly and usually are docile, though when provoked, they will rise up and exhale with a sharp, loud hiss through the open mouth. If tormented, they bite savagely, given the opportunity. After rains that follow a prolonged drought, these skinks emerge at dusk from fig tree hollows to lick up raindrops collected on leaves of their host tree.

FEEDING ECOLOGY AND DIET
This skink is completely herbivorous and consumes a variety of plants, but the bulk of its diet is made up of leaves and flowers of the aroid *Epipremnum pinnatum*.

REPRODUCTIVE BIOLOGY
Live bearers, these skinks give birth to single, large young about one-third the size of adults. Twins are rare. Newborns eat the fecal pellets of adults to establish the gut flora necessary for digestion of plant material.

CONSERVATION STATUS
The species has been overcollected in the Solomons by animal dealers, to supply the foreign pet trade. It is listed on Appendix II of CITES. Thousands are exported from the Solomons each year, seriously threatening the long-term survival of wild populations of this spectacular skink. It is not listed by the IUCN.

SIGNIFICANCE TO HUMANS
Rural Solomon Islanders prize this lizard as food. Many are kept in captivity by herpetoculturists around the world. ◆

Fourteen-lined comb eared skink
Ctenotus quattuordecimlineatus

SUBFAMILY
Lygosominae

TAXONOMY
Ctenotus quattuordecimlineatus Sternfeld, 1919, Hermannsburg Mission, Upper Finke River, Northern Territory.

OTHER COMMON NAMES
None known.

PHYSICAL CHARACTERISTICS
These relatively small, sleek diurnal skinks have 14 pale longitudinal lines on a darker body background. In all species of *Ctenotus*, which translates "comb ear," several scales protrude backward on the anterior edge of the external ear opening.

DISTRIBUTION
The species occurs in central Australia.

HABITAT
These skinks inhabit red, sandy deserts with spinifex grasses.

BEHAVIOR
These alert, wary, active, diurnal skinks are constantly on the move between grass tussocks.

FEEDING ECOLOGY AND DIET
This skink forages widely, constantly moving from tussock to tussock and searching for insect prey, particularly termites.

REPRODUCTIVE BIOLOGY
Females lay two to four eggs. The average clutch size is three eggs.

CONSERVATION STATUS
Not threatened.

SIGNIFICANCE TO HUMANS
None known. ◆

No common name
Cyclodomorphus branchialis

SUBFAMILY
Lygosominae

TAXONOMY
Cyclodomorphus branchialis Gunther, 1867, Champion Bay, Western Australia.

OTHER COMMON NAMES
None known.

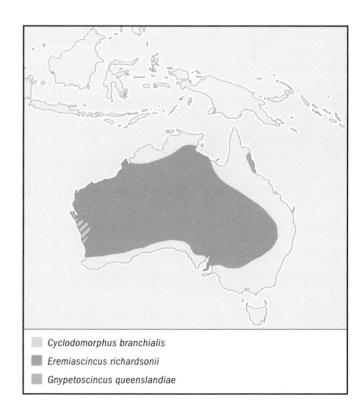

Cyclodomorphus branchialis
Eremiascincus richardsonii
Gnypetoscincus queenslandiae

PHYSICAL CHARACTERISTICS
This is a moderately sized, elongated, slender, short-limbed skink. The color varies from gray to olive brown, sometimes greenish and occasionally dotted with white or cream spots, especially in juveniles. The tail is slightly less than or about equal to the snout to vent length.

DISTRIBUTION
It occurs in Western and South Australia, the southern portion of the Northern Territory, and western Queensland.

HABITAT
This skink inhabits semiarid heaths, woodlands, shrublands, coastal dunes, and red, sandy deserts vegetated with spinifex grasses.

BEHAVIOR
In sandy deserts these secretive lizards spend most of their time within large *Triodia* grass tussocks. In other habitats, they hide in leaf litter or under fallen bushes and trees. They are crepuscular and nocturnal.

FEEDING ECOLOGY AND DIET
The species feeds primarily on a variety of arthropods but also occasionally eats snails and small lizards.

REPRODUCTIVE BIOLOGY
These are live bearers, typically with litter sizes of two to three large young.

CONSERVATION STATUS
Not threatened.

SIGNIFICANCE TO HUMANS
None known. ◆

Menetia greyii
Egernia striata

Night skink
Egernia striata

SUBFAMILY
Lygosominae

TAXONOMY
Egernia striata Sternfeld, 1919, Hermannsburg Mission, Upper Finke River, Northern Territory.

OTHER COMMON NAMES
English: Nocturnal desert skink.

PHYSICAL CHARACTERISTICS
These are moderately large, reddish brown terrestrial skinks with elliptical pupils.

DISTRIBUTION
The species occurs in Central Australia.

HABITAT
The night skink inhabits red, sandy deserts with spinifex grass.

BEHAVIOR
This large nocturnal skink digs elaborate tunnel systems that are used as retreats by many other species of reptiles, both diurnal and nocturnal. These complex burrows are important features of Australian sandy deserts, with several interconnected openings often as far as 3.3 ft (1 m) apart and up to 1.6 ft (0.5 m) deep, vaguely reminiscent of a tiny rabbit warren.

Most sand removed from a night skink burrow is piled up in a large mound outside one "main" entrance.

FEEDING ECOLOGY AND DIET
The major prey is termites, which constitute 76% of the diet by volume. Beetles, ants, cockroaches, and insect larvae also are eaten.

REPRODUCTIVE BIOLOGY
Night skinks are live bearers, giving birth to one to four young. Gravid females with full-term embryos are found from late October through mid-January (with a peak in December). Juveniles stay in the same burrow system with their mothers, as they often contain an adult female plus several newborn young.

CONSERVATION STATUS
Not threatened.

SIGNIFICANCE TO HUMANS
These skinks consume insects. ◆

Broad-banded sand swimmer
Eremiascincus richardsonii

SUBFAMILY
Lygosominae

TAXONOMY
Eremiascincus richardsonii Gray, 1845, Houtman's Abrolhos, Western Australia.

OTHER COMMON NAMES
None known.

PHYSICAL CHARACTERISTICS
These are medium-size yellow-golden brown skinks, with eight to 14 blackish bands on the body and about 20 bands on the tail.

DISTRIBUTION
The species is found throughout most of interior Australia, but not along the eastern, northern, and southern coasts.

HABITAT
They inhabit semihumid to arid areas, including hard or stony substrates with woodlands, shrublands, and hummock grasslands. They also are found throughout the red, sandy deserts of central Australia.

BEHAVIOR
These skinks are crepuscular, nocturnal, and terrestrial. They frequent burrows, often digging their own small burrows off at a right angle to larger burrows, such as a rabbit warrens.

FEEDING ECOLOGY AND DIET
The species feeds on beetles, ants, wasps, and termites.

REPRODUCTIVE BIOLOGY
Little is known about their reproduction, but they lay from three to seven eggs.

CONSERVATION STATUS
Not threatened.

SIGNIFICANCE TO HUMANS
None known. ◆

Prickly forest skink
Gnypetoscincus queenslandiae

SUBFAMILY
Lygosominae

TAXONOMY
Gnypetoscincus queenslandiae De Vis, 1890, Bellendenker and Herberton, Queensland.

OTHER COMMON NAMES
None known.

PHYSICAL CHARACTERISTICS
This small skink is brownish to purple-brown with small keeled scales. The bizarre granular scales and coarse skin appear to help keep the skin evenly moist through capillary spread along the edges of the scales.

DISTRIBUTION
The species is found only in rainforest along the northeastern Queensland coast, from Rossville south to Kirrama.

HABITAT
The prickly forest skink inhabits rainforest.

BEHAVIOR
These skinks shun direct sunlight, and they shelter under rotting logs, stones, and leaf litter deep in rainforest, where damp, shaded conditions prevail. These cryptic skinks have low active body temperatures.

FEEDING ECOLOGY AND DIET
Little is known about the diet, but slugs, snails, and worms are probable food items.

REPRODUCTIVE BIOLOGY
Litters of one to two young are born throughout the year.

CONSERVATION STATUS
Not threatened.

SIGNIFICANCE TO HUMANS
None known. ◆

Striped skink
Mabuya striata

SUBFAMILY
Lygosominae

TAXONOMY
Mabuya striata Peters, 1844, Mozambique. Four subspecies are recognized.

OTHER COMMON NAMES
None known.

PHYSICAL CHARACTERISTICS
The color of this medium-size arboreal skink is dark blackish in the Kalahari, but it varies regionally between subspecies.

DISTRIBUTION
The species is widespread in southern Africa, including extreme southern Angola and Zambia, Namibia, Botswana, Zimbabwe, Mozambique, and parts of central and eastern South Africa.

HABITAT
The habitat is varied; these skinks are found in woodlands, deserts, and mangrove swamps.

BEHAVIOR
This alert climbing skink is found on trees in the Kalahari Desert. It maintains a higher body temperature during the summer than in the winter and basks early and late in the day during summer but at midday in winter.

FEEDING ECOLOGY AND DIET
The diet is composed of a broad variety of insects, including beetles, termites, ants, insect larvae, and spiders.

REPRODUCTIVE BIOLOGY
In northern populations, breeding males have orange-brown heads and yellow-orange throats. The striped skink is live bearing, with an average litter size of 5.4.

CONSERVATION STATUS
Not threatened.

SIGNIFICANCE TO HUMANS
None known. ◆

No common name
Menetia greyii

SUBFAMILY
Lygosominae

TAXONOMY
Menetia greyii Gray, 1845, Australia.

OTHER COMMON NAMES
None known.

PHYSICAL CHARACTERISTICS
These tiny bronze-brown to gray-brown skinks have narrow, broken, dark lines from the neck to the base of the tail. The limbs are short, with four toes on the forelimbs and five toes on the hind limbs.

DISTRIBUTION
The species occurs through most of Australia.

HABITAT
These skinks are versatile, occurring in a wide range of habitats, including sandy spinifex deserts, shrub acacia woodland, mallee, dry sclerophyll forests, and temperate and tropical woodlands. They are largely absent from rainforest.

BEHAVIOR
These tiny diurnal lizards are denizens of leaf litter.

FEEDING ECOLOGY AND DIET
These skinks prey on small termites, spiders, and hemipterans.

REPRODUCTIVE BIOLOGY
Males have an orange throat and yellow venter during the breeding season. Both sexes mature during their first year, and some live to breed again in their second year. Females can produce two clutches of one to three eggs per season. Neonates measure only 0.79 in (2 cm) and weigh only 0.004 oz (0.1 g).

CONSERVATION STATUS
Not threatened.

SIGNIFICANCE TO HUMANS
None known. ◆

Bobtail
Tiliqua rugosa

SUBFAMILY
Lygosominae

TAXONOMY
Tiliqua rugosa Gray, 1825, type locality not specified. There are four subspecies.

OTHER COMMON NAMES
English: Shingleback, sleepy lizard, boggi.

PHYSICAL CHARACTERISTICS
This large lizard is recognized easily by its short, blunt tail and shingle-like scales. The color varies, ranging from white to dark gray to reddish.

DISTRIBUTION
The species occurs in southern Australia, central New South Wales, and southern Queensland.

HABITAT
Bobtails are found in all terrestrial habitats, including sclerophyll forests, hummock grasslands, chenopod shrublands, mallee, and sparsely vegetated coastal dunes, but it is uncommon in dense forests and swamps.

BEHAVIOR
These are slow-moving, lethargic lizards. When threatened, they open their mouths wide and wave their blue tongues.

They often bask on roads, where many are killed by passing vehicles.

FEEDING ECOLOGY AND DIET
Bobtails are omnivorous.

REPRODUCTIVE BIOLOGY
These lizards form long-term pair bonds. Females give birth to two very large young (rarely, one or three young).

CONSERVATION STATUS
Not threatened.

SIGNIFICANCE TO HUMANS
Bobtails may be kept as pets. ◆

Broad-headed skink
Eumeces laticeps

SUBFAMILY
Scincinae

TAXONOMY
Eumeces laticeps Schneider, 1801.

OTHER COMMON NAMES
English: Scorpion, greater five-lined skink; French: Euméces à tête large; German: Breitkopfskink.

PHYSICAL CHARACTERISTICS
These are moderately large, brown skinks. Males have reddish heads during the breeding season.

DISTRIBUTION
The species is widespread in the southeastern United States, from eastern Texas north through eastern Oklahoma and east-

Eumeces laticeps

ern Kansas to southern Missouri. They occur in Illinois, Indiana, and Ohio, east to Maryland, Delaware, and New Jersey and south to northern Florida. They also are found throughout Kentucky, Virginia, Tennessee, North Carolina, South Carolina, Georgia, Alabama, Mississippi, Louisiana, and Arkansas.

HABITAT
Broad-headed skinks inhabit swamps, woodlands, forests, and urban lots strewn with debris.

BEHAVIOR
These skinks are strongly arboreal, adeptly climbing trees and fences. They operate at relatively low body temperatures but are wary, sleek, and slippery lizards.

FEEDING ECOLOGY AND DIET
These active, widely foraging skinks consume a wide variety of arthropods.

REPRODUCTIVE BIOLOGY
In Florida, mating occurs in April and May, but it is somewhat later further north. During the breeding season, males have red heads, which are used as displays in male-male combat as well as in courtship of females. Males face off, moving in a circle; lunge at each other; and aggressively bite the head, neck, and tail of the other male. Tail loss and wounds on the head and neck result, and eventually the loser retreats. Males fight over females, chasing each other up and down trees. Females deposit six to 10 eggs and usually guard their clutches until they hatch.

CONSERVATION STATUS
Not threatened.

SIGNIFICANCE TO HUMANS
None known. ◆

Sandfish
Scincus scincus

SUBFAMILY
Scincinae

TAXONOMY
Scincus scincus Linnaeus, 1759, North Africa. Four subspecies are recognized.

OTHER COMMON NAMES
French: Poisson des sables, Scinque des sables; German: Apothekerskink, Sandfisch.

PHYSICAL CHARACTERISTICS
These medium-size, pale-colored, banded, fusiform skinks have shovel-shaped snouts, countersunk lower jaws, short tails, and enlarged toe lamellae forming fringes along the toes that enhance traction on loose sand.

DISTRIBUTION
The sandfish occurs in the Sahara of northern Africa, from Algeria, Tunisia, and Libya to Egypt and the Mediterranean coast. It also is found in Israel, Jordan, Iraq, and Iran.

HABITAT
These skinks inhabit areas with loose, drifting sand and rich vegetation on the leeward sides of dunes, where sands are not exposed to drying winds. They are found around oases.

BEHAVIOR
These lizards escape from enemies by running along the surface and then suddenly diving into loose sand and swimming a short distance, leaving behind a clear mark where they entered the sand.

FEEDING ECOLOGY AND DIET
The sandfish sometimes swims in loose sand and captures insects on the surface from below. They are omnivorous, preying on scorpions, beetles, other insects, and insect larvae as well as flowers and grains. Occasionally, they eat small *Acanthodactylus* lizards.

REPRODUCTIVE BIOLOGY
Males reach larger sizes than females. Mating occurs in June, and females lay about six eggs shortly thereafter.

CONSERVATION STATUS
Not threatened.

SIGNIFICANCE TO HUMANS
The species once was considered to be a source for a medicinal pharmaceutical against many ailments as well as an aphrodisiac. Dried specimens wrapped in wormwood (*Artemesia*) were imported into Europe via Cairo until the last century. The belief in their medicinal effects seems to have been based on the fact that they feed on wormwood, which is known to have medicinal properties. They are eaten by locals. ◆

Resources

Books
Greer, Allen E. *The Biology and Evolution of Australian Lizards*. Chipping Norton, Australia: Surrey Beatty and Sons, 1989.

Hutchinson, M. N. "Family Scincidae." In *Fauna of Australia*. Vol. 2A, *Amphibia and Reptilia*, edited by C. J. Gasby, C. J. Ross, and P. L. Beesly. Canberra: Australian Biological and Environmental Survey, 1993.

Hutchinson, M. N., and S. C. Donnellan. "Phylogeny and Biogeography of the Squamata." In *Fauna of Australia*. Vol. 2A, *Amphibia and Reptilia*, edited by C. J. Gasby, C. J. Ross, and P. L. Beesly. Canberra: Australian Biological and Environmental Survey, 1993.

Pianka, E. R. *Ecology and Natural History of Desert Lizards: Analyses of the Ecological Niche and Community Structure*. Princeton, NJ: Princeton University Press, 1986.

Pianka, E. R., and L. J. Vitt. *Lizards: Windows to the Evolution of Diversity*. Berkeley: University of California Press, 2003.

Storr, G. M., L. A. Smith, and R. E. Johnstone. *Lizards of Western Australia*. Vol. 1, *Skinks*. Perth: Western Australian Museum, 1999.

Zug, George R., Laurie J. Vitt, and Janalee P. Caldwell. *Herpetology: An Introductory Biology of Amphibians and Reptiles*. 2nd edition. San Diego: Academic Press, 2001.

Resources

Periodicals

Greer, A. E. "Distribution of Maximum Snout-Vent Length Among Species of Scincid Lizards." *Herpetology* 35, no. 3 (2001): 383–395.

Huey, R. B., and E. R. Pianka. "Patterns of Niche Overlap Among Broadly Sympatric Versus Narrowly Sympatric Kalahari Lizards (Scincidae: *Mabuya*)." *Ecology* 58 (1977): 119–128.

———. "Seasonal Variation in Thermoregulatory Behavior and Body Temperature of Diurnal Kalahari Lizards." *Ecology* 58 (1977): 1066–1075.

Huey, R. B., E. R. Pianka, M. E. Egan, and L. W. Coons. "Ecological Shifts in Sympatry: Kalahari Fossorial Lizards (*Typhlosaurus*)." *Ecology* 55 (1974): 304–316.

Pianka, E. R., and W. F. Giles. "Notes on the Biology of Two Species of Nocturnal Skinks, *Egernia inornata* and *Egernia striata*, in the Great Victoria Desert." *Western Australian Naturalist* 15 (1982): 44–49.

Vitt, L. J., and W. E. Cooper Jr. "The Evolution of Sexual Dimorphism in the Skink *Eumeces laticeps*: An Example of Sexual Selection." *Canadian Journal of Zoology* 63 (1985): 995–1002.

———. "Feeding Responses of Broad-Headed Skinks (*Eumeces laticeps*) to Velvet Ants (*Dasymutilla occidentalis*)." *Journal of Herpetology* 22 (1988): 485–488.

———. "The Relationship Between Reproduction and Lipid Cycling in *Eumeces laticeps* with Comments on Brooding Ecology." *Herpetologica* 41 (1985): 419–432.

Other

McCoy, Mike. *Reptiles of the Solomon Islands*. CD-ROM. Kuranda, Australia: ZooGraphics, 2000.

Eric R. Pianka, PhD

Alligator lizards, galliwasps, glass lizards, and relatives

(Anguidae)

Class Reptilia

Order Squamata

Suborder Sauria

Family Anguidae

Thumbnail description
Small to large lizards with elongated bodies; dorsal and ventral scales reinforced by underlying bone (osteoderms); and, in many species, reduced or absent limbs

Size
2.8–55.1 in (70–1,400 mm)

Number of genera, species
14 genera; 112 species

Habitat
Coastal dunes, desert, grasslands, chaparral, pine flatwoods, cloud forest, pine-oak forest, tropical wet forest, and paramo (high-altitude habitat that is cold, wet, and dominated by grasses and shrubs)

Conservation status
Extinct: 1 species; Critically Endangered: 3 species; Endangered: 1 species; Vulnerable: 1 species

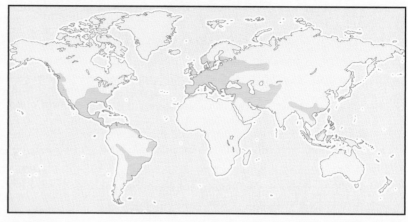

Distribution
North, Central, and South America; the Greater and Lesser Antilles; Europe; North Africa; southwestern Asia; and Southeast Asia, including the islands of the Sunda Shelf

Evolution and systematics

The Anguidae is a monophyletic group within the Anguimorpha and is related most closely to either the Varanoidea (Varanidae, Lanthanotidae, and Helodermatidae) or the Xenosauridae and Shinisauridae. The 14 anguid genera are classified among four subfamilies as follows: Anguinae (glass lizards and slowworm—*Ophisaurus* and *Anguis*), Anniellinae (legless lizards—*Anniella*), Diploglossinae (galliwasps and allies—*Celestus*, *Diploglossus*, *Ophiodes*, *Sauresia*, and *Wetmorena*), and Gerrhonotinae (alligator lizards—*Abronia*, *Barisia*, *Coloptychon*, *Elgaria*, *Gerrhonotus*, and *Mesaspis*).

Phylogenetic analyses of DNA sequence data strongly support the monophyly of the Anguidae and three of its subfamilies, but monophyly of the Diploglossinae requires further evaluation. These analyses also suggest that the Gerrhonotinae and Anguinae are sister taxa and that the Anniellinae is the most basal subfamily. Within the Anguinae, *Ophisaurus* is paraphyletic with respect to *Anguis*; the geographically proximate *A. fragilis* and *O. apodus* are sister taxa. Within the Diploglossinae, *Sauresia* and *Wetmorena* appear to be sister taxa, and *Ophiodes* is related closely to the West Indian *Diploglossus*. Studies with a denser sampling of diploglossine taxa are needed to evaluate these findings further. Within the Gerrhonotinae, available data support a sister relationship between *Elgaria* and the other genera.

Fossils from the late Cretaceous (ca. 75–95 million years ago) of Europe have been assigned to a distinct subfamily (Glyptosaurinae) within the Anguidae. European fossils that may represent anguines or their close relatives are known from the Middle Eocene and later, and, according to J. A. Gauthier, fossils assigned to the *Anguis-Ophisaurus apodus* clade date back to the late Oligocene or early Miocene. On the other hand, *Ophisaurus* fossils are not known from North America before the late Miocene. The fossil record indicates that each anguid subfamily originated at least by the early Eocene (50–55 million years ago).

Physical characteristics

Anguids may be relatively small (e.g., *Celestus macrotus* and *Elgaria parva*) or extremely long (*Anguis fragilis*). Small bony elements called osteoderms are present beneath dorsal and ventral scales. A ventrolateral fold lined with tiny scales allows expansion of the armored skin when food, eggs, or developing young distend the body cavity. This fold is absent in the Anniellinae and most diploglossines. Limb reduction is common in this family, and some species lack limbs altogether. External ear openings are present in most species, but several species lack them. In diploglossines and gerrhonotines, the tail is usually shorter than the body, but in anguines the tail is much longer than the body. Anniellines

A western slender glass lizard (*Ophisaurus a. attenuatus*) in Franklin County, Kansas, USA. (Photo by Suzanne L. Collins and Joseph T. Collins/Photo Researchers, Inc. Reproduced by permission.)

have a very short tail, typically less than two-thirds of the body length.

Caudal autotomy (self-amputation) is widespread among anguids, and fracture planes are present in some caudal vertebrae. Species of *Abronia* are specialized for an arboreal existence and have prehensile tails. Many anguids are some shade of bronze or brown (e.g., *Ophisaurus attenuatus* and *Elgaria kingii*), but some species exhibit more strking coloration (e.g., various shades of green in some *Abronia* and bright bands in some *Diploglossus*). Some species, such as *Diploglossus fasciatus*, may be boldly patterned. L. Vitt has suggested that *D. lessonae* juveniles mimic a toxic species of millipede with their bold markings.

Distribution

The Anniellinae, Diploglossinae, and Gerrhonotinae are restricted to the New World, but the Anguinae has representatives in North America, Europe, Asia, and North Africa. Phylogenetic analyses conducted by J. R. Macey and his colleagues suggest that anguids originated in the Northern Hemisphere (probably in North America) and only later colonized Africa and South America. Anguids are absent from sub-Saharan Africa, Madagascar, and Australia. Anguids occur from sea level to at least 12,470 ft (3,800 m).

Habitat

Anguids occur in a variety of habitats, from coastal dunes and desert scrub to grasslands above the tree line on tropical mountains. They inhabit a variety of forest types, from pine-oak forest to lowland wet forest to cloud forest. While many anguid species are terrestrial, others are fossorial or arboreal.

Behavior

Most anguids easily autotomize (self-amputate) their tails when grasped, and the fragility of this appendage makes the name glass lizards (for *Ophisaurus*) particularly appropriate. Some anguids are active by day (e.g., *Anguis fragilis*), but others are nocturnal (e.g., some *Diploglossus*). Their slow movements and tendency to utilize leaf litter and other surface cover often render these lizards fairly inconspicuous.

Feeding ecology and diet

Anguids are carnivores that rely heavily on arthropod prey. These lizards typically move slowly and deliberately when foraging. In addition to arthropods, they take snails, slugs, nestling rodents, and other small vertebrates.

Reproductive biology

Both live birth and egg laying are widespread in this family. Anniellines give live birth to one to two young. Within the Anguinae, *Anguis* gives live birth, but all species of *Ophisaurus* lay eggs that are attended by the female. Both egg laying and live birth are found in the Diploglossinae and Gerrhonotinae.

Conservation status

The current IUCN listings underestimate the number of imperiled anguid species. The genus *Abronia* is in serious jeopardy from habitat destruction, as discussed thoughtfully by J. Campbell and D. Frost. Habitat destruction is also a grave problem for anguids in the West Indies.

Significance to humans

Anguids are in no way dangerous to humans and do not have commercial value outside the pet trade. As with all lizard groups, some anguids are erroneously believed to be venomous. Some people mistakenly think that the regenerated tail tip in *Ophisaurus* is a stinger.

1. Montane alligator lizard (*Mesaspis monticola*); 2. Texas alligator lizard (*Gerrhonotus liocephalus*); 3. Coban alligator lizard (*Abronia aurita*); 4. *Celestus hylaius*; 5. Moroccan glass lizard (*Ophisaurus koellikeri*); 6. California legless lizard (*Anniella pulchra*). (Illustration by John Megahan)

Species accounts

Moroccan glass lizard
Ophisaurus koellikeri

SUBFAMILY
Anguinae

TAXONOMY
Ophisaurus koellikeri Günther, 1873, North Africa, later restricted to Mogador, Morocco.

OTHER COMMON NAMES
English: Koelliker's glass lizard; French: Ophisaure de Koelliker, l'orvet du Maroc; German: Marokko-schleiche, Türkis-panzerschleiche; Spanish: Lagarto de cristal marroquí.

PHYSICAL CHARACTERISTICS
This slender lizard with a long tail grows to 19.7 in (500 mm) in length. Forelimbs are absent; the greatly reduced hind limbs consist of small flaps near the cloaca. A ventrolateral fold is present, and the tail is extremely fragile. The dorsal coloration is brownish. Over the anterior two-thirds of the body, there are small dark spots (sometimes arranged in transverse rows). Most of the dark spots contain light, iridescent specks.

DISTRIBUTION
This is the only anguid species known from Africa; it is endemic to Morocco at elevations from 160 to 6,560+ ft (50–2,000+ m). Bons and Geniez suggested that it also may occur in Algeria.

HABITAT
The Moroccan glass lizard is known from regions with moderate to high amounts of rainfall; it inhabits open grassy areas near pine and oak forests and also agricultural areas.

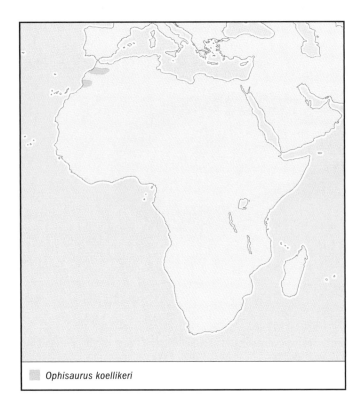

Ophisaurus koellikeri

BEHAVIOR
This species spends much time hidden beneath stones or logs but frequently basks in the morning. Those that live at high elevations hibernate during the cold months. The Moroccan glass lizard rapidly flees from perceived predators and readily autotomizes the tail if it is captured.

FEEDING ECOLOGY AND DIET
This species actively forages for arthropods.

REPRODUCTIVE BIOLOGY
Little is known of the reproductive biology of this species. Ovulation may occur in June, and it is possible that the female guards the eggs.

CONSERVATION STATUS
Although this species has a limited distribution within Morocco, it is not known to be threatened.

SIGNIFICANCE TO HUMANS
None known. ◆

California legless lizard
Anniella pulchra

SUBFAMILY
Anniellinae

TAXONOMY
Anniella pulchra Gray, 1852, 0.5 mi (0.8 km) southeast of Pinnacles National Monument, San Benito County, California, United States (as designated by Murphy and Smith, 1991). Two subspecies have been recognized.

OTHER COMMON NAMES
Spanish: Culebra, Lagartija-sin patas californiana.

PHYSICAL CHARACTERISTICS
The adult California legless lizard has a snout-vent length of 5.9–7.1 in (150–180 mm). This small, serpentiform lizard lacks external ear openings and limbs. The dorsal coloration is silvery gray, dark brown, or black. Three dark longitudinal stripes (one on the center of the back and one on each side of the body) are most distinct on gray individuals. The belly and sides of the body below the lateral stripes are yellow. The ventrolateral fold is absent, and the tail is short.

DISTRIBUTION
This species ranges from the San Francisco Bay Area of California south into northwestern Baja California; it also occurs on several offshore islands. It is known from sea level to 5,085 ft (1,550 m).

HABITAT
The species is found in areas with loose, moist soil (sand or loam) in chaparral, oak woodlands, and, in particular, coastal dunes with sparse vegetation.

BEHAVIOR
The species is primarily fossorial by day but emerges to forage at night.

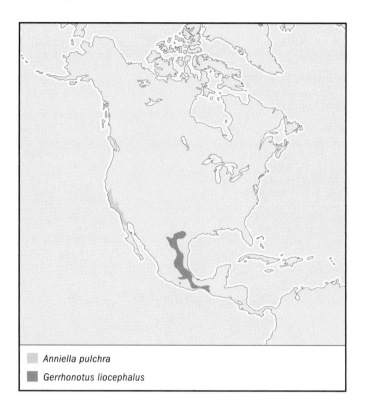

Anniella pulchra

Gerrhonotus liocephalus

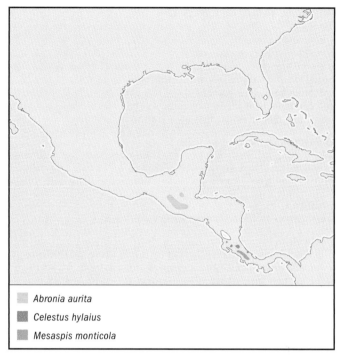

Abronia aurita

Celestus hylaius

Mesaspis monticola

FEEDING ECOLOGY AND DIET
This species actively forages in leaf litter for small arthropods.

REPRODUCTIVE BIOLOGY
From September to November, females typically give birth to one to two young.

CONSERVATION STATUS
The species is not threatened at present, but agriculture has eliminated many California populations.

SIGNIFICANCE TO HUMANS
None known. ◆

No common name
Celestus hylaius

SUBFAMILY
Diploglossinae

TAXONOMY
Celestus hylaius Savage and Lips, 1993, La Selva Biological Station, Canton de Sarapiqui, Heredia, Costa Rica, 130 ft (40 m) above sea level.

OTHER COMMON NAMES
None known.

PHYSICAL CHARACTERISTICS
This is a small lizard, with a total length up to 7.8 in (198 mm). It has short legs and a tail that contributes to slightly more than half of the total length. The dorsal ground color is bronze, with numerous black flecks. Ventrolaterally the black flecks form an irregular stripe. The belly is yellowish green.

DISTRIBUTION
This species is endemic to the lowlands of eastern Costa Rica.

HABITAT
The species has been collected from leaf litter in wet and moist forests.

BEHAVIOR
Although little is known about this infrequently encountered species, it may be diurnal and terrestrial.

FEEDING ECOLOGY AND DIET
It probably actively forages for small arthropods, but the feeding ecology of this species has not been studied.

REPRODUCTIVE BIOLOGY
The species probably bears live young.

CONSERVATION STATUS
Not listed by the IUCN.

SIGNIFICANCE TO HUMANS
None known. ◆

Coban alligator lizard
Abronia aurita

SUBFAMILY
Gerrhonotinae

TAXONOMY
Abronia aurita Cope, 1869, Vera Paz, Guatemala, near Peten and Cobán.

OTHER COMMON NAMES
English: Golden arboreal alligator lizard; Spanish: Escorpión.

PHYSICAL CHARACTERISTICS
This lizard is of moderate size, with snout-vent lengths of adults approximately 3.9–4.9 in (100–125 mm) and total lengths of adults approximately 9.8–12.2 in (250–310 mm). It has well-developed limbs and a prehensile tail. The head is broad posteriorly, narrowing toward the snout. Distinctive spinelike scales are present above the ear opening. The dorsal coloration is pale green or yellowish green with orange or yellow markings or both. The area around the eye is yellow.

DISTRIBUTION
The golden arboreal alligator lizard is endemic to the highlands of Guatemala. It is known from 6,560 to 8,730 ft (2,000–2,660 m) but may have occurred at lower elevations in the past.

HABITAT
The species inhabits pine-oak forest.

BEHAVIOR
This species is diurnal and arboreal, often associated with large bromeliads or Spanish moss growing on oak trees. Males may behave aggressively toward each other.

FEEDING ECOLOGY AND DIET
This species readily takes arthropods in captivity, but little information is available about its feeding habits in the wild.

REPRODUCTIVE BIOLOGY
A captive female gave birth to 12 young on January 26–28, 1992.

CONSERVATION STATUS
The range of this species has been reduced drastically by habitat destruction.

SIGNIFICANCE TO HUMANS
Owing to its spectacular appearance and docile nature, this species is favored by herpetoculturists. ◆

Texas alligator lizard
Gerrhonotus liocephalus

SUBFAMILY
Gerrhonotinae

TAXONOMY
Gerrhonotus liocephalus Wiegmann, 1828, Mexico, later restricted to Tlapancingo, Oaxaca. Up to six subspecies have been recognized.

OTHER COMMON NAMES
Spanish: Culebra con patas, lagartija escorpión de Tejas.

PHYSICAL CHARACTERISTICS
This large lizard grows to 19.7 in (500 mm) in length, though typically it is 9.8–15.7 in (250–400 mm) long. The Texas alligator lizard has strong jaws; four well-developed limbs; and a long, moderately prehensile tail. The dorsal ground color is reddish brown to dull yellow, with irregular transverse bands consisting of dark and light flecks.

DISTRIBUTION
This species occurs in central and southwestern Texas south to San Luis Potosí, Mexico.

HABITAT
It inhabits rocky hillsides with low vegetation.

BEHAVIOR
The Texas alligator lizard is diurnal and primarily terrestrial. A slow-moving species, it may inflate itself with air when disturbed.

FEEDING ECOLOGY AND DIET
This species actively forages for arthropods and small rodents, snakes, and other lizards.

REPRODUCTIVE BIOLOGY
Females attend their clutches of five to 31 eggs, which are laid once per year or more often.

CONSERVATION STATUS
Not threatened.

SIGNIFICANCE TO HUMANS
None known. ◆

Montane alligator lizard
Mesaspis monticola

SUBFAMILY
Gerrhonotinae

TAXONOMY
Mesaspis monticola Cope, 1878, summit of Pico Blanco, Costa Rica

OTHER COMMON NAMES
Spanish: Dragón, lagartija de altura.

PHYSICAL CHARACTERISTICS
These moderately sized lizards (9.3 in [236 mm] in total length) have four rather short limbs and a long tail. A ventrolateral fold is present. Males are bright green or yellowish green with extensive black flecking. Females and juveniles are brownish with black flecking.

DISTRIBUTION
The species occurs in the highlands of Costa Rica and western Panama from 5,900 to 12,470 ft (1,800 to 3,800 m).

HABITAT
The montane alligator lizard inhabits open areas within or at the edges of wet montane forests; it also is found in paramo.

BEHAVIOR
The species is terrestrial and diurnal. It thermoregulates by basking on sunny mornings and may be forced to remain dormant on days without sunshine.

FEEDING ECOLOGY AND DIET
It actively forages for arthropods and juvenile salamanders.

REPRODUCTIVE BIOLOGY
Females give live birth to two to 10 young and may stay with their offspring for a period after birth. It has been suggested that females reproduce only every other year.

CONSERVATION STATUS
Not threatened.

SIGNIFICANCE TO HUMANS
In parts of its range people incorrectly consider this harmless lizard to be capable of delivering potent venom by stinging with its tail. ◆

Resources

Books

Bons, Jacques, and Philippe Geniez. *Amphibiens et Reptiles du Maroc.* Barcelona: Asociación Herpetológica Española, 1996.

Grismer, L. Lee. *Amphibians and Reptiles of Baja California, Including Its Pacific Islands and the Islands in the Sea of Cortés.* Berkeley: University of California Press, 2002.

Savage, Jay M. *The Amphibians and Reptiles of Costa Rica.* Chicago: University of Chicago Press, 2002.

Schleich, H. Hermann, Werner Kästle, and Klaus Kabisch. *Amphibians and Reptiles of North Africa.* Koenigstein, Germany: Koeltz Scientific Publishers, 1996.

Periodicals

Campbell, Jonathan A., and José L. Camarillo. "A New Lizard of the Genus *Diploglossus* (Anguidae: Diploglossinae) from Mexico, with a Review of the Mexican and Northern Central American Species." *Herpetologica* 50 (1994): 193–209.

Campbell, Jonathan A., and Darrel R. Frost. "Anguid Lizards of the Genus *Abronia*: Revisionary Notes, Descriptions of Four New Species, a Phylogenetic Analysis, and Key." *Bulletin of the American Museum of Natural History* 216 (1993): 1–121.

Gauthier, Jacques A. "Fossil Xenosaurid and Anguid Lizards from the Early Eocene Wasatch Formation, Southeast Wyoming, and a Revision of the Anguioidea." *Contributions to Geology (University of Wyoming)* 21 (1982): 7–54.

Good, David A. "Phylogenetic Relationships Among Gerrhonotine Lizards." *University of California Publications in Zoology* 121 (1988): 1–138.

Macey, J. Robert, James A. Schulte II, Allan Larson, Boris S. Tuniyev, Nikolai Orlov, and Theodore J. Papenfuss. "Molecular Phylogenetics, tRNA Evolution, and Historical Biogeography in Anguid Lizards and Related Taxonomic Families." *Molecular Phylogenetics and Evolution* 12 (1999): 250–272.

Savage, Jay M., and Karen R. Lips. "A Review of the Status and Biogeography of the Lizard Genera *Celestus* and *Diploglossus* (Squamata: Anguidae), with Description of Two New Species from Costa Rica." *Revista de Biologia Tropical* 41, no. 3 (1993): 817–842.

Vitt, Laurie J. "Mimicry of Millipedes and Centipedes by Elongate Terrestrial Vertebrates." *Research and Exploration* 8 (1992): 76–95.

Ronald L. Gutberlet, Jr., PhD

Knob-scaled lizards

(*Xenosauridae*)

Class Reptilia
Order Squamata
Suborder Lacertilia
Family Xenosauridae

Thumbnail description
Medium-sized lizards with dorso-ventrally
flattened body and relatively flat, triangular head

Size
Maximum snout-to-vent length between 4.7 and
5.1 in (120 and 130 mm)

Number of genera, species
1 genus; 6 species

Habitat
Tropical scrub, tropical rainforest, cloud forest

Conservation status
Not threatened

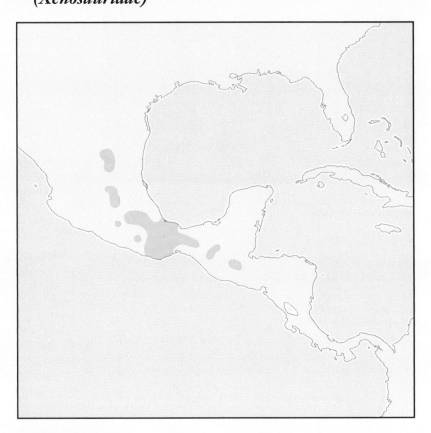

Distribution
Mexico and Guatemala

Evolution and systematics

Originally the family Xenosauridae included two genera, *Xenosaurus* from Mexico and Central America and *Shinisaurus* from China. However, in 1999, J. Robert Macey and colleagues removed *Shinisaurus* from Xenosauridae and placed it into its own family (Shinisauridae) on the basis of DNA evidence. This same study places Xenosauridae in close relationships with the families Shinisauridae, Anniellidae, and Anguidae.

As of 2002, there were six described species of xenosaurs, including one described in 2000 (*X. penai*) and one in 2001 (*X. phalaroantheron*). Given the isolation of many *Xenosaurus* populations and their low mobility, it will not be surprising if further research finds several more undescribed species of *Xenosaurus*. No subfamilies are recognized.

There is no fossil record for the extant genus *Xenosaurus*. Fossils of three extinct xenosaurid genera have been found in North America and Europe.

Physical characteristics

The bodies of xenosaurids are relatively unique, at least compared to the more common lizards, such as *Sceloporus* and *Anolis*, of Central America and Mexico. In fact, the name *Xenosaurus* means "alien lizard." Compared to more typical lizards, xenosaurids have a flattened body and a flat, somewhat triangular head. Presumably the flattened body shape is related to the crevice- or hole-dwelling habit of these lizards.

The flattened head of these crevice-dwellers may influence how hard they can bite. Anthony Herrel and his colleagues found that xenosaurids with taller heads were able to bite more strongly than those with shorter heads. It is not clear whether the reduced bite strength of these flat-headed lizards is important ecologically because the jaws are still strong enough to easily crush their arthropod prey.

Male and female *Xenosaurus* often differ in body and head size, although the extent of sexual dimorphism varies between species. In most cases, females are bigger than males (e.g., *X. newmanorum*, *X. platyceps*), but no difference has been seen in some (e.g., *X. grandis grandis*, *X. rectocollaris*). While

A knob-scaled lizard (*Xenosaurus platyceps*). (Photo by Laurie J. Vitt. Reproduced by permission.)

females are often bigger in body size, males typically have larger heads, perhaps related to aggression between males.

Distribution

Xenosaurids are found in Middle America, from southwestern Tamaulipas, Mexico, at the northern end of its range to central Guatemala at the southern end of its range. Within this broad region, the distribution is far from continuous, with most populations isolated from each other on particular mountain ranges.

Habitat

The habitats in which xenosaurids are found range from cool, tropical cloud forests to fairly dry and hot scrub habitats. All species apparently share a crevice- or hole-dwelling habit. In fact, very few *Xenosaurus* have ever been found outside of a crevice or hole.

Habitat appears to influence the ability of xenosaurids to thermoregulate. Species and populations from rain or cloud forests are unable to raise their body temperatures independently of the environment, whereas those from more open, scrub habitats are able to elevate their body temperatures above the environment's temperature.

Behavior

There appears to be a range in the extent of social interactions in the Xenosauridae. Some species, such as *X. grandis*, are solitary animals that will aggressively interact or even fight when placed in proximity. Such fights may escalate to biting, especially in fights between males, sometimes leaving scars on the lizards' heads. Other species, such as *X. platyceps* and *X. newmanorum*, seem to be much more sociable and often occur in pairs.

Perhaps the most interesting social behavior observed in this family is apparent parental care. In two species (*X.*

newmanorum and *X. platyceps*), adult females have been found in the same crevice as neonates. In each case, the female had recently been pregnant but was no longer. What is particularly interesting about these female/neonate associations is that the adult female is always seen closer to the crevice opening than the neonate, almost as if she were trying to protect the neonate from a predator trying to enter the crevice.

Feeding ecology and diet

Living almost exclusively in crevices or holes, xenosaurids are likely to be sit-and-wait foragers who eat what comes into or near their crevices. Diet analyses of three species by Julio Lemos-Espinal and colleagues found a generalist diet made up primarily of arthropods. However, the diet of *X. newmanorum* included a small amount of mammal and plant material. This diet analysis suggests that xenosaurids are generalists and opportunistic foragers.

Many lizards use their tongues to help detect prey odors or chemicals, and when presented with potential prey items, lizards will typically increase the rate at which they flick their tongues. William E. Cooper Jr. and colleagues tested *X. platyceps* for this response to prey chemicals. Tests of juveniles and adults outside of their crevices found that juveniles responded to prey chemicals but that adults did not. However, when tested in their crevices, both juveniles and adults responded to prey chemicals with elevated tongue-flicking rates, suggesting that the behavior of the adults depends on their location relative to their crevices.

Reproductive biology

All species studied are viviparous with many species having small litter sizes ranging from one to three offspring; however, *X. grandis* from a population in Veracruz, Mexico, has larger litters ranging up to six. Females likely give birth between June and August.

A knob-scaled lizard (*Xenosaurus grandis*). (Photo by R. Wayne Van Devender. Reproduced by permission.)

Conservation status

Because ecological and population studies of xenosaurids are few, their conservation status is unknown. However, given their relatively long lives (*X. newmanorum* can live to be at least seven years), relatively small litters, low mobility, and their specialized way of life (crevice-dwelling), it would seem that these species might be susceptible to any alteration of their habitats by humans. At least one population of *X. newmanorum* does coexist with low levels of human activity: they live in rock walls in a lime and coffee plantation.

Significance to humans

None known.

1. Newman's knob-scaled lizard (*Xenosaurus newmanorum*); 2. Knob-scaled lizard (*Xenosaurus grandis*). (Illustration by Brian Cressman)

Species accounts

Knob-scaled lizard
Xenosaurus grandis

TAXONOMY
Xenosaurus grandis Gray, 1856, Mexico, near Cordova [= Veracruz]. Five subspecies are recognized: *X. g. agrenon*, *X. g. arboreus*, *X. g. grandis*, *X. g. rackhami*, and *X. g. sanmartinensis*.

OTHER COMMON NAMES
None known.

PHYSICAL CHARACTERISTICS
In general form *X. grandis* is similar to other xenosaurids, although it differs in some aspects of scalation and dorsal coloration. *Xenosaurus grandis* also has strikingly red eyes.

DISTRIBUTION
Xenosaurus grandis has the largest geographic range of any xenosaurid, ranging from west-central Veracruz, Mexico, to Guatemala. However, it is possible that *X. grandis* is actually composed of several, as yet undescribed, species.

HABITAT
Tropical forest.

BEHAVIOR
Xenosaurus grandis is one of the more solitary species of xenosaurids, with the vast majority of individuals living alone in their crevices.

FEEDING ECOLOGY AND DIET
Xenosaurus grandis eats a wide range of arthropods.

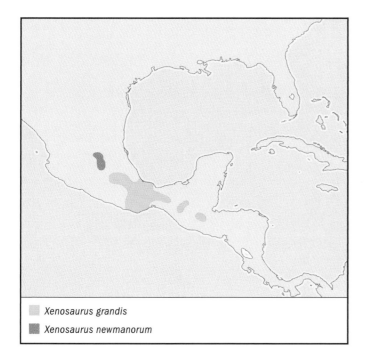

Xenosaurus grandis

Xenosaurus newmanorum

REPRODUCTIVE BIOLOGY
The reproduction of *X. grandis* has been studied in two populations, one in Veracruz, Mexico (*X. grandis grandis*), and one in Oaxaca, Mexico (*X. grandis agrenon*). The Veracruz population has the largest litters in the family (mean = 5.1), whereas the Oaxaca population has a litter size similar to the rest of the family (mean = 3.2).

CONSERVATION STATUS
Unknown, but likely susceptible to habitat changes.

SIGNIFICANCE TO HUMANS
None known. ◆

Newman's knob-scaled lizard
Xenosaurus newmanorum

TAXONOMY
Xenosaurus newmanorum Taylor, 1949, Xilitla region, San Luis Potosi, Mexico. No subspecies are recognized.

OTHER COMMON NAMES
None known.

PHYSICAL CHARACTERISTICS
This species is similar to other xenosaurids except for some elements of scalation. The eyes of *X. newmanorum* are greenish yellow.

DISTRIBUTION
Veracruz and southeastern San Luis Potosi, Mexico.

HABITAT
Cloud forest.

BEHAVIOR
Xenosaurus newmanorum was the first species of xenosaurid found with apparent parental care in the form of females protecting their young. *Xenosaurus newmanorum* is also one of the more gregarious species of xenosaurid, with male-female pairs often seen in the same crevice.

FEEDING ECOLOGY AND DIET
Xenosaurus newmanorum's diet consists primarily of arthropods, but some individuals occasionally eat mammals and vegetation.

REPRODUCTIVE BIOLOGY
Xenosaurus newmanorum has a mean litter size of 2.6.

CONSERVATION STATUS
Unknown, but populations may be susceptible to habitat alteration. However, *X. newmanorum* is known to coexist with limited human activity.

SIGNIFICANCE TO HUMANS
None known. ◆

Resources

Periodicals

Ballinger, Royce E., Julio A. Lemos Espinal, and Geoffrey R. Smith. "Reproduction in Females of Three Species of Crevice-dwelling Lizards (Genus *Xenosaurus*) from Mexico." *Studies on Neotropical Fauna and Environment* 35 (2000): 179–183.

Cooper, William E., Jr., Julio A. Lemos-Espinal, and Geoffrey R. Smith. "Presence and Effect of Defensiveness or Context on Detectability of Prey Chemical Discrimination in the Lizard *Xenosaurus platyceps*." *Herpetologica* 54 (1998): 409–413.

Herrel, Anthony, Ed De Grauw, and Julio A. Lemos-Espinal. "Head Shape and Bite Performance in Xenosaurid Lizards." *Journal of Experimental Zoology* 290 (2001): 101–107.

Lemos-Espinal, Julio A., Geoffrey R. Smith, and Royce E. Ballinger. "Diets of Three Species of Knob-Scaled Lizards (genus *Xenosaurus*) from México." *Southwestern Naturalist* 48 (2003).

———. "Ecology of *Xenosaurus grandis agrenon*, A Knob-Scaled Lizard from Oaxaca, Mexico." *Journal of Herpetology* 37 (2003): In press.

———. "Natural History of the Mexican Knob-scaled Lizard, *Xenosaurus rectocollaris*." *Herpetological Natural History* 4 (1996): 151–154.

———. "Natural History of *Xenosaurus platyceps*, A Crevice-dwelling Lizard from Tamaulipas, Mexico." *Herpetological Natural History* 5 (1997): 181–186.

———. "Neonate-Female Associations in *Xenosaurus newmanorum*: A Case of Parental Care in a Lizard?" *Herpetological Review* 28 (1997): 22–23.

———. "Thermal Ecology of the Crevice-dwelling Lizard, *Xenosaurus newmanorum*." *Journal of Herpetology* 32 (1998): 141–144.

Macey, J. Robert, James A. Schulte II, Allan Larson, Boris S. Tuniyev, Nikolai Orlov, and Theodore J. Papenfuss. "Molecular Phylogenetics, tRNA Evolution, and Historical Biogeography in Anguid Lizards and Related Taxonomic Families." *Molecular Phylogenetics and Evolution* 12 (1999): 250–272.

Nieto-Montes de Oca, A., Jonathan A. Campbell, and O. Flores-Villela. "A New Species of *Xenosaurus* (Squamata: Xenosauridae) from the Sierra Madre del Sur of Oaxaca, Mexico." *Herpetologica* 57 (2001): 32–47.

Pérez Ramos, E., L. Saldaña de la Riva, and Jonathan A. Campbell. "A New Allopatric Species of *Xenosaurus* (Squamata: Xenosauridae) from Guerrero, Mexico." *Herpetologica* 56 (2000): 500–506.

Smith, Geoffrey R., Julio A. Lemos-Espinal, and Royce E. Ballinger. "Sexual Dimorphism in Two Species of Knob-Scaled Lizards (Genus *Xenosaurus*) from Mexico." *Herpetologica* 53 (1997): 200–205.

Geoffrey R. Smith, PhD

Gila monsters and Mexican beaded lizards
(Helodermatidae)

Class Reptilia
Order Squamata
Suborder Scleroglossa
Family Helodermatidae

Thumbnail description
Large, stout, venomous lizards with distinctive, beadlike scales (osteoderms) on the dorsal surfaces of head, limbs, body, and tail

Size
12–18 in (30–45 cm) snout-to-vent length; 14–39 in (35–100 cm) total length; 1.0–4.4 lb (450g–2 kg)

Number of genera, species
1 genus; 2 species

Habitat
Hot desert, tropical deciduous forest

Conservation status
Vulnerable: 2 species

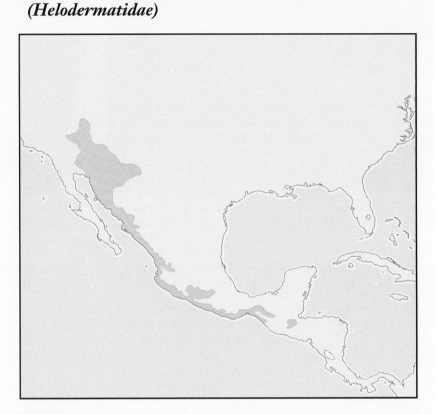

Distribution
Desert regions of the southwestern United States and northwestern Mexico; Pacific drainages along the western slope of Mexico and southern Guatemala; and two Atlantic drainages in Chiapas and eastern Guatemala

Evolution and systematics

The helodermatid clan has a rich and diverse evolutionary history that dates back 98 million years across Europe, Asia, and North America to a time well before many dinosaurs had appeared. The fossil record shows that the remaining species of helodermatid lizards are relics of a more diverse lineage that included at least six other genera inhabiting subtropical desert, forest, and savanna habitats. Family members somehow managed to survive the great Cretaceous extinctions, which vanquished the dinosaurs 65 million years ago. Helodermatid lizards have undergone relatively little gross morphological change over this time, and may appropriately be regarded as living fossils. The genus *Heloderma* has existed since at least the early Miocene (about 23 million years ago).

Today only two species remain: the Gila monster (*Heloderma suspectum*) and the Mexican beaded lizard or escorpíon (*H. horridum*). The two species are readily distinguished from each other by the Mexican beaded lizard's proportionately longer tail (at least 65% of the body length; no more than 55% in the Gila monster). The escorpíon is a longer, lankier, more arboreal lizard than the Gila monster.

Heloderma horridum was first described by Wiegmann in 1829 in Huajintlán, Morelos, Mexico. Four subspecies are recognized. *Heloderma suspectum* was first described by Cope in 1869, on international boundary between the United States and Mexico, Sierra de Moreno, Arizona. Two subspecies are recognized.

No subfamilies are recognized.

Physical characteristics

Helodermatid lizards are named after their distinctively textured skin, consisting of rounded, bony bumps (osteoderms) on their dorsal surfaces. The name *Heloderma* is derived from the Greek for "studded skin." Their lumbering gait, thick forked tongues, robust skull architecture, and venom glands in the lower jaw give them a cumbersome appearance that some consider monsterlike. The body markings can be bright and colorful or faded and cryptic. Juveniles frequently have banded patterns, which break up with age into a variety of adult markings consisting of spots, blotches, or chainlike crossbands of black or yellow on a background of pink, orange, yellow, slate gray, or black. The limbs are rel-

A Gila monster (*Heloderma suspectum*) in Arizona, USA. (Photo by Joe McDonald. Bruce Coleman, Inc. Reproduced by permission.)

A Mexican beaded lizard (*Heloderma horridum*) in western Mexico. (Photo by Animals Animals ©Joe McDonald. Reproduced by permission.)

atively short and strong; the clawed feet are reminiscent of tiny human hands. Fat reserves are stored in the tail, which may be plump in well-fed individuals but is often quite thin in wild-caught lizards. Individuals range in size from barely 6 in (15 cm) total length (hatchling Gila monster) to up to 3.3 ft (1 m) for a large beaded lizard, which can weigh more than 4.4 lb (2 kg).

Distribution

Gila monsters occur from near sea level to about 5,090 ft (1,550 m) from southern Nevada, southwestern Utah, and

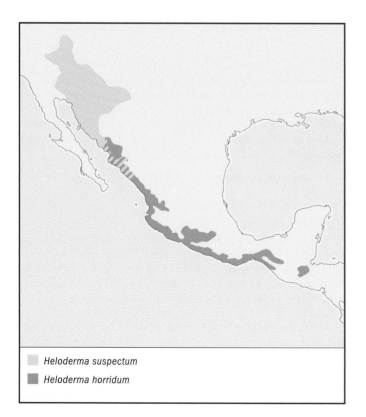

☐ *Heloderma suspectum*
■ *Heloderma horridum*

southeastern California throughout much of Arizona and Sonora, Mexico, and part of southwestern New Mexico. The Mexican beaded lizards occur from sea level to about 5,250 ft (1,600 m) along the Pacific foothills of Mexico from southern Sonora to Chiapas, along Pacific drainages in southern Guatemala, and along two Atlantic drainages in Chiapas and eastern Guatemala.

Habitat

Gila monsters are primarily desert dwellers, although they also inhabit semidesert grassland and woodland communities along mountain foothills. They prefer canyons or adjacent rocky slopes and more rarely open valleys. Their occurrence is strongly influenced by the availability of suitable microenvironments (boulders, burrows, pack rat middens, etc.) used as shelters, where they spend most of their time. Mexican beaded lizards inhabit primarily tropical dry forests and thornscrub, occurring less frequently in lower pine-oak woodlands. They frequent relatively open sandy and rocky arroyos, as well as densely vegetated upland hilly terrain and plateaus.

Behavior

Gila monsters and beaded lizards spend more than 95% of their time hidden within shelters (rocky crevices, burrows, pack rat middens, and trees). When active on the surface, however, they can travel long distances—more than 0.6 mi (1 km) in search of food and mates. Field studies using radiotelemetry have shown that both species are primarily diurnal. The specific timing of activity varies among individuals, seasons, and geographic locations.

During the breeding season, Gila monsters and beaded lizards perform spectacular ritualized male-male combat behaviors that are strikingly similar to those of many monitor lizards (*Varanus*). For the beaded lizards, combat consists of the formation of a high arch posture, with bellies pressed together and snouts, forelimbs, and tail tips forming contact points on the ground. Pressure exerted by the combatants eventually collapses the arch, and the dominant lizard emerges

Three defensive postures of the Gila monster (*Heloderma suspectum*). (Illustration by John Megahan)

on top. Combatants may repeatedly form the arch in matches that may continue for several hours. A typical combat session demands considerable physical effort and leaves both participants exhausted.

The combat of Gila monsters also consists of a series of ritualized wrestling matches, whereby combatants straddle each other, then perform a body twist in an effort to gain the superior position. Gila monsters do not form the arching postures performed by the beaded lizards, probably because their tails are too short. Each bout ends when pressure exerted by the body twist causes the lizards to separate, but bouts can be

repeated many times over several hours. Two fighting males observed in southwestern Utah performed at least 13 individual bouts over nearly three hours of continuous exertion.

Feeding ecology and diet

Gila monsters and beaded lizards are widely searching foragers that feed on the contents of vertebrate nests, primarily reptilian and avian eggs, and juvenile mammals. Among the most common food items are juvenile cottontail rabbits and rodents; snake and lizard eggs, especially those of the spiny-tailed

1. Mexican beaded lizard (*Heloderma horridum*); 2. Gila monster (*Heloderma suspectum*). (Illustration by Brian Cressman)

iguana (*Ctenosaura pectinata*) within the range of the beaded lizard; and quail eggs. Beaded lizards take a greater variety of food items. Gila monsters can fulfill their annual maintenance energy requirements with three large meals. Their relatively large size, low resting metabolic rate, and ability to take large meals make frequent foraging activity unnecessary for helodermatid lizards.

Reproductive biology

In Gila monsters, spermiogenesis, courtship, and mating occur in late April through early June. Eggs are laid in July and August, which coincides with the onset of the summer rains in the southwestern deserts. Hatchling Gila monsters do not emerge until the following April. The clutch size varies from two to 12, with a mean of 5.7. Hatchlings have a snout-to-vent length around 4.5 in (110 mm) and average 6.5 in (165 mm) in total length; they weigh 1.2 oz (33 g).

In beaded lizards, spermiogenesis, courtship, and mating take place in September and October. Eggs are laid between October and December and hatchlings appear in June or July with the onset of the wet season. The clutch size ranges from two to 22 eggs with a mean of seven to nine. Wild hatchlings have a snout-to-vent length of 4.5–5.0 in (115–127 mm) and weigh 0.8–1.0 oz (23–27 g).

Conservation status

Both Gila monsters and beaded lizards are categorized as Vulnerable by the IUCN. In addition, both species are listed by the Convention on International Trade in Endangered Species of Wild Fauna and Flora (CITES) as Appendix II species, which includes those for whom trade must be controlled to avoid overexploitation. They receive full legal state protection from collecting, transport, or killing throughout their ranges. Their greatest threat is from habitat loss, due to the development of their dry forest and desert habitats, and from unscrupulous collectors. The spotty distribution of Gila monsters in the Mojave Desert of the U.S. Southwest, and rapid urban and recreational development in that region, prompted the U.S. Fish and Wildlife Service to classify Gila monsters as a high-priority species that may be vulnerable for listing as a threatened or endangered species in the near future.

Significance to humans

Helodermatids are the only lizards known to be venomous. Their venom apparatus consists of multilobed glands that empty through ducts at the base of grooved, venom-conducting teeth. In contrast to snakes, the venom glands of Gila monsters and beaded lizards are housed in the lower jaw rather than the upper jaw. Their venom is used primarily for defense. A bite from a Gila monster or a beaded lizard causes excruciating pain, swelling, and, in more severe bites, a rapid drop in blood pressure, profuse sweating, and vomiting. Bites to people are rare and almost always result from careless handling. Despite numerous exaggerated accounts before 1950, there has not been a human death reported from a Gila monster bite since 1930. This is attributable more to improved accuracy of reporting and medical record-keeping than to reduced bite frequency or advances in treatment. No first aid measures are recommended aside from carefully cleaning the wound and seeking immediate medical attention. Several important biologically active peptides were discovered in the venom of helodermatid lizards in the 1990s. The best known of these, Exendin 4, is very effective at inducing insulin release in human subjects and has become a promising tool for the treatment of diabetes.

Resources

Books

Brown, David E., and Neil B. Carmony. *Gila Monster: Facts and Folklore of America's Aztec Lizard.* Salt Lake City, UT: University of Utah Press, 1999.

Campbell, Jonathan A., and William W. Lamar. *The Venomous Reptiles of Latin America.* Ithaca, NY: Comstock Publishing Associates, 1989.

Lowe, Charles H., Cecil R. Schwalbe, and Terry B. Johnson. *The Venomous Reptiles of Arizona.* Phoenix: Arizona Game and Fish Department, 1986.

Periodicals

Beck, D. D. "Ecology and Behavior of the Gila Monster in Southwestern Utah." *Journal of Herpetology* 24 (1990): 54–68.

Beck, D. D., et al. "Locomotor Peformance and Activity Energetics of Helodermatid Lizards." *Copeia* (1995): 577–585.

Beck, D. D., and C. H. Lowe. "Ecology of the Beaded Lizard, *Heloderma horridum*, in a Tropical Dry Forest in Jalisco, Mexico." *Journal of Herpetology* 25 (1991): 395–406.

———. "Resting Metabolism of Helodermatid Lizards: Allometric and Ecological Relationships." *Journal of Comparative Physiology B* 164 (1994): 124–129.

Beck, D. D., and A. Ramírez-Bautista. "Combat Behavior of the Beaded Lizard, *Heloderma h. horridum*, in Jalisco, Mexico." *Journal of Herpetology* 25 (1991): 481–484.

Bogert, C. M., and R. M. del Campo. "The Gila Monster and Its Allies: The Relationships, Habits, and Behavior of the Lizards of the Family Helodermatidae." *Bulletin of the American Museum of Natural History* 109 (1956): 1–238.

Doyle, M. E., and J. M. Egan. "Glucagon-like Peptide-1." *Recent Progress in Hormone Research* 56 (2001): 377–399.

Goldberg, S. R., and D. D. Beck. "*Heloderma horridum* (Beaded Lizard): Reproduction." *Herpetological Review* 32 (2001): 255–256.

Goldberg, S. R., and C. H. Lowe. "Reproductive Cycle of the Gila Monster, *Heloderma suspectum*, in Southern Arizona." *Journal of Herpetology* 31 (1997): 161–166.

Resources

Norell, M. A., and K. Gao. "Braincase and Phylogenetic Relationships of *Estesia mongoliensis* from the Late Cretaceous of the Gobi Desert and the Recognition of a New Clade of Lizards." *American Museum Novitates* 3211 (1997): 1–25.

Nydam, R. L. "A New Taxon of Helodermatid-like Lizard from the Albian-Cenomanian of Utah." *Journal of Vertebrate Paleontology* 20, no. 2 (2000): 285–294.

Pregill, G. K., J. A. Gauthier, and H. W. Greene. "The Evolution of Helodermatid Squamates, with Description of a New Taxon and an Overview of Varanoidea." *Transactions of the San Diego Society of Natural History* 21 (1986): 167–202.

Raufman, J. P. "Bioactive Peptides from Lizard Venoms." *Regulatory Peptides* 61 (1996): 1–18.

Organizations

Tucson Herpetological Society. P.O. Box 709, Tucson, Arizona 85702-0709 USA. Web site: <http://tucsonherpsociety.org>

Daniel D. Beck, PhD

Monitors, goannas, and earless monitors
(Varanidae)

Class Reptilia
Order Squamata
Suborder Varanoidei
Family Varanidae

Thumbnail description
Small to very large lizards with long necks and forked tongues

Size
9–122 in (230–3,100 mm) total length

Number of genera, species
2 genera; about 61 species

Habitat
Versatile

Conservation status
Vulnerable: 2 species

Distribution
Africa, Asia, Southeast Asia, and Australia

Evolution and systematics

The Borneon earless monitor *Lanthanotus* is the sister group to all *Varanus*. Recent phylogenetic studies suggest that African monitors are the sister group to all other monitors. There is a large Asian clade and two clades of Australian monitor lizards, one consisting of pygmy monitors and the other of larger species. Monitors may well be more closely related to snakes than to most other lizards.

The varanid lizard body plan has been exceedingly successful as it has been around since the late Cretaceous. Komodo dragons are dwarfed by a closely related, extinct gigantic varanid, *Megalania prisca*, originally placed in the genus *Varanus* (to which it must be returned). This Australian Pleistocene species is estimated to have reached more than 20 ft (6 m) in total length and to have weighed over 1,320 lb (600 kg). *Megalania* fossils have been dated at 19,000–26,000 years before present.

Large monitor lizards evolved large body size in response to availability of very large prey (pygmy elephants in Indonesia for *V. komodoensis* and rhinoceros-sized diprotodont marsupials in Australia for *Megalania*). At the end of the Pleistocene, much of the megafauna went extinct. Without large prey to support them, large predators also died out. Fortunately, *Varanus komodoensis* did not die out when pygmy elephants went extinct at the end of the Pleistocene.

There two living genera: *Varanus* (about 60 species) and *Lanthanotus* (1 species). Two subfamilies are recognized: Varanines and Lanthanotines.

Physical characteristics

Varanus are morphologically conservative but vary widely in size, which makes this genus ideal for comparative studies of the evolution of body size. Small body size has evolved at least twice, once in Australia and another time in an Asian clade. Large body sizes appear to have evolved in several lineages of varanids. These lizards range from the diminutive Australian pygmy monitor *Varanus brevicauda* (only about 6.7–7.9 in or 17–20 cm in total length and 0.28–0.71 oz or 8–20 g in mass) to Indonesian Komodo dragons (*Varanus komodoensis*), which attain lengths of 9.9 ft (3 m) and weights of 331 lb (150 kg). The largest living lizard, *Varanus komodoensis*, was not officially described scientifically as a species until 1912, but Chinese and Dutch travelers must have known of its existence centuries before. Flicking their long forked yellow tongues, surely these great reptiles must have given rise to the very concept of dragons breathing fire. Ancient cartographers marked the Lesser Sunda Islands on their maps of southeastern Asia with an ominous warning: "Here be dragons."

The teeth of most varanids are serrated along the rear edge, which facilitates cutting and tearing skin and flesh of prey as these big lizards pull back on their bite. *V. komodoensis* routinely kills deer and pigs this way, and one Komodo dragon actually eviscerated a water buffalo. Komodo dragons and *Megalania* are/were ecological equivalents of large saber-toothed cats, using their slashing bite to disembowel large mammals.

Varanids have more aerobic capacity and a greater metabolic scope than other lizards and range over larger areas. Because of

An Australian sand goanna (*Varanus gouldii*). (Photo by Animals Animals ©Fritz Prenzel. Reproduced by permission.)

their body size, large individual monitor lizards retain body heat in their nocturnal retreats and can emerge the next morning with body temperatures well above ambient air temperatures. Their mass confers a sort of "inertial homeothermy," which has implications for understanding the evolution of endothermy.

Distribution

About 60 species of *Varanus* are currently recognized worldwide, and all occur in Africa, Asia, Southeast Asia, and Australia (the new world is now sadly impoverished, although fossil varanids are known from North America). The largest adaptive radiations have occurred in Australia, where they are commonly known as "goannas," and where about 30 species are found, including one very interesting clade (subgenus *Odatria*) that has evolved dwarfism. The single species of *Lanthanotus* is found only in Sarawak in northern Borneo.

Habitat

Monitors live in a wide variety of habitats, ranging from mangrove swamps to dense forests to savannas to arid deserts. Some species are aquatic, some semiaquatic, others terrestrial, while still others are saxicolous or semiarboreal or truly arboreal.

Behavior

Monitor lizards adopt characteristic defensive postures, flattening themselves from side to side and extending their gular pouches to make themselves appear as large as possible. Often they hiss loudly and flick their tongues. Big species lash their tails like whips with considerable accuracy. Some species stand up erect on their hind legs during such displays.

Varanids appear to be much more intelligent than most lizards. At the National Zoo in Washington, D.C., individual

Komodo dragons have their own "personalities" and recognize each keeper. Recent experiments on captive *V. albigularis* by John Phillips at the San Diego Zoo suggest that some varanids can actually count. Lizards were conditioned by feeding them groups of four snails in separate compartments with movable partitions, opened one at a time allowing monitors to eat each of the four snails. Upon finishing the fourth snail, lizards were allowed into another chamber containing four more snails. After such conditioning, one snail was removed from some snail groups; lizards searched extensively for the missing fourth snail, even when they could see the next group. Such experiments showed that these varanids can count up to six, but with groups of snails larger than six, the monitors seemed to stop counting and merely classified them as "lots," eating them all before moving on to the next chamber. Such an ability to count probably evolved as a consequence of raiding the nests of reptiles, birds, and mammals, since average clutch or litter size of prey would usually be less than six. A pair of *V. niloticus* were observed to cooperate when raiding a crocodile nest in Africa. After one monitor lured a female crocodile guarding her nest away into the water, another monitor dug into the nest and began eating eggs and hatchlings. Soon it was joined by its accomplice. Similar observations have been made of monitors raiding bird nests.

Some varanids, such as *Varanus niloticus* and *V. mertensi*, are highly aquatic, leaving water only to bask nearby or to dig nest burrows and lay their eggs. Others, such as *V. indicus*, *V. mitchelli*, and the so-called water monitor, *V. salvator*, are more terrestrial but are strong swimmers, very much at home in the

Bengal monitor lizards (*Varanus bengalensis*) engaged in a ritual trial of strength. These fights take place during the mating season; the stronger lizards winning the females. Neither contestant is harmed during the struggle—as soon as one is wrestled to the ground, the battle is over. (Photo by Zingel/Eichhorn, F. L. Bruce Coleman, Inc. Reproduced by permission.)

Komodo dragon (*Varanus komodoensis*) defense includes tail thrashing. (Illustration by Amanda Humphrey)

water. Komodo dragons also swim well, which may facilitate colonization of islands. Even desert monitors swim with ease.

Using high-tech radiotelemetry, Auffenberg followed movements and body temperatures of *Varanus olivaceus* in the Philippines. Surprisingly, he found that this large, rare, previously unstudied, arboreal monitor feeds primarily on fruit. His similar extensive study of *Varanus bengalensis*, primarily in India and Pakistan, showed that this wide-ranging terrestrial monitor feeds primarily on a wide variety of arthropods (earthworms, crustaceans, snails, many other vertebrates, as well as their eggs and young, are also eaten when available).

One of the most arboreal of all monitor lizards is the beautiful green *V. prasinus* from New Guinea and Cape York, Australia. These small climbing monitors have strongly prehensile tails and spend most of their lives above ground in trees. Several new species have recently been described that belong in the *prasinus* group.

In the Australian desert, as many as six or seven species of *Varanus* occur together in sympatry. All are exceedingly wary,

essentially unapproachable, and unobservable lizards. Fortunately, however, they leave fairly conspicuous tracks, and one may deduce quite a lot about their biology from careful study of this spoor. Each species leaves its own distinct track. The largest species, the perentie, *V. giganteus*, reaches 6.6 ft (2 m) or more in total length, whereas some smaller "pygmy goannas," such as the ubiquitous and very important lizard predator *V. eremius*, achieve total lengths of only about 15.8 in (40 cm). Two other species, *V. gouldii* and *V. tristis*, are intermediate in size. Individuals of all four of these species range over extensive areas and consume very large prey items, particularly other vertebrates (especially lizards). Daily forays typically cover distances of a kilometer or more. *V. tristis* and two other little-known small species, *V. caudolineatus* and *V. gilleni*, which have strongly curved very sharp claws, are semiarboreal. Four other species (*V. brevicauda*, *V. eremius*, *V. gouldii*, and *V. giganteus*) are terrestrial.

Feeding ecology and diet

Almost all monitor lizards are active predatory species that raid vertebrate nests and eat large vertebrate prey (some

A juvenile Nile monitor lizard (*Varanus niloticus*) eats a fish in Africa. (Photo by Animals Animals ©Zig Leszczynski. Reproduced by permission.)

smaller species also feed extensively on invertebrates, including centipedes, large insects, earthworms, crustaceans, and snails). Many monitor lizards are top predators in the communities in which they live. Varanids are very active lizards that forage widely, using their forked tongues extensively to locate and discriminate among their prey by scent (vision and sound are also used).

When foraging, *V. gouldii* hunt by smell, swinging their long necks and heads from side-to-side, constantly flicking out their long forked, very snake-like, tongues, searching for scent trails and sweeping over as big an arc as possible so that they cover as much ground as possible. Upon detecting a scent signal, these monitors follow the trail to the source, usually a burrow, and dig up the intended prey. Digging is methodical, using the forearms and sharp claws of the forefeet with the pointed snout, mouth, and sharp teeth right in between ever ready to snatch up prey as they dash to escape. Sand monitors consume many geckos captured in their diurnal retreats, dead-end burrows. Many diurnal species of lizards are also eaten (*Ctenophorus, Ctenotus, Lerista, Lialis, Menetia, Moloch, Pogona*, as well as other *Varanus*, including *V. brevicauda, V. caudolineatus, V. gilleni*, and *V. gouldii*). They also eat reptile eggs, baby mammals, and baby birds. These lizards very probably eat any other lizard that they can catch. Among

specimens examined, the largest relative prey mass was a *Pogona minor* estimated to weigh about 0.88 oz (25 g) eaten by a *V. gouldii* that weighed 6.3 oz (180 g), or about 13.9% of its weight.

Reproductive biology

Male monitor lizards engage in ritualized combat, fighting over females. Larger species wrestle in an upright posture, using their tails for support, grabbing each other with their forelegs and attempting to throw their opponent to the ground. Smaller species grapple each other while lying horizontally, with legs wrapped around each other, rolling over and over on the ground. The victor then courts the female, first flicking his tongue all over her, and if she concurs, climbing on top of her and mating by curling the base of his tail beneath hers and inserting one of his two hemipenes into her cloaca. Male varanids have a cartilaginous or bony support structure in each hemipenis called a hemibaculum.

All monitor lizards lay eggs. Clutch sizes vary widely among species from two to three in the smallest monitors such as *V. brevicauda* to 35–60 in large African species like *V. albigularis* and *V. niloticus*. Some monitors (*V. rosenbergi, V. niloticus, V. prasinus*, and *V. varius*) excavate termitaria for nesting burrows. Termites close off the entrance, sealing the eggs inside in an almost ideal, protected, humid environment of nearly constant temperature. Sometimes, females return to the same termitarium nine months later to open it and free their hatchlings. In some species of monitor lizards, hatchlings are much more vividly marked and colorful than adults (*V. dumerilii, V. griseus, V. rosenbergi, V. tristis*).

Conservation status

The IUCN lists two species as Vulnerable, the Indonesian Komodo dragon and the Philippine Gray's monitor. In Africa and Asia, habitat loss and the skin trade pose serious threats to these magnificent lizards. All varanids are on the CITES endangered list, even though some, such as Australian desert species, are not really in any immediate danger of going extinct.

Significance to humans

Monitor comes from the Latin noun "monition," which means someone who is a warner. According to ancient belief, these lizards were supposed to warn people that crocodiles were in an area. The generic name *Varanus* comes from the Arabic word "waran." Waran is the Egyptian name for the Nile monitor, *Varanus niloticus*, which basically means "monitor" in Arabic. Large monitor lizards, especially *V. salvator*, are hunted for their skins, especially in Southeast Asia. The largest of all lizards, Komodo dragons, are popular centerpieces of reptile exhibits at zoos.

1. Crocodile monitor (*Varanus salvadorii*); 2. Short-tailed monitor (*Varanus brevicauda*); 3. Komodo dragon (*Varanus komodoensis*); 4. Stripe-tailed monitor (*Varanus caudolineatus*); 5. Earless monitor lizard (*Lanthanotus borneensis*). (Illustration by Joseph E. Trumpey)

Species accounts

Earless monitor lizard

Lanthanotus borneensis

SUBFAMILY
Lanthanotines

TAXONOMY
Lanthanotus borneensis Steindachner, 1878, Sarawak, Borneo.

OTHER COMMON NAMES
None known.

PHYSICAL CHARACTERISTICS
One of the least known of all lizards, earless monitor lizards are medium-sized, with adults averaging 16.5–21.6 in (42–55 cm) total length with a relatively long cylindrical body, long neck, and long tail. They have short legs but long, curved, sharp claws. They can wrap their muscular bodies and prehensile tails around a branch in a manner that suggests that they might climb. Most of their scales are small, but six longitudinal rows of enlarged scales run from the head down the back, and two central rows run out on to the tail. Earless monitor lizard tails do not regenerate. They shed their skin in one piece as do some other anguimorphans and snakes. The brain case is similar to snakes in that it is more solidly encased than it is in varanids. The upper temporal arch has been lost and there is a hinge joint in the middle of the lower jaw, as in snakes. Also, like snakes, these lizards have sharp recurved teeth on their premaxillaries, maxillaries, palatines, pterygoids, and dentaries. *Lanthanotus* is the only species among anguimorphans with translucent windows in its lower eyelids, which could be a precursor to the "spectacle" covering the eyes of snakes. Like snakes, *Lanthanotus* (Lanthan = hidden, otus = ear)

have no external ear openings and they have deeply forked tongues. Indeed, the earless monitor lizard could well be closely related to snakes. Viewed from below, males have blunt, rectangular jaws, whereas jaws of females are more pointed. Because of the hemipenes in males, the base of the tail is broader than in females.

DISTRIBUTION
These very secretive lizards are found only in lowland riverine regions of Sarawak on northern Borneo (there are unsubstantiated reports from nearby Kalimantan, Indonesia).

HABITAT
These lizards are found along banks of rivers and ponds.

BEHAVIOR
Earless monitor lizards dig burrows in banks along watercourses and retreat into the water when threatened. By some accounts, the earless monitor is aquatic (individuals have been captured in fish seines and traps), by others, it is a burrower. The species does seem to prefer cool moist habitats. A number of these unusual lizards were collected after severe flooding in Sarawak in 1963. They could have been inactive in underground retreats and emerged when it flooded. They may also climb. Some reports of captive lizards suggest that the earless monitor could be nocturnal. In captivity, earless monitors appear to prefer relatively low ambient temperatures of about 75.2–82.4°F (24–28°C). In captivity, these are sluggish lizards that spend most of their time lying in water, seldom moving. Captives shed their skins very infrequently, less than once per year. Such observations could be mere artifacts of the unusual environmental conditions in captivity and could be largely irrelevant to behaviors of free-ranging wild lizards.

FEEDING ECOLOGY AND DIET
Captives have eaten squid, small bits of fish, earthworms, liver, and even beaten eggs. Earthworm setae have been found in stomachs of museum specimens, but their natural diet remains largely unknown.

REPRODUCTIVE BIOLOGY
Little is known about reproduction in the earless monitor lizard. One lizard was found with six large eggs in September 1976, but these eggs may not have been *Lanthanotus* eggs. Eggs have never been laid in captivity. Clutch sizes from dissected females is three to four. Eggs are large, about 1.2 in (30 mm) long.

CONSERVATION STATUS
The earless monitor lizard is not threatened, but should be considered gravely endangered as it occurs only in lowland riverine areas of Sarawak that have been dramatically altered by human activity.

SIGNIFICANCE TO HUMANS
As the sister species to varanids, this living fossil may offer hints as to what the ancestors of monitor lizards were like. ◆

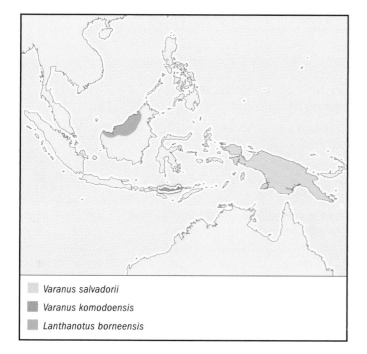

Varanus salvadorii

Varanus komodoensis

Lanthanotus borneensis

Short-tailed monitor
Varanus brevicauda

SUBFAMILY
Varanines

TAXONOMY
Varanus brevicauda Boulenger, 1898, Sherlock River, Nickol Bay, Western Australia.

OTHER COMMON NAMES
English: Short-tailed pygmy monitor.

PHYSICAL CHARACTERISTICS
The short-tailed monitor is the smallest varanid. Adult size is reached at a snout-vent length (SVL) of 3.5–4.3 in (90–110 mm) and a weight of 0.35–0.6 oz (10–17 g). Hatchlings are about 1.8 in (45 mm) SVL and weigh only 0.07–0.1 oz (2–3 g).

DISTRIBUTION
Central and Western Australia.

HABITAT
Red sandy desert dominated by spinifex (*Triodia*) grasses.

BEHAVIOR
Short-tailed monitors are terrestrial, spending most of their time within tussocks of spinifex grass. These small monitors probably climb around within these grass tussocks. Their tails are very muscular and prehensile and they "hang on for dear life" when inside a spinifex grass tussock using their legs as well as their tail. The typical monitor lizard threat posture and behavior has been conserved in the evolution of these diminutive monitors, which hiss and lunge with their throat inflated as if they are a serious threat. This tiny varanid is seldom encountered active above ground; the vast majority of specimens are collected in pit traps. Two were dug up in shallow burrows

during August (one must have been active immediately prior to being exhumed, as crisp, fresh tail lash marks were at the burrow's entrance and the lizard had a body temperature of 95.7°F [35.4°C], 10 degrees above ambient air temperature). Mark-recapture studies show that short-tailed monitors do not move very far. Dozens have been pit trapped on a flat sandplain covered with large, long unburned, clumps of spinifex, possibly the preferred habitat of this monitor. One female weighing 0.32 oz (9.1 g) contained an adult 0.05-oz (1.5-ml) *Ctenotus calurus* skink; this prey item constituted 16.5% of the short-tailed monitor's body weight.

FEEDING ECOLOGY AND DIET
Since these sedentary lizards seldom leave the protective cover of spinifex tussocks, they must forage within tussocks. They eat large insects and, occasionally, small lizards.

REPRODUCTIVE BIOLOGY
In the Great Victoria Desert, the smallest male *V. brevicauda* with enlarged testes was 3.2 in (82 mm) SVL and the smallest gravid female was 3.7 in (94 mm) SVL. In central Australia, sexual maturity is reached in males at about 2.75 in (70 mm) SVL and in females at about 3.3 in (83 mm) SVL. Males may become reproductive at an age of about 10 months, but females probably do not mature until their second spring at an age of about 22 months. A male fell into the same pit only hours after a female was removed from that pit trap, suggesting that males may follow scent trails to find females. Clutch size is usually two to three eggs, although larger clutches have been reported. Relative clutch masses of two females with oviductal eggs was 16.7% of a female's body weight. Mating occurs in the spring (September–October) and eggs are laid in November. Hatchlings emerge in late January to February and are about 1.6–1.8 in (42–45 mm) SVL and weigh only about 0.07 oz (2 g). Incubation takes about 70–84 days at 64.4–77°F (18–25°C).

CONSERVATION STATUS
Not threatened.

SIGNIFICANCE TO HUMANS
As the smallest of all monitors, this diminutive species offers insight into the evolution of small body size. German herpetoculturists have successfully bred this species in captivity. ◆

Stripe-tailed monitor
Varanus caudolineatus

SUBFAMILY
Varanines

TAXONOMY
Varanus caudolineatus Boulenger, 1885, Champion Bay, Western Australia.

OTHER COMMON NAMES
English: Line-tailed pygmy monitor.

PHYSICAL CHARACTERISTICS
Both sexes appear to mature at about 3.6 in (91 mm) SVL, the size of the smallest male with enlarged testes and the smallest gravid female. One specimen with an SVL of 4.4 in (111 mm), estimated to weigh 0.53 oz (15 g), contained an intact 0.1-oz (3-ml) *Gehyra* (20% of its mass).

DISTRIBUTION
Central and interior Western Australia.

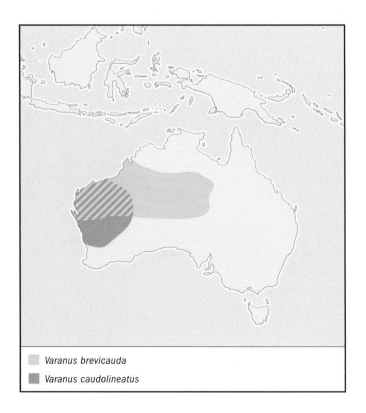

Varanus brevicauda
Varanus caudolineatus

HABITAT

The stripe-tailed monitor is semiarboreal, preferring habitats with mulga trees, which offer small hollows that provide the lizards with tight-fitting safe diurnal and nocturnal retreats.

BEHAVIOR

Movements of stripe-tailed monitors marked with a radioactive tracer were not nearly as extensive as movements observed in other species of varanids, suggesting that these pygmy monitors may be fairly sedentary.

FEEDING ECOLOGY AND DIET

These monitors descend to the ground to forage, evidenced by the fact that three of the 13 active specimens observed were on the ground when first sighted. Moreover, one stomach contained a ground-dwelling *Rhynchoedura ornata* gecko. Others contained tails and intact *Gehyra*, arboreal geckos. Gut contents of another sample consisted largely of scorpions and ground-dwelling spiders. These monitors forage on the ground searching for prey by going down into their burrows. *V. caudolineatus* (and *V. gilleni*) actually "harvest" the exceedingly fragile tails of geckos that are too large to subdue intact.

REPRODUCTIVE BIOLOGY

The male combat ritual for the stripe-tailed monitor is similar to that of *V. gilleni* and most other pygmy varanids in the *Odatria* group. It involves wrestling, longitudinal rolls while embraced with ventral surfaces adjacent, lateral twisting and flexing, and occasional biting on the flank, limbs, and tail. While embraced with both fore and hind limbs, two adversaries form an arch from their snout to their tail.

Mating behavior is similar to that in other varanids. A male will nudge and tongue-flick a female, particularly around the head and neck. Mating occurs when a male lies along the top of a female, using a hind limb and tail to expose her cloaca, enabling him to insert a hemipene. Male testes were largest during July and August, and a female with enlarged oviducts was collected in late December. Smith (1988) reports that a 1.3 oz (37 g) female *V. caudolineatus* laid four eggs (total mass of 0.3 oz or 9 g) on October 23, 1986. *V. caudolineatus* appear to oviposit over an extended period. Clutch size is usually three to four eggs.

CONSERVATION STATUS

Not threatened.

SIGNIFICANCE TO HUMANS

None known. ◆

Komodo dragon

Varanus komodoensis

SUBFAMILY

Varanines

TAXONOMY

Varanus komodoensis Ouwens, 1912, Komodo, Indonesia.

OTHER COMMON NAMES

English: Komodo monitor, ora; French: Dragon des Komodos, Varan de Komodo; German: Komodo-Waran; Spanish: Varano de Komodo.

PHYSICAL CHARACTERISTICS

Indonesian Komodo dragons, which attain lengths of 9.9 ft (3 m) and weights of 331 lb (150 kg), are the largest living lizard.

DISTRIBUTION

Found only on several Indonesian islands in the Lesser Sunda Island chain, Flores, Rinca, and Komodo.

HABITAT

Komodo dragons are found from sea level up to 2,625 ft (800 m), mainly in tropical dry and moist deciduous monsoon forest, savanna, and mangrove forest.

BEHAVIOR

Hatchling Komodo dragons are arboreal. Juveniles and subadults are both arboreal and terrestrial. Adult individuals are strictly terrestrial; their large body size hinders climbing trees. Komodo dragons have been occasionally observed swimming for short distances, close to mangrove forest.

FEEDING ECOLOGY AND DIET

Komodo dragons are versatile predators, and they survived to the present by switching their diet to smaller prey such as large reptiles, mammals, and birds. Only relatively recently, humans introduced the deer and wild pigs that constitute the staple diet of large Komodo dragons today. Komodo monitors have secondarily evolved to become ambush predators, lying in wait along trails for large prey such as small deer or wild pigs. When foraging for smaller prey such as mammals and snakes, Komodo dragons forage widely like other monitors. Juvenile Komodos are active, widely foraging predators. The diet includes insects, small to large vertebrates (lizards, snakes, rodents, monkeys, wild boars, deer, and water buffalo), bird and sea turtle eggs, as well as carrion. Deer make up about 50% of the adult diet. Adults also prey upon the young of their own species. Juvenile Komodos are highly arboreal, which may protect them from being eaten by their larger, less agile brethren. These monitors can detect airborne volatile oils released by decomposition of carcasses, which leads lizards on long feeding excursions.

After following Komodos in the field using radiotelemetry for over a year, Auffenberg summed up their ambush thus: "when these animals decide to attack, nothing can stop them." He followed one lizard for 81 days, during which time it made only two verified successful kills. A 110-lb (50-kg) female consumed a 683-lb (310-kg) boar in just 17 minutes. Auffenberg himself was attacked and treed by a "maverick" Komodo dragon.

REPRODUCTIVE BIOLOGY

In captivity, age of first reproduction is reached at about eight or nine years (no data are available for wild specimens). Courtship and mating occur from May to August. Adult males of similar size engage in ritual combats to gain access to females. Using their tails for support, they wrestle in an upright position, grabbing each other with their forelegs, and try to knock down their opponent. In courtship, a male flicks its tongue on the female's snout and then over her body until he reaches her cloaca. He then presses his snout at the base of her tail, scratches her body with his claws, and eventually crawls on her back.

Females deposit eggs in September in burrows located on hill slopes or in nests of megapode birds. These nests consist of heaps of twigs mixed with earth, up to 5 ft (1.5 m) high and 16.4 ft (5 m) in diameter. A hole is dug into the nest and used to lay eggs. Females sometimes lie on the nest for several months, probably guarding their eggs from predators. Eggs hatch in March–April. In captivity, incubation period averages about 220 days. Parental care has not been observed in Komodo dragons. Average clutch size is 18 eggs. The largest clutch recorded is 36 eggs.

Male Komodo dragons tend to grow bulkier and bigger than females. No obvious morphological differences between sexes exist, except for a specific area of precloacal scales. On Komodo island, sex ratio was estimated at three males per female. Hatchlings gain weight rapidly, and after about five years they may weigh 55 lb (25 kg) and have a total length of over 6.6 ft (2 m). Growth continues slowly throughout life. Males reach a larger size than females. Captive records and field observations suggest a longevity of over 30 years.

CONSERVATION STATUS
The IUCN lists the Komodo dragon as Vulnerable due to its small geographic range, small population size, and loss of habitat due to human encroachment. Populations of these spectacular creatures have now been reduced to only a few thousand lizards on a handful of islands. Komodo dragons have recently been successfully bred in captivity.

SIGNIFICANCE TO HUMANS
Blood plasma of Komodo dragons contains powerful antibacterial substances that could be developed as new useful antibiotics in the ongoing worldwide battle against the evolution of antibiotic-resistant microbes. These giant lizards are often the centerpiece of reptile exhibits at zoos. ◆

Crocodile monitor
Varanus salvadorii

SUBFAMILY
Varanines

TAXONOMY
Varanus salvadorii Peters and Doria, 1878, Southern New Guinea.

OTHER COMMON NAMES
English: Artrellia.

PHYSICAL CHARACTERISTICS
This is an unusual large climbing monitor lizard. Characterized by an extremely long tail and a distinctive arched muzzle and bulbous nose, crocodile monitors could perhaps be allied with the Australian "lace monitor," *V. varius*, also a large climbing monitor. The largest reliable measurements of individual adult crocodile monitors list snout to vent lengths of 30.7–33.5 in (78–85 cm), tails from 60–65 in (153–166 cm), and total lengths of 91–100 in (231–255 cm), although many anecdotal reports of much larger lizards exist.

DISTRIBUTION
Southern New Guinea.

HABITAT
This species is found in lowland rainforest.

BEHAVIOR
Crocodile monitors use their long prehensile tails as counterbalances when climbing in the canopy and as effective whips when threatened. Their teeth are very long and fanglike. Ritualized combat, the clinch-bipedal stance phase, courtship, and copulation have been observed in captive crocodile monitors, which have been successfully bred in Germany several times.

FEEDING ECOLOGY AND DIET
In captivity, these monitors will eat mice, rats, and chickens. In the wild, they probably prey largely on birds.

REPRODUCTIVE BIOLOGY
Hatchlings are large, measuring 20 in (0.5 m) in total length.

CONSERVATION STATUS
Not threatened.

SIGNIFICANCE TO HUMANS
New Guinea natives consider this species to be an "evil spirit that climbs trees, walks upright, breathes fire, and kills men." ◆

Resources

Books
Auffenberg, W. *The Behavioral Ecology of the Komodo Monitor.* Gainesville: University Press of Florida, 1981.

———. *The Bengal Monitor.* Gainesville: University Press of Florida, 1994.

———. *Gray's Monitor Lizard.* Gainesville: University Press of Florida, 1988.

Bennett, D. *Monitor Lizards. Natural History, Biology and Husbandry.* Frankfurt am Main: Edition Chimaira, 1998.

King, D., and B. Green. *Goannas The Biology of Varanid Lizards.* University of New South Wales Press, 1999.

Murphy, J. B., C. Ciofi, C. de la Panouse, and T. Walsh, eds. *Komodo Dragons: Biology and Conservation.* Washington, DC: Smithsonian Institution Press, 2002.

Pianka, E. R., and L. J. Vitt. *Lizards: Windows to the Evolution of Diversity.* Berkeley: University of California Press, 2003.

Zug, G. R., L. J. Vitt, and J. P. Caldwell. *Herpetology: An Introductory Biology of Amphibians and Reptiles.* 2nd edition. San Diego: Academic Press, 2001.

Periodicals
Bayless, M. K. "The Artrellia: Dragon of the Trees." *Reptiles* 6 (1998): 32–47.

Böhme, W., and H. G. Horn. "Advances in Monitor Research." *Mertensiella* 2 (1991): 1–266.

Horn, H.-G., and W. Böhme. "Advances in Monitor Research II." *Mertensiella* 11 (1999): 1–366.

McDowell, S., and C. M. Bogert. "The Systematic Position of *Lanthanotus* and the Affinities of the Anguimorphan Lizards." *Bulletin of the American Museum of Natural History* 105 (1954): 1–142.

Pianka, E. R. "Comparative Ecology of *Varanus* in the Great Victoria Desert." *Australian Journal of Ecology* 19 (1994): 395–408.

Thompson, G. "Daily Movement Patterns and Habitat Preferences of *Varanus caudolineatus* (Reptilia: Varanidae). *Journal of Wildlife Research* 20 (1993): 227–231.

Eric R. Pianka, PhD

Early blindsnakes
(Anomalepididae)

Class Reptilia
Order Squamata
Suborder Serpentes
Family Anomalepididae (Early blindsnakes)

Thumbnail description
Small, fossorial snakes with smooth, uniformly
sized body scales; highly reduced eyes; small,
ventrally placed mouth; rounded snout; and very
short tail

Size
3–16 in (7–41 cm)

Number of genera, species
4 genera; 16 species

Habitat
Soil

Conservation status
Not threatened

Distribution
Southern Central America, northern South America, and east-central South America

Evolution and systematics

The family Anomalepididae was first established in 1939 by Edward H. Taylor after he discovered that snakes of the genus *Anomalepis* exhibit patterns of dentition and scalation that are significantly different from those seen in other blindsnakes. Soon thereafter, anatomical studies revealed that *Helminthophis*, *Liotyphlops*, and *Typhlophis* share many distinctive morphological features with *Anomalepis*, indicating that these three genera also belong in the family Anomalepididae.

Although morphological evidence strongly suggests that Anomalepididae represents a natural, or monophyletic, group, the interrelationships among the four genera of early blindsnakes and their phylogenetic affinities within Serpentes remain poorly understood. Most recent snake systematists have placed Anomalepididae together with the families Leptotyphlopidae (slender blindsnakes) and Typhlopidae (blindsnakes) in the infraorder Scolecophidia. However, the interrelationships of these three families have been controversial. In general, characters relating to cranial morphology, visceral topography, and scalation patterns suggest that, within Scolecophidia, Anomalepididae is most closely related to Typhlopidae. In contrast, characters associated with the morphology of the hyobranchial apparatus support a close relationship between Leptotyphlopidae and Typhlopidae. Adding further confusion to this issue, a small number of morphological and molecular phylogenetic analyses have concluded that Scolecophidia may be polyphyletic (i.e., that the three families of blindsnakes may not share a close common ancestry).

Unfortunately, the fossil record has offered few clues to help resolve this confusion. No fossils have been discovered yet that can be assigned unequivocally to Anomalepididae (although *Suffusio predatrix*, from the Paleocene of Texas, has been identified tentatively as an anomalepidid), and the few leptotyphlopid and typhlopid fossils that have been described are known only from very incomplete remains (usually isolated vertebrae). Thus, the evolutionary origin of early blindsnakes and their relationships to other blindsnakes remain enigmatic.

No subfamilies are recognized.

Physical characteristics

Early blindsnakes are among the smallest serpents in the Neotropics. Although five species (*Helminthophis flavoterminatus*,

1. *Liotyphlops ternetzii*; 2. *Typhlophis squamosus*. (Illustration by Emily Damstra)

Liotyphlops anops, *L. beui*, *L. schubarti*, and *L. ternetzii*) may occasionally reach or exceed 1 ft (31 cm) in total length, most are considerably smaller, reaching adult lengths of only 6–10 in (15–25 cm) and adult weights of less than 0.1 oz (2.8 g). In addition to being short in length, these fossorial snakes are also quite slender, having maximum body widths of only 0.04–10.2 inches (1–5 mm) and aspect ratios (total length divided by body width) ranging from 32 to as high as 86.

Like other blindsnakes (Leptotyphlopidae and Typhlopidae), early blindsnakes have small, ventrally placed mouths; highly reduced eyes that are barely (if at all) visible beneath the overlying head scales; and cylindrical bodies covered with smooth, circular, and uniformly sized scales (they lack the enlarged belly scales present in most other forms of snakes). However, anomalepidids can be distinguished from other sympatric blindsnakes by the presence of two (or rarely three) anal shields (*Leptotyphlops* has only one and *Typhlops* has three to five) and teeth on both the upper and lower jaws (*Leptotyphlops* lacks upper teeth and *Typhlops* lacks lower teeth). In addition, most anomalepidids have more than 20 rows of scales encircling the body, whereas all New World *Leptotyphlops* have 14 and all mainland Neotropical *Typhlops* have 20 or fewer. Finally, most members of the family are characterized by a distinctive color pattern in which the body is uniformly dark in color (black or dark brown), but the head and often at least parts of the tail are light in color (white, yellow, or pink). The four species of *Anomalepis* stand as exceptions to this generalization, however, being uniformly brown or reddish brown in coloration.

In all species of Anomalepididae the snout is bluntly rounded in shape, and in all genera except *Typhlophis* the scales surrounding the snout are somewhat enlarged. These enlarged scales not only reduce friction between the snake's head and the soil during burrowing, they also house numerous pressure-sensitive sensory organs. The tail is universally short (1–3.4% of the snake's total length) and may terminate in a sharp, needle-like apical spine (*T. squamosus* and the seven species of *Liotyphlops*), or end more bluntly as in other snakes (*Anomalepis*, *Helminthophis*, and *T. ayarzaguenai*).

Early blindsnakes are also characterized by a suite of unique internal anatomical features. The most noteworthy of these features are an elaborate, M-shaped hyoid (the skeletal element that supports the tongue, and which in other snakes is V- or Y-shaped), a relatively short tongue, and a pair of peculiar "orbital bones" that are involved in the suspension of the upper jaws. (Though traditionally called "orbital bones," it has recently been established that these bones have nothing to do with the formation of the bony orbit.) Additionally, like many other basal snakes, at least some species of *Liotyphlops* retain a vestigial pelvic girdle. However, no pelvic elements have been found in *Anomalepis*, *Helminthophis*, or *Typhlophis*.

Distribution

Anomalepididae is an exclusively Neotropical family, with an apparently discontinuous distribution in central and northern South America and southern Central America. The genus *Anomalepis* includes four species, which range from northern

Costa Rica southward through northern Peru. *Helminthophis* contains three species, which are distributed from Costa Rica southward through Colombia and Venezuela. In addition, one species of this genus (*H. flavoterminatus*) may have been introduced onto the Indian Ocean island of Mauritius during the nineteenth century. *Liotyphlops* is the most speciose of the four anomalepidid genera, including three species that range from southern Costa Rica southward through Venezuela, Colombia, and Ecuador and four species that are distributed throughout southern Brazil, southeastern Paraguay, and northeastern Argentina. Finally, the genus *Typhlophis* includes only two species, which are distributed from central Venezuela eastward through French Guiana and southward through northeastern Brazil. *T. squamosus* has also been reported to be present on Trinidad, but the validity of this claim remains questionable. The altitudinal distribution of early blindsnakes is poorly known, but available data suggest that these snakes inhabit a relatively wide range of elevations, from sea level to at least 5,968 ft (1,819 m). It should be emphasized, however, that these tiny, secretive snakes inhabit areas that have relatively low human population densities. Furthermore, like other blindsnakes, anomalepidids are probably frequently transported to areas outside of their natural range through human activities (especially through the agricultural trade). Therefore, the known geographical distributions of early blindsnakes may not accurately reflect the natural or complete range of the family.

Liotyphlops ternetzii

Typhlophis squamosus

Habitat

The little information that is available about the natural history of early blindsnakes suggests that these serpents utilize a wide range of macrohabitats. They have been collected in hot, humid rainforests, dry forests, pasturelands, and even rocky, mountainous regions. However, throughout these many macrohabitats, they apparently inhabit a relatively narrow range of microhabitats. They are most commonly encountered in soil, often during the course of digging operations, or hidden beneath logs, stones, or forest debris. It is not known to what depths they may retreat beneath the ground, but one species (*T. ayarzaguenai*) was unearthed by an excavator from a depth of about 1.6 ft (0.5 m) in Venezuela. Although there have been no reports of arboreality for anomalepidids, it is likely that they also occasionally climb trees, as this behavior has been noted in numerous species of leptotyphlopid and typhlopid blindsnakes.

Behavior

Because early blindsnakes occur predominantly in areas with relatively low human population densities, the behavior of these small, secretive serpents is very poorly known. In captivity, *L. beui* behaves similarly to comparably sized typhlopids, exhibiting strong fossorial tendencies and responding to human handling by frantically writhing its body, voiding the contents of its cloaca, and jabbing the sharp terminal spine on the tip of its tail into the skin of its captor. Given their tropical distribution, it is likely that early blindsnakes are active throughout the year. However, like other blindsnakes, they are probably more active at night than during the day.

Feeding ecology and diet

Very little information is available concerning the natural diet or feeding ecology of early blindsnakes. However, both *L. ternetzii* and *T. squamosus* are known to feed on the eggs, larvae, and pupae of ants, suggesting that the trophic ecology of early blindsnakes is similar to that of other blindsnakes (Leptotyphlopidae and Typhlopidae), nearly all of which prey almost exclusively on social insects.

Reproductive biology

The reproductive biology of early blindsnakes is almost completely unknown. All species are presumed to be oviparous, but specific details about reproductive seasonality, courtship and mating behavior, oviposition, clutch size, and incubation time are lacking.

Conservation status

No early blindsnakes are listed as Endangered or Vulnerable by the IUCN. However, six of the 16 species of Anomalepididae are known only from the vicinities of their type localities, and one (*A. aspinosus*) has not been found since 1916. Therefore, the status of many species of early blindsnakes is unknown.

Significance to humans

None known.

Resources

Books

Greene, Harry W. *Snakes: The Evolution of Mystery in Nature.* Berkeley: University of California Press, 1997.

McDiarmid, Roy W., Jonathan A. Campbell, and T'Shaka A. Touré. *Snake Species of the World: A Taxonomic and Geographic Reference.* Vol. 1. Washington, DC: Herpetologists' League, 1999.

Pérez-Santos, Carlos, and Ana G. Moreno. *Ofidios de Colombia.* Museo Regionale di Scienze Naturali, Monografia 6. Turin, Italy: Museo Regionale di Scienze Naturali, 1988.

Peters, James A., and Braulio R. Orejas-Miranda. *Catalogue of the Neotropical Squamata.* Vol. 1, *Snakes.* Washington, DC: Smithsonian Institution Press, 1970.

Periodicals

Cunha, Osvaldo R. da, and Francisco P. do Nascimento. "Ofídios da Amazônia. X.—As cobras da região leste do Pará." *Museu Paraense Emílio Goeldi Publicações Avulsas* 31 (1978): 1–218.

Dixon, James R., and Christopher P. Kofron. "The Central and South American Anomalepid Snakes of the Genus *Liotyphlops.*" *Amphibia-Reptilia* 4 (1983): 241–264.

Hahn, Donald E. "Liste der rezenten Amphibien und Reptilien: Anomalepididae, Leptotyphlopidae, Typhlopidae." *Das Tierreich* 101 (1980): 1–93.

Kofron, Christopher P. "The Central and South American Blindsnakes of the Genus *Anomalepis.*" *Amphibia-Reptilia* 9 (1988): 7–14.

List, James C. "Comparative Osteology of the Snake Families Typhlopidae and Leptotyphlopidae." *Illinois Biological Monographs* 36 (1966): 1–112.

Robb, Joan, and Hobart M. Smith. "The Systematic Position of the Group of Snake Genera Allied to *Anomalepis.*" *Natural History Miscellanea* 184 (1966): 1–8.

Señaris, J. Celsa. "A New Species of *Typhlophis* (Serpentes: Anomalepididae) from Bolívar State, Venezuela." *Amphibia-Reptilia* 19 (1998): 303–310.

Nathan J. Kley, PhD

Slender blindsnakes

(*Leptotyphlopidae*)

Class Reptilia

Order Squamata

Suborder Serpentes

Family Leptotyphlopidae

Thumbnail description
Small, slender, fossorial snakes with smooth, uniformly sized body scales, highly reduced eyes, a small, ventrally placed mouth, rounded or hook-shaped snout, and short tail bearing a sharp terminal spine

Size
2.3–15.3 in (5.8–38.9 cm)

Number of genera, species
2 genera; 93 species

Habitat
Soil

Conservation status
Not classified by the IUCN

Distribution
Africa, southwest Asia, southern North America, Central America, the West Indies, and South America

Evolution and systematics

Most recent phylogenetic analyses have placed Leptotyphlopidae together with Anomalepididae (early blindsnakes) and Typhlopidae (blindsnakes) in Scolecophidia, one of two infraorders recognized within Serpentes (the clade that includes all living snakes). However, the interrelationships among the three groups of blindsnakes are poorly understood. The unusual form and position of the hyoid in Leptotyphlopidae and Typhlopidae are suggestive of a close relationship between these two families. However, similarities in skull structure, visceral anatomy, and scalation patterns suggest that Anomalepididae and Typhlopidae are more closely related to each other than either is to Leptotyphlopidae. Unfortunately, the fossil record for Leptotyphlopidae is exceptionally poor, and the few fossil remains that are known offer little insight into the evolutionary history of the family. Interrelationships within Leptotyphlopidae are also poorly known. Nearly 20 species groups are tentatively recognized, but there have been no large-scale phylogenetic analyses that have addressed interspecific relationships within the family.

No subfamilies are recognized.

Physical characteristics

The family Leptotyphlopidae includes the most highly miniaturized snakes in the world. Although a few species (e.g., *Leptotyphlops humilis, L. melanotermus, L. occidentalis, L. tricolor, L. weyrauchi,* and *Rhinoleptus koniagui*) occasionally grow to lengths of over 1 ft (30 cm), most forms are significantly smaller, ranging between 4 and 10 in (10 and 25 cm) in total length and often weighing less than 0.05 oz (1.4 g). Even more remarkable than their short length, however, is their ex-

tremely narrow build, a characteristic reflected in their common names, "slender blindsnakes," "threadsnakes," and "wormsnakes." Most species attain a maximum body width of only 0.04–0.20 in (0.1–0.5 cm) and exhibit aspect ratios (total length divided by body width) of between 40 and 100. Two exceptionally slender species, *L. macrorhynchus* and *L. occidentalis*, occasionally have aspect ratios exceeding 140, and even the stoutest forms (e.g., *L. broadleyi* and *L. boulengeri*) are more slender than most other snakes, rarely having aspect ratios of less than 30.

Leptotyphlopids bear a strong superficial resemblance to other blindsnakes (Anomalepididae and Typhlopidae) in having cylindrical bodies covered by smooth, equally sized, cycloid scales, short lower jaws countersunk into the ventral surface of the head, and vestigial eyes that are scarcely visible beneath the enlarged head scales. (However, in some species, most notably *Leptotyphlops macrops*, the eyes are larger and more highly developed.) In addition, all forms have numerous tactile organs housed within the anterior head scales, often visible to the naked eye as tiny, light-colored specks on the external surfaces of the scales. However, several morphological characteristics serve to distinguish leptotyphlopids from anomalepidid and typhlopid blindsnakes. In particular, all slender blindsnakes have either 14 (in *Leptotyphlops*) or 16 (in *Rhinoleptus*) rows of scales encircling the body (all anomalepidids and nearly all typhlopids have more than 16 scale rows), a single anal shield (all anomalepidids and nearly all typhlopids have two or more), and a distinctive arrangement of the scales along the upper lip. Moreover, leptotyphlopids are unique among snakes in having teeth only on the lower jaw.

Slender blindsnakes are generally rather dull in appearance. Although a few South American species (e.g., *Leptotyphlops*

The black threadsnake (*Leptotyphlops nigricans*), one of the world's smallest snakes, with rice grain-size eggs that are strung together like sausages. (Photo by Bill Branch. Reproduced by permission.)

alfredschmidti, L. teaguei, L. tricolor) are boldly patterned with multicolored dorsal stripes, most leptotyphlopids are patternless and have a relatively uniform pink, gray, tan, brown, or black dorsal coloration. Those forms that are pinkish in color, such as the two species found in the southwestern United States (*L. dulcis* and *L. humilis*), bear an uncanny superficial resemblance to earthworms, thus giving rise to another common name for these diminutive serpents, "wormsnakes."

The size and shape of both the snout and tail are somewhat variable within Leptotyphlopidae. Most species of *Leptotyphlops* have relatively blunt, rounded snouts. However, several Old World species (e.g., *L. macrorhynchus, L. parkeri, L. rostratus*) have prominent, hook-shaped snouts, and in two Socotran forms (*L. filiformis* and *L. macrurus*), the snout is both hooked and pointed. A pointed snout is also seen in *Rhinoleptus koniagui*. Such highly derived snout morphologies are less common among New World taxa, but they are seen in a few South American species (e.g., *L. borrichianus* and *L. unguirostris*). As in most other blindsnakes, the scales surrounding the snout in leptotyphlopids are somewhat larger than those surrounding the body, and in at least one species (*L. humilis*), the largest of these scales (the rostral) fluoresces under ultraviolet light. In most taxa, the tail constitutes 5–10% of the snake's total length, but this figure may be as low as 2.1% in short-tailed species (e.g., *L. septemstriatus*), or as high as 18.9% in long-tailed species (e.g., *L. macrurus* and *L. wilsoni*). The tail usually terminates in a small needle- or thorn-shaped apical spine.

Leptotyphlopids are also characterized by a number of distinctive internal anatomical features. The most significant of these relate to the structure of the jaws. The upper jaws are toothless and relatively immobile. In contrast, the lower jaw bears teeth and is highly flexible due to the presence of exceptionally well-developed intramandibular joints, which divide the left and right halves of the lower jaw into separate anterior and posterior segments. Also unusual are the form and position of the hyoid apparatus, which is Y-shaped and

located far behind the head (characteristics also seen in typhlopid blindsnakes). The pelvic apparatus is, in general, more complete than that of other snakes, typically consisting of paired ilia, ischia, pubes, and femora (although in some taxa the pelvis is highly reduced [e.g., *Leptotyphlops macrorhynchus*] or absent [e.g., *L. cairi*]). Even in species that possess well-developed femora, however, the horny spurs on the distal ends of the femora rarely protrude through the skin as they commonly do in other basal snakes (e.g., pipesnakes, boas, pythons). Perhaps the most bizarre osteological feature of Leptotyphlopidae is seen in several Old World species of *Leptotyphlops* (e.g., *L. cairi, L. macrorhynchus, L. nursii,* and *L. occidentalis*), in which much of the skull roof has been lost.

Distribution

Slender blindsnakes have a relatively wide geographical distribution, ranging throughout the Ethiopian and Neotropical regions and extending northward into southern portions of the Palearctic and Nearctic regions as well. All but one of the approximately 93 species of Leptotyphlopidae are contained in the genus *Leptotyphlops*. In the Old World, this genus is distributed throughout Africa and the Arabian Peninsula, with two species (*L. blanfordi* and *L. macrorhynchus*) extending eastward as far as northwestern India. Also, three species (*L. filiformis, L. macrurus,* and *L. wilsoni*) are endemic to the island of Socotra in the northwestern Indian Ocean, and a small number of mainland species are known to inhabit several islands off the coast of Africa (e.g., Pemba and Bioco). In the New World, *Leptotyphlops* ranges throughout most of South America (excluding Chile, southern Argentina, and southern Peru) and all of Central America and Mexico, with two species (*L. dulcis* and *L. humilis*) extending northward into the southwestern United States. In addition, six species are endemic to islands of the West Indies, and several mainland species are known from islands along the coasts of Mexico and Central America. The numbers of Old World and New World species of *Leptotyphlops* are approximately equal. The genus *Rhinoleptus* includes only a single species, *R. koniagui*, which is known from Guinea and Senegal in western Africa. The altitudinal distribution of slender blindsnakes is remarkable given the extraordinarily small size of these ectothermic animals. They have been found at elevations ranging from 250 ft (76 m) below sea level (*L. humilis* in Death Valley, California) to 10,660 ft (3,250 m) above sea level (*L. tricolor* in the Peruvian Andes).

Habitat

Slender blindsnakes are known to occur in a relatively broad array of habitats, including deserts, tropical rainforests, dry woodlands, savannas, plantations, and boulder-strewn mountain slopes. Throughout these many macrohabitats, however, they are generally found within a relatively narrow range of microhabitats. They are most frequently found in shallow soil, amidst leaf litter and other surface debris, or beneath stones or logs. They are also occasionally encountered within rotten logs, anthills, and termite nests. The strong preference that these tiny snakes appear to have for such microhabitats is likely related at least in part to their extremely high surface-to-volume ratios, which make the crucial tasks of regulating body temperature and minimizing evaporative water loss especially challenging. Laboratory experiments on captive animals suggest that the hydric environment is especially important to these fossorial snakes. When placed in enclosures containing soils of different moisture levels, they avoid the drier soils, choosing instead to seek out microenvironments having higher moisture levels. One form, *Leptotyphlops natatrix*, may even be semiaquatic or aquatic. This species, known only from the type specimen collected in Gambia in 1931, has a laterally compressed, oarlike tail (like those seen in sea snakes) and was found in a swamp. Several species of *Leptotyphlops* have also been found climbing trees. It is unclear, however, whether arboreality is common among these snakes, or if they merely occasionally pursue their prey (mainly ants and termites) into trees.

Behavior

Leptotyphlopids are predominantly fossorial snakes. They are most commonly encountered by humans either during the course of digging operations (in some cases as far as 49 ft [15 m] beneath the surface) or after heavy rains have flooded them out of their subterranean retreats. No observations have been made on their burrowing behavior, but it is likely that they make extensive use of preexisting animal burrows and root systems when moving about underground. They can quickly burrow into loose soils such as sand, but they appear to lack the strength necessary to construct their own tunnels in compact soils.

Although these secretive snakes spend most of their lives underground, they do occasionally venture above ground during the evening hours to search for food or mates. When disturbed by potential predators during these above-ground excursions, they immediately attempt to escape into the ground. If this fails, however, they have several additional defensive strategies that they may implement. When restrained, they usually thrash about violently in an attempt to escape. If a snake cannot wiggle free from danger, it will jab its captor with its sharp tail spine and void the contents of its cloaca. As a last resort, some species will become rigid and fake death.

Feeding ecology and diet

Slender blindsnakes feed exclusively on small invertebrate prey. Some species consume a relatively wide variety of such animals, including beetles, caterpillars, centipedes, cockroaches, crickets, fly larvae, harvestmen, millipedes, and spiders. However, the bulk of their diet consists mainly of ant brood and termites. Like other snakes, they rely heavily on chemoreception to find their prey. They are able to follow the pheromone trails of ants and termites with relative ease, allowing them to locate large colonies of these abundant social insects in almost any environment. Once the snakes enter these colonies, they go into a feeding frenzy and quickly gorge themselves, often eating hundreds of prey items in a single meal. They ingest their prey using a unique feeding mechanism, in which the front half of the lower jaw is rapidly flexed in and out of the mouth to ratchet prey into the throat. This mandibular raking mechanism allows leptotyphlopids to feed very rapidly, thereby minimizing the time that they are exposed to the attacks of ants and termites defending their nests.

These tiny snakes also have evolved an elaborate defensive behavior to help protect themselves from the bites and stings of ants. When molested, they briefly retreat from their attackers and coil into a ball. They then expel a mixture of glandular secretions and feces from the cloaca and begin to writhe within their coils, deliberately spreading this mixture over their entire bodies. After several minutes of this, the snakes take on a glossy, silvery appearance. More importantly, however, they emerge from their coils with at least a partial immunity to ant attacks. The secret to this defensive strategy is a mixture of chemicals in the snakes' cloacal secretions that has a strong repellent effect on ants. Once the snakes have applied this "ant repellent," they resume feeding, during which time they are generally left unbothered by the ants.

Reproductive biology

The reproductive biology of slender blindsnakes is poorly known. All species are believed to be oviparous, but detailed data are available only for two South African species (*Leptotyphlops conjunctus* and *L. scutifrons*) and two North American species (*L. dulcis* and *L. humilis*). In these subtropical forms, reproduction is highly seasonal, with courtship and mating occurring in the spring and oviposition occurring in the summer. Clutch size typically ranges between two and seven eggs. However, some species occasionally deposit clutches consisting of only a single egg, and one Latin American species (*L. goudotii*) is known to produce clutches of up to 12 eggs. The elongate, thin-shelled eggs are generally 0.6–1 in (1.5–2.5 cm) in length, but measure only 0.08–0.16 in (0.2–0.4 cm) in width. Natural incubation times are unknown, but one clutch of *L. humilis* eggs incubated in captivity at 86°F (30°C) hatched after 94 days. Hatchling size appears to vary widely between species, ranging from less than 2.4 in (6.1 cm) in some small species to over 4.3 in (11 cm) in larger species.

Conservation status

No species are listed by the IUCN.

Significance to humans

Because of their extremely small size and secretive nature, slender blindsnakes are of no economic significance to humans. However, in areas where they are particularly abundant, they may benefit humans by keeping populations of ants and termites in check.

Species accounts

Texas blindsnake
Leptotyphlops dulcis

TAXONOMY
Leptotyphlops dulcis (Baird and Girard, 1853), between San Pedro and Camanche [Comanche] Springs, Texas. Five subspecies are recognized.

OTHER COMMON NAMES
English: Texas threadsnake, Texas wormsnake; French: Leptotyphlops du Texas; German: Texas-Schlankblindschlange; Spanish: Serpiente-lombriz texana.

PHYSICAL CHARACTERISTICS
2.6–10.7 in (6.6–27 cm) in total length. Tail 5–6% of total length. Midbody diameter 0.06–0.22 in (0.15–0.5 cm). Adult aspect ratio of approximately 50. Pink or reddish brown dorsally, light pink or cream-colored ventrally.

Leptotyphlops dulcis

DISTRIBUTION
Southwestern United States (southern Kansas, central and western Oklahoma, central and western Texas, southern New Mexico, and southeastern Arizona) and northeastern Mexico (northeastern Sonora, northeastern Chihuahua, Coahuila, Nuevo Leon, Tamaulipas, northern Veracruz, San Luis Potosi, and northern Zacatecas).

HABITAT
These snakes inhabit deserts, grassy plains, oak and juniper woodlands, and rock-strewn mountain slopes. They are usually found buried in sandy or loamy soil, or beneath stones, logs, or other surface debris, often near some source of water.

BEHAVIOR
Texas blindsnakes are predominantly fossorial. However, they are occasionally encountered above ground at night or after heavy rains. They move somewhat clumsily above ground, using a combination of undulatory, rectilinear, and concertina locomotion. In the case of the latter, the tail spine may be used as an anchor point.

FEEDING ECOLOGY AND DIET
These snakes feed mainly on ant brood and termites. They swallow ant larvae and pupae whole, but their prey-handling strategies vary when they feed on termites. They always attack termites from behind and sometimes swallow them whole. In some cases, they ingest only the abdomen and thorax and break off the head. In still other instances, the snakes merely chew on the termites, draining their abdominal fluids. Less common prey include ant lions, beetles, caterpillars, cockroaches, earwigs, fly larvae, and spiders. Texas blindsnakes are sometimes observed foraging amidst raiding columns of army ants. Eastern screech owls (*Otus asio*) often capture these snakes alive and bring them back to their nests, where the snakes feed on parasitic invertebrates amidst the nest debris.

REPRODUCTIVE BIOLOGY
Courtship and mating occur throughout the spring and often involve aggregations of more than a dozen individuals. Oviposition usually occurs in June or July. Clutch size ranges between two and seven eggs, each measuring approximately 0.59 by 0.16 in (1.5 by 0.4 cm). Following oviposition, females coil around their eggs, in some cases in close proximity to other brooding females. Hatchlings, measuring 2.6–3 in (6.6–7.6 cm) in length, emerge in late summer.

CONSERVATION STATUS
Not threatened.

SIGNIFICANCE TO HUMANS
None known. ◆

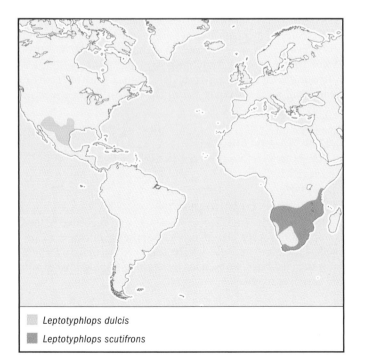

Leptotyphlops dulcis

Leptotyphlops scutifrons

Peters' wormsnake
Leptotyphlops scutifrons

TAXONOMY
Leptotyphlops scutifrons (Peters, 1854), Sena [Mozambique]. Two subspecies are recognized.

OTHER COMMON NAMES
English: Peters' threadsnake, Peters' earthsnake, shielded blindsnake, scaly-fronted wormsnake, glossy wormsnake; German: Glanzende Schlankblindschlange.

PHYSICAL CHARACTERISTICS

2.8–11 in (7–28 cm) in total length. Tail 5–13% of total length. Midbody diameter 0.06–0.16 in (0.15–0.4 cm). Aspect ratio between 40 and 89. Black, dark brown, or reddish brown dorsally (often with pale-edged scales), paler ventrally.

Leptotyphlops scutifrons

DISTRIBUTION

Southern Africa (Republic of South Africa, Swaziland, Namibia, Botswana, Zimbabwe, Mozambique, Angola, Zambia, Malawi, Tanzania, Kenya).

HABITAT

These snakes inhabit mainly savannas, where they are found in soil or beneath stones, logs, and other surface debris.

BEHAVIOR

Peters' wormsnakes are fossorial. They are most frequently found above ground at night after heavy rains.

FEEDING ECOLOGY AND DIET

These snakes feed mainly on the eggs, larvae, and pupae of ants, and occasionally eat termites.

REPRODUCTIVE BIOLOGY

Mating takes place in the spring. Oviposition occurs in the early summer (usually December or January). The eggs, measuring between 0.51 and 0.99 in (1.3–2.5 cm) in length and between 0.09 and 0.16 in (0.2–0.4 cm) in width, are usually deposited in clutches of one to three, although clutches of up to seven eggs have been reported. The elongate eggs are linked together like a string of sausages. The hatchlings, measuring 2.8 in (7.1 cm) or less in length, appear to emerge in late summer or early autumn (February or March).

CONSERVATION STATUS

Not threatened.

SIGNIFICANCE TO HUMANS

None known.

Resources

Books

Greene, H. *Snakes: The Evolution of Mystery in Nature.* Berkeley: University of California Press, 1997.

McDiarmid, R. W., J. A. Campbell, and T. A. Touré. *Snake Species of the World. A Taxonomic and Geographic Reference,* Vol. 1. Washington, DC: Herpetologists' League, 1999.

Shaw, C. E., and S. Campbell. *Snakes of the American West.* New York: Alfred A. Knopf, 1974.

Werler, J. E., and J. R. Dixon. *Texas Snakes: Identification, Distribution, and Natural History.* Austin: University of Texas Press, 2000.

Periodicals

Beebe, W. "Field Notes on the Snakes of Kartabo, British Guiana, and Caripito, Venezuela." *Zoologica* 31 (1946): 11–52.

Brattstrom, B. H., and R. C. Schwenkmeyer. "Notes on the Natural History of the Worm Snake, *Leptotyphlops humilis.*" *Herpetologica* 7 (1951): 193–196.

Broadley, D. G., and S. Broadley. "A Review of the African Worm Snakes from South of Latitude 12°S (Serpentes: Leptotyphlopidae)." *Syntarsus* 5 (1999): 1–36.

Gehlbach, F. R., and R. S. Baldridge. "Live Blind Snakes (*Leptotyphlops dulcis*) in Eastern Screech Owl (*Otus asio*) Nests: A Novel Commensalism." *Oecologica* 71 (1987): 560–563.

Hahn, D. E., and V. Wallach. "Comments on the Systematics of Old World *Leptotyphlops* (Serpentes: Leptotyphlopidae), with Description of a New Species." *Hamadryad* 23 (1998): 50–62.

Hoogmoed, M. S. "On a New Species of *Leptotyphlops* from Surinam, with Notes on the Other Surinam Species of the Genus (Leptotyphlopidae, Serpentes). Notes on the Herpetofauna of Surinam V." *Zoologische Mededelingen* 51 (1977): 99–123.

Klauber, L. M. "The Worm Snakes of the Genus *Leptotyphlops* in the United States and Northern Mexico." *Transactions of the San Diego Society of Natural History* 9 (1940): 87–162.

Kley, N. J., and E. L. Brainerd. "Feeding by Mandibular Raking in a Snake." *Nature* 402 (1999): 369–370.

List, J. C. "Comparative Osteology of the Snake Families Typhlopidae and Leptotyphlopidae." *Illinois Biological Monographs* 36 (1966): 1–112.

Watkins II, J. F., F. R. Gehlbach, and J. C. Kroll. "Attractant-Repellent Secretions of Blind Snakes (*Leptotyphlops dulcis*) and their Army Ant Prey (*Neivamyrmex nigrescens*)." *Ecology* 50 (1969): 1,098–1,102.

Webb, J. K., R. Shine, W. R. Branch, and P. S. Harlow. "Life-History Strategies in Basal Snakes: Reproduction and Dietary Habits of the African Thread Snake *Leptotyphlops scutifrons* (Serpentes: Leptotyphlopidae)." *Journal of Zoology, London* 250 (2000): 321–327.

Nathan J. Kley, PhD

Blindsnakes

(Typhlopidae)

Class Reptilia
Order Squamata
Suborder Serpentes
Family Typhlopidae

Thumbnail description
Small to medium-sized fossorial snakes with smooth, uniformly sized body scales; highly reduced eyes; small, ventrally placed mouth; variably shaped snout; and short tail bearing a sharp terminal spine

Size
2.1–37.4 in (5.3–95 cm)

Number of genera, species
6 genera; 214 species

Habitat
Soil

Conservation status
Endangered: 1 species; Vulnerable: 1 species

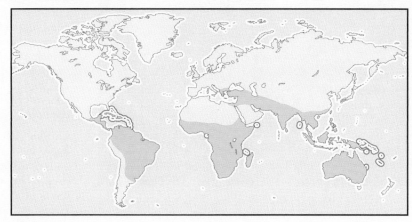

Distribution
Sub-Saharan Africa, Madagascar, southeastern Europe, southern Asia, Australia, southern North America, Central America, the West Indies, and South America

Evolution and systematics

The phylogenetic position of Typhlopidae within the order Squamata has been enigmatic and controversial for more than two centuries. Most early taxonomists grouped typhlopids together with various forms of snake-like lizards, such as amphisbaenians, burrowing skinks, and limbless anguids. This practice continued throughout the early nineteenth century until the publication of the sixth volume of Duméril and Bibron's monumental work, "Erpétologie Général ou Histoire Naturelle Complète des Reptiles" in 1844. Based on extensive anatomical comparisons with other squamates, Duméril and Bibron grouped typhlopids together with leptotyphlopids (to the exclusion of superficially similar limbless lizards) into "Scolécophides," one of five major "sections" that they recognized within Serpentes. This classification scheme soon gained widespread acceptance, and the Latinized name Scolecophidia was introduced in 1864 by Edward Drinker Cope.

Most recent workers have at least tentatively accepted the hypothesis that Scolecophidia represents a natural (monophyletic) group. However, this notion has not gone unchallenged. Several morphological and molecular studies have concluded that Scolecophidia may be polyphyletic, including two influential anatomical studies that suggested that Typhlopidae may be placed improperly within Serpentes (i.e., that typhlopids represent a divergent lineage of limbless lizards). Nevertheless, most cladistic analyses of squamate interrelationships have supported the hypothesis of scolecophidian monophyly. Moreover, the majority of such studies have suggested that within Scolecophidia, Typhlopidae is more closely related to Anomalepididae (early blindsnakes) than to Leptotyphlopidae (slender blindsnakes).

The fossil record for Typhlopidae is extremely poor. Only two fossil taxa have been described. *Typhlops grivensis* is known from three trunk vertebrae found in Middle Miocene deposits in France, and *T. cariei* is known from seven trunk vertebrae discovered in Subrecent deposits on Mauritius. Given that these species were described solely on the basis of vertebral morphology, their placement in the genus *Typhlops* must be considered tentative. Unidentified scolecophidian fossils have also been found in Tertiary deposits in Europe, Africa, Australia, and South America, and it is likely that many of these fragmentary fossils represent typhlopid remains. Unfortunately, the very incomplete scolecophidian fossils that have been unearthed thus far have shed little light on the evolution of blindsnakes. However, some of these fossil remains have improved our understanding of the biogeographical history of blindsnakes.

No subfamilies are recognized.

Physical characteristics

Perhaps the most conspicuous physical characteristic of blindsnakes is their small size. Although Typhlopidae includes a small number of "giant" species that may reach over 2 ft (61 cm) in total length (e.g., *Ramphotyphlops proximus, R. unguirostris, Rhinotyphlops acutus, R. schlegelii, Typhlops angolensis, T. lineolatus,* and *T. punctatus*), most species are much smaller, and many reach a maximum adult length of less than 1 ft (31 cm). The smallest blindsnakes known are hatchling *Ramphotyphlops braminus,* which measure little more than 2 in (5 cm) in length, less than 0.04 in (1 mm) in width, and weigh only 0.005 oz (0.13 g). In contrast, the largest blindsnake, *Rhinotyphlops schlegelii,* reaches a maximum length of 3.1 ft (95 cm),

A blindsnake above ground after a heavy rain in the Solomon Islands. (Photo by Animals Animals ©W. Cheng. Reproduced by permission.)

a width of almost 1.2 in (3 cm), and a weight of nearly 1.1 lb (0.5 kg).

In general, typhlopids are somewhat stouter than other blindsnakes (Anomalepididae and Leptotyphlopidae). However, "aspect ratios" (total length divided by body width) vary widely within this family, ranging from 16 (in the stocky Brazilian species *Typhlops paucisquamus*) to at least 130 (in the extremely slender Australian species *Ramphotyphlops grypus*). In most forms the body is relatively round in cross-section and is nearly uniform in width from the head to the base of the tail. In some larger species, however, the rear part of the body may become markedly distended in adults due to the deposition of fat throughout the posterior portions of the abdominal cavity.

The body scales of typhlopids are smooth, shiny, and strongly overlapping. As in Anomalepididae and Leptotyphlopidae, these scales are more or less circular in shape (at least along their posterior margins) and uniform in size. The only enlarged scales are the anal shields (numbering between one and five) and the scales surrounding the head. The number of longitudinal scale rows encircling the body ranges from 16 to 44, and the number of transverse scale rows along the length of the body ranges from 169 to 709. The body scales are remarkably thick, providing these snakes with an effective line of defense against the bites and stings of the insects on which they feed. This thickening of the epidermis also has a profound effect on ecdysis (shedding of skin). The shed skins of typhlopids (especially those of larger species of *Ramphotyphlops*, *Rhinotyphlops*, and *Typhlops*) are much thicker than those of other snakes and often have a rubbery consistency. In addition, the skin is usually shed in a series of rings rather than in a single piece as in most other snakes.

Like other blindsnakes, typhlopids have small, ventrally placed mouths and highly reduced eyes that are covered by the enlarged scales surrounding the head. However, the general shape of the head is much more variable in Typhlopi-

dae than in Anomalepididae and Leptotyphlopidae. The snout may be bluntly rounded (*Cyclotyphlops*, most *Typhlops*, and many *Ramphotyphlops*), flattened and anteroventrally sloping (*Xenotyphlops* and some *Rhinotyphlops*), acutely conical (*Acutotyphlops*), hooked (e.g., *Ramphotyphlops grypus*), or trilobulate (*Ramphotyphlops bituberculatus*) in shape. In most species epidermal glands are visible along the edges of the anterior head scales. In addition, the head shields house numerous tactile organs, and in some taxa (e.g., *Xenotyphlops* and some *Rhinotyphlops*) flexible, papilla-like structures project externally from the anterior-most scales surrounding the snout.

The tail is relatively short in most typhlopids, and in many species of *Rhinotyphlops* and *Typhlops*, this structure accounts for less than 1% of the snake's total length. However, in some species of *Acutotyphlops* and *Ramphotyphlops* the tail is often somewhat longer (5% or more of total length), and in one species that appears to have strong arboreal tendencies (*Ramphotyphlops cumingii*), the tail may account for as much as 10% of the snake's total length. The tail terminates in a sharp, narrow apical spine in most species, but this feature is absent in some taxa (e.g., *Xenotyphlops* and the two Asian species of *Rhinotyphlops*), and in others it may become somewhat enlarged (e.g., *Typhlops depressiceps* and *Acutotyphlops subocularis*).

Coloration and patterning are highly variable within Typhlopidae. The majority of species are essentially unpatterned, and most are relatively dark in color (some shade of black, gray, or brown). A wider array of colors and patterns are seen in the genera *Rhinotyphlops* and *Typhlops*. Although both of these genera include rather drab, unicolored forms, they also include species boldly patterned with speckles (e.g., *R. schlegelii*), blotches (e.g., *T. congestus*), or stripes (e.g., *R. unitaeniatus*). In such taxa, bright colors (blue, orange, yellow, white) are often incorporated into these elaborate patterns. In most blindsnakes the ventral scales are somewhat lighter in color than the dorsal scales, and in some species (e.g., *T. reticulatus*) this difference is especially striking. Finally, a small number of typhlopids appear to lack pigmentation entirely (e.g., *Xenotyphlops grandidieri*).

The most distinctive internal anatomical features of typhlopids relate to their jaw apparatus. In contrast to other snakes, the lower jaw in Typhlopidae is rigid and toothless. The only teeth in the skull are located on the maxillae, which are suspended from the rest of the skull almost entirely by muscles and ligaments. Also peculiar is the orientation of the maxillae, which are positioned horizontally in the mouth. As a result, the transversely oriented maxillary tooth rows point directly posteriorly toward the throat. As in Leptotyphlopidae, the hyoid (tongue skeleton) is Y-shaped and positioned far behind the head. The pelvic apparatus is highly reduced. In most species only the rod-like ischia become ossified, and the pubes and ilia (when present) are generally fused and remain cartilaginous. Femora are absent. In addition to these skeletal characteristics, several unique soft anatomical features have also been documented in Typhlopidae. Perhaps the most noteworthy among these is the solid distal portion of the hemipenis in males of the genus *Ramphotyphlops*, a feature that is unique among squamates.

Distribution

Typhlopidae is the largest and most widely distributed of the three families of blindsnakes. It includes six genera and approximately 214 species and has a predominantly tropicopolitan distribution. The genus *Acutotyphlops* includes four species that are restricted to easternmost Papua New Guinea, the Bismarck Archipelago, and the Solomon Islands. *Cyclotyphlops* includes a single species, *C. deharvengi*, which is known only from the type specimen collected in southeastern Sulawesi. The genus *Ramphotyphlops* contains approximately 57 species that are distributed throughout Southeast Asia, Australia, New Guinea, the Bismarck Archipelago, the Solomon Islands, New Caledonia, and perhaps other Pacific Islands as well. One species (*R. exocoeti*) is endemic to Christmas Island in the eastern Indian Ocean. In addition, the parthenogenetic flowerpot blindsnake, *R. braminus*, has been introduced throughout much of the world through the horticultural trade and is now established in Africa, Madagascar, the Seychelles, Mexico, Central America, the Hawaiian Islands, Florida, and even Boston, Massachusetts. The genus *Rhinotyphlops* includes 30 species, most of which are restricted to mainland Africa. However, *R. feae* and *R. newtoni* are endemic to the islands of São Tomé and Príncipe in the Gulf of Guinea; *R. simoni* is native to Syria, Jordan, and Israel; and *R. acutus* is endemic to the Indian subcontinent. *Typhlops* is the largest of the six typhlopid genera, containing approximately 121 species. In the Old World, this genus is distributed from southeastern Europe through Southeast Asia (including Sri Lanka and the Indian subcontinent), and is also found throughout subequatorial Africa, Madagascar, and the Middle East. In addition, a small number of species are known from Príncipe Island in the Gulf of Guinea (*T. elegans*) and Socotra (*T. socotranus*), the Comoro Islands, (*T. comorensis*), and the Andaman Islands (*T. andamanesis* and *T. oatesii*) in the Indian Ocean. In the New World, *Typhlops* is confined to the Neotropical region, ranging from southern Mexico southward to northeastern Argentina. In addition, 26 species are endemic to the islands of the West Indies. Finally, the genus *Xenotyphlops* includes a single species, *X. grandidieri*, which is known only from two specimens believed to have been collected in Madagascar.

From a standpoint of species diversity, the worldwide distribution of Typhlopidae is highly uneven; nearly 85% of all recognized species are restricted to the Old World. In addition, approximately 45% of typhlopid species are known only from islands. Typhlopids are known to occur throughout a wide range of elevations, ranging from sea level to at least 7,497 ft (2,285 m) above sea level (*T. meszoelyi* in the Indian Himalayas).

Habitat

Blindsnakes occur in a wide variety of habitats, including tropical rainforests, arid deserts, savannas, dry woodlands, cultivated farmlands, and even coastal beaches. They are most commonly found underground (often during the course of digging operations), or beneath stones, logs, or other surface debris (e.g., bark, leaf litter, coconut husks, etc.). They are also frequently discovered in rotting logs, ant hills, and termitaria. Although most typhlopids are believed to be pre-

A blindsnake (*Typhlops* sp.) on a leaf in a forest in Ecuador. The *Typhlops* are very primitive snakes that spend most of their time underground, where they hunt insects such as ants and termites. Their eyes are poorly developed and can only distiguish between light and dark. The animal burrows through the ground using its head (center right). (Photo by Dr. Morley Read/Science Photo Library/Photo Researchers, Inc. Reproduced by permission.)

dominantly fossorial, some forms are occasionally encountered above ground (often following heavy rains), and several species of *Ramphotyphlops* are commonly found in trees. In fact, one Philippine species (*R. cumingii*) has frequently been discovered in the root masses of bird's nest ferns (*Asplenium nidus*) growing near the top of tall forest trees. It is not known whether such species are truly arboreal, or if they only climb trees occasionally in search of food.

Behavior

The natural behavior of typhlopids is poorly known. Because these small snakes spend most of their lives hidden underground, much of what is known about their behavior derives from observations of the defensive strategies that they implement when they are unearthed by humans. When first exposed, most blindsnakes immediately endeavor to burrow into the soil. If captured, they attempt to escape by frantically thrashing their bodies back and forth, defecating, and voiding the contents of their anal glands. In addition, in many instances the apical tail spine is jabbed forcefully into the captor's skin. No blindsnakes are known to bite in defense, but some will gape widely when restrained, and a small number of *Rhinotyphlops* and *Ramphotyphlops* have been reported to emit faint squeaking sounds when handled roughly.

One additional aspect of typhlopid behavior that has been noted with some frequency is the tendency of these snakes to congregate in relatively large groups. This phenomenon, which has also been reported for several leptotyphlopid taxa, has been documented in numerous species of *Ramphotyphlops* and *Typhlops*, and may be common among other typhlopids as well. In some instances, more than 20 snakes have been found coiled together beneath a single stone. These congregations do not appear to be related to reproduction, as they usually include both juveniles and adults. Instead, they appear to result from multiple individuals seeking out and utilizing the same favorable microhabitat.

Feeding ecology and diet

Like other scolecophidian snakes, typhlopids feed exclusively on small invertebrate prey. Most taxa feed predominantly on the larvae and pupae of ants, although many species frequently eat termites as well. Several Australian species of *Ramphotyphlops* are also known to occasionally include earthworms in their diets, and at least one species of *Acutotyphlops* (*A. subocularis*) is believed to feed exclusively on earthworms.

Typhlopids feed with astonishing rapidity, in some cases ingesting nearly 100 prey items per minute. They transport their prey using a peculiar feeding mechanism in which the toothed maxillae are rotated in and out of the mouth as many as five times per second. This rapid maxillary raking mechanism allows typhlopids to quickly gorge themselves once they have located an abundant source of food, such as an ant nest or termite mound. Large, heavy-bodied species of *Ramphotyphlops*, *Rhinotyphlops*, and *Typhlops* may consume hundreds of ant larvae and pupae within a single nest before becoming satiated.

Reproductive biology

Like many other aspects of their natural history, the reproductive biology of blindsnakes is poorly understood. Studies of preserved museum specimens suggest that most typhlopids are oviparous. In the few instances in which clutches of eggs have been deposited in captivity, incubation periods have generally ranged between one and two months. However, the eggs of one African species (*Typhlops bibronii*) have extraordinarily thin shells and hatch within a week of deposition, and one specimen of *T. diardii* from Vietnam was found to contain 14 full-term embryos, suggesting that this species may be viviparous throughout at least some parts of its range. The most divergent reproductive strategy among typhlopids is that of *Ramphotyphlops braminus*, an all-female, triploid species that reproduces parthenogenetically.

Clutch size is highly variable within Typhlopidae. Small individuals of relatively slender species (e.g., *Ramphotyphlops braminus*, *R. wiedii*) may deposit only a single egg per clutch. In contrast, clutches of more than 50 eggs have been reported

for the "giant" African species *Rhinotyphlops schlegelii*. Egg size is similarly variable, although precise measurements have been reported only rarely. In the slender species *Ramphotyphlops braminus*, eggs average approximately 0.16 in (4 mm) in width, 0.59 in (15 mm) in length, and 0.006 oz (0.18 g) in mass. However, in the more heavy-bodied Australian species *Ramphotyphlops nigrescens*, eggs may be as large as 0.47 in (12 mm) in width, 0.95 in (24 mm) in length, and more than 0.07 oz (2 g) in mass.

Reproduction appears to be highly seasonal in the subtropical species of Typhlopidae that have been studied in detail. In Australian *Ramphotyphlops*, South African *Rhinotyphlops* and *Typhlops*, and a Japanese population of *Ramphotyphlops braminus*, mating usually occurs in late spring (except in the parthenogenetic *R. braminus*) and oviposition occurs in the summer. *Ramphotyphlops braminus* appears to reproduce throughout the year in the Seychelles, suggesting that tropical populations and species may not exhibit such seasonal reproductive cycles.

Mating behavior has been observed only in *Typhlops vermicularis*. In this species, the male wraps several tight coils around the posterior portion of the female's body during copulation. This behavior presumably allows the male to keep the female's cloaca positioned properly during mating. In some instances, the male and female become intertwined with one another during copulation. On other occasions, the snakes may face in opposite directions while mating.

Conservation status

Only two of the approximately 214 species of blindsnakes are listed by the IUCN. The Mona Island blindsnake (*Typhlops monensis*) is listed as Endangered, and the Christmas Island blindsnake (*Ramphotyphlops exocoeti*) is listed as Vulnerable due to its severely limited range. In addition, *T. cariei*, known only from Subrecent fossil material from Mauritius, is listed as Extinct. However, nearly 20% of the recognized species of blindsnakes are known only from their type localities, and thus the status of many species is uncertain.

Significance to humans

Although some species of Typhlopidae have become incorporated into the legends and folklore of many African and Asian cultures, blindsnakes are of no economic significance to humans.

Species accounts

Schlegel's blindsnake
Rhinotyphlops schlegelii

TAXONOMY
Rhinotyphlops schlegelii Bianconi, 1847, Inhambane, Mozambique. Four subspecies are recognized (although some or all of these may represent distinct species).

OTHER COMMON NAMES
English: Schlegel's beaked snake, giant blindsnake; French: Typhlops de Schlegel; German: Afrikanische Blindschlange.

PHYSICAL CHARACTERISTICS
This species ranges between 4.5–37.4 in (11.5–95.0 cm) in total length and between 0.14–1.1 in (3.5–28.1 mm) in midbody diameter. The tail is short, usually 1–2% of total length. Aspect ratios range from less than 20 to more than 50, but average approximately 30. Females grow longer and heavier than males, but have

Rhinotyphlops schlegelii

slightly shorter tails. The large rostral scale is strongly angled and heavily keratinized, forming a prominent snout with a sharp, horizontally oriented cutting edge. The eyes are relatively distinct, usually lying beneath the suture between the preocular and ocular scales. There are 30–44 longitudinal scale rows at midbody and 307–624 scales along the dorsal midline. Dorsal color and pattern are highly variable (especially in *R. s. mucruso*). Blotched, speckled, lineolate, and unicolor morphs are known, exhibiting a relatively wide array of colors (e.g., brown, black, gray, blue, yellow, white). The belly is generally unpatterned and is usually some shade of yellow or white.

DISTRIBUTION
Schlegel's blindsnake ranges widely throughout sub-Saharan Africa (Sudan, Ethiopia, Somalia, Kenya, Uganda, Democratic Republic of Congo, Angola, Zambia, Malawi, Tanzania, Mozambique, Zimbabwe, Botswana, Namibia, Republic of South Africa, Swaziland).

HABITAT
This blindsnake inhabits mainly wooded savannas. It is usually found amid soil or underneath stones, logs, and other surface debris.

BEHAVIOR
Schlegel's blindsnake is fossorial. Smaller individuals are frequently seen above ground after heavy rains. Large adults are encountered less frequently, suggesting that they may dig deeper underground than juveniles.

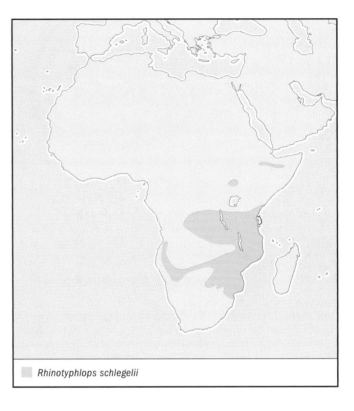

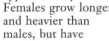
Rhinotyphlops schlegelii

FEEDING ECOLOGY AND DIET
This blindsnake feeds mainly on the larvae and pupae of ants, although termites are also frequently eaten.

REPRODUCTIVE BIOLOGY
Mating and oviposition apparently occur most frequently in late spring and early summer, respectively. However, this species may breed throughout the year in at least some areas of its range. Clutch size ranges from fewer than 10 to more than 50 eggs, each of which may be as large as 0.47 × 0.83 in (12 × 21 mm). The eggs hatch after a relatively short incubation period of four to six weeks.

CONSERVATION STATUS
Not listed by the IUCN.

SIGNIFICANCE TO HUMANS
None known. ◆

Blackish blindsnake
Ramphotyphlops nigrescens

TAXONOMY
Ramphotyphlops nigrescens Gray, 1845, Australia, Paramatta [Parramatta, New South Wales]. No subspecies are recognized.

OTHER COMMON NAMES
English: Eastern blindsnake.

PHYSICAL CHARACTERISTICS

This blindsnake has a total length of 3.8–22.7 in (9.7–57.6 cm). Aspect ratios typically range between 30 and 60. Females grow significantly longer and heavier than males. The snout is bluntly rounded and the eyes are distinctly visible beneath the ocular scales. There are 22 longitudinal scale rows at midbody. The color is pinkish-brown, purple, or black dorsally, and pink or cream-colored ventrally. The tail often has dark patches on either side of the vent.

Ramphotyphlops nigrescens

DISTRIBUTION

Eastern Australia, from southeastern Queensland to Victoria.

HABITAT

The blackish blindsnake occurs in a variety of habitats, from coastal forests to cultivated farmlands. It is relatively common in areas of high human population densities (e.g., Sydney and Brisbane) and is frequently discovered beneath rocks and logs in such areas.

BEHAVIOR

Like other blindsnakes, the blackish blindsnake is fossorial. It frequently congregates in relatively large groups of up to 30 or more individuals. Occasionally, individuals are seen crawling above ground after prolonged drenching rains. One specimen was found 16.4 ft (5 m) above ground in a she-oak (*Casuarina*).

FEEDING ECOLOGY AND DIET

The blackish blindsnake feeds almost exclusively on ant brood, although other small invertebrates (e.g., earthworms, leeches) may also be eaten. Ant nests are located by following pheromone trails laid down by adult ants. As many as 1,500 or more ant larvae and pupae may be consumed in a single meal. The species generally feeds only during the spring and summer.

REPRODUCTIVE BIOLOGY

Reproduction in this southern species is highly seasonal. Mating occurs most commonly in late spring, and eggs are deposited during the summer. Clutch size ranges from five to 20 eggs, each weighing 0.04–0.07 oz (1–2 g) and measuring 0.75–1.3 in (19–34 mm) in length and 0.32–0.47 in (8–12 mm) in width. Artificially incubated eggs have taken between 30 and 72 days to hatch (depending on temperature), but natural incubation periods are unknown.

CONSERVATION STATUS

Not listed by the IUCN.

SIGNIFICANCE TO HUMANS

None known. ◆

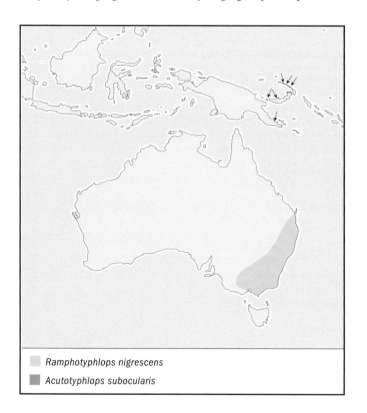

⬜ *Ramphotyphlops nigrescens*

⬛ *Acutotyphlops subocularis*

Bismarck blindsnake
Acutotyphlops subocularis

TAXONOMY

Acutotyphlops subocularis Waite, 1897, Duke of York Island (Bismarck Archipelago). No subspecies are recognized.

OTHER COMMON NAMES

None known.

PHYSICAL CHARACTERISTICS

This species has a total length of 7.5–15.5 in (19.1–39.4 cm). The tail is 3.0–4.5% of total length in females, but may be as long as 6.3% of total length in males. Males also have a greater number of subcaudal scales (22–31) than females (14–23). In both sexes, the tail termi-

Acutotyphlops subocularis

nates in a relatively large, thornlike apical spine. Aspect ratios range from 23 to 44, but average approximately 32. The snout is somewhat pointed, especially in lateral view (but not as acutely conical as in *A. infralabialis* and *A. kunuaensis*). The lower jaw is V-shaped in ventral view. The small eyes are visible beneath the ocular scale. There are 32–36 longitudinal scale rows at midbody, and there are 363–472 middorsal scales. The color is dark brown dorsally, and golden-yellow ventrally.

DISTRIBUTION
Bismarck Archipelago (New Britain, New Ireland, Umboi, and Duke of York Islands) and eastern Papua New Guinea (Morobe Province).

HABITAT
Not known.

BEHAVIOR
Little is known about the behavior of this species. However, numerous morphological features (e.g., pointed head, solidly constructed bony snout, thick neck) suggest that these snakes may be particularly powerful burrowers.

FEEDING ECOLOGY AND DIET
This is the only species of blindsnake known to feed exclusively (or at least predominantly) on earthworms. Based on the condition of prey found in the stomachs of preserved museum specimens, it appears that worms are swallowed whole.

REPRODUCTIVE BIOLOGY
The reproductive biology is not known, but the closely related *A. kunuaensis* from Bougainville Island is believed to deposit relatively small clutches (often consisting of only one or two eggs) in August or September.

CONSERVATION STATUS
Not listed by the IUCN.

SIGNIFICANCE TO HUMANS
None known. ◆

Resources

Books

Cogger, H. G. *Reptiles and Amphibians of Australia.* Fifth edition. Ithaca, NY: Comstock Publishing Associates/Cornell University Press, 1994.

Ehmann, H., and M. J. Bamford. "Family Typhlopidae." In *Fauna of Australia.* Vol. 2A, *Amphibia & Reptilia,* edited by C. J. Glasby, G. J. B. Ross, and P. L. Beesley. Canberra: Australian Government Publishing Service, 1993.

FitzSimons, V. F. M. *Snakes of Southern Africa.* Cape Town and Johannesburg: Purnell and Sons, 1962.

Greer, A. E. *The Biology and Evolution of Australian Snakes.* Chipping Norton, New South Wales, Australia: Surrey Beatty & Sons, 1997.

McDiarmid, R. W., J. A. Campbell, and T. A. Touré. *Snake Species of the World: A Taxonomic and Geographic Reference.* Volume 1. Washington, DC: Herpetologists' League, 1999.

Schwartz, A., and R. W. Henderson. *Amphibians and Reptiles of the West Indies: Descriptions, Distributions, and Natural History.* Gainesville: University of Florida Press, 1991.

Periodicals

Dixon, J. R., and F. S. Hendricks. "The Wormsnakes (Family Typhlopidae) of the Neotropics, Exclusive of the Antilles." *Zoologische Verhandelingen* 173 (1979): 1–39.

Kley, N. J. "Prey Transport Mechanisms in Blindsnakes and the Evolution of Unilateral Feeding Systems in Snakes." *American Zoologist* 41 (2001): 1321–1337.

List, J. C. "Comparative Osteology of the Snake Families Typhlopidae and Leptotyphlopidae." *Illinois Biological Monographs* 36 (1966): 1–112.

McDowell, S. B. "A Catalogue of the Snakes of New Guinea and the Solomons, with Special Reference to Those in the Bernice P. Bishop Museum. Part I. Scolecophidia." *Journal of Herpetology* 8 (1974): 1–57.

Richter, R. "Aus dem Leben der Wurmschlangen." *Natur und Volk* 85 (1955): 360–363.

Roux-Estève, R. "Révision systématique des Typhlopidae d'Afrique Reptilia-Serpentes." *Mémoires du Muséum National D'Histoire Naturelle, Série A, Zoologie* 87 (1974): 1–313.

Shine, R. and J. K. Webb. "Natural History of Australian Typhlopid Snakes." *Journal of Herpetology* 24 (1990): 357–363.

Wallach, V. "The Status of the Indian Endemic *Typhlops acutus* (Duméril & Bibron) and the Identity of *Typhlops psittacus* Werner (Reptilia, Serpentes, Typhlopidae)." *Bulletin De L'Institut Royal Des Sciences Naturelles De Belgique, Biologie* 64 (1994): 209–229.

———. "A New Genus for the *Ramphotyphlops subocularis* Species Group (Serpentes: Typhlopidae), with Description of a New Species." *Asiatic Herpetological Research* 6 (1995): 132–150.

Wallach, V., and I. Ineich. "Redescription of a Rare Malagasy Blind Snake, *Typhlops grandidieri* Mocquard, with Placement in a New Genus (Serpentes: Typhlopidae)." *Journal of Herpetology* 30 (1996): 367–376.

Webb, J. K., W. R. Branch, and R. Shine. "Dietary Habits and Reproductive Biology of Typhlopid Snakes from Southern Africa." *Journal of Herpetology* 35 (2001): 558–567.

Nathan J. Kley, PhD

False blindsnakes

(*Anomochilidae*)

Class Reptilia
Order Squamata
Suborder Serpentes
Family Anomochilidae

Thumbnail description
Superficially like small pipesnakes
(Cylindrophiidae) that have a short mouth, tiny
eyes, and lack a mental groove

Size
8–14 in (20–36 cm)

Number of genera, species
1 genus; 2 species

Habitat
Probably leaf litter, loose soils, but little known

Conservation status
Data Deficient: 2 species

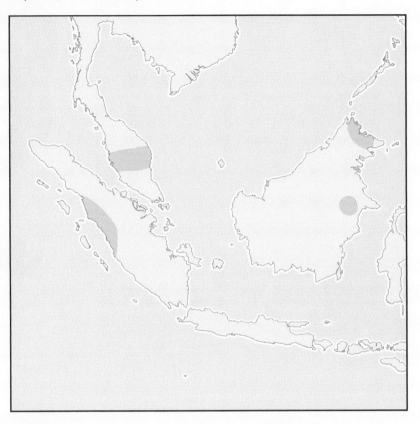

Distribution
Sumatra, Malaysian Peninsula, and Borneo

Evolution and systematics

Although the genus *Anomochilus* is known from very few specimens, none of which have been kept alive, a detailed anatomical study of a single specimen of *A. weberi* suggested

Close-up of a false blindsnake head (*Anomochilus leonardi*) from Sabah, East Malaysia. (Photo by Dr. Indraneil Das. Reproduced by permission.)

that this genus is the most primitive living member of the group (clade) to which all of the large-jawed ("true") snakes (Alethinophidia) belong. Current systematic treatments therefore place the genus among the most basal members, or as the most basal member, of the group. No subfamilies are recognized.

Physical characteristics

These small snakes have blunt heads, short tails, and cylindrical bodies of uniform diameter. The mouth is short and slightly subterminal, the eyes are small, and the spectacle is reduced or absent. Both species have 17–19 scale rows (scale row formulae of 17–19–17, 17–17–17, 17–17–15 reported) and the number of ventrals ranges from 222–252 in *A. leonardi* and 236–248 in *A. weberi*. In keeping with its short tail, there are only six to eight subcaudal scales.

Distribution

These snakes are found in Sumatra, Malaysian Peninsula, and Borneo. However, the few known specimens have a

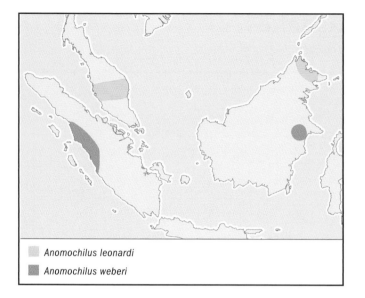

Anomochilus leonardi

Anomochilus weberi

The false blindsnake *Anomochilus leonardi*. (Illustration by Emily Damstra)

strange pattern of distribution. Specimens of *A. weberi* come from west-central Sumatra and south-central Kalimantan, Borneo, and the specimens of *A. leonardi* were caught in central peninsular Malaysia and in Sabah, near the northeast coast of Borneo. It seems likely that these specimens represent either a single species or a number of species different from current accounts.

Habitat

Very little is known about this family's habitat, but it probably includes leaf litter and loose soils.

Behavior

Not known.

Feeding ecology and diet

Both species are assumed to feed on invertebrates (worms, insect larvae) but no stomach contents are known.

Reproductive biology

The presence of shelled eggs in one female suggests that the genus is probably oviparous.

Conservation status

Listed as Data Deficient by the IUCN.

Significance to humans

None known.

The false blindsnake *Anomochilus weberi*. (Illustration by Emily Damstra)

Resources

Books

David, P., and G. Vogel. *The Snakes of Sumatra*. Frankfurt am Main, Germany: Édition Chimaira, 1996.

Greene, H. W. *Snakes: The Evolution of Mystery in Nature.* Berkeley: University of California Press, 1999.

Periodicals

Cundall, D., and D. A. Rossman. "Cephalic Anatomy of the Rare Indonesian Snake *Anomochilus weberi*." *Zoological Journal of the Linnean Society* 109 (1993): 235–273.

Cundall, D., V. Wallach, and D. A. Rossman. "The Systematic Relationships of the Snake Genus *Anomochilus*." *Zoological Journal of the Linnean Society* 109 (1993): 275–299.

Steubing, R. B., and R. Goh. "A New Record of Leonard's Pipe Snake *Anomochilus leonardi* Smith (Serpentes: Uropeltidae: Cylindrophinae) from Sabah, Northwestern Borneo." *Raffles Bulletin of Zoology* 41 (1993): 311–314.

David Cundall, PhD

Shieldtail snakes

(*Uropeltidae*)

Class Reptilia
Order Squamata
Suborder Serpentes
Family Uropeltidae

Thumbnail description
Small, cylindrical, smooth-scaled snakes; many
species have conical, pointed heads smaller
than the anterior trunk and a blunt or even an
obliquely flattened tail covered with one or more
enlarged scales decorated with keels or spines

Size
7–23 in (18–58 cm) total length

Number of genera, species
8 genera; 47 species

Habitat
Moist soils of lowland and upland forests, under
logs and loose cover, and among roots of trees
and shrubs; many species occur in relatively
compacted clay soils in which they actively
burrow

Conservation status
Not classified by IUCN

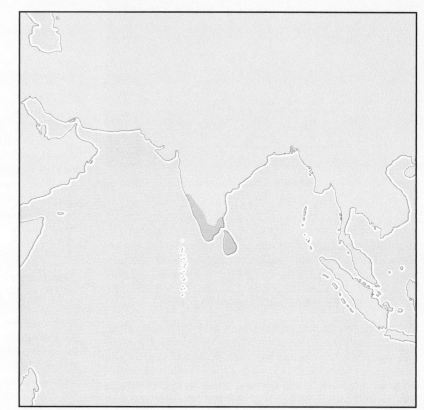

Distribution
Sri Lanka and southern India

Evolution and systematics

No fossil uropeltids are known. Their current distribution
suggests a relict pattern except that their phylogenetic relationships appear to be with the most basal of living
alethinophidian ("true," or large-jawed) snakes. Both molecular and morphological features support affinity with cylindrophiids. No subfamilies are recognized.

Physical characteristics

Uropeltids display a surprising diversity of external features despite their superficial resemblance. All are relatively
small snakes, most having adult sizes below 12 in (30 cm) total length. A few species (e.g., *Rhinophis oxyrhynchus*, *Uropeltis
ocellatus*) attain sizes nearly twice this length. In all uropeltids
the eye is covered by an ocular shield with no separate spectacle. Instead, the region of the ocular shield overlying the
eye is transparent. In species other than *Platyplectrurus*, the
head tapers from a wider anterior trunk, and the snout is flattened either horizontally (dorsoventrally), as in *Melanophidium* and *Platyplectrurus*, or vertically (mediolaterally), as in
most species of *Rhinophis*. Species with pointed snouts have

modified rostral scales with thickened keratinized layers. In
Rhinophis, one of the largest members of the family, the rostral scale is greatly enlarged and forms a prominent ridge over
the dorsal surface of the snout. At the caudal end of these
snakes, the tail has a variety of scale shapes, from pointed
(*Teretrurus sanguineus*, *Brachyophidium rhodogaster*) to an
obliquely flattened, very blunt end (most species of *Rhinophis*
and *Uropeltis*). Uropeltids are designated "shieldtail" snakes
because many species have a greatly enlarged terminal scale
that has numerous spines or keels. The terminal scale in many
is preceded by a region of thickened, keeled scales that form
an oval, obliquely flattened surface. The keels and spines on
these scales become encrusted with soil particles and, in wet
soils, an appreciable plug of soil several millimeters thick may
tightly adhere to the modified scales of the tail tip. Carl Gans
has suggested that the soil plug may serve in defense against
some types of burrowing predators.

Uropeltid skin colors vary from browns, grays, and black,
often with light yellow or white scale edges on the ventral surface (many species), to dark, iridescent blue dorsally and a
bright yellow with darker spots ventrally (*Uropeltis myhendrae*).
A number of species are brown with dark bands rising from the

A large-scaled shieldtail (*Uropeltis macrolepis macrolepis*). Photo by A. Captain/R. Kulkarni/S. Thakur. Reproduced by permission.

ventral surface and appear superficially similar to some species of large centipedes. Gans suggested that some Sri Lankan species may mimic small snakes of the family Elapidae, and tests with domestic chickens suggested that some ground-feeding birds, like jungle fowl, may avoid exposed uropeltids.

Among the more unusual features of uropeltids is the modification of their trunk musculature into an anterior region of red fibers rich in myoglobin and mitochondria, whereas the remainder of the trunk muscles are white. These snakes burrow by forcing their heads through the soil, anchoring the anterior trunk with tight bends of the vertebral column, then straightening the trunk anterior to these bends. Much of the work during burrowing is thus apparently done by the anterior trunk and its fatigue-resistant red muscles.

Although uropeltids retain many primitive features in skeletal and muscle arrangements, they have no pelvic vestiges, no premaxillary teeth, and reduced left lungs. Among uropeltid genera considered derived for the group (*Rhinophis* and *Uropeltis*), the maxilla is firmly anchored to the premaxilla, and many of the bones at the rear of the skull are fused. In more basal uropeltids, the maxilla is free of the premaxilla, and there is less fusion of skull elements.

Distribution

Shieldtail snakes are found in western and eastern Ghats, and the Anaimalai and Nilgiri Hills of southern and west-central India; some species reach the lowlands on the western coast of India near Bombay, and the lowlands and hills of Sri Lanka. Three of the eight genera have representatives in both India and Sri Lanka, but only one of the 47 species (*Platyplectrurus madurensis*) is reported from both.

Habitat

Shieldtail snakes are assumed to have originally inhabited moist soils of montane forests. Most species now occur in soils

of tea plantations, kitchen gardens, and irrigated agricultural lands. It has been suggested that ranges of most species have been diminished by widespread deforestation of southern India and Sri Lanka. Although a few species are widespread and occur in lowlands, most appear to be limited to montane areas in which the temperature rarely exceeds 77°F (25°C). Many species are active at temperatures below 68°F (20°C) and appear to be intolerant of both temperature and humidity fluctuations.

Behavior

These small snakes are most often seen after rains when they emerge from their burrows. When caught, they tend to coil the anterior trunk around whatever is holding them and wave the caudal trunk around as though it were a head. When placed on the ground, they move rapidly with the head pointed downward until a soft area is found. They then burrow rapidly into the soil.

Feeding ecology and diet

All species appear to feed primarily on worms; a few species also eat caterpillars and termites. At least one species, *Uropeltis rubromaculata*, has been seen to emerge during rain to feed on earthworms. Although some species feed readily in captivity, there are no published details on how they feed, although different species of *Uropeltis* have been seen catching worms either at one end or in the middle. Worms caught in the middle were folded in half as they were dragged backward into a burrow. Feeding trials in captive specimens of seven or eight species showed that all ate earthworms and one, *U. maculatus*, also ate termites.

Reproductive biology

Shieldtail snakes are viviparous, producing relatively small numbers of relatively large young. The largest clutch size recorded is nine for oviductal eggs in *Uropeltis*, but litter sizes for most species range from two to five, depending on the size of the female. Reproductive seasons and reproductive behavior are poorly known. Young for different species have been recorded as being found anywhere from March to September, and females with embryos have been collected in most of the intervening months. In most species embryos are found in the right oviduct, but in a few species (e.g., *U. phipsonii*) embryos have been found in the left oviduct.

Conservation status

Many species would appear to be threatened by habitat destruction, but there are no detailed population studies.

Significance to humans

None known.

Species accounts

Nilgiri burrowing snake
Plectrurus perrotetii

TAXONOMY
Plectrurus perrotetii Dumeril, 1851, Nilgiri Hills, Tamil Nadu State, southwestern India. No subspecies are recognized.

OTHER COMMON NAMES
None known.

PHYSICAL CHARACTERISTICS

Plectrurus perrotetii

Moderate to large uropeltids, reaching a length of 17.3 in (44 cm), either uniformly brown or purplish above and yellowish or paler brown below, or with paler spots in the center of every scale. Head slightly compressed dorsoventrally, tail compressed laterally. Tail tip formed by a laterally compressed, cup-shaped single scale with two larger spines one above the other and smaller spines on the lateral surfaces. The body is cylindrical, with ventrals about one and one-half times the width of the dorsal scales. Like species of *Platyplectrurus* and *Teretrurus*, *Plectrurus perrotetii* has supraocular scales separating the ocular shield from the frontal.

DISTRIBUTION
Nilgiri and Anamalai Hills at high elevations.

Plectrurus perrotetii
Uropeltis phipsonii

HABITAT
Most specimens have come from cultivated areas or gardens. Common in heavily manured soils at depths of 4–6 in (10–15 cm).

BEHAVIOR
Rajendran reports that *P. perrotetii* will move out of their burrows into heaps of dung left on the soil by farmers, particularly when air temperatures fall.

FEEDING ECOLOGY AND DIET
This species feeds primarily on worms.

REPRODUCTIVE BIOLOGY
Viviparous, producing three to six young born in July or August.

CONSERVATION STATUS
Not listed by IUCN.

SIGNIFICANCE TO HUMANS
None known. ◆

Phipson's shieldtail snake
Uropeltis phipsonii

TAXONOMY
Uropeltis phipsonii Mason, 1888, "Bombay Ghats." No subspecies are recognized.

OTHER COMMON NAMES
None known.

PHYSICAL CHARACTERISTICS
One of the larger species of uropeltids, reaching over 11.8 in (30 cm) in length. These snakes are dark brown above, paler below, with yellow stripes on the tail that meet ven-

Uropeltis phipsonii

trally at the anal scales and a variable number of yellowish triangles that extend upward from the ventral scales in the anterior trunk. The ventral and adjacent scale rows have dark brown bases (anterior half) and yellow edges. The ventrals are nearly twice the width of the adjacent scales. The head is distinctly smaller than the neck (anterior trunk). The rostral scale is large, caps the entire front of the snout, and extends caudally between the nasal scales. The eye is relatively large, occupying half or more of the ocular shield, which meets the frontal scale. There are no supraocular scales. The caudal shield is large and either flat or depressed, composed of about 40 strongly bicarinate scales and a large terminal scute.

DISTRIBUTION
From sea level to about 1,640 ft (500 m) in the Western Ghats of southern India.

HABITAT

Originally thick forest but replaced by rubber plantations. The soils these snakes live in become extremely hard during the dry season (which lasts about three months).

BEHAVIOR

Uropeltis phipsonii burrows well and has been seen in captivity to move both forward and backward in its tunnels.

FEEDING ECOLOGY AND DIET

Stomach contents of wild-caught specimens contained only earthworms and humus, but captive specimens may eat earwigs as well. In captivity, this species will emerge from tunnels at night to forage on the surface. Earthworms thrown on the surface are immediately grasped and pulled back into the tunnel.

REPRODUCTIVE BIOLOGY

Gives birth in March or April. Largest number of oviductal embryos found in a uropeltid was nine, in this species.

CONSERVATION STATUS

Not listed by IUCN.

SIGNIFICANCE TO HUMANS

None known.

Resources

Books

Deraniyagala, P. E. P. *A Colored Atlas of Some Vertebrates from Ceylon.* Vol. 3, *Serpentoid Reptilia.* Colombo, Sri Lanka: Government Press, 1955.

Frank, N., and E. Ramus. *A Complete Guide to Scientific and Common Names of Reptiles and Amphibians of the World.* Pottsville, PA: NG Publishing, 1996.

McDiarmid, R. W., J. A. Campbell, and T. A. Toure. *Snake Species of the World,* Vol. 1. Washington, DC: The Herpetologists' League, 1999.

Pough, F. H., R. M. Andrews, J. E. Cadle, M. L. Crump, A. H. Savitzky, and K. D. Wells. *Herpetology,* 2nd ed. Upper Saddle River, NJ: Prentice Hall, 2001.

Rajendran, M. V. *Studies in Uropeltid Snakes.* Madurai, India: Madurai Kamaraj University Publications, 1985.

Smith, M. A. *The Fauna of British India, Ceylon and Burma, Including the Whole of the Indo-Chinese Sub-Region. Reptilia and Amphibia.* Vol. 3, *Serpentes.* London: Taylor and Francis, 1943.

Periodicals

Cadle, J. E., H. C. Dessauer, C. Gans, and D. F. Gartside. "Phylogenetic Relationships and Molecular Evolution in Uropeltid Snakes (Serpentes: Uropeltidae): Allozymes and Albumin Immunology." *Biological Journal of the Linnean Society* 40 (1990): 293–320.

Cundall, D., V. Wallach, and D. A. Rossman. "The Systematic Relationships of the Snake Genus *Anomochilus*." *Zoological Journal of the Linnean Society* 109 (1993): 275–299.

Gans, Carl. "Aspects of the Biology of Uropeltid Snakes." *Linnean Society Symposia* 3 (1976): 191–204.

Gans, C., H. C. Dessauer, and D. Baic. "Axial Differences in the Musculature of Uropeltid Snakes: The Freight-Train Approach to Burrowing." *Science* 199 (1978): 189–192.

David Cundall, PhD

Pipe snakes
(*Cylindrophiidae*)

Class Reptilia
Order Squamata
Suborder Serpentes
Family Cylindrophiidae

Thumbnail description
Small to moderate-sized, southeast Asian cylindrical-bodied snakes with ventral scales barely wider than the first dorsal scale row, short tails, and heads not demarcated from the trunk

Size
1–3 ft (0.4–1 m) total length

Number of genera, species
1 genus; 9 species

Habitat
Lowland forests, rice fields, but also urban areas; live predominately in leaf litter or loose soil and may be more common near water

Conservation status
Not classified by the IUCN

Distribution
Sri Lanka, Indonesia, Borneo, Myanmar, southern China, Thailand, Laos, Cambodia, Vietnam

Evolution and systematics

There are no fossils definitely assignable to this family and hence there is no evidence of its geological age. Molecular and morphological data suggest that cylindrophiids are distantly related to other basal alethinophidians (uropeltids, aniliids) as well as to basal macrostomatans. Current phylogenies place cylindrophiids near uropeltids and aniliids. No subfamilies are recognized.

Physical characteristics

Color patterns among the species vary, but most have a dark ground color (black or dark brown) with 40–60 yellow or red half-bands on the ventral surface that extend dorsally to different scale rows in different species. Some species have either longitudinal stripes (*Cylindrophis lineatus*) or a row of small light spots on either side of the dorsal midline (*C. engkariensis*). They differ from more basal alethinophidians (anomochilids and uropeltids) in having a fully toothed palate, using constriction to restrain prey, and in exploiting small vertebrates as prey. All species have a distinct mental groove in the lower jaw, no loreal or preocular scales (the prefrontals

meet the supralabial scales), nostril located within a single nasal scale, 17–23 dorsal scale rows (most species have 19, 21, or 23; the most recently described species from Sarawak, Borneo, *C. engkariensis*, has 17 scale rows, the same number found in anomochilids), short tails (five to seven subcaudals), and a

Blotched pipe snake (*Cylindrophis maculatus*). (Illustration by Bruce Worden)

Red-tailed pipe snake (*Cylindrophis ruffus*). (Illustration by Bruce Worden)

spectacle covering a very small eye with a round or vertically subelliptical pupil. The skull is marked by relatively small numbers of robust teeth on the maxillae, dentaries, palatines, and pterygoids. There are no premaxillary teeth. A postorbital bone is present but typically floats in connective tissue and may be lost in many prepared skulls. The lower jaw has a highly mobile intramandibular joint but tight attachment between the tips of the mandibles.

Distribution

The distribution of the family is unusual. One species, *C. maculatus*, occurs in the lowlands of Sri Lanka. With the exception of *C. ruffus*, all of the other species occur on Indonesian islands and Borneo (shared by Indonesia and Malaysia). The rest of the range of the family is accounted for by *C. ruffus*, which is found from Myanmar east through southern China (including Hainan Island), Thailand, Laos, Cambodia, Vietnam, and south through the Malaysian peninsula, Sumatra, Borneo, Java, Sulawesi, Sula, and Buton. The most easterly distributed species is *C. aruensis* from Aru Island, south of Irian Jaya.

Habitat

These snakes inhabit lowland forests and rice fields, as well as urban areas. They live predominately in leaf litter or loose soil and may be more common near water. Specimens are often found after heavy rains.

Behavior

Very little is recorded of the behavior of most species of this family. One remarkable behavioral feature they have when threatened is to flatten the body and curl the tail backward over the trunk to expose bright red or yellow bands on the ventral tail. This display is often combined with continued movement of the tail and hiding the head under part of the trunk. Thus the hind part of the body gives the appearance of a small cobra in threatening display. If touched, the snakes may eject a fluid from the vent which seems to be a mixture of excretory and scent gland products. Captives usually cease this display within a few weeks. Observations on burrowing in captive *C. ruffus* show that these snakes are capable of burrowing quickly in loose soils and form tunnels approximately twice the cross sectional area of the body. When in the tunnel, snakes could move in either direction with nearly equal speed and were also capable of turning around in the tunnel and passing back along their own bodies. The feeding behavior of pipe snakes is unusual in that it depends on two distinct mechanisms for moving prey through the mouth. The initial swallowing is done by tiny movements of the upper jaws that also involve slight movements of the lower snout bones. Once the prey reaches the rear of the head, the snakes switch to a mechanism in which the anterior trunk compresses by tight curves of the vertebral column within the skin while the mouth is held closed. The mouth then opens and the vertebral column straightens, shooting the snake's head forward over the prey. This mechanism is very similar to one proposed for burrowing in uropeltids, and it is possible that cylindrophiids use this mechanism during burrowing (although it has not yet been reported).

Feeding ecology and diet

Most observations of feeding have been made in *C. ruffus*, but probably apply to the other species. Cylindrophiids appear to feed on a variety of small vertebrates, predominantly elongate vertebrates such as other snakes, elongate lizards, and eels. In captivity *C. ruffus* readily take small mice and fish. Prey are restrained or killed by constriction or crushed by the jaws.

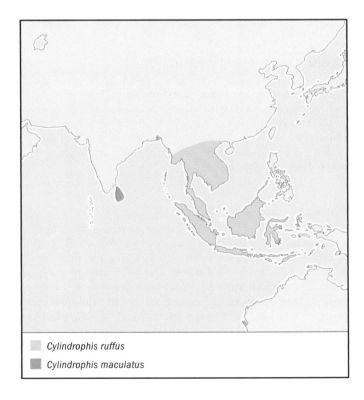

Cylindrophis ruffus
Cylindrophis maculatus

The pipe snake (*Cylindrophis ruffus*) tail display draws attention away from its head (right). (Photo by Animals Animals ©David Dennis. Reproduced by permission.)

Reproductive biology

Pipe snakes are viviparous, with the number of young correlated with the size of the mother. Litters tend to be small (two to three, possibly five) and the young are relatively large, approximately one-third to nearly one-half the length of the mother.

Conservation status

These snakes are not commonly encountered, and, like most fossorial groups, the status of their populations remains very poorly known. No species are listed by the IUCN.

Significance to humans

None known.

Resources

Books

Mattison, C. *The Encyclopedia of Snakes.* New York: Facts on File, 1995.

Zug, G. R., L. J. Vitt, and J. P. Caldwell. *Herpetology: An Introductory Biology of Amphibians and Reptiles.* 2nd ed. San Diego: Academic Press, 2001.

Periodicals

Cundall, D. "Feeding Behaviour in *Cylindrophis* and Its Bearing on the Evolution of Alethinophidian Snakes." *Journal of Zoology* (London) 237 (1995): 353–376.

Cundall, D., D. A. Rossman, and V. Wallach. "The Systematic Relationships of the Snake Genus *Anomochilus*." *Zoological Journal of the Linnean Society* 109 (1993): 275–299.

McDowell S. B., Jr. "A Catalogue of the Snakes of New Guinea and the Solomons, with Special Reference to Those in the Bernice P. Bishop Museum. Part 2, Anilioidea and Pythoninae." *Journal of Herpetology* 9 (1975): 1–79.

Steubing, R. "A New Species of *Cylindrophis* (Serpentes: Cylindrophiidae) from Sarawak, Western Borneo." *Raffles Bulletin of Zoology* 42 (1994): 967–973.

David Cundall, PhD

False coral snakes

(*Aniliidae*)

Class Reptilia
Order Squamata
Suborder Serpentes
Family Aniliidae

Thumbnail description
Moderately sized Amazonian snakes with a
cylindrical body of uniform diameter, very short
tail and a blunt, slightly flattened head not
demarcated from the body; body has smooth
scales, a ground color of pink or red with 50–60
black bands or half bands, each two to four
scales long, distributed evenly down the trunk

Size
2–3 ft (0.6–1 m) total length

Number of genera, species
1 genus; 1 species

Habitat
Rainforest and riparian lowland areas

Conservation status
Data Deficient

Distribution
Amazon Basin of eastern Peru and Ecuador, southern Colombia, northern Bolivia,
and Brazil

Evolution and systematics

The evolutionary origin of aniliids remains obscure. Fossil vertebrae ascribed to this family have been recovered from
Cretaceous deposits in Wyoming and New Mexico. These
fossils look superficially like the vertebrae of living aniliids except for two features of the neural arch that are more similar
to those found in living scolecophidian snakes. Much of the
older literature uses the term "aniliid" to refer to all basal
alethinophidian lineages. Hence, it is unclear whether many
of these data actually apply to Aniliidae *sensu stricto*. Recent
phylogenetic analyses have produced conflicting results with
respect to the relationships of aniliids, although most analyses
place aniliids closest to the anomochilid-uropeltid-cylindrophiid
lineages. Aniliids are therefore one of a group of relict, basal
alethinophidian taxa.

The family contains a single species that has been divided
into two subspecies based on ventral counts and the relative
lengths of black and red rings on the body. The northern
(Venezuelan) subspecies *A. scytale phelpsorum* is purported to
have fewer than 225 ventrals and have black bands longer than
the red bands, whereas the subspecies occupying the remainder
of the family's range, *A. scytale scytale*, has more than 225 ven-

trals and has black bands shorter than the red bands. Specimens
with characteristics of both subspecies have been found in northern Brazil and French Guiana. No subfamilies are recognized.

False coral snake (*Anilius scytale*). (Illustration by Jonathan Higgins)

A false coral snake (*Anilius scytale*) from Colombia, South America. (Photo by W. W. Lamar/GreenTracks. Reproduced by permission.)

Physical characteristics

Among the features denoting the basal position of anilids are the tiny eyes covered by one of the larger head scales (as in uropeltids) rather than a differentiated spectacle, as in most other snakes. Externally, these snakes have a superficial resemblance to the pipe snakes of Southeast Asia and to some of the venomous coral snake species of the Amazon Basin, being red or pink with about 50 black bands distributed evenly down the trunk. They have smooth, shiny scales in 15 (precloacal) to 21 (midbody) rows and the ventrals are one and one-half to two times the width of the adjacent scale rows. Internally, they retain teeth on the premaxilla, a vestigial pelvic girdle capped by a cloacal spur, a small left lung, and paired carotid arteries. Their skulls have massive jaws with small numbers of relatively large, slightly recurved conical teeth.

Distribution

False coral snakes are found in the Amazon Basin of eastern Peru and Ecuador, southern Colombia, northern Bolivia, and Brazil. In the east and north their range includes French Guiana, Suriname, Guyana, and the southern Orinoco Basin of southeastern Venezuela.

Habitat

Rainforest and riparian lowland areas.

Behavior

Very little is known.

Feeding ecology and diet

Anilius feeds primarily on small, elongate vertebrates such as eels, amphisbaenians, caecilians, and other snakes.

Reproductive biology

Females give birth to up to 15 young.

Conservation status

This species is listed as Data Deficient by the IUCN, but it is probably threatened in many areas of its range by habitat destruction or modification.

Significance to humans

None known.

Resources

Books

Chippaux, J.-P. *Les Serpents de la Guyane Francaise*. Paris: Éditions de l'ORSTOM, 1986.

Greene, H. W. *Snakes: The Evolution of Mystery in Nature*. Berkeley: University of California Press, 1997.

Peters, J. A., R. Donoso-Barros, and P. E. Vanzolini. *Catalogue of the Neotropical Squamata*. Part 1, *Snakes*. Washington, DC: Smithsonian Institution Press, 1986.

Roze, J. A. *La Taxonomia y Zoogeografia de los Ofidios de Venezuela*. Caracas: Ediciones de la Biblioteca, Univ. Central de Venezuela, 1966.

Zug, G. R., L. J. Vitt, and J. P. Caldwell. *Herpetology: An Introductory Biology of Amphibians and Reptiles*, 2nd ed. San Diego: Academic Press, 2001.

David Cundall, PhD

Sunbeam snakes

(Xenopeltidae)

Class Reptilia
Order Squamata
Suborder Serpentes
Family Xenopeltidae

Thumbnail description
Common semifossorial snakes with smooth, shiny, iridescent scales. Although dark brown on top and pale gray or yellow beneath, the scales reflect the colors of the spectrum, hence their common name, sunbeam snakes

Size
2–3 ft (approximately 1 m)

Number of genera, species
1 genus; 2 species

Habitat
Agricultural and settled areas and along the edges of forests

Conservation status
Not threatened

Distribution
Southeast Asia

Evolution and systematics

Xenopeltids appear to be a relict lineage basal to the Macrostomata, although some recent authors have included the family within Macrostomata on the assumption that these snakes possess macrostomatan properties. The two recognized species of *Xenopeltis* are very similar to each other but differ in a number of structural features from other snakes. Current phylogenetic analyses place the family between anilids and booids (pythonids, boids, and related snakes). No subfamilies are recognized.

Physical characteristics

Adult *Xenopeltis* have bodies that appear slightly flattened, but otherwise similar to pipe snakes. The dorsal half of the body is a dark, purplish brown, while the ventral half is white, light gray, or light yellow. On the head the yellow extends onto the upper labial scales. The young have a distinct light ring around the rear of the head and anterior neck. The head is wedge-shaped and flattened. The tail comprises about a tenth of the total length and has paired subcaudal scales. Scales on the body are large, typically in 15 rows throughout the length of the trunk, head plates are reduced in number and large, and the eye is

small. These snakes have a number of unusual anatomical features, including teeth on the premaxilla, mobile attachment of teeth (hinged teeth) on all of the toothed bones, a left lung about half the length of the right lung, no pelvic vestiges, a palate

Common sunbeam snake (*Xenopeltis unicolor*). (Illustration by Jonathan Higgins)

The common sunbeam snake's (*Xenopeltis unicolor*) scales appear iridescent as it moves along the forest floor in China. (Photo by Animals Animals ©Joe Mc Donald. Reproduced by permission.)

tightly attached to the snout, but a snout and facial region (maxilla, prefrontals) that can move up and down on the braincase, and an extraordinary mobile toothed process on the dentary bone that extends backward more than half the length of the lower jaw. Their quadrate is short and vertically oriented and is attached dorsally to a supratemporal partially embedded in the bones roofing the ear. Unlike macrostomatan snakes, the anterior tips of the lower jaw in *Xenopeltis* are tightly bound, as are the maxillae to the premaxilla. The lower jaw also has long, splintlike coronoids and a mobile intramandibular joint.

Distribution

From Myanmar south to the Nicobar and Andaman Islands, east through Thailand, Laos, Cambodia, Vietnam, and southern China, south through Malaysia and Indonesia east to Borneo, Java, Sulawesi, and the southwestern Philippine Islands.

Habitat

These snakes are common in agricultural and settled areas and along the edges of forests, but either rare or more difficult to find in deep forest. Most specimens are found in litter, under trash, or in disturbed areas.

Behavior

When not burrowing, xenopeltids move rapidly and nervously with their head against the substrate and with rapid tongue flicking. If touched, they jerk stiffly but rapidly in unpredictable directions. They have also been reported to vibrate their tail rapidly in a manner similar to rattlesnakes.

Feeding ecology and diet

Most information available is known to apply only to the common sunbeam snake (*Xenopeltis unicolor*). Stomach contents of wild-caught snakes contain predominantly lizards (particularly skinks), snakes, and frogs, although small mammals and birds have also been recorded. In captivity, sunbeam snakes will usually eat mice, including adult mice that they kill by constriction. Prey capture in captivity is unusual in appearing accidental and undirected, although snakes will pursue mice. Swallowing is extremely rapid despite the limited mobility of the upper jaws.

Reproductive biology

Sunbeam snakes are oviparous. However, remarkably little is known, and captive breeding has not been reported. Females have been found with up to 17 eggs.

Conservation status

Not threatened.

Significance to humans

None known.

Resources

Books

Campden-Main, S. M. *A Field Guide to the Snakes of South Vietnam.* Washington, DC: Smithsonian Institution, 1970.

Cox, M. J. *The Snakes of Thailand and Their Husbandry.* Malabar, FL: Krieger Publishing Company, 1991.

Deuve, J. *Serpents du Laos.* Paris: ORSTOM, 1970.

Zug, G. R., L. J. Vitt, and J. P. Caldwell. *Herpetology: An Introductory Biology of Amphibians and Reptiles,* 2nd ed. San Diego: Academic Press, 2001.

Periodicals

Frazzetta, T. H. "Adaptations and Significance of the Cranial Feeding Apparatus of the Sunbeam Snake (*Xenopeltis unicolor*): Part 1. Anatomy of the Skull." *Journal of Morphology* 239 (1999): 27–43.

David Cundall, PhD

Neotropical sunbeam snakes

(Loxocemidae)

Class Reptilia
Order Squamata
Suborder Serpentes
Family Loxocemidae

Thumbnail description
A medium-sized snake with cylindrical, muscular body; small eyes with vertically elliptical pupils; small mouth; and short tail

Size
The maximum recorded total length is 5 ft (1.53 m), but most adults are less than 3 ft (1 m); the tail accounts for 10–14% of the snake's total length

Number of genera, species
1 genus, 1 species

Habitat
Tropical moist and dry forests, including coastal beaches

Conservation status
Not threatened

Distribution
Southwestern Mexico, Guatemala, Honduras, El Salvador, Nicaragua, and northwestern Costa Rica

Evolution and systematics

Loxocemus represents a lineage that probably arose independently and early in snake evolution. The genus was variously placed in the families Pythonidae (pythons), Boidae (boas and sand boas), or Xenopeltidae (Asian sunbeam snakes), but now systematists agree that it comprises a distinct family, Loxocemidae. Previously, some herpetologists recognized the form *sumichrasti* as either a distinct species or a subspecies of *L. bicolor*, but detailed studies of morphological and geographic variation demonstrated that no distinction between *bicolor* and *sumichrasti* is possible, and therefore a single, monotypic species is currently recognized, *L. bicolor*.

The taxonomy for this species is *Loxocemus bicolor* Cope, 1861, La Unión, San Salvador [El Salvador].

English common names for this species include Neotropical sunbeam snake, Mexican burrowing python, Mesoamerican python, New World python, ground python, and burrowing boa; Spanish common names include chatilla, pitón excavador, and boa excavadora.

Physical characteristics

Neotropical sunbeam snakes have a prominently upturned snout. The scales on top of the head are large, whereas the

Neotropical sunbeam snake (*Loxocemus bicolor*). (Illustration by Bruce Worden)

scales on the body are smaller, smooth, and slightly iridescent. Pelvic spurs are present in both sexes but usually are not visible in females. Individuals are uniformly brown or have a white or cream belly sharply set off from the dark dorsum. In addition to this variation in body coloration, many individuals have small, highly irregular white blotches.

Distribution

This species is found in low and moderate elevations in southwestern Mexico, Guatemala, Honduras, El Salvador, Nicaragua, and northwestern Costa Rica.

Habitat

This species inhabits tropical moist and dry forests, including coastal beaches. Specimens have been found in banks of arroyos, rock piles, holes in the ground, under leaf litter and logs, behind the bark of logs, and inside ant nests.

Behavior

The species is relatively uncommon or infrequently seen throughout its distribution. As a result, its behavior in the wild is poorly known. The snakes are semifossorial, burrowing in loose soil or rotting foliage with their upturned snouts during the day and moving on the surface at night or on rainy days. Males engage in physical combat during the breeding season and can inflict severe wounds on their rivals by biting them.

Feeding ecology and diet

Loxocemus bicolor is a constrictor snake that forages widely for prey, relying on chemical and visual cues to locate prey outright or to restrict its attention to specific microhabitats.

Loxocemus bicolor has scales that are slightly iridescent. (Photo by R. Wayne Van Devender. Reproduced by permission.)

Food items of three snakes from Mexico and one from Costa Rica included two whip-tailed lizards (*Cnemidophorus* sp.) and two small rodents. Costa Rican specimens also take eggs and hatchlings of black iguanas (*Ctenosaura similis*) and eggs of green iguanas (*Iguana iguana*) and olive ridley sea turtles (*Lepidochelys olivacea*). A 54 in (138 cm) (total body length) snake consumed 32 *C. similis* eggs, whereas a 55 in (140 cm) specimen ate 23 *I. iguana* and four *C. similis* eggs. *Loxocemus* apparently enters the nesting tunnels of lizards and sea turtles in search of eggs, which probably constitute seasonally important prey for the snake. Hatchlings are seized as they emerge from their nests. When feeding on sea turtle eggs, *Loxocemus* loops its body around the eggs before swallowing them whole; the coils are made with the venter toward the snake's head. In captivity, *C. similis* eggs are first bitten, pushed against the body with the mouth, and then ingested. Some eggs may be broken in the process, but most are swallowed intact.

Reproductive biology

The species is oviparous. One captive female laid four eggs, which averaged 3 in (78.8 mm) in length, 1.4 in (37 mm) in width, and 2.16 oz (61.5 g). In the wild, neonates are reportedly found in May and take four to five years to reach sexual maturity.

Conservation status

The species is not considered to be endangered or threatened but is listed on CITES Appendix II, which includes species whose trade must be controlled in order to avoid use that is incompatible with their survival.

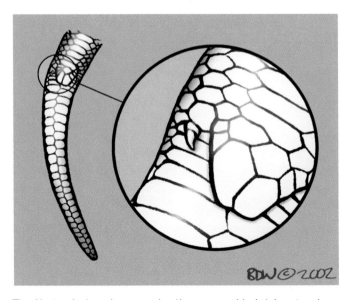

The Neotropical sunbeam snake (*Loxocemus bicolor*) has two bony spurs (vestigial femurs) projecting from its ventral surface (its underside). (Illustration by Bruce Worden)

Significance to humans

None known.

Resources

Books

Greene, H. W. *Snakes: The Evolution of Mystery in Nature.* Berkeley: University of California Press, 1997.

McDiarmid, R. W., J. A. Campbell, and T. Touré. *Snake Species of the World.* Vol. 1, *A Taxonomic and Geographic Reference.* Washington, DC: The Herpetologists' League, 1999.

Savage, J. M. *The Amphibians and Reptiles of Costa Rica.* Chicago: University of Chicago Press, 2002.

Wilson, L. D., and J. R. Meyer. *The Snakes of Honduras,* 2nd ed. Milwaukee: Milwaukee Public Museum, 1985.

Periodicals

Cundall, D., V. Wallach, and D. A. Rossman. "The Systematic Relationships of the Snake Genus *Anomochilus.*" *Zoological Journal of the Linnean Society* 109 (1993): 275–299.

Greene, H. W. "Dietary Correlates of the Origin and Radiation of Snakes." *American Zoologist* 23 (1983): 431–441.

Mora, J. M. "Natural History Notes: *Loxocemus bicolor* (Burrowing Python). Feeding Behavior." *Herpetological Review* 22 (1991): 61.

———. "Predation by *Loxocemus bicolor* on the Eggs of *Ctenosaura similis* and *Iguana iguana.*" *Journal of Herpetology* 21 (1987): 334–335.

Mora, J. M., and D. C. Robinson. "Predation of Sea Turtle Eggs (*Lepidochelys*) by the Snake *Loxocemus bicolor* Cope." *Revista de Biología Tropical* 32 (1984): 161–162.

Nelson, C. E., and J. R. Meyer. "Variation and Distribution of the Middle American Snake Genus, *Loxocemus* Cope (Boidae?)." *Southwestern Naturalist* 12 (1967): 439–453.

Odinchenko, V. I., and V. A. Latyshev. "Keeping and Breeding in Captivity the Mexican Burrowing Python *Loxocemus bicolor* (Cope, 1961) at Moscow Zoo." *Russian Journal of Herpetology* 3 (1996): 95–96.

Willard, D. E. "Constricting Methods of Snakes." *Copeia* 1977 (1977): 379–382.

Woodbury, A. M., and D. M. Woodbury. "Notes on Mexican Snakes from Oaxaca." *Journal of the Washington Academy of Sciences* 34 (1944): 360–373.

Other

"Convention on International Trade in Endangered Species of Wild Fauna and Flora." [cited September 24, 2002]. <http://www.cites.org>

Javier A. Rodríguez-Robles, PhD

Boas
(Boidae)

Class Reptilia
Order Squamata
Suborder Serpentes
Family Boidae

Thumbnail description
Small to giant constricting snakes possessing paired lungs, cloacal spurs, and toothless premaxilla; most species are viviparous

Size
1.2–25 ft (0.37–7.7 m); 0.2–320+ lb (0.1–145+ kg)

Number of genera, species
7 genera; 41 species

Habitat
Loose sand, burrows, grasslands, savanna, forest, various freshwater habitats

Conservation status
Endangered: 1 species; Vulnerable: 4 species; Lower Risk/Near Threatened: 2 species

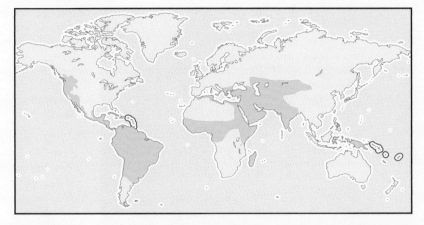

Distribution
South America, Central America, Mexico, southwestern Canada, western United States, and West Indies; southeastern Europe and Asia Minor; sub-Saharan western Africa east to Tanzania, north through Egypt, and Mediterranean coast from Egypt to eastern Morocco; Madagascar and Reunion Island; Arabian Peninsula; southwestern and central Asia; Indian subcontinent and Sri Lanka; Sulawesi, Moluccan Islands; New Guinea; Bismarck Archipelago; and Melanesia east to American Samoa

Evolution and systematics

Fossil snake skulls are rarely recovered because the bones of the skull are small and loosely connected and typically become separated soon after death. Vertebrae are the most common snake fossil. In general, few snake families can be identified incontrovertibly by vertebrae and ribs. Living erycine snakes, however, have several unique vertebral characters, and on this basis numerous fossil vertebrae have been identified as erycine (in the subfamily Erycinae).

The modern Boidae is believed to have descended from basal macrostomatans; it is one of several snake lineages that diverged from the primitive alethinophidians (true snakes) near the end of the Cretaceous. Macrostomatan snakes are distinguished by characters of the skull and musculature that allow them increased jaw flexibility, a greater gape, and the ability to consume larger prey.

Boid snakes share many characters with other basal macrostomatan snakes, including fully functional paired lungs, smooth scales (with some exceptions) vestiges of a pelvic girdle, and cloacal spurs. The cloacal spurs of boas are two claw-like structures that are located one on each side of the anal scale. They are usually larger in male boas than in females; the females of some species may not have apparent cloacal spurs. Characters shared with the Pythonidae, the sister taxon of the Boidae, include elliptical pupils and pitted lip scales. The pits in the lips are associated with thermoreception, the ability to detect differences in temperature.

Boas differ from pythons in numerous characters, including: Boid snakes do not have a supraorbital bone (with one

exception) while all pythons have a supraorbital bone. Not all boas have labial pits; when present, the labial pits are located between the labial scales while the labial pits of pythons are centered in the labial scales. Two premaxilla are fused together to form a small bone across the front of the upper jaw; the premaxilla of boid snakes is without teeth while the premaxilla of most pythons is toothed. Most boas are viviparous, meaning they bear live young; all pythons lay eggs. Three taxa of boid snakes, *Charina reinhardtii*, *Eryx muelleri*, and *Eryx jayakari*, are oviparous and lay eggs.

Undoubtedly, the Boidae is more speciose than is recognized. An analysis of the geographic variation and systematic relationships of populations of *Boa constrictor*, the most widely distributed boid species, has yet to be accomplished. Likewise, the systematic relationships of the insular populations of all three species of *Candoia* remain to be investigated fully. There is little doubt that there will be further taxonomic changes within the Boidae.

As of 2002, science recognizes 41 species in seven genera and two subfamilies in the Boidae. The Boinae, the larger subfamily, includes the boas and anacondas, 27 species in five genera. The Erycinae includes the sandboa, rubber boa, rosy boa, and Calabar boa, 14 species in two genera. The division of the Boidae into these two subfamilies is based primarily on osteological characters.

Physical characteristics

The Boidae is distributed widely on five continents and countless islands and occurs in many different habitats.

An Amazon tree boa (*Corallus hortulanus*) eating, Brazil. (Photo by Joe McDonald. Bruce Coleman, Inc. Reproduced by permission.)

rostrals (scales on the end of the snout) are strong and broad, and the lower jaws are underslung and close tightly. The heads are narrow and the necks thick, the bodies are round and usually smooth scaled, and the tails are short with thickened skin on the upper surfaces. Several species have blunt tails. No erycine snakes have labial pits or have been shown to have well-developed temperature-sensing abilities. The largest erycine snake is the brown sandboa, *Eryx johnii*; this species reaches a maximum length slightly exceeding 4 ft (1.3 m). There are several small species—the smallest is probably the Arabian sandboa, *E. jayakari*, with an average adult size of 0.9 ft (28 cm).

The boine snakes are mostly medium sized, athletic, terrestrial, and arboreal. The smallest species is the Abaco boa, *Epicrates exsul*, with a maximum length of 31.5 in (810 mm). The best-known arboreal species is the emerald tree boa, *Corallus caninus*. The anacondas are the largest and most aquatic of the boas; they have soft, loose skin that can withstand long periods of immersion, and their eyes and nostrils are directed upward, so that they can see and breathe with most of the head submerged. Most boine snakes have large heads that are distinctly wider than their necks, large eyes, laterally compressed bodies (to varying degrees), and long tails. Most have temperature-sensing labial pits, and several species in the genera *Boa* and *Eunectes* have temperature-sensing abilities even without labial pits.

The boas include many beautiful species. Several species have a polymorphic appearance; one such species is the Amazon tree boa, *Corallus hortulanus*, with patterned and unpatterned appearances that vary in color from gray to brown to yellow to orange to red. The skin of many species exhibits a beautiful iridescence. Some populations of the boa constrictor, *B. constrictor*, and the rainbow boa, *Epicrates cenchria*, have a remarkable ability to change the color of their skin; they typically appear darkest during the day and much paler at night. The Fiji Island boa, *Candoia bibroni*, has been seen to change from black to pale pink in a period of six hours.

Among the species, great variation in size, scalation, diet, habitat, and many other characters can be seen. There are several species of erycine snakes that are not known to exceed 3 ft (1 m) in length. The Haitian vine boa, *Epicrates gracilis*, is one of the most elongate and slender of all snakes. The boa constrictor is a large species, approaching 15 ft (4.6 m) in maximum length.

The green anaconda, *Eunectes murinus*, is the largest boid snake. Its maximum size is a topic of controversy. There are stories and reports in the literature of anacondas measuring 33–45 ft (10–14 m). Murphy and Henderson list 25 ft (7.7 m) as the longest specimen whose length actually was measured, not just estimated. This anaconda is probably the heaviest snake species in the world. While all of the giant snake species attain great weight in captivity, there are numerous records of wild specimens of green anacondas exceeding 300 lb (136 kg).

With the exception of the rosy boa, *Charina trivirgata*, the bodies of erycine snakes are modified for burrowing. The eyes are small and often set high on the sides of their heads, the

Distribution

The Boidae has one of the most extensive distributions of a snake family. Several genera have disjunct distributions. In the Boinae, *Boa* occurs in northwestern and northeastern Mexico south through Central and South America, the Lesser Antilles, Madagascar, and Reunion. *Eunectes* is found in tropical South America from Colombia to Argentina. *Corallus* occurs in Central America from Honduras south to the Amazon drainage in eastern Bolivia and Brazil; it is found on numerous islands in the West Indies. One species of *Epicrates* is widespread from southern Central America and South America to Argentina; the other species are distributed throughout the West Indies. *Candoia* is distributed in the Indo-Pacific region from Sulawesi and the Moluccas, New Guinea, Bismarck Archipelago, Melanesia, and Polynesia east to American Samoa.

Of the Erycinae, *Charina* is found in western North America, including southwestern Canada, western United States, and northwestern Mexico. *Charina* also occurs in tropical central Africa, from Liberia east to Cameroon, Gabon, and

Congo and into Zaire. In Africa, *Eryx* occurs in the Sahel region from Mauritania and Senegal to Kenya and Tanzania, northeastern Africa, coastal northern Africa, and the Arabian Peninsula; it also occurs in southeastern Europe, the Middle East, and Asia Minor to central Asia, India, and Sri Lanka.

Habitat

Boas can be found in nearly every habitat known to host snakes of any sort, except marine habitats. Many of the sandboas, such as the Arabian sandboa, are well adapted to live in extremely hot and dry habitats; in contrast, *Eryx tataricus* is found in a very cold climate in southern Mongolia. The range of the viper boa, *Candoia aspera*, includes New Ireland in the Bismarck Archipelago, one of the rainiest locales on Earth. The wide-ranging boa constrictor can be found in the Sonoran Desert in northwestern Mexico, the rainforest in Brazil, and the temperate grasslands of northern Argentina.

Behavior

Boid snakes tend to be nocturnal, but they are often encountered moving or basking during the day. Faced with a perceived threat, the larger boid snakes typically defend themselves with cloacal discharge, hissing, striking, and biting. When threatened, many sandboas are reluctant to bite and instead roll into a tight ball with their heads in the center; several of the blunt-tailed species then will use their tails to mimic their heads. *C. bibroni* has been seen to flatten the head and the anterior half of the body, much in the manner of a cobra.

Feeding ecology and diet

Boid snakes are primarily ambush hunters that consume vertebrate prey. Ambush techniques range from that of sandboas, which lie buried in wait for lizards or small mammals, to that of Amazon tree boas perching in trees over watercourses waiting for birds to fly by; to that of Puerto Rican boas, *Epicrates inornatus*, which sit high in cave entrances to intercept bats. Boid snakes can and do incorporate active foraging behavior as well.

Many small boid snakes consume lizards, both small taxa and the young of larger taxa. Anoline lizards are the favored food of many of the West Indian *Epicrates*. Mammals become an increasingly significant percentage of the diet as boas grow in size. Most species will consume birds whenever the opportunity presents itself. Snakes are included in the diet of anacondas, but in general snake-eating appears to be rare in the Boidae. Boas in the genera *Boa* and *Corallus* are known to caudal lure by wriggling the distal portion of their tails to attract prey. Although it does not appear to be well documented, there are a sufficient number of published reports that green anacondas kill and occasionally consume humans to assume that it does happen. It is not a common occurrence.

Reproductive biology

In most species the female is the larger sex. In some boid species, males will fight when competing to breed with a fe-

male, often incorporating wrestling and biting to achieve dominance. Male anacondas apparently do not engage in combat; groups of males are sometimes observed simultaneously courting one female.

All boine snakes and most erycine snakes are live-bearers. Litters of well-formed live young are born, typically all delivered within a short time. There is one record of a green anaconda delivering 82 young in one litter. A captive East African sandboa, *Eryx colubrinus*, produced more than 250 young, breeding 14 times in a 16-year period. The Arabian sandboa (*E. jayakari*), Sahara sandboa (*E. muelleri*), and the Calabar boa (*C. reinhardtii*) are oviparous and reproduce by laying eggs.

Conservation status

The rarest boa in the world is *Corallus cropanii*, known only from its type locality in southeastern Brazil. It receives no formal protection. More than 40 years have passed since the last specimen was collected, and though it has not been formally declared extinct, many authorities believe this is the case.

The Mona boa, *Epicrates monensis*, including both subspecies, *E. m. granti* and *E. m. monensis*, is listed as Endangered by the IUCN, owing to the fragmentation of habitat and populations, a low number of adults that is declining, habitat degradation, and introduced predators. A captive-breeding program for the species was begun in 1985, and there is now a self-sustaining captive population of several hundred individuals maintained in zoos. Reintroduction to former habitats was started in 1993, after rats and cats were eliminated from those areas. The reintroduced populations appear to be breeding.

The four species listed as Vulnerable by the IUCN are the Jamaican boa, *Epicrates subflavus*; Dumeril's boa, *Boa dumerili*; the Madagascar boa, *Boa madagascariensis*; and the Madagascar tree boa, *Boa mandrita*. At the time of this writing, all are believed to be stable.

Significance to humans

In general, few species in the Boidae are persecuted actively by humans. Most are too small to be of value in the skin trade. In the recent past, many boa species certainly were considered an important natural resource by indigenous peoples, but today it seems that most boas escape much human attention. Some of the larger boas still may be hunted for meat in remote areas, and body parts are used in folk medicine in some areas. There is commerce in the skins of anacondas and boa constrictors, but it does not approach the magnitude of the trade in python skins. With the exception of *C. cropanii* and *Eryx somalicus*, all of the boid species are kept in captivity. Boa constrictors, rosy boas, and East African sandboas are among the most commonly kept snake species; thousands are bred and born in captivity every year. As of 2002 all but three or four boid species have been reproduced in captivity.

1. Emerald tree boa (*Corallus caninus*); 2. Viper boa (*Candoia aspera*); 3. Green anaconda (*Eunectes murinus*); 4. East African sandboa (*Eryx colubrinus*); 5. Rosy boa (*Charina trivirgata*); 6. Cuban boa (*Epicrates angulifer*); 7. Calabar boa (*Charina reinhardtii*); 8. Boa constrictor (*Boa constrictor*). (Illustration by Marguette Dongvillo)

Species accounts

Boa constrictor
Boa constrictor

SUBFAMILY
Boinae

TAXONOMY
Boa constrictor Linnaeus, 1758, "Indiis" (erroneous).

OTHER COMMON NAMES
English: Boa constrictor, boa, redtail boa; French: Boa constricteur; German: Konigsboa, Sbgottshlangen; Spanish: Mazacuata, travaganado, macuarel, darura; Portuguese: Jibóia.

PHYSICAL CHARACTERISTICS
This is a medium-size to large species with a large head distinct from the neck, a laterally compressed body, and a long prehensile tail. Throughout the extensive range, there is considerable variation in pattern and color, but most boas are brown snakes with dark brown markings on the back that expand to become red, reddish brown, or dark brown blotches on the tail.

DISTRIBUTION
This species is found within 150 mi (240 km) of the U.S. border in northwestern and northeastern Mexico. The range includes Mexico, Central America, most of South America north of 35° south latitude. It also occurs on Dominica and Saint Lucia in the Lesser Antilles and on many small islands along the coasts of Mexico, Central America, and South America.

HABITAT
This adaptable species can be found in desert, grasslands, and forest.

◻ *Boa constrictor*
◼ *Charina trivirgata*
◼ *Epicrates angulifer*

BEHAVIOR
Boas often spend much of their time in trees. Large specimens are probably more terrestrial in their habits, but even very large boas are known to climb. Boas seek shelter in the burrows of agoutis, pacas, and armadillos.

FEEDING ECOLOGY AND DIET
Boas are primarily ambush hunters. There are records of large specimens consuming ocelots and porcupines, but the typical prey includes rodents, such as rats, squirrels, agoutis, and pacas, as well as birds, monkeys, and bats. They eat large lizards, including ameivas, tegus, and iguanas. In captivity boa constrictors typically are fed mice and rats.

REPRODUCTIVE BIOLOGY
Little is known about reproduction in the wild. In captivity boas can become sexually mature in their second year, but maturity more typically comes in the third or fourth year. Babies usually are born 120–145 days after ovulation.

CONSERVATION STATUS
There are no baseline data regarding wild boa populations, but the species appears to be holding its own throughout its range.

SIGNIFICANCE TO HUMANS
In agricultural areas boa constrictors are important predators of rodents. Boa constrictors are commonly kept in captivity, especially in North America and Europe. ◆

Viper boa
Candoia aspera

SUBFAMILY
Boinae

TAXONOMY
Candoia aspera Günther, 1877, Duke of York Island, Bismarck Archipelago.

OTHER COMMON NAMES
English: New Guinea ground boa, Papuan ground boa; French: Boa nain; German: Pazifik Boa.

PHYSICAL CHARACTERISTICS
This is a short, heavy-bodied boa with a very short tail. The head is triangular and distinct from the neck. All of the dorsal scales are keeled with a prominent, raised, longitudinal ridge that runs down the center of each scale. The maximum length of this species approaches 3 ft (1 m), but most adults are only about half that size.

DISTRIBUTION
The species inhabits New Guinea and the Bismarck Archipelago.

HABITAT
Viper boas have been encountered in coconut husk piles, on the coast under driftwood, in trees, and in leaf litter on forest floors. They often are found in swampy areas and mudflats.

Candoia aspera

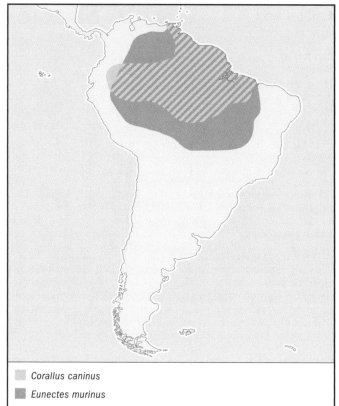

Corallus caninus

Eunectes murinus

BEHAVIOR
This snake is nocturnal and secretive and is rarely encountered during the day except after rain. It is known to coil in a ball for defense. It also can deliver a painful bite. It is believed that this snake mimics the death adder, *Acanthophis* sp., in areas where the two are sympatric.

FEEDING ECOLOGY AND DIET
Viper boas commonly feed on frogs and lizards. Small mammals also are taken.

REPRODUCTIVE BIOLOGY
Little is known about reproduction in nature. Captured gravid females have delivered litters of five to 15 babies.

CONSERVATION STATUS
Not threatened.

SIGNIFICANCE TO HUMANS
None known. ◆

Emerald tree boa
Corallus caninus

SUBFAMILY
Boinae

TAXONOMY
Corallus caninus Linnaeus, 1758, "Americae."

OTHER COMMON NAMES
French: Boa canin, boa émeraude; German: Grüner Hundskopfshlinger; Spanish: Falsa mapanare verde, boa esmeralda; Portuguese: Arara bóia, cobra verde.

PHYSICAL CHARACTERISTICS
This is a beautiful green snake highly adapted for an arboreal life. The emerald tree boa has a large head with prominently

pitted labial scales. The body is elongate and very laterally compressed, and the tail is long and very prehensile. This species has the longest teeth of any nonvenomous snake; the front teeth of a large specimen can be 1.5 in (3.7 cm) in length.

DISTRIBUTION
This species is known from the tropical rainforests in and surrounding the Amazon Basin of South America.

HABITAT
This species is associated with primary and secondary forest. It is often found in trees overhanging or near water courses.

BEHAVIOR
The resting pose of this species appears to be a flat coil that is folded over a branch, with the head in the center pointing down as if to watch below.

FEEDING ECOLOGY AND DIET
Rodents, parrots, passerines, and small monkeys have all been recorded as prey.

REPRODUCTIVE BIOLOGY
Babies are red, orange, or green. They change to the adult green color as they approach sexual maturity.

CONSERVATION STATUS
Many Latin American countries control or forbid the export of native boid snakes; emerald tree boas thereby receive some protection from commercial collecting in most of the countries where the species occurs. The species appears to be stable and not threatened, but its future is tied to the forests it inhabits.

SIGNIFICANCE TO HUMANS
This attractive snake is popular among keepers of boas. The species is hardy in captivity and is regularly bred in captivity. ◆

Cuban boa
Epicrates angulifer

SUBFAMILY
Boinae

TAXONOMY
Epicrates angulifer Bibron, 1840, Cuba.

OTHER COMMON NAMES
French: Boa de Cuba; German: Kuba-Schlankboa; Spanish: Majá de Santa Maria.

PHYSICAL CHARACTERISTICS
This is the largest species in the genus *Epicrates*, with a maximum size approaching 13 ft (4 m).

DISTRIBUTION
E. angulifer is found throughout Cuba at altitudes up to 1,000 ft (310 m); the species also occurs on nearby smaller islands.

HABITAT
This is an adaptable snake most often found in wooded areas, often off the ground in trees but also on rocky hillsides, in caves, and in talus.

BEHAVIOR
Captive juvenile Cuban boas tend to be excitable and irritable, often biting their keepers when given the chance; adults tend to be calm and placid animals that do well in captivity. Like most boas in the genus *Epicrates*, the Cuban boa often excretes viscous liquid uric acid (white insoluble nitrogenous waste products) when excited or frightened.

FEEDING ECOLOGY AND DIET
Cuban boas are both ambush predators and active foragers. They are known to eat bats, rodents, chickens, native birds, and iguanas.

REPRODUCTIVE BIOLOGY
Males fight during the breeding season. Females deliver litters of two to 10 large babies that are up to 24 in (61 cm) in length.

CONSERVATION STATUS
Not threatened. Cuban boas are protected by Cuban law, and are believed to be common throughout Cuba. The species seems to coexist well with humans when it is not persecuted.

SIGNIFICANCE TO HUMANS
This species occasionally preys on domestic fowl. ◆

Green anaconda
Eunectes murinus

SUBFAMILY
Boinae

TAXONOMY
Eunectes murinus Linnaeus, 1758, "America."

OTHER COMMON NAMES
English: Water boa; huilla, huilia, camoudi; French: Anaconda commun; German: Grosse Anakonda; Portuguese: Arigbóia, boiuna, boicu, boiguacu, sucuri, sucuriju, sucurijuba, Spanish: Culebra de agua.

PHYSICAL CHARACTERISTICS
This is a giant, heavy-bodied, dark green boa with black spots. Even an average specimen, 10–15 ft (3–4.6 m) in length, appears immense because of its girth.

DISTRIBUTION
This species occurs in the Amazonian and Orinoco drainages from Columbia and Venezuela to eastern Bolivia and central Brazil. It also is known from Trinidad.

HABITAT
The green anaconda is associated strongly with watercourses, swamps, and other freshwater habitats.

BEHAVIOR
Anacondas are rarely found far from water. Feeding usually takes place in the water.

FEEDING ECOLOGY AND DIET
Anacondas typically lie in wait for prey at the water's edge. They are known to eat a wide variety of vertebrate prey, including monkeys, deer, peccaries, pacas, agoutis, birds, fish, caiman, and turtles.

REPRODUCTIVE BIOLOGY
Breeding usually occurs during the dry season. A group of males will court a receptive female, competing peacefully to copulate.

CONSERVATION STATUS
Not threatened. So far as is known, the green anaconda exists in reasonable numbers throughout its range.

SIGNIFICANCE TO HUMANS
Anacondas seem to be largely unmolested by humans. They are not commonly harvested for meat, and throughout most of the range there is little or no harvest of skins. Small specimens are collected and exported for the live trade, but the species is not particularly popular in captivity, largely because of its size but in part because many specimens are foul-tempered. ◆

Calabar boa
Charina reinhardtii

SUBFAMILY
Erycinae

TAXONOMY
Charina reinhardtii Schlegel, 1848, originally designated as "Old Calabar, West Africa" and now annotated to "Gold Coast."

OTHER COMMON NAMES
English: Burrowing python, Calabar ground python; German: Erdpython.

PHYSICAL CHARACTERISTICS
The Calabar boa is a small species that only rarely grows longer than 30 in (80 cm). The head is small and not distinguished from the neck. The body is round, the skin is soft, the scales are smooth, and the tail is blunt. Individuals are dark brown or black with red or orange scales randomly scattered on the body. Hatchlings have white markings on their tails that disappear with age.

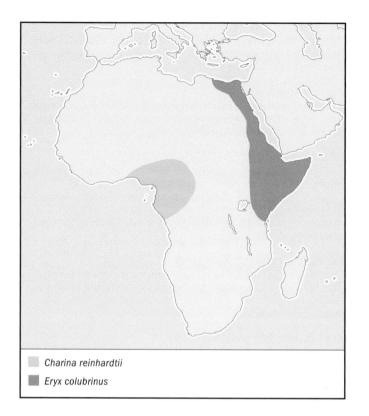

■ *Charina reinhardtii*
■ *Eryx colubrinus*

DISTRIBUTION
The species occurs in west and central Africa from Guinea and Liberia east to Cameroon, Gabon, Congo, and into Zaire.

HABITAT
The Calabar boa is associated with forest, soft soil, and leaf litter. It most often is seen out and moving after rains, both during the day and at night.

BEHAVIOR
Calabar boas form a tight ball with the head in the middle and use the tail as a decoy. The tail of wild adults usually is scarred. So strong is the instinct for this defensive behavior that even captive-hatched and raised animals rarely will uncoil when handled. This species never bites in defense.

FEEDING ECOLOGY AND DIET
In nature the diet includes rodents and insects. In captivity the species eats rodents at all ages.

REPRODUCTIVE BIOLOGY
The Calabar boa is oviparous; relative to body size, the eggs are immense. The eggs are delicate and thin-shelled. Clutch size ranges from one to 12 eggs; most clutches number two to four eggs.

CONSERVATION STATUS
Not known.

SIGNIFICANCE TO HUMANS
It is reported that this species is feared by some local people, who believe that it has two heads. ◆

Rosy boa
Charina trivirgata

SUBFAMILY
Erycinae

TAXONOMY
Charina trivirgata Cope, 1861, Cape San Lucas, Baja California Sur, Mexico.

OTHER COMMON NAMES
French: Boa à trois bandes; German: Deistreifen-Rosenboa.

PHYSICAL CHARACTERISTICS
The rosy boa is a small, heavy-bodied snake with a small head that is barely distinct from the neck. The eyes are small, and the pupils are vertical. The tail is relatively long and thick, coming to a blunt point. The scales are smooth and shiny. Large specimens attain considerable bulk and girth. Adult females are about 28–36 in (71–95 cm) in total length; most adult males are 18–26 in (46–67 cm). The maximum size of this form approaches 4 ft (1.3 m).

DISTRIBUTION
The species ranges across southern California, southwestern Arizona, and northwestern Mexico.

HABITAT
This is a saxicolous species, strongly associated with rocky canyons and rocky ridges and hills.

BEHAVIOR
Rosy boas are usually docile snakes that are deliberate in their actions and reluctant to bite in defense.

FEEDING ECOLOGY AND DIET
Small rodents and lizards make up the bulk of the diet.

REPRODUCTIVE BIOLOGY
A gravid female will complete a shed 16–20 days after ovulation. Usually babies are born 100–120 days after that shed. Litters typically are born from mid-August to early October. Clutch size is reported to vary from one to 13; most litters number four or five.

CONSERVATION STATUS
Not threatened.

SIGNIFICANCE TO HUMANS
This is a very popular snake species in captivity; every year, thousands of captive-bred rosy boas are born. ◆

East African sandboa
Eryx colubrinus

SUBFAMILY
Erycinae

TAXONOMY
Anguis colubrinus Linnaeus, 1758, "Egypto."

OTHER COMMON NAMES

French: Boa des éscailles rugueuses; German: Aegyptische Sandboa, Kenyan Sandboa.

PHYSICAL CHARACTERISTICS

This is a heavy-bodied sandboa with a maximum length that approaches 36 in (90 cm). The northern populations are smaller and patterned in tan and yellow, whereas the southern populations are larger and darker with an orange pattern. The undersurface is an immaculate white. The dorsal surface of the tail is armored with thick skin and hooked keeled scales.

DISTRIBUTION

The species is distributed in northeast Africa from central Tanzania north to Egypt and from there west into Niger.

HABITAT

In the northern part of the range this species is more likely found in arid sandy and rocky areas; to the south these sandboas are found in rocky hills and in burrows in soil. They often occupy agricultural fields.

BEHAVIOR

This sandboa rarely strikes forward in defense, but it defends itself with quick backward-directed slashes when its body is touched.

FEEDING ECOLOGY AND DIET

In captivity the larger animals from southern populations feed on rodents at all ages; the northern animals more often feed on geckos and skinks when they are young.

REPRODUCTIVE BIOLOGY

Most litters contain 12–20 babies, but litters of 32 babies are reported.

CONSERVATION STATUS

Nothing is known about this species in nature.

SIGNIFICANCE TO HUMANS

This is a very popular snake to keep in captivity; thousands of babies are born in captivity annually. The captive specimens constitute a viable self-sustaining population; very few East African sandboas are taken from the wild.

Resources

Books

de Vosjoli, Philippe, Roger Klingenberg, and Jeff Ronne. *The Boa Constrictor Manual*. Santee, CA: Advanced Vivarium Systems, 1998.

Greene, Harry W. *Snakes: The Evolution of Mystery in Nature*. Berkeley: University of California Press, 1997.

Minton, Sherman A., and Madge Rutherford Minton. *Giant Reptiles*. New York: Charles Scribner's Sons, 1973.

Murphy, John C., and Robert W. Henderson. *Tales of Giant Snakes: A Historical Natural History of Anacondas and Pythons*. Malabar, FL: Krieger Publishing Company, 1997.

O'Shea, Mark. *A Guide to the Snakes of Papua New Guinea*. Port Moresby, Papua New Guinea: Independent Publishing, 1996.

Pope, Clifford Millhouse. *The Giant Snakes: The Natural History of the Boa Constrictor, the Anaconda, and the Largest Pythons, Including Comparative Facts About Other Snakes and Basic Information on Reptiles in General*. New York: Alfred A. Knopf, 1961.

Stafford, Peter J., and Robert W. Henderson. *Kaleidoscopic Tree Boas: The Genus* Corallus *of Tropical America*. Malabar, FL: Krieger, 1996.

Stebbins, Robert C. *A Field Guide to the Western Reptiles and Amphibians: Field Marks of All Species in Western North America, Including Baja California*. 2nd edition. Boston: Houghton Mifflin, 1985.

Tolson, P. J., and R. W. Henderson. *The Natural History of West Indian Boas*. Taunton, England: R & A Publishing Limited, 1993.

Periodicals

Kluge, Arnold G. "Boine Snake Phylogeny and Research Cycles." *Miscellaneous Publications, Museum of Zoology, University of Michigan*, no. 178 (1991): 1–58.

———. "*Calabaria* and the Phylogeny of Erycine Snakes." *Zoological Journal of the Linnean Society* 107 (1993): 293–351.

McDowell, S. B. "A Catalogue of the Snakes of New Guinea and the Solomons, with Special Reference to Those in the Bernice P. Bishop Museum. Part III. Boinae and Acrochordoidea (Reptilia, Serpentes)." *Journal of Herpetology* 13, no. 1 (1979): 1–92.

David G. Barker, MS
Tracy M. Barker, MS

Pythons
(Pythonidae)

Class Reptilia

Order Squamata

Suborder Serpentes

Family Pythonidae

Thumbnail description
Small to giant, finely scaled, oviparous, constricting snakes possessing paired lungs, cloacal spurs, and supraorbital bone

Size
1.5–33 ft (0.5–10.1 m) in length, with a weight of 0.3–320 lb (0.14–145 kg)

Number of genera, species
8 genera; 32 species

Habitat
Rocky and sandy deserts, savanna, open woodlands, and forest

Conservation status
Endangered: 1 species; Vulnerable: 31 species

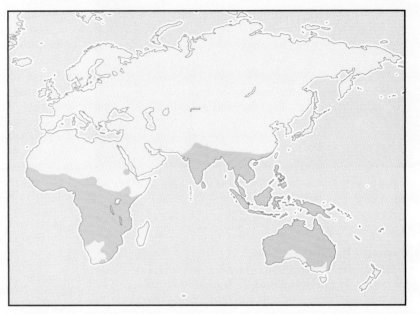

Distribution
Sub-Saharan Africa, southern Asia and Southeast Asia, southeastern China, Indonesia, Philippines, Papua New Guinea, and Australia

Evolution and systematics

There are few snake fossils that can be incontrovertibly identified as pythons; these fossils are relatively recent, and all have been assigned to extant genera. Snake fossils from the mid- to late Miocene in Australia have been identified as *Liasis* and *Morelia* (formerly *Montypythonoides*). Fossils assigned to the genus *Python* are known from the Pliocene of eastern Africa.

Pythons are considered to be basal macrostomatans, one of several ophidian lineages that diverged from the primitive alethinophidians near the end of the Cretaceous. Macrostomatan snakes are distinguished by characters of the skull and musculature that allow them increased jaw flexibility, a greater gape, and the ability to consume larger prey. The pythons likely evolved in the Australo-Papuan region where today they exist with the greatest diversity. Seven of the eight genera of pythons occur in Australia and New Guinea. The genus *Aspidites*, an Australian endemic, is the sister taxon to all other pythons. It is assumed that *Python*, the most widespread genus, evolved in the Indo-Papuan region and dispersed north to Southeast Asia and then west into Africa.

Pythons share many characters with other basal macrostomatan snakes, including fully functional paired lungs, smooth scales (with some exceptions), vestiges of a pelvic girdle, and cloacal spurs. A cloacal spur is a claw-like structure used in courtship that is found on either side of the anal scale;

male pythons tend to have larger cloacal spurs than female pythons. Characters shared with the Boidae, the sister taxon of the Pythonidae, include elliptical pupils and pitted lip scales that are associated with thermoreception. Thermoreception is the ability to sense differences in temperatures.

Pythons differ from boas in numerous characters, including: The supraorbital bone of pythons is in contact with the prefrontal. The supraorbital bone is a small bone located at the top of the eye socket. Only one taxon of boa possesses a supraorbital bone, and it does not contact the prefrontal. Python labial pits are located in the centers of the lip scales; when present in boas, pits are located between the lip scales. Two premaxilla are fused together to make the small bone at the front of the upper jaw: the premaxilla of most adult pythons is toothed, while the premaxilla of boid snakes is without teeth. All pythons are oviparous and lay eggs, while all but three taxa of boid snakes give live birth.

The Pythonidae is a widely distributed lineage that is undoubtedly more species-rich than is recognized currently. Mainly because of the practical problems involved in preserving and caring for large specimens, there is a surprising paucity of representative material from most python populations in museum collections. In 2000 and 2001, taxonomic revisions resulted in an increase in the number of python species from 25 species to the currently recognized 32 species. There is little doubt that future investigations will identify more taxa. No subfamilies are recognized.

Burmese python (*Python molurus bivittatus*) laying eggs. (Photo by Animals Animals ©Jim Tuten. Reproduced by permission.)

Physical characteristics

While pythons are widely known for their great size, only three of the 32 species exceed 20 ft (6.1 m) in length. The smallest of the three is the Indian python, *Python molurus*, with records of wild specimens ranging from 19 to 22 ft (5.8 to 6.7 m). As of 2002, there were numerous living captive specimens that are purported to exceed 22 ft (6.7 m) and 300 lb (136 kg). Records for the African rock python, *Python sebae*, range from 28 to 32 ft (8.5 to 9.8 m). The most commonly cited maximum length for any snake is 33 ft (10.1 m); that length is based on a reticulated python *P. reticulatus* killed in Sulawesi in 1912. The largest snake ever kept in captivity was a reticulated python named Colossus, kept at the Pittsburgh Zoo from 1949 until 1956; that snake was measured reliably as 28 ft (8.5 m), 6 in (15 cm) and weighed 320 lb (145 kg).

Most pythons are cryptically patterned with blotches, bands, or rings. About one third of the species are patternless, nearly patternless, or have a patternless appearance. Most species undergo a color change as they mature. Many pythons have the ability to change color to a slight degree, but three taxa, *Morelia oenpelliensis*, *Morelia carinata*, and *Apodora papuana*, can change the hue and intensity of their color dramatically and exhibit about the highest development of this ability among all snakes. Many pythons display iridescent colors that reflect off their skins. The ringed python (*Bothrochilus boa*), the white-lipped python (*Leiopython albertisii*), and the black python (*Morelia boeleni*) are among the most iridescent snakes in the world.

Distribution

Pythons occur in sub-Saharan Africa, southern Asia, Southeast Asia, southeastern China, Indonesia, the Philippines, Papua New Guinea, and Australia. There are no pythons in the Western Hemisphere. The genus *Python* is the most widespread, occurring in Africa, Asia, Philippines, and western Indonesia in the Greater Sundas, Lesser Sundas, and Sulawesi. The genus *Morelia* is distributed extensively throughout Australia, New Guinea, and the Indonesian province of Maluku. *Liasis* occurs in northern Australia, central-southern

New Guinea, and islands in the Indonesia province of Nusa Tenggara. *Aspidites* is endemic to Australia. *Antaresia*, long considered to be endemic to Australia, was discovered in 2001 to occur in south-central New Guinea. *Apodora* and *Leiopython* occur through the lower elevations of New Guinea and nearby islands. *Bothrochilus* is restricted to islands in the Bismarck Archipelago of Papua New Guinea.

Habitat

Pythons can be found in a variety of habitats, including harsh deserts, wetlands, gum tree forest, open woodlands, savanna, rocky slopes, and rainforests. For about six months of the year, water pythons (*Liasis fuscus*) are essentially aquatic, living in the vast monsoon-flooded plains of northern Australia. The Lesser Sundas python, *Python timoriensis*, is a terrestrial species adapted to the rolling grasslands of Flores and nearby smaller islands. The green python, *Morelia viridis*, has obvious adaptations for an arboreal life in forests, including green coloration, a laterally compressed body, and a long tail adapted for grasping.

Desert-adapted python species, such as the woma, *Aspidites ramsayi*, of central Australia and the Angolan python, *Python anchietae*, found in the rocky escarpment along the eastern margin of the Namib desert in Angola and Namibia, survive in areas that receive little or no precipitation in some years. Contrast that to the ringed python living on New Ireland with more than 400 in (more than 10 m) of annual precipitation.

Most python species occurring in New Guinea can be found at elevations from sea level to at least 5,000 ft (1,500 m). The African rock python, *P. sebae*, has been recorded at elevations up to 7,500 ft (2,300 m). The black python, *M. boeleni*, endemic to the New Guinea highland, holds the elevation record for pythons; it is encountered most commonly at elevations of 5,500–8,000 ft (1,700–2,400 m), living on eroded karst slopes overgrown with low heather and scrub brush.

Behavior

Little is known about the behavior of pythons in nature. The 1990s saw the first radio-tracking studies documenting the behavior, natural history, and ecology of several python taxa, mostly in Australia. Pythons tend to be nocturnal, but they often are encountered moving or basking during the day. Faced with a perceived threat, a python will display an instinctive and stereotypic defensive behavior. Most python species incorporate hissing, striking, and biting into their defense, along with the release of musk and feces. The ball python, *Python regius*, rarely actively defends itself, preferring to coil the body into a ball, with the head pulled into the center of the coils. The rough-scaled python, *Morelia carinata*, exhibits an open-mouth threat display, extending its open mouth toward the perceived threat and exhibiting its extraordinarily long teeth while waving the head from side to side.

Most pythons occur in temperate or tropical climates. There is anecdotal evidence that suggests that most taxa are inactive for a portion of the year, during which time breeding takes place. The timing of this inactive period usually is dictated by seasonal heat, drought, or cool weather. The di-

Green python (*Morelia viridis*) wrapped around a tree branch. (Photo by JLM Visuals. Reproduced by permission.)

amond python, *Morelia s. spilota*, of southeastern Australia hibernates during the winter months, often choosing rock shelters with northern and western exposures.

Feeding ecology and diet

Pythons are primarily ambush hunters that consume vertebrate prey. Pythons can and do incorporate active foraging behavior as well. Lizards make up a large percentage of the diet of small pythons, both small taxa and the young of larger taxa. With an increase in size, mammals become an increasingly significant percentage of the diet. Many pythons consume birds when given the opportunity, but birds make up a small portion of the diet of pythons in nature. Pythons occasionally consume snakes; the genera *Aspidites*, *Antaresia*, *Apodora*, and *Bothrochilus* regularly include snakes in their diets. There are records of attacks on humans by the three largest species, dating back to the early eighteenth century. While such predation is rare, it is undisputed that reticulated pythons, African pythons, and Indian pythons grab and constrict humans from time to time and then attempt to consume them; they sometimes succeed.

Reproductive biology

All pythons reproduce by laying eggs. Python eggs range from the size of a grape in the case of the pygmy python,

Antaresia perthensis, to the size of a medium white potato in the case of the reticulated python, *P. reticulatus*. Python eggs are white, and when freshly laid the shells are taut and leathery to the touch. Typically, all or most of the eggs in a clutch adhere together for the duration of incubation.

When eggs are laid, female pythons tightly wrap coils around the eggs and remain with their clutches until they hatch. During incubation, the females of many species of pythons are capable of elevating their body temperatures. A female accomplishes this thermal feat by increasing her metabolic rate either through rhythmic muscle contractions that give the impression that she is shivering or by isotonic muscle contractions that allow her to remain motionless. Some pythons are observed to supplement their thermal exertions during incubation by briefly leaving their clutches to bask and then returning to the task of incubating the eggs when their bodies have been warmed by the sun.

Conservation status

All pythons are listed as Vulnerable by the IUCN and are on CITES Appendix II. A subspecies of the Indian python, *P. m. molurus*, is listed as Endangered by IUCN and is on CITES Appendix I. It also is considered Endangered by the U.S. Endangered Species Act. Pythons receive varying degrees of protection in the countries in which they occur. Australia, India, Papua New Guinea, South Africa, and Namibia are notable for the degree of protection they afford to their indigenous pythons. Little is known about the status of any python species in the wild. No baseline population density studies are published. There is anecdotal evidence to suggest that populations of several of the larger species, notably reticulated pythons and Indian pythons, are in decline or have been locally extirpated. At this time there is no evidence that any species is endangered by human activities.

Significance to humans

Throughout their range, pythons are considered an important natural resource by indigenous peoples. Pythons are consumed as food, and python parts are used in folk medicine. Python skins from the larger species are a valuable commodity. During the 1990s about a million python skin exports were reported annually to CITES, but authorities believe that the actual number is higher. Many of the larger python species are considered to be livestock predators. Throughout many rural areas of Indonesia and Southeast Asia, reticulated pythons are feared as human predators. Conversely, pythons are revered by some cultures. They are kept in captivity all over the world. All species but one have been reproduced in captivity. The captive breeding of pythons for pets is a multimillion-dollar annual enterprise in the United States and Europe.

1. Pygmy python (*Antaresia perthensis*); 2. Green python (*Morelia viridis*); 3. Halmahera python (*Morelia tracyae*); 4. Papuan python (*Apodora papuana*); 5. Blood python (*Python brongersmai*); 6. Reticulated python (*Python reticulatus*); 7. Ball python (*Python regius*); 8. Black-headed python (*Aspidites melanocephalus*). (Illustration by Brian Cressman)

Species accounts

Pygmy python
Antaresia perthensis

TAXONOMY

Antaresia perthensis Stull, 1932, Perth, western Australia (erroneous).

OTHER COMMON NAMES

English: Ant hill python.

PHYSICAL CHARACTERISTICS

This is considered to be the smallest python species; most adults measure 18.5–22 in (48–56 cm).

DISTRIBUTION

Pygmy pythons are found in the Pilbara region of Western Australia.

HABITAT

The species most often is encountered in areas of rocky hills vegetated in shrubs and spinifex. In some areas they shelter in termite mounds.

BEHAVIOR

Little is known of the pygmy python in the wild. It is most often encountered crossing roads at night. In captivity the pygmy python is known to be a hardy species that does well for its keepers.

FEEDING ECOLOGY AND DIET

Pygmy pythons are primarily lizard eaters; small mammals are included in the diet of adults.

REPRODUCTIVE BIOLOGY

A pygmy python usually becomes sexually mature in its third year. Females are larger than males, though the difference is not great in older animals. The eggs of this species measure about 1.5 in (38 mm) in length. Clutch size ranges up to 10 eggs. Hatchlings are about 7–8 in (18–20 cm) in length. Babies have a more defined pattern than the adults.

CONSERVATION STATUS

Not threatened. Found only in the state of Western Australia, pygmy pythons are afforded strict protection.

SIGNIFICANCE TO HUMANS

The pygmy python exists largely unseen and undisturbed by humans. ◆

Papuan python
Apodora papuana

TAXONOMY

Apodora papuana Peters and Doria, 1878, Ramoi, Sorong Peninsula, Irian Jaya, Indonesia.

OTHER COMMON NAMES

English: New Guinea olive python.

PHYSICAL CHARACTERISTICS

This is a large, elongate, athletic python with a large head,

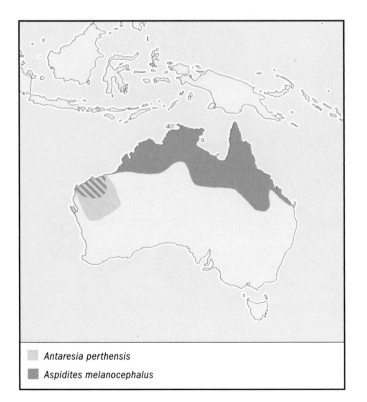

Antaresia perthensis
Aspidites melanocephalus

Apodora papuana

large eyes, and a long, deeply forked tongue. The lining of the mouth is dark. Adults are 10–14 ft (3–4.3 m) in length. This is the largest New Guinea python, with a record size of 16 ft (4.9 m), 10 in (25 cm).

DISTRIBUTION
The species is found throughout the lower elevations of New Guinea and on nearby islands, including Misool, Biak, and Fergusson.

HABITAT
The Papuan python is widespread through a variety of habitats but is associated strongly with river gallery forest.

BEHAVIOR
Papuan pythons have a remarkable ability to change color; their heads turn from pale gray to dark brown and their bodies from mustard yellow to dark brown.

FEEDING ECOLOGY AND DIET
They are recorded to eat a wide variety of vertebrates, including snakes, wallabies, flying fox, bandicoots, and rodents.

REPRODUCTIVE BIOLOGY
A Papuan python usually becomes sexually mature in its fifth or sixth year. Females have a larger average size than males. The eggs of this species measure about 4 in (11 cm) in length. Clutch size ranges up to 28 eggs. Hatchlings are about 24 in (61 cm) in length. Babies are darker in color than adults.

CONSERVATION STATUS
This python is rarely encountered; nothing is known about its status in the wild.

SIGNIFICANCE TO HUMANS
The Papuan python is largely undisturbed by humans. Some indigenous cultures do hunt the species for its meat and skin. ◆

Black-headed python
Aspidites melanocephalus

TAXONOMY
Aspidites melanocephalus Kreft, 1864, Bowen, Queensland, Australia.

OTHER COMMON NAMES
French: Pythons à tête noire; German: Schwarzkopfpythons.

PHYSICAL CHARACTERISTICS
This is a slender, muscular snake with a long head that is indistinct from the neck. The head and neck of this python look as if they have just been dipped into shiny black lacquer.

DISTRIBUTION
This species is found across the northern third of Australia.

HABITAT
These pythons are most commonly encountered in wooded savanna and open forest, but they are reported from tropical forest, grasslands, scrub lands, open sclerophyll forest, and open rocky habitats.

BEHAVIOR
Black-headed pythons often inhabit the burrows of other animals, but they are capable of excavating burrows on their own.

FEEDING ECOLOGY AND DIET
Lizards and snakes make up most of the natural diet of this species, including venomous snakes. Mammals and birds occasionally are taken.

REPRODUCTIVE BIOLOGY
A black-headed python usually becomes sexually mature in its forth or fifth year. Females are usually larger than males, but some older males do attain similar size. The eggs of this species measure about 3.5 in (8.9 cm) in length. Clutch size ranges up to 18 eggs. Hatchlings are about 24 in (61 cm) in length. Babies have brighter colors and a more contrasting pattern than adults.

CONSERVATION STATUS
Little is known about the status of wild populations, but the species has a large range, most of it unoccupied and unmodified by humans.

SIGNIFICANCE TO HUMANS
The species exists largely unseen and undisturbed by humans. The common knowledge that this python species eats venomous snakes does afford it some protection when it encounters people. ◆

Halmahera python
Morelia tracyae

TAXONOMY
Morelia tracyae Harvey Barker, Ammerman, and Chippindale, 2000, "near Tobelo", Halmahera, Maluku, Indonesia.

OTHER COMMON NAMES
None known.

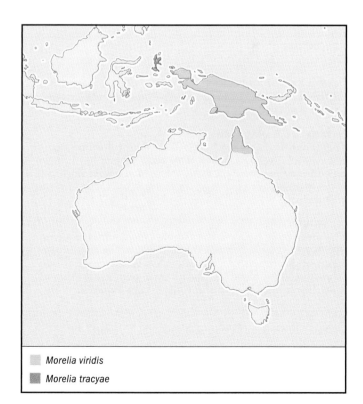

Morelia viridis
Morelia tracyae

PHYSICAL CHARACTERISTICS

This is a long, slender animal with a large head, large eyes, and long, straight teeth. This species of *Morelia* has large head plates. It is one of the least-known pythons; most adults of record are 8–11 ft (2.4–3.4 m) in length.

DISTRIBUTION

This python is endemic to Halmahera, a large island in the Maluku Province of eastern Indonesia.

HABITAT

The Halmahera python is found in primary and secondary forest; it is equally at home on the ground or in the trees.

BEHAVIOR

Halmahera pythons are generally docile and deliberate in their actions.

FEEDING ECOLOGY AND DIET

Little is known about the diet of this species. Local people say that it is often found at fruit bat rookeries.

REPRODUCTIVE BIOLOGY

Not known. As of August 2002, this is the only python species that has not been bred in captivity.

CONSERVATION STATUS

Halmahera python skins are harvested, but the numbers taken are not known, because until recently the species has been identified as *Morelia amethistina*.

SIGNIFICANCE TO HUMANS

Halmahera pythons are eaten by some peoples in Halmahera, and the species is persecuted as a predator of domestic chickens. ◆

Green python
Morelia viridis

TAXONOMY

Morelia viridis Schlegel, 1872, Aru Islands, Indonesia.

OTHER COMMON NAMES

English: Green tree python; French: Python vert; German: Grüner Baumpython.

PHYSICAL CHARACTERISTICS

This is one of the most identifiable of the pythons because of its beautiful green coloration. This small python has a finely scaled head; long, straight teeth; a tented vertebral ridge; and a strongly prehensile tail. Most adults are 4.5–6 ft (1.4–1.8 cm) in length; the maximum size approaches 7.1 ft (2.2 m).

DISTRIBUTION

The green python ranges throughout New Guinea from sea level up to elevations of 6,000 ft (1,800 m). The species also occurs on nearby islands, including Aru Islands, Biak, Misool, Salawati, and Normanby Island. A small population exists in Australia on the east side of the Cape York Peninsula.

HABITAT

Green pythons are found in primary and secondary forest.

BEHAVIOR

Green pythons often exhibit caudal-luring behavior when they are hungry. They seductively wriggle the distal portion of their tails; lizards are attracted to the motion.

FEEDING ECOLOGY AND DIET

Youngsters feed primarily on skinks and geckos. Adults apparently feed on the ground—their diet is made up largely of terrestrial rodents. Birds are rarely consumed.

REPRODUCTIVE BIOLOGY

A green python usually becomes sexually mature in its third year. Females have a larger average size than males. The eggs of this species measure about 1.6 in (4 cm) in length. Clutch size ranges up to 30 eggs. Hatchlings vary from about 11–14 in (28–36 cm) in length. Babies are very differently colored than adults; many babies are yellow, some are brick red or dark red.

CONSERVATION STATUS

This is the most widespread and common python species in New Guinea.

SIGNIFICANCE TO HUMANS

Indigenous people are known to eat green pythons. ◆

Blood python
Python brongersmai

TAXONOMY

Python brongersmai Stull, 1938, Singapore.

OTHER COMMON NAMES

English: Short python, short-tailed python; French: Python malais; German: Buntpython.

PHYSICAL CHARACTERISTICS

Average adults of this heavy-bodied species are 4–4.7 ft (1.2–1.4 m) in length. The maximum length approaches 8.5 ft (2.6 m). In most populations there are several color phases, including red, brown, yellow, and orange, with red being the most common.

Python reticulatus

Python brongersmai

DISTRIBUTION

The species occurs in the lowlands of eastern Sumatra; on islands in the Strait of Malacca, including Bangka, Riau, and the Lingga islands; and throughout Western Malaysia, barely entering southwestern Thailand. The species is not believed to occur in Singapore, despite that being the type locality.

HABITAT

Blood pythons are found in primary and secondary forest, open woodlands, and dense bamboo. The species appears to thrive in the palm-oil plantations of eastern Sumatra. They are encountered under piles of leaves and fronds that are trimmed from the palm-oil palms, and they are seen seeking shelter in burrows.

BEHAVIOR

When forced to defend itself, a blood python will face its attacker and strike. If pressed, it may eject feces and musk while thrashing its tail. Its physical presence is emphasized by flattening its body and moving in quick jerks, thereby drawing attention to its large girth and increasing the impression of its size.

FEEDING ECOLOGY AND DIET

At all ages this species feeds predominantly on rodents.

REPRODUCTIVE BIOLOGY

A blood python usually becomes sexually mature in its third year. Females are usually the larger sex, but older males sometimes attain equal size. The eggs of this species measure about 3–3.5 in (7.6–8.9 cm) in length. Clutch size ranges up to 29 eggs. Hatchlings are about 18 in (46 cm) in length. Babies are patterned as adults and usually tan in color.

CONSERVATION STATUS

Not threatened, but about 50,000 skins a year are reported to CITES by Indonesia, the largest producer.

SIGNIFICANCE TO HUMANS

This species is commonly kept in captivity. The source of most live juveniles exported to captivity is eggs harvested from gravid females brought to skinning businesses. ◆

Ball python
Python regius

TAXONOMY

Python regius Shaw, 1802, no type locality specified.

OTHER COMMON NAMES

English: Royal python, West African python; French: Python royal; German: Königspython.

PHYSICAL CHARACTERISTICS

This is a small, heavy-bodied species with a slender neck and large head. Most adults measure 3.5–5 ft (1.1–1.5 m). The maximum length approaches 6.5 ft (2 m).

DISTRIBUTION

Ball pythons are found in sub-Saharan central Africa, from Senegal to Liberia and east through Nigeria, Cameroon, and Chad to Sudan and Uganda.

HABITAT

They are most commonly found in grasslands, savannas, open woodlands, and agricultural areas. Ball pythons often shelter in rodent burrows.

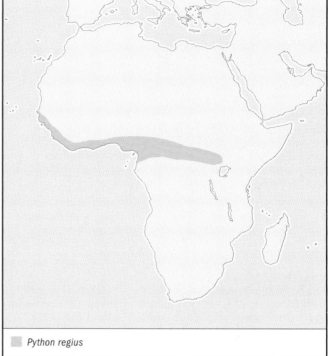

Python regius

BEHAVIOR

Ball pythons are named for their defensive behavior of coiling into a ball, protecting the head in the center of the coils.

FEEDING ECOLOGY AND DIET

At all ages, this species feeds predominantly on rodents.

REPRODUCTIVE BIOLOGY

A ball python female usually becomes sexually mature in its third year; males have been know to successfully breed at less than a year of age. Females are generally larger than males. The eggs of this species measure about 3.5 in (8.9 cm) in length. Clutch size ranges up to 16 eggs. Hatchlings are about 16 in (41 cm) in length. Babies are similar to adults in color and pattern.

CONSERVATION STATUS

Not threatened. There is anecdotal evidence that ball python populations have benefited from the loss of forest that has occurred throughout their distribution.

SIGNIFICANCE TO HUMANS

This is the most common python in captivity. About one million live specimens were exported to captivity in the 1990s. They breed readily in captivity. This species holds the longevity record for all snakes, based on a specimen in the Philadelphia Zoo that was captive for more than 49 years. ◆

Reticulated python
Python reticulatus

TAXONOMY

Python reticulatus Schneider, 1801, no type locality specified in original description but later designated as "Java".

OTHER COMMON NAMES
English: Regal python; French: Python réticulé; German: Netzpython.

PHYSICAL CHARACTERISTICS
This is a giant python, one of the largest snake species. Hatchlings measure 18–35 in (46–89 cm) in length. Most adults are 12–15 ft (3.7–4.6 m), and specimens of 20 ft (6.1 m) are not uncommon.

DISTRIBUTION
This species occurs on the Nicobar Islands in India and throughout most of Southeast Asia from southeastern Bangladesh east to Vietnam and south through western Malaysia to Singapore. The species is widespread throughout the Philippines and Indonesia.

HABITAT
Throughout their extensive range, reticulated pythons can be found in a variety of habitats, including dense forest, open woodlands, rocky areas, caves, swamps, rivers, and lakes. This species is seldom found far from fresh water.

BEHAVIOR
The keepers of reticulated pythons report that there is geographic variation of the temperament of this species. For example, the reticulated pythons of central Thailand and of the Lesser Sundas Islands of Indonesia can be expected to be calm and docile snakes in captivity; from other areas, such as the island of Sulawesi in Indonesia, reticulated pythons are typically irritable and defensive.

FEEDING ECOLOGY AND DIET
Taking advantage of their large size, reticulated pythons are known to consume a wide variety of prey, including primates, pangolins, rodents, canids, felids, waterfowl, pigs, and cervids.

REPRODUCTIVE BIOLOGY
A reticulated python usually becomes sexually mature in its third or fourth year. At the onset of maturity, males are usually 7–10 ft (2.1–3.1 m) in length. Females become mature at 10–13 ft (3.1–4 m). The eggs of this species measure 4–5 in (10–13 cm) in length. Clutch size can exceed 100 eggs. Hatchlings are 24–35 in (61–89 cm) in length. Babies are similar to adults in color and pattern.

CONSERVATION STATUS
Nothing is known about the numbers in the wild. More than half a million skins of reticulated pythons are harvested officially each year, and the actual numbers are likely greater. There is anecdotal evidence and reports that populations are in decline in some areas where there are active skinning businesses and in areas of dense human population, but throughout most of the range the species is believed to be holding its own.

SIGNIFICANCE TO HUMANS
Reticulated pythons are hunted for meat, skin, and parts for folk medicine. They also are persecuted as predators of domestic livestock and feared as predators of humans. They are common in captivity, but the large size of the species makes them unsuitable for most keepers. ◆

Resources

Books

Barker, David G., and Tracy M. Barker. *Pythons of the World.* Vol. 1, *Australia.* Lakeside, CA: Advanced Vivarium Systems, Inc., 1994.

Cogger, Harold, G. *Reptiles and Amphibians of Australia.* Sydney, Australia: Reed New Holland, 2000.

Greene, Harry W. *Snakes: The Evolution of Mystery in Nature.* Berkeley: University of California Press, 1997.

Minton, Sherman A., and Madge Rutherford Minton. *Giant Reptiles.* New York: Charles Scribner's Sons, 1973.

Murphy, John C., and Robert W. Henderson. *Tales of Giant Snakes: A Historical Natural History of Anacondas and Pythons.* Malabar, FL: Krieger Publishing Company, 1997.

O'Shea, Mark. *A Guide to the Snakes of Papua New Guinea.* Port Moresby, Papua New Guinea: Independent Publishing Group, 1996.

Pope, Clifford Millhouse. *The Giant Snakes: The Natural History of the Boa Constrictor, the Anaconda, and the Largest Pythons, Including Comparative Facts About Other Snakes and Basic Information on Reptiles in General.* New York: Alfred A. Knopf, 1961.

Torr, Geordie. *Pythons of Australia: A Natural History.* Sydney, Australia: University of New South Wales Press, 2000.

Periodicals

Kluge, Arnold G. "*Aspidites* and the Phylogeny of the Pythonine Snakes." *Records of the Australian Museum Supplement* 19 (1993): 1–78.

McDowell, S. B. "A Catalogue of the Snakes of New Guinea and the Solomons, with Special Reference to Those in the Bernice P. Bishop Museum. Part 2. Anilioidae and Pythonidae." *Journal of Herpetology* 9, no. 1 (1975): 1–79.

Weigel, J., and T. Russell. "A Record of a Third Specimen of the Rough-Scaled Python, *Morelia carinata.*" *Herpetofauna* 23, no. 2 (1993): 1–5.

David G. Barker
Tracy M. Barker

Splitjaw snakes

(Bolyeriidae)

Class Reptilia
Order Squamata
Family Bolyeriidae

Thumbnail description
A monotypic family consisting of a species of small, slender snakes characterized by a divided and hinged maxilla (upper jaw bone)

Size
Maximum size approaches 4 ft (1.3 m); largest specimen caught in the wild, collected by the Jersey Wildlife Preservation Trust in 1977, weighed 17 oz (510 g)

Number of genera, species
1 genus; 1 species

Habitat
Found throughout Round Island, most often in or near the remnants of forest

Conservation status
Endangered

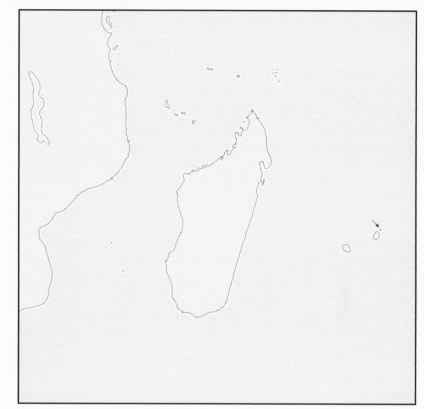

Distribution
Restricted to Round Island, a small island north of Mauritius in the Indian Ocean

Evolution and systematics

Two genera of bolyeriid snakes survived into the late twentieth century on Mauritius and several closely associated small islands. This family is believed to be the modern descendant of an early lineage of macrostomatan snakes that diverged from the alethinophidians at the end of the Cretaceous. It is distinguished by a character that is unique among all vertebrate animals: the maxilla is divided into separate anterior and posterior parts by a movable joint. The joint is located beneath the eye; it allows the front of the upper jaw to bend up or down independently of the rear portion of the jaw.

For many years the Bolyeriidae family was classified as a subfamily within the Boidae. For that reason bolyeriid snakes most often were referred to as "boas." It now is recognized that the Bolyeriidae is a unique lineage not closely allied with the Boidae. In recognition of the uniqueness of the modified maxilla of the Bolyeriidae, the snakes in this family are identified as "splitjaw snakes." The sole member of the Bolyeriidae to survive into the twenty-first century is the keel-scaled splitjaw, *Casarea dussumieri*.

The smooth-scaled splitjaw, *Bolyeria multocarinata*, is the second bolyeriid species known to science. One specimen, identified by a unique scar and believed to be the last survivor of the species, was found in faunal surveys of Round Island in the 1960s and early 1970s. It was last seen in 1974, and the species is considered extinct.

Splitjaw snake (*Casarea dussumieri*). (Illustration by Marguette Dongvillo)

No fossils of this family exist. Other common names of the splitjaw snake include Round Island boa, splitjaw boa, and keel-scaled boa. No subfamilies are recognized.

The taxonomy of the keel-scaled splitjaw is *Casarea dussumieri* Schlegel, 1837, Round Island, Mauritius.

Physical characteristics

The splitjaw snake has a long and flat head that is distinct from the neck. The eye has an elliptical pupil. The snake is slender and has a long tail that equals or exceeds 25% of the total body length. The dorsal body scales are hexagonal, with little overlap. There is no tracheal lung, and the left lung is small and poorly developed, more similar to the condition of colubrid snakes than that of boid snakes. Vestiges of the pelvic bones and cloacal spurs are absent.

Distribution

The Bolyeriidae was endemic to Mauritius and several nearby small islands including Flat Island, Gunner Quoin, and Round Island. The Bolyeriidae are the only snakes native to this region. Today the sole surviving species is restricted to Round Island, a 374-acre (151-ha) island of volcanic origin that lies approximately 13 mi (20 km) north and slightly west of Mauritius.

Habitat

Round Island was once a verdant island, forested with a hardwood scrub plant community that included many endemic and unique species of palms. Introduction of goats and rabbits in the early nineteenth century caused rampant ecological damage, including the near total lost of the forest. The loss of the dominant plant community resulted in dramatic erosion; it is estimated that 90% of the soil was lost, and the island is now crisscrossed with deep gullies. The flora and fauna of Round Island rank among the world's most threatened.

Round Island acted as a final refuge for many Mauritian plants and animals, because rats were never introduced. The plight of Round Island came to the attention of conservationists in the 1970s, in part through the efforts of Gerald Durrell and the Jersey Wildlife Preservation Trust in the United Kingdom. Goats were removed from the island in 1979, and rabbits were exterminated in 1986, protecting some of the remaining plant species from grazing. Keel-scaled splitjaws are found throughout the island, but they are associated strongly with the sparse remaining forest.

Behavior

The splitjaw snake has been encountered at all times of the day, but it is primarily nocturnal. Most often the species is found on the ground, but it has been seen in vegetation up to 8 ft (2.5 m) above the ground. It is believed to shelter in humid burrows.

Feeding ecology and diet

This species is a visually oriented predator of lizards. These snakes have been observed feeding in daylight on a day gecko, *Phelsuma ornata;* the Round Island skink, *Leiolopisma telfairii;* and another skink, *Gongylomorphus bojerii.* It is believed that the snake employs ambush-hunting techniques during the day and forages actively at night. The keel-scaled splitjaw employs an unusual method to stalk lizards at night. The snake raises its head several inches (centimeters) above the substrate and, while continually flicking its tongue, slowly approaches its prey using rectilinear motion. In captivity, adult splitjaws accept mice as prey; the species is not known to feed on mammals in nature.

Reproductive biology

The splitjaw snake is oviparous (egg-laying). Nothing is known about its reproduction in the wild. The species was first bred in captivity in 1982 at the Jersey Wildlife Preservation Trust. In the following years, the species has been bred in numbers in captivity. It has been determined that copulations most likely to result in fertile eggs occur between March and July, with oviposition then taking place between May and October. Clutch size varies from three to 11 eggs.

Conservation status

The species is considered Endangered by the IUCN. In the late 1980s, surveys of Round Island indicated that the population of splitjaws there numbered several hundred animals and the species was apparently stable and holding its own. Between 1977 and 1984, 11 specimens were taken from Round Island to the Jersey Wildlife Preservation Trust in the United Kingdom. It was not until two young were hatched there in 1982 that it was realized that the snake is oviparous. From this nucleus, there is now a stable, self-sustaining captive population that is distributed among several zoos. This provides a buffer against the extinction of the species, should something go wrong on Round Island. Keel-scaled splitjaws breed well in captivity, but hatchlings have proved very difficult to raise; they rarely feed voluntarily and must be force-fed until they are large enough to feed on young mice.

Smooth-scaled splitjaw (*Bolyeria multocarinata*). (Photo by Virtual Museum of Natural History at curator.org. Photo reproduced by permission.)

Because of the presence of rats and other introduced animals on the large island of Mauritius, it is unlikely that keel-scaled splitjaws will be reestablished there. Efforts are underway, however, to remove introduced species from the other small islands included in the original range of the splitjaw snake, and there are plans to reintroduce endemic animals and plants, including splitjaws, when the habitats are recovered sufficiently.

Significance to humans

None known.

Resources

Books

Day, David. *The Doomsday Book of Animals.* London: London Editions Limited, 1981.

Greene, Harry W. *Snakes: The Evolution of Mystery in Nature.* Berkeley: University of California Press, 1997.

Periodicals

Bloxam, Q. M. C., and S. J. Tonge. "The Round Island Boa *Casarea dussumieri* Breeding Program at the Jersey Wildlife Preservation Trust." *Dodo, Journal of the Jersey Wildlife Preservation Trust* 23 (1986): 101–107.

Cundall, David, and Frances Irish. "Aspects of Locomotor and Feeding Behavior in the Round Island Boa *Casarea dussumieri.*" *Dodo, Journal of the Jersey Wildlife Preservation Trust* 23 (1986): 108–111.

Frazzetta, T. H. "From Hopeful Monsters to Bolyerine Snakes?" *American Naturalist* 104, no. 935 (January–February 1970): 55–72.

Guibe, J. "Les Serpents de Madagascar." *Mémoires de L'Institut Scientifique de Madagascar* série A., tome 12 (1958): 190–260.

Kluge, Arnold G. "Boine Snake Phylogeny and Research Cycles." *Miscellaneous Publications, Museum of Zoology, University of Michigan,* no. 178 (1991): 1–58.

McAlpine, Donald F. "Activity Patterns of the Keel-Scaled Boa (*Casarea dussumieri*) at the Jersey Wildlife Preservation Trust." *Dodo, Journal of the Jersey Wildlife Preservation Trust* 18 (1981): 74–78.

McDowell, S. B. "A Catalogue of the Snakes of New Guinea and the Solomons, with Special Reference to Those in the Bernice P. Bishop Museum. Part 2. Anilioidae and Pythonidae." *Journal of Herpetology* 9, no. 1 (1975): 1–79.

David G. Barker, MS
Tracy M. Barker, MS

Woodsnakes and spinejaw snakes

(Tropidophiidae)

Class Reptilia
Order Squamata
Suborder Serpentes
Family Tropidophiidae

Thumbnail description
Small constricting snakes possessing tracheal lung and without functional left lung; in most species males possess cloacal spurs and vestige of pelvic girdle

Size
4–41 in (10–106 cm); 1–16 oz (30–450 g)

Number of genera, species
5 genera; 25 species

Habitat
Open woodland, forest, cloud forest, palm groves, dry scrub forest, rocky hillsides, cliff faces, and caves

Conservation status
No species listed by IUCN

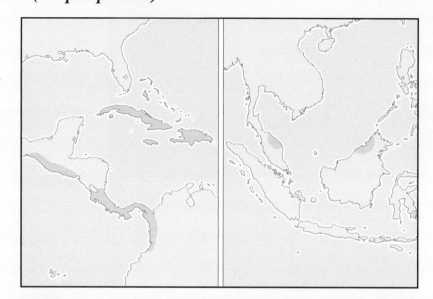

Distribution
Northwestern South America in Colombia and Ecuador; Amazonian Ecuador, Peru, and northwestern Brazil; southeastern Brazil; Central America; southern Mexico; West Indies; peninsular Malaysia and northern Borneo

Evolution and systematics

There are few snake fossils that can be incontrovertibly identified as woodsnakes. The fossil genus *Boavus* dating from the Eocene may belong to the Tropidophiidae. More recent fossils identified as *Tropidophis* are known from Pleistocene deposits in caves.

Woodsnakes are believed to have diverged from basal macrostomatan stock sometime after the divergence of the macrostomatans from alethinophidians at the end of the Cretaceous. Macrostomatan snakes are distinguished by characters of the skull and musculature that allow them increased jaw flexibility, a greater gape, and the ability to consume larger prey.

The woodsnakes are believed to have originated in northern South America and from there spread into Central America and the West Indies. The greatest diversity of woodsnakes exists in Cuba, where in the relative absence of other snakes the tropidophiid lineage became the dominant family of snakes on the large island. The origin and relationships of the two Asian species of spinejaw snakes in the genus *Xenophidion* are unclear at this time.

The phylogenetic relationship of the Tropidophiidae to other snake families is unclear. Traditionally, the sister taxon has been identified as the Boidae, but in recent years many authors have come to identify the Bolyeriidae as the most likely sister taxon.

As is typical of most macrostomatan snakes, most tropidophiid species have the vestiges of a pelvic girdle and cloacal spurs. However, in the Tropidophiidae, only male snakes have the vestigial pelvic girdle and cloacal spurs; *Tropidophis semicinctus* and both species of *Xenophidion* do not have the pelvic girdle and spurs.

Tropidophiid snakes do not have a functional left lung, as is characteristic of all caenophidian snakes. They do have a well-developed tracheal lung, a characteristic of many colubridoid snakes.

The Tropidophiidae has been well investigated and is relatively well-known. However, it is a widely distributed lineage that may be more speciose than is currently recognized. It seems likely that future phylogenetic analyses based on genetic characters will recognize some subspecies and some disjunct populations as new species. Two species, *Tropidophis celiae* and *Tropidophis spiritus*, were described as recently as 1999. The Asian genus *Xenophidion* was recognized in 1996, and tentatively placed in the Tropidophiidae in 2001.

Some authors recognize three distinct lineages within the Tropidophiidae, namely the Tropidophiinae comprised of the genera *Tropidophis* and *Trachyboa*; the Ungaliophiinae comprised of *Ungaliophis* and *Exiliboa*; and the Xenophidiinae with *Xenophidion*. It has been proposed that each of the three lineages should be recognized as a family. Future investigations will undoubtedly address and resolve the relationships of these genera.

Physical characteristics

The woodsnakes and spinejaw snakes are small, inoffensive, boa-like snakes. The bones of the lower jaw and the ves-

Cuban ground boa (*Tropidophis melanurus*) is the largest of the Neotropical woodsnake species. (Photo by Animals Animals ©David M. Dennis. Reproduced by permission.)

tigial pelvic girdle are similar to that of boas, while the absence of a functional left lung and the presence of a tracheal lung are more typical of colubrids. The hyoid apparatus and the condition of the contact between the prefrontal and internasal bones of the snout also are more similar to colubrids than boas.

The smallest woodsnake species is probably *Tropidophis fuscus* of Cuba; it is not known to exceed 12 in (30 cm). The largest species is *Tropidophis melanurus* of Cuba, with a reported maximum length of 41 in (106 cm).

Most woodsnakes and the spinejaw snakes are colored in shades of browns and grays, patterned with muted small blotches or stripes. There is a wide range of variation in the keels on dorsal scales in the genus *Tropidophis*; some species have individuals with smooth dorsal scales and others with strongly keeled dorsal scales. The incredibly rough, keeled, and spiky scalation of *Trachyboa boulengeri* makes it one of the most unusual snakes in the world.

Tropidophis feicki has the ability to change color to a slight degree, turning darker during the day and paler at night. The skin of several smooth-scaled taxa of woodsnakes exhibits iridescence; the large flat scales of *Exiliboa placata* are both shiny and iridescent, while the smaller ruggedly keeled scales of *Trachyboa gularis* give a drab and dusty appearance.

Distribution

The Tropidophiidae is distributed from southeastern Brazil to southern Mexico and the West Indies. The genus *Tropidophis* is the most widespread, with three species found in South America and 15 species in the West Indies. The genus *Trachyboa* occurs from the Choco region of Panama south into Ecuador. *Ungaliophis* is found from the state of Chiapas in southern Mexico south to the Pacific coast of Colombia. *Exiliboa* is known only from two mountain ranges in Oaxaca, Mexico. *Xenophidion* is known from peninsular Malaysia and northern Borneo.

Habitat

Woodsnakes can be found in a variety of habitats, including open woodlands, forest, cloud forest, dry scrub forest, rocky slopes, rainforests, palm groves, agricultural areas, caves, and cliffs. Woodsnakes in general are mesophilic in their habitat preferences, avoiding extremes and existing in environments that are neither too wet nor too dry.

Most tropidophiid species are found at lower elevations from sea level to 2,000 ft (less than 600 m). Two species are found at high elevation: *Exiliboa placata* is found at elevations of 7,500 ft (2,300 m) and *Tropidophis taczanowskyi* occurs in the northern Andes at elevations of 6,500–10,000 ft (2,000–3,000 m).

Behavior

Woodsnakes tend to be nocturnal, but they are often encountered moving or basking during the day. Faced with a perceived threat, most woodsnakes meekly coil into a ball; *Trachyboa* coils into a flat disk with its head in the center. If physically molested, a woodsnake may release odiferous anal secretions. It is rare that any woodsnake bites in defense. Several species of *Tropidophis* are reported to autohemorrhage, spontaneously bleeding from the mouth, nostrils, and eyes when severely stressed.

Most tropidophiid snakes are terrestrial, but many are occasionally observed to climb into bushes, vines, and low trees. The bromeliad woodsnakes, *Ungaliophis panamensis* and *Ungaliophis continentalis*, are probably the most arboreal of the woodsnakes; both species are known to live high in trees, burrowing in the epiphytic growth on large limbs; *U. panamensis* has been accidentally shipped to Europe and the United States in bunches of bananas. *Tropidophis paucisquamis* also has been observed climbing in vegetation 3–10 ft (1–3 m) above the ground.

Feeding ecology and diet

The feeding behavior of woodsnakes is not well known. They have been observed actively foraging, and it is likely that

Dwarf boa (*Ungaliophis continentalis*) is found in Mexico, Guatemala, and Honduras. (Photo by Animals Animals ©David M. Dennis. Reproduced by permission.)

they also incorporate ambush techniques. Anoline lizards comprise a large percentage of the diet of West Indian *Tropidophis* species. Most woodsnakes will accept eleutherodactylid frogs as prey; in captivity, *E. placata* readily accepts small eleutherodactylid frogs as prey, but refuses similar-sized hylid and ranid frogs. Small salamanders and frog eggs have been found in the stomachs of wild *Exiliboa*. Both *Trachyboa* species feed on fishes and amphibians. Large individuals of *T. melanurus* are known to feed on small mammals and birds. Larger adult specimens of many species of *Tropidophis* and both species of *Ungaliophis* are known to accept newborn mice in captivity.

Reproductive biology
All woodsnakes bear live young.

Conservation status
No species of Tropidophiidae is listed on the 2002 IUCN Red List. In the 1990s, the Navassa woodsnake, *Tropidophis bucculentus*, was reported as likely extinct. Habitat disturbance and mongoose predation are reported to be significant factors in the demise of this species.

Little is known about the status of any woodsnake species in the wild. No base-line population density studies are published.

Significance to humans
Woodsnakes in nature exist largely unseen and unbothered by humans. Some species are kept and bred in captivity.

Species accounts

Banded woodsnake
Tropidophis feicki

TAXONOMY
Tropidophis feicki Schwartz, 1957, Cuéva de los Índios, San Ví-
cente, Pínar del Río Province, Cuba.

OTHER COMMON NAMES
English: Banded dwarf boa, Feick's dwarf boa; French: Boa
forestier de Feick; Spanish: Majá.

**PHYSICAL
CHARACTERISTICS**
This is a medium-
sized woodsnake with
a reported maximum
length of 20 in (50.5
cm). The head of
this slender snake is
distinctly wider than
the neck. Banded
woodsnakes are
boldly marked with
17–26 dark bands on
the body. They are
smooth-scaled, shiny, and iridescent.

Tropidophis feicki

DISTRIBUTION
This species is found in the western third of Cuba.

HABITAT
The banded woodsnake is associated with wooded areas. It has
been collected on cliff faces and in caves, as well. While the

species is considered to be predominantly terrestrial, specimens
have been found climbing in trees and vines.

BEHAVIOR
Like most woodsnakes, the banded woodsnake is a calm and
docile snake that coils into a ball when threatened. This beautiful snake tends to be slow and deliberate in its movements.

FEEDING ECOLOGY AND DIET
The banded woodsnake feeds primarily on anoline lizards. It is
reported that in captivity all ages accept appropriately sized
Anolis carolinensis as suitable prey. Large adults will feed on
pink mice in captivity.

REPRODUCTIVE BIOLOGY
This species is viviparous. Little is known about its reproduction in nature.

The San Antonio Zoo reported on the birth of two litters
of *T. feicki*, both born in September 1999. The eight neonates
ranged in length from 5.6 in to 7.3 in (14.4 cm to 18.5 cm)
and in weight from 0.07 oz to 0.12 oz (2.1g to 3.5 g). Prior to
breeding, the two pairs of adult parents were subjected to
both daily and seasonal temperature fluctuations. Throughout
most of the year, temperatures were kept fairly constant, varying only 78–80°F (26–27°C); from December through February the daily temperatures varied from 62.6°F to 80.6°F (17°C
to 27°C).

CONSERVATION STATUS
Nothing is known about the numbers in the wild.

SIGNIFICANCE TO HUMANS
The banded woodsnake exists in nature largely unseen and unmolested by humans. It is one of the most attractive species of
woodsnakes, but few specimens have come from Cuba and the
species is rarely seen in captivity. ◆

Southern bromeliad woodsnake
Ungaliophis panamensis

TAXONOMY
Ungaliophis panamensis Schmidt, 1933, Cérro Brujo, Colón
Province, Panamá.

OTHER COMMON NAMES
English: Bromeliad boa, bromeliad dwarf boa, banana boa;
French: Boa nain; German: Bananenboa; Spanish: Boa enana.

PHYSICAL CHARACTERISTICS
This is a medium-sized woodsnake with a reported maximum
length approaching 30 in (76 cm). This is a slender, smooth-
scaled snake, pale gray or tan with a distinct pattern of black triangular blotches on the back. There is a single large prefrontal
scale, the scale on top of the snout; this character distinguishes
this genus from other tropidophiids. Females do not have cloacal

Tropidophis feicki
Ungaliophis panamensis

spurs, but males have prominent large spurs.

DISTRIBUTION
The southern bromeliad woodsnake occurs at low to moderate elevations in southeastern Nicaragua, Costa Rica, Panama, and western Colombia.

HABITAT
Southern bromeliad woodsnakes are associated with primary and secondary forest. The species has been encountered on the ground and has been collected in the verdant epiphytic growth of large trees when they are felled.

ESD©2003

Ungaliophis panamensis

BEHAVIOR
This is a very pleasant snake to handle, being inoffensive by nature and deliberate and docile in actions. The southern bromeliad woodsnake does not bite in defense. When threatened or molested, it coils into a ball. Only rarely does this species discharge its odiferous anal secretions when molested.

FEEDING ECOLOGY AND DIET
In nature, it is believed that this species feeds primarily on small lizards and frogs. In captivity, all ages will usually accept appropriately sized *Anolis sagrei* and *Anolis carolinensis* lizards as prey; adults usually feed on appropriately sized rodents.

REPRODUCTIVE BIOLOGY
This species is viviparous. Very little is known about the reproduction of this species in nature or captivity. Neonates are about 6 in (15 cm) in length.

CONSERVATION STATUS
Nothing is known about the numbers in the wild.

SIGNIFICANCE TO HUMANS
This species is rarely kept in captivity. In nature the southern bromeliad woodsnake is rarely observed. The species is largely unseen and unmolested by humans. ◆

Resources

Books

Crother, Brian I., ed. *Caribbean Amphibians and Reptiles.* San Diego: Academic Press, 1999.

Duellman, William E., ed. *The South American Herpetofauna: Its Origin, Evolution and Dispersal.* Monograph of the Museum of Natural History, Number 7. Lawrence: The University of Kansas, 1979.

Greene, Harry W. *Snakes: The Evolution of Mystery in Nature.* Berkeley: University of California Press, 1997.

McDiarmid, Roy W., Jonathan A. Campbell, and T'Shaka A. Touré. *Snake Species of the World.* Washington, DC: The Herpetologists' League, 1999.

Schwartz, Albert, and Robert W. Henderson. *Amphibians and Reptiles of the West Indies.* Gainesville: University of Florida Press, 1991.

Tolson, P. J., and R. W. Henderson. *The Natural History of West Indian Boas.* Taunton: R & A Publishing Limited, 1993.

Zug, G. R., L. J. Vitt, and J. P. Caldwell. *Herpetology.* 2nd ed. San Diego: Academic Press, 2001.

Periodicals

Barbour, Thomas, and Charles T. Ramsden. "The Herpetology of Cuba." *Memoirs of the Museum of Comparative Zoology* XLVII, no. 2 (May 1919): 1–213 plus 15 plates.

Bogert, Charles M. "Variations and Affinities of the Dwarf Boas of the Genus *Ungaliophis.*" *American Museum Novitates* no. 2340 (August 9, 1968): 1–26.

———. "A New Genus and Species of Dwarf Boa from Southern Mexico." *American Museum Novitates* no. 2354 (December 18, 1968): 1–38.

Burger, R. Michael. "The Arboreal Burrower: The Dwarf Boa *Ungaliophis.*" *Vivarium* 7, no. 2 (June 1998): 46–49.

———. "Observations on Courtship Behavior in *Ungaliophis* Mueller." *Bulletin of the Chicago Herpetological Society* 31, no. 4 (April 1996): 57–59.

———. "The Bromeliad Boa (*Ungaliophis continentalis*)." *Reptiles* 6, no. 6 (June 1998): 12–14.

McDowell, S. B. "A Catalogue of the Snakes of New Guinea and the Solomons, with Special Reference to Those in the Bernice P. Bishop Museum. Part 2. Anilioidae and Pythoninae." *Journal of Herpetology* 9, no. 1 (1975): 1–80.

Schwartz, Albert. "A New Species of Boa (genus *Tropidophis*) from Western Cuba." *American Museum Novitates,* no. 1839 (August 19, 1957): 1–8.

Schwartz, Albert, and Robert J. Marsh. "A Review of the *Pardalis-maculatus* Complex of the Boid Genus *Tropidophis* of the West Indies." *Bulletin of the Museum of Comparative Zoology* 123, no. 2 (1960): 49–89.

Stull, Olive Griffith. "A Revision of the Genus *Tropidophis.*" *Occasional Papers of the Museum of Zoology,* no. 135 (October 1, 1928): 1–51.

Wallach, V., and R. Günther. "Visceral Anatomy of the Malaysian Snake Genus *Xenophidion,* Including a Cladistic Analysis and Allocation to a New Fmaily." *Amphibia Reptilia* 19, no. 4 (1998): 385–404.

Resources

Wilcox, T. P., Derrick J. Zwickl, Tracy A. Heath, and David M. Hillis. "Phylogenetic Relationships of the Dwarf Boas and a Comparison of Bayesian and Bootstrap Measures of Phylogenetic Support." *Molecular Phylogenetics and Evolution* 25, no. 2 (2002): 361–371.

David G. Barker, MS
Tracy M. Barker, MS

File snakes
(Acrochordidae)

Class Reptilia
Order Squamata
Suborder Serpentes
Family Acrochordidae

Thumbnail description
Small to medium-sized aquatic snakes with blunt heads not distinct from the neck, dorsally directed small eyes and nares, flabby and roughened skin bearing small, spinate scales. Body stout and capable of lateral compression for swimming. Sluggish behavior.

Size
20–76 inches (0.5–2 m)

Number of genera, species
1 genus, 3 species

Habitat
Shallow tropical waters associated with coastal mangroves, estuaries, and river systems

Conservation status
Data Deficient, but some populations may be locally threatened

Distribution
Waters of Indo-Australian region, from western India to Southeast Asia, Malaysia, Indonesia, east to Solomon Islands and south to northern Australia

Evolution and systematics

The Acrochordidae, called wart snakes or file snakes, is an unusual family of strictly aquatic snakes consisting of a single genus and three species. The little file snake (*Acrochordus granulatus*) is the smallest of the three species and has a largely marine distribution. It originally was thought to be venomous and was placed in a separate genus (*Chersydrus*). It seems best to include this species in the genus *Acrochordus*, although some systematists have maintained the two separate genera. The Arafura file snake (*A. arafurae*) and the Java file snake (*A. javanicus*) are roughly twice the size of the little file snake and are largely freshwater in distribution. A fourth extinct species is known that existed in the Upper Miocene and Lower Pliocene epochs of Pakistan.

The phylogenetic relationships between acrochordids and other snakes are unclear. The snakes in this family retain some primitive characteristics but have evolved numerous specialized traits. Some morphologic features are so different from those of other snakes that file snakes have been joined in a separate superfamily, the Acrochordoidea. They appear to be related most closely to advanced snakes: colubrids (a large

family including many common and familiar snakes such as garter snakes, rat snakes, and king snakes), elapids (cobras and their relatives), viperids (vipers and pit vipers), and a group of African snakes known as Atractaspididae.

Taxonomy for these species: *Acrochordus arafurae* McDowell, 1979, Papua New Guinea, Western Province, Lake Daviumbo; commonly known as the Arafura file snake. *Acrochordus granulatus* Schneider, 1799, "India"; commonly known as the little file snake. *Acrochordus javanicus* Hornstedt, 1787, Java; commonly known as the Java file snake, wart snake, or elephant trunk snake.

Physical characteristics

File snakes exhibit many fascinating and novel features of morphology, physiology, and behavior. The little file snake is the smallest of the three species. Adults average 20–28 in (50–70 cm) and grow to a maximum length of approximately 40 in (1 m). The other species are nearly twice this size, the Arafura file snake reaching a maximum length of approximately 67 inches (1.7 m) and the Java file snake reaching a maximum length of nearly 80 inches (2 m). All three species

Skin of the Java file snake (*Acrochordus javanicus*). (Photo by Al Savitsky. Reproduced by permission.)

of file snake show sexual dimorphism, females having larger heads, shorter tails, and generally heavier bodies than males of the same species.

A number of the unusual features of anatomy can be interpreted as adaptations for aquatic life and a specialized diet of fishes. File snakes are distinguished by a very loose skin and supple musculature that allow strong lateral compression and enable these snakes to seize and hold struggling fish. The ventral scutes are very small and project downward at the midline to form a compressed ventral keel during swimming. The flabby skin enhances mobility beneath water but sags noticeably when a snake is out of water. The skin is prominent in being roughened by spines or tubercles that project from each of the numerous small scales covering the body. These scales enable grasping of fish and are sensory. A bundle of bristle-like structures is present in the dome of the tubercles, and the base of this structure is richly supplied with nerves. The skin between the small scales may be developed into smaller bristle-bearing tubercles. These are presumed to be sense organs that detect mechanical stimuli and aid in movement, orientation, and the capture of fish in waters where visibility can be extremely limited.

The tail is laterally compressed to assist swimming, although this feature is not as pronounced as it is in sea snakes. The nostrils are valved and located at the dorsal aspect of the snout. This feature enables these snakes to periodically breathe atmospheric air while the remainder of the body remains underwater. The vertebrae are relatively short and have a small condyle that is partially freed such that flexibility is enhanced for both swimming and constriction of fishes. The skull is flexible, quadrate bones are elongated, and features of articulation are well adapted for swallowing fishes.

File snakes have a heart similar to that of other snakes, but the position is more central (mid-body) than it is in many terrestrial species of snakes, in which the heart is closer to the head. Unlike that of terrestrial snakes, the lung of file snakes contains vascularized tissue for respiratory gas exchange that extends almost the entire length of the body cav-

ity. The veins are capacious and accommodate a relatively large volume of blood. This characteristic is presumed to be adaptive with respect to storage of oxygen in support of prolonged dives.

Some interesting metabolic, respiratory, and cardiovascular adaptations are related to aquatic habits. The metabolic rate is relatively low compared with that of other snake species, and the low rate of energy use appears related to the generally sluggish lifestyle of file snakes. Laboratory studies of the Arafura file snake indicate that the capacity for generating metabolic energy is low and cannot sustain vigorous activity for more than a few minutes. These snakes are lung breathers but can remain submerged for several hours. The skin functions as an accessory respiratory organ and exchanges a considerable fraction of oxygen and carbon dioxide when snakes are in well-oxygenated water. Relatively long submergence times are related to the low metabolic rate, cutaneous gas exchange, sluggish behavior, and large oxygen store attributed to the elongated lung and to the presence of a large volume of circulating blood, which contains large amounts of red blood cells and hemoglobin.

File snakes prefer the high body temperatures achieved in shallow tropical waters. Body temperature typically is 77–86°F (25–30°C). These snakes largely conform to the temperatures that prevail in surrounding water, but there is evidence that the Arafura file snake selects specific thermal microhabitats where variation of body temperature is minimized. The little file snake can tolerate a range of water salinity from freshwater to seawater, and the other species tolerate water conditions ranging from fresh to brackish. The little file snake has a sublingual salt gland that is presumed to function in osmoregulation. Little is known, however, about the importance of this gland. Marine populations of this snake need fresh water, which they obtain from surface lenses of fresh water that form temporarily during rainstorms.

A little file snake (*Acrochordus granulatus*) showing knotting behavior. (Photo by H. Lillywhite. Reproduced by permission.)

Acrochordus granulatus

Distribution

Species of Acrochordidae are distributed in tropical waters, but there are important habitat distinctions. The range of the little file snake extends from the western coast of India through tropical Asia to the Philippines, south to Timor, and east to Papua New Guinea, northern Australia, the Bismarck Archipelago, and the Solomon Islands. The Java file snake ranges from Thailand through Malaysia and the Greater Sunda Islands of Indonesia. The range of the Arafura file snake appears confined to the freshwater drainages of Papua New Guinea and those of Australia connected to the Gulf of Carpentaria and the Arafura Sea.

Habitat

The smallest species, the little file snake, is primarily marine but tolerates water of varying salinity, including freshwater. These snakes are found in the sea but live more typically in mangroves or other areas of shallow coastal waters, including estuaries. Little file snakes have been captured at depths of 13–66 ft (4–20 m) as far as 1–6 mi (2–10 km) offshore, but shallow waters of a few feet (meters) or less are more typical of primary habitat. Populations enter rivers throughout the range, and a few populations are known to live in freshwater lakes in Papua New Guinea and the Philippines. The inland distribution of this species is probably limited by waterfalls rather than by lack of tolerance for freshwater.

The Java file snake is largely an inhabitant of lagoons and streams as well as other areas of permanent freshwater. The species also enters estuaries and the sea, but permanent occupation of marine habitats is unlikely. The Java file snake should be considered a freshwater species.

The Arafura file snake is a freshwater species that occupies tropical rivers and billabongs (dry streambeds that fill

only seasonally), reaching high population densities in some of the drainage systems of northern Australia. Much of this habitat is subject to periodic fluctuation in water level caused by seasonal aridity in parts of the range. The snakes live in billabongs during the dry season but disperse into inundated grassland with the onset of wet-season flooding.

Behavior

All three species of file snake are aquatic and appear to be nocturnal. Nighttime activity seems related largely to foraging, and snakes have been observed patrolling shallow tidal pools, where fish tend to become concentrated. During daylight hours these snakes are very reclusive, hiding among tangled mangrove roots, river edges, or in holes and burrows. In captivity, these snakes bury themselves in mud. Java file snakes have been observed to burrow in riverbanks beneath the roots of trees, where they are sometimes found in large aggregations. Java file snakes are occasionally seen swimming, mostly because of the periodic need to breathe air at the water's surface. Although they are adept swimmers, file snakes often move by crawling sluggishly over muddy substrates at the bottom of streams or swamps. Younger, smaller snakes are seldom seen, and very little is known about their ecology.

Although movements of file snakes are generally described as sluggish, data on the Arafura file snake indicate these snakes sometimes make extensive daily excursions. Occasional specimens of the little file snake have been found on tidal mud flats, and it seems likely that file snakes might occasionally leave water to travel between bodies of water during times of tidal or seasonal fluctuation of water level. In at least some instances, however, file snakes are known to remain within a limited area year after year. Because file snakes have low energy requirements and exist in areas where fishes tend to be concentrated, population densities may be very high and are reported to be at least 100 snakes for every 2 acres (1 hectare) in some Australian billabongs.

Little is known about predation on file snakes. They may be eaten by crocodiles, various birds, and other animals and be captured by humans. It seems likely that the physiological and behavioral adaptations for prolonged submergence are related to avoidance of predation in shallow-water habitats.

Feeding ecology and diet

All three species of file snake feed almost exclusively on fishes, including carrion and a large variety of species. The little file snake appears to specialize on gobiid and goby-like fishes. Stomach contents indicate that crustaceans may be eaten occasionally. Australian Arafura file snakes consume a diversity of fish species and act as scavengers as well as predators in watercourse systems where they have been studied. Sleepy cod and barramundi are important prey. Java file snakes specialize on freshwater eels and catfishes in Malaysia. The spines of catfishes occasionally perforate the digestive tracts of these snakes and can cause wounds and abscesses.

File snakes feed by seizing fish rapidly with the mouth or by swiftly ensnaring fish in coils of the body or tail. In either

Acrochordus arafurae
Acrochordus javanicus

matic variability and the aridity of the Australian habitat. Litter size averages five for the little file snake, 17 for the Arafura file snake, and 26 for the Java file snake. Information on Java file snakes indicates that larger females produce both larger litters and larger offspring.

Conservation status

Although no species are listed in the IUCN Red Book, there are two areas of concern in relation to the conservation status of acrochordid snakes. One is commercial exploitation, largely in relation to the skin trade, and the other is depletion of habitat and its quality. Commercial hunting of file snakes and sale of the products are prohibited by law in some countries where these snakes exist, but such protection is either absent or ineffectively enforced throughout much of Asia. Because of slow growth and low reproductive frequency, file snakes may be quite sensitive to harvesting in areas where such exploitation is intensive. On the other hand, much of the habitat of these snakes is relatively inaccessible to large numbers of human hunters. Other potential threats to file snakes are related to physical disturbances of rivers, estuaries, and wetlands and to various forms of water pollution. The effect of such factors on the distribution and abundance of file snakes is almost totally unknown and is in need of evaluation.

Significance to humans

Female Arafura file snakes are hunted by indigenous peoples and used as food in parts of Australia. Because of the pronounced sexual dimorphism that characterizes this species, the larger females are more easily found and captured by aboriginal hunters. Such subsistence use of these snakes by relatively small numbers of people appears to have little effect on populations of this species.

The other two species of file snake are harvested commercially for their skins. The skin of the Java file snake is used for *karung*, ornamental leather, and is heavily harvested in the Indonesian archipelago. It has been difficult to estimate how widespread or intensive such exploitation has become.

Both the Java file snake and the little file snake have made occasional appearances in the pet trade, but use of these snakes as pets appears to be uncommon. Given the difficulties associated with maintaining these snakes in captivity, exploitation related to the pet trade seems to have little importance and may likely disappear altogether.

case, the body is wrapped quickly around the fish to subdue and to hold it. These snakes have been labeled constrictors, but the body coils function to hold and immobilize prey rather than to cause death. Captured fish are swallowed very quickly, thus little water is ingested with prey. File snakes have been observed to forage nocturnally in areas of shallow water, and captive specimens feed more readily and capture fish more easily in shallow than in deep water. Stomach contents indicate file snakes feed infrequently and appear to grow slowly.

Reproductive biology

All three species of file snake appear to have seasonal reproductive cycles, even though they are active year-round in tropical habitats. These snakes are viviparous, and their young are born alive. Gestation begins in the middle of the year, and birth usually occurs in December. Evidence suggests that female snakes do not give birth every year, so the reproductive frequency is less than annual. Little file snakes and Java file snakes reproduce, on average, once every two years. Arafura file snakes reproduce less frequently because of greater cli-

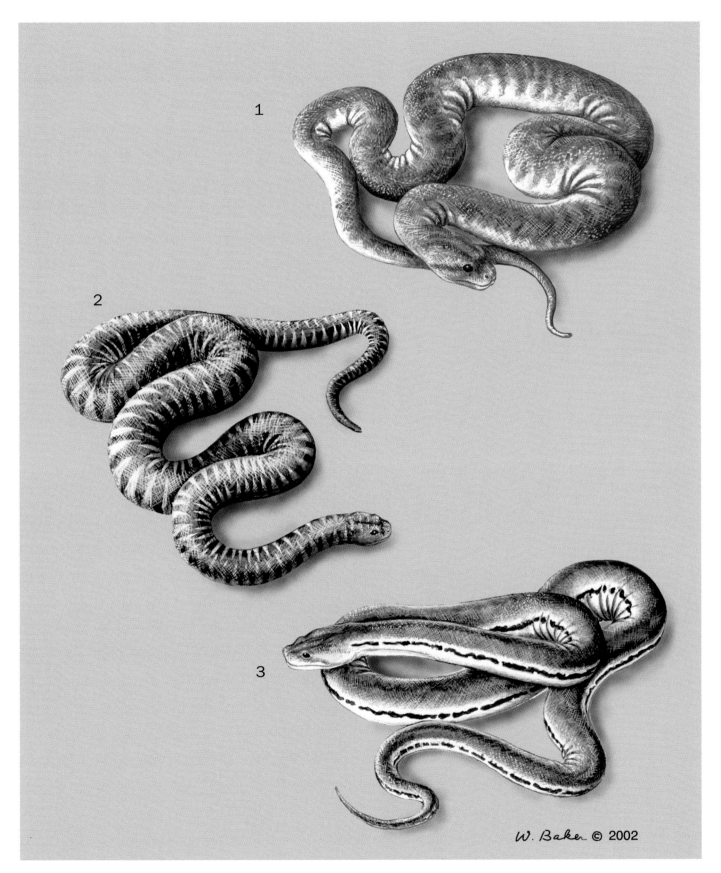

1. Arafura file snake (*Acrochordus arafurae*); 2. Little file snake (*Acrochordus granulatus*); 3. Java file snake (*Acrochordus javanicus*). (Illustration by Wendy Baker)

Resources

Periodicals

Dowling, Herndon.G. "The Curious Feeding Habits of the Java Wart Snake." *Animal Kingdom* 63 (1960): 13–15.

Lillywhite, Harvey B. "The Biology and Conservation of Acrochordid Snakes." *Hamadryad* 16 (1991): 1–9.

———. "Husbandry of the Little File Snake, *Acrochordus granulatus.*" *Zoo Biology* 15 (1996): 315–327.

Lillywhite, Harvey B., and Tamir M. Ellis. "Ecophysiological Aspects of the Coastal-Estuarine Distribution of Acrochordid Snakes." *Estuaries* 17 (1994): 53–61.

McDowell, S. B. "A Catalogue of the Snakes of New Guinea and the Solomons, with Special Reference to Those in the Bernice P. Bishop Museum: Part III, Boinae and Acrochordoidea (Reptilia, Serpentes)." *Journal of Herpetology* 13 (1979): 1–92.

Seymour, Roger S., G. P. Dobson, and John Baldwin. "Respiratory and Cardiovascular Physiology of the Aquatic Snake, *Acrochordus arafurae.*" *Journal of Comparative Physiology* 144 (1981): 215–227.

Shine, Richard. "Ecology of a Low-Energy Specialist: Food Habits and Reproductive Biology of the Arafura Filesnake (Acrochordidae)." *Copeia* 1986 (1986): 424–437.

Shine, Richard, and R. Lambeck. "A Radiotelemetric Study of Movements, Thermoregulation and Niche Utilization in an Aquatic Snake, *Acrochordus arafurae.*" *Herpetologica* 41 (1985): 351–361.

Voris, Harold K., and G. S. Glodek. "Habitat, Diet and Reproduction of the File Snake, *Acrochordus granulatus,* in the Straits of Malacca." *Journal of Herpetology* 14 (1980): 105–108.

Harvey B. Lillywhite, PhD

Vipers and pitvipers
(Viperidae)

Class Reptilia
Order Squamata
Suborder Serpentes
Family Viperidae

Thumbnail description
Small to large venomous snakes with hollow
fangs attached to shortened, movable maxillary
bones

Size
ca. 1–11.8 ft (30–360 cm)

Number of genera, species
36 genera; 256 species

Habitat
Deserts, steppes, mountains, forests,
meadows, and savannas

Conservation status
Critically Endangered: 7 species; Endangered: 4
species; Vulnerable: 7 species; Data Deficient:
1 species

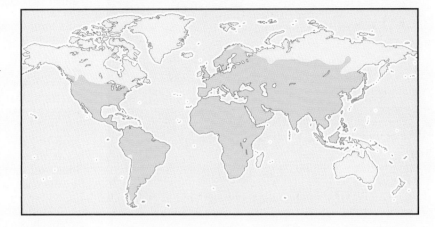

Distribution
Africa, Europe, Asia, North America, Central America, and South America

Evolution and systematics

Researchers have devoted considerable effort to reconstructing the phylogenetic history of the Viperidae, and, consequently, extensive revision of the classification of these snakes was done throughout the end of the twentieth century and has continued into the twenty-first century. Monophyly of the Viperidae is well supported by molecular and anatomical data. Available evidence suggests that the Viperidae is the most basal family of the Colubroidea, which also includes the Colubridae, Elapidae, and Atractaspididae. Four subfamilies are recognized.

The Causinae includes a single genus, *Causus*, with six species; the Viperinae includes 12 genera with 75 species; the Azemiopinae contains a single genus and species (*Azemiops feae*); and the pitvipers are classified in the Crotalinae, in which 22 genera and 174 species are recognized. With the possible exception of the Viperinae, the monophyly of each viperid subfamily is well supported by anatomical and molecular data. A potential synapomorphy of the Viperinae is the ventral course of the facial carotid artery, providing at least some support for the monophyly of this subfamily. The Causinae is thought to be sister to either all other viperids or only the Viperinae. The Azemiopinae is sister to the Crotalinae. As of 2002 almost all genera and subgenera recognized are monophyletic, although relationships among genera are not yet resolved fully.

Viperinae consists of *Vipera*, *Macrovipera*, *Pseudocerastes*, *Eristicophis*, *Daboia*, *Echis*, *Cerastes*, *Atheris*, *Bitis*, *Proatheris*, *Adenorhinos*, and *Montatheris*. The Eurasian genus *Vipera* is divided into the subgenera *Vipera*, *Pelias*, *Montivipera*, and

Acridophaga, each with an evolutionary history that can be traced back to the Miocene through the fossil record or "molecular clock" calibrations. It is the same with the African genus *Bitis*, which is divided into the subgenera *Bitis*, *Macrocerastes*, *Calechidna*, and *Keniabitis*, each with its own evolutionary history, as inferred from molecular data. Unfortunately, there are no viper fossils known from tropical Africa. The genera of Old World pitvipers are *Calloselasma*, *Deinagkistrodon*, *Ermia*, *Gloydius*, *Hypnale*, *Ovophis*, *Protobothrops*, *Triceratolepidophis*, *Trimeresurus*, and *Tropidolaemus*, and the genera of New World pitvipers are *Agkistrodon*, *Atropoides*, *Bothriechis*, *Bothriopsis*, *Bothrocophias*, *Bothrops*, *Cerrophidion*, *Crotalus*, *Lachesis*, *Ophryacus*, *Porthidium*, and *Sistrurus*.

The region of origin for viperids remains undetermined. The earliest fossil specimens of this family are known from the Lower Miocene (ca. 20 million years ago) of Europe and West Asia. These findings are well documented, but there are no fossil records from the Oligocene, and there is a general opinion that the viperines arose somewhere else. The origin of African viperines has been dated to at least 50 million years before the present based on molecular evidence, and there is much support for a tropical African origin of the Viperidae.

The striking examples of convergent evolution between the Asian and American pitviper radiations had long obscured attempts to understand the phylogeny of these snakes; however, data from mitochondrial DNA sequences strongly suggest that pitvipers originated in Asia, and these findings are consistent with other lines of evidence. For example, the closest living relative of all pitvipers, *Azemiops feae*, is an Asian species. The earliest known fossil pitviper from the New World is from the Miocene. All New World pitvipers are descended

Pope's pitviper (*Trimeresurus popeorum*) in the montane rainforest of Malaysia. (Photo by Fletcher & Baylis/Photo Researchers, Inc. Reproduced by permission.)

from a single pitviper species that extended its range across the Bering Land Bridge into North America. Phylogenetic studies suggest that the original American pitviper resembled *Gloydius blomhoffi* in many respects.

Physical characteristics

In vipers or adders (Viperinae) and pitvipers (Crotalinae), the head is roughly triangular and is distinct from the rest of the body. The head bears nine symmetrical plates (as in *Agkistrodon* and *Gloydius*), fragmented head plates (as in many *Vipera*), or numerous small scales (as in *Bitis* and *Ophryacus*). The pupils are usually vertical and elliptical. All the face bones are movable. Each of the two shortened, retractile upper jawbones (maxillae) bears only the tubular venom fang (which can be activated only for a short period of time) and often one to several significantly smaller reserve teeth of various sizes, none being a firmly positioned poison fang. The tail is short, and the male copulatory organ is bifid (forked).

Viperines are generally compact, sturdy snakes, and their length ranges from 11.8 in (30 cm: dwarf puff adder) to 5.9 ft (180 cm: Gaboon adder). Coloration is usually drab, and in the genus *Vipera* it often includes a dark zigzag pattern or a rhomboid band along the back. Desert species are sand yellow, whereas jungle vipers often have a colorful carpet marking.

Pitvipers exhibit great variety of size, shape, and color, but any pitviper can be recognized easily by the deep, heat-sensitive pits, one on each side of the head, between the eye and the nostril. Color variation within the Crotalinae reflects the diversity of habitats occupied by these snakes: invariably, pitvipers are cryptic in their native haunts. In general, pitvipers are relatively stout, although some arboreal species may be more slender. Tail length varies from quite short in rattlesnakes to

relatively long in most arboreal species. The tail is prehensile in species that are adapted fully to an arboreal existence. The longest viperids are pitvipers of the genus *Lachesis*, some specimens of which are known to reach 11.8 ft (3.6 m).

One of the most well-known and interesting features of pitvipers is the rattle, which is unique to the genera *Crotalus* and *Sistrurus*. A neonatal rattlesnake has a keratinized "button" at the tip of its tail; each time the snake sheds, a rattle segment is added. Segments of the rattle fit loosely together, so that a clearly audible sound is produced when the tail is vibrated. Interspecific variation in rattle size has considerable effect on the sound, with large species producing particularly loud and effective warnings. Because most rattlesnakes shed more than once a year and because the ends of rattles tend to break off in wild snakes, the number of rattle segments cannot be used as a direct indication of age.

Sometimes known as night adders, species of *Causus* are relatively small, terrestrial snakes with a stout body covered by weakly keeled scales. Nine plates cover the top of the head. The pupils are round, and the fangs are relatively short. Despite the short fangs, some species have very long venom glands. The sole living species of Azemiopinae is characterized by smooth scales and nine large plates on top of the head.

The most important distinguishing characteristic in all viperids is the venom apparatus. Their poison fangs have no sign of grooves; they actually have enclosed canals within the fangs that transmit venom out of the body, very much like a hypodermic needle. The two upper jawbones (maxillae), which bear the fangs, are very short. Each maxilla has a special joint that permits this bone, along with the fang anchored firmly within it, to rotate 90°. When the viper closes its mouth, the fangs lay back, tip inward, and are covered by a fold in the mucous membrane. When the mouth is opened, a lifting mechanism is activated, putting the fangs into a vertical position by means of the rod-shaped ectopterygoid bones and the pterygoid. The fangs are then in position to bite and inject venom. The fangs (or, more precisely, the maxillae) are laid back with the same action as when a pocketknife is snapped together. The adaptation of folding back the fangs (referred to as solenoglyphous dentition) permits them to be extremely long, far exceeding the length of those in such snakes as cobras, which bear fangs that are fixed in the down, or vertical, position. The fangs of the giant king cobra are not much longer than those of the rather small adder. The long fangs enable vipers to bite deeply into the tissues and cause the victim to suffer severe necrosis. The fangs fold back into the mouth after they are withdrawn from the victim.

Viper venom contains primarily hematoxic material (i.e., substances injurious to the blood and the blood vessels). Thus, a viper bite typically has a very different effect from a cobra or mamba bite (their venom being primarily neurotoxic, that is, injurious to the nervous system). Viper bites are accompanied by prominent local irritation and symptoms of severe blood poisoning, with burning pain, inflamed swellings, pronounced discoloration, sudden drop in blood pressure, internal bleeding, degeneration of the tissues, and the formation of an abscess. Death ensues because the heart stops, not as the result of respiratory arrest, as in cobra bites. Some vipers, whose venom contains neurotoxic as well as hematoxic substances, are especially dangerous.

Distribution

Viperines, *Causus*, and *Azemiops* are found only in the Old World (Europe, Asia, and Africa), but crotalines inhabit both the Eastern and Western Hemispheres. Australia lacks viperids, suggesting that this family evolved after the Australian continent became a separate landmass. The distribution of *Causus* is restricted to sub-Saharan Africa. *Azemiops* occurs in southern China, Burma (Myanmar), Laos, and Vietnam. Viperines are distributed in Africa, Europe, and Asia. Crotalines are found in North America, Central America, and South America as well as East Asia and Central Asia. One species (*Gloydius halys*) enters Europe north of the Caspian Sea.

Habitat

Viperines occur in both tropical and temperate environments, and species have adapted to numerous microhabitats. In temperate regions, species also migrate between different habitats during their active season. The common adder, *Vipera berus*, moves to meadow habitats with populations of rodents during the summer feeding period, whereas it often occurs on south-facing rocky slopes during the spring and mating periods. Distinct groups of vipers have different connections to special habitats. *Macrovipera* species occur in dry steppe habitats, whereas the subgenus *Montivipera* is affiliated with rocky habitats in mountains, the subgenus *Pelias* with moist grasslands, the subgenus *Acridophaga* with dry grasslands, and so on. In tropical habitats most species are terrestrial, but *Atheris* species are arboreal. Of the terrestrial species, some are connected to wet forests (*Bitis gabonica* and *B. nasicornis*), and others are savanna inhabitants (*B. arietans*).

Pitvipers also occupy a wide variety of habitats in temperate and tropical regions. These habitats include temperate forests (*Gloydius caliginosus* and *Agkistrodon contortrix*), tropical wet forests (*Hypnale hypnale* and *Bothrocophias hyoprora*), tropical deciduous forests (*Calloselasma rhodostoma* and *Porthidium ophryomegas*), montane pine-oak forests (*Crotalus willardi*), cloud forests (*Atropoides nummifer*), deserts (*Crotalus cerastes*), and grasslands (*Sistrurus catenatus*). Several genera of tropical pitvipers (*Bothriechis, Bothriopsis, Trimeresurus,* and *Tropidolaemus*) are specialized for an arboreal existence, but most pitviper species are primarily terrestrial. Even terrestrial species occasionally are encountered in trees or shrubs, however. The cottonmouth (*Agkistrodon piscivorus*) is semiaquatic.

Behavior

Some viperids move over large areas in search of prey during their active season, whereas others are more sedentary. In temperate areas vipers and pitvipers hibernate for several months, and some species (*Vipera dinniki, V. darevskii,* and *Gloydius monticola*) at high elevations can hibernate for two-thirds of the year. There is no real territoriality, but in some species, such as *V. berus*, the males actively protect areas around reproductive females during the mating period.

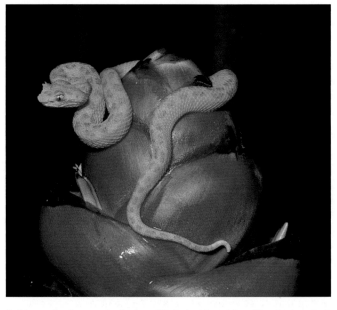

Gold morph of an eyelash viper (*Bothriechis schlegelii*) coiled on heliconia (*Heliconia imbricata*). (Photo by Michael Fogden. Bruce Coleman, Inc. Reproduced by permisson.)

Some vipers inflate their bodies into sausage shapes when they are excited. Almost all vipers also can assume a plate-shaped coiled position as a threat gesture, in which they lift up the neck and hold it in an S shape. Other threat behaviors include loud hissing and rapid forward jerks of the head. Some sand dwellers, such as the saw-scaled vipers, create a particularly impressive sound by rubbing their scales together. Many species of pitvipers vibrate their tails when disturbed, and the evolution of the rattle resulted in amplifiction of the sound produced by this behavior. Several pitviper species (e.g., *Agkistrodon piscivorus* and *Bothriechis schlegelii*) give a silent but effective warning by gaping to reveal the bright white lining of the mouth.

Feeding ecology and diet

Viperids feed chiefly on small vertebrates, particularly rats, mice, and lizards, and less often on frogs and birds, paralyzing or killing their prey by biting it. Some of the smallest vipers prefer locusts, and various other vertebrate and invertebrate prey types are known. Many viperids are ambushers and generally lie in wait for their prey, sitting quietly in one spot for long periods of time. Other species may forage actively or employ a combination of active and ambush foraging. The most specialized ambushers may feed quite infrequently. Many species show an ontogenetic shift in prey preference, feeding on small ectotherms (e.g., lizards and frogs) as juveniles and taking endotherms (e.g., rodents) when they are adults. Caudal luring has been observed in juveniles of several viperid species (e.g., *Bitis peringueyi* and *Agkistrodon contortrix*), and this behavior is retained in adults of *Bothriopsis bilineata*. Associated with caudal luring is a distinctive coloration of the tail tip—from black to brown to bright yellow or bright green, depending on the species. Many species withdraw the head immediately after striking prey and subsequently locate the prey by using their Jacobson's organ. Other species, including many arboreal species, hold the prey until it is immobilized and then swallow it. Many vipers are useful for controlling rodent pests.

Reproductive biology

Most viperids have an annual or biennial breeding cycle. A biennial cycle normally is encountered in females in temperate regions, as in Europe, but in some species (e.g., *Crotalus horridus*) females may reproduce much less frequently. Males always have annual reproduction, which means that reproductive males always outnumber reproductive females in the local population. In *Vipera berus* the production of sperm takes place during the fall and the following spring, whereas in *V. aspis* ripe sperm is ready in late fall. In the former species mating is a spring event, triggered by the first molt for the season in males; in the latter species mating can take place during both fall and spring, even if spring mating is most common. In this case the spring molt is not involved in the start of mating activities.

In some genera, such as *Vipera*, *Bitis*, *Agkistrodon*, and *Crotalus*, males engage in ritualized fights that once were interpreted falsely as mating ceremonies. They lift their heads and approach each other in this vertical position, wrap their bodies around each other, and push with their fore bodies, head to head. In *Vipera berus* a male hierarchy becomes established, and the dominant male is the first to copulate with a female. By activating a sphincter muscle in the female genital tract, a copulatory plug effect is created temporarily. This lessens the possibility that additional males will fertilize the eggs. In some tropical genera a more lengthy breeding season is reported.

Most viperids are ovoviviparous, bearing live young that hatch from membranous eggs. Pregnancy time, meaning the time from ovulation to birth, is normally about 2.5 months for European vipers. Some viperids (e.g., *Lachesis* and *Deinagkistrodon*) lay eggs. The night adders (Causinae) of Africa and the rare Fea's viper (Azemiopinae) of Asia also lay eggs. Incubation has not been noted in any viperine species, nor has any kind of parental care, but egg or neonate attendance has been documented in various pitviper species.

Conservation status

Four viperine species are listed in CITES: Indian *Daboia russelii* is in Appendix III; *Vipera wagneri* is in Appendix II, and *V. latifii* and populations of *V. ursinii* from regions outside former Soviet territories are in Appendix I. Honduras has seven pitviper species listed in CITES Appendix III: *Agkistrodon bilineatus*, *Atropoides nummifer*, *Bothriechis schlegelii*, *Bothrops asper*, *Crotalus durissus*, *Porthidium nasutum*, and *P. ophryomegas*. The IUCN lists 19 viperid taxa: 7 as Critically Endangered, 4 as Endangered, 7 as Vulnerable, and 1 as Data Deficient. Habitat destruction is a serious threat to viperid species throughout their range, and several species are threatened by intentional persecution by humans. Many species have extremely limited distributions (e.g., *Bothrops insularis* and *B. alcatraz*), leaving them vulnerable to local disturbances. Details of the natural history (including geographic distribution and species limits) of most viperids are not well known, and the number of species listed by CITES and the IUCN probably represents a serious underestimate of the conservation problems affecting this family.

Significance to humans

Bites from viperids are problems at the local level. Snake bites kill 30,000–40,000 people in the world each year. The majority of snakebite cases occur in Asia, with India having the highest number, 10,000–15,000 deaths per year. Vipers such as *Daboia russellii* and *Echis carinatus* contribute to a large number of these incidents. Viperids have figured prominently in the legends and religious ceremonies of many cultures. For example, in the first edition of *Grzimek's Animal Life Encyclopedia*, H.-G. Petzold provided an account of the Hopi snake dance. Certain Christian groups in the southeastern United States "take up serpents" such as *Crotalus horridus* and *Agkistrodon contortrix*, as part of their religious ceremonies, grasping the snakes at midbody and dancing with them. The pharmaceutical ancrod (an anticoagulant) is derived from the venom of *Calloselasma rhodostoma*. In the Old World, viperids are hunted for use in traditional medicine, and in the New World they are persecuted during organized events called rattlesnake roundups. Viperids play critical roles in food webs that affect humans.

1. Russel's viper (*Daboia russelii*); 2. Fea's viper (*Azemiops feae*); 3. Horned viper (*Cerastes cerastes*); 4. Green bush viper (*Atheris squamigera*); 5. Common adder (*Vipera berus*); 6. Saw-scaled viper (*Echis carinatus*); 7. Rhombic night adder (*Causus rhombeatus*); 8. Gaboon adder (*Bitis gabonica*); 9. Levantine viper (*Macrovipera lebetina*). (Illustration by Dan Erickson)

1. Cottonmouth (*Agkistrodon piscivorus*); 2. Tibetan pitviper (*Gloydius strauchi*); 3. Sri Lankan hump-nosed pitviper (*Hypnale nepa*); 4. Hundred-pace pitviper (*Deinagkistrodon acutus*); 5. Patagonian lancehead (*Bothrops ammodytoides*); 6. Yellow-blotched palm pitviper (*Bothriechis aurifer*); 7. Timber rattlesnake (*Crotalus horridus*); 8. Black-headed bushmaster (*Lachesis melanocephala*). (Illustration by Dan Erickson)

Species accounts

Fea's viper
Azemiops feae

SUBFAMILY
Azemiopinae

TAXONOMY
Azemiops feae Boulenger, 1888, "Kakhien Hills (Kachin Hills)" Myanmar (Burma).

OTHER COMMON NAMES
German: Fea-Vipern.

PHYSICAL CHARACTERISTICS
Fea's viper has no facial pit between the nostrils and eyes. The head is white in color and covered by large symmetrical shields. The body and tail are black with about 18 short transverse orange to yellow bands laterally along each side.

DISTRIBUTION
Fea's viper occurs in central and southern China from western Yunnan and Sichuan east to Zhejiang and south to Guangxi. It also inhabits northern Burma and northern Vietnam (Tonkin).

HABITAT
Fea's viper inhabits bamboo and tree fern forest alternating with open-light sites. It prefers ground covered with soft layers of deciduous leaves, decomposed trunks of tree ferns, and vigorous outcrops of the karst formation, permanently permeated by numerous open and subterranean streams.

BEHAVIOR
Fea's viper moves very slowly and shows no aggressive behavior.

FEEDING ECOLOGY AND DIET
A shrew was found in the stomach of one specimen, and in captivity mice and lizards have been accepted.

REPRODUCTIVE BIOLOGY
Nikolai Orlov describes the mating behavior as similar to that of other vipers. The male courts the female by twitching the head along the female's body. The short copulation lasts about 10 minutes. The species is oviparous. Little else is known about the reproductive biology.

CONSERVATION STATUS
Although not listed by the IUCN, this is a very rare snake.

SIGNIFICANCE TO HUMANS
There are no conflicts with humans. ◆

Rhombic night adder
Causus rhombeatus

SUBFAMILY
Causinae

TAXONOMY
Causus rhombeatus Lichtenstein, 1823, type locality not specified.

OTHER COMMON NAMES
English: Common night adder, demon adder; French: Vipére nocturne, vipère-démon; German: Krötenotter, Pfeilotter, Nachtadder, Nachtotter.

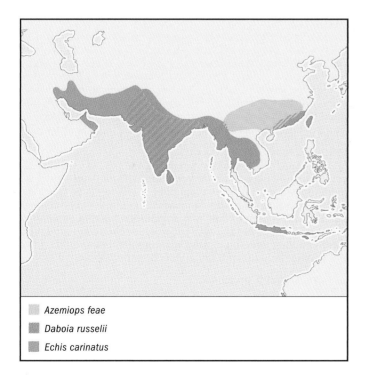

Azemiops feae
Daboia russelii
Echis carinatus

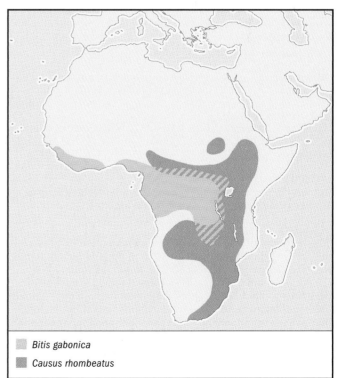

Bitis gabonica
Causus rhombeatus

PHYSICAL CHARACTERISTICS
The night adder is a rather small, stout viper with a distinct rounded head covered by nine larger shields. The pupil is round. The dorsal scales are weakly keeled, and the body is cylindrical. The tail is short. The maximum size of this viper is less than 3.3 ft (1 m), but the average size is only about 11.8–23.6 in (30–60 cm). The normal color pattern is brownish gray with shades of green. Along the back there is a series of about 20–30 dark, light-edged rhombic blotches. On the head there is a characteristic arrow-shaped pattern. The dorsal pattern can vary locally, and occasionally there are patternless specimens.

DISTRIBUTION
The rhombic night adder is distributed widely in southern and eastern Africa and reaches the Cape of Good Hope in the south. It also has a scattered distribution in Zimbabwe, Angola, Zambia, Congo, Kenya, Nigeria, Ethiopia, and Sudan.

HABITAT
It occurs in open habitats, such as savanna, grasslands, forest edges, and swamps.

BEHAVIOR
The viper hisses ferociously when disturbed.

FEEDING ECOLOGY AND DIET
Frogs and toads are the most important prey items.

REPRODUCTIVE BIOLOGY
The species is oviparous and can lay from seven to 26 eggs.

CONSERVATION STATUS
Not threatened.

SIGNIFICANCE TO HUMANS
Very few bites are recorded. ◆

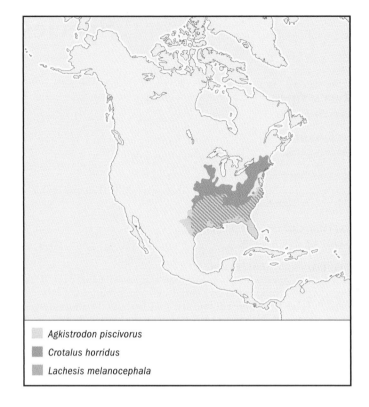

Agkistrodon piscivorus

Crotalus horridus

Lachesis melanocephala

Cottonmouth
Agkistrodon piscivorus

SUBFAMILY
Crotalinae

TAXONOMY
Agkistrodon piscivorus Lacépède 1789 Carolina, later restricted to the vicinity of Charleston, South Carolina. Three subspecies are recognized.

OTHER COMMON NAMES
English: Water moccasin; French: Mocassin d'eau; German: Wassermokassinschlange, Wassermokassinotter.

PHYSICAL CHARACTERISTICS
The cottonmouth is a large, robust snake, occasionally exceeding 5.9 ft (1.8 m) in length and 10 lb (4.6 kg) in weight. One well-fed captive specimen reached a weight of about 23 lb (10.4 kg). In older adults, the dorsum typically is a uniform dark brown, black, or olive. Younger specimens have a lighter ground color with dark brown or reddish brown bands. Juveniles have a bright yellow or greenish tail tip. The loreal scale is absent (fused to the upper preocular scale). For this species, 23–27 (25 in most individuals) midbody scale rows have been recorded, along with 6–9 (8 in most individuals) supralabial scales, 128–145 ventral scales, and 36–53 subcaudal scales.

DISTRIBUTION
The cottonmouth inhabits the southeastern United States from southeastern Virginia south to Florida and west to central Texas and Oklahoma.

HABITAT
This snake inhabits swamps, streams, rivers, ponds, lakes, coastal marshes, rice fields, and several offshore islands. Occasionally it is found away from water.

BEHAVIOR
Cottonmouths are semiaquatic and can be found coiled in the open during the day or night. The name cottonmouth describes the bright white lining of the mouth, which is shown as a warning to creatures that approach too closely. It may be active during any month in the southern part of the range, but more northern populations hibernate. Most individuals are not aggressive, but numerous exceptions are known. This dangerous snake should not be handled.

FEEDING ECOLOGY AND DIET
The cottonmouth frequently forages by ambush, but it also may search actively for prey or feed on carrion. An interesting population on Seahorse Key, Florida, relies heavily on dead fish and young birds that drop from the nests of cormorants, herons, and egrets. Known prey include fish, amphibians, reptiles (among them, small alligators and turtles and even other cottonmouths), birds and their eggs, mammals, snails, insects, and crayfish. Juveniles attract prey by undulating their brightly colored tail tips.

REPRODUCTIVE BIOLOGY
This species gives live birth to 1–16 young, usually in August or September. Male combat has been documented.

CONSERVATION STATUS
The cottonmouth is still an abundant snake in many parts of its range, although local populations are threatened frequently

or extirpated by habitat alteration (e.g., draining of wetlands, channeling of streams, and building of dams).

SIGNIFICANCE TO HUMANS
The cottonmouth is a dangerously venomous snake; bites can cause severe trauma and even death. In the past, this species has been killed systematically through organized snake hunts. ◆

Yellow-blotched palm-pitviper
Bothriechis aurifer

SUBFAMILY
Crotalinae

TAXONOMY
Bothriechis aurifer Salvin, 1860, Cobá, Alta Verapaz, Guatemala.

OTHER COMMON NAMES
Spanish: Cantil loro, cantil verde.

PHYSICAL CHARACTERISTICS
The yellow-blotched palm-pitviper is a small, relatively slender snake with a prehensile tail. Specimens longer than 3.3 ft (1 m) are known, but the length of most adults is less than 2.3 ft (70 cm). The dorsum is green with small yellow blotches with black borders. A black postocular stripe is almost always present. A few specimens may be uniformly green dorsally. For this species, 18–21 midbody (most individuals have 19) scale rows have been recorded, along with 8–12 supralabial scales, 148–167 ventral scales, and 48–64 subcaudal scales.

DISTRIBUTION
The species occurs in southern Mexico and Guatemala.

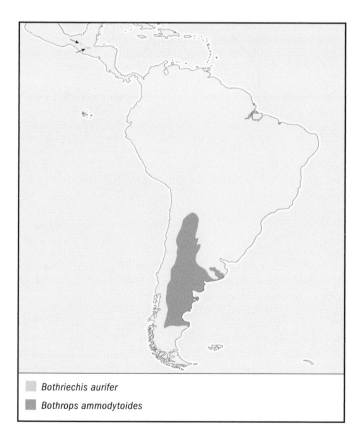

■ *Bothriechis aurifer*
■ *Bothrops ammodytoides*

HABITAT
The yellow-blotched palm-pitviper inhabits cloud forest and pine-oak forest, usually at elevations of 3,940–7,550 ft (1,200–2,300 m).

BEHAVIOR
The species is arboreal and diurnal.

FEEDING ECOLOGY AND DIET
On many occasions, the yellow-blotched palm-pitviper has been observed actively foraging in terrestrial bromeliads. Known prey include treefrogs and small mammals.

REPRODUCTIVE BIOLOGY
It gives live birth to five to eight young.

CONSERVATION STATUS
This species is not listed by the IUCN. However, extensive destruction of cloud forest throughout the range of this snake continues to extirpate local populations and threatens the entire species.

SIGNIFICANCE TO HUMANS
This is a dangerously venomous snake; at least one human fatality has resulted from the bite of this species. ◆

Patagonian lancehead
Bothrops ammodytoides

SUBFAMILY
Crotalinae

TAXONOMY
Bothrops ammodytoides Leybold, 1873, Mendoza Province, northern Argentina.

OTHER COMMON NAMES
German: Argentinische Jararaca; Spanish: Cenicienta.

PHYSICAL CHARACTERISTICS
Adults range from 1.5 ft (45 cm) to 3.3 ft (1 m) in total length. The dorsal ground color is light brown or gray with dark brown blotches. The snout is elevated. Eight to 11 supralabial, 147–160 ventral, 30–41 subcaudal, and 23–25 midbody scale rows have been recorded for this species.

DISTRIBUTION
The species occurs in Argentina, from sea level to 6,560 ft (2,000 m).

HABITAT
It inhabits temperate to subtropical savannas and steppes.

BEHAVIOR
The species is terrestrial. In parts of its range, the Patagonian lancehead may use the tunnels of the chinchillid mammal *Lagostomus maximus*. The species has been characterized as an irritable snake. It is inactive during the austral winter.

FEEDING ECOLOGY AND DIET
It feeds primarily on lizards.

REPRODUCTIVE BIOLOGY
The Patagonian lancehead gives live birth, but little else is known.

CONSERVATION STATUS
Not threatened.

SIGNIFICANCE TO HUMANS
This snake is dangerously venomous. ◆

Timber rattlesnake
Crotalus horridus

SUBFAMILY
Crotalinae

TAXONOMY
Crotalus horridus Linnaeus, 1758, "America," later restricted to the vicinity of New York City, United States. Two subspecies have been recognized, but this taxonomy has been questioned.

OTHER COMMON NAMES
English: Canebrake (for southern populations); French: Crotale des bois; German: Waldklapperschlange.

PHYSICAL CHARACTERISTICS
This large, heavy-bodied snake reaches almost 5 ft (1.5 m) in total length; examples exceeding 6 ft (1.8 m) are known. The ground color of the dorsum may be yellow, gray, tan, or brown to black with dark chevron-shaped blotches. An orange to rust-colored vertebral stripe is present in many individuals; the stripe is especially prominent in snakes from the southern and western parts of the range. The tail is black with a large rattle at the tip. A postocular stripe may be present or absent. Ten to 17 supralabial, 158–183 ventral,13–30 subcaudal, and 21–26 midbody scale rows have been recorded for this species.

DISTRIBUTION
The snake occurs in the eastern United States from New Hampshire, Vermont, and New York south to Florida and west to Minnesota, Nebraska, Kansas, Oklahoma, and Texas.

HABITAT
Forested hillsides with rock outcrops are preferred den sites for northern populations of timber rattlesnakes, but males and nongravid females move into more densely canopied forest during the active season. Southern populations occur in hardwood forests of bottomland floodplains, wet pine flat woods, upland woodlands, and canebrakes (thickets of cane).

BEHAVIOR
Communal dens are used for hibernation in the north, but in the south, snakes hibernate individually or in very small groups. During the active season, some individuals travel great distances from the den site; a migration of more than 4.3 mi (7 km) was documented for one male in a New York population.

FEEDING ECOLOGY AND DIET
Timber rattlesnakes forage mainly by ambush and may consume only 6–20 meals per year. They feed mainly on mammals but also take birds, lizards, snakes, anurans, and insects.

REPRODUCTIVE BIOLOGY
This species gives live birth to three to 19 (usually six to 10) young. Females may give birth every other year, every third year, or every fourth year; even longer intervals have been reported. Females do not produce their first litters until they have reached an age of four to nine years. Male combat has been observed (usually in April or May), but reports from

Louisiana indicate that combat there occurs in the fall.

CONSERVATION STATUS
This species is not listed by the IUCN. However, as of August 2002 the species is listed as threatened or endangered by eight U.S. states, and it is believed to be extirpated from Maine and Rhode Island and from Ontario, Canada. With its low reproductive rate and typically long-lived adults, this species is extremely vulnerable to human disturbance.

SIGNIFICANCE TO HUMANS
The bites of this dangerously venomous snake are potentially life-threatening. In *Landscape with Reptile*, Thomas Palmer recounts the long history of interactions between humans and timber rattlesnakes. There are elements of hope in this story but also much to regret about the toll that humans have taken on this beautiful and unusual species. ◆

Hundred-pace pitviper
Deinagkistrodon acutus

SUBFAMILY
Crotalinae

TAXONOMY
Deinagkistrodon acutus Günther, 1888, Wusueh, Hupeh Province, China. No subspecies are recognized.

OTHER COMMON NAMES
English: Long-nosed pitviper, sharp-nosed pitviper; German: Chinesische Nasenotter.

PHYSICAL CHARACTERISTICS
This is a large, stout-bodied pitviper, sometimes exceeding 5 ft (1.5 m) in total length. Distinctive features of this species include a protuberant snout and tuberculate keels on the dorsal scales. Nine symmetrical plates cover the crown, although some

☐ *Deinagkistrodon acutus*

▨ *Gloydius strauchi*

■ *Hypnale nepa*

fragmentation of these plates is evident in many specimens. Typically, there are seven supralabial and 21 middorsal scale rows and 157–174 ventral and 51–61 subcaudal scale rows. The top of the head is dark brown with a thin, darker postocular stripe. The dorsal ground color is pale gray or brownish gray with dark crossbands that are triangular in lateral view.

DISTRIBUTION
The species occurs in southeastern China, Taiwan, and northern Vietnam.

HABITAT
It inhabits forested hills and mountains at elevations of 330–4,920 ft (100–1,500 m). It often is found near streams and in rocky areas.

BEHAVIOR
The hundred-pace pitviper often is found coiled in the open during the day but also is frequently active at night. It usually raises the head, vibrates the tail, and then strikes if disturbed. In northern Fukien Province, China, these snakes are known to hibernate from late December to early March, but probably they do not hibernate in some warmer parts of the range.

FEEDING ECOLOGY AND DIET
This snake is known to prey on amphibians, reptiles, birds, and mammals; rodents and anurans are the most common prey.

REPRODUCTIVE BIOLOGY
Courtship has been observed from March through May and also September through December. The female lays five to 32 eggs between June and early September. Eggs, which are attended by the female, hatch in approximately 20–30 days.

CONSERVATION STATUS
This species is not listed by the IUCN. However, with the widespread collection of these snakes for the traditional medicine trade, their status warrants investigation.

SIGNIFICANCE TO HUMANS
In their classic monograph on the *Agkistrodon* complex, Howard Gloyd and Roger Conant summarize the extensive use of the hundred-pace pitviper by various Asian cultures. According to the legends and myths of the Paiwan, a tribe indigenous to Taiwan, their first leaders came from the eggs of the hundred-pace pitviper, and the snake features prominently in their art. It is exploited heavily for the traditional medicine trade; raw gallbladders of these snakes are especially prized, but the skin, flesh, eyes, and bones also are used. Snake soup sometimes is prepared from this species; alternatively, entire snakes are preserved in wine, and the wine is used as medicine. Although this is a dangerously venomous snake, its most common English name, hundred-pacer (one can walk only 100 paces before dying), exaggerates the danger; with prompt and proper treatment, bites are rarely life-threatening. ◆

Tibetan pitviper
Gloydius strauchi

SUBFAMILY
Crotalinae

TAXONOMY
Gloydius strauchi Bedriaga, 1912, Tungngolo, Szechwan Province, China (as restricted by Pope in his 1935 work on the reptiles of China). No subspecies are recognized.

OTHER COMMON NAMES
None known.

PHYSICAL CHARACTERISTICS
This is a relatively small snake that typically does not exceed 19.7–21.7 in (500–550 mm) in total length. The largest male recorded was 20 in (510 mm) in total length, and the largest female was 21.5 in (547 mm) in total length. The ground color of the dorsum may be brown or greenish brown with four dark longitudinal stripes. These stripes are usually incomplete and curved such that individuals may appear spotted or exhibit a zigzag pattern on parts of the dorsum. A dark postocular stripe is present, and there are dark markings on the top of the head. Some adults are uniformly dark in color. The crown has nine symmetrical plates. Most specimens have seven supralabial and 21 middorsal scale rows. Ventral scales range from 145 to 175, and subcaudal scales range from 34 to 44.

DISTRIBUTION
The Tibetan pitviper occurs in southern China in Szechuan and Tsinghai Provinces.

HABITAT
The species inhabits the Tibetan Plateau at elevations between 9,470 ft (2,886 m) and 14,000 ft (4,267 m).

BEHAVIOR
Little is known about this terrestrial species, but it must hibernate for extended periods, owing to the climatic conditions within its range.

FEEDING ECOLOGY AND DIET
One specimen was reported to contain a young pika (*Lagomys*).

REPRODUCTIVE BIOLOGY
The Tibetan pitviper is thought to give live birth; seven partially developed embryos were found in one specimen.

CONSERVATION STATUS
Not listed by the IUCN. Additional study of this snake is needed; at present too little is known of its biology to evaluate its status.

SIGNIFICANCE TO HUMANS
There are no records of human envenomation, nor is this species known to be exploited by humans to any appreciable degree. ◆

Sri Lankan hump-nosed pitviper
Hypnale nepa

SUBFAMILY
Crotalinae

TAXONOMY
Hypnale nepa Laurenti, 1768, Sri Lanka. No subspecies are recognized.

OTHER COMMON NAMES
None known.

PHYSICAL CHARACTERISTICS
The Sri Lankan hump-nosed pitviper is a small snake. Adults usually are 11.8–13.8 in (300–350 mm) in total length, but slightly larger specimens are known. The ground color of the dorsum is light or dark brown with 17–26 pairs of small dark

brown blotches. In some individuals, the thin, dark postocular stripe is bordered dorsally by a thin white line. The snout is upturned and bears a small hump covered by tiny scales. As in other members of the genus, the five posterior head plates are mainly intact, but the anterior plates (internasals and prefrontals) are replaced by numerous small scales. Seven to nine supralabial, 124–142 ventral, 33–41 subcaudal, and 17 midbody scale rows have been recorded for this species.

DISTRIBUTION
The species occurs in southwestern Sri Lanka.

HABITAT
Their habitat is wet forest from near sea level to 6,000 ft (1,830 m). Additional study is needed to assess the lower elevation limits of this species; it is possible that specimens of *Hypnale nepa* and *H. walli* have been confused in some reports.

BEHAVIOR
The Sri Lankan hump-nosed pitviper is terrestrial and mainly nocturnal, but it sometimes is encountered during the day. It is not aggressive and generally does not attempt to bite unless it is molested.

FEEDING ECOLOGY AND DIET
Known prey include lizards, snakes, frogs, and reptile eggs.

REPRODUCTIVE BIOLOGY
Little is known about the reproduction of this species. It gives birth to live young. Gravid females containing four to six immature eggs have been reported.

CONSERVATION STATUS
This species is not listed by the IUCN. However, with its limited distribution and preference for areas of dense forest, the Sri Lankan hump-nosed pitviper is vulnerable to habitat destruction. Fortunately, there are national parks and forest reserves within its range.

SIGNIFICANCE TO HUMANS
This is a dangerously venomous snake, but reports of contact with humans are few. ◆

Black-headed bushmaster
Lachesis melanocephala

SUBFAMILY
Crotalinae

TAXONOMY
Lachesis melanocephala Solórzano and Cerdas, 1986, 5.6 mi (9 km) north of Ciudad Neily, Puntarenas Province, Costa Rica.

OTHER COMMON NAMES
German: Schwarzkopf-Buschmeister; Spanish: Cascabel muda, matabuey, plato y negro.

PHYSICAL CHARACTERISTICS
This extremely large snake commonly reaches 6.6 ft (2 m), and specimens of 7.5–7.9 ft (2.3–2.4 m) have been documented. The snout is distinctively rounded. A vertebral ridge is present; and the dorsal scales bear tubercular keels such that the body appears rough. The dorsal ground color is yellow, tan, or brown with black diamond-shaped blotches. The top of the head is black. Seven to nine supralabial, 209–222 ventral,

35–54 subcaudal, and 36–41 midbody scale rows have been recorded for this species.

DISTRIBUTION
The species occurs in southwestern Costa Rica and possibly adjacent parts of Panama.

HABITAT
It inhabits lowland wet and moist forest and premontane wet forest.

BEHAVIOR
The black-headed bushmaster is terrestrial. Some use rodent or armadillo burrows as refuges. It may vibrate the tail when disturbed.

FEEDING ECOLOGY AND DIET
This species forages by ambush. Rodents and marsupials are probably the main prey.

REPRODUCTIVE BIOLOGY
Females may lay up to 16 eggs; two clutches laid in captivity included nine and 13 eggs, respectively. Females remain with the eggs until they hatch.

CONSERVATION STATUS
Although not listed by the IUCN, this species is relatively uncommon and has a restricted distribution.

SIGNIFICANCE TO HUMANS
The black-headed bushmaster is dangerously venomous; any bite should be considered life-threatening. Other species of bushmaster (e.g., *Lachesis muta*) are used as food. Bushmasters, in general, feature prominently in forest lore, probably owing mainly to their formidable size and lethal venom. ◆

Green bush viper
Atheris squamigera

SUBFAMILY
Viperinae

TAXONOMY
Atheris squamigera Hallowell, 1856, "near the river Gabon, Guinea" (Gabon). Two subspecies are recognized.

OTHER COMMON NAMES
English: Common bush viper, leaf viper, variable bush viper; French: Vipère d'arbre; German: Blattgrüne Buschviper.

PHYSICAL CHARACTERISTICS
The green bush viper is a species that grows up to 31 in (78 cm) in size, with the females larger than the males. The body is covered with strongly keeled scales. The tail is comparatively long and prehensile. The color varies, with yellow, reddish, and gray specimens. The majority of individuals are greenish.

DISTRIBUTION
The species occurs through the tropical belt of western and central Africa, from Ghana in the west to Uganda and western Kenya in the east. Southward it is distributed down to northern Angola.

HABITAT
This is a rainforest species, often preferring low, thick, flowering bushes.

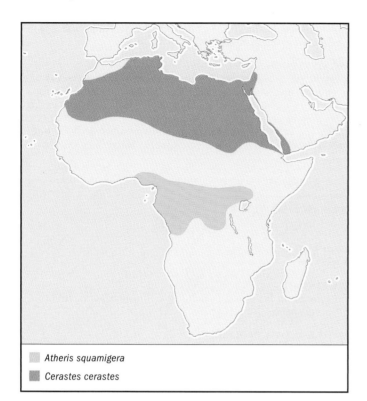

Atheris squamigera

Cerastes cerastes

BEHAVIOR
During the daytime it can be found up in the vegetation, while in some investigated populations it moves down toward the ground at night to prey on ground rodents.

FEEDING ECOLOGY AND DIET
Rodents are the main prey items for adult bush vipers.

REPRODUCTIVE BIOLOGY
Mating takes place in October, and the live young are born during March and April. The clutch size is about seven to nine young.

CONSERVATION STATUS
Not threatened.

SIGNIFICANCE TO HUMANS
Few bites are recorded, but fatal cases are known. ◆

Gaboon adder
Bitis gabonica

SUBFAMILY
Viperinae

TAXONOMY
Bitis gabonica Duméril Bibron, and Duméril, 1854, "Gabon." Two subspecies are recognized.

OTHER COMMON NAMES
English: Gaboon viper; French: Vipère du Gabon; German: Gabunviper, Gabunotter.

PHYSICAL CHARACTERISTICS
This species reaches an average size of about 4 ft (120 cm), but it can grow to more than 6.6 ft (2 m). The fangs can be longer

than 2 in (50 mm), and the head and body are very broad. The West African subspecies, *B. g. rhinoceros*, bears a pair of horn-like knobs on the top of the snout. The Gaboon adder has a geometric pattern with rich pastel, brown and whitish colors.

DISTRIBUTION
The nominate subspecies of the Gaboon adder is distributed in central, eastern and southern Africa. The western subspecies, *B. g. rhinoceros*, occurs in West Africa.

HABITAT
This is a tropical forest species, which is well camouflaged on the forest floor.

BEHAVIOR
The Gaboon adder is a placid species that very rarely strikes. The protection behavior is a loud hissing sound.

FEEDING ECOLOGY AND DIET
Like most vipers, they forage from an ambush position on the forest floor. The prey consists mostly of rodents, even when larger prey, such as mongoose, hares, and monkeys, can be taken.

REPRODUCTIVE BIOLOGY
This species gives birth to live young. Clutch sizes can be very large, as many as 60 young in a single clutch. Normally, the clutch size is smaller (16–30 in the western subspecies).

CONSERVATION STATUS
Not threatened. The species is considered very common in prime forest habitats.

SIGNIFICANCE TO HUMANS
Bites are very rare, as the species is very docile and lives in the rainforest. When bites occur, however, fatalities are common. ◆

Horned viper
Cerastes cerastes

SUBFAMILY
Viperinae

TAXONOMY
Cerastes cerastes Linnaeus, 1758, "Egypt." Two subspecies are recognized.

OTHER COMMON NAMES
English: Desert horned viper, Saharan horned viper; French: Vipère à cornes; German: Hornviper.

PHYSICAL CHARACTERISTICS
The horned viper is rather short and stout in body shape, and the tail is short. The head is triangular and covered with small scales. Many specimens have a raised horn over each eye. Hornless specimens have a raised brow ridge. The horn is made up of a single elongated scale. The neck is thin, and the body is covered with pronouncedly keeled scales, in 25–35 rows. The size typically is between 11.8 in (30 cm) and 23.6 in (60 cm), but occasionally specimens can grow to 2.8 ft (85 cm). The ground color is mostly grayish, yellowish, or reddish with a series of rectangular brown blotches along the back.

DISTRIBUTION
The horned viper occurs in northern Africa in sandy habitats. It also occurs in eastern Sinai.

HABITAT
To a large extent, this is a sand-dwelling species, but it often may live in habitats mixed with rocks and stones. Occasionally, it occurs in ground with more clay, typically mixed with some sandy soils.

BEHAVIOR
During the daytime this nocturnal species hides under rocks or in rodent burrows, and it will even dig down into the sand.

FEEDING ECOLOGY AND DIET
The prey consist of rodents, lizards, and birds.

REPRODUCTIVE BIOLOGY
This viper is oviparous and can produce between 10 and 23 eggs.

CONSERVATION STATUS
The species has a rather wide distribution and is not threatened.

SIGNIFICANCE TO HUMANS
Bites do occur, but the species is not thought to be particularly dangerous. ◆

Russel's viper
Daboia russelii

SUBFAMILY
Viperinae

TAXONOMY
Daboia russelii Shaw and Nodder, 1797, India (Coromandel Coast). Two subspecies are recognized.

OTHER COMMON NAMES
English: Chain viper; French: Vipère de Russel; German: Kettenviper.

PHYSICAL CHARACTERISTICS
This species is characterized by a pattern of three longitudinal rows of reddish brown, black-edged, oval or circular spots on a mostly brownish yellow or brownish gray ground color.

DISTRIBUTION
Russel's viper occurs in Southeast Asia, from India, Pakistan, and Sri Lanka to Taiwan in the east and Java, Komodos, and Flores in the south. The distribution is not continuous within this large area.

HABITAT
The species prefers plains, savannas, or hills, and it is encountered regularly in agricultural areas.

BEHAVIOR
It is mainly a nocturnal snake; if encountered during the daytime, it is rather sluggish and, unless attacked, is not aggressive.

FEEDING ECOLOGY AND DIET
It mainly preys on rodents but also takes frogs, birds, and lizards.

REPRODUCTIVE BIOLOGY
This large ovoviviparous species can give birth to 20–63 live young in each clutch.

CONSERVATION STATUS
This species is not listed by the IUCN. However, some populations of Russel's viper are endangered, primarily because of collection for leather production or as food. The Indian populations are listed in Appendix III of CITES.

SIGNIFICANCE TO HUMANS
Russel's viper is the most dangerous viper in Southeast Asia and is of great medical importance. This species is responsible for a majority of cases of snakebite injury and death within its range, especially in densely populated areas of the Indian subcontinent. In Sri Lanka and Burma (Myanmar), approximately 2,000 bites and 900 deaths per year are attributed to Russel's viper. ◆

Saw-scaled viper
Echis carinatus

SUBFAMILY
Viperinae

TAXONOMY
Echis carinatus Schneider, 1801, "Arni," near Madras, India. Four subspecies are recognized.

OTHER COMMON NAMES
English: Carpet viper; French: Échide carénée; German: Sandrasselotter.

PHYSICAL CHARACTERISTICS
The saw-scaled viper is a comparatively small snake, 15–23.6 in (38–60 cm) in length, with an oval head and strongly keeled scales in 27–36 rows around the body. It produces its characteristic sound by coiling its body in parallel loops and rubbing the body sides, with its serrated lateral body scales, together.

DISTRIBUTION
It occurs as several subspecies from India and Sri Lanka in the east to the United Arab Emirates and Oman in the west. In northern Africa and other parts of Asia there are numerous similar species that until recently were considered subspecies of *E. carinatus* but which today are separated at the species level.

HABITAT
The saw-scaled viper prefers dry, often rocky habitats but also may occur on dry, grassy slopes or even sandy habitats.

BEHAVIOR
This species is a good sidewinder and can move quickly. When threatened, it coils up in a horseshoe fashion and rubs the sides of its body together in opposite directions while producing a sharp sizzling sound. This species is mostly nocturnal.

FEEDING ECOLOGY AND DIET
This viper forages broadly and can eat many different kinds of small animals, such as centipedes, scorpions, birds, rodents, frogs, lizards, and even other snakes.

REPRODUCTIVE BIOLOGY
The saw-scaled viper is oviparous and can produce up to 20 eggs.

CONSERVATION STATUS
Not threatened.

SIGNIFICANCE TO HUMANS
Saw-scaled vipers are very aggressive and quick to strike. They produce highly virulent hematoxic venom. In addition, they are often abundant in heavily populated regions, which makes them some of the most dangerous snakes in the world. ◆

Levantine viper
Macrovipera lebetina

SUBFAMILY
Viperinae

TAXONOMY
Macrovipera lebetina Linnaeus, 1758, Cyprus. Five subspecies are recognized.

OTHER COMMON NAMES
English: Blunt-nosed viper; French: Vipère du Levant; German: Levante-Otter.

PHYSICAL CHARACTERISTICS
This is a large viper with a head covered with small scales. In some eastern subspecies the supraocular scale can be larger but almost always is semidivided. All body scales are keeled. The color typically is grayish with numerous dark brown or dark gray crossbands, which may be more or less indistinct. The maximum size of this species is 7 ft (214 cm).

DISTRIBUTION
Numerous subspecies occur in dry habitats in western Asia from Cyprus and central Turkey to Pakistan and Kashmir. One subspecies (*M. l. transmediterranea*) is distributed in North Africa (Tunisia and Algeria).

HABITAT
The species occurs in several kinds of dry habitats, such clay steppes, rocky mountain slopes, grass meadows, and dry deciduous forests. It often is seen near natural water streams or human-made channels. It is well adapted to agriculture areas.

BEHAVIOR
The Levantine viper is mainly nocturnal and is largely responsible for snake bites in the Near East and Middle East, which occur when people walk around at night. When it is disturbed, it typically makes a loud hissing sound.

FEEDING ECOLOGY AND DIET
It is an opportunistic species and feeds on various kinds of rodents, birds, and lizards.

REPRODUCTIVE BIOLOGY
Like all *Macrovipera*, the Levantine viper is oviparous, but the hatching time varies between different populations and areas. In some cases the embryos are rather well developed when the eggs are deposited. A single clutch can consist of up to 35 eggs.

CONSERVATION STATUS
Not threatened. The species is locally abundant within its range.

SIGNIFICANCE TO HUMANS
The Levantine viper generally is considered dangerously venomous. Bites have been fatal to humans, horses, cows, and camels. ◆

Common adder
Vipera berus

SUBFAMILY
Viperinae

TAXONOMY
Vipera berus Linnaeus, 1758, Uppsala, Sweden (terra typica restricta). Three subspecies are recognized.

OTHER COMMON NAMES
French: Vipér peliad; German: Kreuzotter.

PHYSICAL CHARACTERISTICS
Males are 21.7–25.6 in (55–65 cm), and females are 23.6–27.6 in (60–70 cm) in length. This is a medium-sized viper with sexual dimorphism. Males have a gray ground color with a black dorsal zigzag band; females are brown or reddish brown, with a dark brown zigzag band along the back. Locally, up to 50% of the snakes in the population are totally melanistic.

DISTRIBUTION
The species is distributed in Europe from the Russian Kola Peninsula in the north to northern Italy and Greece in the south. Eastward it is distributed through Asia to the island of Sakhalin north of Japan and to North Korea.

HABITAT
The species prefers mesic habitats with meadows and moorland. It can be found at the edges of forests.

BEHAVIOR
Males defend a small territory around reproductive females during a few spring weeks (May) and engage in pronounced combat behavior during the mating period.

FEEDING ECOLOGY AND DIET
Small rodents and especially voles (*Microtus*) are the main food item for adults. Occasionally birds, frogs, and lizards are eaten. Juvenile adders take frogs and lizards as prey.

REPRODUCTIVE BIOLOGY
Females usually reproduce every second year, but occasionally less often. Males reproduce every year. The mating period is in May, and the young are born (ovoviviparity) in August or September. The clutch size is, on average, 10 young, but it can vary between two and 18 young.

CONSERVATION STATUS
Although not listed by the IUCN, this species has a fragmented distribution in central Europe and is locally endangered.

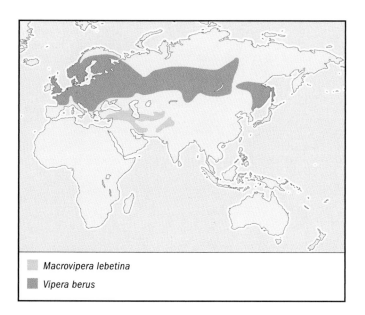

■ *Macrovipera lebetina*
■ *Vipera berus*

It is more abundant in the northern parts of its distribution (Scandinavia and Russia). It is protected in most European countries.

SIGNIFICANCE TO HUMANS
The common adder causes numerous bites throughout its range every year, but very few are fatal. ◆

Resources

Books

Campbell, Jonathan A., and Edmund D. Brodie Jr. *Biology of the Pitvipers.* Tyler, TX: Selva, 1992.

Campbell, Jonathan A., and William W. Lamar. *The Venomous Reptiles of Latin America.* Ithaca, NY: Cornell University Press, 1989.

———. *The Venomous Reptiles of the Western Hemisphere.* Ithaca, NY: Cornell University Press, 2003.

Ernst, Carl H. *Venomous Reptiles of North America.* Washington, DC: Smithsonian Institution Press, 1992.

Gloyd, Howard K., and Roger Conant. *Snakes of the Agkistrodon Complex: A Monographic Review.* Vol. 6, *Contributions to Herpetology.* Oxford, OH: Society for the Study of Amphibians and Reptiles, 1990.

Jena, I. *Snakes of Medical Importance and Snake-bite Treatment.* New Delhi: Ashish Publishing House, 1985.

Mallow, D., D. Ludwig, and G. Nilson, *True Vipers: Natural History and Toxinology of Old World Vipers.* Melbourne, FL: Krieger Publishing Company, 2003.

Nilson, G., and C. Andrén. "Evolution, Systematics and Biogeography of Palearctic Vipers." In *Venomous Snakes: Ecology, Evolution and Snakebite,* edited by R. S. Thorpe, W. Wüster, and A. Malhotra. Symposia of the Zoological Society of London. London: Oxford University Press, 1997.

Palmer, T. *Landscape with Reptile: Rattlesnakes in an Urban World.* New York: Ticknor and Fields, 1992.

Schuett, Gordon W., Mats Höggren, Michael E. Douglas, and Harry W. Greene, eds. *Biology of the Vipers.* Eagle Mountain, UT: Eagle Mountain Publishing, 2002.

Spawls S., and B. Branch. *The Dangerous Snakes of Africa.* Sanibel Island, FL: Ralph Curtis Books, 1995.

Periodicals

Lenk P., S. Kalyabina, M. Wink, and U. Joger. "Evolutionary Relationships Among the True Vipers (Viperinae) Inferred from Mitochondrial DNA Sequences." *Molecular Phylogenetics and Evolution* 19, no. 1 (2001): 94–104.

Parkinson, C. L. "Molecular Systematics and Biogeographical History of Pitvipers as Determined by Mitochondrial Ribosomal DNA Sequences." *Copeia* (1999): 576–586.

Parkinson, C. L., K. M. Zamudio, and H. W. Greene. "Phylogeography of the Pitviper Clade *Agkistrodon:* Historical Ecology, Species Status, and Conservation of Cantils." *Molecular Ecology* 9 (2000): 411–420.

Warrell D. "Tropical Snakebite: Clinical Studies in South East Asia." *Toxicon* 23 (1985): 543.

Göran Nilson, PhD
Ronald L. Gutberlet, Jr., PhD

African burrowing snakes

(Atractaspididae)

Class Reptilia

Order Squamata

Suborder Serpentes

Family Atractaspididae

Thumbnail description
These snakes have small heads, smooth scales without apical pits, small to minute eyes with round pupils, and no loreal scale; one group, burrowing asps (*Atractaspis*), has unusual erectile front fangs

Size
12–40 in (30–102 cm)

Number of genera, species
12 genera; 62 species

Habitat
Forest, savanna, and near desert

Conservation status
Not threatened

Distribution
Sub-Saharan Africa, with one genus entering the Near East

Evolution and systematics

These unique snakes have caused great taxonomic confusion. Early classifications placed undue emphasis on the presence of erectile front fangs in burrowing asps. Due to this sole feature, for many years they were called mole vipers or burrowing adders and mistakenly were placed in the family Viperidae, even though they did not look like other vipers. It is now known that the fang erection mechanism of burrowing asps is unique and unlike that of true vipers.

Burrowing asps look very similar to and have similar lifestyles to other African burrowers, including the exotic-sounding purple-glossed snakes (*Amblyodipsas*), quill-snouted snakes (*Xenocalamus*), and centipede eaters (*Aparallactus*). They are now placed together, although confusion remains over the relationships and status of some genera within the family. Many have been combined into the snake eaters (*Polemon*), while at least one other genus (*Elapotinus*) is known from only a single specimen that may not even be African. The affinities of African burrowing snakes with other snakes also remain obscure. They share some intriguing similarities with primitive elapid snakes, e.g., the harlequin snakes (*Homoroselaps*) from South Africa, and may be

close to the basal stock from which the important and venomous elapid snakes arose.

Two subfamilies are recognized. The Aparallactinae contains 11 genera, 25 species, of small- to medium-sized burrowers that are mainly back-fanged. The Atractaspidinae contains only a single genus (*Atractaspis*) with 17 species, distinguished by its unusual method of fang erection.

Physical characteristics

A wide range of fang types and associated glands is found in these unusual burrowing snakes. They all have similar skulls, vertebrae, and hemipenes. They are usually back-fanged, but one genus has long, hollow fangs that can rotate forward and inject venom. The small head is not distinct from the neck, and the snout is usually bluntly rounded (but very pointed in quill-snouted snakes, *Xenocalamus*). To aid in burrowing, the head shields often fuse and are thus reduced in number. The body is cylindrical and sometimes very long and thin. Many of these snakes are uniform black or brown, often with a light or dark collar. A few species are brightly striped, but blotched patterns are rare.

Distribution

Restricted to sub-Saharan Africa, with the small-scaled burrowing asps (*Atractaspis microlepidota*) extending into Israel and another (*Micrelaps muelleri*) into Jordan.

Habitat

Most species burrow in sand soils in savannas or desert scrublands. Others live in moist, humid soils of lowland forests.

Behavior

All live underground, some utilizing existing animal tunnels and others pushing through loose sand or leaf litter. Most species are gentle in the hand, but burrowing asps are irascible (easily provoked) and strike quickly and often. Only one fang is used and the mouth remains closed. The straight, thin fang protrudes from the side of the mouth and is hooked sideways and backwards into the prey or victim. This unusual behavior has led to another common name, the side-stabbing snake. They may also be called stiletto snakes. Unlike other venomous snakes, they cannot be safely held behind the head.

Yellow and black burrowing snakes (*Chilorhinophis*) have blunt, tiny tails that mimic the head color. When disturbed by a predator, it hides its head beneath body coils and slowly waves the raised tail to deflect attack away from the delicate head.

Feeding ecology and diet

Most species eat other burrowing reptiles, including blindsnakes, wormlizards, or legless lizards. Others have more specialized diets. As their name implies, centipede eaters (*Aparallactus*) feed almost exclusively on centipedes, which they grab and chew. The prey quickly succumbs to the venom and is swallowed head first. The very elongate quill-snouted snakes feed almost exclusively on wormlizards (*Monopeltis* sp.). Larger species, such as the purple-glossed snakes (*Amblyodipsas*), eat small rodents as well as snakes, while the Natal black snakes (*Macrelaps microlepidotus*) also eat rain frogs (*Breviceps* sp.). The unusual fang erection of burrowing asps allows them to crawl past prey within the narrow confines of a burrow and envenomate the prey by stabbing backward.

Reproductive biology

All but one species are oviparous and lay small clutches (two to 15) of elongate eggs. These are laid in a dead termite nest or in moist soil beneath a rotting log or boulder, and take six to eight weeks to develop. Jackson's centipede eater (*Aparallactus jacksonii*) gives birth in June to two to three minute babies (about 4 in [10 cm]).

Conservation status

No species are threatened, although some have very restricted ranges.

Significance to humans

Although most aparallactines are harmless, the Natal black snakes require care. Its bite has caused nausea and loss of consciousness, although not death. All burrowing asps, however, are venomous and a few are dangerously so. The venom glands are large and in some species extend into the neck. The venom is unique among snakes and causes severe pain, swelling, local blistering, nausea, vomiting, and diarrhea. Bites from most species are mild, but deaths have been caused by three species. There is no effective antivenom. Most bites occur at night when the victim either treads on the snake or rolls over onto it while asleep. They are easily confused with harmless snakes.

Species accounts

Variable quill-snouted snake
Xenocalamus bicolor

SUBFAMILY
Aparallactinae

TAXONOMY
Xenocalamus bicolor Günther, 1868, Zambezi (later corrected to Damaraland), Namibia.

OTHER COMMON NAMES
English: Striped quill-snouted snake.

PHYSICAL CHARACTERISTICS
The female is 28.5 in (72 cm) and the male 22.5 in (57 cm). This very thin, elongate snake has smooth scales in 17 midbody rows. The snout is extremely pointed with an underslung mouth. It is pale above with paired dark blotches or stripes; the belly is white.

Xenocalamus bicolor

DISTRIBUTION
From the southern parts of the Democratic Republic of the Congo, through Angola and western Zambia, to central South Africa and Mozambique.

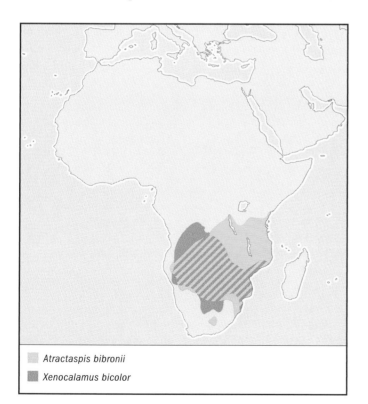

◻ *Atractaspis bibronii*
◼ *Xenocalamus bicolor*

HABITAT
Deep sandy soils with mixed savanna.

BEHAVIOR
These bizarre burrowing snakes tunnel deep into sandy soils searching for their prey. They come to the surface only when searching for mates or after heavy rain. Gentle in the hand, they rarely bite but may give a sharp prick with their hard snout.

FEEDING ECOLOGY AND DIET
This snake is a specialist feeder on large wormlizards (e.g., *Monopeltis* sp.). How quill snouts manage to kill and overpower such large prey within the confines of their tight burrows remains a mystery—perhaps they stab them to death.

REPRODUCTIVE BIOLOGY
Oviparous, they lay a few (up to four) elongate eggs in a chamber in moist sand. The young measure about 8 in (20 cm).

CONSERVATION STATUS
Not threatened.

SIGNIFICANCE TO HUMANS
None known. ◆

Southern burrowing asp
Atractaspis bibronii

SUBFAMILY
Atractaspidinae

TAXONOMY
Atractaspis bibronii A. Smith, 1849, eastern districts of the Cape Colony, South Africa.

OTHER COMMON NAMES
English: Bibron's burrowing asp, side-stabbing snake.

PHYSICAL CHARACTERISTICS
The female is 24.4 in (62 cm) and the male 26 in (66 cm). This short, stocky snake has smooth, close-fitting scales in 19–25 rows at midbody. It is uniformly purple-brown to black above. The belly is usually uniform dark gray, but sometimes white to cream with scattered dark blotches.

Atractaspis bibronii

DISTRIBUTION
From the Northern Cape of South Africa, through Botswana, Namibia, and Angola to Tanzania and coastal Kenya.

HABITAT
Varied, ranging from savanna and semidesert to coastal thicket.

BEHAVIOR
At night they may emerge on the surface, particularly after rain. Underground they use the side of the head to excavate a chamber beneath a sun-warmed stone. They have a peculiar "aromatic" smell, the function of which is unknown. On hard surfaces the neck is flexed, with the nose pointing down vertically, as they try to burrow.

FEEDING ECOLOGY AND DIET
The diet mainly includes other burrowing reptiles, but nestling rodents and small frogs are also eaten.

REPRODUCTIVE BIOLOGY
Oviparous, they lay a few (usually four to seven, but up to 11) elongate eggs (1.1–1.4 in [27–36 mm] long by 0.4–0.5 in [10–12 mm] wide) in summer. The young measure about 6 in (15 cm).

CONSERVATION STATUS
Not threatened.

SIGNIFICANCE TO HUMANS
These asps are responsible for many bites in rural areas. The glands yield minute amounts of straw-colored venom (1.3–7.4 mg), which causes immediate pain and local swelling. Mild neurotoxic symptoms (e.g., nausea, dry throat, and vertigo) may be present in the early stages, but necrosis is rare and no fatalities are known. Polyvalent antivenom unfortunately is ineffective.

Resources

Books

Branch, Bill. *Field Guide to Snakes and Other Reptiles of Southern Africa*. South Africa, 1998.

Spawls, Stephen, and Bill Branch. *The Dangerous Snakes of Africa: Natural History, Species Directory, Venoms, and Snakebite*. Sanibel Island, FL: Ralph Curtis Books, 1995.

Spawls, Stephen, et al. *A Field Guide to the Reptiles of East Africa: Kenya, Tanzania, Uganda, Rwanda, and Burundi*. San Diego: Academic Press, 2002.

Periodicals

Underwood, G., and E. Kochva. "On the Affinities of the Burrowing Asps *Atractaspis* (Serpentes: Atractaspididae)." *Zoological Journal of the Linnean Society* 107 (1993): 3–64.

Organizations

Herpetological Association of Africa. P.O. Box 20142, Durban North, 4016 South Africa.
Web site: <http://www.wits.ac.za/haa>

Bill Branch, PhD

Colubrids
(Colubridae)

Class Reptilia
Order Squamata
Suborder Serpentes
Family Colubridae

Thumbnail description
Highly variable in size, body form, and color
pattern; ranges from short, stout, drab species
to large, slender, boldly marked forms

Size
6 in–12 ft (160 mm–3.7 m)

Number of genera, species
Approximately 300 genera; approximately 1,700
species

Habitat
The family includes terrestrial, fossorial,
arboreal, and aquatic species

Conservation status
Extinct: 1 species; Critically Endangered: 6
species; Endangered: 7 species; Vulnerable: 8
species; Lower Risk/Near Threatened: 4
species; Data Deficient: 10 species

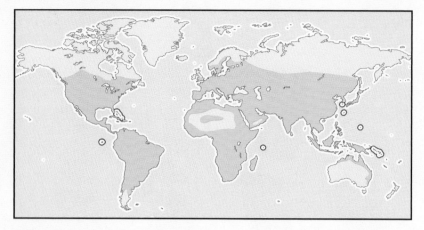

Distribution
Worldwide except Antarctica, extremely high latitudes of Eurasia and North America,
and central and western Australia

Evolution and systematics

The Colubridae comprise by far the largest and most diverse family of snakes, containing about 70% of all snake
species. For that reason, few useful generalizations apply to
this expansive family. Indeed, it is the ability of colubrids to
adapt to widely different habitats, diets, and life history
modes that above all characterizes this extraordinary lineage.
Although some colubrids are generalists, many exhibit
strong specialization for a particular environment and/or
specific prey.

Fossils attributed to the Colubridae first appear in the
Lower Oligocene, about 35 million years ago. However, the
group seems to have radiated rapidly during the Miocene
(5–25 million years ago), and by the end of that period a fauna
previously dominated by boa-like species had largely been replaced by colubrids, elapids, and viperids. However, our
knowledge of fossil snake faunas is drawn primarily from temperate remains, so inferences for tropical regions should be
drawn with caution.

The relationships among genera of colubrids remain
poorly understood and highly controversial, despite numerous attempts to bring order to this complex group. Early classifications were based largely upon similarities in scale
characteristics, dentition, and the form of the hemipenes (the
paired copulatory organs of male squamates). In some cases
these features yielded groupings that have stood the test of
time, but in many instances the failure to distinguish between
ancestral and derived conditions resulted in unnatural groups.

The application of both phylogenetic systematic (cladistic)
and molecular methods has helped to clarify the relationships
within many groups of colubrid snakes. However, even when
well-defined clusters of genera are confirmed, the relationships among those clusters often remain unclear, and widely
accepted phylogenetic hypotheses concerning colubrid relationships have remained elusive. Furthermore, the relationship between colubrids and two other colubroid families, the
Elapidae and Atractaspididae, is unclear, and it is possible that
the Colubridae itself is paraphyletic relative to one or both of
those families. That is, some colubrids may be more closely
related to members of one of those families than they are to
certain other colubrids.

That said, several subfamilies of Colubridae are widely recognized, if not universally accepted. Seven are recognized
here. The Xenodermatinae is a small group of six genera and
about 15 species from southern and eastern Asia. Little is
known of the biology of these strange colubrids, many of
which have unusual, protuberant scales on the body or head.
Most occupy terrestrial habitats in moist tropical forests, and
some are known to prey on amphibians. The Pareatinae is a
small but well-defined lineage of three genera and about 18
species of Southeast Asian snakes that are highly specialized
to prey on terrestrial mollusks (snails and/or slugs). A slightly
larger but still well-defined lineage of Asian colubrids is the
Homalopsinae, with about 10 genera and 35 species. All are
strongly aquatic and some, such as the tentacled snake (*Erpeton tentaculatus*), rarely leave the water. Several occupy mud
flats and mangrove forests, including the dog-faced watersnake

A sample of the wide variety of scale pattern and color in the milksnake subspecies of *Lampropeltis triangulum*. From left to right: *L. t. blanchardi*; *L. t. campbelli*; *L. t. triangulum*; *L. t. elapsoides*; *L. t. gaigeae*; *L. t. hondurensis* (Tangarine morph); *L. t. taylori*. (Illustration by Barbara Duperron)

(*Cerberus rynchops*) and the white-bellied mangrove snake (*Fordonia leucobalia*), while the keel-bellied watersnake (*Bitia hydroides*) occupies coastal marine waters. Many feed on fishes, but some prey on frogs and *Fordonia* feeds on crustaceans. All are rear-fanged.

Four other subfamilies are much larger, and their monophyly is uncertain. The Natricinae is the most cohesive of these, including many species with aquatic tendencies, including the familiar North American gartersnakes (*Thamnophis*), North American watersnakes (*Nerodia*), and the grass snake (*Natrix natrix*) of Europe. The subfamily includes about 38 genera and almost 200 species distributed throughout Europe, Asia, Africa, and North and Central America. Especially large radiations occur in southern and eastern Asia and in North America. The American members constitute a well-defined tribe, the Thamnophiini, all of which are viviparous, in contrast to most of the Old World forms. In addition to the many aquatic species of natricines, some are small cryptozoic or fossorial forms.

Another subfamily with many northern representatives is the Colubrinae, with over 100 genera and roughly 650 species. Also included in this subfamily are many tropical species and several clusters of genera that are sometimes recognized as tribes or even subfamilies in their own right. Worldwide in distribution, the Colubrinae include significant radiations in North America, Eurasia, Africa, and Southeast Asia. Many are fast-moving terrestrial species, although arboreal and fossorial members are numerous. Many are rear-fanged, although others lack enlarged rear maxillary teeth and some have evolved constricting behavior. Among the more familiar groups are the North American racers (*Coluber*), kingsnakes (*Lampropeltis*), and ratsnakes (*Elaphe*). Other important members of the Colubrinae include a large group of tropical Asian

burrowing snakes, including the reedsnakes (*Calamaria*) and related genera, and a group of African genera that include the dangerously venomous boomslang (*Dispholidus*) and twigsnake (*Theletornis*). Few alethinophidian snakes prey on arthropods, but the colubrines include the largest radiation of arthropod-eating snakes, the sonorines, with about a dozen genera, including the North American groundsnakes (*Sonora*), black-headed snakes (*Tantilla*), and shovel-nosed snakes (*Chionactis*). Sonorines are broadly distributed throughout the Americas, but they have their greatest diversity in the dry regions of southwestern North America. Another important cluster of genera includes the diverse African sandsnakes (*Psammophis*) and related species, including the Montpellier snake (*Malpolon monspessulanus*) of the Mediterranean region. Several genera of arboreal colubrines have independently evolved the so-called vinesnake morphology, with very narrow heads, slender bodies, and even behavioral modifications that allow them to blend into the surrounding vegetation. These include genera in the Neotropics (*Oxybelis*), Africa (*Theletornis*), and Southeast Asia (*Ahaetulla*). (Other vinesnakes have evolved in the subfamilies Xenodontinae and Lamprophiinae.) In addition to many small, burrowing species, the Colubrinae includes some giants among colubrids, such as the indigo snake (*Drymarchon*), tiger ratsnake (*Spilotes*), and bird snake (*Pseustes*) of the Americas, and the banded ratsnake (*Ptyas*) of Asia, which approaches or exceeds 10 ft (about 3 m) in length.

The Xenodontinae comprise another large colubrid subfamily, with a distribution limited to the New World. With about 90 genera and over 500 species, this group dominates the colubrid fauna of the Neotropics, where most of its diversity occurs. Two major groups have been identified, one centered in Central America and one in South America, although the two groups broadly overlap geographically, and a number of genera cannot be assigned to either of those two major clades. Some familiar North American species belong to the Xenodontinae, including the ring-necked snakes (*Diadophis*), wormsnakes (*Carphophis*), and mudsnakes (*Farancia*). Even more than the colubrines, the Xenodontinae encompass an extraordinary range of natural histories. Among the major themes are repeated invasions of aquatic, arboreal, and fossorial habitats, as well as specialization on any of a wide range of prey, including fishes, amphibians, earthworms, and terrestrial mollusks. A number of well-defined clades are recognized within the Xenodontinae. Among those in the South American group are the false pitvipers (*Xenodon*) and such related genera as the Neotropical hog-nosed snakes (*Lystrophis*) and *Liophis*. Many members of this group feed on frogs, including such highly toxic species as toads (*Bufo*) and poison frogs (Dendrobatidae). The North American hog-nosed snakes (*Heterodon*) are also xenodontines, but apparently represent an independent evolution of toad-eating habits. Other primarily South American groups include the pseudoboines, which include the mussurana (*Clelia*) and related genera, the diverse terrestrial to arboreal species of *Philodryas*, and the highly aquatic members of the genera *Helicops* and *Hydrops*. A substantial radiation from the South American clade occurs in the West Indies, and a few species are found on the Galápagos Islands. Some have evolved the vinesnake morphology, including *Uromacer* and *Xenoxybelis*. The Central American clade includes a very important radiation of predators on worms or terrestrial mollusks. Among them are worm-eaters, such as *Geophis* and *Atractus*, and the so-called snail-suckers, such as *Dipsas* and *Sibon*, which prey on snails and/or slugs. The snail-eating xenodontines strongly resemble the pareatines of Asia, both externally and in details of their skulls and jaw muscles, and they apparently use similar feeding mechanics to extract snails from their shells. Many xenodontines of both major clades are rear-fanged. A number of xenodontine species mimic venomous coralsnakes (*Micrurus*), as do several colubrines.

The final subfamily of colubrids is the Lamprophiinae, a more modest radiation in both numbers and distribution. The 44 genera and roughly 200 species of lamprophiines are limited to sub-Saharan Africa. Most are small to moderate-sized snakes, and many prey upon reptiles, including other snakes. Some are more specialized, however, including aquatic specialists such as *Lycodonomorphus* and *Grayia*, and prey specialists such as the slug-eaters (*Duberria*). An important radiation occurs on Madagascar, where members of this group dominate the snake fauna. Among that distinctive fauna are the Madagascan hog-nosed snakes (*Lioheterodon*) and the bizarre arboreal Madagascan vinesnakes (*Langaha*), which have a scaly proboscis on their snout.

Physical characteristics

Again, generalizations are difficult in light of the extraordinary diversity encompassed within this family. Compared with basal families and with boas and pythons, colubrids are characterized by the loss or simplification of many features. Notably, all vestiges of the hind limbs are absent in colubrids,

and the coronoid bones of the lower jaws have been lost. Externally, most colubrids have relatively wider ventral scales and fewer rows of dorsal scales than boas and pythons. In addition, most colubrids possess a standard complement of enlarged scales on the head, including nine scales on the top of the head. That feature, like other external attributes, is shared with members of the Atractaspididae and Elapidae and even with some Viperidae, and at the same time is not seen in all members of the Colubridae. The vertebrae of colubrids are, in general, more slender and lightly built than those of basal snakes, boas, and pythons. In addition, the interconnecting chains of muscles that effect locomotion in snakes are longer in colubrids than in those other groups.

One important feature that distinguishes most colubrids is the presence of Duvernoy's glands, a pair of glands located on either side of the head behind the eye. These are the homologues (the evolutionary counterparts) of the venom glands of vipers, elapids, and *Atractaspis*. Although they are often described as modified salivary glands, they in fact are very different in both their tissue characteristics and their origin. The Duvernoy's glands are associated with the rear teeth of the maxillary bones, the major toothed bones of the upper jaws, and they secrete a complex mix of chemicals whose composition is still poorly understood. In some colubrids the secretion is known to serve as a slow-acting venom or as a digestive adjunct. The rear pair of maxillary teeth are often modified to assist with the delivery of secretion from Duvernoy's glands, and may be either enlarged, grooved, or both. That is generally known as the rear-fanged, or ophisthoglyphous, condition, although a variety of terms has been applied to specific conditions of the rear teeth. In most cases the secretion of Duvernoy's gland has no apparent effect on humans, but in some species it can cause local swelling. A few taxa, including the boomslang (*Dispholidus*) and twigsnake (*Theletornis*) of Africa and the Yamakagashi (*Rhabdophis*) of Asia, are capable of delivering a lethal bite to humans. The xenodontine genera *Apostolepis* and *Elapomorphus* resemble some aparallactine atractaspidids in approaching the front-fanged condition of elapid snakes, with large Duvernoy's glands and few teeth preceding the greatly enlarged rear fangs. Although Duvernoy's gland is found, to some degree, in most colubrids, it is not universally present. In some colubrids it may have been lost as a consequence of the evolution of constricting behavior, which constitutes an alternative mechanism for immobilizing prey. In other colubrids the enlarged rear maxillary teeth may serve a purely mechanical function. In some toad-eating snakes, such as hog-nosed snakes (*Heterodon*), those teeth may be used to deflate the prey, which in others, such as kukrisnakes (*Oligodon*), they are used to slice open the eggs of lizards and snakes.

Distribution

Members of the Colubridae occur worldwide, except Antarctica, the highest latitudes of North America and Eurasia, and the central and western regions of Australia. Only vipers extend farther north, in Scandanavia, than do colubrids. In Australia, unlike other regions, colubrids account for a minority of the snake fauna, with members of the Elapidae predominating.

Habitat

Colubrids occupy a wide array of habitats, and independent lineages have repeatedly entered the same habitat. It is impossible to know the ancestral habitat of the Colubridae. Although the earliest fossils have been allied with living terrestrial lineages, the vertebrae of most colubrids do not unambiguously reflect the species' habitat. Terrestrial species occur on all occupied continents and in all subfamilies except the Homalopsinae. Likewise, aquatic species abound in most subfamilies, and their morphologies range from minimally specialized to highly modified. Examples of the former include many members of the Natricinae, including the North American watersnakes (*Nerodia*) and Eurasian watersnakes (*Natrix*), as well as the African watersnakes (*Lycodonomorphus*) of the Lamprophiinae. The Xenodontinae include many aquatic taxa, from the relatively generalized Neotropical watersnakes (*Helicops*) to the more highly modified Neotropical swampsnake (*Tretanorhinus*), which has dorsally placed eyes and valvular nostrils. Some aquatic xenodontines, such as mudsnakes (*Farancia*) of North America and *Hydrops* and *Pseudoeryx* of South America, have rounded heads and smooth, shiny scales, and seem to be adapted for burrowing in aquatic habitats. The entire subfamily Homalopsinae is aquatic, and its members range from moderately specialized forms such as *Enhydris* and the puff-faced watersnake (*Homalopsis*) to extreme specialists such as the keel-bellied watersnake (*Bitia*) and the bizarre tentacled snake (*Erpeton*), which rarely leave the water. Homalopsines have valvular nostrils and mouths similar to those of seasnakes.

Some primarily terrestrial colubrids occasionally climb small shrubs, and even some relatively unmodified taxa, such as the ratsnakes (*Elaphe*) of the Colubrinae, are adept climbers of trees. More specialized arboreal species include the various vinesnakes of the Colubrinae, Xenodontinae, and Lamprophiinae, described above, and the nearly as specialized rough greensnake (*Opheodrys aestivus*), a North American colubrine. The Pareatinae are primarily arboreal, and the Colubrinae and Xenodontinae include many arboreal species in the tropics. Such species usually exhibit slender bodies, which may be laterally compressed. The most extreme condition occurs in the blunt-headed vinesnakes (*Imantodes*), Neotropical xenodontines that have extremely slender bodies and chunky heads. The Southeast Asian flyingsnakes (*Chrysopelea*) are arboreal colubrines that can flatten their bodies and glide from tree to tree.

Likewise, cryptozoic (hiding) and fossorial (burrowing) forms abound in most subfamilies. Aside from being relatively small, cryptozoic species are sometimes little modified morphologically, as in the colubrine North American groundsnakes (*Sonora*) and the xenodontine coffeesnakes (*Ninia*). Others are highly modified for burrowing, including the colubrine shovel-nosed snakes (*Chionactis*) of North America and shovel-snouted snakes (*Prosymna*) of Africa. Fossorial species abound in the Xenodontinae, including the Neotropical burrowing snakes (*Geophis*) of Central America and *Apostolepis* of South America. Such species often have pointed snouts, fused head scales, compact cranial bones, and smooth, shiny scales on the body. Even a few members of the Natricinae exhibit fossorial adaptations, such as the lined snake (*Tropidoclonion*) and rough earthsnake (*Virginia striatula*) of North America.

Behavior

As expected, behaviors are extremely diverse among Colubridae and generally poorly known, with the exception of a few common or unusual species. The behavior of a species often reflects its environment, and arboreal, cryptozoic, and fossorial species are especially difficult to study behaviorally. Defensive behavior, in particular, seems to reflect both coloration and habitat. Many species are cryptically colored (camouflaged) and, in the various vinesnakes, similarity to the snake's surroundings is enhanced by the narrow body and pointed head. More active defenses include threat displays, which may involve flattening the body, as in many North American watersnakes (*Nerodia*); spreading of a horizontal hood reminiscent of cobras, as seen in the xenodontine false water-cobras (*Hydrodynastes*); or vertical inflation of the neck, as in the colubrine tiger ratsnake (*Spilotes*). In general, horizontal displays characterize terrestrial species, whereas arboreal ones have vertical displays. Tail displays also occur in some species, such as the xendontine ring-necked snakes (*Diadophis*), in which the tail is upturned to show the bright ventral coloration and is moved in a distinctive spiral action. Other species gape when confronted by a predator and many bite, although others simply attempt to escape. Asian natricine snakes of the genera *Rhabdophis* and *Balanophis* have glands on the dorsal surface of their neck or along their entire back that exude a noxious compound when the snakes are threatened. Tail loss is rare among snakes, but a few taxa apparently lose the tail freely to escape predators, including the neck-banded snake (*Scaphiodontophis*), a Neotropical colubrine. Unlike the tails of some lizards, those of snakes do not regenerate if lost. Mimicry is common among snakes. Some species, such as the false pitvipers (*Xenodon*), mimic vipers, while in the Neotropics many species of xenodontines and a few colubrines mimic the elapid American coral snakes (*Micrurus*).

Colubrid snakes do not appear to be territorial. Instead, the males of some species, especially among the colubrines, are known to engage in ritual combat to obtain access to mates. Such combat generally resembles that of vipers and elapids, with the individuals intertwining their bodies and attempting to topple each other. That behavior may have served as the inspiration for the caduceus, the staff with entwining snakes that was said to have been carried by the god Hermes and that now is emblematic of the medical profession. In some species biting may also occur during combat. Species with male combat generally exhibit sexual size dimorphism in which males are larger than females, reflecting the fact that larger males generally are more successful in combat and therefore are favored to reproduce. Some other species exhibit the opposite pattern of sexual size dimorphism, in which females are larger than males. That condition reflects selection in which larger female size is associated with greater numbers of offspring.

Courtship behavior is known in a number of colubrid species, and it generally involves the male tongue-flicking and pressing his head along the back of the female. If the female is receptive, the male inserts one of his paired hemipenes into the female's cloaca to inseminate her. Males often locate females by following pheromone trails, a behavior mediated by the vomeronasal organ located in the roof of the mouth. That

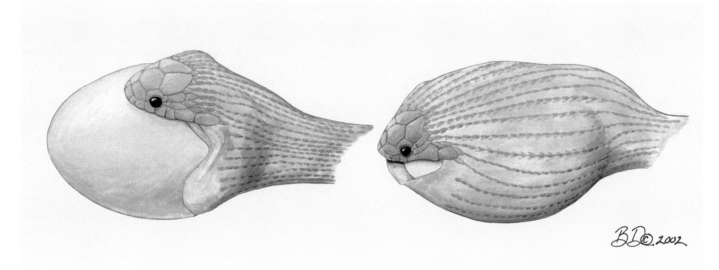

The common egg-eater snake (*Dasypeltis scabra*) must unhinge its jaw to eat an egg. (Illustration by Barbara Duperron)

organ receives cues from the environment by means of the tongue tips. In some species courting males aggregate around females and compete for matings, as in certain Canadian populations of the common gartersnake (*Thamnophis sirtalis*).

Feeding ecology and diet

Colubrids feed on a wide variety of prey. Some species are prey generalists, feeding on virtually any animals within an appropriate size range. The racer (*Coluber constrictor*), for example, is known to feed on a wide variety of prey, including mammals, lizards, frogs, baby turtles, and insects. Some wide-ranging species, such as the ring-necked snake (*Diadophis punctatus*), consume different prey in different parts of their range, while in others with varied diets, such as the common garter snake (*Thamnophis sirtalis*), prey preference is influenced by early experience with a particular prey. Location of prey frequently involves the use of chemical cues, which are picked up by the tongue and delivered to the vomeronasal organ. However, many terrestrial and arboreal snakes are also strongly dependent upon vision to locate prey.

More often, however, colubrids are prey specialists, preferring either a general class of prey, such as fishes, frogs, or mammals, or a very specific diet, such as lizard eggs or skinks. Unlike lizards, few snakes feed regularly on insects. Among colubrids, only a few lineages take large numbers of insects or other terrestrial arthropods, of which the sonorines are the largest. Even within that group, some (such as *Sonora*) take a variety of arthropods, whereas others specialize on particular arthropods, such as centipedes (*Tantilla*). *Stenorrhina* preys primarily upon scorpions, tarantulas, and grasshoppers. Some snakes specialize on prey of a certain shape. Often these are fossorial snakes, whose narrow heads limit them to relatively elongate prey. For smaller species these may be worms, whereas large species may feed on snakes or elongate lizards. Within the aquatic genus *Farancia*, the rainbow snake (*F. erytrogramma*) feeds mainly on eels, whereas the red-bellied mudsnake (*F. abacura*) feeds primarily on elongate aquatic amphibians, such as sirens (*Siren*). Eggs are a rich, if seasonal, source of food for many snakes. The arboreal cat-eyed snakes (*Leptodeira*), of the Neotropics, feed both on frogs and on frog eggs that are laid on vegetation overhanging water, such as those of the red-eyed treefrog (*Agalychnis callidryas*). The soft-shelled eggs of lizards and snakes are eaten by a variety of snakes, including the scarletsnake, *Cemophora*, a North American colubrine, the Asian kukrisnakes (*Oligodon*), and several Neotropical xenodontine genera, such as *Umbrivaga* and *Enulius*. An extreme case of specialization involves the egg-eaters (*Dasypeltis*), African colubrines that feed only on bird eggs. They consume eggs that may be several times the diameter of their own heads, forcing their nearly toothless jaws around the egg. Once in the esophagus, the eggs are cracked on ventral projections of the vertebrae; the shell is regurgitated and the liquid contents are swallowed. A few snakes possess hinged teeth, which fold back when their hard-bodied prey are consumed. Among these are several lineages of skink-eating snakes, including the neck-banded snake (*Scaphiodontophis*), a Neotropical colubrine, and some that feed on hard-shelled crayfish, including the striped crayfish snake (*Regina alleni*), a natricine from southeastern North America.

Reproductive biology

Although most species of colubrid snakes are oviparous, a large number of viviparous species are also known. Eggs may be laid in a hole or burrow in the ground or within rotting vegetation, where the heat generated by decomposition presumably speeds development. Viviparous species are especially prevalent at high latitudes and at high elevations, where females apparently can overcome the adverse effects of cold environmental temperatures on embryonic development by actively thermoregulating their own temperature and thus warming their embryos.

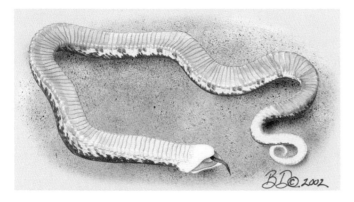

Death feign (faking death), a defensive posture, in *Heterodon platirhinos*. (Illustration by Barbara Duperron)

The size of a clutch or litter varies widely among colubrids. Small, fossorial oviparous species often lay only a few eggs (the wormsnakes, *Carphophis*, average about three per clutch), whereas larger terrestrial species often lay several times as many (for example, common kingsnakes, *Lampropeltis getula*, average about 10). In some species clutch size varies geographically, as in the Eastern racer (*Coluber constrictor*). Some species lay many more, such as the Eastern hog-nosed snake (*Heterodon platirhinos*), which averages more than 20 eggs per clutch, and the mudsnakes (*Farancia*), which average over 30. The females of several oviparous species of colubrids are reported to attend their clutches, and others are known to lay their eggs in communal oviposition sites. Likewise, litter size varies greatly among viviparous species. All of the thamnophiine natricines are viviparous, with the average number of young varying from about seven in the North American earthsnakes (*Virginia*) to almost 30 in the Plains garter snake (*Thamnophis radix*) and almost 50 in the diamond-backed watersnake (*Nerodia rhombifera*).

Most colubrid snakes reproduce annually, even in tropical regions, and reproduction may be timed to the seasonal patterns of temperature or rainfall. The embryos of both oviparous and viviparous snakes are well provisioned with yolk, although in some viviparous species a placenta permits the transfer of additional nutrients from mother to embryo. Even in oviparous colubrids, embryonic development usually begins well before the eggs are laid, and the embryos are already about one-third of the way through development by the time of oviposition.

Conservation status

The 2002 IUCN Red List includes 36 colubrid species. One species, *Alsophis sanctaecrucis*, is categorized as Extinct. In addition, 6 species are listed as Critically Endangered; 7 as Endangered; 8 as Vulnerable; 4 as Lower Risk/Near Threatened; and 10 as Data Deficient.

Very few colubrids are legally protected at either the national or international levels, although more regional regulations protect a few more. No colubrid species appear on Appendix I of CITES, only four appear on Appendix II, and three appear on Appendix III. As with most organisms, habitat destruction is probably the leading threat to colubrid snakes throughout most of their range. In addition, however, collection for the pet trade has contributed to the decline of some species, such as the Eastern indigo snake (*Drymarchon corais couperi*), and some species may be at risk due to their use in the skin trade and other commerce.

Overwhelmingly, it is humans who pose an ongoing threat to many colubrid snakes. However, one species, the brown treesnake (*Boiga irregularis*), has itself become a severe threat to the native fauna of the Pacific island of Guam. The accidental introduction of this adaptable predator to the island following World War II, apparently as stowaways on military transports, has resulted in the decimation of the native lizards and birds of the island and has resulted in the collapse of the natural food web. Efforts are under way to control this species and to recover the populations of those birds that have not already been completely extirpated.

Significance to humans

Colubrid snakes appear to figure in human commerce in two primary ways. First, many species are sold in the pet trade, including large numbers of wild-caught individuals of many species. Several groups of colubrids have become sufficiently popular that they now are bred in captivity in large numbers. Among the latter are the cornsnake (*Elaphe guttata*) and several species of kingsnakes (*Lampropeltis*). Although most species are too small to be useful as leather, a few colubrids are exploited in the skin trade, including the large Asian ratsnakes (*Ptyas*). Recently a large trade in aquatic colubrids of the subfamily Homalopsinae has been documented in Cambodia, and much of that trade involves the preparation of skins for leather. In addition, however, snakes are used as food for humans and are also fed to farm-raised crocodiles that are in turn used for leather. The magnitude of the trade was enormous, with an estimated 4,000–8,500 snakes per day traded at one port during peak periods. That catch is likely to be unsustainable.

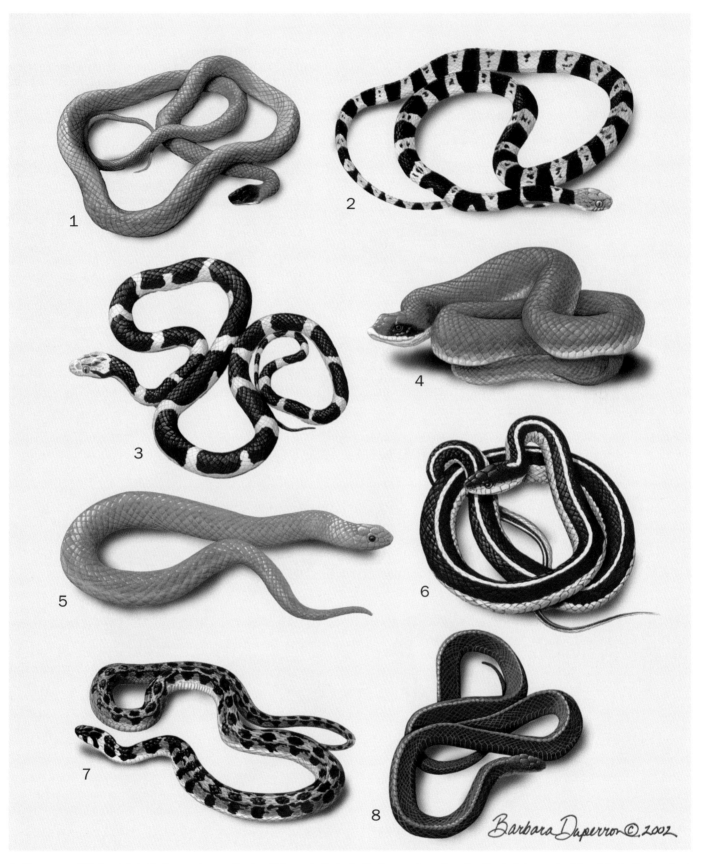

1. Mussurana (*Clelia clelia*); 2. Amazonian snail-eater (*Dipsas indica*); 3. Northern cat-eyed snake (*Leptodeira septentrionalis*); 4. Eastern hog-nosed snake (*Heterodon platirhinos*); 5. Common slug-eater (*Duberria lutrix*); 6. Common garter snake (*Thamnophis sirtalis*); 7. Yamakagashi (*Rhabdophis tigrinus*); 8. Cape filesnake (*Mehelya capensis*). (Illustration by Barbara Duperron)

1. Brown treesnake (*Boiga irregularis*); 2. Smooth snake (*Coronella austriaca*); 3. Boomslang (*Dispholidus typus*); 4. Common egg-eater (*Dasypeltis scabra*); 5. Milksnake (*Lampropeltis triangulum*); 6. Indigo snake (*Drymarchon corais*); 7. Tentacled snake (*Erpeton tentaculatus*); 8. Dog-faced watersnake (*Cerberus rynchops*). (Illustration by Barbara Duperron)

Species accounts

Brown treesnake
Boiga irregularis

SUBFAMILY
Colubrinae

TAXONOMY
Coluber irregularis Merrem, 1802, Bechstein.

OTHER COMMON NAMES
None known.

PHYSICAL CHARACTERISTICS
This snake is about 4.5–6.5 ft (1.4–2 m) in length. It has a slender body, very narrow neck, rounded head, and large eyes. It is tan to brown in color with irregular, darker brown, transverse markings.

DISTRIBUTION
The brown treesnake is found in Indonesia, New Guinea, Solomon Islands, and northern Australia. It has been introduced on Guam.

HABITAT
This snake is primarily arboreal, but it also forages terrestrially. It occurs in a range of habitats, including forests, swamps, and coastal regions.

BEHAVIOR
The brown treesnake is primarily nocturnal. It strikes defensively when threatened.

FEEDING ECOLOGY AND DIET
This species feeds on a variety of vertebrates, primarily lizards, birds, and mammals.

REPRODUCTIVE BIOLOGY
The brown treesnake is oviparous, laying about a dozen eggs per clutch.

CONSERVATION STATUS
Not threatened. On Guam, where the species has been introduced, it is implicated in the extinction of several species of vertebrates, including most of the native birds.

SIGNIFICANCE TO HUMANS
There is concern that this species may spread to other Pacific Islands, prompting security measures on some islands, including Hawaii, in an attempt to prevent its introduction. ◆

Smooth snake
Coronella austriaca

SUBFAMILY
Colubrinae

TAXONOMY
Coronella austriaca Laurenti, 1768. Three subspecies are recognized.

OTHER COMMON NAMES
French: Couleuvre lisse; German: Glattnatter.

PHYSICAL CHARACTERISTICS
Adults reach about 24 in (60 cm) in length. This is a shiny, grayish or reddish brown snake with ill-defined, paired blotches down the back.

DISTRIBUTION
Europe (including southern England) and western Asia, from Scandinavia to northern Spain and east to Kazakhstan and northern Iran.

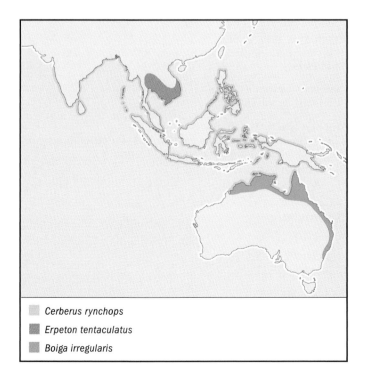

☐ *Cerberus rynchops*
◼ *Erpeton tentaculatus*
▨ *Boiga irregularis*

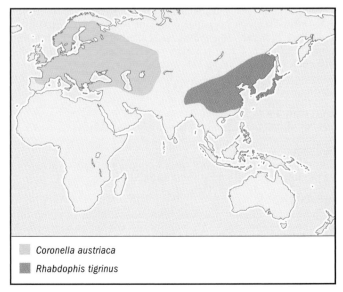

☐ *Coronella austriaca*
◼ *Rhabdophis tigrinus*

HABITAT

Fairly dry, open, rocky habitats are favored, including stone walls. In England this species favors heathlands, finding shelter under rocks and logs.

BEHAVIOR

The smooth snake forages diurnally.

FEEDING ECOLOGY AND DIET

This snake feeds primarily on lizards, but other small vertebrates are also taken. The prey is constricted.

REPRODUCTIVE BIOLOGY

This species is viviparous, giving birth to about four to 15 young.

CONSERVATION STATUS

Numbers reportedly have declined greatly in England and elsewhere in its European range. Although not listed by the IUCN, the species is fully protected in Europe, being listed in Appendix II of the Bern Convention of European Wildlife and Natural Habitats and on Annex IV of the European Union Habitat and Species Directive.

SIGNIFICANCE TO HUMANS

The smooth snake is apparently a sensitive indicator of the health of heathlands in southern England. ◆

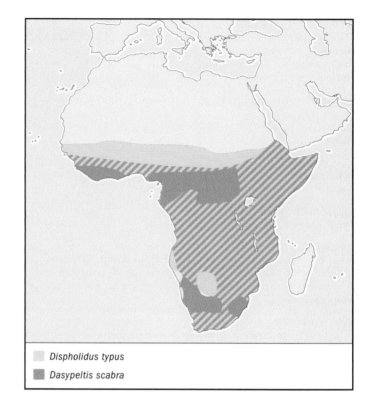

☐ *Dispholidus typus*
■ *Dasypeltis scabra*

Common egg-eater

Dasypeltis scabra

SUBFAMILY

Colubrinae

TAXONOMY

Coluber scaber Linnaeus, 1758, Indiis. Two subspecies are sometimes recognized.

OTHER COMMON NAMES

English: African egg-eater, rhombic egg-eater.

PHYSICAL CHARACTERISTICS

This snake is about 20–35 in (50–90 cm) in length. It is a very slender snake with a narrow head; the dorsal scales are keeled, and the keels are serrated on the lower lateral scales. The color varies but is usually gray or brown, with prominent dark blotches down the center of the back and narrower ones on the sides. The jaws are nearly toothless, the skin of the neck is capable of an astonishing degree of stretching, and the anterior vertebrae are modified to crack the shells of bird eggs.

DISTRIBUTION

Widely distributed throughout sub-Saharan Africa.

HABITAT

The common egg-eater occupies a range of habitats, exclusive of the driest deserts and wettest tropical forests.

BEHAVIOR

The common egg-eater mimics various species of vipers in different parts of its range. Mimicry of the saw-scaled viper (*Echis carinatus*) involves not only appearance but behavior. Like that viper, the egg-eater throws its body into a series of curved loops and rubs the serrated lateral scales against each other to produce a hissing sound.

FEEDING ECOLOGY AND DIET

This snake feeds exclusively on bird eggs, and its morphology is highly modified for that diet. The head is forced over an egg, which may be several times the diameter of the head, and the egg is cracked using modified ventral processes (hypapophyses) of the anterior vertebrae, which actually penetrate the esophagus. The liquid contents of the egg are swallowed, and the shell is crushed and regurgitated. The egg-eater feeds heavily in the spring and summer, when birds are nesting.

REPRODUCTIVE BIOLOGY

The common egg-eater is oviparous, with a clutch size of six to 25 eggs.

CONSERVATION STATUS

Not listed by the IUCN.

SIGNIFICANCE TO HUMANS

The common egg-eater may be mistaken as a venomous snake, due to its mimicry of several species of vipers ◆

Boomslang

Dispholidus typus

SUBFAMILY

Colubrinae

TAXONOMY

Bucephalus typus Smith, 1829, Old Latakoo, South Africa. Two or three subspecies are recognized.

OTHER COMMON NAMES

French: Serpent d'arbre du Cap; German: Boomslang.

PHYSICAL CHARACTERISTICS
The boomslang is about 4 ft (1.25 m) in length. It is a long, slender snake with distinctive oblique dorsal scales. Its color is highly variable, ranging from nearly black to bright green and even reddish; some individuals have black scales with a bright yellow spot in the center. Males are more brightly colored than females. The head is large and the eyes are prominent.

DISTRIBUTION
Ranges widely throughout sub-Saharan Africa.

HABITAT
This species occurs in a wide variety of habitats, including forest and savanna.

BEHAVIOR
The boomslang is highly arboreal.

FEEDING ECOLOGY AND DIET
This snake forages diurnally, feeding on a variety of arboreal prey, especially birds and chameleons.

REPRODUCTIVE BIOLOGY
The boomslang is oviparous, with a clutch size of about a dozen eggs.

CONSERVATION STATUS
Not listed by the IUCN.

SIGNIFICANCE TO HUMANS
The boomslang is one of the few colubrid snakes capable of producing a lethal bite in humans. The venom, which is injected through very long rear fangs, acts slowly to impair the blood's ability to clot, resulting in death by hemorrhage. The prominent herpetologist Karl P. Schmidt died from the bite of a boomslang in 1957. ◆

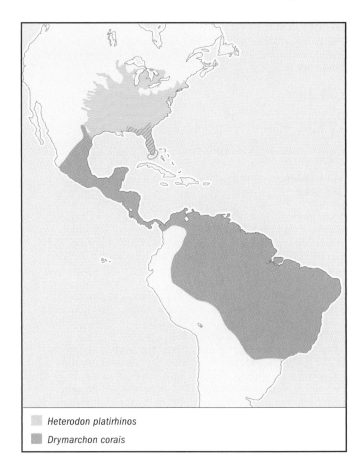

Heterodon platirhinos
Drymarchon corais

Indigo snake
Drymarchon corais

SUBFAMILY
Colubrinae

TAXONOMY
Coluber corais Boie, 1827, America. Eight subspecies are recognized, and some workers believe that several separate species should be recognized.

OTHER COMMON NAMES
English: Blacktail, cribo; Spanish: Cola sucia.

PHYSICAL CHARACTERISTICS
This very large colubrid may reach almost 10 ft (3 m) in length. It has large, shiny, slightly oblique dorsal scales. Its color varies geographically. *D. c. couperi*, of the southeastern United States, is jet black with a reddish throat. Tropical subspecies may be black, gray, brown, or yellow, sometimes with a sharply contrasting tail.

DISTRIBUTION
Ranges from the southeastern United States to northern Argentina.

HABITAT
In the southeastern United States, this species prefers environments with sandy soils, such as pine savanna and scrub. Tropi-

cal populations occupy a wider range of habitats, from moist forest to savanna and thorn forest. The species is frequently found near water.

BEHAVIOR
The indigo snake is diurnal, foraging actively over large areas.

FEEDING ECOLOGY AND DIET
A wide variety of prey are consumed, including other fishes, frogs, turtles, birds, and mammals. Other snakes appear to comprise a substantial proportion of the diet, including pitvipers. It does not constrict its prey but simply overpowers it.

REPRODUCTIVE BIOLOGY
The indigo snake is oviparous, with a clutch size of about four to 12. Hatchlings are large, sometimes over 2 ft (60 cm).

CONSERVATION STATUS
This species is not listed by the IUCN. However, in the southeastern United States, the eastern indigo snake (*D. c. couperi*) is listed as Threatened by the U.S. Fish and Wildlife Service; it is also listed by several states. The subspecies is highly prized in the pet trade, and illegal collecting continues to pose a threat. The species also is a frequent inhabitant of the burrows dug by gopher tortoises (*Gopherus polyphemus*), as is the eastern diamondback rattlesnake (*Crotalus adamanteus*). Rattlesnakes are collected in some regions for "rattlesnake round-ups" or for their skins, and collection often involves pouring gasoline down a tortoise burrow. That action can incidentally harm indigo snakes. Finally, this subspecies is strongly affected by the rapid residential and commercial development of its habitat in the southeastern United States.

SIGNIFICANCE TO HUMANS
Indigo snakes are highly valued as pets, especially *D. c. couperi*. Some individuals now are being produced through captive breeding. ◆

Milksnake
Lampropeltis triangulum

SUBFAMILY
Colubrinae

TAXONOMY
Coluber triangulum Lacépède, 1788, America. Twenty-five subspecies are recognized, and it is likely that this species ultimately will be divided into several separate species.

OTHER COMMON NAMES
English: Scarlet kingsnake (for subspecies *L. t. elapsoides*); French: Couleuvre tachetée; Spanish: Culebra-real coralillo.

PHYSICAL CHARACTERISTICS
This snake varies greatly in size and appearance across its broad geographic range, as reflected in the large number of subspecies recognized. Adult length varies from about 20–60 in (50–152 cm), depending upon the population. The usual color pattern is one of alternating bands of red-black-yellow/white-black-red. However, some populations have red blotches rather than bands, and one subspecies is melanistic (virtually all black). The scales are smooth and shiny.

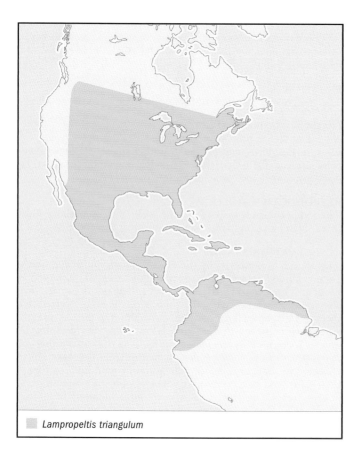

Lampropeltis triangulum

DISTRIBUTION
This species has a wide range, from southeastern Canada to western Ecuador and northern Venezuela.

HABITAT
Habitat varies across the wide range of this species. It generally occupies forested environments, but in some regions it can also be found in open prairies. In various parts of its range this species frequently occupies rocky slopes.

BEHAVIOR
The milksnake is essentially terrestrial (although *L. t. elapsoides* sometimes takes shelter beneath the bark of standing dead pine trees), and activity is largely nocturnal.

FEEDING ECOLOGY AND DIET
The diet varies by region but often includes lizards (especially skinks), snakes, and small mammals.

REPRODUCTIVE BIOLOGY
Milksnakes are oviparous, laying an average of about 10 eggs per clutch, although that number may vary geographically.

CONSERVATION STATUS
Not listed by the IUCN, although in some regions there may be substantial pressure from collection for the pet trade.

SIGNIFICANCE TO HUMANS
The common name of the species derives from an early myth that the species sucks milk from cows. That erroneous idea presumably derived from the discovery of milksnakes near barns, where they may have been feeding on mice. This species is highly valued in the pet trade, and many subspecies are now being bred in captivity for sale. ◆

Dog-faced watersnake
Cerberus rynchops

SUBFAMILY
Homalopsinae

TAXONOMY
Hydrus rynchops Schneider, 1799, Ganjam, India. Two subspecies are recognized.

OTHER COMMON NAMES
English: Bockadam.

PHYSICAL CHARACTERISTICS
This snake is about 24 in (60 cm) in length. It is a slender, grayish to reddish snake with a narrowly triangular head and dorsal eyes and nostrils.

DISTRIBUTION
This is a coastal species found from western India through Southeast Asia, the Philippines, and Indonesia, to northern Australia.

HABITAT
This species occupies coastal mangrove forests and estuaries, where it is often found on mud flats.

BEHAVIOR
The dog-faced watersnake forages in the water and on mud flats, where it moves by sidewinding. It is reported to release a foul-smelling musk for defense.

FEEDING ECOLOGY AND DIET
This species feeds primarily upon fishes.

REPRODUCTIVE BIOLOGY
This snake is viviparous, giving birth to about eight young.

CONSERVATION STATUS
Not listed by the IUCN, although coastal environments in general are subject to increasing pressure from expanding human populations.

SIGNIFICANCE TO HUMANS
None known. ◆

Tentacled snake
Erpeton tentaculatus

SUBFAMILY
Homalopsinae

TAXONOMY
Erpeton tentaculatus Lacépède, 1800, locality unknown.

OTHER COMMON NAMES
None known.

PHYSICAL CHARACTERISTICS
One of the most distinctive snakes, this species is 20 in (50 cm) in length and is readily identifiable by the paired, scaly appendages that project from the front of the snout. The head and body are extremely flat, the nostrils are valvular and dorsal in position, and the color may be light brown with darker stripes or very dark with lighter brown mottling.

DISTRIBUTION
Central and southern Thailand, and southern Cambodia and Vietnam.

HABITAT
Tentacled snakes are found in ponds and slow-moving streams.

BEHAVIOR
This snake is entirely aquatic, lying quietly among submerged vegetation waiting for prey. The flattened body is very stiff, capable of bending laterally but exhibiting little vertical flexibility.

FEEDING ECOLOGY AND DIET
These snakes are ambush predators of fishes.

REPRODUCTIVE BIOLOGY
This species is viviparous.

CONSERVATION STATUS
Not listed by the IUCN.

SIGNIFICANCE TO HUMANS
Tentacled snakes appear from time to time in the pet trade. In some regions they are erroneously regarded as dangerously venomous. ◆

Common slug-eater
Duberria lutrix

SUBFAMILY
Lamprophiinae

TAXONOMY
Coluber lutrix Linnaeus, 1758, Indiis. Six subspecies recognized.

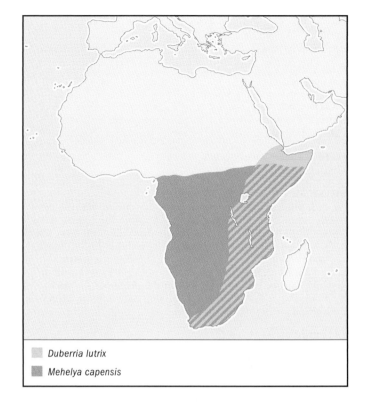

◻ *Duberria lutrix*
◼ *Mehelya capensis*

OTHER COMMON NAMES
None known.

PHYSICAL CHARACTERISTICS
The common slug-eater is a small snake, about 14 in (35 cm) in length, with a thick body and a narrow head. The scales are smooth, and the dorsal surface is brown or reddish; the ventral color is lighter.

DISTRIBUTION
Eastern Africa from Ethiopia to the Cape.

HABITAT
This snake occupies savannas and bush habitats.

BEHAVIOR
This species coils tightly when threatened.

FEEDING ECOLOGY AND DIET
The common slug-eater forages in moist situations, preying primarily on slugs. Land snails are also eaten.

REPRODUCTIVE BIOLOGY
This species is viviparous, giving birth to about 10 relatively large young per litter.

CONSERVATION STATUS
Not threatened.

SIGNIFICANCE TO HUMANS
None known. ◆

Cape filesnake
Mehelya capensis

SUBFAMILY
Lamprophiinae

TAXONOMY
Heterolepis capensis Smith, 1847, Eastern districts of Cape Province, South Africa. Three subspecies are recognized.

OTHER COMMON NAMES
German: Kap-Feilennatter.

PHYSICAL CHARACTERISTICS
As with other filesnakes, the body of this species has the distinctive triangular cross-section that inspired the common name. The shape results from highly modified vertebrae, which have long processes that project dorsally and laterally. Reaching about 60 in (1.5 m) in length, this species has narrow, keeled dorsal scales with patches of bare skin between them and an enlarged row of scales down the center of the back. The dorsal color is gray or dark brown, with a white or yellow mid-dorsal stripe.

DISTRIBUTION
Ranges through most of sub-Saharan Africa, from Cameroon and Somalia to the Cape.

HABITAT
The Cape filesnake occupies a variety of habitats but is especially prevalent in savannas.

BEHAVIOR
This is a terrestrial species that is active nocturnally.

FEEDING ECOLOGY AND DIET
Although the Cape filesnake feeds on a variety of vertebrates, including lizards and mammals, snakes are the preferred prey. This species constricts its prey and may consume snakes that are very long relative to its own body.

REPRODUCTIVE BIOLOGY
This species is oviparous, usually laying about six eggs. The hatchlings are large, about 15.8 in (40 cm) in length.

CONSERVATION STATUS
Not threatened.

SIGNIFICANCE TO HUMANS
None known. ◆

Yamakagashi
Rhabdophis tigrinus

SUBFAMILY
Natricinae

TAXONOMY
Tropidonotus tigrinus Boie, 1826. Two subspecies are recognized.

OTHER COMMON NAMES
English: Tiger watersnake.

PHYSICAL CHARACTERISTICS
This is a rather slender snake that attains a length of about 40 in (1 m). Color varies, but most specimens are drab green to yellow, with red skin between the scales of the neck. Young have a bright light ring around the neck. The dorsal scales are keeled, and enlarged scales on the neck overlie special nuchal glands.

DISTRIBUTION
Widely distributed in eastern Asia, including much of China, Korea, and Japan.

HABITAT
This snake is terrestrial but is generally found in association with water, including rice paddies.

BEHAVIOR
The defensive behavior of this snake is very unusual. When threatened by a predator, the snake arches its neck toward the attacker and releases the contents of paired nuchal glands that lie in the dorsal skin. The product of those glands is distasteful and irritating to the eyes and contains compounds similar to those found in the skin glands of toads.

FEEDING ECOLOGY AND DIET
This species feeds almost entirely on frogs and toads, although it occasionally eats fishes.

REPRODUCTIVE BIOLOGY
This snake is oviparous, laying about a dozen eggs per clutch.

CONSERVATION STATUS
Not listed by the IUCN. At one site in Japan, this species is reported to have declined due to changes in agricultural practices, including the use of pesticides and chemical fertilizers, which may have reduced the numbers of frogs.

SIGNIFICANCE TO HUMANS
Although this snake is reluctant to bite defensively, the bite has been known to cause fatalities in humans. The venom acts very slowly, inhibiting the ability of the blood to clot and causing death by hemorrhage. ◆

Common garter snake
Thamnophis sirtalis

SUBFAMILY
Natricinae

TAXONOMY
Coluber sirtalis Linnaeus, 1758, Canada. Eleven subspecies are recognized.

OTHER COMMON NAMES
French: Couleuvre rayèe, serpent-jarretière; German: Gewöhnliche Strumpfbandnatter; Spanish: Culebra-listonada común.

PHYSICAL CHARACTERISTICS
The common garter snake is a relatively slender snake generally reaching about 28 in (70 cm) in length. Color varies greatly across the wide geographic range of this species, but in most populations the snake exhibits three longitudinal yellow stripes. The ground color may be dark brown, olive green, or red, sometimes with black patches.

DISTRIBUTION
This species is widely distributed across North America, including the eastern half of the continent, the northern Great Plains and northern Rocky Mountains, and the Pacific Coast.

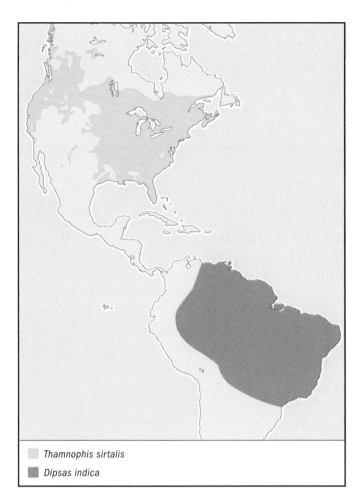

Thamnophis sirtalis

Dipsas indica

emerge earlier, so females are courted by large numbers of waiting males.

CONSERVATION STATUS
Not listed by the IUCN. The San Francisco garter snake (*T. s. infernalis*, previously known as *T. s. tetrataenia*) is the only snake in the United States listed as Endangered by the U.S. Fish and Wildlife Service. Habitat loss is primarily responsible for its decline. In Manitoba, Canada, where the red-sided garter snake (*T. s. parietalis*) aggregates by the thousands in communal hibernacula, the populations were once threatened by overcollection for commercial purposes. A ban on export of garter snakes from Manitoba was imposed in 1991.

SIGNIFICANCE TO HUMANS
Because of their abundance and ease of maintenance in captivity, common garter snakes are used for a variety of laboratory studies and are also maintained as pets. ◆

Mussurana
Clelia clelia

SUBFAMILY
Xenodontinae

TAXONOMY
Coluber clelia Daudin, 1826, Surinam. Three or four subspecies are recognized.

OTHER COMMON NAMES
Spanish: Culebrera, sumbadora.

PHYSICAL CHARACTERISTICS
This is a large species, reaching over 6 ft (2 m) in length. Adults are uniform shiny black or dark gray dorsally, and

Some Canadian populations extend far north, one reaching the southern Northwest Territories.

HABITAT
Common garter snakes are found in a wide variety of habitats, including marshes, fields, woodlands, and forest edges. This species is often found in association with water.

BEHAVIOR
In general this snake is diurnally active, with a long activity season. It is active year round in the southern part of the range, and even at extreme northern localities it has been reported to be active for five months out of the year. Individuals in northern populations may travel long distances from their hibernation sites to their summer foraging habitats.

FEEDING ECOLOGY AND DIET
This species forages actively on a wide variety of prey, including both vertebrates and invertebrates. Common prey include earthworms, fishes, and frogs, but a host of other prey also are eaten, occasionally including small mammals or birds.

REPRODUCTIVE BIOLOGY
This snake is viviparous, with litter size varying geographically. The usual litter size is about 10–15 young, although lower numbers are common in some northern populations, and an average of over 30 was recorded for a population in Maryland. Populations in Manitoba, Canada, are well known for the enormous aggregations of males seeking to mate upon emergence from hibernation. The males, which are smaller,

Clelia clelia

Leptodeira septentrionalis

lighter ventrally. Juveniles are very different, having a bright red body with black tips on the dorsal scales, a broad cream collar, and a black snout.

DISTRIBUTION
Central and South America, from Guatemala and Belize to northern Argentina.

HABITAT
The mussurana is an occupant of both forested and open habitats, including marshes.

BEHAVIOR
This active species largely forages nocturnally, but it can also be found abroad during the day.

FEEDING ECOLOGY AND DIET
Snakes are the preferred prey of this species, including pitvipers of the genus *Bothrops*, the lanceheads. Lizards and mammals are also eaten. This powerful snake combines constriction with rear-fanged envenomation.

REPRODUCTIVE BIOLOGY
This snake is oviparous, laying clutches of about 10–20 eggs.

CONSERVATION STATUS
Not listed by the IUCN. This species is listed on Appendix II of CITES, one of the few colubrids to receive such protection.

SIGNIFICANCE TO HUMANS
The mussurana is highly regarded within its range for its tendency to prey on venomous snakes. Protection under CITES presumably reflects a belief that this species might be desirable in the pet trade, although similar protection is not afforded countless species of other tropical colubrids. ◆

Amazonian snail-eater
Dipsas indica

SUBFAMILY
Xenodontinae

TAXONOMY
Dipsas indica Laurenti, 1768, Ceylon (in error). Five subspecies are recognized.

OTHER COMMON NAMES
None known.

PHYSICAL CHARACTERISTICS
This is a slender snake about 28 in (70 cm) in length, with a laterally compressed body and a row of enlarged mid-dorsal scales. The body is gray to brown with large, dark brown blotches arranged in pairs on the sides. In some populations the edges of blotches may be somewhat irregular.

DISTRIBUTION
Found in tropical South America, from the Amazon Basin to Bolivia and northern Argentina.

HABITAT
This species occurs in both primary and secondary forest.

BEHAVIOR
This snake is highly arboreal, foraging at night.

FEEDING ECOLOGY AND DIET
The Amazonian snail-eater feeds on snails, which it extracts from the shell. After the snake seizes the exposed body of a snail, the slender lower jaws of the snake are drawn into the shell as the snail retracts. The snake then slowly pulls on the soft body of the snail with its lower jaws, eventually tearing the body away from the shell, which is discarded.

REPRODUCTIVE BIOLOGY
This oviparous species presumably resembles other species of *Dipsas*, which apparently lay very small clutches, as do some other slender arboreal snakes.

CONSERVATION STATUS
Not listed by the IUCN, although it is clearly dependent upon tropical forest habitat, which is being cleared for timber and agriculture in many regions.

SIGNIFICANCE TO HUMANS
None known. ◆

Eastern hog-nosed snake
Heterodon platirhinos

SUBFAMILY
Xenodontinae

TAXONOMY
Heterodon platirhinos Latreille, 1801, vicinity of Philadelphia.

OTHER COMMON NAMES
English: Hissing adder, puff adder, spreading adder; French: Hétérodon commun; German: Gewöhnliche hakennatter.

PHYSICAL CHARACTERISTICS
This is a stout-bodied snake, usually about 30 in (75 cm) in length, with a broad head and a distinctive, upturned snout. The keeled dorsal scales are arranged in oblique rows. The coloration is highly variable, often yellow, orange, or olive green, with dark brown dorsal and lateral blotches. Some individuals become melanistic (entirely black), especially in the southeastern United States.

DISTRIBUTION
Widely distributed across eastern and central North America, including southern Ontario, Canada.

HABITAT
This species occurs in a range of environments, including drier forests, but it prefers open habitats, generally with well-drained, sandy soils.

BEHAVIOR
The Eastern hog-nosed snake has a diverse repertoire of defensive behaviors, some of which have given rise to colorful common names. When first approached by a predator, this species flattens its head and spreads the ribs of its neck to form a broad hood, while hissing loudly. The snake may raise its head and even strike, although the species almost never bites. If that display of bravado fails and the predator persists, the snake rolls on its back, usually with its mouth agape and its tongue hanging out, apparently feigning death. If turned right side up, the snake immediately rolls over on its back again, remaining in that position until the threat has passed.

FEEDING ECOLOGY AND DIET

Most populations of these snakes feed overwhelmingly on toads (*Bufo*), although other frogs may also be eaten, as well as occasional salamanders and even mammals. The snakes have been reported to use the upturned snout to excavate toads from sandy soils. The maxillary bone is highly mobile and has a pair of greatly enlarged rear teeth that apparently help to deflate toads that have puffed themselves up with air to resist being swallowed.

REPRODUCTIVE BIOLOGY

This snake is oviparous, laying relatively large clutches of eggs (generally about 20, although much larger clutches have been reported).

CONSERVATION STATUS

Not listed by the IUCN, although loss of habitat, declines in amphibian populations, and the threatening defensive behavior of this species (its stout body form also resembles that of pitvipers) probably make its populations vulnerable to human impact.

SIGNIFICANCE TO HUMANS

None known. ◆

Northern cat-eyed snake

Leptodeira septentrionalis

SUBFAMILY

Xenodontinae

TAXONOMY

Dipsas septentrionalis Kennicott in Baird, 1859, Matamoros, Tamaulipas, Mexico and Brownsville, Texas. Five subspecies are recognized.

OTHER COMMON NAMES

Spanish: Culebra destinida, culebra nocturna.

PHYSICAL CHARACTERISTICS

This is a slender snake about 33 in (85 cm) long, with a moderately wide head and large eyes with vertical pupils. In general, the color pattern is brown with irregular, darker brown spots. However, the pattern varies considerably over the range of the species. In some regions the dark blotches extend fully across the body, whereas in others the dark markings are limited to small, irregular spots down the middle of the back.

DISTRIBUTION

Distributed widely through the northern Neotropics, from southern Texas and Sinaloa, Mexico, to western Venezuela and northwestern Peru.

HABITAT

The northern cat-eyed snake is primarily an inhabitant of moist forests, but it also occupies drier habitats in western Mexico and Peru.

BEHAVIOR

This species is primarily arboreal, but it also engages in terrestrial activity, especially in drier portions of its range. Although it is rear-fanged, this species typically does not bite defensively when captured.

FEEDING ECOLOGY AND DIET

Frogs are the preferred prey of this species, which often forages among large choruses of calling males. Where frogs lay eggs on vegetation overhanging water, as in the case of the leaf frogs (*Agalychnis*), cat-eyed snakes also consume the eggs, pushing their faces into the mass of egg jelly to reach the developing embryos. In addition, lizards and even a snake have been reported as occasional prey.

REPRODUCTIVE BIOLOGY

This snake is oviparous, laying about 10 eggs per clutch.

CONSERVATION STATUS

Although not listed by the IUCN, this species is listed as Threatened by the Texas Parks and Wildlife Department, reflecting its very limited range at the southern tip of that state.

SIGNIFICANCE TO HUMANS

Before the advent of effective fumigation, snakes of the genus *Leptodeira* frequently stowed away in bunches of bananas, arriving alive at ports such as New York City.

Resources

Books

Branch, Bill. *Field Guide to Snakes and Other Reptiles of Southern Africa.* 3rd edition. Sanibel Island, FL: Ralph Curtis Books, 1998.

Cadle, John E., and Harry W. Greene. "Phylogenetic Patterns, Biogeography, and the Ecological Structure of Neotropical Snake Assemblages." In *Species Diversity in Ecological Communities: Historical and Geographical Perspectives,* edited by Robert E. Ricklefs and Dolf Schluter. Chicago: University of Chicago Press, 1993: 281–293.

Greene, Harry W. *Snakes: The Evolution of Mystery in Nature.* Berkeley: University of California Press, 1997.

Lee, Julian C. *The Amphibians and Reptiles of the Yucatán Peninsula.* Ithaca, NY: Cornell University Press, 1996.

Pough, F. Harvey, Robin M. Andrews, John E. Cadle, Martha L. Crump, Alan H. Savitzky, and Kentwood D. Wells.

Herpetology. 2nd edition. Upper Saddle River, NJ: Prentice Hall, 2001.

Rossman, Douglas A., Neil B. Ford, and Richard A. Seigel. *The Garter Snakes: Evolution and Ecology.* Norman: University of Oklahoma Press, 1996.

Shine, Richard. "The Evolution of Viviparity in Reptiles: An Ecological Analysis." 605–694. In *Biology of the Reptilia, Volume 15, Development B,* edited by Carl Gans and Frank Billett. New York: John Wiley & Sons, 1985.

———. "Parental Care in Reptiles." In *Biology of the Reptilia.* Vol. 16, *Ecology B: Defense and Life History,* edited by Carl Gans and Raymond B. Huey. New York: Alan R. Liss, 1988: 275–329.

———. *Australian Snakes: A Natural History.* Ithaca, NY: Cornell University Press, 1991.

Resources

Zug, George R., Laurie J. Vitt, and Janalee P. Caldwell. *Herpetology: An Introductory Biology of Amphibians and Reptiles.* 2nd edition. San Diego: Academic Press, 2001.

Periodicals

Cadle, John E. "Molecular Systematics of Neotropical Xenodontine Snakes. III. Overview of Xenodontine Phylogeny and the History of New World Snakes." *Copeia* 1984 (1984): 641–652.

Cadle, John E. "Phylogenetic Relationships Among Advanced Snakes: A Molecular Perspective." *University of California Publications in Zoology* 119 (1988): 1–70.

———. "The Colubrid Radiation in Africa (Serpentes: Colubridae): Phylogenetic Relationships and Evolutionary Patterns Based on Immunological Data." *Zoological Journal of the Linnean Society* 110 (1994): 103–140.

Fritts, Thomas H., and Gordon H. Rodda. "The Role of Introduced Species in the Degradation of Island Ecosystems: A Case History of Guam." *Annual Review of Ecology and Systematics* 29 (1998): 113–140.

Gyi, Ko Ko. "A Revision of Colubrid Snakes of the Subfamily Homalopsinae." *University of Kansas Publications, Museum of Natural History* 20 (1970): 47–223.

Karns, Daryl R., Harold K. Voris, Tanya Chan-ard, Jeffrey C. Goodwin, and John C. Murphy. "The Spatial Ecology of the Rainbow Water Snake, *Enhydris enhydris* (Homalopsinae) in Southern Thailand." *Herpetological Natural History* 7 (2000): 97–115.

Kraus, Fred, and Wesley M. Brown. "Phylogenetic Relationships of Colubrid Snakes Based on Mitochondrial DNA Sequences." *Zoological Journal of the Linnean Society* 122 (1998): 455–487.

Lopez, T. J., and L. Maxson. "Mitochondrial DNA Sequence Variation and Genetic Differentiation Among Colubrine Snakes (Reptilia: Colubridae: Colubrinae)." *Biochemical Systematics and Ecology* 23 (1995): 487–505.

Shine, Richard. "Strangers in a Strange Land: Ecology of Australian Colubrid Snakes." *Copeia* 1991 (1991): 120–131.

———. "Sexual Size Dimorphism in Snakes Revisited." *Copeia* 1994 (1994): 326–346.

Stuart, Bryan L., Jady Smith, Kate Davey, Prom Din, and Steven G. Platt. "Homalopsine Watersnakes: The Harvest and Trade from Tonle Sap, Cambodia." *TRAFFIC Bulletin* 18 (2000): 115–124.

Zaher, Hussam. "Hemipenial Morphology of the South American Xenodontine Snakes, with a Proposal for a Monophyletic Xenodontinae and a Reappraisal of Colubroid Hemipenes." *Bulletin of the American Museum of Natural History* 240 (2000): 1–168.

Other

European Molecular Biology Laboratory. The EMBL Reptile Database. Family Colubridae (Colubrids). June 1, 2002 [cited October 31, 2002]. <http://www.embl-heidelberg.de/uetz/families/Colubridae.html>

Alan H. Savitzky, PhD

Cobras, kraits, seasnakes, death adders, and relatives

(Elapidae)

Class Reptilia

Order Squamata

Suborder Serpentes

Family Elapidae

Thumbnail description
Small to large venomous snakes

Size
7–200 in (18–500 cm)

Number of genera, species
60 genera; more than 300 species

Habitat
Highly variable depending on species; desert, savanna, rainforest, fully arboreal to fully marine

Conservation status
Vulnerable: 7 species; Lower Risk/Near Threatened: 2 species

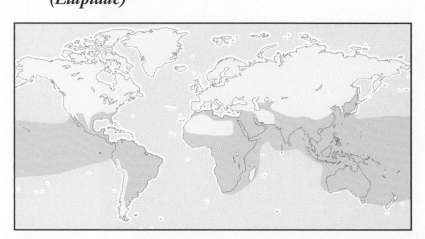

Distribution
Southern United States to Central and South America, Africa, Asia, Australia, and the Pacific and Indian Oceans

Evolution and systematics

The two major families of venomous snakes are Elapidae, or the elapid snakes, and Viperidae, the vipers and pitvipers. The snakes in these families are similar in that they have fangs in the front of the mouth. The two groups arose independently from nonvenomous snake ancestors, however, so there are important differences between them. The overall appearance of elapids is much more like that of the primarily nonvenomous colubrids than it is of the generally plump, short-tailed vipers. The main differences between elapids and vipers are in the structure of the venom delivery apparatus and the nature of the venom. Elapids have proteroglyphic dentition, which means "fixed front fangs." Vipers have solenoglyphic dentition, which means "movable front fangs." In elapids, the fangs are fixed in one position and are relatively short to avoid puncturing the snake's bottom lip. Vipers have long fangs that are hinged and fold back into the mouth. The venom of elapids is quite different from that of vipers. Elapids have neurotoxic venom (nerve poison), which acts mainly on the central nervous system. The venom affects heart function and breathing but causes little or no damage at the bite site. Vipers have primarily hemotoxic and myotoxic venom, which produces severe damage at the bite site, including complete necrosis of the surrounding tissue.

The venomous elapid snakes include 60 genera and more than 300 species. Because elapids represent approximately 10% of living snake species and more than 50% of species of venomous snakes, they are of considerable medical importance. The elapids are fantastically diverse in size, shape, color, ecology, and behavior, but they can be classified as follows according to size and distribution: cobras and mambas; coral snakes; terrestrial kraits; Australo-Papuan elapids, which include brown snakes, taipans, and death adders; sea kraits; and seasnakes.

Little is known about the origin of elapid snakes except that they are related to some African forms that seem to have "protoelapid" fangs. For example, the African and Middle East members of the genus *Atractaspis* are venomous and have front fangs, but they also have a number of characteristics that differentiate them from elapids and unite them with primarily nonvenomous species. The southern African genus *Homoroselaps* is confusing in that it has elapid fangs and venom but seems to have characteristics of *Atractaspis*.

Despite the confusion, elapids seem to form a monophyletic group, as does each of the major elapid lineages. Together elapids are primarily defined by the presence of a venom delivery system comprising two small permanently erect front fangs. Diverse data sets have been used to elucidate relationships among and within elapid lineages, including various aspects of morphology, protein albumins, karyotypes, allozymes, venom protein sequences, and DNA sequences. There is still some disagreement about the relationships between the major groups of elapids.

The number and content of elapid families and subfamilies have varied widely. Depending on perceived levels of differentiation, various authors have recognized either a single family, Elapidae, with two to six subfamilies or two families: the Elapidae, terrestrial elapids, and the Hydrophiidae, seasnakes. Evidence from studies of morphology and DNA sequences shows that seasnakes are most closely related to Australo-Papuan elapids and thus are part of elapid radiation.

An Egyptian banded cobra (*Naja naja annulifera*). (Photo by Animals Animals ©Ralph Reinhold. Reproduced by permission.)

The fully marine seasnakes evolved from terrestrial live-bearing Australian ancestors, and the partially marine sea kraits seem to be most closely related to terrestrial elapids in Asia and Melanesia. Most authorities recognize a single family, Elapidae, that has two subfamilies: the Elapinae, including coral snakes, cobras, mambas, and terrestrial kraits, and the Hydrophiinae, including all the Australo-Papuan elapids, sea kraits, and seasnakes.

Definite elapid snake fossils are rare but have been found in Miocene deposits in Europe, North America, Africa, and Australia. Because there are so few, these fossils have contributed little to the understanding of elapid evolution.

Physical characteristics

Elapids are generally slender, highly agile snakes with a colubrid-like head that is not very distinct from the neck and bears large, colubrid-like scales or scutes. Elapids lack the loreal scute that separates the nasal scute from the preorbital scutes (most nonvenomous colubrid snakes have this scute). Because the fangs are short, the mouth does not have to open wide when the snake strikes. The length of these snakes varies from 7 in (18 cm) (the rare Fijian, *Ogmodon vitianus*) to more than 200 in (5 m) (king cobra, *Ophiophagus hannah*). The body often has stripes that may be very colorful. Many cobras flatten when excited, and cobras are famous for the ability to spread their neck ribs to form a hood.

The coral snakes of the Americas can be unicolored (no bands), but most species are famous for having a bright series of alternating color bands. The snakes may be bicolored, tricolored, or even quadricolored. The bands serve as a warning to potential predators. Also famous is the diverse radiation of nonvenomous snake mimics of the coral snakes. Many species of nonvenomous snakes that live in the same regions as coral snakes have evolved coloration almost identical to that of coral snakes. It has been estimated that 18% of all snakes found in the Americas are coral snake mimics. There are twice the number of mimics as there are coral snake species.

Seasnakes have evolved many adaptations, from the partially marine existence of the sea kraits (*Laticauda*) to the fully marine existence of the seasnakes. The nostrils of all seasnakes have valves that form a tight seal around the mouth when the snake dives. Fully marine seasnakes move sinusoidally as do land snakes, but they propel themselves through the water with a paddle-shaped tail rather than by grabbing the substrate with wide belly scales as land snakes do. The belly scales of fully marine seasnakes are almost the same size as their other body scales.

Distribution

Elapids are found in the southern United States to Central and South America, Africa except for Madagascar, southern Asia, Australia, and the intervening Pacific and Indian Oceans. They are most diverse in equatorial regions. Although widely distributed, each of the major elapid groups tends to occupy a particular region. For example, the elapid fauna of the Americas includes only the diverse coral snake lineage, which has approximately 60 species. Several coral snake species exist in the United States from North Carolina to Florida and west to Arizona. Coral snake diversity increases greatly in Mexico and Central and South America. The cobra group occupies almost all of Africa, the Middle East, and all of southern Asia. Cobras reach to Java in the Indonesian archipelago. Mambas are found in southern and central Africa, and terrestrial kraits are found from India through Southeast Asia. The Australo-Papuan elapids are the most diverse in terms of species number. They are found throughout Australia and New Guinea. A few species of elapids are found on the Solomon Islands. The unusual *Ogmodon vitianus* is the sole species in Polynesia, where it lives deep underground on Fiji. Sea kraits are found in coastal areas of southern Asia through Southeast Asia, Melanesia, and Polynesia. Seasnakes are abundant on coral reefs in the warm waters around northern Australia, New Guinea, Indonesia, the Philippines, and all of Southeast Asia. A few species exist as far west as the Persian Gulf and as far east as French Polynesia. Only one species, the yellow-bellied seasnake (*Pelamis platurus*), extends beyond this region, and it is the only open-water or pelagic species. This snake is found in warm waters from the east coast of Africa to the west coast of North and Central America in the Pacific and Indian Oceans. It is almost certainly the most widely distributed snake species. No seasnakes are known to exist in the Atlantic Ocean.

Habitat

Elapid snakes have diverse habitats. Most are ground dwellers, found everywhere from rainforest to savanna to grassy plains to harsh desert. Some species have a preferred

Monocled cobras (*N. naja kaouthia*) hatching in Thailand. (Photo by Animals Animals ©Davud M. Dennis. Reproduced by prmission.)

habitat; others are generalists. Some elapids seek shelter under rocks or in rodent burrows; others burrow into loose soil. Most cobras are terrestrial, but some are mostly arboreal or aquatic. African mambas spend most of their time in trees, where they are exceptionally graceful and fast. Almost all of the fully marine seasnakes and the partially marine sea kraits inhabit coral reefs, where they forage for prey, mostly fish and eels. Sea kraits come onto beaches and the surrounding rocks when they need to rest or lay their eggs.

Behavior

Many elapid snakes are active at dusk and at night. Others are active daytime foragers. Because elapids, like all snakes, are ectoderms and therefore must thermoregulate, the time at which they are active depends on the temperature. In cooler regions, such as southern Africa and southern Australia, elapid activity follows the seasons. Peak activity occurs during the warmer months, and hibernation during the colder months, although many species emerge on sunny winter days to bask. During the heat of summer, diurnal snakes are most active in the morning, late afternoon, and early evening, when it is cooler. In the spring and autumn, these species are active throughout the day because they do not become overheated.

Because snakes can be difficult to find, surprisingly little research has been conducted on the behavior of elapid snakes and snakes in general. The introduction of radio transmitters small enough to be surgically implanted into snakes has allowed researchers to follow snakes and document their daily activity through the seasons. The findings have shown that many elapids once thought sedentary are actually highly mobile, such as Australian death adders (genus *Acanthophis*) and the Australian broad-headed snake (*Hoplocephalus bungaroides*).

A cobra emerging from a woven basket and "dancing" to a snake charmer's flute is a familiar image. Egyptian, Asian, and Indian cobras are used for these demonstrations. Contrary to popular belief, the snakes are not being charmed or hypnotized. The snake is collected and placed in a woven basket,

where it is secure. The charmer may reach into the basket and grab the snake at mid body but is careful to keep the snake off balance. When the charmer lifts the lid of the basket, the snake rises in a vertical defensive posture with hood spread. Because he knows cobras strike from a vertical posture downward, the charmer stays out of reach and sways from side to side as he plays. Snakes lack external ears and pick up only low-frequency airborne sounds, therefore the music has no influence on the cobra. The charmer's flute is only a prop; the cobra follows the charmer's movements. Some charmers use snakes immobilized by cooling, and some use unaltered cobras. There is evidence, however, that some charmers provoke cobras to strike a stick or a piece of rough cloth, which is forcefully pulled from the snake's mouth, taking the fangs with it.

Feeding ecology and diet

Elapids are diverse in both diet and method of obtaining food. These snakes use envenomation rather than constriction to subdue prey. The chief prey are small vertebrates (rats, mice, birds, snakes, lizards, frogs, and fishes) and sometimes eggs. Some snakes specialize. The southern African Rinkhal's cobra (*Hemachatus haemachatus*) has a special fondness for toads.

In Australia, only death adders (genus *Acanthophis*), brown snakes (genus *Pseudonaja*), black snakes (genus *Pseudechis*), and taipans (genus *Oxyuranus*) eat small mammals as a large part of the diet, but they also eat other prey. Many of Australia's diverse terrestrial elapid fauna specialize on small reptiles, mostly scincid lizards, which the snakes find by searching under cover or by active foraging. Other elapids specialize on frogs, which they find at water's edge or under cover.

Both the partially marine sea kraits (*Laticauda*) and the diverse fully marine seasnakes obtain all their food from the aquatic habitat. Sea kraits specialize on eels they find among the reefs. Seasnakes have diverse diets. Most eat relatively sedentary fish that are easy to catch, but they tend to specialize on one or a few fish shapes, ranging from short gobies to long eels to squid. Three species of seasnake eat only the egg masses of fishes.

A spotted harlequin snake (*Homoroselaps lacteus*), a very primitive elapid, eating its main diet, black threadsnakes (*Leptotyphlops nigricans*). (Photo by Bill Branch. Reproduced by permission.)

A sea krait (*Laticauda colubrina*). (Photo by A. Power. Bruce Coleman, Inc. Reproduced by permission.)

King cobras eat other snakes, including venomous species. Australian bandy-bandy snakes (genus *Vermicella*) eat nothing but blindsnakes. Many coral snakes specialize on other snakes. Some species of Australian sand-swimming snakes of the genus *Simoselaps* eat nothing but the eggs of other reptiles. They ingest the small eggs whole and then, it is thought, regurgitate the empty shells.

Most elapids are active foragers. The Australian death adder (genus *Acanthophis*), however, stays in position and undulates the tip of its tail (which in contrast to the rest of the tail is yellowish white, resembling a larval insect) to lure prey. Australian whipsnakes (genus *Demansia*) have large eyes and are very active and visual daytime hunters. African mambas (genus *Dendroaspis*) also have large eyes to help them locate small mammals.

Reproductive biology

Elapids tend to reproduce once a year in spring, often after bouts of male combat over females. All coral snakes, mambas, terrestrial kraits, sea kraits, almost all cobras, and approximately half of the Australo-Papuan elapids are egg layers. Most snakes lay eggs, but viviparity (live-bearing) has evolved multiple times independently. Live-bearing is more common in species that live in cool climates because it is thought that mothers are able to control the developmental temperature of their offspring by behavioral thermoregulation. This ability is an important advantage in a short summer. The only cobra to evolve live-bearing is the southern African Rinkhal's cobra, which is reported to have litters of as many as 60 offspring. In Australia there is a diverse radiation of live-bearing elapids. Approximately half of the 20 genera and more than 90 species in Australia are live-bearers.

There is dispute about how many times live-bearing has evolved in the Australian elapid radiation. It is known that live-bearing has evolved at least twice independently, once in the main live-bearing radiation and once in the red-bellied black snake (*Pseudechis porphyriacus*). The other members of this genus are egg layers. The fully marine seasnakes also are live-bearers.

Most elapids do not take care of their eggs or young. In egg-laying species, females find suitable spots to lay eggs—under a rock, in or under a log, or in a crevice—and vacate the site. The eggs incubate for approximately three months, and the young hatch and are immediately on their own. In live-bearing species, the mother goes through a three-month pregnancy and gives birth in a secluded spot. Like the hatchlings, the liveborn young are immediately on their own. An exception is king cobras, which form a pair bond and build a nest from leaves and soil. King cobra pairs protect their nests and their eggs and can be very aggressive during breeding season.

Conservation status

Nine species are listed on the IUCN Red List. Seven of these are categorized as Vulnerable: *Austrelaps labialis*, *Denisonia maculata*, *Echiopsis atriceps*, *E. curta*, *Furina dunmalli*, *Hoplocephalus bungaroides*, and *Ogmodon vitianus*. Two species are categorized as Lower Risk/Near Threatened: *Elapognathus minor* and *Simoselaps calonotus*.

Conservation of snakes is relatively rare in most parts of the world, partly because little is known about most species. The best-studied elapid snake is the Australian broad-headed snake. This snake is distributed only in the sandstone country that surrounds greater Sydney. It is now rare and considered Vulnerable. Over the course of more than 10 years, researchers from the University of Sydney have documented the movement, behavior, and habitat preference of these snakes. It has long been known that broad-headed snakes over-winter under rocks on the edges of cliffs. The snakes, however, seemed to disappear in summer, so radio tracking was used to follow their

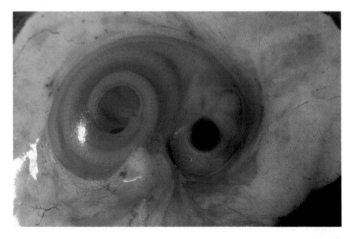

A night adder (*Causus rhombeatus*) embryo at 40-days-old in South Africa. (Photo by John Visser. Bruce Coleman, Inc. Reproduced by permission.)

venom gland

Spitting cobra mechanism. (Illustration by Dan Erickson)

movements. It was found that the snakes spend the summer far up in the forest canopy, where they hide in hollows, but that they use only large trees. The investigators also found that during the winter the snakes are very particular about the size of the rocks they use—too thin, and the snake becomes too hot; too thick, and the snake is not warm enough. Unfortunately for the snakes, the rocks are the same size that landscapers sell for gardens. Both large trees and appropriate-sized rocks must be preserved for the species to survive.

Much less is known about the conservation status of other elapid species. Hundreds of thousands of cobras are collected from the wild in Indonesia and other parts of Asia for the reptile skin trade. The cobra skins are turned into belts, wallets, and other pieces of apparel. There is little in-

formation about the effect of this practice on cobra populations. Similarly, degradation of the marine habitats of sea kraits and seasnakes is probably having an effect, but the effect has not been quantified. Loss of habitat is a primary concern for many elapid species because many of them are habitat specialists.

Significance to humans

Elapid snakes are one of the two major groups of venomous snakes. Many species are of special importance to humans because of the danger they represent. Many of the most venomous snakes are elapids. People are killed by elapid snakebites, but the danger of a snake has just as much to do with behavior as it does level of toxicity. For example, the

Australian inland taipan, or fierce snake (*Oxyuranus microlepidotus*), has the most potent venom. Few people have been killed by this snake, however, because it inhabits inhospitable areas where people tend not to live. The taipan also is very shy and always retreats if it can. Similarly, seasnakes are highly venomous, but most are not inclined to bite, so the incidence of snake bites from seasnakes is extremely low. In contrast, some species of Asian cobra are less toxic but are common in densely populated regions, so people tend to encounter them more than they do more venomous snakes. Thus the incidence of fatal snake bites can be high. Australia has the greatest diversity of elapid snakes in terms of species number. Death from elapid bites is rare in Australia because of access to antivenin and widespread knowledge of the Sutherland pressure-immobilization first-aid technique (wrapping of the bitten area and splinting of the affected extremity). In parts of Africa, India, and southern Asia, death from elapid bite is a major medical problem.

1. Red-bellied black snake (*Pseudechis porphyriacus*); 2. Half-girdled snake (*Simoselaps semifasciatus*); 3. Brown snake (*Pseudonaja textilis*); 4. Death adder (*Acanthophis antarcticus*); 5. Yellow-bellied seasnake (*Pelamis platurus*); 6. Olive seasnake (*Aipysurus laevis*); 7. Bandy-bandy snake (*Vermicella annulata*). (Illustration by Dan Erickson)

1. North American coral snake (*Micrurus fulvius*); 2. Sea krait (*Laticauda colubrina*); 3. Turtle-headed seasnake (*Emydocephalus annulatus*); 4. Forest cobra (*Naja melanoleuca*); 5. Black-necked spitting cobra (*Naja nigricollis*); 6. King cobra (*Ophiophagus hannah*); 7. Tiger snake (*Notechis scutatus*); 8. Taipan (*Oxyuranus scutellatus*); 9. Black mamba (*Dendroaspis polylepis*). (Illustration by Dan Erickson)

Species accounts

Black mamba
Dendroaspis polylepis

SUBFAMILY
Elapinae

TAXONOMY
Dendroaspis polylepis Günther, 1864, Zambezi River, Mozambique.

OTHER COMMON NAMES
French: Mamba noir; German: Schwarze Mamba.

PHYSICAL CHARACTERISTICS
The black mamba has a length of 78–118 in (2–3 m). It is a dark olive, gray, or gunmetal color, with a large head and eyes.

DISTRIBUTION
Central and southern Africa.

HABITAT
The black mamba's habitat is highly arboreal but, unlike the green mamba, the species is equally at home on the ground, where it can move very quickly. It prefers low-lying savanna.

BEHAVIOR
The black mamba is generally diurnal but crepuscular in some parts of its range. It often uses a semipermanent home base in holes or cracks in trees or termite mounds for many years. Although considered aggressive, black mambas usually flee if given the opportunity.

FEEDING ECOLOGY AND DIET
This elapid has a varied diet of small mammals but also feeds on birds and other snakes.

REPRODUCTIVE BIOLOGY
The female lays six to 17 eggs.

CONSERVATION STATUS
Not threatened.

SIGNIFICANCE TO HUMANS
The black mamba is highly venomous. As recently as the 1960s, bites from the black mamba were almost always fatal. Bites are still dangerous but are treatable if appropriate first aid is initiated and antivenin is administered quickly. ◆

North American coral snake
Micrurus fulvius

SUBFAMILY
Elapinae

TAXONOMY
Coluber fulvius Linnaeus, 1766, Carolina.

OTHER COMMON NAMES
English: Northern coral snake; Spanish: Serpiente-coralillo arlequin.

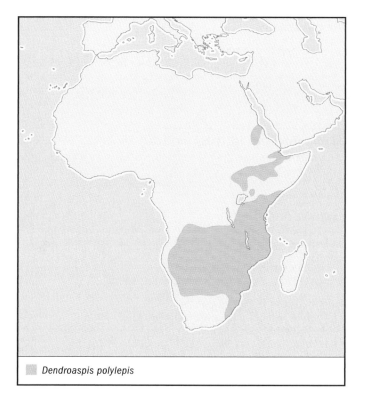

Dendroaspis polylepis

Micrurus fulvius

PHYSICAL CHARACTERISTICS
This slender snake reaches a length of 18–28 in (45–70 cm), but one specimen was recorded at 51 in (130 cm). It has thick red and black bands and thin yellow bands in an alternating pattern with yellow between black and red.

DISTRIBUTION
Eastern and southeastern United States from North Carolina to the southern tip of Florida, west to eastern and southern Texas and south to central Mexico.

HABITAT
The North American coral snake habitat is highly variable, ranging from forest to desert.

BEHAVIOR
Little is known.

FEEDING ECOLOGY AND DIET
This snake eats mainly small lizards, but it also consumes other snakes. It is an active forager that seems to be able to follow odor trails left by potential prey.

REPRODUCTIVE BIOLOGY
The female snake lays as many as 13 eggs but generally fewer than nine.

CONSERVATION STATUS
Not threatened.

SIGNIFICANCE TO HUMANS
This is a venomous species. Bites can be fatal, but fatal bites now are rare because of habitat encroachment and the introduction of antivenin. ◆

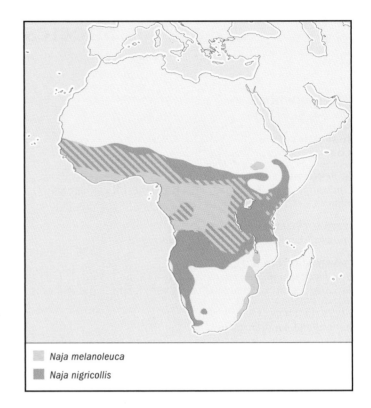

Naja melanoleuca
Naja nigricollis

FEEDING ECOLOGY AND DIET
This is an active forager that feeds on a wide variety of prey, including mammals, birds, reptiles, and amphibians.

REPRODUCTIVE BIOLOGY
The female lays 15–26 eggs.

CONSERVATION STATUS
Not threatened.

SIGNIFICANCE TO HUMANS
This species is venomous, but little is known about the effects of a bite because the snake is successful at avoiding humans. ◆

Forest cobra
Naja melanoleuca

SUBFAMILY
Elapinae

TAXONOMY
Naja melanoleuca Hallowell, 1857, Gaboon.

OTHER COMMON NAMES
None known.

PHYSICAL CHARACTERISTICS
The forest cobra has a length of 79–118 in (2–3 m). With its large, thick body, it is Africa's largest cobra. Its color is variable by region but usually is dark with crossbars or blotches.

DISTRIBUTION
Western and central Africa and eastern coastal parts of southern Africa.

HABITAT
This species generally inhabits forest and woodland, but it can also be found in open savanna and grassland in some parts of its range.

BEHAVIOR
The forest cobra is fast, very active, and an agile climber. It is active both day and night and can be aggressive if not able to retreat.

Black-necked spitting cobra
Naja nigricollis

SUBFAMILY
Elapinae

TAXONOMY
Naja nigricollis Reinhardt, 1843, Guinea. Genus probably includes several species.

OTHER COMMON NAMES
French: Cobra à cou noir; German: Speikobra.

PHYSICAL CHARACTERISTICS
This snake reaches a length of approximately 79 in (2 m). The species is highly variable in color, ranging from dull brown to contrasting black and white bands to jet black. Its fangs are like hypodermic needles in that each fang has an opening at the tip that points outward, thus allowing the cobra to "spit" venom a considerable distance.

DISTRIBUTION
Western, central, and southern Africa.

HABITAT
The black-necked spitting cobra generally prefers open savanna but can be found in all types of terrestrial habitat, including urban areas.

BEHAVIOR
This cobra is terrestrial but is a good climber. It is generally active at night but sometimes during the day as well.

FEEDING ECOLOGY AND DIET
This cobra has a varied diet, including reptiles and amphibians (even toads), birds, and eggs.

REPRODUCTIVE BIOLOGY
The female lays eight to 20 eggs.

CONSERVATION STATUS
Not threatened.

SIGNIFICANCE TO HUMANS
The snake squirts venom from its fangs, aiming for the eyes of the target. A large snake can spit a jet of venom up to 118 in (3 m). The venom causes extreme pain and can cause temporary blindness if treatment is not initiated immediately. The bite can be fatal. ◆

Ophiophagus hannah

King cobra
Ophiophagus hannah

SUBFAMILY
Elapinae

TAXONOMY
Hamadryas hannah Cantor, 1836, Calcutta.

OTHER COMMON NAMES
French: Cobra hannah; German: Königskobra.

PHYSICAL CHARACTERISTICS
The king cobra has an average length of approximately 118 in (3 m), but has been reported to reach a length of more than 197 in (5 m). It is the longest venomous snake. Its color varies greatly in combinations of brown, yellow, olive green, and black. Some snakes are banded with a yellow venter.

DISTRIBUTION
India to southeastern China; Philippines and Indonesian archipelago.

HABITAT
The king cobra is found in dense, high jungle, often near water.

BEHAVIOR
This is a shy species that avoids humans but can become aggressive when cornered. It attacks to guard its nest.

FEEDING ECOLOGY AND DIET
This species preys mainly on other snakes, which it finds through active foraging.

REPRODUCTIVE BIOLOGY
The king cobra is one of the only snake species to construct a nest, which is made of dead vegetation and soil. Bamboo thickets are the preferred nesting site. Both the male and female remain with the eggs and protect them until hatching.

CONSERVATION STATUS
Not threatened.

SIGNIFICANCE TO HUMANS
This species is considered highly dangerous because of its large size and potent venom. It is revered in Indian and Southeast Asian societies because of these traits. ◆

Death adder
Acanthophis antarcticus

SUBFAMILY
Hydrophiinae

TAXONOMY
Boa antarctica Shaw, 1794, Australia.

OTHER COMMON NAMES
German: Todesotter.

PHYSICAL CHARACTERISTICS
This species has an average length of 20–39 in (0.5–1 m). Unlike other elapids, it is thick bodied and thus resembles vipers. It is variable in color, ranging from light brown to nearly black, usually with a banding pattern.

DISTRIBUTION
Eastern and southern Australia.

HABITAT
The death adder inhabits grasslands to desert.

BEHAVIOR
Unlike most elapids, death adders wait for prey. They bury themselves in substrate and attract prey with a worm-like tail. The species is nocturnal and secretive.

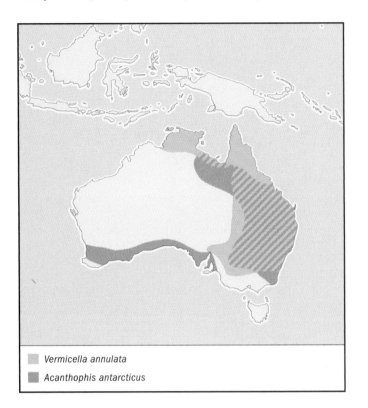

Vermicella annulata
Acanthophis antarcticus

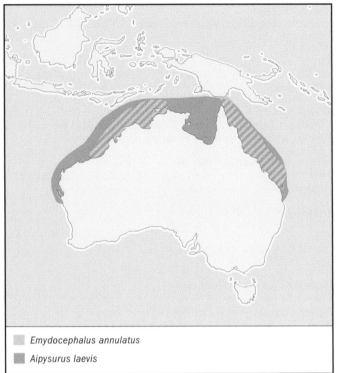

Emydocephalus annulatus
Aipysurus laevis

FEEDING ECOLOGY AND DIET
The death adder feeds mostly on small reptiles, but it also preys on frogs and small mammals.

REPRODUCTIVE BIOLOGY
This species is live-bearing, with females giving birth to up to 20 young.

CONSERVATION STATUS
Not threatened.

SIGNIFICANCE TO HUMANS
The death adder is considered one of the most dangerous snakes because it is so highly venomous. The death rate due to bites has declined because of the availability of antivenin and widespread knowledge of the Sutherland pressure-immobilization first-aid technique. This snake is sometimes found in urban areas. ◆

Olive seasnake
Aipysurus laevis

SUBFAMILY
Hydrophiinae

TAXONOMY
Aipysurus laevis Lacepède, 1804, Arafura Sea.

OTHER COMMON NAMES
None known.

PHYSICAL CHARACTERISTICS
This species has an average length of 47–79 in (1.2–2 m). One of the largest of the fully marine seasnakes, it is thick bodied and highly variable in color and pattern, ranging from light brown to dark with speckles.

DISTRIBUTION
Warm tropical waters of northern Australia and New Guinea.

HABITAT
This species inhabits coral reefs.

BEHAVIOR
This seasnake is unusually curious. Many scuba divers and snorkelers have experienced close encounters with olive seasnakes.

FEEDING ECOLOGY AND DIET
Olive seasnakes eat a wide variety of fish.

REPRODUCTIVE BIOLOGY
This live-bearing snake gives birth to two to six young.

CONSERVATION STATUS
Not threatened.

SIGNIFICANCE TO HUMANS
Because of curiosity, abundance, and large size, the olive seasnake is considered dangerous to humans. ◆

Turtle-headed seasnake
Emydocephalus annulatus

SUBFAMILY
Hydrophiinae

TAXONOMY
Emydocephalus annulatus Krefft, 1869, Loyalty Islands.

OTHER COMMON NAMES
None known.

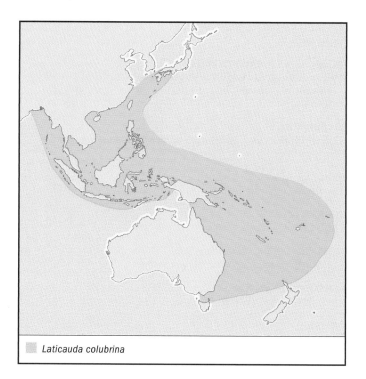

Laticauda colubrina

PHYSICAL CHARACTERISTICS
This species has an average length of approximately 30 in (75 cm). It is highly variable in color and pattern, ranging from a strong pattern with light and dark bands to almost uniform black or gray.

DISTRIBUTION
Warm tropical waters of Australia.

HABITAT
This species is found in shallow coral reefs.

BEHAVIOR
The turtle-headed seasnake lives in very high densities on some reefs.

FEEDING ECOLOGY AND DIET
This seasnake specializes on egg masses of fishes, particularly the eggs of small gobies and blennies.

REPRODUCTIVE BIOLOGY
This species is live-bearing, but little else is known about its reproduction.

CONSERVATION STATUS
Not threatened.

SIGNIFICANCE TO HUMANS
This species is venomous but of little threat to humans. ◆

Sea krait
Laticauda colubrina

SUBFAMILY
Hydrophiinae

TAXONOMY
Hydrus colubrinus Schneider, 1799, type locality unknown.

OTHER COMMON NAMES
None known.

PHYSICAL CHARACTERISTICS
The sea krait has an average length of approximately 39 in (1 m) but can reach 55 in (1.4 m). It has a striking banded pattern with alternating blue or bluish gray and black bands. Adaptations for marine life include valved nostrils and a paddle-shaped tail. Well-developed ventrals facilitate terrestrial locomotion after emergence from water.

DISTRIBUTION
Coastal regions from India through Southeast Asia to New Guinea and many Pacific islands.

HABITAT
This snake forages on coral reefs and comes onto beaches and nearby rocky areas to rest and lay eggs. It occasionally enters mangrove areas.

BEHAVIOR
The sea krait is mainly nocturnal but sometimes forages during the day.

FEEDING ECOLOGY AND DIET
This snake feeds almost exclusively on eels in coral reefs.

REPRODUCTIVE BIOLOGY
The female lays up to 18 eggs on land.

CONSERVATION STATUS
Not threatened

SIGNIFICANCE TO HUMANS
The sea krait is highly venomous but has a gentle nature and rarely bites humans. ◆

Tiger snake
Notechis scutatus

SUBFAMILY
Hydrophiinae

TAXONOMY
Naja scutata Peters, 1861, Java (in error). The taxonomy is confused and in dispute. Two species are recognized, but considerable DNA evidence suggests that tiger snakes are a single highly variable species.

OTHER COMMON NAMES
English: Tropical rat snake; French: Serpents tigrés; German: Tigerottern.

PHYSICAL CHARACTERISTICS
Average length is approximately 47 in (1.2 m) for most mainland individuals. Some island populations have giants that can reach nearly 79 in (2 m) and dwarfs that are shorter than 28 in (70 cm). The tiger snake is highly variable in color and pattern, ranging from light gray to brown to black with or without a banded pattern.

DISTRIBUTION
Southern and southeastern Australia.

HABITAT
This snake's habitat is highly variable, but it is often found around moist areas near creeks and other bodies of water.

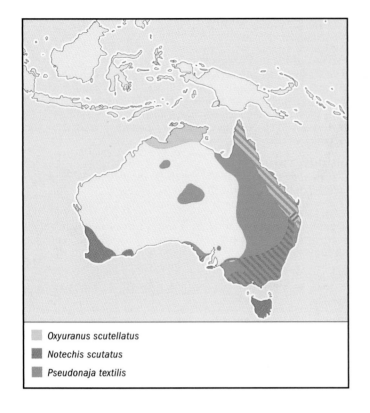

Oxyuranus scutellatus

Notechis scutatus

Pseudonaja textilis

BEHAVIOR
The tiger snake is generally active during the day but becomes nocturnal on warm nights. Males sometimes engage in combat over females.

FEEDING ECOLOGY AND DIET
This snake has a varied diet. It preys on both reptiles and mammals, but frogs make up a large part of the diet of many populations. The island giants, such as the Chappell Island tiger snake, eat large prey, including mutton bird chicks and stick-nest rats.

REPRODUCTIVE BIOLOGY
This species is live-bearing. Females give birth to up to 30 young.

CONSERVATION STATUS
Not threatened.

SIGNIFICANCE TO HUMANS
The tiger snake is highly venomous. It is dangerous because it is often found in or near urban areas. The bite can be fatal, but the death rate has decreased owing to the availability of an-tivenin and widespread knowledge of the Sutherland pressure-immobilization technique. ◆

Taipan
Oxyuranus scutellatus

SUBFAMILY
Hydrophiinae

TAXONOMY
Oxyuranus scutellatus Peters, 1867, Rockhampton.

OTHER COMMON NAMES
None known.

PHYSICAL CHARACTERISTICS
This species often reaches a length more than 79 in (2 m). Its coloration is generally light to dark brown with no obvious pattern.

DISTRIBUTION
Eastern Queensland, northeastern Western Australia, and Northern Territory.

HABITAT
The taipan inhabits forest to open savanna.

BEHAVIOR
This snake is mainly active during the day.

FEEDING ECOLOGY AND DIET
The taipan specializes on small mammals.

REPRODUCTIVE BIOLOGY
The female lays up to 20 eggs.

CONSERVATION STATUS
Not threatened.

SIGNIFICANCE TO HUMANS
The taipan is considered the second most venomous snake. Only its close relative, the inland taipan, or fierce snake, is more venomous. Few fatal bites occur, however, because taipans inhabit inhospitable areas where people tend not to live. The taipan also is very shy and always retreats if it can. ◆

Yellow-bellied seasnake
Pelamis platurus

SUBFAMILY
Hydrophiinae

TAXONOMY
Anguis platurus Linnaeus, 1766, no type locality.

OTHER COMMON NAMES
German: Plättchen-Seeschlange; Spanish: Serpiente-marina pelágica.

PHYSICAL CHARACTERISTICS
This species has an average length of approximately 28 in (70 cm). Yellow and black stripes run the length of the body.

DISTRIBUTION
Equatorial waters of the Pacific and Indian Oceans from the east coast of Africa to the west coast of the Americas.

HABITAT
This species is fully pelagic, inhabiting open waters.

BEHAVIOR
This seasnake may use migratory routes. It reaches high density in warm tropical waters. To molt, the snake coils itself into a ball, rubbing the skin of one area of its body against the skin of another area.

FEEDING ECOLOGY AND DIET
This species preys on surface-active fish in slicks, areas of calm water where two ocean currents meet.

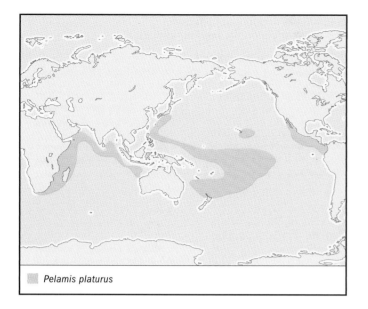

Pelamis platurus

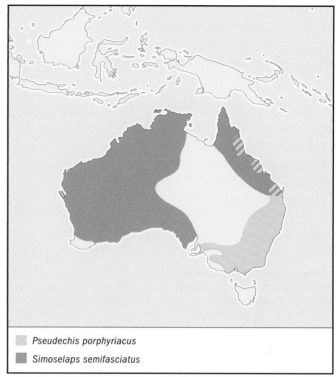

Pseudechis porphyriacus

Simoselaps semifasciatus

REPRODUCTIVE BIOLOGY
This snake is live-bearing, with females giving birth to two to six young.

CONSERVATION STATUS
Not threatened.

SIGNIFICANCE TO HUMANS
This species is venomous but of little threat to humans. ◆

Red-bellied black snake
Pseudechis porphyriacus

SUBFAMILY
Hydrophiinae

TAXONOMY
Coluber porphyriacus Shaw, 1794, Australia.

OTHER COMMON NAMES
None known.

PHYSICAL CHARACTERISTICS
This snake's average length is approximately 59 in (1.5 m). It is shiny black with a bright red venter.

DISTRIBUTION
Southeastern Australia; isolated populations in coastal Queensland.

HABITAT
This snake inhabits creeks, ponds, and swamps.

BEHAVIOR
This species is active during the day. It is gentle and calm and slow to react. Males engage in combat over females.

FEEDING ECOLOGY AND DIET
The red-bellied black snake's diet consists primarily of frogs, but it also preys on reptiles and small mammals.

REPRODUCTIVE BIOLOGY
The species is live-bearing; the female gives birth to eight to 40 young.

CONSERVATION STATUS
Not threatened.

SIGNIFICANCE TO HUMANS
This snake is highly venomous. Fatal bites have occurred, but this snake has a calm disposition and tries to avoid human contact. However, it is often found in urban areas. ◆

Brown snake
Pseudonaja textilis

SUBFAMILY
Hydrophiinae

TAXONOMY
Furina textilis Duméril Bibron, and Duméril, 1854, New South Wales.

OTHER COMMON NAMES
English: DeKay's brown snake; French: Couleuvre brune, serpents jaunes; German: Braunschlangen; Spanish: Culebra-parda de Kay.

PHYSICAL CHARACTERISTICS
This snake reaches an average length of approximately 59 in (1.5 m). It is variable in color and pattern. The young often are heavily banded in alternating black and brown, but the bands fade as the snakes mature. Adults usually are a uniform light brown color but can be almost black.

DISTRIBUTION
Much of the eastern half of Australia.

HABITAT
This species is found from desert to savanna to thick forest. It can be found in urban areas.

BEHAVIOR
The brown snake is very fast moving and is active during the day.

FEEDING ECOLOGY AND DIET
This species' diet is variable, but it feeds mostly on small mammals and reptiles. It is an active hunter.

REPRODUCTIVE BIOLOGY
The female lays 10–35 eggs.

CONSERVATION STATUS
Not threatened.

SIGNIFICANCE TO HUMANS
The brown snake is considered highly dangerous because of its abundance and highly toxic venom. Fatal bites used to be common, but the death rate has decreased because of the availability of antivenin and widespread knowledge of the Sutherland pressure-immobilization technique. ◆

Half-girdled snake
Simoselaps semifasciatus

SUBFAMILY
Hydrophiinae

TAXONOMY
Brachyurophis semifasciatus Günther, 1863, Western Australia.

OTHER COMMON NAMES
English: Sand swimmer.

PHYSICAL CHARACTERISTICS
This snake has an average length of 12 in (30 cm). It has an alternating pattern of varying light and dark blotches and bars but is highly variable in color and pattern. This snake is also known by the common name sand swimmer because of its use of an upturned snout for burrowing.

DISTRIBUTION
Western Australia and much of the Northern Territory, South Australia, and Queensland.

HABITAT
This snake is found in arid and semiarid habitats, including grasslands and coastal dunes.

BEHAVIOR
The half-girdled snake is highly secretive, and its behavior is not well known.

FEEDING ECOLOGY AND DIET
This snake preys on the eggs of other reptiles.

REPRODUCTIVE BIOLOGY
This species lays eggs, but little else is known.

CONSERVATION STATUS
Not threatened.

SIGNIFICANCE TO HUMANS
This snake is venomous but poses almost no threat to humans because of its small size and distribution in generally uninhabited areas. ◆

Bandy-bandy snake
Vermicella annulata

SUBFAMILY
Hydrophiinae

TAXONOMY
Calamaria annulata Gray, 1849, Australia.

OTHER COMMON NAMES
None known.

PHYSICAL CHARACTERISTICS
This snake has an average length of 24–39 in (0.6–1 m). It has an alternating pattern of black and white bands.

DISTRIBUTION
Queensland, New South Wales, and parts of Victoria and South Australia.

HABITAT
This snake can be found in almost any habitat.

BEHAVIOR
The bandy-bandy snake is a secretive and shy burrower. Usually seen only after heavy rains, it is well known for a defensive display in which it alternates between thrashing and contorting itself into one or more large, vertically oriented coils, apparently to frighten predators.

FEEDING ECOLOGY AND DIET
This snake feeds exclusively on blindsnakes nearly equal in size to itself.

REPRODUCTIVE BIOLOGY
The female lays two to 13 eggs.

CONSERVATION STATUS
Not threatened.

SIGNIFICANCE TO HUMANS
This species is venomous but poses almost no threat to humans because of its small size and distribution in generally uninhabited areas.

Resources

Books

Branch, B. *Field Guide to Snakes and Other Reptiles of Southern Africa.* Cape Town: Struik Publishers, 1998.

Broadley, D. G. *FitzSimons' Snakes of Southern Africa.* Johannesburg: Delta Books, 1983.

Campbell, J. A., and W. W. Lamar. *The Venomous Reptiles of Latin America.* Ithaca: Cornell University Press, 1989.

Cogger, H. G. *Reptiles and Amphibians of Australia.* 6th edition. Sydney: Reed New Holland, 2000.

Greene, Harry W. *Snakes: The Evolution of Mystery in Nature.* Berkley: University of California Press, 1997.

Greer, A. *The Biology and Evolution of Australian Snakes.* Chipping Norton, New South Wales: Surrey Beatty and Sons, 1997.

Heatwole, H. *Sea Snakes.* Sydney: University of New South Wales Press, 1999.

Roze, J. A. *Coral Snakes of the Americas: Biology, Identification, and Venoms.* Malabar, FL: Krieger Publishing, 1998.

Shine, R. *Australian Snakes: A Natural History.* Ithaca: Cornell University Press, 1991.

Spawls, S., and B. Branch. *The Dangerous Snakes of Africa.* Halfway House, South Africa: Southern Book Publishers, 1995.

Periodicals

Keogh, J. S. "Molecular Phylogeny of Elapid Snakes and a Consideration of Their Biogeographic History." *Biological Journal of the Linnean Society* 63 (1998): 177–203.

Shine, R. "Allometric Patterns in the Ecology of Australian Snakes." *Copeia* 1994 (1994): 851–867.

———. "Sexual Size Dimorphism in Snakes Revisited." *Copeia* 1994 (1994): 326–346.

Slowinski, J., and J. S. Keogh. "Phylogenetic Relationships of Elapid Snakes Based on Cytochrome b mtDNA Sequences." *Molecular Phylogenetics and Evolution* 15 (2000): 157–164.

J. Scott Keogh, PhD

For further reading

Ackerman, Lowell, ed. *The Biology, Husbandry, and Health Care of Reptiles.* 3 vol. Neptune City, NJ: T. F. H. Publications, Inc., 1997.

Adler, Kraig K. *A Brief History of Herpetology in North America before 1900.* Milwaukee, WI: Society for the Study of Amphibians and Reptiles, 1979.

———, ed. *Herpetology: Current Research on the Biology of Amphibians and Reptiles: Proceedings of the First World Congress of Herpetology.* [England]: Society for the Study of Amphibians and Reptiles, 1992.

———, ed. *Early Herpetological Studies and Surveys in the Eastern United States.* New York: Arno Press, 1979.

———, ed. *Herpetological Explorations of the Great American West.* New York: Arno Press, 1978.

Alderton, David. *Crocodiles & Alligators of the World.* New York: Facts on File, 1991.

Arnold, E. N., and J. A. Burton. *A Field Guide to the Reptiles and Amphibians of Britain and Europe.* London: Collins, 1978.

Avila-Pires, T. C. S. *Lizards of Brazilian Amazonia (Reptilia: Squamata).* Leiden, Germany: Zoologische Verhandelingen, 1995.

Aymar, Brandt., ed. *Treasury of Snake Lore.* New York: Greenberg Publishers, 1956.

Bambaradeniya, Channa N. B., and Vidhisha N. Samarasekara, eds. *An Overview of the Threatened Herpetofauna of South Asia.* Colombo, Sri Lanka: IUCN Sri Lanka and Asia Regional Biodiversity Programme, 2001.

Barker, David G., and Tracy M. Barker. *Pythons of the World.* Vol. 1, *Australia.* Lakeside, CA: Advanced Vivarium Systems, Inc., 1994.

Bartlett, Richard D. *In Search of Reptiles and Amphibians.* New York: E. J. Brill, 1988.

Bauer, Aaron M., and Ross A. Sadlier. *The Herpetofauna of New Caledonia.* Ithaca, NY: Society for the Study of Amphibians and Reptiles, in cooperation with the Institut de Recherche pur le Développment, 2000.

Behler, John L., and F. Wayne King. *The Audubon Society Field Guide to North American Reptiles and Amphibians.* New York: Knopf; Distributed by: Random House, 1979.

Bellairs, Angus d'A. *Reptiles.* London: Hutchinson, 1970.

Bennett, Daniel. *Monitor Lizards. Natural History, Biology and Husbandry.* Frankfurt: Edition Chimaira, 1998.

Benton, Michael J. *Vertebrate Paleontology.* 2nd edition. Oxford: Blackwell Science, 2000.

Benton, Michael J., and D. A. T. Harper. *Basic Palaeontology.* London: Addison Wesley Longman, 1997.

Bjorndal, Karen A., ed. *Biology and Conservation of Sea Turtles.* Washington, DC: Smithsonian Institution Press, 1995.

Boos, Hans E. A. *The Snakes of Trinidad and Tobago.* College Station, TX: Texas A&M University Press, 2001.

Boycott, R. C., and O. Bourquin. *The Southern African Tortoise Book: A Guide to Southern African Tortoises, Terrapins and Turtles.* Privately published, KwaZulu-Natal, South Africa, 2000.

Branch, William R. *Field Guide to Snakes and Other Reptiles of Southern Africa.* South Africa, 1998.

———, ed. *South African Red Data Book: Reptiles and Amphibians.* Report no. 151. Pretoria: South African National Scientific Programmes, 1988.

Brazaitis, Peter, and Myrna E. Watanabe. *Snakes of the World.* New York: Crescent Books, 1992.

Broadley, Donald G. *FitzSimons' Snakes of Southern Africa.* Johannesburg: Delta Books, 1983.

Brodmann, Peter. *Die Giftschlangen Europas: und die Gattung Vipera in Afrika und Asien.* Bern: Kümmery & Frey, 1987.

Brown, Philip R., and John W. Wright. *Herpetology of the North American Deserts: Proceedings of a Symposium.* Van Nuys, CA: Southwestern Herpetologists Society; Excelsior, MN: Trade distribution by Serpent's Tale Books, 1994.

Burghardt, Gordon M., and A. Stanley Rand, eds. *Iguanas of the World: Their Behavior, Ecology, and Conservation.* Park Ridge, NJ: Noyes Publications, 1982.

Burton, Maurice. *The World of Reptiles & Amphibians.* London: Orbis Publishing, 1973.

Campbell, Jonathan A. *Amphibians and Reptiles of Northern Guatemala, the Yucatan and Belize.* Animal Natural History Series, Vol. 4. Norman: University of Oklahoma Press, 1998.

Campbell, Jonathan A., and Edmund D. Brodie, Jr. *Biology of the Pitvipers*. Tyler, TX: Selva, 1992.

Campbell, Jonathan A., and William W. Lamar. *The Venomous Reptiles of Latin America*. Ithaca: Cornell University Press, 1989.

Cann, John. *Australian Freshwater Turtles*. Singapore: Beaumont Publishing, 1998.

Carr, Archie Fairly. *The Windward Road: Adventures of a Natualist on Remote Caribbean Shores*. Tallahassee, FL: University Presses of Florida, 1979.

———. *Handbook of Turtles: The Turtles of the United States, Canada, and Baja California*. Ithaca, NY: Comstock Publishing Associates, 1952.

Carroll, Robert L. *Patterns and Processes in Vertebrate Evolution*. Cambridge: Cambridge University Press, 1997.

Cloudsley-Thompson, John L. *The Diversity of Amphibians and Reptiles: An Introduction*. New York: Springer, 1999.

Coborn, John. *The Atlas of Snakes of the World*. Neptune City, NJ: T.H.F. Publications, 1991.

Cogger, Harold G. *Reptiles and Amphibians of Australia*. Ithaca, NY: Cornell University Press, 1992.

Cogger, Harold G., E. E. Cameron, R. A. Sadlier, and P. Eggler, eds. *The Action Plan for Australian Reptiles*. Endangered Species Unit, project number 124. Sydney: Australian Nature Conservation Agency, 1993.

Cogger, Harold, and Richard Zweifel, eds. *Encyclopedia of Reptiles and Amphibians*. San Diego: Academic Press, 1998.

Colbert, Edwin H. *The Age of Reptiles*. New York: Norton, 1965.

Colbert, Edwin H., Michael Morales, and Eli C. Minkoff. *Colbert's Evolution of the Vertebrates*. 5th edition. New York: John Wiley and Sons, Inc., 2001.

Committee on Sea Turtle Conservation. *Decline of the Sea Turtles: Causes and Prevention*. Washington, DC: National Academy Press, 1990.

Conant, Roger, Joseph T. Collins, Isabelle H. Conant, and Thomas R. Johnson. *A Field Guide to Reptiles & Amphibians of Eastern & Central North America*. Peterson Field Guide Series. Boston: Houghton Mifflin Co., 1998.

Corbett, Keith. *Conservation of European Reptiles and Amphibians*. London: Christopher Helm, 1989.

Cowan, Richard. *History of Life*. 3rd edition. Malden, Mass.: Blackwell Science, 2000.

Crother, Brian I., ed. *Caribbean Amphibians and Reptiles*. San Diego: Academic Press, 1999.

Curran, Charles Howard, and Carl Kauffeld. *Snakes and Their Ways*. New York: Harper & Brothers, 1937.

Daniel, J. C. *The Book of Indian Reptiles*. Bombay, India: Bombay Natural History Society, Oxford University Press, 1983.

Day, David. *The Doomsday Book of Animals*. London: London Editions Limited, 1981.

Deraniyagala, P. E. P. *A Colored Atlas of Some Vertebrates From Ceylon*. Colombo: Ceylon Government Press, 1953.

Diaz-Bolio, Jose. *The Geometry of the Maya and Their Rattlesnake Art*. Merida, Mexico: Area Maya-Mayan Area, 1987.

Ditmars, Raymond L. *The Book of Living Reptiles*. Philadelphia: J. B. Lippincott Company, 1936.

———. *Reptiles of the World: Tortoises and Turtles, Crocodilians, Lizards and Snakes of the Eastern and Western Hemispheres*. New York: The Macmillan Company, 1928.

Dixon, James R., and Pekka Soini. *The Reptiles of the Upper Amazon River Basin, Iquitos Region, Peru*, 2nd rev. ed. Milwaukee: Milwaukee Public Museum, 1986.

Duellman, William E. *The Biology of an Equatorial Herpetofauna in Amazonian Ecuador*. Lawrence, KS: University of Kansas, Museum of Natural History, 1978.

Engelmann, Wolf-Eberhard, and Fritz Jürgen Obst. *Snakes: Biology, Behavior and Relationship to Man*. New York: Exeter Books, 1982.

Ernst, Carl H., and Roger W. Barbour. *Turtles of the World*. Washington, DC: Smithsonian Institution Press, 1989.

Ernst, Carl H., Jeffrey E. Lovich, and Roger W. Barbour. *Turtles of the United States and Canada*. Washington, DC: Smithsonian Institution Press, 1994.

Ernst, C. H., George R. Zug. *Snakes in Question: The Smithsonian Answer Book*. Washington, DC: Smithsonian Institution Press, 1996.

Fitch, Henry S. *A Kansas Snake Community: Composition and Changes over 50 Years*. Malabar, FL: Krieger Publishing Co., 1999.

———. *Reproductive Cycles in Lizards and Snakes*. Lawrence, KS: University of Kansas Natural History Museum, 1970.

———. *Sexual Size Differences in Reptiles*. Lawrence, KS: University of Kansas, Museum of Natural History, 1981.

———. *Variation in Clutch and Litter Size in New World Reptiles*. Lawrence: University of Kansas, 1985.

Fizgerald, Sarah. *International Wildlife Trade: Whose Business Is It?* Washington, DC: World Wildlife Fund, 1989.

FitzSimons, F. W., assisted by V. F. M. FitzSimons. *Snakes and the Treatment of Snake Bite*. Cape Town, South Africa: Specialty Press of S. A., 1929.

FitzSimons, V. F. M. *Snakes of Southern Africa*. Cape Town and Johannesburg: Purnell and Sons, 1962.

Frank, Norman, and Erica Ramus. *A Complete Guide to Scientific and Common Names of Reptiles and Amphibians of the World*. Pottsville, PA.: NG Publishing Inc., 1996.

Franke, Joseph, and T. M. Telecky. *Reptiles as Pets: An Examination of the Trade in Live Reptiles in the United States*. Washington DC: Humane Society, 2001.

Frye, Fredric L. *Reptile Care: An Atlas of Diseases and Treatments.* Neptune City, NJ: T. F. H. Publications, Inc., 1991.

———. *Reptile Clinician's Handbook: A Compact Clinical and Surgical Reference.* Malabar, FL: Krieger Publishing, Co., 1995.

Frye, Fredric L., and David L. Williams. *Self-Assessment Color Review of Reptiles and Amphibians.* Ames, IA: Iowa State University Press, 1995.

Frye, Fredric L., and Wendy Townsend. *Iguanas: A Guide To Their Biology and Captive Care.* Malabar, FL: Krieger Publishing, Co., 1993.

Gadow, Hans. *Amphibia and Reptiles.* New York: Macmillan and Co., Ltd., 1901.

Gans, Carl, ed. *Biology of the Reptilia.* 19 vols. London, New York: Academic Press, 1998.

———. *Biomechanics: An Approach to Vertebrate Biology.* Philadelphia, PA: J. B. Lippincott Company, 1974.

Gibbons, Whit. *Their Blood Runs Cold: Adventures with Reptiles and Amphibians.* University, AL: University of Alabama Press, 1983.

Glaw, Frank. *A Fieldguide to the Amphibians and Reptiles of Madagascar.* Köln, Germany: M. Vences & F. Glaw Verlags, 1994.

Gloyd, Howard K., and Roger Conant. *Snakes of the Agkistrodon Complex: A Monographic Review.* St. Louis, MO: Society for the Study of Amphibians and Reptiles, 1990.

Goin, Coleman J., Olive B. Goin, George R. Zug. *Introducton to Herpetology.* San Francisco, CA: W. H. Freeman, 1978.

Graham, A. D. *Eyelids of Morning: The Mingled Destinies of Crocodiles and Men.* Greenwich, CT: New York Graphic Society, 1973.

Greene, Harry W. *Snakes: The Evolution of Mystery in Nature.* Berkeley: University of California Press, 1997.

Greer, Allen E. *The Biology and Evolution of Australian Lizards.* Chipping Norton, New South Wales, Australia: Surrey Beatty & Sons, 1989.

———. *The Biology and Evolution of Australian Snakes.* Chipping Norton, New South Wales, Australia: Surrey Beatty & Sons, 1997.

Grismer, L. Lee, and Harry W. Green. *Amphibians and Reptiles of Baja California.* Berkeley: University of California Press, 2002.

Guggisberg, C.A.W. *Crocodiles: Their Natural History, Folklore and Conservation.* Harrisburg, PA: Stackpole Books, 1972.

Günther, Albert C. L. G. *The Gigantic Land-Tortoises (Living and Extinct) in the Collection of the British Museum.* London: Printed by order of the Trustees, British Museum (Natural History), 1877.

Hallam, Arthur, and P. P. Wignall. *Mass Extinctions and Their Aftermath.* New York: Oxford University Press, 1997.

Halliday, Tim, and Kraig Adler. *The Encyclopedia of Reptiles and Amphibians.* New York: Facts on File, 1986.

Hambly, Wilfrid D. *Serpent Worship in Africa.* Chicago: Field Museum of Natural History, 1931.

Harding, Keith A., and Kenneth R. G. Welch. *Venomous Snakes of The World: A Checklist.* New York: Pergamon Press, 1980.

Harrison, Hal H. *The World of the Snake.* Philadelphia: Lippincott, 1971.

Heatwole, Harold. *Reptile Ecology.* St. Lucia, Australia: University of Queensland Press, 1976.

———. *Sea Snakes.* Sydney: University of New South Wales Press, 1999.

Holman, J. Alan. *Vertebrate Life of the Past.* Dubuque, Iowa: William C. Brown Publishers, 1994.

———. *Fossil Snakes of North America: Origin, Evolution, Distribution, Paleoecology.* Bloomington and Indianapolis: Indiana University Press, 2000.

Hopley, Catherine C. *Snakes: Curiosity and Wonders of Serpent Life.* London: Griffith & Farran; New York: E. P. Dutton & Co., 1882.

Hoser, Raymond. *Smuggled: The Underground Trade in Australia's Wildlife.* Sydney, Australia: Apollo Books, 1993.

Huey, Raymond B., Eric R. Pianka, and Thomas W. Schoener, eds. *Lizard Ecology: Studies of a Model Organism.* Cambridge, MA: Harvard University Press, 1983.

Iverson, John B. *A Checklist with Distribution Maps of the Turtles of the World.* Richmond, IN: Paust Printing, 1986.

Kauffeld, Carl. *Snakes: The Keeper and the Kept.* New York: Doubleday, 1969.

———. *Snakes and Snake Hunting.* Garden City, NY: Hanover House, 1957.

King, Dennis, and Brian Green. *Goanna: The Biology of Varanid Lizards.* Kensington, NSW, Australia: New South Wales University Press; Portland, OR: Available in North America through International Specialized Book Services, 1993.

King, Dennis, and Brian Green. *Monitors: The Biology of Varanid Lizards.* Malabar, FL: Krieger, 1999.

King, Wayne, and Russell Burke, eds. *Crocodilian, Tuatara, and Turtle Species of the World: A Taxonomic and Geographic Reference.* Washington, DC: Association of Systematics Collections, 1989.

Klauber, Laurence Monroe. *Rattlesnakes: Their Habits, Life Stories, and Influence on Mankind.* Berkeley, CA: Published for the Zoological Society of San Diego by the University of California Press, 1972.

Klemens, Michael W., ed. *Turtle Conservation.* Washington, DC: Smithsonian Institution Press, 2000.

Lamar, William W. *The World's Most Spectacular Reptiles & Amphibians.* Tampa, FL: World Publications, 1997.

FOR FURTHER READING

Langton, T., and J. A. Burton. *Amphibians and Reptiles: Conservation Management of Species and Habitats.* Strasbourg, France: Council of Europe Publishing, 1998.

Lee, Julian C. *The Amphibians and Reptiles of the Yucatán Peninsula.* Ithaca, NY: Comstock, 1996.

———. *A Field Guide to the Amphibians and Reptiles of the Maya World: The Lowlands of Mexico, Northern Guatemala, and Belize.* Ithaca, NY: Comstock Publishing Associates, 1995.

Levell, John. P. *A Field Guide to Reptiles and the Law.* 2nd revised edition. Lanesboro, MN: Serpent's Tale Books, 1997.

Liat, Lim Boo, and Indraneil Das. *Turtles of Borneo and Peninsular Malaysia.* Kota Kinabalu, Malaysia: Natural History Publications (Borneo), 1999.

Lowe, Charles H., Cecil R. Schwalbe, and Terry B. Johnson. *The Venomous Reptiles of Arizona.* Phoenix: Arizona Game and Fish Department, 1986.

Mader, Douglas R, ed. *Reptile Medicine and Surgery.* Philadelphia, PA: W. B. Saunders, 1996.

Maki, Moichirō. *Monograph of the Snakes in Japan.* Tokyo: Dai-ichi Shobō, [1931?].

Martin, James. *Masters of Disguise: A Natural History of Chameleons.* New York: Facts on File, 1992.

Mattison, C. *The Encyclopedia of Snakes.* New York: Facts on File, 1995.

———. *Lizards of the World.* New York: Facts on File, 1989.

———. *The Care of Reptiles and Amphibians in Captivity.* Poole, UK: Blandford Press; New York: Distributed in the U.S. by Sterling Pub. Co., 1987.

———. *The Encyclopedia of Snakes.* New York: Facts on File, 1995.

McDiarmid, Roy W., Jonathan A. Campbell, and T'Shakar A. Touré. *Snake Species of the World: A Taxonomic and Geographic Reference.* Volume 1. Washington, DC: Herpetologists' League, 1999.

Medem, Federico M. *Los Crocodylia de Sur América.* Bogotá, Colombia: Ministerio de Educación Nacional, Fondo Colombiano de Investigaciones Científicas y Proyectos Especiales "Francisco José de Caldas", 1981–1983.

Mellado, J., L. Gimenez, J. J. Gomez, et al. *El Camaleón en Andalucía: Distribución actual y amenazas para su supervivencia.* Rota: Fundacion Alcalde Zoilo Ruiz-Mateos, 2001.

Mehrtens, John M. *Living Snakes of the World in Color.* New York: Sterling Publishing Company, 1987.

Milstead, William, ed. *Lizard Ecology: A Symposium.* Columbia: University of Missouri Press, 1967.

Minton, Sherman A. *Venom Diseases.* Springfield, IL: Thomas, [1974].

Minton, Sherman A., and Madge Rutherford Minton. *Giant Reptiles.* New York: Charles Scribner's Sons, 1973.

Moriarty, John J., and Aaron M. Bauer. *State and Provincial Amphibian and Reptile Publications for the United States and Canada.* St. Louis, MO: Society for the Study of Amphibians and Reptiles, 2000.

Morris, Ramona, and Desmond Morris. *Men and Snakes.* New York: McGraw-Hill, 1965.

Müller-Schwarze, D., and R. M. Silverstein, eds. *Chemical Signals: Vertebrates and Aquatic Invertebrates.* New York: Plenum Press, 1980.

Murphy, James B., Kraig Adler, and Joseph T. Collins, eds. *Captive Management and Conservation of Amphibians and Reptiles.* Ithaca, NY: Society for the Study of Amphibians and Reptiles, 1994.

Murphy, James B., C. Ciofi, C. de la Panouse, and T. Walsh, eds. *Komodo Dragons: Biology and Conservation.* Washington, DC: Smithsonian Institution Press, 2002.

Murphy, John C. *Amphibians and Reptiles of Trinidad and Tobago.* Malabar, FL: Krieger, 1997.

Murphy, John C., and Robert W. Henderson. *Tales of Giant Snakes: A Historical Natural History of Anacondas and Pythons.* Malabar, FL: Krieger Publishing Company, 1997.

Nečas, P. *Chameleons: Nature's Hidden Jewels.* Malabar, FL: Krieger Publishing Co., 1999.

Nečas, P., David Modrý, Vít Savadil. *Czech Recent and Fossil Amphibians and Reptiles: An Atlas and Field Guide.* Frankfurt am Main, Germany: Edition Chimaira, 1997.

Neill, Wilfred T. *The Last of the Ruling Reptiles: Alligators, Crocodiles, and Their Kin.* New York: Columbia University Press, 1971.

———. *Reptiles and Amphibians in the Service of Man.* Indianapolis, IN: Pegasus, 1974.

Newman, Don. *Tuatara.* Endangered New Zealand Wildlife Series. Dunedin: John McIndoe, 1987.

Noble, G. K., and H. T. Bradley. *The Mating Behavior of Lizards: Its Bearing on the Theory of Sexual Selection. New York Academy of Sciences Annals* 35 (1933): 25–100.

Obst, Fritz Jürgen. *Turtles, Tortoises, and Terrapins.* New York: St. Martin's Press, 1988.

Oliver, James Arthur. *The Natural History of North American Amphibians and Reptiles.* Princeton, NJ: Van Nostrand, [1955].

———. *Snakes, in Fact and Fiction.* New York: Macmillian, 1958.

O'Shea, Mark. *A Guide to the Snakes of Papua New Guinea.* Port Moresby, Papua New Guinea: Independent Publishing, 1996.

Parker, H. W. *Snakes.* New York: Norton, 1963.

———. *Snakes: A Natural History.* London: British Museum (Natural History); Ithaca, NY: Cornell University Press, 1977.

Peters, James A. *Dictionary of Herpetology; A Brief and Meaningful Definition of Words and Terms Used in Herpetology.* New York: Hafner, 1964.

Peters, James A., and Roberto Donoso-Barros. *Catalogue of the Neotropical Squamata.* Part 1, *Snakes.* Washington, DC: Smithsonian Institution Press, 1970.

Peters, James A., and Braulio Orejas-Miranda. *Catalogue of the Neotropical Squamata.* Part 2, *Lizards and Amphisbaenians.* Washington, DC: Smithsonian Institution Press, 1970.

Pianka, Eric R. *Ecology and Natural History of Desert Lizards: Analyses of the Ecological Niche and Community Structure.* Princeton, NJ: Princeton University Press, 1986.

Pianka, Eric R. , and Laurie J. Vitt. *Lizards: Windows to the Evolution of Diversity.* Berkeley: University of California Press, 2003.

Pickwell, George V. *Amphibians and Reptiles of the Pacific States.* New York: Dover Publications, 1972.

Pope, Clifford Millhouse. *The Giant Snakes: The Natural History of the Boa Constrictor, the Anaconda, and the Largest Pythons, Including Comparative Facts About Other Snakes and Basic Information on Reptiles in General.* New York: Alfred A. Knopf, 1961.

Pough, F. Harvey, Robin M. Andrews, John E. Cadle, Martha L. Crump, Alan H. Savitzky, and Kentwood D. Wells. *Herpetology.* Upper Saddle River, NJ: Prentice Hall, 1998.

Powell, Robert, and Robert W. Henderson, eds. *Contributions to West Indian Herpetology: A Tribute to Albert Schwartz.* Ithaca, NY: Society for the Study of Amphibians and Reptiles, 1996.

Pritchard, Peter C. H. *Encyclopedia of Turtles.* Neptune, NJ: T. F. H. Publications, Inc., 1979.

————. *Living Turtles of the World.* Neptune, NJ: T. F. H. Publications, Inc., 1967.

Rhodin, Anders G. J., and Kenneth Miyata, eds. *Advances in Herpetology and Evolutionary Biology: Essays in Honor of Ernest E. Williams.* Cambridge, MA: Museum of Comparative Zoology, 1983.

Rivero, Juan A. *Los Anfibios y Reptiles de Puerto Rico.* San Juan, Puerto Rico: Universidad de Puerto Rico, Editorial Universitaria, 1998.

Romer, Alfred Sherwood. *Osteology of the Reptiles.* Chicago: University of Chicago Press, 1956.

Ross, James Perran, ed. *Crocodiles: Status Survey and Conservation Action Plan.* 2nd edition. Gland, Switzerland: IUCN/SSC Crocodile Specialist Group, 1998.

Rossman, Douglas A., Neil B. Ford, and Richard A. Seigel. *The Garter Snakes: Evolution and Ecology.* Norman, OK: University of Oklahoma Press, 1996.

Roze, Janis A. *Coral Snakes of the Americas: Biology, Identification, and Venoms.* Malabar, FL: Krieger Publishing, 1998.

————. *La Taxonomia y Zoogeografia de los Ofidios de Venezuela.* Caracas: Ediciones de la Biblioteca, Univ. Central de Venezuela, 1966.

Ruckdeschel, C., C. R. Shoop, G. R. Zug. *Sea Turtles of the Georgia Coast.* St. Mary's, GA: Cumberland Island Museum, 2000.

Russell, Findlay E. *Snake Venom Poisoning.* Great Neck, NY: Scholium International, 1983.

Russell, Findlay E., and Richard S. Scharffenberg. *Bibliography of Snake Venoms and Venomous Snakes.* West Covina, CA: 1964.

Savage, Jay M., Michael Fogden, and Patricia Fogden. *The Amphibians and Reptiles of Costa Rica: A Herpetofauna Between Two Continents, Between Two Seas.* Chicago: University of Chicago Press, 2002.

Saville-Kent, W. *The Naturalist in Australia.* London, 1897.

Schleich, Hans-Hermann, Werner Kästle, eds. *Contributions to the Herpetology of South-Asia (Nepal, India).* Wuppertal, Germany: Fuhlrott Museum, 1998.

Schleich, Hans-Hermann, Werner Kästle, and Klaus Kabisch. *Amphibians and Reptiles of North Africa: Biology, Systematics, Field Guide.* Koenigstein, Germany: Koeltz Books, 1996.

Schmidt, Karl P. *A Check List of North American Amphibians and Reptiles.* 6th edition. Chicago: University of Chicago Press, 1953.

Schmidt, Karl P., and Dwight Davis. *Field Book of Snakes of the United States and Canada.* New York: G. P. Putnam's Sons, 1941.

Schmidt, Karl P., and Robert F. Inger. *Living Reptiles of the World.* Garden City, NY: Hanover House, 1957.

Schuett, Gordon W., Mats Höggren, Michael E. Douglas, and Harry W. Greene, eds. *Biology of the Vipers.* Eagle Mountain, CO: Eagle Mountain Publishing, 2002.

Schwartz, Albert, and Robert W. Henderson. *Amphibians and Reptiles of the West Indies: Descriptions, Distributions, and Natural History.* Gainesville: University of Florida Press, 1991.

Seigel, Richard A., and Joseph T. Collins, eds. *Snakes: Ecology and Behavior.* New York: McGraw-Hill, 1993.

Seigel, Richard A., Joseph T. Collins, and Susan S. Novak, eds. *Snakes: Ecology and Evolutionary Biology,* New York: McGraw-Hill, 1987.

Sharell, Richard. *The Tuatara, Lizards and Frogs of New Zealand.* London: Collins, 1975.

Shine, Richard. *Australian Snakes: A Natural History.* Ithaca: Cornell University Press, 1991.

Smith, Hobart M. *Handbook of Lizards: Lizards of the United States and Canada.* Ithaca, NY: Comstock Publishing Co., 1946.

Smith, Hobart M., and Edmund D. Brodie, Jr. *Reptiles of North America: A Guide to Field Identification.* New York: Golden Press, 1982.

Smith, Malcolm A. *The British Amphibians & Reptiles*. London: Collins, 1951.

―――. *Monograph of the Sea-snakes (Hydrophiidae)*. London: Printed by Order of the Trustees of the British Museum, 1926.

―――. *Reptilia and Amphibia*. London: Taylor and Francis, 1931.

―――. *The Reptilia and Amphibia of the Malay Peninsula*. Singapore: Printed at the Govt. Print. Off., 1930.

Shaw, Charles E., and Sheldon Campbell. *Snakes of the American West*. New York: Alfred A. Knopf, 1974.

Spawls, Stephen, and Bill Branch. *The Dangerous Snakes of Africa: Natural History, Species Directory, Venoms, and Snakebite*. Sanibel Island, FL: Ralph Curtis Books, 1995.

Spawls, Stephen, Kim Howell, Robert Drewes, and James Ashe. *A Field Guide to the Reptiles of East Africa*. San Diego, CA: Academic Press, 2002.

Spellerberg, Ian F. *Biology of Reptiles: An Ecological Approach*. Glasgow: Blackie; New York: Distributed in the U.S. by Chapman and Hall, 1982.

Sprackland, Robert George. *Giant Lizards*. Neptune City, NJ: T. F. H. Publications, Inc., 1992.

Stafford, Peter J. *Snakes*. Washington, DC: Smithsonian Institution Press in association with the Natural History Museum, London, 2000.

Stebbins, Robert. *Amphibians and Reptiles of California*. Berkeley, CA: University of California Press, 1972.

―――. *A Field Guide to Western Reptiles and Amphibians*. Boston: Houghton Mifflin Co., 1985.

Stejneger, Leonhard. *Herpetology of Japan and Adjacent Territory*. Washington, DC: U.S. Government Printing Office, 1907.

―――. *Herpetology of Porto Rico*. Washington, DC: U.S. Government Printing Office, 1904.

―――. *The Poisonous Snakes of North America*. Seattle, WA: Shorey Book Store, 1971.

Steward, J. W. *The Snakes of Europe*. Rutherford, NJ: Fairleigh Dickinson University Press [1971].

Street, Donald. *The Reptiles of Northern and Central Europe*. London: B. T. Batsford, 1979.

Swingland, Ian R., and Michael W. Klemens, ed. *The Conservation Biology of Tortoises*. Gland, Switzerland: IUCN, 1989.

Thorpe, R. S., W. Wüster, and Anita Malhotra, eds. *Venomous Snakes: Ecology, Evolution, and Snakebite*. Oxford: Published for the Zoological Society of London by Clarendon Press; New York: Oxford University Press, 1997.

Tolson, Peter J., and Robert W. Henderson. *The Natural History of West Indian Boas*. Taunton, England: R & A Publishing Limited, 1993.

Trueb, Linda. *Catalogue of Publications in Herpetology*. Lawrence, KS: University of Kansas, Museum of Natural History, 1976.

Truntnau, Ludwig. *Nonvenomous Snakes*. Woodbury, NY: Barron's, 1986.

Tu, Anthony T. *Reptile Venoms and Toxins*. New York: M. Dekker, 1991.

Van Devender, Thomas R. *The Sonoran Desert Tortoise: Natural History, Biology, and Conservation*. Tucson: University of Arizona Press, 2002.

Van Dijk, Peter Paul, Bryan L. Stuart, and Anders G. J. Rhodin, eds. *Asian Turtle Trade: Proceedings of a Workshop on Conservation and Trade of Freshwater Turtles and Tortoises in Asia—Phnom Penh, Cambodia, 1–4 December, 1999*. Lunenburg, MA: Chelonian Research Foundation, 2000.

Visser, John. *Poisonous Snakes of Southern Africa, and the Treatment of Snakebite*. Cape Town: H. Timmins, 1966.

Webb, Grahame J. W., S. Charlie Manolis, and Peter J. Whitehead, eds. *Wildlife Management: Crocodiles and Alligators*. Chipping Norton, Australia: Surrey Beatty & Sons, 1987.

Worrell, Eric. *Dangerous Snakes of Australia and New Guinea*. Sydney: Angus and Robertson, 1961.

Wright, Albert Hansen, and Anna Allen Wright. *Handbook of Snakes of the United States and Canada*. Ithaca, NY: Comstock Pub. Associates, 1957.

Zappalorti, Robert T. *The Amateur Zoologist's Guide to Turtles and Crocodilians*. Harrisburg, PA: Stackpole Books, 1976.

Zhao, Ermi, and Kraig Adler. *Herpetology of China*. Oxford, OH: Society for the Study of Amphibians and Reptiles, in cooperation with Chinese Society for the Study of Amphibians and Reptiles, 1993.

Zug, George R. *Age Determination in Turtles*. Ithaca, NY: Society for the Study of Amphibians and Reptiles, 1991.

―――. *Herpetology: Introducing the Biology of Amphibians and Reptiles*. San Diego, CA: Academic Press, 1993.

―――. *The Lizards of Fiji: Natural History and Systematics*. Honolulu, HI: Bishop Museum Press, 1991.

Zug, George R., Laurie J. Vitt, and Janalee P. Caldwell. *Herpetology: An Introductory Biology of Amphibians and Reptiles*. 2nd edition. San Diego: Academic Press, 2001.

· · · · ·

Organizations

American Society of Ichthyologists and Herpetologists
donnelly@fiu.edu
Phone: (305) 919-5651
<http://199.245.200.110/>

American Zoo and Aquarium Association
8403 Colesville Road, Suite 710
Silver Spring, MD 20910
<http://www.aza.org>

Asociación Herpetológica Española
Apartado de Correos 191
28911 Leganés
Madrid
Spain
<http://elebo.fbiolo.uv.es/zoologia/AHE/>

Australian Society of Herpetologists
c/- CSIRO Wildlife and Ecology
PO Box 84
Lyneham, ACT2602
Australia
J.Wombey@dwe.csiro.au
<http://www.gu.edu.au/school/asc/ppages/academic/jmhero/
ash/frameintro.html>

The Center for North American Herpetology
<http://www.naherpetology.org/>

The Chameleon Information Network
13419 Appalachian Way,
San Diego,, California 92129
United States
Phone: 858-484-2669
Fax: 858- 484-4757
E-mail: chamnet1@aol.com
<http://www.animalarkshelter.org/cin>

Chicago Herpetological Society
2430 North Cannon Drive
Chicago, Illinois60614
Phone: 773-281-1800
<http://www.chicagoherp.org/>

Deutschen Gesellschaft für Herpetologie
Postfach 1421
D-53351
Rheinbach
Germany
Phone: 49-2225-703333
Fax: 49-2225-703338
gs@dght.de
<http://www.dght.de/>

Global Gecko Association
4920 Chester Street Spencer, Oklahoma
73084–2560
United States
<http://www.gekkota.com>

Herpetological Association of Africa
P.O. Box 20142
Durban North, 4016
South Africa
<http://www.wits.ac.za/haa/>

Iguana Specialist Group of the World Conservation Union
(IUCN)
<http://www.iucn-isg.org/index.php>

The International Iguana Society
<http://www.iguanasociety.com/>
133 Steele Rd.
West Hartford, CT 06119
United States

IUCN: The World Conservation Union
Rue Mauverney 28
1196
Gland
Switzerland
Phone: 41-22-999-0000
mail@hq.iucn.org
<http://www.iucn.org>

The League of Florida Herpetological Societies
<http://www.calusaherp.org/league.htm>
<http://www.flmnh.ufl.edu/natsci/herpetology/fhl.htm>

New York Herpetological Society
  P.O. Box 1245
  New York, NY10163-1245
  Phone: 212-740-3580
  <http://www.nyhs.org/>

Societas Europaea Herpetologica
  Natural Resources Institute, University of Greenwich
  Central Avenue, Chatham Maritime
  Kent
  ME4 4TB
  United Kingdom
  mike.lambert@nri.org
  <http://www.gli.cas.cz/SEH/>

Society for Research on Amphibians and Reptiles in New
Zealand (SRARNZ)
  SBS, Victoria University of Wellington
  PO Box 600
  Wellington
  New Zealand

Society for the Study of Amphibians and Reptiles
  gpisani@ku.edu
  <http://www.ukans.edu/ssar/>

Tucson Herpetological Society
  P.O. Box 709
  Tucson, Arizona85702-0709
  United States
  <http://tucsonherpsociety.org/>

The Tuatara Recovery Group, c/o Department of Conservation
  P.O. Box 10 420
  Wellington,
  New Zealand
  Phone: 64-4-471-0726
  <http://www.doc.govt.nz>

Contributors to the first edition

The following individuals contributed chapters to the original edition of Grzimek's Animal Life Encyclopedia, *which was edited by Dr. Bernhard Grzimek, Professor, Justus Liebig University of Giessen, Germany; Director, Frankfurt Zoological Garden, Germany; and Trustee, Tanzanian National Parks, Tanzania.*

Dr. Michael Abs
Curator, Ruhr University
Bochum, Germany

Dr. Salim Ali
Bombay Natural History Society
Bombay, India

Dr. Rudolph Altevogt
Professor, Zoological Institute,
University of Münster
Münster, Germany

Dr. Renate Angermann
Curator, Institute of Zoology,
Humboldt University
Berlin, Germany

Edward A. Armstrong
Cambridge University
Cambridge, England

Dr. Peter Ax
Professor, Second Zoological Institute
and Museum, University of Göttingen
Göttingen, Germany

Dr. Franz Bachmaier
Zoological Collection of the State of
Bavaria
Munich, Germany

Dr. Pedru Banarescu
Academy of the Roumanian Socialist
Republic, Trajan Savulescu Institute of
Biology
Bucharest, Romania

Dr. A. G. Bannikow
Professor, Institute of Veterinary
Medicine
Moscow, Russia

Dr. Hilde Baumgärtner
Zoological Collection of the State of
Bavaria
Munich, Germany

C. W. Benson
Department of Zoology, Cambridge
University
Cambridge, England

Dr. Andrew Berger
Chairman, Department of Zoology,
University of Hawaii
Honolulu, Hawaii, U.S.A.

Dr. J. Berlioz
National Museum of Natural History
Paris, France

Dr. Rudolf Berndt
Director, Institute for Population
Ecology, Hiligoland Ornithological
Station
Braunschweig, Germany

Dieter Blume
Instructor of Biology, Freiherr-vom-
Stein School
Gladenbach, Germany

Dr. Maximilian Boecker
Zoological Research Institute and A.
Koenig Museum
Bonn, Germany

Dr. Carl-Heinz Brandes
Curator and Director, The Aquarium,
Overseas Museum
Bremen, Germany

Dr. Donald G. Broadley
Curator, Umtali Museum
Mutare, Zimbabwe

Dr. Heinz Brüll
Director; Game, Forest, and Fields
Research Station
Hartenholm, Germany

Dr. Herbert Bruns
Director, Institute of Zoology and the
Protection of Life
Schlangenbad, Germany

Hans Bub
Heligoland Ornithological Station
Wilhelmshaven, Germany

A. H. Chrisholm
Sydney, Australia

Herbert Thomas Condon
Curator of Birds, South Australian
Museum
Adelaide, Australia

Dr. Eberhard Curio
Director, Laboratory of Ethology,
Ruhr University
Bochum, Germany

Dr. Serge Daan
Laboratory of Animal Physiology,
University of Amsterdam
Amsterdam, The Netherlands

Dr. Heinrich Dathe
Professor and Director, Animal Park
and Zoological Research Station,
German Academy of Sciences
Berlin, Germany

Dr. Wolfgang Dierl
Zoological Collection of the State of
Bavaria
Munich, Germany

Dr. Fritz Dieterlen
Zoological Research Institute, A.
Koenig Museum
Bonn, Germany

Dr. Rolf Dircksen
Professor, Pedagogical Institute
Bielefeld, Germany

Josef Donner
Instructor of Biology
Katzelsdorf, Austria

Dr. Jean Dorst
Professor, National Museum of
Natural History
Paris, France

Dr. Gerti DÜcker
Professor and Chief Curator,
Zoological Institute, University of
Münster
Münster, Germany

Dr. Michael Dzwillo
Zoological Institute and Museum,
University of Hamburg
Hamburg, Germany

Dr. Irenäus Eibl-Eibesfeldt
Professor and Director, Institute of
Human Ethology, Max Planck
Institute for Behavioral Physiology
Percha/Starnberg, Germany

Dr. Martin Eisentraut
Professor and Director, Zoological
Research Institute and A. Koenig
Museum
Bonn, Germany

Dr. Eberhard Ernst
Swiss Tropical Institute
Basel, Switzerland

R. D. Etchecopar
Director, National Museum of
Natural History
Paris, France

Dr. R. A. Falla
Director, Dominion Museum
Wellington, New Zealand

Dr. Hubert Fechter
Curator, Lower Animals, Zoological
Collection of the State of Bavaria
Munich, Germany

Dr. Walter Fiedler
Docent, University of Vienna, and
Director, Schönbrunn Zoo
Vienna, Austria

Wolfgang Fischer
Inspector of Animals, Animal Park
Berlin, Germany

Dr. C. A. Fleming
Geological Survey Department of
Scientific and Industrial Research
Lower Hutt, New Zealand

Dr. Hans Frädrich
Zoological Garden
Berlin, Germany

Dr. Hans-Albrecht Freye
Professor and Director, Biological
Institute of the Medical School
Halle a.d.S., Germany

Günther E. Freytag
Former Director, Reptile and
Amphibian Collection, Museum of
Cultural History in Magdeburg
Berlin, Germany

Dr. Herbert Friedmann
Director, Los Angeles County
Museum of Natural History
Los Angeles, California, U.S.A.

Dr. H. Friedrich
Professor, Overseas Museum
Bremen, Germany

Dr. Jan Frijlink
Zoological Laboratory, University of
Amsterdam
Amsterdam, The Netherlands

Dr. Dr. H.C. Karl Von Frisch
Professor Emeritus and former
Director, Zoological Institute,
University of Munich
Munich, Germany

Dr. H. J. Frith
C.S.I.R.O. Research Institute
Canberra, Australia

Dr. Ion E. Fuhn
Academy of the Roumanian Socialist
Republic, Trajan Savulescu Institute of
Biology
Bucharest, Romania

Dr. Carl Gans
Professor, Department of Biology,
State University of New York at
Buffalo
Buffalo, New York, U.S.A.

Dr. Rudolf Geigy
Professor and Director, Swiss Tropical
Institute
Basel, Switzerland

Dr. Jacques Gery
St. Genies, France

Dr. Wolfgang Gewalt
Director, Animal Park
Duisburg, Germany

Dr. Dr. H.C. Dr. H.C. Viktor
Goerttler
Professor Emeritus, University of Jena
Jena, Germany

Dr. Friedrich Goethe
Director, Institute of Ornithology,
Heligoland Ornithological Station
Wilhelmshaven, Germany

Dr. Ulrich F. Gruber
Herpetological Section, Zoological
Research Institute and A. Koenig
Museum
Bonn, Germany

Dr. H. R. Haefelfinger
Museum of Natural History
Basel, Switzerland

Dr. Theodor Haltenorth
Director, Mammalology, Zoological
Collection of the State of Bavaria
Munich, Germany

Barbara Harrisson
Sarawak Museum, Kuching, Borneo
Ithaca, New York, U.S.A.

Dr. Francois Haverschmidt
President, High Court (retired)
Paramaribo, Suriname

Dr. Heinz Heck
Director, Catskill Game Farm
Catskill, New York, U.S.A.

Dr. Lutz Heck
Professor (retired), and Director,
Zoological Garden, Berlin
Wiesbaden, Germany

Dr. Dr. H.C. Heini Hediger
Director, Zoological Garder
Zurich, Switzerland

Dr. Dietrich Heinemann
Director, Zoological Garden, Münster
Dörnigheim, Germany

Dr. Helmut Hemmer
Institute for Physiological Zoology,
University of Mainz
Mainz, Germany

Dr. W. G. Heptner
Professor, Zoological Museum,
University of Moscow
Moscow, Russia

Dr. Konrad Herter
Professor Emeritus and Director
(retired), Zoological Institute, Free
University of Berlin
Berlin, Germany

Dr. Hans Rudolf Heusser
Zoological Museum, University of
Zurich
Zurich, Switzerland

Dr. Emil Otto Höhn
Associate Professor of Physiology,
University of Alberta
Edmonton, Canada

Dr. W. Hohorst
Professor and Director, Parasitological
Institute, Farbwerke Hoechst A.G.
Frankfurt-Höchst, Germany

Dr. Folkhart Hückinghaus
Director, Senckenbergische Anatomy,
University of Frankfurt a.M.
Frankfurt a.M., Germany

Francois Hüe
National Museum of Natural History
Paris, France

Dr. K. Immelmann
Professor, Zoological Institute,
Technical University of Braunschweig
Braunschweig, Germany

Dr. Junichiro Itani
Kyoto University
Kyoto, Japan

Dr. Richard F. Johnston
Professor of Zoology, University of
Kansas
Lawrence, Kansas, U.S.A.

Otto Jost
Oberstudienrat, Freiherr-vom-Stein
Gymnasium
Fulda, Germany

Dr. Paul Kähsbauer
Curator, Fishes, Museum of Natural
History
Vienna, Austria

Dr. Ludwig Karbe
Zoological State Institute and
Museum
Hamburg, Germany

Dr. N. N. Kartaschew
Docent, Department of Biology,
Lomonossow State University
Moscow, Russia

Dr. Werner Kästle
Oberstudienrat, Gisela Gymnasium
Munich, Germany

Dr. Reinhard Kaufmann
Field Station of the Tropical Institute,
Justus Liebig University, Giessen,
Germany
Santa Marta, Colombia

Dr. Masao Kawai
Primate Research Institute, Kyoto
University
Kyoto, Japan

Dr. Ernst F. Kilian
Professor, Giessen University and
Catedratico Universidad Austral,
Valdivia-Chile
Giessen, Germany

Dr. Ragnar Kinzelbach
Institute for General Zoology,
University of Mainz
Mainz, Germany

Dr. Heinrich Kirchner
Landwirtschaftsrat (retired)
Bad Oldesloe, Germany

Dr. Rosl Kirchshofer
Zoological Garden, University of
Frankfort a.M.
Frankfurt a.M., Germany

Dr. Wolfgang Klausewitz
Curator, Senckenberg Nature
Museum and Research Institute
Frankfurt a.M., Germany

Dr. Konrad Klemmer
Curator, Senckenberg Nature
Museum and Research Institute
Frankfurt a.M., Germany

Dr. Erich Klinghammer
Laboratory of Ethology, Purdue
University
Lafayette, Indiana, U.S.A.

Dr. Heinz-Georg Klös
Professor and Director, Zoological
Garden
Berlin, Germany

Ursula Klös
Zoological Garden
Berlin, Germany

Dr. Otto Koehler
Professor Emeritus, Zoological
Institute, University of Freiburg
Freiburg i. BR., Germany

Dr. Kurt Kolar
Institute of Ethology, Austrian
Academy of Sciences
Vienna, Austria

Dr. Claus König
State Ornithological Station of Baden-
Württemberg
Ludwigsburg, Germany

Dr. Adriaan Kortlandt
Zoological Laboratory, University of
Amsterdam
Amsterdam, The Netherlands

Dr. Helmut Kraft
Professor and Scientific Councillor,
Medical Animal Clinic, University of
Munich
Munich, Germany

Dr. Helmut Kramer
Zoological Research Institute and A.
Koenig Museum
Bonn, Germany

Dr. Franz Krapp
Zoological Institute, University of
Freiburg
Freiburg, Switzerland

Dr. Otto Kraus
Professor, University of Hamburg,
and Director, Zoological Institute and
Museum
Hamburg, Germany

Dr. Dr. Hans Krieg
Professor and First Director (retired),
Scientific Collections of the State of
Bavaria
Munich, Germany

Dr. Heinrich Kühl
Federal Research Institute for
Fisheries, Cuxhaven Laboratory
Cuxhaven, Germany

Dr. Oskar Kuhn
Professor, formerly University
Halle/Saale
Munich, Germany

Dr. Hans Kumerloeve
First Director (retired), State
Scientific Museum, Vienna
Munich, Germany

Dr. Nagamichi Kuroda
Yamashina Ornithological Institute,
Shibuya-Ku
Tokyo, Japan

Dr. Fred Kurt
Zoological Museum of Zurich
University, Smithsonian Elephant
Survey
Colombo, Ceylon

Dr. Werner Ladiges
Professor and Chief Curator,
Zoological Institute and Museum,
University of Hamburg
Hamburg, Germany

Leslie Laidlaw
Department of Animal Sciences,
Purdue University
Lafayette, Indiana, U.S.A.

Dr. Ernst M. Lang
Director, Zoological Garden
Basel, Switzerland

Dr. Alfredo Langguth
Department of Zoology, Faculty of
Humanities and Sciences, University
of the Republic
Montevideo, Uruguay

Leo Lehtonen
Science Writer
Helsinki, Finland

Bernd Leisler
Second Zoological Institute,
University of Vienna
Vienna, Austria

Dr. Kurt Lillelund
Professor and Director, Institute for
Hydrobiology and Fishery Sciences,
University of Hamburg
Hamburg, Germany

R. Liversidge
Alexander MacGregor Memorial
Museum
Kimberley, South Africa

Dr. Dr. Konrad Lorenz
Professor and Director, Max Planck
Institute for Behavioral Physiology
Seewiesen/Obb., Germany

Dr. Dr. Martin Lühmann
Federal Research Institute for the
Breeding of Small Animals
Celle, Germany

Dr. Johannes Lüttschwager
Oberstudienrat (retired)
Heidelberg, Germany

Dr. Wolfgang Makatsch
Bautzen, Germany

Dr. Hubert Markl
Professor and Director, Zoological
Institute, Technical University of
Darmstadt
Darmstadt, Germany

Basil J. Marlow , B.SC. (Hons)
Curator, Australian Museum
Sydney, Australia

Dr. Theodor Mebs
Instructor of Biology
Weissenhaus/Ostsee, Germany

Dr. Gerlof Fokko Mees
Curator of Birds, Rijks Museum of
Natural History
Leiden, The Netherlands

Hermann Meinken
Director, Fish Identification Institute,
V.D.A.
Bremen, Germany

Dr. Wilhelm Meise
Chief Curator, Zoological Institute
and Museum, University of Hamburg
Hamburg, Germany

Dr. Joachim Messtorff
Field Station of the Federal Fisheries
Research Institute
Bremerhaven, Germany

Dr. Marian Mlynarski
Professor, Polish Academy of
Sciences, Institute for Systematic and
Experimental Zoology
Cracow, Poland

Dr. Walburga Moeller
Nature Museum
Hamburg, Germany

Dr. H.C. Erna Mohr
Curator (retired), Zoological State
Institute and Museum
Hamburg, Germany

Dr. Karl-Heinz Moll
Waren/Müritz, Germany

Dr. Detlev Müller-Using
Professor, Institute for Game
Management, University of Göttingen
Hannoversch-Münden, Germany

Werner Münster
Instructor of Biology
Ebersbach, Germany

Dr. Joachim Münzing
Altona Museum
Hamburg, Germany

Dr. Wilbert Neugebauer
Wilhelma Zoo
Stuttgart-Bad Cannstatt, Germany

Dr. Ian Newton
Senior Scientific Officer, The Nature
Conservancy
Edinburgh, Scotland

Dr. Jürgen Nicolai
Max Planck Institute for Behavioral
Physiology
Seewiesen/Obb., Germany

Dr. Günther Niethammer
Professor, Zoological Research
Institute and A. Koenig Museum
Bonn, Germany

Dr. Bernhard Nievergelt
Zoological Museum, University of
Zurich
Zurich, Switzerland

Dr. C. C. Olrog
Institut Miguel Lillo San Miguel de
Tucuman
Tucuman, Argentina

Alwin Pedersen
Mammal Research and aRctic
Explorer
Holte, Denmark

Dr. Dieter Stefan Peters
Nature Museum and Senckenberg
Research Institute
Frankfurt a.M., Germany

Dr. Nicolaus Peters
Scientific Councillor and Docent,
Institute of Hydrobiology and
Fisheries, University of Hamburg
Hamburg, Germany

Dr. Hans-Günter Petzold
Assistant Director, Zoological Garden
Berlin, Germany

Dr. Rudolf Piechocki
Docent, Zoological Institute,
University of Halle
Halle a.d.S., Germany

Dr. Ivo Poglayen-Neuwall
Director, Zoological Garden
Louisville, Kentucky, U.S.A.

Dr. Egon Popp
Zoological Collection of the State of
Bavaria
Munich, Germany

Dr. Dr. H.C. Adolf Portmann
Professor Emeritus, Zoological
Institute, University of Basel
Basel, Switzerland

Hans Psenner
Professor and Director, Alpine Zoo
Innsbruck, Austria

Dr. Heinz-Siburd Raethel
Oberveterinärrat
Berlin, Germany

Dr. Urs H. Rahm
Professor, Museum of Natural History
Basel, Switzerland

Dr. Werner Rathmayer
Biology Institute, University of
Konstanz
Konstanz, Germany

Walter Reinhard
Biologist
Baden-Baden, Germany

Dr. H. H. Reinsch
Federal Fisheries Research Institute
Bremerhaven, Germany

Dr. Bernhard Rensch
Professor Emeritus, Zoological
Institute, University of Münster
Münster, Germany

Dr. Vernon Reynolds
Docent, Department of Sociology,
University of Bristol
Bristol, England

Dr. Rupert Riedl
Professor, Department of Zoology,
University of North Carolina
Chapel Hill, North Carolina, U.S.A.

Dr. Peter Rietschel
Professor (retired), Zoological
Institute, University of Frankfurt a.M.
Frankfurt a.M., Germany

Dr. Siegfried Rietschel
Docent, University of Frankfurt;
Curator, Nature Museum and
Research Institute Senckenberg
Frankfurt a.M., Germany

Herbert Ringleben
Institute of Ornithology, Heligoland
Ornithological Station
Wilhelmshaven, Germany

Dr. K. Rohde
Institute for General Zoology, Ruhr
University
Bochum, Germany

Dr. Peter Röben
Academic Councillor, Zoological
Institute, Heidelberg University
Heidelberg, Germany

Dr. Anton E. M. De Roo
Royal Museum of Central Africa
Tervuren, South Africa

Dr. Hubert Saint Girons
Research Director, Center for
National Scientific Research
Brunoy (Essonne), France

Dr. Luitfried Von Salvini-Plawen
First Zoological Institute, University
of Vienna
Vienna, Austria

Dr. Kurt Sanft
Oberstudienrat, Diesterweg-
Gymnasium
Berlin, Germany

Dr. E. G. Franz Sauer
Professor, Zoological Research
Institute and A. Koenig Museum,
University of Bonn
Bonn, Germany

Dr. Eleonore M. Sauer
Zoological Research Institute and A.
Koenig Museum, University of Bonn
Bonn, Germany

Dr. Ernst Schäfer
Curator, State Museum of Lower
Saxony
Hannover, Germany

Dr. Friedrich Schaller
Professor and Chairman, First
Zoological Institute, University of
Vienna
Vienna, Austria

Dr. George B. Schaller
Serengeti Research Institute, Michael
Grzimek Laboratory
Seronera, Tanzania

Dr. Georg Scheer
Chief Curator and Director,
Zoological Institute, State Museum of
Hesse
Darmstadt, Germany

Dr. Christoph Scherpner
Zoological Garden
Frankfurt a.M., Germany

Dr. Herbert Schifter
Bird Collection, Museum of Natural
History
Vienna, Austria

Dr. Marco Schnitter
Zoological Museum, Zurich
University
Zurich, Switzerland

Dr. Kurt Schubert
Federal Fisheries Research Institute
Hamburg, Germany

Eugen Schuhmacher
Director, Animals Films, I.U.C.N.
Munich, Germany

Dr. Thomas Schultze-Westrum
Zoological Institute, University of
Munich
Munich, Germany

Dr. Ernst Schüt
Professor and Director (retired), State
Museum of Natural History
Stuttgart, Germany

Dr. Lester L. Short , Jr.
Associate Curator, American Museum
of Natural History
New York, New York, U.S.A.

Dr. Helmut Sick
National Museum
Rio de Janeiro, Brazil

Dr. Alexander F. Skutch
Professor of Ornithology, University
of Costa Rica
San Isidro del General, Costa Rica

Dr. Everhard J. Slijper
Professor, Zoological Laboratory,
University of Amsterdam
Amsterdam, The Netherlands

Bertram E. Smythies
Curator (retired), Division of Forestry
Management, Sarawak-Malaysia
Estepona, Spain

Dr. Kenneth E. Stager
Chief Curator, Los Angeles County
Museum of Natural History
Los Angeles, California, U.S.A.

Dr. H.C. Georg H.W. Stein
Professor, Curator of Mammals,
Institute of Zoology and Zoological
Museum, Humboldt University
Berlin, Germany

Dr. Joachim Steinbacher
Curator, Nature Museum and
Senckenberg Research Institute
Frankfurt a.M., Germany

Dr. Bernard Stonehouse
Canterbury University
Christchurch, New Zealand

Dr. Richard Zur Strassen
Curator, Nature Museum and
Senckenberg Research Institute
Frandfurt a.M., Germany

Dr. Adelheid Studer-Thiersch
Zoological Garden
Basel, Switzerland

Dr. Ernst Sutter
Museum of Natural History
Basel, Switzerland

Dr. Fritz Terofal
Director, Fish Collection, Zoological
Collection of the State of Bavaria
Munich, Germany

Dr. G. F. Van Tets
Wildlife Research
Canberra, Australia

Ellen Thaler-Kottek
Institute of Zoology, University of
Innsbruck
Innsbruck, Austria

Dr. Erich Thenius
Professor and Director, Institute of
Paleontolgy, University of Vienna
Vienna, Austria

Dr. Niko Tinbergen
Professor of Animal Behavior,
Department of Zoology, Oxford
University
Oxford, England

Alexander Tsurikov
Lecturer, University of Munich
Munich, Germany

Dr. Wolfgang Villwock
Zoological Institute and Museum,
University of Hamburg
Hamburg, Germany

Zdenek Vogel
Director, Suchdol Herpetological
Station
Prague, Czechoslovakia

Dieter Vogt
Schorndorf, Germany

Dr. Jiri Volf
Zoological Garden
Prague, Czechoslovakia

Otto Wadewitz
Leipzig, Germany

Dr. Helmut O. Wagner
Director (retired), Overseas Museum,
Bremen
Mexico City, Mexico

Dr. Fritz Walther
Professor, Texas A & M University
College Station, Texas, U.S.A.

John Warham
Zoology Department, Canterbury
University
Christchurch, New Zealand

Dr. Sherwood L. Washburn
University of California at Berkeley
Berkeley, California, U.S.A.

Eberhard Wawra
First Zoological Institute, University
of Vienna
Vienna, Austria

Dr. Ingrid Weigel
Zoological Collection of the State of
Bavaria
Munich, Germany

Dr. B. Weischer
Institute of Nematode Research,
Federal Biological Institute
Münster/Westfalen, Germany

Herbert Wendt
Author, Natural History
Baden-Baden, Germany

Dr. Heinz Wermuth
Chief Curator, State Nature Museum,
Stuttgart
Ludwigsburg, Germany

Dr. Wolfgang Von Westernhagen
Preetz/Holstein, Germany

Dr. Alexander Wetmore
United States National Museum,
Smithsonian Institution
Washington, D.C., U.S.A.

Dr. Dietrich E. Wilcke
Röttgen, Germany

Dr. Helmut Wilkens
Professor and Director, Institute of
Anatomy, School of Veterinary
Medicine
Hannover, Germany

Dr. Michael L. Wolfe
Utah, U.S.A.

Hans Edmund Wolters
Zoological Research Institute and A.
Koenig Museum
Bonn, Germany

Dr. Arnfrid Wünschmann
Research Associate, Zoological Garden
Berlin, Germany

Dr. Walter Wüst
Instructor, Wilhelms Gymnasium
Munich, Germany

Dr. Heinz Wundt
Zoological Collection of the State of
Bavaria
Munich, Germany

Dr. Claus-Dieter Zander
Zoological Institute and Museum,
University of Hamburg
Hamburg, Germany

Dr. Dr. Fritz Zumpt
Director, Entomology and
Parasitology, South African Institute
for Medical Research
Johannesburg, South Africa

Dr. Richard L. Zusi
Curator of Birds, United States
National Museum, Smithsonian
Institution
Washington, D.C., U.S.A.

CONTRIBUTORS TO THE FIRST EDITION

Glossary

Acrodont—Having teeth attached to the edge of the jawbone without sockets.

Anterior—The front or head of an animal.

Apical—Of, relating to, or situated at an apex.

Arribada—A massive, coordinated arrival of seaturtles, and some freshwater species, at a nesting beach.

Aspect ratios—Total length divided by body width

Autotomy—Self-amputation of a body part, typically a tail; used as a defense mechanism.

Axillary—Of, relating to, or located near the axilla; the cavity beneath the junction of a forelimb and the body.

Barbels—Fleshy, tubular extension of skin, usually on the head or neck.

Basal—Arising from the base of a stem; of, or relating to the foundation, base or essence; of or relating to, or being essential for maintaining the fundamental vital activities of an organism.

Carapace—The dorsal (upper) shell of a turtle.

Caudal luring—A specific movement of the tail which is meant to attract or lure prey; an adaption that evolved to help increase the animal's chance of getting food and therefore allowing it to survive.

Ceratobranchials—Principal paired derivatives from the 3rd through 6th visceral arches found in the hyoid apparatus (the lateral and posterior paired parts of the hyoid).

Chemoreception—The sensory reception of chemical stimuli; capacity to detect and differentiate certain chemicals in the surrounding environment.

Chromosome—Thread-like structure consisting mostly of genetic material (DNA) in the nucleus of cells.

Clade—An evolutionary lineage of organisms that includes the most recent common ancestor of all those organisms and all the descendants of that common ancestor.

Cladogram—Graphic, tree-like representation that shows the evolutionary relationships of organisms.

Concertina—A mode of locomotion in snakes, characterized by sequential extension and contraction of the body from one anchored and stationary site to the next as the animal moves with accordion-like appearance in one direction.

Congeneric—Of or belonging to the same genus.

Conspecific—Of or belonging to the same species.

Cruciform—Forming or arranged in a cross.

Crypsis—Reference to camouflage or matching between the color, pattern, or shape of an animal and a random sample of the background it is viewed against, as perceived by another animal.

Diapause—A period of physiologically enforced dormancy between periods of activity.

Dimorphism—The existence of two different forms (color, size, sex) of a species in the same population.

Distal—Toward the tip of a body extremity, such as a phalange.

Diurnal—Active by day.

Dorsal—Relating to or situated near or on the back of an animal.

Dorsolaterally—Relating to, or involving both the back and the sides.

Dorsoventrally—Relating to or involving, or extending along the axis joining the dorsal and ventral sides.

Ecdysis—The act of molting or shedding an outer cuticular or epidermal layer.

Ectoderm—The outer cellular membrane of a diploblastic animal (one whose embryo has two germ layers) or the outermost of the three primary germ layers of an embryo.

Ectopterygoid bones—Dermal bone in the palate of most reptiles.

Ectotherm—An animal whose body temperature is determined principally by the environment.

Emarginated—To deprive of a margin; having a margin notched.

Endoderm—The innermost of the germ layers of an embryo that is the source of the epithelium of the digestive tract and its derivatives.

Endotherm—An animal whose body temperature is determined by internal metabolic heat production.

Extant—The state of a taxonomic group being alive in the present; not extinct.

External fertilization—The joining of sperm and eggs outside of the female's body.

Extinct—The state of a taxonomic group being lost; no longer in existence.

Femoral—Of, or relating to the femur or thigh.

Fenestra—A small anatomical opening.

Fertilization—The penetration of an egg by sperm.

Fossorial—Adapted for or used in burrowing or digging.

Fusiform—Tapering towards each end.

Gravid—Female carrying young or eggs.

Gular pouch—A pouch of bare skin between the lower mandibles (jaws).

Hatchling—A young animal that has just emerged from an egg.

Heliothermic—Basks in the heat of the sun (or other overhead bright heat source). Commonly occurs in many temperate reptiles including squamates, crocodilians, and semiaquatic turtles. Snakes use a combination of heliothermic and thigmothermic strategies.

Hemipene—The bi-lobed male reproductive organs in most reptiles, kept inverted in the tail until needed.

Heterospecific—Members of a different species.

Hybrid—Individual resulting from mating of parents that belong to different species.

Hyoid bone—A bone or complex of bones situated at the base of the tongue and supporting the tongue and its muscles.

Inframarginal—Lamina (thin plate or scale) lying between the marginals of the carapace and the lateral margin of the normal plastral laminae.

Inguinal fat body—Body fat relating to, or situated on the groin or in either of the lowest lateral regions of the abdomen.

Internal fertilization—Penetration of eggs by sperm inside the female's body.

Intrascalar—Within or underneath the skin.

Intromission—The act of insertion.

Jacobson's organ—A convoluted blind sac opening into the roof of the mouth and lined with sensory structures that detect odorant particles brought into the mouth on the tongue. This organ is absent in adult crocodilians. Also known as vomeronasal organ.

Keeled—A scale with a raised ridge running down its midline.

Keratinized—Tissue that is hardened with keratin, such as human fingernails or toenails.

Lamella—A thin, flat scale, membrane, or layer.

Lateral—Related to the side of an animal.

Loreal scale—The scale between the preocular and nasal scales.

Lotic—Of, relating to, or living in actively moving water.

Marginal—A lamina in the outer series on the carapace of a turtle, visible both from above and below and characterized by a sharp angle marking the edge of the bony part of the carapace.

Maxillary—Pertaining to a maxilla, or jaw.

Monophyletic—Developed from a single common ancestral form or line.

Monophyly—A single common ancestral form or line.

Morph—A unique body form or coloration.

Morphology—The study of an animal's shape or form.

Necrosis—Localized death of living tissue.

Nocturnal—Active at night.

Ocelli—A minute simple eye or eyespot of an invertebrate; an eyelike colored spot.

Ontogeny—The development or course of development of an individual organism.

Oviparous—Producing eggs that develop and hatch outside the maternal body.

Ovoviviparous—Producing eggs that develop within the maternal body and hatch within or immediately after extrusion from the parent.

Paramo—High-altitude habitat that is cold, wet, and dominated by grasses and shrubs.

Parietal—Of or relating to the walls of a part or cavity; of, relating to, or forming the upper posterior wall of the head.

Phalange—One of the digits in the hand or foot.

Phylogenetic—Relating to the evolutionary history of an organism or group of organisms.

Pit organ—Specialized and highly sensitive infrared receptors that detect rapid changes in infrared radiation. These have evolved independently in boid and crotaline snakes and provide very precise discrimination of the direction and distance of an infrared source. They are located in upper and lower labial scales of boid snakes and occur as single structures between the eye and nostril on either side of the head of crotaline snakes.

Plastron—The ventral (underside) part of a turtle or tortoise shell.

Pleurodont—Having the teeth attached by their sides to the inner side of the jaw.

Posterior—Related to the rear or tail end of an animal.

Postocular—That part of the head behind the eye

Prefrontals—1) Either of a pair of large scales directly anterior to the frontal scales. 2) A dermal bone forming part of the roof of the orbit.

Prehensile—Adapted for seizing or grasping especially by wrapping around an item.

Preocular—The part of the head in front of or before the eye.

Pterygoid—Of, relating to, or lying in the region of the inferior part of the sphenoid bone of the vertebrate skull.

Rectilinear locomotion—A mode of locomotion used primarily by heavy-bodied snakes that move in a straight line. Alternating sections of the ventral skin are lifted clear of the substrate and are pulled forward by muscles originating on ribs and connecting to the ventral skin.

Relict—A persistent remnant of an otherwise extinct organism.

Rostrals—Situated toward the oral or nasal region.

Sexual dimorphism—Difference of physical form (shape) or coloration between the sexes; any consistent difference between males and females beyond the basic functional portions of the sex organs.

Scute—An external bony or horny plate or a large scale.

Serpentine—Of or relating to snakes; the type of limbless movements used by snakes.

Sidewinding—A mode of locomotion in which a snake raises its body in loops and rests its weight on two or three points that are the only places where the body contacts the ground. The loops are swung forward through the air while the points of contact are moved smoothly along the body length. This locomotion is characteristic of snakes that live in deserts where sandy substrates have particles that slip and induce sliding during serpentine locomotion.

Speciose—A taxonomic group with many species.

Subcaudals—Scales on the ventral side of the tail of reptiles, commonly arranged in either a single series or two series in snakes.

Supralabials—Enlarged scales on the edge of the lip of the upper jaw.

Supramarginal—Any one of the plates situated between the mid-lateral and marginal plates of the carapace in some species of turtles.

Supraocular—Either of a pair of shields that lie dorsal to the eyes of snakes.

Suprapygals—Any of the bones (usually two) that lie across the midline of the turtle carapace, just anterior to the hindmost plate of the shell

Supranasals—Any scale located directly above the nasal scale of squamate reptiles.

SVL—Abbreviation for snout-vent length. Measures from the tip of the animal's snout to the end of its vent.

Synapomorphy—An apomorphy (derived or specialised character) shared by two or more groups which originated in their last common ancestor

Tail autotomy—A defense mechanism whereby an organism can drop its tail if it feels threatened or is grabbed by the tail. Tails have also been dropped as a result of stress.

Thigmothermic—Absorbs heat by contact with warm surfaces, for example, by lying on a warm rock. Snakes use this strategy.

Tubercles—A small projection or nodule that grows on a plant or animal.

Tuberculate—A plant or animal that has nodules or tubercles growing on them.

Tympanic membrane—Thin membrane that closes externally the cavity of the middle ear and functions in the mechanical reception of sound waves and in their transmission to the site of sensory reception.

Undulatory—A type of limbless movement in which the body moves through a series of curves while at least three points on the body slide past irregular points of contact where reaction forces are generated to move the body in the forward direction.

Ventral—Of or relating to the belly or abdomen of an animal.

Viviparous—Giving birth to live young rather than eggs.

Warts—Any hard or cornified prominence on the skin, sometimes glandular in substructure.

GLOSSARY

Reptiles species list

Testudines [Order]
Carettochelyidae [Family]
  Carettochelys [Genus]
    C. insculpta

Chelidae [Family]
  Acanthochelys [Genus]
    A. chacoensis [Species]
    A. macrocephala
    A. pallidipectoris
    A. radiolata
    A. spixii
  Chelodina [Genus]
    C. canni [Species]
    C. expansa
    C. longicollis
    C. mccordi
    C. novaeguineae
    C. oblonga
    C. parkeri
    C. pritchardi
    C. reimanni
    C. rugosa
    C. siebenrocki
    C. steindachneri
  Chelus [Genus]
    C. fimbriata [Species]
  Elseya [Genus]
    E. dentata [Species]
    E. georgesi
    E. irwini
    E. latisternum
    E. lavarackorum
    E. novaeguineae
    E. purvisi
  Elusor [Genus]
    E. macrurus [Species]
  Emydura [Genus]
    E. australis [Species]
    E. krefftii
    E. macquarrii
    E. signata
    E. subglobosa
    E. tanybaraga
    E. victoriae
  Hydromedusa [Genus]
    H. maximiliani [Species]
    H. tectifera

  Phrynops [Genus]
    P. dahli [Species]
    P. geoffroanus
    P. gibbus
    P. heliostemma
    P. hilarii
    P. hogei
    P. nasutus
    P. raniceps
    P. rufipes
    P. tuberculatus
    P. vanderhaegei
    P. wermuthi
    P. williamsi
    P. zuliae
  Platemys [Genus]
    P. platycephala [Species]
  Pseudemydura [Genus]
    P. umbrina [Species]
  Rheodytes [Genus]
    R. leukops [Species]

Cheloniidae [Family]
  Caretta [Genus]
    C. caretta [Species]
  Chelonia [Genus]
    C. mydas [Species]
  Eretmochelys [Genus]
    E. imbricata [Species]
  Lepidochelys [Genus]
    L. kempii [Species]
    L. olivacea
  Natator [Genus]
    N. depressa [Species]

Chelydridae [Family]
  Chelydra [Genus]
    C. serpentina [Species]
  Macrochelys [Genus]
    M. temminckii [Species]

Dermatemydidae [Family]
  Dermatemys [Genus]
    D. mawii [Species]

Dermochelyidae [Family]
  Dermochelys [Genus]
    D. coriacea [Species]

Emydidae [Family]
  Chrysemys [Genus]
    C. picta [Species]
  Clemmys [Genus]
    C. guttata [Species]
    C. insculpta
    C. marmorata
    C. muhlenbergii
  Deirochelys [Genus]
    D. reticularia [Species]
  Emydoidea [Genus]
    E. blandingii [Species]
  Emys [Genus]
    E. orbicularis [Species]
  Graptemys [Genus]
    G. barbouri [Species]
    G. caglei
    G. ernsti
    G. flavimaculata
    G. geographica
    G. gibbonsi
    G. nigrinoda
    G. oculifera
    G. ouachitensis
    G. pseudogeographica
    G. pulchra
    G. versa
  Malaclemys [Genus]
    M. terrapin [Species]
  Pseudemys [Genus]
    P. alabamensis [Species]
    P. concinna
    P. floridana
    P. gorzugi
    P. nelsoni
    P. peninsularis
    P. rubriventris
    P. suwanniensis
    P. texana
  Terrapene [Genus]
    T. carolina [Species]
    T. coahuila
    T. nelsoni
    T. ornata
  Trachemys [Genus]
    T. adiutrix [Species]

T. decorata
T. decussata
T. dorbigni
T. gaigeae
T. ornata
T. scripta
T. stejnegeri
T. terrapen

Geoemydidae [Family]
  Batagur [Genus]
    B. baska [Species]
  Callagur [Genus]
    C. borneoensis [Species]
  Chinemys [Genus]
    C. megalocephala [Species]
    C. nigricans
    C. reevesii
  Cistoclemmys [Genus]
    C. flavomarginata [Species]
    C. galbinifrons
  Cuora [Genus]
    C. amboinensis [Species]
    C. aurocapitata
    C. flavomarginata
    C. galbinifrons
    C. hainanensis
    C. mccordi
    C. pani
    C. trifasciata
    C. yunnanensis
    C. zhoui
  Cyclemys [Genus]
    C. atripons [Species]
    C. dentata
    C. oldhami
    C. tcheponensis
  Geoclemys [Genus]
    G. hamiltonii [Species]
  Geoemyda [Genus]
    G. depressa [Species]
    G. japonica
    G. leytensis
    G. silvatica
    G. spengleri
  Hardella [Genus]
    H. thurjii [Species]
  Heosemys [Genus]
    H. depressa [Species]
    H. grandis
    H. leytensis
    H. silvatica
    H. spinosa
  Hieremys [Genus]
    H. annandalii [Species]
  Kachuga [Genus]
    K. dhongoka [Species]
    K. kachuga
    K. smithii
    K. sylhetensis
    K. tecta

K. tentoria
K. trivittata
Malayemys [Genus]
  M. subtrijuga [Species]
Mauremys [Genus]
  M. annamensis [Species]
  M. caspica
  M. iversoni
  M. japonica
  M. leprosa
  M. mutica
  M. pritchardi
  M. rivulata
Melanochelys [Genus]
  M. tricarinata [Species]
  M. trijuga
Morenia [Genus]
  M. ocellata [Species]
  M. petersi
Notochelys [Genus]
  N. platynota [Species]
Ocadia [Genus]
  O. glyphistoma [Species]
  O. philippeni
  O. sinensis
Orlitia [Genus]
  O. borneensis [Species]
Pyxidea [Genus]
  P. mouhotii [Species]
Rhinoclemmys [Genus]
  R. annulata [Species]
  R. areolata
  R. diademata
  R. funerea
  R. melanosterna
  R. nasuta
  R. pulcherrima
  R. punctularia
  R. rubida
Sacalia [Genus]
  S. bealei [Species]
  S. pseudocellata
  S. quadriocellata
Siebenrockiella [Genus]
  S. crassicollis [Species]
Leucocephalon [Genus]
  L. yuwonoi [Species]

Kinosternidae [Family]
  Claudius [Genus]
    C. angustatus [Species]
  Kinosternon [Genus]
    K. acutum [Species]
    K. alamosae
    K. angustipons
    K. baurii
    K. chimalhuaca
    K. creaseri
    K. cruentatum
    K. dunni
    K. flavescens

K. herrerai
K. hirtipes
K. integrum
K. leucostomum
K. oaxacae
K. postinguinale
K. scorpioides
K. sonoriense
K. spurrelli
K. subrubrum
Staurotypus [Genus]
  S. salvinii [Species]
  S. triporcatus
Sternotherus [Genus]
  S. carinatus [Species]
  S. depressus
  S. minor
  S. odoratus

Pelomedusidae [Family]
  Pelomedusa [Genus]
    P. subrufa [Species]
  Pelusios [Genus]
    P. adansonii [Species]
    P. bechuanicus
    P. broadleyi
    P. carinatus
    P. castaneus
    P. castanoides
    P. chapini
    P. gabonensis
    P. marani
    P. nanus
    P. niger
    P. rhodesianus
    P. seychellensis
    P. sinuatus
    P. subniger
    P. upembae
    P. williamsi

Platysternidae [Family]
  Platysternon [Genus]
    P. megacephalum [Species]

Podocnemidae [Family]
  Erymnochelys [Genus]
    E. madagascariensis [Species]
  Peltocephalus [Genus]
    P. dumerilianus [Species]
  Podocnemis [Genus]
    P. erythrocephala [Species]
    P. expansa
    P. lewyana
    P. sextuberculata
    P. unifilis
    P. vogli

Testudinidae [Family]
  Chersina [Genus]
    C. angulata [Species]
  Dipsochelys [Genus]
    D. arnoldi [Species]

D. dussumieri
D. hololissa
Geochelone [Genus]
  G. carbonaria [Species]
  G. chilensis
  G. denticulata
  G. elegans
  G. gigantea
  G. nigra
  G. pardalis
  G. platynota
  G. radiata
  G. sulcata
  G. yniphora
Gopherus [Genus]
  G. agassizii [Species]
  G. berlandieri
  G. flavomarginatus
  G. polyphemus
Homopus [Genus]
  H. areolatus [Species]
  H. bergeri
  H. boulengeri
  H. femoralis
  H. signatus
Indotestudo [Genus]
  I. elongata [Species]
  I. forstenii
  I. travancorica
Kinixys [Genus]
  K. belliana [Species]
  K. erosa
  K. homeana
  K. natalensis
  k. spekii
Malacochersus [Genus]
  M. tornieri [Species]
Manouria [Genus]
  M. emys [Species]
  M. impressa
Psammobates [Genus]
  P. geometricus [Species]
  P. oculifera
  P. tentorius
Pyxis [Genus]
  P. arachnoides [Species]
  P. planicauda
Testudo [Genus]
  t. antakyensis [Species]
  T. graeca
  T. hermanni
  T. horsfieldii
  T. kleinmanni
  T. marginata
  T. weissingeri
  T. werneri

Trionychidae [Family]
  Amyda [Genus]
    A. cartilaginea [Species]
    A. nakornsrithammarajensis

Apalone [Genus]
  A. ferox [Species]
  A. mutica
  A. spinifera
Aspideretes [Genus]
  A. gangeticus [Species]
  A. hurum
  A. leithii
  A. nigricans
Chitra [Genus]
  C. indica [Species]
Cyclanorbis [Genus]
  C. elegans [Species]
  C. senegalensis
Cycloderma [Genus]
  C. aubryi [Species]
  C. frenatum
Dogania [Genus]
  D. subplana [Species]
Lissemys [Genus]
  L. punctata [Species]
  L. scutata
Nilssonia [Genus]
  N. formosa [Species]
Palea [Genus]
  P. steindachneri [Species]
Pelochelys [Genus]
  P. bibroni [Species]
  P. cantori
Pelodiscus [Genus]
  P. parviformis [Species]
  P. sinensis
Rafetus [Genus]
  R. euphraticus [Species]
  R. swinhoei
Trionyx [Genus]
  T. axenaria [Species]
  T. triunguis

Crocodylia [Order]

Gavialidae [Family]
  Gavialis [Genus]
    Gavialis gangeticus [Species]

Alligatoridae [Family]
  Alligator [Genus]
    A. mississippiensis [Species]
    A. sinensis
  Caiman [Genus]
    C. crocodilus [Species]
    C. latirostris
    C. yacare
  Melanosuchus [Genus]
    M. niger [Species]
  Paleosuchus [Genus]
    P. palpebrosus [Species]
    P. trigonatus

Crocodylidae [Family]
  Crocodylus [Genus]
    C acutus [Species]

    C. cataphractus
    C. intermedius
    C. johnsoni
    C. mindorensis
    C. moreletii
    C. niloticus
    C. novaeguineae
    C. palustris
    C. porosus
    C. rhombifer
    C. siamensis
  Osteolaemus [Genus]
    O. tetraspis [Species]
  Tomistoma [Genus]
    T. schlegelii [Species]

Rhynchocephalia [Order]

Sphenodontidae [Family]
  Sphenodon [Genus]
    S. guntheri [Species]
    S. punctatus

Squamata [Order]

Agamidae [Family]
  Acanthocercus [Genus]
    A. Acanthocercus [Species]
    A. cyanogaster
    A. phillipsii
    A. trachypleurus
    A. yemensis
    A. zonurus
  Acanthosaura [Genus]
    A. armata [Species]
    A. capra
    A. crucigera
    A. lepidogaster
  Agama [Genus]
    A. aculeata [Species]
    A. agama
    A. anchietae
    A. armata
    A. atra
    A. bocourti
    A. bottegi
    A. boueti
    A. boulengeri
    A. caudospinosa
    A. cornii
    A. doriae
    A. etoshae
    A. gracilimembris
    A. hartmanni
    A. hispida
    A. impalearis
    A. insularis
    A. kirkii
    A. mehelyi
    A. montana
    A. mossambica
    A. mwanzae

A. paragama
A. persimilis
A. planiceps
A. robecchii
A. rueppelli
A. sankaranica
A. spinosa
A. weidholzi
Amphibolurus [Genus]
A. muricatus [Species]
A. nobbi
A. norrisi
Aphaniotis [Genus]
A. acutirostris [Species]
A. fusca
A. ornata
Brachysaura [Genus]
B. minor [Species]
Bronchocela [Genus]
B. celebensis [Species]
B. cristatella
B. danieli
B. hayeki
B. jubata
B. marmorata
B. smaragdina
Bufoniceps [Genus]
B. laungwalaensis [Species]
Caimanops [Genus]
C. amphiboluroides [Species]
Calotes [Genus]
C. andamanensis [Species]
C. bhutanensis
C. calotes
C. ceylonensis
C. ellioti
C. emma
C. grandisquamis
C. jerdoni
C. kingdonwardi
C. liocephalus
C. liolepis
C. maria
C. medogensis
C. mystaceus
C. nemoricola
C. nigrigularis
C. nigrilabris
C. nigriplicatus
C. rouxii
C. versicolor
Ceratophora [Genus]
C. aspera [Species]
C. erdeleni
C. karu
C. stoddartii
C. tennentii
Chelosania [Genus]
C. brunnea [Species]
Chlamydosaurus [Genus]
C. kingii [Species]

Cophotis [Genus]
C. ceylanica [Species]
Cryptagama [Genus]
C. aurita [Species]
Ctenophorus [Genus]
C. caudicinctus [Species]
C. clayi
C. cristatus
C. decresii
C. femoralis
C. fionni
C. fordi
C. gibba
C. isolepis
C. maculatus
C. maculosus
C. mckenziei
C. nuchalis
C. ornatus
C. pictus
C. reticulatus
C. rufescens
C. salinarum
C. scutulatus
C. tjantjalka
C. vadnappa
C. yinnietharra
Dendragama [Genus]
D. boulengeri [Species]
Diporiphora [Genus]
D. albilabris [Species]
D. arnhemica
D. australis
D. bennettii
D. bilineata
D. convergens
D. lalliae
D. linga
D. magna
D. margaretae
D. pindan
D. reginae
D. superba
D. valens
D. winneckei
Draco [Genus]
D. affinis [Species]
D. biaro
D. bimaculatus
D. blanfordii
D. caerulhians
D. cornutus
D. cristatellus
D. cyanopterus
D. dussumieri
D. fimbriatus
D. guentheri
D. haematopogon
D. jareckii
D. lineatus
D. maculatus

D. maximus
D. melanopogon
D. mindanensis
D. norvillii
D. obscurus
D. ornatus
D. palawanensis
D. quadrasi
D. quinquefasciatus
D. reticulatus
D. spilopterus
D. taeniopterus
D. volans
Gonocephalus [Genus]
G. bellii [Species]
G. beyschlagi
G. borneensis
G. chamaeleontinus
G. doriae
G. grandis
G. interruptus
G. klossi
G. kuhlii
G. lacunosus
G. liogaster
G. megalepis
G. mjoebergi
G. robinsonii
G. semperi
G. sophiae
Harpesaurus [Genus]
H. beccarii [Species]
H. borneensis
H. ensicauda
H. modigliani
H. thescelorhinos
H. tricinctus
Hydrosaurus [Genus]
H. amboinensis [Species]
H. pustulatus
H. weberi
Hypsicalotes [Genus]
H. kinabaluensis [Species]
Hypsilurus [Genus]
H. auritus [Species]
H. binotatus
H. boydii
H. bruijnii
H. dilophus
H. geelvinkianus
H. godeffroyi
H. modestus
H. nigrigularis
H. papuensis
H. schoedei
H. schultzewestrumi
H. spinipes
Japalura [Genus]
J. andersoniana [Species]
J. brevipes
J. chapaensis

REPTILES SPECIES LIST

J. dymondi
J. fasciata
J. flaviceps
J. grahami
J. hamptoni
J. kaulbacki
J. kumaonensis
J. luei
J. major
J. makii
J. micangshanensis
J. planidorsata
J. polygonata
J. sagittifera
J. splendida
J. swinhonis
J. tricarinata
J. varcoae
J. variegata
J. yunnanensis
Laudakia [Genus]
   L. adramitanus [Species]
   L. agrorensis
   L. atricollis
   L. badakhshana
   L. bochariensis
   L. caucasia
   L. dayana
   L. erythrogastra
   L. himalayana
   L. kirmanensis
   L. lehmanni
   L. melanura
   L. microlepis
   L. nupta
   L. nuristanica
   L. pakistanica
   L. papenfussi
   L. sacra
   L. stellio
   L. stoliczkana
   L. tuberculata
   L. wui
Leiolepis [Genus]
   L. belliana [Species]
   L. boehmei
   L. guentherpetersi
   L. guttata
   L. peguensis
   L. reevesii
   L. triploida
Lophocalotes [Genus]
   L. ludekingi [Species]
Lophognathus [Genus]
   L. gilberti [Species]
   L. longirostris
   L. maculilabris
   L. temporalis
Lyriocephalus [Genus]
   L. scutatus [Species]

Mictopholis [Genus]
   M. austeniana [Species]
Moloch [Genus]
   M. horridus [Species]
Oreodeira [Genus]
   O. gracilipes [Species]
Oriocalotes [Genus]
   O. paulus [Species]
Otocryptis [Genus]
   O. beddomii [Species]
   O. wiegmanni
Phoxophrys [Genus]
   P. borneensis [Species]
   P. cephalum
   P. nigrilabris
   P. spiniceps
   P. tuberculata
Phrynocephalus [Genus]
   P. affinis [Species]
   P. albolineatus
   P. alticola
   P. arabicus
   P. arcellazzii
   P. axillaris
   P. birulai
   P. clarkorum
   P. elegans
   P. euptilopus
   P. forsythii
   P. frontalis
   P. geckoides
   P. golubewii
   P. guttatus
   P. helioscopus
   P. hongyuanensis
   P. interscapularis
   P. lidskii
   P. luteoguttatus
   P. maculatus
   P. melanurus
   P. moltschanowi
   P. mystaceus
   P. nasatus
   P. ornatus
   P. parvulus
   P. parvus
   P. przewalskii
   P. pylzowi
   P. raddei
   P. reticulatus
   P. roborowskii
   P. rossikowi
   P. salenskyi
   P. scutellatus
   P. sogdianus
   P. steindachneri
   P. strauchi
   P. theobaldi
   P. versicolor
   P. vlangalii

   P. zetanensis
Physignathus [Genus]
   P. cocincinus [Species]
   P. lesueurii
Pogona [Genus]
   P. barbata [Species]
   P. henrylawsoni
   P. microlepidota
   P. minima
   P. minor
   P. nullarbor
   P. vitticeps
Pseudocalotes [Genus]
   P. brevipes [Species]
   P. dringi
   P. flavigula
   P. floweri
   P. larutensis
   P. microlepis
   P. poilani
   P. saravacensis
   P. sumatrana
   P. tympanistriga
Psammophilus [Genus]
   P. blanfordanus [Species]
   P. dorsalis
Pseudotrapelus [Genus]
   P. sinaitus [Species]
Ptyctolaemus [Genus]
   P. gularis [Species]
   P. phuwuanensis
Rankinia [Genus]
   R. adelaidensis [Species]
   R. diemensis
Salea [Genus]
   S. anamallayana [Species]
   S. gularis
   S. horsfieldii
   S. kakhienensis
Sitana [Genus]
   S. fusca [Species]
   S. ponticeriana
   S. sivalensis
Trapelus [Genus]
   T. agilis [Species]
   T. blanfordi
   T. flavimaculatus
   T. jayakari
   T. lessonae
   T. megalonyx
   T. microtympanum
   T. mutabilis
   T. pallidus
   T. persicus
   T. rubrigularis
   T. ruderatus
   T. sanguinolentus
   T. savignii
   T. tournevillei
Tympanocryptis [Genus]

T. cephalus [Species]
T. intima
T. lineata
T. parviceps
T. tetraporophora
T. uniformis
Uromastyx [Genus]
U. acanthinura [Species]
U. aegyptia
U. alfredschmidti
U. asmussi
U. benti
U. dispar
U. geyri
U. hardwickii
U. leptieni
U. loricata
U. macfadyeni
U. occidentalis
U. ocellata
U. ornata
U. princeps
U. thomasi
Xenagama [Genus]
X. batillifera [Species]
X. taylori

Chamaeleonidae [Family]
Bradypodion [Genus]
B. adolfifriderici [Species]
B. caffer
B. carpenteri
B. damaranum
B. dracomontanum
B. excubitor
B. fischeri
B. gutturale
B. karrooicum
B. melanocephalum
B. mlanjense
B. nemorale
B. occidentale
B. oxyrhinum
B. pumilum
B. setaroi
B. spinosum
B. taeniabronchum
B. tavetanum
B. tenue
B. thamnobates
B. transvaalense
B. uthmoelleri
B. ventrale
B. xenorhinum
Brookesia [Genus]
B. ambreensis [Species]
B. antakarana
B. bekolosy
B. betschi
B. bonsi
B. brygooi

B. decaryi
B. dentata
B. ebenaui
B. exarmata
B. griveaudi
B. karchei
B. lambertoni
B. lineata
B. lolontany
B. minima
B. nasus
B. perarmata
B. peyrierasi
B. stumpffi
B. superciliaris
B. therezieni
B. thieli
B. tuberculata
B. vadoni
B. valerieae
Calumma [Genus]
C. andringitraensis [Species]
C. boettgeri
C. brevicornis
C. capuroni
C. cucullata
C. fallax
C. furcifer
C. gallus
C. gastrotaenia
C. glawi
C. globifer
C. guibei
C. guillaumeti
C. hilleniusi
C. linota
C. malthe
C. marojezensis
C. nasuta
C. oshaughnessyi
C. parsonii
C. peyrierasi
C. tigris
C. tsaratananensis
Chamaeleo [Genus]
C. affinis [Species]
C. africanus
C. anchietae
C. arabicus
C. balebicornatus
C. bitaeniatus
C. calcaricarens
C. calyptratus
C. camerunensis
C. chamaeleon
C. chapini
C. conirostratus
C. cristatus
C. deremensis
C. dilepis

C. eisentrauti
C. ellioti
C. etiennei
C. feae
C. fuelleborni
C. goetzei
C. gracilis
C. harennae
C. hoehnelii
C. incornutus
C. ituriensis
C. jacksonii
C. johnstoni
C. kinetensis
C. laevigatus
C. laterispinis
C. marsabitensis
C. melleri
C. monachus
C. montium
C. namaquensis
C. oweni
C. pfefferi
C. quadricornis
C. quilensis
C. roperi
C. rudis
C. ruspolii
C. schoutedeni
C. schubotzi
C. senegalensis
C. sternfeldi
C. tempeli
C. tremperi
C. werneri
C. wiedersheimi
C. zeylanicus
Furcifer [Genus]
F. angeli [Species]
F. antimena
F. balteatus
F. belalandaensis
F. bifidus
F. campani
F. cephalolepis
F. labordi
F. lateralis
F. minor
F. monoceras
F. nicosiai
F. oustaleti
F. pardalis
F. petteri
F. polleni
F. rhinoceratus
F. tuzetae
F. verrucosus
F. willsii
Rhampholeon [Genus]
R. boulengeri [Species]

REPTILES SPECIES LIST

R. *brachyurus*
R. *brevicaudatus*
R. *chapmanorum*
R. *kerstenii*
R. *marshalli*
R. *moyeri*
R. *nchisiensis*
R. *platyceps*
R. *spectrum*
R. *temporalis*
R. *uluguruensis*

Iguanidae [Family]
  Amblyrhynchus [Genus]
    A. *cristatus* [Species]
  Anisolepis [Genus]
    A. *grilli* [Species]
    A. *longicauda*
    A. *undulatus*
  Anolis [Genus]
    A. *achilles* [Species]
    A. *acutus*
    A. *adleri*
    A. *aeneus*
    A. *aequatorialis*
    A. *agassizi*
    A. *agueroi*
    A. *alayoni*
    A. *albimaculatus*
    A. *alfaroi*
    A. *aliniger*
    A. *allisoni*
    A. *altavelensis*
    A. *alumina*
    A. *alutaceus*
    A. *andianus*
    A. *anfiloquioi*
    A. *angusticeps*
    A. *antioquiae*
    A. *apollinaris*
    A. *argenteolus*
    A. *argillaceus*
    A. *armouri*
    A. *attenuatus*
    A. *bahorucoensis*
    A. *baleatus*
    A. *baracoae*
    A. *barahonae*
    A. *barbatus*
    A. *barbouri*
    A. *bartschi*
    A. *bimaculatus*
    A. *binotatus*
    A. *blanquillanus*
    A. *boettgeri*
    A. *bonairensis*
    A. *brevirostris*
    A. *brunneus*
    A. *calimae*
    A. *caquetae*
    A. *carolinensis*

A. *casildae*
A. *caudalis*
A. *centralis*
A. *chamaeleonides*
A. *chloris*
A. *chlorocyanus*
A. *chocorum*
A. *christophei*
A. *chrysolepis*
A. *clivicola*
A. *coelestinus*
A. *cooki*
A. *cristatellus*
A. *cristifer*
A. *cupeyalensis*
A. *cuvieri*
A. *cyanopleurus*
A. *cybotes*
A. *danieli*
A. *darlingtoni*
A. *deltae*
A. *desechensis*
A. *dissimilis*
A. *distichus*
A. *dolichocephalus*
A. *eewi*
A. *equestris*
A. *ernestwilliamsi*
A. *etheridgei*
A. *eugenegrahami*
A. *eulaemus*
A. *evermanni*
A. *extremus*
A. *fairchildi*
A. *fasciatus*
A. *ferreus*
A. *festae*
A. *fitchi*
A. *fowleri*
A. *fraseri*
A. *frenatus*
A. *fugitivus*
A. *garridoi*
A. *gemmosus*
A. *gingivinus*
A. *gorgonae*
A. *greyi*
A. *griseus*
A. *guamuhaya*
A. *gundlachi*
A. *haetianus*
A. *hendersoni*
A. *huilae*
A. *impetigosus*
A. *incredulus*
A. *inexpectatus*
A. *insignis*
A. *insolitus*
A. *isolepis*
A. *jacare*

A. *juangundlachi*
A. *koopmani*
A. *krugi*
A. *laevis*
A. *lamari*
A. *latifrons*
A. *lividus*
A. *longicauda*
A. *longiceps*
A. *longitibialis*
A. *loysianus*
A. *luciae*
A. *lucius*
A. *luteogularis*
A. *luteosignifer*
A. *macilentus*
A. *maculigula*
A. *malkini*
A. *marcanoi*
A. *marmoratus*
A. *marron*
A. *maynardi*
A. *megalopithecus*
A. *menta*
A. *microtus*
A. *mirus*
A. *monensis*
A. *monticola*
A. *nannodes*
A. *nasofrontalis*
A. *nelsoni*
A. *nigrolineatus*
A. *nigropunctatus*
A. *noblei*
A. *nubilis*
A. *occultus*
A. *oculatus*
A. *oligaspis*
A. *olssoni*
A. *oporinus*
A. *palmeri*
A. *parilis*
A. *paternus*
A. *peraccae*
A. *philopunctatus*
A. *phyllorhinus*
A. *pigmaequestris*
A. *placidus*
A. *poncensis*
A. *porcatus*
A. *porcus*
A. *princeps*
A. *proboscis*
A. *propinquus*
A. *pseudotigrinus*
A. *pulchellus*
A. *pumilus*
A. *punctatus*
A. *purpurescens*
A. *quaggulus*

A. radulinus
A. rejectus
A. richardii
A. ricordi
A. rimarum
A. roosevelti
A. roquet
A. ruizi
A. rupinae
A. sabanus
A. santamartae
A. scriptus
A. semilineatus
A. sheplani
A. shrevei
Basiliscus [Genus]
B. basiliscus [Species]
A. shrevei
A. singularis
A. smallwoodi
A. smaragdinus
A. solitarius
A. spectrum
A. squamulatus
A. strahmi
A. stratulus
A. tigrinus
A. toldo
A. transversalis
A. trinitatis
A. vanidicus
A. vaupesianus
A. ventrimaculatus
A. vermiculatus
A. vescus
A. wattsi
A. websteri
A. whitemani
A. williamsii
Basiliscus [Genus]
B. basiliscus [Species]
B. galeritus
B. plumifrons
B. vittatus
Brachylophus [Genus]
B. fasciatus [Species]
B. vitiensis
Callisaurus [Genus]
C. draconoides [Species]
Chalarodon [Genus]
C. madagascariensis [Species]
Conolophus [Genus]
C. pallidus [Species]
C. subcristatus
Cophosaurus [Genus]
C. texanus [Species]
Corytophanes [Genus]
C. cristatus [Species]
C. hernandezi
C. percarinatus
Crotaphytus [Genus]

C. antiquus [Species]
C. collaris
C. grismeri
C. insularis
C. nebrius
C. reticulatus
C. vestigium
Ctenosaura [Genus]
C. acanthura [Species]
C. alfredschmidti
C. bakeri
C. clarki
C. defensor
C. flavidorsalis
C. hemilopha
C. melanosterna
C. oaxacana
C. oedirhina
C. palearis
C. pectinata
C. quinquecarinata
C. similis
Cyclura [Genus]
C. carinata [Species]
C. collei
C. cornuta
C. cychlura
C. nubila
C. pinguis
C. ricordi
C. rileyi
Diplolaemus [Genus]
D. bibronii [Species]
D. darwinii
D. leopardinus
Dipsosaurus [Genus]
D. dorsalis [Species]
Enyalioides [Genus]
E. cofanorum [Species]
E. heterolepis
E. laticeps
E. microlepis
E. oshaughnessyi
E. palpebralis
E. praestabilis
Enyalius [Genus]
E. bibronii [Species]
E. bilineatus
E. brasiliensis
E. catenatus
E. iheringii
E. leechii
E. perditus
E. pictus
Gambelia [Genus]
G. copeii [Species]
G. sila
G. wislizenii
Holbrookia [Genus]
H. lacerata [Species]
H. maculata

H. propinqua
H. subcaudalis
Hoplocercus [Genus]
H. spinosus [Species]
Iguana [Genus]
I. delicatissima [Species]
I. iguana
Laemanctus [Genus]
L. longipes [Species]
L. serratus
Leiocephalus [Genus]
L. anonymous [Species]
L. apertosulcus
L. barahonensis
L. carinatus
L. cubensis
L. cuneus
L. endomychus
L. eremitus
L. etheridgei
L. greenwayi
L. herminieri
L. inaguae
L. jamaicensis
L. loxogrammus
L. lunatus
L. macropus
L. melanochlorus
L. onaneyi
L. personatus
L. pratensis
L. psammodromus
L. punctatus
L. raviceps
L. rhutidira
L. schreibersii
L. semilineatus
L. stictigaster
L. vinculum
Leiosaurus [Genus]
L. bellii [Species]
L. catamarcensis
L. paronae
Liolaemus [Genus]
L. abaucan [Species]
L. albiceps
L. alticolor
L. andinus
L. anomalus
L. archeforus
L. atacamensis
L. audituvelatus
L. austromendocinus
L. baguali
L. bellii
L. bibronii
L. bisignatus
L. bitaeniatus
L. boulengeri
L. buergeri
L. calchaqui

REPTILES SPECIES LIST

L. canqueli
L. capillitas
L. ceii
L. chacoensis
L. chiliensis
L. coeruleus
L. constanzae
L. copiapensis
L. cranwelli
L. cristiani
L. curicensis
L. curis
L. cuyanus
L. cyanogaster
L. darwinii
L. disjunctus
L. donosobarrosi
L. dorbignyi
L. duellmani
L. eleodori
L. elongatus
L. erroneus
L. escarchadosi
L. etheridgei
L. exploratorum
L. fabiani
L. famatinae
L. fittkaui
L. fitzgeraldi
L. fitzingerii
L. forsteri
L. foxi
L. fuscus
L. gallardoi
L. gracilis
L. gravenhorstii
L. griseus
L. grosseorum
L. hatcheri
L. heliodermis
L. hellmichi
L. hernani
L. huacahuasicus
L. insolitus
L. irregularis
L. isabelae
L. islugensis
L. jamesi
L. josephorum
L. juanortizi
L. kingii
L. koslowskyi
L. kriegi
L. kuhlmanni
L. laurenti
L. lemniscatus
L. leopardinus
L. lineomaculatus
L. lorenzmuelleri
L. lutzae
L. magellanicus

L. maldonadae
L. melanogaster
L. melanops
L. montanus
L. monticola
L. multicolor
L. multimaculatus
L. nigriceps
L. nigromaculatus
L. nigroventrolateralis
L. nigroviridis
L. nitidus
L. occipitalis
L. olongasta
L. orientalis
L. ornatus
L. ortizii
L. pagaburoi
L. pantherinus
L. patriciaiturrae
L. paulinae
L. petrophilus
L. pictus
L. platei
L. pleopholis
L. polystictus
L. pseudoanomalus
L. pseudolemniscatus
L. pulcherrimus
L. quilmes
L. rabinoi
L. ramirezae
L. ramonensis
L. reichei
L. riojanus
L. robertmertensi
L. robustus
L. rosenmanni
L. rothi
L. ruibali
L. salinicola
L. sanjuanensis
L. sarmientoi
L. saxatilis
L. scapularis
L. schmidti
L. schroederi
L. signifer
L. silvai
L. silvanae
L. somuncurae
L. tacnae
L. tari
L. telsen
L. tenuis
L. thermarum
L. thomasi
L. tristis
L. uspallatensis
L. valdesianus
L. vallecurensis

L. variegatus
L. velosoi
L. walkeri
L. wiegmannii
L. williamsi
L. xanthoviridis
L. zapallarensis
L. zullyi
Microlophus [Genus]
 M. albemarlensis [Species]
 M. atacamensis
 M. bivittatus
 M. delanonis
 M. duncanensis
 M. grayii
 M. habelii
 M. heterolepis
 M. koepckeorum
 M. occipitalis
 M. pacificus
 M. peruvianus
 M. quadrivittatus
 M. stolzmanni
 M. tarapacensis
 M. theresiae
 M. theresioides
 M. thoracicus
 M. tigris
 M. yanezi
Morunasaurus [Genus]
 M. annularis [Species]
 M. groi
Oplurus [Genus]
 O. cuvieri [Species]
 O. cyclurus
 O. fierinensis
 O. grandidieri
 O. quadrimaculatus
 O. saxicola
Petrosaurus [Genus]
 P. mearnsi [Species]
 P. thalassinus
Phrynosoma [Genus]
 P. asio [Species]
 P. braconnieri
 P. cerroense
 P. cornutum
 P. coronatum
 P. ditmarsi
 P. douglassii
 P. hernandesi
 P. mcallii
 P. modestum
 P. orbiculare
 P. platyrhinos
 P. solare
 P. taurus
Phymaturus [Genus]
 P. antofagastensis [Species]
 P. indistinctus
 P. mallimaccii

P. nevadoi
P. palluma
P. patagonicus
P. payunae
P. punae
P. somuncurensis
P. zapalensis
Plica [Genus]
    P. lumaria [Species]
    P. plica
    P. umbra
Polychrus [Genus]
    P. acutirostris [Species]
    P. femoralis
    P. gutturosus
    P. liogaster
    P. marmoratus
    P. peruvianus
Pristidactylus [Genus]
    P. achalensis [Species]
    P. alvaroi
    P. araucanus
    P. casuhatiensis
    P. fasciatus
    P. nigroiugulus
    P. scapulatus
    P. torquatus
    P. valeriae
    P. volcanensis
Phenacosaurus [Genus]
    P. bellipeniculus [Species]
    P. carlostoddi
    P. euskalerriari
    P. heterodermus
    P. inderenae
    P. neblininus
    P. nicefori
    P. orcesi
    P. tetarii
    P. vanzolinii
Sauromalus [Genus]
    S. ater [Species]
    S. hispidus
    S. klauberi
    S. obesus
    S. slevini
    S. varius
Sceloporus [Genus]
    S. acanthinus [Species]
    S. adleri
    S. aeneus
    S. anahuacus
    S. arenicolus
    S. asper
    S. bicanthalis
    S. bulleri
    S. carinatus
    S. cautus
    S. chaneyi
    S. chrysostictus
    S. clarkii

S. couchii
S. cozumelae
S. cryptus
S. dugesii
S. edwardtaylori
S. exsul
S. formosus
S. gadovae
S. goldmani
S. graciosus
S. grammicus
S. heterolepis
S. horridus
S. hunsakeri
S. insignis
S. jalapae
S. jarrovii
S. licki
S. lineatulus
S. lundelli
S. macdougalli
S. maculosus
S. magister
S. malachiticus
S. megalepidurus
S. melanorhinus
S. merriami
S. monserratensis
S. mucronatus
S. nelsoni
S. occidentalis
S. ochoterenae
S. olivaceus
S. orcutti
S. ornatus
S. palaciosi
S. parvus
S. poinsettii
S. pyrocephalus
S. rufidorsum
S. salvini
S. samcolemani
S. scalaris
S. serrifer
S. siniferus
S. slevini
S. smaragdinus
S. smithi
S. spinosus
S. squamosus
S. stejnegeri
S. subniger
S. subpictus
S. taeniocnemis
S. tanneri
S. teapensis
S. torquatus
S. undulatus
S. utiformis
S. vandenburgianus
S. variabilis

S. virgatus
S. woodi
S. zosteromus
Stenocercus [Genus]
    S. aculeatus [Species]
    S. angel
    S. apurimacus
    S. arenarius
    S. azureus
    S. boettgeri
    S. bolivarensis
    S. caducus
    S. carrioni
    S. chlorostictus
    S. chota
    S. chrysopygus
    S. crassicaudatus
    S. cupreus
    S. doellojuradoi
    S. dumerilii
    S. empetrus
    S. erythrogaster
    S. eunetopsis
    S. festae
    S. fimbriatus
    S. formosus
    S. guentheri
    S. haenschi
    S. huancabambae
    S. humeralis
    S. imitator
    S. iridescens
    S. ivitus
    S. lache
    S. latebrosus
    S. limitaris
    S. marmoratus
    S. melanopygus
    S. modestus
    S. nigromaculatus
    S. nubicola
    S. ochoai
    S. orientalis
    S. ornatissimus
    S. ornatus
    S. pectinatus
    S. percultus
    S. praeornatus
    S. rhodomelas
    S. roseiventris
    S. scapularis
    S. simonsii
    S. stigmosus
    S. trachycephalus
    S. tricristatus
    S. variabilis
    S. varius
Tropidurus [Genus]
    T. amathites [Species]
    T. bogerti
    T. callathelys

T. catalanensis
T. chromatops
T. cocorobensis
T. divaricatus
T. erythrocephalus
T. etheridgei
T. guarani
T. helenae
T. hispidus
T. hygomi
T. insulanus
T. itambere
T. melanopleurus
T. montanus
T. mucujensis
T. nanuzae
T. oreadicus
T. panstictus
T. pinima
T. psammonastes
T. semitaeniatus
T. spinulosus
T. torquatus
T. xanthochilus
Uma [Genus]
U. exsul [Species]
U. inornata
U. notata
U. paraphygas
U. scoparia
Uracentron [Genus]
U. azureum [Species]
U. flaviceps
Uranoscodon [Genus]
U. superciliosus [Species]
Urosaurus [Genus]
U. auriculatus [Species]
U. bicarinatus
U. clarionensis
U. gadovi
U. graciosus
U. irregularis
U. lahtelai
U. nigricaudus
U. ornatus
Urostrophus [Genus]
U. gallardoi [Species]
U. vautieri
Uta [Genus]
U. concinna [Species]
U. encantadae
U. lowei
U. nolascensis
U. palmeri
U. squamata
U. stansburiana
U. stejnegeri
U. tumidarostra

Gekkonidae [Family]
Aeluroscalabotes [Genus]
A. felinus [Species]

Afroedura [Genus]
A. africana [Species]
A. amatolica
A. bogerti
A. hawequensis
A. karroica
A. nivaria
A. pondolia
A. tembulica
A. transvaalica
Afrogecko [Genus]
A. ansorgii [Species]
A. porphyreus
A. swartbergensis
Agamura [Genus]
A. femoralis [Species]
A. gastropholis
A. misonnei
A. persica
Ailuronyx [Genus]
A.seychellensis [Species]
A.tachyscopaeus
A.trachygaster
Alsophylax [Genus]
A. boehmei [Species]
A. laevis
A. loricatus
A. pipiens
A. przewalskii
A. tadjikiensis
A. tokobajevi
Aristelliger [Genus]
A. barbouri [Species]
A. cochranae
A. expectatus
A. georgeensis
A. hechti
A. lar
A. praesignis
Asaccus [Genus]
A. caudivolvulus [Species]
A. elisae
A. gallagheri
A. griseonotus
A. kermanshahensis
A. montanus
A. platyrhynchus
Bavayia [Genus]
B. crassicollis [Species]
B. cyclura
B. exsuccida
B. geitaina
B. madjo
B. montana
B. ornata
B. pulchella
B. robusta
B. sauvagii
B. septuiclavis
B. validiclavis

Blaesodactylus [Genus]
B. antongilensis [Species]
B. boivini
B. sakalava
Bogertia [Genus]
B. lutzae [Species]
Briba [Genus]
B. brasiliana [Species]
Bunopus [Genus]
B. blanfordii [Species]
B. crassicauda
B. spatalurus
B. tuberculatus
Calodactylodes [Genus]
C. aureus [Species]
C. illingworthi
Carinatogecko [Genus]
C. aspratilis [Species]
C. heteropholis
Carphodactylus [Genus]
C. laevis [Species]
Chondrodactylus [Genus]
C. angulifer [Species]
Christinus [Genus]
C. guentheri [Species]
C. marmoratus
Cnemaspis [Genus]
C. affinis [Species]
C. africana
C. argus
C. assamensis
C. barbouri
C. beddomei
C. boiei
C. boulengerii
C. chanthaburiensis
C. dickersoni
C. dilepis
C. dringi
C. flavolineata
C. gigas
C. goaensis
C. gordongekkoi
C. indica
C. jerdonii
C. kandianus
C. kendallii
C. koehleri
C. kumpoli
C. littoralis
C. nairi
C. nigridius
C. occidentalis
C. ornata
C. otai
C. petrodroma
C. podihuna
C. quattuorseriata
C. siamensis
C. sisparensis

C. spinicollis
C. timoriensis
C. tropidogaster
C. uzungwae
C. wynadensis
C. yercaudensis
Coleodactylus [Genus]
    C. amazonicus [Species]
    C. brachystoma
    C. meridionalis
    C. natalensis
    C. septentrionalis
Coleonyx [Genus]
    C. brevis [Species]
    C. elegans
    C. fasciatus
    C. mitratus
    C. reticulatus
    C. switaki
    C. variegatus
Colopus [Genus]
    C. wahlbergii [Species]
Cosymbotus [Genus]
    C. craspedotus [Species]
    C. platyurus
Crenadactylus [Genus]
    C. ocellatus [Species]
Crossobamon [Genus]
    C. eversmanni [Species]
    C. orientalis
Cryptactites [Genus]
    C. peringueyi [Species]
Cyrtodactylus [Genus]
    C. adleri [Species]
    C. agusanensis
    C. angularis
    C. annulatus
    C. aravallensis
    C. baluensis
    C. basoglui
    C. biordinis
    C. brevipalmatus
    C. cavernicolus
    C. collegalensis
    C. condorensis
    C. consobrinoides
    C. consobrinus
    C. darmandvillei
    C. deccanensis
    C. derongo
    C. deveti
    C. elok
    C. feae
    C. fraenatus
    C. gubernatoris
    C. ingeri
    C. interdigitalis
    C. intermedius
    C. irianjayaensis
    C. irregularis
    C. jarujini

C. jellesmae
C. khasiensis
C. laevigatus
C. lateralis
C. loriae
C. louisiadensis
C. malayanus
C. malcomsmithi
C. mansarulus
C. marmoratus
C. matsuii
C. mimikanus
C. nebulosus
C. novaeguineae
C. oldhami
C. papilionoides
C. papuensis
C. peguensis
C. philippinicus
C. pubisulcus
C. pulchellus
C. quadrivirgatus
C. redimiculus
C. rubidus
C. sadleiri
C. sermowaiensis
C. stoliczkai
C. sworderi
C. tibetanus
C. tiomanensis
C. variegatus
C. walli
C. wetariensis
C. yoshii
Cyrtopodion [Genus]
    C. agamuroides [Species]
    C. amictopholis
    C. battalensis
    C. baturensis
    C. brevipes
    C. caspius
    C. chitralensis
    C. dattanensis
    C. elongatus
    C. fasciolatus
    C. fedtschenkoi
    C. fortmunroi
    C. heterocercus
    C. himalayanus
    C. indusoani
    C. kachhensis
    C. kirmanensis
    C. kohsulaimanai
    C. kotschyi
    C. lawderanus
    C. longipes
    C. medogensis
    C. mintoni
    C. montiumsalsorum
    C. narynensis
    C. potoharensis

C. rhodocaudus
C. rohtasfortai
C. russowii
C. sagittifer
C. scaber
C. spinicaudus
C. turcmenicus
C. voraginosus
C. watsoni
Diplodactylus [Genus]
    D. alboguttatus [Species]
    D. assimilis
    D. byrnei
    D. ciliaris
    D. conspicillatus
    D. elderi
    D. fulleri
    D. furcosus
    D. galeatus
    D. granariensis
    D. immaculatus
    D. intermedius
    D. jeanae
    D. kenneallyi
    D. klugei
    D. maini
    D. mcmillani
    D. michaelseni
    D. mitchelli
    D. occultus
    D. ornatus
    D. polyophthalmus
    D. pulcher
    D. rankini
    D. robinsoni
    D. savagei
    D. spinigerus
    D. squarrosus
    D. steindachneri
    D. stenodactylus
    D. strophurus
    D. taeniatus
    D. taenicauda
    D. tessellatus
    D. vittatus
    D. wellingtonae
    D. williamsi
    D. wilsoni
    D. wombeyi
Dixonius [Genus]
    D. melanostictus [Species]
    D. siamensis
Dravidogecko [Genus]
    D. anamallensis [Species]
Ebenavia [Genus]
    E. inunguis [Species]
    E. maintimainty
Eublepharis [Genus]
    E. angramainyu [Species]
    E. fuscus
    E. hardwickii

E. macularius
E. turcmenicus
Euleptes [Genus]
   E. europaea [Species]
Eurydactylodes [Genus]
   E. agricolae [Species]
   E. symmetricus
   E. vieillardi
Geckolepis [Genus]
   G. anomala [Species]
   G. maculata
   G. petiti
   G. polylepis
   G. typica
Geckonia [Genus]
   G. chazaliae [Species]
Gehyra [Genus]
   G. angusticaudata [Species]
   G. australis
   G. baliola
   G. barea
   G. borroloola
   G. brevipalmata
   G. butleri
   G. catenata
   G. dubia
   G. fehlmanni
   G. fenestra
   G. intermedia
   G. interstitialis
   G. kimberleyi
   G. lacerata
   G. lampei
   G. leopoldi
   G. marginata
   G. membranacruralis
   G. minuta
   G. montium
   G. mutilata
   G. nana
   G. occidentalis
   G. oceanica
   G. pamela
   G. papuana
   G. pilbara
   G. punctata
   G. purpurascens
   G. robusta
   G. variegata
   G. vorax
   G. xenopus
Gekko [Genus]
   G. athymus [Species]
   G. auriverrucosus
   G. badenii
   G. chinensis
   G. gecko
   G. gigante
   G. grossmanni
   G. hokouensis

G. japonicus
G. kikuchii
G. mindorensis
G. monarchus
G. palawanensis
G. palmatus
G. petricolus
G. porosus
G. romblon
G. scabridus
G. siamensis
G. similignum
G. smithii
G. subpalmatus
G. swinhonis
G. taibaiensis
G. tawaensis
G. taylori
G. ulikovskii
G. verreauxi
G. vittatus
G. yakuensis
Goggia [Genus]
   G. braacki [Species]
   G. essexi
   G. gemmula
   G. hewitti
   G. hexapora
   G. lineata
   G. microlepidota
   G. rupicola
Gonatodes [Genus]
   G. albogularis [Species]
   G. annularis
   G. antillensis
   G. atricucullaris
   G. caudiscutatus
   G. ceciliae
   G. concinnatus
   G. eladioi
   G. falconensis
   G. hasemani
   G. humeralis
   G. ocellatus
   G. petersi
   G. seigliei
   G. taniae
   G. tapajonicus
   G. vittatus
Goniurosaurus [Genus]
   G. araneus [Species]
   G. bawanglingensis
   G. kuroiwae
   G. lichtenfelderi
   G. luii
Gonydactylus [Genus]
   G. markuscombaii [Species]
   G. martinstolli
   G. nepalensis
   G. paradoxus

Gymnodactylus [Genus]
   G. darwinii [Species]
   G. geckoides
   G. guttulatus
Haemodracon [Genus]
   H. riebeckii [Species]
   H. trachyrhinus
Hemidactylus [Genus]
   H. agrius [Species]
   H. albopunctatus
   H. aporus
   H. arnoldi
   H. barodanus
   H. bavazzanoi
   H. bayonii
   H. bouvieri
   H. bowringii
   H. brookii
   H. citernii
   H. curlei
   H. depressus
   H. dracaenacolus
   H. echinus
   H. fasciatus
   H. flaviviridis
   H. forbesii
   H. frenatus
   H. funaiolii
   H. garnotii
   H. giganteus
   H. gracilis
   H. granchii
   H. granti
   H. greefii
   H. haitianus
   H. homoeolepis
   H. intestinalis
   H. isolepis
   H. jubensis
   H. kamdemtohami
   H. karenorum
   H. klauberi
   H. laevis
   H. laticaudatus
   H. lemurinus
   H. leschenaultii
   H. longicephalus
   H. mabouia
   H. macropholis
   H. maculatus
   H. mahendrai
   H. marmoratus
   H. matschiei
   H. megalops
   H. mercatorius
   H. modestus
   H. muriceus
   H. newtoni
   H. ophiolepis
   H. ophiolepoides

H. oxyrhinus
H. palaichthus
H. persicus
H. platycephalus
H. porbandarensis
H. prashadi
H. puccionii
H. pumilio
H. reticulatus
H. richardsonii
H. ruspolii
H. scabriceps
H. sinaitus
H. smithi
H. somalicus
H. squamulatus
H. stejnegeri
H. subtriedrus
H. tanganicus
H. tasmani
H. taylori
H. triedrus
H. tropidolepis
H. turcicus
H. vietnamensis
H. yerburyi
Hemiphyllodactylus [Genus]
H. aurantiacus [Species]
H. larutensis
H. typus
H. yunnanensis
Hemitheconyx [Genus]
H. caudicinctus [Species]
H. taylori
Heteronotia [Genus]
H. binoei [Species]
H. planiceps
H. spelea
Holodactylus [Genus]
H. africanus [Species]
H. cornii
Homonota [Genus]
H. andicola [Species]
H. borellii
H. darwinii
H. fasciata
H. gaudichaudii
H. horrida
H. penai
H. underwoodi
H. uruguayensis
H. whitii
Homopholis [Genus]
H. fasciata [Species]
H. mulleri
H. walbergii
Hoplodactylus [Genus]
H. chrysosireticus [Species]
H. delcourti
H. duvaucelii

H. granulatus
H. kahutarae
H. maculatus
H. nebulosus
H. pacificus
H. rakiurae
H. stephensi
Lepidoblepharis [Genus]
L. buchwaldi [Species]
L. colombianus
L. duolepis
L. festae
L. grandis
L. heyerorum
L. hoogmoedi
L. intermedius
L. microlepis
L. miyatai
L. montecanoensis
L. oxycephalus
L. peraccae
L. ruthveni
L. sanctaemartae
L. williamsi
L. xanthostigma
Lepidodactylus [Genus]
L. aureolineatus [Species]
L. balioburius
L. browni
L. christiani
L. euaensis
L. flaviocularis
L. gardineri
L. guppyi
L. herrei
L. intermedius
L. listeri
L. lombocensis
L. lugubris
L. magnus
L. manni
L. moestus
L. mutahi
L. novaeguineae
L. oortii
L. orientalis
L. paurolepis
L. planicaudus
L. pulcher
L. pumilus
L. pusillus
L. ranauensis
L. shebae
L. vanuatuensis
L. woodfordi
L. yami
Lucasium [Genus]
L. damaeum [Species]
Luperosaurus [Genus]
L. brooksii [Species]

L. browni
L. cumingii
L. iskandari
L. joloensis
L. macgregori
L. palawanensis
L. yasumai
Lygodactylus [Genus]
L. angolensis [Species]
L. angularis
L. arnoulti
L. bernardi
L. blancae
L. blanci
L. bradfieldi
L. broadleyi
L. capensis
L. chobiensis
L. conradti
L. conraui
L. decaryi
L. depressus
L. expectatus
L. fischeri
L. grandisonae
L. graniticolus
L. gravis
L. guibei
L. gutturalis
L. heterurus
L. howelli
L. inexpectatus
L. insularis
L. intermedius
L. keniensis
L. kimhowelli
L. klemmeri
L. klugei
L. lawrencei
L. luteopicturatus
L. madagascariensis
L. manni
L. methueni
L. miops
L. mirabilis
L. montanus
L. nigropunctatus
L. ocellatus
L. ornatus
L. pauliani
L. picturatus
L. pictus
L. praecox
L. rarus
L. rex
L. scheffleri
L. scorteccii
L. septemtuberculatus
L. somalicus
L. stevensoni

L. thomensis
L. tolampyae
L. tuberosus
L. verticillatus
L. waterbergensis
L. wetzeli
L. williamsi
Matoatoa [Genus]
  M. brevipes [Species]
  M. spannringi
Microscalabotes [Genus]
  M. bivittis [Species]
Nactus [Genus]
  N. cheverti [Species]
  N. coindemirensis
  N. eboracensis
  N. galgajuga
  N. multicarinatus
  N. pelagicus
  N. serpensinsula
  N. vankampeni
Narudasia [Genus]
  N. festiva [Species]
Naultinus [Genus]
  N. elegans [Species]
  N. gemmeus
  N. grayii
  N. manukanus
  N. poecilochlorus
  N. rudis
  N. stellatus
  N. tuberculatus
Nephrurus [Genus]
  N. amyae [Species]
  N. asper
  N. deleani
  N. laevissimus
  N. levis
  N. sheai
  N. stellatus
  N. vertebralis
  N. wheeleri
Oedura [Genus]
  O. castelnaui [Species]
  O. coggeri
  O. filicipoda
  O. gemmata
  O. gracilis
  O. lesueurii
  O. marmorata
  O. monilis
  O. obscura
  O. reticulata
  O. rhombifer
  O. robusta
  O. tryoni
Pachydactylus [Genus]
  P. amoenus [Species]
  P. austeni
  P. barnardi

P. bibronii
P. bicolor
P. capensis
P. caraculicus
P. fasciatus
P. formosus
P. gaiasensis
P. geitje
P. haackei
P. kladaroderma
P. kobosensis
P. kochii
P. labialis
P. laevigatus
P. maculatus
P. mariquensis
P. monticolus
P. namaquensis
P. oculatus
P. oreophilus
P. oshaughnessyi
P. punctatus
P. rugosus
P. sansteyni
P. scherzi
P. scutatus
P. serval
P. tetensis
P. tigrinus
P. tsodiloensis
P. tuberculosus
P. turneri
P. vansoni
P. weberi
Palmatogecko [Genus]
  P. rangei [Species]
  P. vanzyli
Paragehyra [Genus]
  P. gabriellae [Species]
  P. petiti
Paroedura [Genus]
  P. androyensis [Species]
  P. bastardi
  P. gracilis
  P. homalorhinus
  P. karstophila
  P. lohatsara
  P. maingoka
  P. masobe
  P. oviceps
  P. picta
  P. sanctijohannis
  P. stumpffi
  P. tanjaka
  P. vahiny
  P. vazimba
Perochirus [Genus]
  P. ateles [Species]
  P. guentheri
  P. scutellatus

Phelsuma [Genus]
  P. abbotti [Species]
  P. andamanense
  P. antanosy
  P. astriata
  P. barbouri
  P. berghofi
  P. borbonica
  P. breviceps
  P. cepediana
  P. comorensis
  P. dubia
  P. edwardnewtoni
  P. flavigularis
  P. gigas
  P. guentheri
  P. guimbeaui
  P. guttata
  P. hielscheri
  P. inexpectata
  P. klemmeri
  P. laticauda
  P. lineata
  P. madagascariensis
  P. malamakibo
  P. masohoala
  P. modesta
  P. mutabilis
  P. nigristriata
  P. ocellata
  P. ornata
  P. parkeri
  P. pronki
  P. pusilla
  P. quadriocellata
  P. robertmertensi
  P. seippi
  P. serraticauda
  P. standingi
  P. sundbergi
  P. v-nigra
Phyllodactylus [Genus]
  P. angelensis [Species]
  P. angustidigitus
  P. apricus
  P. barringtonensis
  P. baurii
  P. bordai
  P. bugastrolepis
  P. clinatus
  P. darwini
  P. davisi
  P. delcampoi
  P. dixoni
  P. duellmani
  P. galapagensis
  P. gerrhopygus
  P. gilberti
  P. heterurus
  P. homolepidurus

P. inaequalis
P. insularis
P. interandinus
P. johnwrighti
P. julieni
P. kofordi
P. lanei
P. leei
P. lepidopygus
P. martini
P. microphyllus
P. muralis
P. nocticolus
P. palmeus
P. partidus
P. paucituberculatus
P. pulcher
P. pumilius
P. reissii
P. rutteni
P. santacruzensis
P. sentosus
P. tinklei
P. transversalis
P. tuberculosus
P. unctus
P. ventralis
P. wirshingi
P. xanti
Phyllopezus [Genus]
P. periosus [Species]
P. pollicaris
Phyllurus [Genus]
P. amnicola [Species]
P. caudiannulatus
P. championae
P. isis
P. nepthys
P. ossa
P. platurus
Pristurus [Genus]
P. abdelkuri [Species]
P. carteri
P. celerrimus
P. collaris
P. crucifer
P. flavipunctatus
P. gasperetti
P. guichardi
P. insignis
P. insignoides
P. minimus
P. obsti
P. ornithocephalus
P. phillipsii
P. popovi
P. rupestris
P. saada
P. samhaensis
P. simonettai

P. sokotranus
P. somalicus
Pseudogekko [Genus]
P. brevipes [Species]
P. compressicorpus
P. labialis
P. smaragdinus
Pseudogonatodes [Genus]
P. barbouri [Species]
P. furvus
P. gasconi
P. guinanensis
P. lunulatus
P. manessi
P. peruvianus
Pseudothecadactylus [Genus]
P. australis [Species]
P. cavaticus
P. lindneri
Ptenopus [Genus]
P. carpi [Species]
P. garrulus
P. kochi
Ptychozoon [Genus]
P. horsfieldii [Species]
P. intermedium
P. kuhli
P. lionotum
P. rhacophorus
P. trinotaterra
Ptyodactylus [Genus]
P. guttatus [Species]
P. hasselquistii
P. homolepis
P. oudrii
P. puiseuxi
P. ragazzii
Quedenfeldtia [Genus]
Q. moerens [Species]
Q. trachyblepharus
Rhacodactylus [Genus]
R. auriculatus [Species]
R. chahoua
R. ciliatus
R. leachianus
R. sarasinorum
R. trachyrhynchus
Rhoptropus [Genus]
R. afer [Species]
R. barnardi
R. biporosus
R. boultoni
R. braconnieri
R. bradfieldi
R. taeniostictus
Rhynchoedura [Genus]
R. ornata [Species]
Saltuarius [Genus]
S. cornutus [Species]
S. occultus

S. salebrosus
S. swaini
S. wyberba
Saurodactylus [Genus]
S. fasciatus [Species]
S. mauritanicus
Sphaerodactylus [Genus]
S. altavelensis [Species]
S. argivus
S. argus
S. ariasae
S. armasi
S. armstrongi
S. asterulus
S. beattyi
S. becki
S. bromeliarum
S. caicosensis
S. callocricus
S. celicara
S. cinereus
S. clenchi
S. cochranae
S. copei
S. corticola
S. cricoderus
S. cryphius
S. darlingtoni
S. difficilis
S. docimus
S. dunni
S. elasmorhynchus
S. elegans
S. elegantulus
S. epiurus
S. fantasticus
S. gaigeae
S. gilvitorques
S. glaucus
S. goniorhynchus
S. graptolaemus
S. heliconiae
S. homolepis
S. inaguae
S. intermedius
S. kirbyi
S. klauberi
S. ladae
S. lazelli
S. leucaster
S. levinsi
S. lineolatus
S. macrolepis
S. mariguanae
S. microlepis
S. micropithecus
S. millepunctatus
S. molei
S. monensis
S. nicholsi

S. nigropunctatus
S. notatus
S. nycteropus
S. ocoae
S. oliveri
S. omoglaux
S. oxyrhinus
S. pacificus
S. parkeri
S. parthenopion
S. perissodactylius
S. pimienta
S. plummeri
S. ramsdeni
S. randi
S. rhabdotus
S. richardi
S. richardsonii
S. roosevelti
S. rosaurae
S. ruibali
S. sabanus
S. samanensis
S. savagei
S. scaber
S. scapularis
S. schuberti
S. schwartzi
S. semasiops
S. shrevei
S. sommeri
S. sputator
S. storeyae
S. streptophorus
S. thompsoni
S. torrei
S. townsendi
S. underwoodi
S. vincenti
S. williamsi
S. zygaena
Stenodactylus [Genus]
S. affinis [Species]
S. arabicus
S. doriae
S. grandiceps
S. khobarensis
S. leptocosymbotus
S. petrii
S. pulcher
S. slevini
S. sthenodactylus
S. yemenensis
Tarentola [Genus]
T. albertschwartzi [Species]
T. americana
T. angustimentalis
T. annularis
T. bischoffi
T. boehmei

T. boettgeri
T. caboverdianus
T. darwini
T. delalandii
T. deserti
T. ephippiata
T. gigas
T. gomerensis
T. mauritanica
T. mindiae
T. neglecta
T. parvicarinata
T. rudis
Teratolepis [Genus]
T. albofasciatus [Species]
T. fasciata
Teratoscincus [Genus]
T. bedriagai [Species]
T. microlepis
T. przewalskii
T. roborowskii
T. scincus
T. toksunicus
Thecadactylus [Genus]
T. rapicauda [Species]
Tropiocolotes [Genus]
T. helenae [Species]
T. latifi
T. nattereri
T. nubicus
T. persicus
T. scortecci
T. steudneri
T. tripolitanus
Underwoodisaurus [Genus]
U. milii [Species]
U. sphyrurus
Urocotyledon [Genus]
U. inexpectata [Species]
U. palmata
U. weileri
U. wolterstorffi
Uroplatus [Genus]
U. alluaudi [Species]
U. ebenaui
U. fimbriatus
U. guentheri
U. henkeli
U. lineatus
U. malahelo
U. malama
U. phantasticus
U. sikorae

Pygopodidae [Family]
Aprasia [Genus]
A. aurita [Species]
A. haroldi
A. inaurita
A. parapulchella
A. picturata

A. pseudopulchella
A. pulchella
A. repens
A. rostrata
A. smithi
A. striolata
Delma [Genus]
D. australis [Species]
D. borea
D. butleri
D. elegans
D. fraseri
D. grayii
D. impar
D. inornata
D. labialis
D. mitella
D. molleri
D. nasuta
D. pax
D. plebeia
D. tincta
D. torquata
Lialis [Genus]
L. burtonis [Species]
L. jicari
Ophidiocephalus [Genus]
O. taeniatus [Species]
Paradelma [Genus]
P. orientalis [Species]
Pletholax [Genus]
P. gracilis [Species]
Pygopus [Genus]
P. lepidopodus [Species]
P. nigriceps
P. steelescotti

Dibamidae [Family]
Anelytropsis [Genus]
A. papillosus [Species]
Dibamus [Genus]
D. alfredi [Species]
D. bogadeki
D. bourreti
D. celebensis
D. deharvengi
D. greeri
D. kondaoensis
D. leucurus
D. montanus
D. nicobaricum
D. novaeguineae
D. seramensis
D. smithi
D. somsaki
D. taylori

Amphisbaenidae [Family]
Amphisbaena [Genus]
A. alba [Species]
A. anaemariae

A. angustifrons
A. arenaria
A. bakeri
A. barbouri
A. blanoides
A. bolivica
A. caeca
A. camura
A. carlgansi
A. carvalhoi
A. caudalis
A. cegei
A. crisae
A. cubana
A. cunhai
A. darwinii
A. dubia
A. fenestrata
A. frontalis
A. fuliginosa
A. gonavensis
A. gracilis
A. hastata
A. heathi
A. hogei
A. hugoi
A. hyporissor
A. ignatiana
A. innocens
A. leeseri
A. leucocephala
A. lumbricalis
A. manni
A. medemi
A. mensae
A. mertensii
A. minuta
A. miringoera
A. mitchelli
A. munoai
A. myersi
A. neglecta
A. nigricauda
A. occidentalis
A. palirostrata
A. pericensis
A. plumbea
A. polygrammica
A. pretrei
A. prunicolor
A. ridleyi
A. rozei
A. sanctaeritae
A. schmidti
A. silvestrii
A. slateri
A. slevini
A. spurelli
A. stejnegeri
A. talisiae
A. tragorrhectes

A. vanzolinii
A. vermicularis
A. xera
Ancylocranium [Genus]
A. barkeri [Species]
A. ionidesi
A. somalicum
Anops [Genus]
A. bilabialatus [Species]
A. kingii
Aulura [Genus]
A. anomala [Species]
Baikia [Genus]
B. africana [Species]
Blanus [Genus]
B. cinereus [Species]
B. mettetali
B. strauchi
B. tingitanus
Bronia [Genus]
B. bedai [Species]
B. brasiliana
B. kraoh
Cercolophia [Genus]
C. bahiana [Species]
C. borelli
C. roberti
C. steindachneri
Chirindia [Genus]
C. ewerbecki [Species]
C. langi
C. mpwapwaensis
C. rondoensis
C. swynnertoni
Cynisca [Genus]
C. bifrontalis [Species]
C. degrysi
C. feae
C. gansi
C. haughi
C. kigomensis
C. kraussi
C. leonina
C. leucura
C. liberiensis
C. muelleri
C. nigeriensis
C. oligopholis
C. rouxae
C. schaeferi
C. senegalensis
C. williamsi
Dalophia [Genus]
D. angolensis [Species]
D. ellenbergeri
D. gigantea
D. longicauda
D. luluae
D. pistillum
D. welwitschii
Geocalamus [Genus]

G. acutus [Species]
G. modestus
Leposternon [Genus]
L. infraorbitale [Species]
L. microcephalum
L. octostegum
L. polystegum
L. scutigerum
L. wuchereri
Loveridgea [Genus]
L. ionidesii [Species]
L. phylofiniens
Mesobaena [Genus]
M. huebneri [Species]
Monopeltis [Genus]
M. adercae [Species]
M. anchietae
M. capensis
M. decosteri
M. galeata
M. guentheri
M. infuscata
M. jugularis
M. kabindae
M. leonhardi
M. luandae
M. perplexus
M. remaclei
M. rhodesiana
M. scalper
M. schoutedeni
M. sphenorhynchus
M. vanderysti
M. zambezensis
Zygaspis [Genus]
Z. ferox [Species]
Z. kafuensis
Z. niger
Z. quadrifrons
Z. vandami
Z. violacea

Bipedidae [Family]
Bipes [Genus]
B. alvarezi [Species]
B. biporus
B. canaliculatus
B. tridactylus

Rhineauridae [Family]
Rhineura [Genus]
R. floridana [Species]

Trogonophidae [Family]
Agamodon [Genus]
A. anguliceps [Species]
A. arabicus
A. compressus
Diplometopon [Genus]
D. zarudnyi [Species]
Pachycalamus [Genus]
P. brevis [Species]

Trogonophis [Genus]
    T. wiegmanni [Species]

Xantusiidae [Family]
  Cricosaura [Genus]
    C. typica [Species]
  Lepidophyma [Genus]
    L. alvarezi [Species]
    L. chicoasensis
    L. dontomasi
    L. flavimaculatum
    L. gaigeae
    L. lipetzi
    L. lowei
    L. mayae
    L. micropholis
    L. obscurum
    L. occulor
    L. pajapanensis
    L. radula
    L. reticulatum
    L. sawini
    L. smithii
    L. sylvaticum
    L. tarascae
    L. tuxtlae
  Xantusia [Genus]
    X. bezyi [Species]
    X. bolsonae
    X. henshawi
    X. riversiana
    X. sanchezi
    X. vigilis

Lacertidae [Family]
  Acanthodactylus [Genus]
    A. arabicus [Species]
    A. aureus
    A. bedriagai
    A. beershebensis
    A. blanci
    A. blanfordii
    A. boskianus
    A. boueti
    A. busacki
    A. cantoris
    A. dumerilii
    A. erythrurus
    A. felicis
    A. gongrorhynchatus
    A. grandis
    A. guineensis
    A. haasi
    A. harranensis
    A. lineomaculatus
    A. longipes
    A. maculatus
    A. masirae
    A. micropholis
    A. nilsoni
    A. opheodurus

    A. orientalis
    A. pardalis
    A. robustus
    A. savignyi
    A. schmidti
    A. schreiberi
    A. scutellatus
    A. spinicauda
    A. taghitensis
    A. tilburyi
    A. tristrami
    A. yemenicus
  Adolfus [Genus]
    A. africanus [Species]
    A. alleni
    A. jacksoni
    A. vaureselli
  Algyroides [Genus]
    A. fitzingeri [Species]
    A. marchi
    A. moreoticus
    A. nigropunctatus
  Australolacerta [Genus]
    A. australis [Species]
    A. rupicola
  Darevskia [Genus]
    D. alpina [Species]
    D. armeniaca
    D. bendimahiensis
    D. brauneri
    D. caucasica
    D. clarkorum
    D. daghestanica
    D. dahli
    D. derjugini
    D. lindholmi
    D. mixta
    D. parvula
    D. portschinskii
    D. praticola
    D. raddei
    D. rostombekovi
    D. rudis
    D. sapphirina
    D. saxicola
    D. unisexualis
    D. uzzelli
    D. valentini
  Eremias [Genus]
    E. acutirostris [Species]
    E. afghanistanica
    E. andersoni
    E. argus
    E. arguta
    E. aria
    E. brenchleyi
    E. buechneri
    E. fasciata
    E. grammica
    E. intermedia

    E. lalezharica
    E. lineolata
    E. montanus
    E. multiocellata
    E. nigrocellata
    E. nigrolateralis
    E. nikolskii
    E. persica
    E. pleskei
    E. przewalskii
    E. quadrifrons
    E. regeli
    E. scripta
    E. strauchi
    E. suphani
    E. velox
    E. vermiculata
    E. cincus fasciolatus
    E. cincus richardsonii
  Gallotia [Genus]
    G. atlantica [Species]
    G. caesaris
    G. galloti
    G. gomerana
    G. intermedia
    G. simonyi
    G. stehlini
  Gastropholis [Genus]
    G. echinata [Species]
    G. prasina
    G. tropidopholis
    G. vittata
  Holaspis [Genus]
    H. guentheri [Species]
    H. laevis
  Heliobolus [Genus]
    H. lugubris [Species]
    H. neumanni
    H. nitida
    H. spekii
  Ichnotropis [Genus]
    I. bivittata [Species]
    I. capensis
    I. chapini
    I. grandiceps
    I. microlepidota
    I. squamulosa
    I. tanganicana
  Lacerta [Genus]
    L. agilis [Species]
    L. anatolica
    L. andreanskyi
    L. aranica
    L. aurelioi
    L. bedriagae
    L. bilineata
    L. bonnali
    L. brandtii
    L. cappadocica
    L. chlorogaster

L. cyanisparsa
L. cyanura
L. danfordi
L. defilippii
L. dryada
L. fraasii
L. graeca
L. herseyi
L. horvathi
L. jayakari
L. kulzeri
L. laevis
L. media
L. monticola
L. mosorensis
L. mostoufii
L. oertzeni
L. oxycephala
L. pamphylica
L. parva
L. schreiberi
L. steineri
L. strigata
L. trilineata
L. viridis
L. vivipara
L. zagrosica
Latastia [Genus]
L. boscai [Species]
L. carinata
L. cherchii
L. doriai
L. johnstonii
L. lanzai
L. longicaudata
L. ornata
L. siebenrocki
L. taylori
Meroles [Genus]
M. anchietae [Species]
M. ctenodactylus
M. cuneirostris
M. knoxii
M. micropholidotus
M. reticulatus
M. suborbitalis
Mesalina [Genus]
M. adramitana [Species]
M. ayunensis
M. balfouri
M. brevirostris
M. ercolinii
M. guttulata
M. martini
M. olivieri
M. pasteuri
M. rubropunctata
M. simoni
M. watsonana
Nucras [Genus]
N. boulengeri [Species]

N. caesicaudata
N. intertexta
N. lalandii
N. livida
N. scalaris
N. taeniolata
N. tessellata
Ophisops [Genus]
O. beddomei [Species]
O. elbaensis
O. elegans
O. jerdonii
O. leschenaultii
O. microlepis
O. minor
O. occidentalis
Pedioplanis [Genus]
P. benguelensis [Species]
P. breviceps
P. burchelli
P. gaerdesi
P. husabensis
P. laticeps
P. lineoocellata
P. namaquensis
P. rubens
P. undata
Philochortus [Genus]
P. hardeggeri [Species]
P. intermedius
P. lhotei
P. neumanni
P. phillipsi
P. spinalis
P. zolii
Podarcis [Genus]
P. bocagei [Species]
P. dugesii
P. erhardii
P. filfolensis
P. gaigeae
P. hispanica
P. lilfordi
P. melisellensis
P. milensis
P. muralis
P. peloponnesiaca
P. perspicillata
P. pityusensis
P. raffonei
P. sicula
P. taurica
P. tiliguerta
P. wagleriana
Poromera [Genus]
P. fordii [Species]
Psammodromus [Genus]
P. algirus [Species]
P. blanci
P. hispanicus
P. microdactylus

Pseuderemias [Genus]
P. brenneri [Species]
P. erythrosticta
P. mucronata
P. savagei
P. septemstriata
P. smithii
P. striatus
Takydromus [Genus]
T. amurensis [Species]
T. dorsalis
T. formosanus
T. hani
T. haughtonianus
T. hsuehshanensis
T. intermedius
T. khasiensis
T. kuehnei
T. sauteri
T. septentrionalis
T. sexlineatus
T. smaragdinus
T. stejnegeri
T. sylvaticus
T. tachydromoides
T. toyamai
T. wolteri
Timon [Genus]
T. lepidus [Species]
T. pater
T. princeps
Tropidosaura [Genus]
T. cottrelli [Species]
T. essexi
T. gularis
T. montana

Gymnophthalmidae [Family]
Alopoglossus [Genus]
A. andeanus [Species]
A. angulatus
A. atriventris
A. buckleyi
A. copii
A. festae
A. lehmanni
Amapasaurus [Genus]
A. tetradactylus [Species]
Anadia [Genus]
A. altaserrania [Species]
A. bitaeniata
A. blakei
A. bogotensis
A. brevifrontalis
A. hobarti
A. marmorata
A. metallica
A. ocellata
A. pamplonensis
A. petersi
A. pulchella

A. rhombifera
A. steyeri
A. vittata
Anotosaura [Genus]
A. brachylepis [Species]
A. collaris
Arthrosaura [Genus]
A. guianensis [Species]
A. kockii
A. reticulata
A. synaptolepis
A. testigensis
A. tyleri
A. versteegii
Arthroseps [Genus]
A. fluminensis [Species]
A. werneri
Bachia [Genus]
B. barbouri [Species]
B. bicolor
B. bresslaui
B. cacerensis
B. cuvieri
B. dorbignyi
B. flavescens
B. guianensis
B. heteropa
B. huallagana
B. intermedia
B. pallidiceps
B. panoplia
B. peruana
B. pyburni
B. scolecoides
B. talpa
B. trisanale
Calyptommatus [Genus]
C. leiolepis [Species]
C. nicterus
C. sinebrachiatus
Cercosaura [Genus]
C. ocellata [Species]
Colobodactylus [Genus]
C. dalcyanus [Species]
C. taunayi
Colobosaura [Genus]
C. kraepelini [Species]
C. mentalis
C. modesta
Colobosauroides [Genus]
C. carvalhoi [Species]
C. cearensis
Echinosaura [Genus]
E. horrida [Species]
E. orcesi
Ecpleopus [Genus]
E. Ecpleopus [Species]
Euspondylus [Genus]
E. acutirostris [Species]
E. guentheri
E. maculatus

E. monsfumus
E. phelpsorum
E. rahmi
E. simonsii
E. spinalis
E. stenolepis
Gymnophthalmus [Genus]
G. cryptus [Species]
G. leucomystax
G. lineatus
G. pleii
G. speciosus
G. underwoodi
G. vanzoi
Heterodactylus [Genus]
H. imbricatus [Species]
H. lundii
Iphisa [Genus]
I. elegans [Species]
Leposoma [Genus]
L. baturitensis [Species]
L. guianense
L. hexalepis
L. ioanna
L. nanodactylus
L. osvaldoi
L. parietale
L. percarinatum
L. rugiceps
L. scincoides
L. snethlageae
L. southi
Macropholidus [Genus]
M. ataktolepis [Species]
M. ruthveni
Micrablepharus [Genus]
M. atticolus [Species]
M. maximiliani
Neusticurus [Genus]
N. apodemus [Species]
N. bicarinatus
N. cochranae
N. ecpleopus
N. juruazensis
N. medemi
N. ocellatus
N. racenisi
N. rudis
N. strangulatus
N. tatei
Nothobachia [Genus]
N. ablephara [Species]
Opipeuter [Genus]
O. xestus [Species]
Pantodactylus [Genus]
P. quadrilineatus [Species]
P. schreibersii
Pholidobolus [Genus]
P. affinis [Species]
P. annectens
P. anomalus

P. huancabambae
P. macbrydei
P. montium
P. prefrontalis
Placosoma [Genus]
P. cipoense [Species]
P. cordylinum
P. glabella
Prionodactylus [Genus]
P. ampuedae [Species]
P. argulus
P. dicrus
P. eigenmanni
P. goeleti
P. manicatus
P. nigroventris
P. oshaughnessyi
P. vertebralis
Procellosaurinus [Genus]
P. erythrocercus [Species]
P. tetradactylus
Proctoporus [Genus]
P. achlyens [Species]
P. anatoloros
P. balneator
P. bogotensis
P. bolivianus
P. cashcaensis
P. colomaromani
P. columbianus
P. guentheri
P. hyposticus
P. labionis
P. laevis
P. luctuosus
P. meleagris
P. oculatus
P. orcesi
P. pachyurus
P. petrorum
P. raneyi
P. serranus
P. shrevei
P. simoterus
P. stigmatoral
P. striatus
P. unicolor
P. ventrimaculatus
P. vespertinus
P. vietus
Psilophthalmus [Genus]
P. paeminosus [Species]
Ptychoglossus [Genus]
P. bicolor [Species]
P. bilineatus
P. brevifrontalis
P. danieli
P. eurylepis
P. festae
P. gorgonae
P. grandisquamatus

P. kugleri
P. myersi
P. nicefori
P. plicatus
P. romaleos
P. stenolepis
P. vallensis
Riolama [Genus]
R. leucostictus [Species]
Stenolepis [Genus]
R. ridleyi [Species]
Teuchocercus [Genus]
T. keyi [Species]
Tretioscincus [Genus]
T. agilis [Species]
T. bifasciatus
T. oriximinensis
Vanzosaura [Genus]
V. rubricauda [Species]

Teiidae [Family]
Ameiva [Genus]
A. ameiva [Species]
A. anomala
A. auberi
A. bifrontata
A. bridgesii
A. chaitzami
A. chrysolaema
A. cineracea
A. corax
A. corvina
A. dorsalis
A. edracantha
A. erythrocephala
A. exsul
A. festiva
A. fuscata
A. griswoldi
A. leberi
A. leptophrys
A. lineolata
A. major
A. maynardi
A. niceforoi
A. orcesi
A. plei
A. pluvianotata
A. polops
A. quadrilineata
A. septemlineata
A. taeniura
A. undulata
A. vittata
A. wetmorei
Callopistes [Genus]
C. flavipunctatus [Species]
C. maculatus
Cnemidophorus [Genus]
C. angusticeps [Species]
C. arenivagus

C. arizonae
C. arubensis
C. burti
C. calidipes
C. ceralbensis
C. communis
C. costatus
C. cozumelae
C. cryptus
C. deppei
C. dixoni
C. exsanguis
C. flagellicaudus
C. gramivagus
C. gularis
C. guttatus
C. gypsi
C. hyperythrus
C. inornatus
C. labialis
C. lacertoides
C. laredoensis
C. leachei
C. lemniscatus
C. lineattissimus
C. littoralis
C. longicaudus
C. marmoratus
C. martyris
C. maximus
C. mexicanus
C. motaguae
C. murinus
C. nativo
C. neomexicanus
C. neotesselatus
C. nigricolor
C. ocellifer
C. opatae
C. pai
C. parvisocius
C. pseudolemniscatus
C. rodecki
C. sackii
C. scalaris
C. septemvittatus
C. serranus
C. sexlineatus
C. sonorae
C. tesselatus
C. tigris
C. uniparens
C. vacariensis
C. vanzoi
C. velox
Crocodilurus [Genus]
C. lacertinus [Species]
Dicrodon [Genus]
D. guttulatum [Species]
D. heterolepis
D. holmbergi

Dracaena [Genus]
D. guianensis [Species]
D. paraguayensis
Kentropyx [Genus]
K. altamazonica [Species]
K. borckiana
K. calcarata
K. intermedius
K. paulensis
K. pelviceps
K. striatus
K. vanzoi
K. viridistriga
Teius [Genus]
T. oculatus [Species]
T. suquiensis
T. teyou
Tupinambis [Genus]
T. duseni [Species]
T. longilineus
T. merianae
T. quadrilineatus
T. rufescens
T. teguixin

Cordylidae [Family]
Angolosaurus [Genus]
A. skoogi [Species]
Chamaesaura [Genus]
C. aenea [Species]
C. anguina
C. macrolepis
Cordylosaurus [Genus]
C. subtessellatus [Species]
Cordylus [Genus]
C. angolensis [Species]
C. aridus
C. beraduccii
C. campbelli
C. cataphractus
C. cloetei
C. coeruleopunctatus
C. cordylus
C. giganteus
C. imkeae
C. jonesii
C. jordani
C. lawrenci
C. macropholis
C. mclachlani
C. minor
C. namaquensis
C. niger
C. nyikae
C. oelofseni
C. peersi
C. polyzonus
C. pustulatus
C. rhodesianus
C. rivae

C. tasmani
C. tropidosternum
C. ukingensis
C. vittifer
C. warreni
Gerrhosaurus [Genus]
G. flavigularis [Species]
G. major
G. multilineatus
G. nigrolineatus
G. typicus
G. validus
Platysaurus [Genus]
P. broadleyi [Species]
P. capensis
P. guttatus
P. imperator
P. intermedius
P. lebomboensis
P. maculatus
P. minor
P. mitchelli
P. monotropis
P. ocellatus
P. orientalis
P. pungweensis
P. relictus
P. torquatus
Pseudocordylus [Genus]
P. capensis [Species]
P. langi
P. melanotus
P. microlepidotus
P. nebulosus
P. spinosus
Tracheloptychus [Genus]
T. madagascariensis [Species]
T. petersi
Tetradactylus [Genus]
T. africanus [Species]
T. breyeri
T. eastwoodae
T. ellenbergeri
T. seps
T. tetradactylus
Zonosaurus [Genus]
Z. aeneus [Species]
Z. anelanelany
Z. bemaraha
Z. boettgeri
Z. brygooi
Z. haraldmeieri
Z. karsteni
Z. laticaudatus
Z. madagascariensis
Z. maximus
Z. ornatus
Z. quadrilineatus
Z. rufipes
Z. subunicolor

Z. trilineatus
Z. tsingy

Scincidae [Family]
Ablepharus [Genus]
A. bivittatus [Species]
A. chernovi
A. darvazi
A. deserti
A. grayanus
A. kitaibelii
A. pannonicus
Acontias [Genus]
A. breviceps [Species]
A. gracilicauda
A. lineatus
A. litoralis
A. meleagris
A. percivali
A. plumbeus
A. poecilus
Acontophiops [Genus]
A. lineatus [Species]
Afroblepharus [Genus]
A. duruarus [Species]
A. seydeli
A. tancredi
A. wilsoni
Amphiglossus [Genus]
A. alluaudi [Species]
A. andranovahensis
A. ankodabensis
A. anosyensis
A. ardouini
A. astrolabi
A. crenni
A. decaryi
A. elongatus
A. frontoparietalis
A. gastrostictus
A. igneocaudatus
A. intermedius
A. johannae
A. macrocercus
A. macrolepis
A. mandady
A. mandokava
A. melanopleura
A. melanurus
A. minutus
A. mouroundavae
A. nanus
A. ornaticeps
A. poecilopus
A. polleni
A. praeornatus
A. punctatus
A. reticulatus
A. spilostichus
A. splendidus
A. stumpffi

A. stylus
A. tanysoma
A. tsaratananensis
A. valhallae
A. waterloti
Androngo [Genus]
A. trivittatus [Species]
Anomalopus [Genus]
A. brevicollis [Species]
A. gowi
A. leuckartii
A. mackayi
A. pluto
A. swansoni
A. verreauxi
Apterygodon [Genus]
A. vittatum [Species]
Asymblepharus [Genus]
A. alaicus [Species]
A. mahabharatus
A. nepalensis
A. tragbulense
Ateuchosaurus [Genus]
A. chinensis [Species]
A. pellopleurus
Barkudia [Genus]
B. insularis [Species]
B. melanosticta
Bartleia [Genus]
B. jigurru [Species]
Bassiana [Genus]
B. duperreyi [Species]
B. platynota
B. trilineata
Brachymeles [Genus]
B. apus [Species]
B. bicolor
B. bonitae
B. boulengeri
B. cebuensis
B. elerae
B. gracilis
B. hilong
B. minimus
B. pathfinderi
B. samarensis
B. schadenbergi
B. talinis
B. tridactylus
B. vermis
B. wrighti
Caledoniscincus [Genus]
C. aquilonius [Species]
C. atropunctatus
C. auratus
C. austrocaledonicus
C. chazeaui
C. cryptos
C. festivus
C. haplorhinus
C. orestes

C. renevieri
C. terma
Calyptotis [Genus]
    C. lepidorostrum [Species]
    C. ruficauda
    C. scutirostrum
    C. temporalis
    C. thorntonensis
Carlia [Genus]
    C. amax [Species]
    C. bicarinata
    C. coensis
    C. dogare
    C. fusca
    C. gracilis
    C. jarnoldae
    C. johnstonei
    C. longipes
    C. munda
    C. mundivensis
    C. parrhasius
    C. pectoralis
    C. prava
    C. rhomboidalis
    C. rimula
    C. rostralis
    C. rubrigularis
    C. rufilatus
    C. schlegelii
    C. schmeltzii
    C. scirtetis
    C. storri
    C. tetradactyla
    C. triacantha
    C. vivax
Cautula [Genus]
    C. zia [Species]
Chabanaudia [Genus]
    C. boulengeri [Species]
Chalcides [Genus]
    C. armitagei [Species]
    C. bedriagai
    C. chalcides
    C. colosii
    C. ebneri
    C. guentheri
    C. lanzai
    C. levitoni
    C. manueli
    C. mauritanicus
    C. minutus
    C. mionecton
    C. montanus
    C. ocellatus
    C. parallelus
    C. pentadactylus
    C. pistaciae
    C. polylepis
    C. pseudostriatus
    C. pulchellus

C. ragazzii
C. sexlineatus
C. striatus
C. thierryi
C. viridanus
Chalcidoseps [Genus]
    C. thwaitesi [Species]
Coeranoscincus [Genus]
    C. frontalis [Species]
    C. reticulatus
Cophoscincopus [Genus]
    C. durus [Species]
    C. greeri
    C. simulans
Corucia [Genus]
    C. zebrata [Species]
Cryptoblepharus [Genus]
    C. africanus [Species]
    C. aldabrae
    C. ater
    C. balinensis
    C. bitaeniatus
    C. boutonii
    C. burdeni
    C. carnabyi
    C. caudatus
    C. cursor
    C. degrijsi
    C. egeriae
    C. eximius
    C. fuhni
    C. gloriosus
    C. keiensis
    C. leschenault
    C. litoralis
    C. megastictus
    C. mohelicus
    C. novaeguineae
    C. novocaledonicus
    C. pallidus
    C. plagiocephalus
    C. poecilopleurus
    C. renschi
    C. rutilus
    C. sumbawanus
    C. virgatus
Cryptoscincus [Genus]
    C. minimus [Species]
Ctenotus [Genus]
    C. agrestis [Species]
    C. alacer
    C. alleni
    C. allotropis
    C. angusticeps
    C. aphrodite
    C. arcanus
    C. ariadnae
    C. arnhemensis
    C. astarte
    C. astictus

C. atlas
C. australis
C. borealis
C. brachyonyx
C. brooksi
C. burbidgei
C. calurus
C. capricorni
C. catenifer
C. coggeri
C. colletti
C. decaneurus
C. delli
C. dux
C. ehmanni
C. essingtonii
C. eurydice
C. eutaenius
C. fallens
C. gagudju
C. gemmula
C. grandis
C. greeri
C. hanloni
C. hebetior
C. helenae
C. hilli
C. iapetus
C. impar
C. ingrami
C. inornatus
C. joanae
C. kurnbudj
C. labillardieri
C. lancelini
C. lateralis
C. leae
C. leonhardii
C. maryani
C. mastigura
C. militaris
C. mimetes
C. monticola
C. nasutus
C. nigrilineatus
C. nullum
C. olympicus
C. pallescens
C. pantherinus
C. piankai
C. pulchellus
C. quattuordecimlineatus
C. quinkan
C. rawlinsoni
C. regius
C. rimacolus
C. robustus
C. rubicundus
C. rufescens
C. rutilans

C. saxatilis
C. schevilli
C. schomburgkii
C. septenarius
C. serotinus
C. serventyi
C. severus
C. spaldingi
C. spec.
C. storri
C. strauchii
C. striaticeps
C. stuarti
C. taeniolatus
C. tanamiensis
C. tantillus
C. terrareginae
C. uber
C. vertebralis
C. xenopleura
C. youngsoni
C. zastictus
C. zebrilla
Cyclodina [Genus]
C. aenea [Species]
C. alani
C. lichenigera
C. macgregori
C. oliveri
C. ornata
C. whitakeri
Cyclodomorphus [Genus]
C. branchialis [Species]
C. casuarinae
C. celatus
C. maxima
C. melanops
C. michaeli
C. praealtus
C. venustus
Dasia [Genus]
D. griffini [Species]
D. grisea
D. haliana
D. nicobarensis
D. olivacea
D. semicincta
D. subcaerulea
Davewakeum [Genus]
D. miriamae [Species]
Egernia [Genus]
E. arnhemensis [Species]
E. carinata
E. coventryi
E. cunninghami
E. depressa
E. douglasi
E. formosa
E. frerei
E. hosmeri

E. inornata
E. kingii
E. kintorei
E. luctuosa
E. major
E. margaretae
E. mcpheei
E. modesta
E. multiscutata
E. napoleonis
E. pilbarensis
E. pulchra
E. richardi
E. rugosa
E. saxatilis
E. slateri
E. stokesii
E. striata
E. striolata
E. whitii
Emoia [Genus]
E. adspersa [Species]
E. aenea
E. ahli
E. aneityumensis
E. arnoensis
E. atrocostata
E. aurulenta
E. battersbyi
E. bismarckensis
E. boettgeri
E. bogerti
E. brongersmai
E. caeruleocauda
E. callisticta
E. campbelli
E. coggeri
E. concolor
E. cyanogaster
E. cyanura
E. cyclops
E. digul
E. erronan
E. flavigularis
E. guttata
E. impar
E. irianensis
E. isolata
E. jakati
E. jamur
E. kitcheneri
E. klossi
E. kordoana
E. kuekenthali
E. laobaoense
E. lawesi
E. longicauda
E. loveridgei
E. loyaltiensis
E. maculata

E. maxima
E. mivarti
E. mokosariniveikau
E. montana
E. nativittatis
E. nigra
E. nigromarginata
E. obscura
E. oribata
E. pallidiceps
E. paniai
E. parkeri
E. physicae
E. physicina
E. ponapea
E. popei
E. pseudocyanura
E. pseudopallidiceps
E. reimschisseli
E. rennellensis
E. ruficauda
E. rufilabialis
E. samoensis
E. sanfordi
E. schmidti
E. similis
E. slevini
E. sorex
E. submetallica
E. taumakoensis
E. tetrataenia
E. tongana
E. tropidolepis
E. trossula
E. veracunda
Eremiascincus [Genus]
E. cincus fasciolatus [Species]
E. cincus richardsonii
Eroticoscincus [Genus]
E. graciloides [Species]
Eugongylus [Genus]
E. albofasciolatus [Species]
E. microlepis
E. rufescens
E. sulaensis
E. unilineatus
Eulamprus [Genus]
E. amplus [Species]
E. brachyosoma
E. frerei
E. heatwolei
E. kosciuskoi
E. leuraensis
E. luteilateralis
E. martini
E. murrayi
E. quoyii
E. sokosoma
E. tenuis
E. tigrinus

E. tryoni
E. tympanum
Eumeces [Genus]
    E. anthracinus [Species]
    E. barbouri
    E. brevirostris
    E. callicephalus
    E. capito
    E. chinensis
    E. colimensis
    E. copei
    E. coreensis
    E. dugesii
    E. egregius
    E. elegans
    E. fasciatus
    E. gilberti
    E. inexpectatus
    E. kishinouyei
    E. lagunensis
    E. laticeps
    E. latiscutatus
    E. liui
    E. longirostris
    E. lynxe
    E. marginatus
    E. multilineatus
    E. multivirgatus
    E. obsoletus
    E. obtusirostris
    E. ochoterenae
    E. okadae
    E. parviauriculatus
    E. parvulus
    E. popei
    E. quadrilineatus
    E. septentrionalis
    E. skiltonianus
    E. stimpsonii
    E. sumichrasti
    E. tamdaoensis
    E. tetragrammus
    E. tunganus
Eumecia [Genus]
    E. anchietae [Species]
    E. johnstoni
Euprepes [Genus]
    E. chaperi [Species]
Eurylepis [Genus]
    E. indothalensis [Species]
    E. poonaensis
    E. taeniolatus
Evesia [Genus]
    E. bellii [Species]
Feylinia [Genus]
    F. boulengeri [Species]
    F. currori
    F. elegans
    F. grandisquamis
    F. polylepis

Fojia [Genus]
    F. bumui [Species]
Geomyersia [Genus]
    G. coggeri [Species]
    G. glabra
Geoscincus [Genus]
    G. haraldmeieri [Species]
Glaphyromorphus [Genus]
    G. antoniorum [Species]
    G. brongersmai
    G. butlerorum
    G. cracens
    G. darwiniensis
    G. douglasi
    G. emigrans
    G. fuscicaudis
    G. gracilipes
    G. isolepis
    G. mjobergi
    G. nigricaudis
    G. pardalis
    G. pumilus
    G. punctulatus
    G. timorensis
Gnypetoscincus [Genus]
    G. queenslandiae [Species]
Gongylomorphus [Genus]
    G. bojerii [Species]
Gongylus [Genus]
    G. androvandii [Species]
    G. brachypoda
    G. crassicaudum
Graciliscincus [Genus]
    G. shonae [Species]
Haackgreerius [Genus]
    H. miopus [Species]
Hemiergis [Genus]
    H. decresiensis [Species]
    H. initialis
    H. millewae
    H. peronii
    H. quadrilineatum
Hemisphaeriodon [Genus]
    H. gerrardii [Species]
Isopachys [Genus]
    I. anguinoides [Species]
    I. borealis
    I. gyldenstolpei
    I. roulei
Janetaescincus [Genus]
    J. braueri [Species]
    J. veseyfitzgeraldi
Lacertoides [Genus]
    L. pardalis [Species]
Lamprolepis [Genus]
    L. leucosticta [Species]
    L. nieuwenhuisi
    L. smaragdina
    L. vyneri
Lampropholis [Genus]

L. adonis [Species]
L. amicula
L. caligula
L. coggeri
L. colossus
L. couperi
L. delicata
L. elongata
L. guichenoti
L. mirabilis
L. robertsi
Lankascincus [Genus]
    L. deignani [Species]
    L. deraniyagalae
    L. fallax
    L. gansi
    L. taprobanensis
    L. taylori
Leiolopisma [Genus]
    L. alazon [Species]
    L. bardensis
    L. eulepis
    L. fasciolare
    L. lioscincus
    L. mauritiana
    L. paronae
    L. telfairii
Leptoseps [Genus]
    L. osellai [Species]
    L. poilani
Leptosiaphos [Genus]
    L. aloysiisabaudiae [Species]
    L. amieti
    L. blochmanni
    L. dewittei
    L. fuhni
    L. gemmiventris
    L. graueri
    L. hackarsi
    L. hylophilus
    L. ianthinoxantha
    L. kilimensis
    L. lepesmei
    L. luberoensis
    L. meleagris
    L. pauliani
Lerista [Genus]
    L. aericeps [Species]
    L. allanae
    L. allochira
    L. ameles
    L. apoda
    L. arenicola
    L. axillaris
    L. baynesi
    L. bipes
    L. borealis
    L. bougainvillii
    L. bunglebungle
    L. carpentariae

L. chalybura
L. christinae
L. cinerea
L. colliveri
L. connivens
L. desertorum
L. distinguenda
L. dorsalis
L. edwardsae
L. elegans
L. elongata
L. emmotti
L. eupoda
L. flammicauda
L. fragilis
L. frosti
L. gascoynensis
L. gerrardii
L. greeri
L. griffini
L. haroldi
L. humphriesi
L. ingrami
L. ips
L. kalumburu
L. karlschmidti
L. kendricki
L. kennedyensis
L. labialis
L. lineata
L. lineopunctulata
L. macropisthopus
L. maculosa
L. microtis
L. muelleri
L. neander
L. nichollsi
L. onsloviana
L. orientalis
L. petersoni
L. picturata
L. planiventralis
L. praefrontalis
L. praepedita
L. punctatovittata
L. puncticauda
L. quadrivincula
L. robusta
L. separanda
L. simillima
L. speciosa
L. stictopleura
L. storri
L. stylis
L. taeniata
L. talpina
L. terdigitata
L. tridactyla
L. uniduo
L. varia

L. vermicularis
L. viduata
L. vittata
L. walkeri
L. wilkinsi
L. xanthura
L. yuna
L. zonulata
Lioscincus [Genus]
  L. greeri [Species]
  L. maruia
  L. nigrofasciolatum
  L. novaecaledoniae
  L. steindachneri
  L. tillieri
Lipinia [Genus]
  L. auriculata [Species]
  L. cheesmanae
  L. leptosoma
  L. longiceps
  L. macrotympanum
  L. miangensis
  L. noctua
  L. nototaenia
  L. occidentalis
  L. pulchella
  L. pulchra
  L. quadrivittata
  L. rabori
  L. relicta
  L. rouxi
  L. semperi
  L. septentrionalis
  L. subvittata
  L. venemai
  L. vittigera
  L. vulcania
  L. zamboangensis
Lobulia [Genus]
  L. brongersmai [Species]
  L. elegans
Lygisaurus [Genus]
  L. aeratus [Species]
  L. foliorum
  L. laevis
  L. macfarlani
  L. rococo
  L. sesbrauna
  L. tanneri
  L. zuma
Lygosoma [Genus]
  L. afrum [Species]
  L. albopunctata
  L. angeli
  L. anguinum
  L. ashwamedhi
  L. bowringii
  L. carinatum
  L. corpulentum
  L. frontoparietale

L. goaensis
L. grandisonianum
L. guentheri
L. haroldyoungi
L. isodactylum
L. koratense
L. laeviceps
L. lanceolatum
L. lineata
L. lineolatum
L. mabuiiforme
L. mafianum
L. mocquardi
L. muelleri
L. paedocarinatum
L. pembanum
L. popae
L. productum
L. pruthi
L. punctata
L. quadrupes
L. simonettai
L. singha
L. somalicum
L. tanae
L. tersum
L. vinciguerrae
L. vosmaeri
Mabuya [Genus]
  M. acutilabris [Species]
  M. affinis
  M. albilabris
  M. allapallensis
  M. andamanensis
  M. angolensis
  M. arajara
  M. atlantica
  M. aurata
  M. aureopunctata
  M. bayonii
  M. beddomii
  M. bensonii
  M. betsileana
  M. bibronii
  M. binotata
  M. bistriata
  M. bocagii
  M. boettgeri
  M. bontocensis
  M. boulengeri
  M. brauni
  M. brevicollis
  M. breviparietalis
  M. buettneri
  M. caissara
  M. capensis
  M. carinata
  M. carvalhoi
  M. chimbana
  M. clivicola

M. cochabambae
M. comorensis
M. croizati
M. cumingi
M. darevskii
M. delalandii
M. dissimilis
M. dorsivittata
M. dumasi
M. elegans
M. englei
M. falconensis
M. ferrarai
M. fogoensis
M. frenata
M. gansi
M. geisthardti
M. gravenhorstii
M. guaporicola
M. heathi
M. hemmingi
M. hildae
M. hildebrandtii
M. hoeschi
M. homalocephala
M. indeprensa
M. infralineata
M. innotata
M. irregularis
M. ivensii
M. lacertiformis
M. laevis
M. lavarambo
M. lineolata
M. longicaudata
M. mabouya
M. macleani
M. macrorhyncha
M. macularia
M. maculata
M. maculilabris
M. madagascariensis
M. margaritifera
M. megalura
M. mekuana
M. mlanjensis
M. multicarinata
M. multifasciata
M. nagarjuni
M. nancycoutuae
M. nigropalmata
M. nigropunctata
M. novemcarinata
M. occidentalis
M. pendeana
M. perrotetii
M. planifrons
M. polytropis
M. punctulata
M. quadratilobus

M. quadricarinata
M. quinquetaeniata
M. rodenburgi
M. rudis
M. rugifera
M. seychellensis
M. socotrana
M. spilogaster
M. spinalis
M. stangeri
M. stanjorgeri
M. striata
M. sulcata
M. tandrefana
M. tavaratra
M. tessellata
M. trivittata
M. tytleri
M. unimarginata
M. vaillantii
M. varia
M. variegata
M. vato
M. vezo
M. vittata
M. volamenaloha
M. wingati
M. wrightii
Macroscincus [Genus]
   M. coctei [Species]
Marmorosphax [Genus]
   M. euryotis [Species]
   M. montana
   M. tricolor
Melanoseps [Genus]
   M. ater [Species]
   M. loveridgei
   M. occidentalis
   M. rondoensis
Menetia [Genus]
   M. alanae [Species]
   M. amaura
   M. concinna
   M. greyii
   M. koshlandae
   M. maini
   M. sadlieri
   M. surda
   M. timlowi
Mesoscincus [Genus]
   M. altamirani [Species]
   M. managuae
   M. schwartzei
Mochlus [Genus]
   M. brevicaudis [Species]
   M. fernandi
   M. guineensis
   M. sundevalli
Morethia [Genus]
   M. adelaidensis [Species]

M. boulengeri
M. butleri
M. lineoocellata
M. obscura
M. ruficauda
M. storri
M. taeniopleura
Nangura [Genus]
   N. spinosa [Species]
Nannoscincus [Genus]
   N. exos [Species]
   N. gracilis
   N. greeri
   N. hanchisteus
   N. humectus
   N. maccoyi
   N. mariei
   N. rankini
   N. slevini
Neoseps [Genus]
   N. reynoldsi [Species]
Nessia [Genus]
   N. bipes [Species]
   N. burtonii
   N. deraniyagalai
   N. didactyla
   N. hickanala
   N. layardi
   N. monodactyla
   N. sarasinorum
Niveoscincus [Genus]
   N. coventryi [Species]
   N. greeni
   N. metallicus
   N. microlepidotus
   N. ocellatus
   N. orocryptus
   N. palfreymani
   N. pretiosus
Notoscincus [Genus]
   N. butleri [Species]
   N. ornatus
Novoeumeces [Genus]
   N. algeriensis [Species]
   N. blythianus
   N. schneideri
Oligosoma [Genus]
   O. acrinasum [Species]
   O. chloronoton
   O. fallai
   O. gracilicorpus
   O. grande
   O. homalonotum
   O. inconspicuum
   O. infrapunctatum
   O. lineoocellatum
   O. longipes
   O. maccanni
   O. microlepis
   O. moco

O. nigriplantare
O. notosaurus
O. otagense
O. smithi
O. stenotis
O. striatum
O. suteri
O. waimatense
O. zelandicum
Ophiomorus [Genus]
　O. blanfordi [Species]
　O. brevipes
　O. chernovi
　O. latastii
　O. nuchalis
　O. persicus
　O. punctatissimus
　O. raithmai
　O. streeti
　O. tridactylus
Ophioscincus [Genus]
　O. cooloolensis [Species]
　O. ophioscincus
　O. truncatus
Pamelaescincus [Genus]
　P. gardineri [Species]
Panaspis [Genus]
　P. africana [Species]
　P. annobonensis
　P. breviceps
　P. burgeoni
　P. cabindae
　P. chriswildi
　P. helleri
　P. kitsoni
　P. maculicollis
　P. megalurus
　P. nimbaensis
　P. quattuordigitata
　P. reichenowi
　P. rohdei
　P. thomasi
　P. togoensis
　P. wahlbergi
Papuascincus [Genus]
　P. buergersi [Species]
　P. morokanus
　P. phaeodes
　P. stanleyanus
Parachalcides [Genus]
　P. socotranus [Species]
Paracontias [Genus]
　P. brocchii [Species]
　P. hafa
　P. hildebrandti
　P. holomelas
　P. manify
　P. milloti
　P. rothschildi
　P. tsararano

Paralipinia [Genus]
　P. rara [Species]
Parvoscincus [Genus]
　P. palawanensis [Species]
　P. sisoni
Phoboscincus [Genus]
　P. bocourti [Species]
　P. garnieri
Prasinohaema [Genus]
　P. flavipes [Species]
　P. parkeri
　P. prehensicauda
　P. semoni
　P. virens
Proablepharus [Genus]
　P. kinghorni [Species]
　P. reginae
　P. tenuis
Proscelotes [Genus]
　P. aenea [Species]
　P. arnoldi
　P. eggeli
Pseudoacontias [Genus]
　P. angelorum [Species]
　P. madagascariensis
　P. menamainty
Pseudemoia [Genus]
　P. baudini [Species]
　P. cryodroma
　P. entrecasteauxii
　P. pagenstecheri
　P. rawlinsoni
　P. spenceri
Pygomeles [Genus]
　P. braconnieri [Species]
　P. petteri
Riopa [Genus]
　R. bampfyldei [Species]
　R. herberti
　R. opisthorhodum
Ristella [Genus]
　R. beddomii [Species]
　R. guentheri
　R. rurkii
　R. travancorica
Saiphos [Genus]
　S. equalis [Species]
Saproscincus [Genus]
　S. basiliscus [Species]
　S. challengeri
　S. czechurai
　S. hannahae
　S. lewisi
　S. mustelinus
　S. oriarius
　S. rosei
　S. spectabilis
　S. tetradactylus
Scelotes [Genus]
　S. anguina [Species]

S. arenicola
S. bicolor
S. bidigittatus
S. bipes
S. bourquini
S. caffer
S. capensis
S. duttoni
S. fitzsimonsi
S. gronovii
S. guentheri
S. inornatus
S. insularis
S. kasneri
S. limpopoensis
S. mirus
S. mossambicus
S. poensis
S. schebeni
S. sexlineatus
S. uluguruensis
S. vestigifer
Scincella [Genus]
　S. barbouri [Species]
　S. bilineata
　S. capitanea
　S. caudaequinae
　S. doriae
　S. forbesora
　S. formosensis
　S. gemmingeri
　S. huanrenensis
　S. inconspicua
　S. ladacensis
　S. lateralis
　S. macrotis
　S. melanosticta
　S. modesta
　S. monticola
　S. przewalskii
　S. punctatolineata
　S. reevesii
　S. sikimmensis
　S. silvicola
　S. travancorica
　S. tsinlingensis
　S. vandenburghi
　S. victoriana
Scincopus [Genus]
　S. fasciatus [Species]
Scincus [Genus]
　S. hemprichii [Species]
　S. mitranus
　S. scincus
Scolecoseps [Genus]
　S. acontias [Species]
　S. boulengeri
　S. litipoensis
Sepsina [Genus]
　S. alberti [Species]

S. angolensis
S. bayoni
S. copei
S. tetradactyla
Sigaloseps [Genus]
    S. deplanchei [Species]
    S. ruficauda
Simiscincus [Genus]
    S. aurantiacus [Species]
Sphenomorphus [Genus]
    S. abdictus [Species]
    S. acutus
    S. aesculeticola
    S. amblyplacodes
    S. annectens
    S. anotus
    S. arborens
    S. assatus
    S. atrigularis
    S. beauforti
    S. beyeri
    S. bignelli
    S. biparietalis
    S. bruneus
    S. buenloicus
    S. buettikoferi
    S. celebense
    S. cherriei
    S. cinereus
    S. concinnatus
    S. consobrinus
    S. courcyanum
    S. coxi
    S. cranei
    S. crassa
    S. cumingi
    S. cyanolaemus
    S. darlingtoni
    S. decipiens
    S. derroyae
    S. diwata
    S. dorsicatenatus
    S. dussumieri
    S. fasciatus
    S. florensis
    S. forbesi
    S. fragosus
    S. grandisonae
    S. granulatus
    S. haasi
    S. hallieri
    S. helenae
    S. incertus
    S. incognitus
    S. indicus
    S. jagori
    S. jobiensis
    S. kinabaluensis
    S. kitangladensis
    S. knollmanae

S. kuehnei
S. laterimaculatus
S. lawtoni
S. leptofasciatus
S. leucospilos
S. lineopunctulatus
S. llanosi
S. luzonense
S. maculatus
S. maculicollus
S. maindroni
S. malayanum
S. megalops
S. microtympanus
S. mimicus
S. mimikanum
S. mindanensis
S. minutus
S. modigliani
S. muelleri
S. multisquamatus
S. murudensis
S. necopinatus
S. neuhaussi
S. nigrolabris
S. nigrolineata
S. nitens
S. oligolepis
S. praesignis
S. pratti
S. puncticentralis
S. rarus
S. rufocaudatus
S. sabanus
S. sanctus
S. sarasinorus
S. schultzei
S. scotophilus
S. scutatus
S. shelfordi
S. simus
S. solomonis
S. steerei
S. stellatus
S. stickeli
S. striatopunctatum
S. striolatus
S. tagapayo
S. taiwanensis
S. tanahtinggi
S. tanneri
S. taylori
S. temmincki
S. tenuiculus
S. transversus
S. tropidonotus
S. undulatus
S. vanheurni
S. variegatus
S. victoria

S. wolfi
S. woodfordi
S. wrighti
Sphenops [Genus]
    S. delislei [Species]
    S. sepsoides
    S. sphenopsiformis
Tachygia [Genus]
    T. microlepis [Species]
Tiliqua [Genus]
    T. adelaidensis [Species]
    T. gigas
    T. multifasciata
    T. nigrolutea
    T. occipitalis
    T. scincoides
Trachydosaurus [Genus]
    T. rugosus [Species]
Tribolonotus [Genus]
    T. annectens [Species]
    T. blanchardi
    T. brongersmai
    T. gracilis
    T. novaeguineae
    T. ponceleti
    T. pseudoponceleti
    T. schmidti
Tropidophorus [Genus]
    T. assamensis [Species]
    T. baviensis
    T. beccarii
    T. berdmorei
    T. brookei
    T. cocincinensis
    T. davaoensis
    T. grayi
    T. guangxiensis
    T. hainanus
    T. iniquus
    T. laotus
    T. latiscutatus
    T. matsuii
    T. microlepis
    T. micropus
    T. misaminius
    T. mocquardi
    T. murphyi
    T. partelloi
    T. perplexus
    T. robinsoni
    T. sinicus
    T. thai
Tropidoscincus [Genus]
    T. aubrianus [Species]
    T. boreus
    T. variabilis
Typhlacontias [Genus]
    T. brevipes [Species]
    T. gracilis
    T. johnsonii

T. ngamiensis
T. punctatissimus
T. rohani
T. rudebecki
Typhlosaurus [Genus]
  T. aurantiacus [Species]
  T. braini
  T. caecus
  T. cregoi
  T. gariepensis
  T. lineatus
  T. lomii
  T. meyeri
  T. vermis
Voeltzkowia [Genus]
  V. fierinensis [Species]
  V. lineata
  V. mira
  V. petiti
  V. rubrocaudata

Anguidae [Family]
  Abronia [Genus]
    A. anzuetoi [Species]
    A. aurita
    A. bogerti
    A. campbelli
    A. chiszari
    A. deppii
    A. fimbriata
    A. frosti
    A. fuscolabialis
    A. gaiophantasma
    A. graminea
    A. leurolepis
    A. lythrochila
    A. matudai
    A. meledona
    A. mitchelli
    A. mixteca
    A. montecristoi
    A. oaxacae
    A. ochoterenai
    A. ornelasi
    A. ramirezi
    A. reidi
    A. salvadorensis
    A. smithi
    A. taeniata
  Anguis [Genus]
    A. cephalonnicus [Species]
    A. fragilis
  Anniella [Genus]
    A. geronimensis [Species]
    A. pulchra
  Barisia [Genus]
    B. herrerae [Species]
    B. imbricata
    B. levicollis
    B. rudicollis

Celestus [Genus]
  C. agasepsoides [Species]
  C. anelpistus
  C. badius
  C. barbouri
  C. carraui
  C. costatus
  C. crusculus
  C. curtissi
  C. cyanochloris
  C. darlingtoni
  C. duquesneyi
  C. enneagrammus
  C. fowleri
  C. haetianus
  C. hewardi
  C. hylaius
  C. macrotus
  C. marcanoi
  C. microblepharis
  C. montanus
  C. occiduus
  C. orobius
  C. rozellae
  C. scansorius
  C. sepsoides
  C. stenurus
  C. warreni
Coloptychon [Genus]
  C. rhombifer [Species]
Diploglossus [Genus]
  D. atitlanensis [Species]
  D. bilobatus
  D. bivittatus
  D. delasagra
  D. fasciatus
  D. garridoi
  D. legnotus
  D. lessonae
  D. maculatus
  D. microcephalus
  D. microlepis
  D. millepunctatus
  D. monotropis
  D. montisilvestris
  D. montisserrati
  D. nigropunctatus
  D. owenii
  D. pleii
Elgaria [Genus]
  E. coerulea [Species]
  E. kingii
  E. multicarinata
  E. panamintina
  E. parva
  E. paucicarinata
  E. velazquezi
Gerrhonotus [Genus]
  G. liocephalus [Species]
  G. lugoi

Mesaspis [Genus]
  M. antauges [Species]
  M. gadovii
  M. juarezi
  M. monticola
  M. moreletii
  M. viridiflava
Ophiodes [Genus]
  O. intermedius [Species]
  O. striatus
  O. vertebralis
  O. yacupoi
Ophisaurus [Genus]
  O. attenuatus [Species]
  O. buettikoferi
  O. ceroni
  O. compressus
  O. formosensis
  O. gracilis
  O. hainanensis
  O. harti
  O. incomptus
  O. koellikeri
  O. mimicus
  O. sokolovi
  O. ventralis
  O. wegneri
Pseudopus [Genus]
  P. apodus

Xenosauridae [Family]
  Shinisaurus [Genus]
    S. crocodilurus [Species]
  Xenosaurus [Genus]
    X. grandis [Species]
    X. newmanorum
    X. penai
    X. phalaroantheron
    X. platyceps
    X. rectocollaris

Helodermatidae [Family]
  Heloderma [Genus]
    H. horridum [Species]
    H. suspectum

Varanidae [Family]
  Lanthanotus [Genus]
    L. borneensis [Species]
  Varanus [Genus]
    V. acanthurus [Species]
    V. albigularis
    V. auffenbergi
    V. baritji
    V. beccarii
    V. bengalensis
    V. bogerti
    V. brevicauda
    V. caerulivirens
    V. caudolineatus
    V. cerambonensis

V. doreanus
V. dumerilii
V. eremius
V. exanthematicus
V. finschi
V. flavescens
V. flavirufus
V. giganteus
V. gilleni
V. glauerti
V. glebopalma
V. gouldii
V. griseus
V. indicus
V. jobiensis
V. juxtindicus
V. keithhornei
V. kingorum
V. komodoensis
V. mabitang
V. macraei
V. melinus
V. mertensi
V. mitchelli
V. niloticus
V. olivaceus
V. ornatus
V. panoptes
V. pilbarensis
V. prasinus
V. primordius
V. rosenbergi
V. rudicollis
V. salvadorii
V. salvator
V. scalaris
V. semiremex
V. spenceri
V. spinulosus
V. storri
V. telenesetes
V. timorensis
V. tristis
V. varius
V. yemenensis
V. yuwonoi

Anomalepidae [Family]
  Anomalepis [Genus]
    A. aspinosus [Species]
    A. colombia
    A. flavapices
    A. mexicanus
  Helminthophis [Genus]
    H. flavoterminatus [Species]
    H. frontalis
    H. praeocularis
  Liotyphlops [Genus]
    L. albirostris [Species]
    L. anops

L. argaleus
L. beui
L. schubarti
L. ternetzii
L. wilderi
Typhlophis [Genus]
  T. ayarzaguenai [Species]
  T. squamosus

Leptotyphlopidae [Family]
  Leptotyphlops [Genus]
    L. adleri [Species]
    L. affinis
    L. albifrons
    L. albipuncta
    L. albiventer
    L. alfredschmidti
    L. algeriensis
    L. anthracinus
    L. asbolepis
    L. australis
    L. bicolor
    L. bilineatus
    L. blanfordi
    L. borapeliotes
    L. borrichianus
    L. boueti
    L. boulengeri
    L. brasiliensis
    L. bressoni
    L. brevissimus
    L. broadleyi
    L. burii
    L. cairi
    L. calypso
    L. collaris
    L. columbi
    L. conjunctus
    L. cupinensis
    L. debilis
    L. diaplocius
    L. dimidiatus
    L. dissimilis
    L. distanti
    L. drewesi
    L. dugandi
    L. dulcis
    L. emini
    L. filiformis
    L. goudotii
    L. gracilior
    L. guayaquilensis
    L. humilis
    L. jacobseni
    L. joshuai
    L. koppesi
    L. labialis
    L. leptipilepta
    L. longicaudus
    L. macrolepis

L. macrops
L. macrorhynchus
L. macrurus
L. maximus
L. melanotermus
L. melanurus
L. munoai
L. narirostris
L. nasalis
L. natatrix
L. nicefori
L. nigricans
L. nursii
L. occidentalis
L. parkeri
L. pembae
L. perreti
L. peruvianus
L. phillipsi
L. pungwensis
L. pyrites
L. reticulatus
L. rostratus
L. rubrolineatus
L. rufidorsus
L. salgueiroi
L. scutifrons
L. septemstriatus
L. signatus
L. striatula
L. subcrotillus
L. sundewalli
L. sylvicolus
L. teaguei
L. telloi
L. tenellus
L. tesselatus
L. tricolor
L. undecimstriatus
L. unguirostris
L. vellardi
L. weyrauchi
L. wilsoni
Rhinoleptus [Genus]
  R. koniagui [Species]

Typhlopidae [Family]
  Acutotyphlops [Genus]
    A. infralabialis [Species]
    A. kunuaensis
    A. solomonis
    A. subocularis
  Cyclotyphlops [Genus]
    C. deharvengi [Species]
  Ramphotyphlops [Genus]
    R. acuticauda [Species]
    R. affinis
    R. albiceps
    R. angusticeps
    R. aspina

REPTILES SPECIES LIST

R. australis
R. batillus
R. bituberculatus
R. braminus
R. broomi
R. centralis
R. chamodracaena
R. cumingii
R. depressus
R. diversus
R. endoterus
R. erycinus
R. exocoeti
R. flaviventer
R. ganei
R. grypus
R. guentheri
R. hamatus
R. howi
R. kimberleyensis
R. leptosoma
R. leucoproctus
R. ligatus
R. lineatus
R. longissimus
R. lorenzi
R. margaretae
R. melanocephalus
R. micromma
R. minimus
R. multilineatus
R. nema
R. nigrescens
R. nigroterminatus
R. olivaceus
R. pilbarensis
R. pinguis
R. polygrammicus
R. proximus
R. robertsi
R. silvia
R. similis
R. splendidus
R. suluensis
R. supranasalis
R. tovelli
R. troglodytes
R. unguirostris
R. waitii
R. wiedii
R. willeyi
R. yampiensis
R. yirrikalae
Rhinotyphlops [Genus]
R. acutus [Species]
R. anomalus
R. ataeniatus
R. boylei
R. caecus
R. crossii

R. debilis
R. episcopus
R. erythraeus
R. feae
R. gracilis
R. graueri
R. kibarae
R. lalandei
R. leucocephalus
R. lumbriciformis
R. newtoni
R. nigrocandidus
R. pallidus
R. praeocularis
R. rufescens
R. schinzi
R. schlegelii
R. scortecci
R. simonii
R. somalicus
R. stejnegeri
R. sudanensis
R. unitaeniatus
R. wittei
Typhlops [Genus]
T. ahsanai [Species]
T. albanalis
T. andamanesis
T. angolensis
T. annae
T. arenarius
T. ater
T. beddomi
T. bibronii
T. biminiensis
T. bisubocularis
T. blanfordii
T. bothriorhynchus
T. brongersmianus
T. caecatus
T. canlaonensis
T. capensis
T. capitulatus
T. cariei
T. castanotus
T. catapontus
T. caymanensis
T. collaris
T. comorensis
T. congestus
T. conradi
T. costaricensis
T. cuneirostris
T. decorosus
T. decorsei
T. depressiceps
T. diardii
T. disparilis
T. domerguei
T. dominicanus

T. elegans
T. epactius
T. exiguus
T. filiformis
T. fletcheri
T. floweri
T. fornasinii
T. fredparkeri
T. fuscus
T. giadinhensis
T. gierrai
T. gonavensis
T. granti
T. hectus
T. hedraeus
T. hypogius
T. hypomethes
T. hypsobothrius
T. inornatus
T. jamaicensis
T. jerdoni
T. khoratensis
T. klemmeri
T. koekkoeki
T. koshunensis
T. kraali
T. lankaensis
T. lehneri
T. leucomelas
T. leucostictus
T. lineolatus
T. longissimus
T. loveridgei
T. lumbricalis
T. luzonensis
T. mackinnoni
T. madagascariensis
T. madgemintonai
T. malcolmi
T. manilae
T. manni
T. marxi
T. mcdowelli
T. meszoelyi
T. microcephalus
T. microstomus
T. minuisquamus
T. mirus
T. monastus
T. monensis
T. mucronatus
T. muelleri
T. oatesii
T. obtusus
T. ocularis
T. oligolepis
T. pammeces
T. paucisquamus
T. platycephalus
T. platyrhynchus

T. porrectus
T. punctatus
T. pusillus
T. reticulatus
T. reuteri
T. richardi
T. rondoensis
T. rostellatus
T. ruber
T. ruficaudus
T. schmidti
T. schmutzi
T. schwartzi
T. siamensis
T. socotranus
T. stadelmani
T. steinhausi
T. sulcatus
T. syntherus
T. tasymicris
T. tenebrarum
T. tenuicollis
T. tenuis
T. tetrathyreus
T. thurstoni
T. tindalli
T. titanops
T. trangensis
T. trinitatus
T. uluguruensis
T. unilineatus
T. veddae
T. vermicularis
T. verticalis
T. violaceus
T. wilsoni
T. yonenagae
T. zenkeri
Xenotyphlops [Genus]
  X. grandidieri [Species]

Anomochilidae [Family]
Anomochilus [Genus]
  A. leonardi [Species]
  A. weberi

Uropeltidae [Family]
Brachyophidium [Genus]
  B. rhodogaster [Species]
Melanophidium [Genus]
  M. bilineatum [Species]
  M. punctatum
  M. wynaudense
Platyplectrurus [Genus]
  P. madurensis [Species]
  P. trilineatus
Plectrurus [Genus]
  P. aureus [Species]
  P. canaricus
  P. guentheri
  P. perroteti

Pseudotyphlops [Genus]
  P. philippinus [Species]
Rhinophis [Genus]
  R. blythii [Species]
  R. dorsimaculatus
  R. drummondhayi
  R. fergusonianus
  R. oxyrhynchus
  R. philippinus
  R. porrectus
  R. punctatus
  R. sanguineus
  R. travancoricus
  R. trevelyana
  R. tricolorata
Teretrurus [Genus]
  T. sanguineus [Species]
Uropeltis [Genus]
  U. arcticeps [Species]
  U. beddomii
  U. broughami
  U. ceylanicus
  U. dindigalensis
  U. ellioti
  U. liura
  U. macrolepis
  U. macrorhynchus
  U. maculatus
  U. melanogaster
  U. myhendrae
  U. nitidus
  U. ocellatus
  U. petersi
  U. phillipsi
  U. phipsonii
  U. pulneyensis
  U. rubrolineatus
  U. rubromaculatus
  U. ruhunae
  U. smithi
  U. woodmasoni

Cylindrophiidae [Family]
Cylindrophis [Genus]
  C. aruensis [Species]
  C. boulengeri
  C. engkariensis
  C. isolepis
  C. lineatus
  C. maculatus
  C. melanotus
  C. opisthorhodus
  C. ruffus
  C. yamdena

Aniliidae [Family]
Anilius [Genus]
  A. scytale [Species]

Xenopeltidae [Family]
Xenopeltis [Genus]

X. hainanensis [Species]
X. unicolor

Loxocemidae [Family]
Loxocemus [Genus]
  L. bicolor [Species]

Boidae [Family]
Antaresia [Genus]
  A. childreni [Species]
  A. maculosa
  A. perthensis
  A. stimsoni
Apodora [Genus]
  A. papuana [Species]
Boa [Genus]
  B. constrictor [Species]
Bothrochilus [Genus]
  B. boa [Species]
Calabaria [Genus]
  C. reinhardtii [Species]
Candoia [Genus]
  C. aspera [Species]
  C. bibroni
  C. carinata
Charina [Genus]
  C. bottae [Species]
  C. trivirgata
Corallus [Genus]
  C. annulatus [Species]
  C. caninus
  C. cookii
  C. cropanii
  C. hortulanus
  C. ruschenbergerii
Epicrates [Genus]
  E. angulifer [Species]
  E. cenchria
  E. chrysogaster
  E. exsul
  E. fordii
  E. gracilis
  E. inornatus
  E. monensis
  E. striatus
  E. subflavus
Eryx [Genus]
  E. elegans [Species]
  E. jaculus
  E. jayakari
  E. johnii
  E. miliaris
  E. somalicus
  E. tataricus
  E. whitakeri
Eunectes [Genus]
  E. beniensis [Species]
  E. deschauenseei
  E. murinus
  E. notaeus

Gongylophis [Genus]
    G. colubrinus [Species]
    G. conicus
    G. muelleri
Leiopython [Genus]
    L. albertisii [Species]

Pythonidae [Family]
    Aspidites [Genus]
        A. melanocephalus [Species]
        A. ramsayi
    Liasis [Genus]
        L. fuscus [Species]
        L. mackloti
        L. olivaceus
    Morelia [Genus]
        M. amethistina [Species]
        M. boeleni
        M. bredli
        M. carinata
        M. clastolepis
        M. kinghorni
        M. nauta
        M. oenpelliensis
        M. spilota
        M. tracyae
        M. viridis
    Python [Genus]
        P. anchietae [Species]
        P. breitensteini
        P. brongersmai
        P. curtus
        P. molurus
        P. natalensis
        P. regius
        P. reticulatus
        P. sebae
        P. timoriensis

Bolyeridae [Family]
    Bolyeria [Genus]
        B. multocarinata [Species]
    Casarea [Genus]
        C. dussumieri [Species]

Tropidophiidae [Family]
    Exiliboa [Genus]
        E. placata [Species]
    Trachyboa [Genus]
        T. boulengeri [Species]
        T. gularis
    Tropidophis [Genus]
        T. battersbyi [Species]
        T. canus
        T. caymanensis
        T. celiae
        T. feicki
        T. fuscus
        T. greenwayi
        T. haetianus
        T. hendersoni

        T. maculatus
        T. melanurus
        T. morenoi
        T. nigriventris
        T. pardalis
        T. paucisquamis
        T. pilsbryi
        T. semicinctus
        T. spiritus
        T. taczanowskyi
        T. wrighti
    Ungaliophis [Genus]
        U. continentalis [Species]
        U. panamensis
    Xenophidion [Genus]
        X. acanthognathus [Species]
        X. schaeferi

Acrochordidae [Family]
    Acrochordus [Genus]
        A. arafurae [Species]
        A. granulatus
        A. javanicus

Viperidae [Family]
    Adenorhinos [Genus]
        A. barbouri [Species]
    Agkistrodon [Genus]
        A. bilineatus [Species]
        A. contortrix
        A. piscivorus
    Atheris [Genus]
        A. acuminata [Species]
        A. broadleyi
        A. ceratophora
        A. chlorechis
        A. desaixi
        A. hirsuta
        A. hispida
        A. katangensis
        A. nitschei
        A. rungweensis
        A. squamigera
        A. subocularis
    Atropoides [Genus]
        A. nummifer [Species]
        A. olmec
        A. picadoi
    Azemiops [Genus]
        A. feae [Species]
    Bitis [Genus]
        B. albanica [Species]
        B. arietans
        B. armata
        B. atropos
        B. caudalis
        B. cornuta
        B. gabonica
        B. heraldica
        B. inornata
        B. nasicornis

        B. parviocula
        B. peringueyi
        B. rubida
        B. schneideri
        B. worthingtoni
        B. xeropaga
    Bothriechis [Genus]
        B. aurifer [Species]
        B. bicolor
        B. lateralis
        B. marchi
        B. nigroviridis
        B. rowleyi
        B. schlegelii
        B. thalassinus
    Bothriopsis [Genus]
        B. bilineata [Species]
        B. medusa
        B. oligolepis
        B. peruviana
        B. pulchra
        B. punctata
        B. taeniata
    Bothrops [Genus]
        B. alcatraz [Species]
        B. alternatus
        B. ammodytoides
        B. andianus
        B. asper
        B. atrox
        B. barnetti
        B. brazili
        B. campbelli
        B. caribbaeus
        B. colombianus
        B. colombiensis
        B. cotiara
        B. erythromelas
        B. fonsecai
        B. hyoprora
        B. iglesiasi
        B. insularis
        B. itapetiningae
        B. jararaca
        B. jararacussu
        B. jonathani
        B. lanceolatus
        B. leucurus
        B. lojanus
        B. marajoensis
        B. microphthalmus
        B. moojeni
        B. muriciensis
        B. neuwiedi
        B. pictus
        B. pirajai
        B. sanctaecrucis
        B. venezuelensis
    Calloselasma [Genus]
        C. rhodostoma [Species]

Causus [Genus]
  C. bilineatus [Species]
  C. defilippii
  C. lichtensteinii
  C. maculatus
  C. resimus
  C. rhombeatus
Cerastes [Genus]
  C. cerastes [Species]
  C. gasperettii
  C. vipera
Cerrophidion [Genus]
  C. barbouri [Species]
  C. godmani
  C. petlalcalensis
  C. tzotzilorum
Crotalus [Genus]
  C. adamanteus [Species]
  C. aquilus
  C. atrox
  C. basiliscus
  C. catalinensis
  C. cerastes
  C. durissus
  C. enyo
  C. exsul
  C. horridus
  C. intermedius
  C. lannomi
  C. lepidus
  C. mitchelli
  C. molossus
  C. polystictus
  C. pricei
  C. pusillus
  C. ruber
  C. scutulatus
  C. stejnegeri
  C. tigris
  C. tortugensis
  C. transversus
  C. triseriatus
  C. unicolor
  C. vegrandis
  C. viridis
  C. willardi
Daboia [Genus]
  D. russelii [Species]
Deinagkistrodon [Genus]
  D. acutus [Species]
Echis [Genus]
  E. carinatus [Species]
  E. coloratus
  E. hughesi
  E. jogeri
  E. leucogaster
  E. megalocephalus
  E. ocellatus
  E. pyramidum

Eristicophis [Genus]
  E. macmahoni [Species]
Ermia [Genus]
  E. mangshanensis [Species]
Gloydius [Genus]
  G. blomhoffi [Species]
  G. halys
  G. himalayanus
  G. intermedius
  G. monticola
  G. saxatilis
  G. shedaoensis
  G. strauchi
  G. tsushimaensis
  G. ussuriensis
Hypnale [Genus]
  H. hypnale [Species]
  H. nepa
  H. walli
Lachesis [Genus]
  L. melanocephala [Species]
  L. muta
  L. stenophrys
Macrovipera [Genus]
  M. deserti [Species]
  M. lebetina
  M. mauritanica
  M. schweizeri
Montatheris [Genus]
  M. hindii [Species]
Ophryacus [Genus]
  O. melanurus [Species]
  O. undulatus
Ovophis [Genus]
  O. chaseni [Species]
  O. monticola
  O. okinavensis
  O. tonkinensis
Porthidium [Genus]
  P. dunni [Species]
  P. hespere
  P. lansbergii
  P. nasutum
  P. ophryomegas
  P. volcanicum
  P. yucatanicum
Proatheris [Genus]
  P. superciliaris [Species]
Protobothrops [Genus]
  P. elegans [Species]
  P. flavoviridis
  P. jerdonii
  P. kaulbacki
  P. mucrosquamatus
  P. strigatus
  P. tokarensis
Pseudocerastes [Genus]
  P. fieldi [Species]
  P. persicus

Sistrurus [Genus]
  S. catenatus [Species]
  S. miliarius
  S. ravus
Trimeresurus [Genus]
  T. albolabris [Species]
  T. andersonii
  T. borneensis
  T. brongersmai
  T. cantori
  T. cornutus
  T. erythrurus
  T. fasciatus
  T. flavomaculatus
  T. gracilis
  T. gramineus
  T. gumprechti
  T. hageni
  T. kanburiensis
  T. karanshahi
  T. labialis
  T. macrolepis
  T. macrops
  T. malabaricus
  T. malcolmi
  T. medoensis
  T. popeiorum
  T. puniceus
  T. purpureomaculatus
  T. schultzei
  T. stejnegeri
  T. sumatranus
  T. tibetanus
  T. trigonocephalus
  T. vogeli
  T. xiangchengensis
  T. yunnanensis
Tropidolaemus [Genus]
  T. huttoni [Species]
  T. wagleri
Vipera [Genus]
  V. albizona [Species]
  V. ammodytes
  V. aspis
  V. barani
  V. berus
  V. bornmuelleri
  V. darevskii
  V. dinniki
  V. eriwanensis
  V. kaznakovi
  V. latastei
  V. latifii
  V. lotievi
  V. magnifica
  V. monticola
  V. nikolskii
  V. orlovi
  V. palaestinae
  V. pontica

V. raddei
V. renardi
V. sachalinensis
V. seoanei
V. transcaucasiana
V. ursinii
V. wagneri
V. xanthina

Atractaspididae [Family]
Amblyodipsas [Genus]
  A. concolor [Species]
  A. dimidiata
  A. katangensis
  A. microphthalma
  A. polylepis
  A. rodhaini
  A. teitana
  A. unicolor
  A. ventrimaculata
Aparallactus [Genus]
  A. capensis [Species]
  A. guentheri
  A. jacksonii
  A. lineatus
  A. lunulatus
  A. modestus
  A. moeruensis
  A. niger
  A. nigriceps
  A. turneri
  A. werneri
Atractaspis [Genus]
  A. aterrima [Species]
  A. battersbyi
  A. bibronii
  A. boulengeri
  A. coalescens
  A. congica
  A. corpulenta
  A. dahomeyensis
  A. duerdeni
  A. engaddensis
  A. engdahli
  A. fallax
  A. irregularis
  A. leucomelas
  A. microlepidota
  A. micropholis
  A. reticulata
  A. scorteccii
Brachyophis [Genus]
  B. revoili [Species]
Chilorhinophis [Genus]
  C. butleri [Species]
  C. carpenteri
  C. gerardi
Elapotinus [Genus]
  E. picteti [Species]
Hypoptophis [Genus]
  H. wilsoni [Species]

Macrelaps [Genus]
  M. microlepidotus [Species]
Micrelaps [Genus]
  M. bicoloratus [Species]
  M. boettgeri
  M. muelleri
  M. vaillanti
Polemon [Genus]
  P. acanthias [Species]
  P. barthii
  P. bocourti
  P. christyi
  P. collaris
  P. fulvicollis
  P. gabonensis
  P. gracilis
  P. griseiceps
  P. leopoldi
  P. neuwiedi
  P. notatum
  P. robustus
Xenocalamus [Genus]
  X. bicolor [Species]
  X. mechowii
  X. michellii
  X. sabiensis
  X. transvaalensis

Colubridae [Family]
Achalinus [Genus]
  A. ater [Species]
  A. formosanus
  A. hainanus
  A. jinggangensis
  A. meiguensis
  A. niger
  A. rufescens
  A. spinalis
  A. werneri
Adelophis [Genus]
  A. copei [Species]
  A. foxi
Adelphicos [Genus]
  A. daryi [Species]
  A. ibarrorum
  A. latifasciatus
  A. nigrilatum
  A. quadrivirgatus
  A. veraepacis
Aeluroglena [Genus]
  A. cucullata [Species]
Afronatrix [Genus]
  A. anoscopus [Species]
Ahaetulla [Genus]
  A. dispar [Species]
  A. fasciolata
  A. fronticincta
  A. mycterizans
  A. nasuta
  A. perroteti
  A. prasina

  A. pulverulenta
Alluaudina [Genus]
  A. bellyi [Species]
  A. mocquardi
Alsophis [Genus]
  A. anomalus [Species]
  A. antiguae
  A. antillensis
  A. ater
  A. biserialis
  A. cantherigerus
  A. elegans
  A. melanichnus
  A. portoricensis
  A. rijersmai
  A. rufiventris
  A. sanctaecrucis
  A. vudii
Amastridium [Genus]
  A. veliferum [Species]
Amphiesma [Genus]
  A. atemporale [Species]
  A. beddomei
  A. bitaeniatum
  A. boulengeri
  A. celebicum
  A. concelarum
  A. craspedogaster
  A. deschauenseei
  A. flavifrons
  A. frenatum
  A. groundwateri
  A. inas
  A. ishigakiense
  A. johannis
  A. khasiense
  A. metusium
  A. miyajimae
  A. modestum
  A. monticola
  A. nicobariense
  A. octolineatum
  A. optatum
  A. parallelum
  A. pealii
  A. petersii
  A. platyceps
  A. popei
  A. pryeri
  A. sanguinea
  A. sarasinorum
  A. sarawacense
  A. sauteri
  A. sieboldii
  A. stolatum
  A. venningi
  A. vibakari
  A. viperinum
  A. xenura
Amphiesmoides [Genus]
  A. ornaticeps [Species]

Amplorhinus [Genus]
    A. multimaculatus [Species]
Anoplohydrus [Genus]
    A. aemulans [Species]
Antillophis [Genus]
    A. andreae [Species]
    A. parvifrons
Aplopeltura [Genus]
    A. boa [Species]
Apostolepis [Genus]
    A. ambinigra [Species]
    A. arenarius
    A. assimilis
    A. breviceps
    A. cearensis
    A. coronata
    A. dimidiata
    A. dorbignyi
    A. flavotorquata
    A. gaboi
    A. goiasensis
    A. intermedia
    A. longicaudata
    A. multicincta
    A. niceforoi
    A. nigroterminata
    A. phillipsi
    A. polylepis
    A. pymi
    A. quinquelineata
    A. quirogai
    A. sanctaeritae
    A. tenuis
    A. vittata
Argyrogena [Genus]
    A. fasciolata [Species]
Arizona [Genus]
    A. elegans [Species]
    A. occidentalis
Arrhyton [Genus]
    A. ainictum [Species]
    A. callilaemum
    A. dolichura
    A. exiguum
    A. funereum
    A. landoi
    A. polylepis
    A. procerum
    A. redimitum
    A. supernum
    A. taeniatum
    A. tanyplectum
    A. vittatum
Aspidura [Genus]
    A. brachyorrhos [Species]
    A. copei
    A. deraniyagalae
    A. drummondhayi
    A. guentheri
    A. trachyprocta

Atractus [Genus]
    A. albuquerquei [Species]
    A. alphonsehogei
    A. andinus
    A. arangoi
    A. badius
    A. balzani
    A. biseriatus
    A. bocki
    A. bocourti
    A. boettgeri
    A. boulengerii
    A. canedii
    A. carrioni
    A. clarki
    A. collaris
    A. crassicaudatus
    A. duidensis
    A. dunni
    A. ecuadorensis
    A. elaps
    A. emigdioi
    A. emmeli
    A. erythromelas
    A. favae
    A. flammigerus
    A. fuliginosus
    A. gaigeae
    A. guentheri
    A. indistinctus
    A. insipidus
    A. iridescens
    A. lancinii
    A. lasallei
    A. latifrons
    A. lehmanni
    A. limitaneus
    A. loveridgei
    A. maculatus
    A. major
    A. manizalesensis
    A. mariselae
    A. melanogaster
    A. melas
    A. micheli
    A. microrhynchus
    A. modestus
    A. multicinctus
    A. nebularis
    A. nicefori
    A. nigricaudus
    A. nigriventris
    A. obesus
    A. obtusirostris
    A. occidentalis
    A. occipitoalbus
    A. oculotemporalis
    A. pamplonensis
    A. pantostictus
    A. paravertebralis

    A. paucidens
    A. pauciscutatus
    A. peruvianus
    A. poeppigi
    A. potschi
    A. punctiventris
    A. resplendens
    A. reticulatus
    A. riveroi
    A. roulei
    A. sanctaemartae
    A. sanguineus
    A. schach
    A. serranus
    A. snethlageae
    A. steyermarki
    A. subbicinctum
    A. taeniatus
    A. taphorni
    A. torquatus
    A. trihedrurus
    A. trilineatus
    A. trivittatus
    A. turikensis
    A. univittatus
    A. variegatus
    A. ventrimaculatus
    A. vertebralis
    A. vertebrolineatus
    A. vittatus
    A. wagleri
    A. werneri
    A. zidoki
Atretium [Genus]
    A. schistosum [Species]
    A. yunnanensis
Balanophis [Genus]
    B. ceylonensis [Species]
Bitia [Genus]
    B. hydroides [Species]
Blythia [Genus]
    B. reticulata [Species]
Bogertophis [Genus]
    B. rosaliae [Species]
    B. subocularis
Boiga [Genus]
    B. andamanensis [Species]
    B. angulata
    B. barnesii
    B. beddomei
    B. blandingii
    B. ceylonensis
    B. cyanea
    B. cynodon
    B. dendrophila
    B. dightoni
    B. drapiezii
    B. forsteni
    B. gokool
    B. guangxiensis

B. irregularis
B. jaspidea
B. kraepelini
B. multifasciata
B. multomaculata
B. nigriceps
B. nuchalis
B. ocellata
B. ochracea
B. philippina
B. pulverulenta
B. quincunciata
B. saengsomi
B. schultzei
B. trigonata
B. wallachi
Boiruna [Genus]
  B. sertaneja [Species]
Bothrolycus [Genus]
  B. ater [Species]
Bothrophthalmus [Genus]
  B. lineatus [Species]
Brachyorrhos [Genus]
  B. albus [Species]
Brygophis [Genus]
  B. coulangesi [Species]
Buhoma [Genus]
  B. depressiceps [Species]
  B. procterae
  B. vauerocegae
Calamaria [Genus]
  C. abstrusa [Species]
  C. acutirostris
  C. albiventer
  C. alidae
  C. apraeocularis
  C. battersbyi
  C. bicolor
  C. bitorques
  C. boesemani
  C. borneensis
  C. brongersmai
  C. buchi
  C. ceramensis
  C. crassa
  C. curta
  C. doederleini
  C. eiselti
  C. everetti
  C. forcarti
  C. gervaisii
  C. grabowskyi
  C. gracillima
  C. griswoldi
  C. hilleniusi
  C. javanica
  C. joloensis
  C. lateralis
  C. leucogaster
  C. linnaei

C. lowii
C. lumbricoidea
C. lumholtzi
C. margaritophora
C. mecheli
C. melanota
C. modesta
C. muelleri
C. nuchalis
C. palavanensis
C. pavimentata
C. pfefferi
C. prakkei
C. rebentischi
C. schlegelii
C. schmidti
C. septentrionalis
C. suluensis
C. sumatrana
C. ulmeri
C. virgulata
C. yunnanensis
Calamodontophis [Genus]
  C. ipaucidens [Species]
Calamorhabdium [Genus]
  C. acuticeps [Species]
  C. kuekenthali
Cantoria [Genus]
  C. annulata [Species]
  C. violacea
Carphophis [Genus]
  C. amoenus [Species]
  C. vermis
Cemophora [Genus]
  C. coccinea [Species]
Cerberus [Genus]
  C. microlepis [Species]
  C. rynchops
Cercaspis [Genus]
  C. carinatus [Species]
Cercophis [Genus]
  C. auratus [Species]
Chamaelycus [Genus]
  C. christyi [Species]
  C. fasciatus
  C. parkeri
  C. werneri
Chapinophis [Genus]
  C. xanthocheilus [Species]
Chersodromus [Genus]
  C. liebmanni [Species]
  C. rubriventris
Chilomeniscus [Genus]
  C. savagei [Species]
  C. stramineus
Chionactis [Genus]
  C. occipitalis [Species]
  C. palarostris
  C. saxatilis
Chironius [Genus]

C. bicarinatus [Species]
C. carinatus
C. exoletus
C. flavolineatus
C. fuscus
C. grandisquamis
C. laevicollis
C. laurenti
C. monticola
C. multiventris
C. quadricarinatus
C. scurrulus
C. vincenti
Chrysopelea [Genus]
  C. ornata [Species]
  C. paradisi
  C. pelias
  C. rhodopleuron
  C. taprobanica
Clelia [Genus]
  C. bicolor [Species]
  C. clelia
  C. equatoriana
  C. errabunda
  C. montana
  C. quimi
  C. rustica
  C. scytalina
Clonophis [Genus]
  C. kirtlandii [Species]
Collorhabdium [Genus]
  C. williamsoni [Species]
Coluber [Genus]
  C. algirus [Species]
  C. atayevi
  C. bholanathi
  C. brevis
  C. caspius
  C. constrictor
  C. cypriensis
  C. dorri
  C. elegantissimus
  C. florulentus
  C. gemonensis
  C. gracilis
  C. gyarosensis
  C. hippocrepis
  C. insulanus
  C. jugularis
  C. karelini
  C. largeni
  C. manseri
  C. messanai
  C. mormon
  C. najadum
  C. nummifer
  C. ravergieri
  C. rhodorhachis
  C. rogersi
  C. rubriceps

C. schmidti
C. schmidtleri
C. scortecci
C. sinai
C. smithi
C. socotrae
C. somalicus
C. spinalis
C. taylori
C. thomasi
C. variabilis
C. ventromaculatus
C. viridiflavus
C. vittacaudatus
C. zebrinus
Compsophis [Genus]
   C. albiventris [Species]
Coniophanes [Genus]
   C. alvarezi [Species]
   C. andresensis
   C. bipunctatus
   C. dromiciformis
   C. fissidens
   C. imperialis
   C. joanae
   C. lateritius
   C. longinquus
   C. meridanus
   C. piceivittis
   C. quinquevittatus
   C. schmidti
Conophis [Genus]
   C. lineatus [Species]
   C. pulcher
   C. vittatus
Conopsis [Genus]
   C. amphisticha [Species]
   C. biserialis
   C. conica
   C. lineata
   C. megalodon
   C. nasus
Contia [Genus]
   C. tenuis [Species]
Coronella [Genus]
   C. austriaca [Species]
   C. brachyura
   C. girondica
Crisantophis [Genus]
   C. nevermanni [Species]
Crotaphopeltis [Genus]
   C. barotseensis [Species]
   C. braestrupi
   C. degeni
   C. hippocrepis
   C. hotamboeia
   C. tornieri
Cryophis [Genus]
   C. hallbergi [Species]
Cyclocorus [Genus]

C. lineatus [Species]
C. nuchalis
Cyclophiops [Genus]
   C. doriae [Species]
   C. major
   C. multicinctus
   C. semicarinatus
Darlingtonia [Genus]
   D. haetiana [Species]
Dasypeltis [Genus]
   D. atra [Species]
   D. fasciata
   D. inornata
   D. medici
   D. scabra
Dendrelaphis [Genus]
   D. bifrenalis [Species]
   D. calligastra
   D. caudolineatus
   D. cyanochloris
   D. formosus
   D. gastrostictus
   D. gorei
   D. grandoculis
   D. humayuni
   D. inornatus
   D. lorentzi
   D. ngansonensis
   D. oliveri
   D. papuensis
   D. pictus
   D. punctulata
   D. salomonis
   D. striatus
   D. subocularis
   D. tristis
Dendrolycus [Genus]
   D. elapoides [Species]
Dendrophidion [Genus]
   D. bivittatus [Species]
   D. boshelli
   D. brunneus
   D. dendrophis
   D. nuchalis
   D. paucicarinatus
   D. percarinatus
   D. vinitor
Diadophis [Genus]
   D. punctatus [Species]
Diaphorolepis [Genus]
   D. laevis [Species]
   D. wagneri
Dinodon [Genus]
   D. flavozonatum [Species]
   D. gammiei
   D. orientale
   D. rosozonatum
   D. rufozonatum
   D. semicarinatum
   D. septentrionalis

Dipsadoboa [Genus]
   D. aulica [Species]
   D. brevirostris
   D. duchesnei
   D. elongata
   D. flavida
   D. shrevei
   D. underwoodi
   D. unicolor
   D. viridis
   D. weileri
   D. werneri
Dipsas [Genus]
   D. albifrons [Species]
   D. articulata
   D. bicolor
   D. boettgeri
   D. brevifacies
   D. catesbyi
   D. chaparensis
   D. copei
   D. elegans
   D. gaigeae
   D. gracilis
   D. incerta
   D. indica
   D. infrenalis
   D. latifasciata
   D. latifrontalis
   D. maxillaris
   D. neivai
   D. oreas
   D. pavonina
   D. perijanensis
   D. peruana
   D. polylepis
   D. pratti
   D. sanctijoannis
   D. schunkii
   D. temporalis
   D. tenuissima
   D. variegata
   D. vermiculata
   D. viguieri
Dipsina [Genus]
   D. multimaculata [Species]
Dispholidus [Genus]
   D. typus [Species]
Ditaxodon [Genus]
   D. taeniatus [Species]
Ditypophis [Genus]
   D. vivax [Species]
Drepanoides [Genus]
   D. anomalus [Species]
Dromicodryas [Genus]
   D. bernieri [Species]
   D. quadrilineatus
Dromophis [Genus]
   D. lineatus [Species]
   D. praeornatus

Dryadophis [Genus]
    D. cliftoni [Species]
    D. dorsalis
    D. melanolomus
Drymarchon [Genus]
    D. caudomaculatus [Species]
    D. corais
    D. couperi
Drymobius [Genus]
    D. chloroticus [Species]
    D. margaritiferus
    D. melanotropis
    D. reissi
    D. rhombifer
Drymoluber [Genus]
    D. brazili [Species]
    D. dichrous
Dryocalamus [Genus]
    D. davisonii [Species]
    D. gracilis
    D. nympha
    D. philippinus
    D. subannulatus
    D. tristrigatus
Dryophiops [Genus]
    D. philippina [Species]
    D. rubescens
Duberria [Genus]
    D. lutrix [Spccics]
    D. variegata
Echinanthera [Genus]
    E. affinis [Species]
    E. amoena
    E. bilineata
    E. brevirostris
    E. cephalomaculata
    E. cephalostriata
    E. cyanopleura
    E. melanostigma
    E. occipitalis
    E. persimilis
    E. poecilopogon
    E. undulata
Eirenis [Genus]
    E. africana [Species]
    E. aurolineatus
    E. barani
    E. collaris
    E. coronella
    E. decemlineata
    E. eiselti
    E. hakkariensis
    E. levantinus
    E. lineomaculata
    E. mcmahoni
    E. medus
    E. modestus
    E. punctatolineatus
    E. rechingeri
    E. rothii

    E. thospitis
Elachistodon [Genus]
    E. westermanni [Species]
Elaphe [Genus]
    E. bairdi [Species]
    E. bella
    E. bimaculata
    E. carinata
    E. climacophora
    E. conspicillata
    E. davidi
    E. dione
    E. emoryi
    E. erythrura
    E. flavirufa
    E. flavolineata
    E. gloydi
    E. guttata
    E. helena
    E. hohenackeri
    E. leonardi
    E. lineata
    E. longissima
    E. maculata
    E. mandarina
    E. moellendorffi
    E. obsoleta
    E. perlacea
    E. persica
    E. porphyracea
    E. prasina
    E. quadrivirgata
    E. quatuorlineata
    E. radiata
    E. rufodorsata
    E. scalaris
    E. schrenckii
    E. situla
    E. slowinksii
    E. subradiata
    E. taeniura
    E. vulpina
Elapoidis [Genus]
    E. fusca [Species]
Elapomorphus [Genus]
    E. lemniscatus [Species]
    E. lepidus
    E. quinquelineatus
    E. spegazzinii
    E. wuchereri
Emmochliophis [Genus]
    E. fugleri [Species]
    E. miops
Enhydris [Genus]
    E. albomaculata [Species]
    E. alternans
    E. bennettii
    E. bocourti
    E. chinensis
    E. doriae

    E. dussumieri
    E. enhydris
    E. indica
    E. innominata
    E. jagorii
    E. longicauda
    E. maculosa
    E. matannensis
    E. pahangensis
    E. pakistanica
    E. plumbea
    E. polylepis
    E. punctata
    E. sieboldi
    E. smithi
Enuliophis [Genus]
    E. sclateri [Species]
Enulius [Genus]
    E. bifoveatus [Species]
    E. flavitorques
    E. oligostichus
    E. roatanensis
Eridiphas [Genus]
    E. slevini [Species]
Erpeton [Genus]
    E. tentaculatum [Species]
Erythrolamprus [Genus]
    E. aesculapii [Species]
    E. bizonus
    E. guentheri
    E. mimus
    E. ocellatus
    E. pseudocorallus
Etheridgeum [Genus]
    E. pulchrum [Species]
Exallodontophis [Genus]
    E. albignaci [Species]
Farancia [Genus]
    F. abacura [Species]
    F. erytrogramma
Ficimia [Genus]
    F. hardyi [Species]
    F. olivacea
    F. publia
    F. ramirezi
    F. ruspator
    F. streckeri
    F. variegata
Fimbrios [Genus]
    F. klossi [Species]
Fordonia [Genus]
    F. leucobalia [Species]
Gastropyxis [Genus]
    G. smaragdina [Species]
Geagras [Genus]
    G. redimitus [Species]
Geodipsas [Genus]
    G. boulengeri [Species]
    G. infralineata
    G. laphystia

G. vinckei
G. zeny
Geophis [Genus]
  G. anocularis [Species]
  G. betaniensis
  G. bicolor
  G. blanchardi
  G. brachycephalus
  G. cancellatus
  G. carinosus
  G. chalybeus
  G. championi
  G. damiani
  G. downsi
  G. dubius
  G. duellmani
  G. dugesii
  G. dunni
  G. fulvoguttatus
  G. godmani
  G. hoffmanni
  G. immaculatus
  G. incomptus
  G. isthmicus
  G. juliai
  G. laticinctus
  G. laticollaris
  G. latifrontalis
  G. maculiferus
  G. mutitorques
  G. nasalis
  G. nigrocinctus
  G. omiltemanus
  G. petersii
  G. pyburni
  G. rhodogaster
  G. russatus
  G. ruthveni
  G. sallaei
  G. semidoliatus
  G. sieboldi
  G. talamancae
  G. tarascae
  G. zeledoni
Gerarda [Genus]
  G. prevostiana [Species]
Gomesophis [Genus]
  G. brasiliensis [Species]
Gongylosoma [Genus]
  G. baliodeirus [Species]
  G. longicauda
  G. nicobariensis
Gonionotophis [Genus]
  G. brussauxi [Species]
  G. grantii
  G. klingi
Gonyophis [Genus]
  G. margaritatus [Species]
Gonyosoma [Genus]
  G. cantoris [Species]

G. frenatum
G. hodgsoni
G. jansenii
G. oxycephalum
Grayia [Genus]
  G. caesar [Species]
  G. ornata
  G. smythii
  G. tholloni
Gyalopion [Genus]
  G. canum [Species]
  G. quadrangulare
Haplocercus [Genus]
  H. ceylonensis [Species]
Hapsidophrys [Genus]
  H. lineatus [Species]
Helicops [Genus]
  H. angulatus [Species]
  H. carinicaudus
  H. danieli
  H. gomesi
  H. hagmanni
  H. hogei
  H. infrataeniatus
  H. leopardinus
  H. modestus
  H. pastazae
  H. petersi
  H. polylepis
  H. scalaris
  H. trivittatus
  H. yacu
Helophis [Genus]
  H. schoutedeni [Species]
Hemirhagerrhis [Genus]
  H. hildebrandtii [Species]
  H. kelleri
  H. nototaenia
  H. viperina
Heterodon [Genus]
  H. nasicus [Species]
  H. platirhinos
  H. simus
Heteroliodon [Genus]
  H. lava [Species]
  H. occipitalis
Heurnia [Genus]
  H. ventromaculata [Species]
Hologerrhum [Genus]
  H. dermali [Species]
  H. philippinum
Homalopsis [Genus]
  H. buccata [Species]
Hormonotus [Genus]
  H. modestus [Species]
Hydrablabes [Genus]
  H. periops [Species]
  H. praefrontalis
Hydraethiops [Genus]
  H. laevis [Species]

H. melanogaster
Hydrodynastes [Genus]
  H. bicinctus [Species]
  H. gigas
Hydromorphus [Genus]
  H. concolor [Species]
  H. dunni
Hydrops [Genus]
  H. martii [Species]
  H. triangularis
Hypsiglena [Genus]
  H. tanzeri [Species]
  H. torquata
Hypsirhynchus [Genus]
  H. ferox [Species]
Ialtris [Genus]
  I. agyrtes [Species]
  I. dorsalis
  I. parishi
Iguanognathus [Genus]
  I. werneri [Species]
Imantodes [Genus]
  I. cenchoa [Species]
  I. gemmistratus
  I. inornatus
  I. lentiferus
  I. phantasma
  I. tenuissimus
Internatus [Genus]
  I. laevis [Species]
  I. malaccanus
  I. tropidonotus
Ithycyphus [Genus]
  I. blanci [Species]
  I. goudoti
  I. miniatus
  I. oursi
  I. perineti
Lampropeltis [Genus]
  L. alterna [Species]
  L. calligaster
  L. getula
  L. mexicana
  L. pyromelana
  L. ruthveni
  L. triangulum
  L. zonata
Lamprophis [Genus]
  L. abyssinicus [Species]
  L. aurora
  L. erlangeri
  L. fiskii
  L. fuliginosus
  L. fuscus
  L. geometricus
  L. guttatus
  L. inornatus
  L. lineatus
  L. maculatus
  L. olivaceus

REPTILES SPECIES LIST

L. swazicus
L. virgatus
Langaha [Genus]
    L. alluaudi [Species]
    L. madagascariensis
    L. pseudoalluaudi
Leioheterodon [Genus]
    L. geayi [Species]
    L. madagascariensis
    L. modestus
Leptodeira [Genus]
    L. annulata [Species]
    L. bakeri
    L. frenata
    L. maculata
    L. nigrofasciata
    L. punctata
    L. rubricata
    L. septentrionalis
    L. splendida
Leptodrymus [Genus]
    L. pulcherrimus [Species]
Leptophis [Genus]
    L. ahaetulla [Species]
    L. cupreus
    L. depressirostris
    L. diplotropis
    L. mexicanus
    L. modestus
    L. nebulosus
    L. riveti
    L. santamartensis
    L. stimsoni
Lepturophis [Genus]
    L. albofuscus [Species]
    L. borneensis
Limnophis [Genus]
    L. bicolor [Species]
Lioheterophis [Genus]
    L. iheringi [Species]
Liopeltis [Genus]
    L. calamaria [Species]
    L. frenatus
    L. herminae
    L. philippinus
    L. rappi
    L. scriptus
    L. stoliczkae
    L. tricolor
Liophidium [Genus]
    L. apperti [Species]
    L. chabaudi
    L. mayottensis
    L. rhodogaster
    L. therezieni
    L. torquatum
    L. trilineatum
    L. vaillanti
Liophis [Genus]
    L. albiceps [Species]
    L. almadensis

L. amarali
L. andinus
L. angustilineatus
L. anomalus
L. atraventer
L. boursieri
L. breviceps
L. carajasensis
L. ceii
L. cobellus
L. cursor
L. dilepis
L. elegantissimus
L. epinephelus
L. festae
L. flavifrenatus
L. frenatus
L. guentheri
L. jaegeri
L. janaleeae
L. juliae
L. leucogaster
L. lineatus
L. longiventris
L. maryellenae
L. melanauchen
L. melanotus
L. meridionalis
L. miliaris
L. ornatus
L. paucidens
L. perfuscus
L. poecilogyrus
L. problematicus
L. reginae
L. sagittifer
L. steinbachi
L. subocularis
L. tachymenoides
L. taeniurus
L. torrenicolus
L. triscalis
L. tristriatus
L. typhlus
L. vanzolinii
L. viridis
L. vitti
L. williamsi
Liopholidophis [Genus]
    L. dolichocercus [Species]
    L. epistibes
    L. grandidieri
    L. infrasignatus
    L. lateralis
    L. rhadinaea
    L. sexlineatus
    L. stumpffi
    L. varius
Lycodon [Genus]
    L. alcalai [Species]
    L. aulicus

L. bibonius
L. butleri
L. capucinus
L. cardamomensis
L. chrysoprateros
L. dumerili
L. effraenis
L. fasciatus
L. fausti
L. ferroni
L. flavomaculatus
L. jara
L. kundui
L. laoensis
L. mackinnoni
L. muelleri
L. osmanhilli
L. paucifasciatus
L. ruhstrati
L. solivagus
L. stormi
L. striatus
L. subcinctus
L. tessellatus
L. tiwarii
L. travancoricus
L. zawi
Lycodonomorphus [Genus]
    L. bicolor [Species]
    L. laevissimus
    L. leleupi
    L. rufulus
    L. subtaeniatus
    L. whytii
Lycodryas [Genus]
    L. maculatus [Species]
    L. sanctijohannis
Lycognathophis [Genus]
    L. seychellensis [Species]
Lycophidion [Genus]
    L. acutirostre [Species]
    L. albomaculatum
    L. capense
    L. depressirostre
    L. hellmichi
    L. irroratum
    L. laterale
    L. meleagris
    L. namibianum
    L. nanus
    L. nigromaculatum
    L. ornatum
    L. pygmaeum
    L. semiannule
    L. semicinctum
    L. taylori
    L. uzungwense
    L. variegatum
Lystrophis [Genus]
    L. dorbignyi [Species]
    L. histricus

L. matogrossensis
L. pulcher
L. semicinctus
Lytorhynchus [Genus]
  L. diadema [Species]
  L. gasperetti
  L. kennedyi
  L. maynardi
  L. paradoxus
  L. ridgewayi
Macrocalamus [Genus]
  M. jasoni [Species]
  M. lateralis
  M. schulzi
  M. tweediei
Macropisthodon [Genus]
  M. flaviceps [Species]
  M. plumbicolor
  M. rhodomelas
  M. rudis
Macroprotodon [Genus]
  M. cucullatus [Species]
Madagascarophis [Genus]
  M. citrinus [Species]
  M. colubrinus
  M. meridionalis
  M. ocellatus
Malpolon [Genus]
  M. moilensis [Species]
  M. monspessulanus
Manolepis [Genus]
  M. putnami [Species]
Masticophis [Genus]
  M. anthonyi [Species]
  M. aurigulus
  M. bilineatus
  M. flagellum
  M. lateralis
  M. lineolatus
  M. mentovarius
  M. schotti
  M. taeniatus
Mastigodryas [Genus]
  M. amarali [Species]
  M. bifossatus
  M. boddaerti
  M. bruesi
  M. danieli
  M. heathii
  M. pleei
  M. pulchriceps
  M. sanguiventris
Mehelya [Genus]
  M. capensis [Species]
  M. crossi
  M. egbensis
  M. guirali
  M. laurenti
  M. nyassae
  M. poensis

M. riggenbachi
M. stenophthalmus
M. vernayi
Meizodon [Genus]
  M. coronatus [Species]
  M. krameri
  M. plumbiceps
  M. regularis
  M. semiornatus
Micropisthodon [Genus]
  M. ochraceus [Species]
Mimophis [Genus]
  M. mahfalensis [Species]
Montaspis [Genus]
  M. gilvomaculata [Species]
Myersophis [Genus]
  M. alpestris [Species]
Myron [Genus]
  M. richardsonii [Species]
Natriciteres [Genus]
  N. fuliginoides [Species]
  N. olivacea
  N. variegata
Natrix [Genus]
  N. flavifrons [Species]
  N. maura
  N. megalocephala
  N. natrix
  N. tessellata
Nerodia [Genus]
  N. clarkii [Species]
  N. cyclopion
  N. erythrogaster
  N. fasciata
  N. floridana
  N. harteri
  N. paucimaculata
  N. rhombifera
  N. sipedon
  N. taxispilota
Ninia [Genus]
  N. atrata [Species]
  N. celata
  N. diademata
  N. espinali
  N. hudsoni
  N. maculata
  N. pavimentata
  N. psephota
  N. sebae
Nothopsis [Genus]
  N. rugosus [Species]
Oligodon [Genus]
  O. affinis [Species]
  O. albocinctus
  O. ancorus
  O. annulifer
  O. arnensis
  O. barroni
  O. bitorquatus

O. brevicauda
O. catenata
O. chinensis
O. cinereus
O. cruentatus
O. cyclurus
O. dorsalis
O. durheimi
O. eberhardti
O. erythrogaster
O. erythrorhachis
O. everetti
O. forbesi
O. formosanus
O. hamptoni
O. inornatus
O. joynsoni
O. juglandifer
O. kunmingensis
O. lacroixi
O. lungshenensis
O. macrurus
O. maculatus
O. mcdougalli
O. melaneus
O. melanozonatus
O. meyerinkii
O. modestum
O. mouhoti
O. multizonatus
O. nikhili
O. ningshaanensis
O. ocellatus
O. octolineatus
O. ornatus
O. perkinsi
O. petronellae
O. planiceps
O. praefrontalis
O. pulcherrimus
O. purpurascens
O. rhombifer
O. semicinctus
O. signatus
O. splendidus
O. subcarinatus
O. sublineatus
O. taeniatus
O. taeniolatus
O. templetoni
O. theobaldi
O. torquatus
O. travancoricus
O. trilineatus
O. unicolor
O. venustus
O. vertebralis
O. waandersi
O. woodmasoni
Opheodrys [Genus]

O. aestivus [Species]
O. vernalis
Opisthotropis [Genus]
O. alcalai [Species]
O. andersonii
O. balteatus
O. boonsongi
O. daovantieni
O. guangxiensis
O. jacobi
O. kikuzatoi
O. kuatunensis
O. lateralis
O. latouchii
O. maxwelli
O. premaxillaris
O. rugosa
O. spenceri
O. typica
Oreocalamus [Genus]
O. hanitschi [Species]
Oxybelis [Genus]
O. aeneus [Species]
O. brevirostris
O. fulgidus
O. wilsoni
Oxyrhabdium [Genus]
O. leporinum [Species]
O. modestum
Oxyrhopus [Genus]
O. clathratus [Species]
O. doliatus
O. fitzingeri
O. formosus
O. guibei
O. leucomelas
O. marcapatae
O. melanogenys
O. occipitalis
O. petola
O. rhombifer
O. trigeminus
O. venezuelanus
Parahelicops [Genus]
P. annamensis [Species]
Pararhabdophis [Genus]
P. chapaensis [Species]
Pararhadinaea [Genus]
P. melanogaster [Species]
Pareas [Genus]
P. boulengeri [Species]
P. carinatus
P. chinensis
P. formosensis
P. hamptoni
P. iwasakii
P. macularius
P. margaritophorus
P. monticola
P. nuchalis

P. stanleyi
P. vertebralis
Phalotris [Genus]
P. bilineatus [Species]
P. concolor
P. cuyanus
P. lativittatus
P. mertensi
P. multipunctatus
P. nasutus
P. nigrilatus
P. punctatus
P. tricolor
Philodryas [Genus]
P. aestivus [Species]
P. arnaldoi
P. baroni
P. bolivianus
P. chamissonis
P. cordatus
P. hoodensis
P. inca
P. laticeps
P. livida
P. mattogrossensis
P. nattereri
P. olfersii
P. oligolepis
P. patagoniensis
P. psammophideus
P. simonsii
P. tachymenoides
P. trilineatus
P. varius
P. viridissimus
Philothamnus [Genus]
P. angolensis [Species]
P. battersbyi
P. bequaerti
P. carinatus
P. dorsalis
P. girardi
P. heterodermus
P. heterolepidotus
P. hoplogaster
P. hughesi
P. irregularis
P. macrops
P. natalensis
P. nitidus
P. ornatus
P. punctatus
P. semivariegatus
P. thomensis
Phimophis [Genus]
P. chui [Species]
P. guerini
P. guianensis
P. iglesiasi
P. scriptorcibatus

P. vittatus
Phyllorhynchus [Genus]
P. browni [Species]
P. decurtatus
Pituophis [Genus]
P. catenifer [Species]
P. deppei
P. lineaticollis
P. melanoleucus
P. ruthveni
Plagiopholis [Genus]
P. blakewayi [Species]
P. delacouri
P. nuchalis
P. styani
P. unipostocularis
Pliocercus [Genus]
P. euryzonus [Species]
P. wilmarai
Poecilopholis [Genus]
P. cameronensis [Species]
Prosymna [Genus]
P. ambigua [Species]
P. angolensis
P. bivittata
P. frontalis
P. janii
P. meleagris
P. ornatissima
P. pitmani
P. ruspolii
P. semifasciata
P. somalica
P. sundevalli
P. visseri
Psammodynastes [Genus]
P. pictus [Species]
P. pulverulentus
Psammophis [Genus]
P. aegyptius [Species]
P. angolensis
P. ansorgii
P. biseriatus
P. condanarus
P. crucifer
P. elegans
P. jallae
P. leightoni
P. leithii
P. lineolatus
P. longifrons
P. notostictus
P. phillipsi
P. pulcher
P. punctulatus
P. rukwae
P. schokari
P. sibilans
P. subtaeniatus
P. tanganicus

P. trigrammus
Psammophylax [Genus]
   P. rhombeatus [Species]
   P. tritaeniatus
   P. variabilis
Pseudablabes [Genus]
   P. agassizii [Species]
Pseudaspis [Genus]
   P. cana [Species]
Pseudoboa [Genus]
   P. coronata [Species]
   P. haasi
   P. neuwiedii
   P. nigra
   P. serrana
Pseudoboodon [Genus]
   P. boehmei [Species]
   P. gascae
   P. lemniscatus
Pseudocyclophis [Genus]
   P. persicus [Species]
Pseudoeryx [Genus]
   P. plicatilis [Species]
Pseudoficimia [Genus]
   P. frontalis [Species]
Pseudoxyrhopus [Genus]
   P. ambreensis [Species]
   P. analabe
   P. ankafinaensis
   P. heterurus
   P. imerinae
   P. kely
   P. microps
   P. oblectator
   P. quinquelineatus
   P. sokosoko
   P. tritaeniatus
Pseudoleptodeira [Genus]
   P. latifasciata [Species]
   P. uribei
Pseudorabdion [Genus]
   P. albonuchalis [Species]
   P. ater
   P. collaris
   P. eiselti
   P. longiceps
   P. mcnamarae
   P. montanum
   P. oxycephalum
   P. sarasinorum
   P. saravacense
   P. talonuran
   P. taylori
Pseudotomodon [Genus]
   P. trigonatus [Species]
Pseudoxenodon [Genus]
   P. bambusicola [Species]
   P. baramensis
   P. inornatus
   P. karlschmidti

P. macrops
P. stejnegeri
Pseustes [Genus]
   P. cinnamomeus [Species]
   P. poecilonotus
   P. sexcarinatus
   P. shropshirei
   P. sulphureus
Psomophis [Genus]
   P. genimaculatus [Species]
   P. joberti
   P. obtusus
Ptyas [Genus]
   P. carinatus [Species]
   P. dhumnades
   P. dipsas
   P. fuscus
   P. korros
   P. luzonensis
   P. mucosus
   P. nigromarginatus
Ptychophis [Genus]
   P. flavovirgatus [Species]
Pythonodipsas [Genus]
   P. carinata [Species]
Rabdion [Genus]
   R. forsteni [Species]
Regina [Genus]
   R. alleni [Species]
   R. grahami
   R. rigida
   R. septemvittata
Rhabdophis [Genus]
   R. adleri [Species]
   R. angeli
   R. auriculata
   R. barbouri
   R. callichroma
   R. chrysargoides
   R. chrysargos
   R. conspicillatus
   R. himalayanus
   R. leonardi
   R. lineatus
   R. murudensis
   R. nigrocinctus
   R. nuchalis
   R. spilogaster
   R. subminiatus
   R. swinhonis
   R. tigrinus
Rhabdops [Genus]
   R. bicolor [Species]
   R. olivaceus
Rhachidelus [Genus]
   R. brazili [Species]
Rhadinaea [Genus]
   R. anachoreta [Species]
   R. bogertorum
   R. calligaster

R. cuneata
R. decorata
R. flavilata
R. forbesi
R. fulvivittis
R. gaigeae
R. godmani
R. hannsteini
R. hempsteadae
R. hesperia
R. kanalchutchan
R. kinkelini
R. lachrymans
R. laureata
R. macdougalli
R. marcellae
R. montana
R. montecristi
R. myersi
R. omiltemana
R. pilonaorum
R. posadasi
R. pulveriventris
R. quinquelineata
R. rogerromani
R. sargenti
R. schistosa
R. serperastra
R. stadelmani
R. taeniata
R. tolpanorum
R. vermiculaticeps
Rhadinophanes [Genus]
   R. monticola [Species]
Rhamnophis [Genus]
   R. aethiopissa [Species]
   R. batesii
Rhamphiophis [Genus]
   R. acutus [Species]
   R. maradiensis
   R. oxyrhynchus
   R. rubropunctatus
Rhinobothryum [Genus]
   R. bovallii [Species]
   R. lentiginosum
Rhinocheilus [Genus]
   R. lecontei [Species]
Rhynchocalamus [Genus]
   R. arabicus [Species]
   R. melanocephalus
Rhynchophis [Genus]
   R. boulengeri [Species]
Salvadora [Genus]
   S. bairdi [Species]
   S. deserticola
   S. grahamiae
   S. hexalepis
   S. intermedia
   S. lemniscata
   S. mexicana

Saphenophis [Genus]
　S. antioquiensis [Species]
　S. atahuallpae
　S. sneiderni
Scaphiodontophis [Genus]
　S. annulatus [Species]
　S. venustissimus
Scaphiophis [Genus]
　S. albopunctatus [Species]
　S. raffreyi
Scolecophis [Genus]
　S. atrocinctus [Species]
Seminatrix [Genus]
　S. pygaea [Species]
Senticolis [Genus]
　S. triaspis [Species]
Sibon [Genus]
　S. annulata [Species]
　S. annulifera
　S. anthracops
　S. argus
　S. carri
　S. dimidiata
　S. dunni
　S. fischeri
　S. longifrenis
　S. nebulatus
　S. sanniola
　S. sartorii
Sibynomorphus [Genus]
　S. inaequifasciatus [Species]
　S. lavillai
　S. mikanii
　S. neuwiedi
　S. oligozonatus
　S. oneilli
　S. petersi
　S. turgidus
　S. vagrans
　S. vagus
　S. ventrimaculatus
　S. williamsi
Sibynophis [Genus]
　S. bistrigatus [Species]
　S. bivittatus
　S. chinensis
　S. collaris
　S. geminatus
　S. melanocephalus
　S. sagittarius
　S. subpunctatus
　S. triangularis
Simophis [Genus]
　S. rhinostoma [Species]
　S. rohdei
Sinonatrix [Genus]
　S. aequifasciata [Species]
　S. annularis
　S. percarinata
　S. yunnanensis

Siphlophis [Genus]
　S. cervinus [Species]
　S. compressus
　S. leucocephalus
　S. longicaudatus
　S. pulcher
　S. worontzowi
Sonora [Genus]
　S. aemula [Species]
　S. michoacanensis
　S. semiannulata
Sordellina [Genus]
　S. punctata [Species]
Spalerosophis [Genus]
　S. arenarius [Species]
　S. diadema
　S. dolichospilus
　S. josephscorteccii
　S. microlepis
Spilotes [Genus]
　S. pullatus [Species]
Stegonotus [Genus]
　S. batjanensis [Species]
　S. borneensis
　S. cucullatus
　S. diehli
　S. dumerilii
　S. florensis
　S. guentheri
　S. heterurus
　S. modestus
　S. parvus
Stenophis [Genus]
　S. arctifasciatus [Species]
　S. betsileanus
　S. capuroni
　S. carleti
　S. citrinus
　S. gaimardi
　S. granuliceps
　S. guentheri
　S. iarakaensis
　S. inopinae
　S. inornatus
　S. jaosoloa
　S. pseudogranuliceps
　S. tulearensis
　S. variabilis
Stenorrhina [Genus]
　S. degenhardtii [Species]
　S. freminvillei
Stilosoma [Genus]
　S. extenuatum [Species]
Stoliczkaia [Genus]
　S. borneensis [Species]
　S. khasiensis
Storeria [Genus]
　S. dekayi [Species]
　S. hidalgoensis
　S. occipitomaculata

　S. storerioides
Symphimus [Genus]
　S. leucostomus [Species]
　S. mayae
Sympholis [Genus]
　S. lippiens [Species]
Synophis [Genus]
　S. bicolor [Species]
　S. calamitus
　S. lasallei
　S. plectovertebralis
Tachymenis [Genus]
　T. affinis [Species]
　T. attenuata
　T. chilensis
　T. elongata
　T. peruviana
　T. surinamensis
　T. tarmensis
Taeniophallus [Genus]
　T. nicagus [Species]
Tantalophis [Genus]
　T. discolor [Species]
Tantilla [Genus]
　T. albiceps [Species]
　T. alticola
　T. andinista
　T. armillata
　T. atriceps
　T. bairdi
　T. bocourti
　T. brevicauda
　T. briggsi
　T. calamarina
　T. capistrata
　T. cascadae
　T. coronadoi
　T. coronata
　T. cucullata
　T. cuesta
　T. cuniculator
　T. deppei
　T. deviatrix
　T. equatoriana
　T. flavilineata
　T. fraseri
　T. gracilis
　T. hobartsmithi
　T. impensa
　T. insulamontana
　T. jani
　T. johnsoni
　T. lempira
　T. longifrontalis
　T. melanocephala
　T. mexicana
　T. miyatai
　T. moesta
　T. nigra
　T. nigriceps

T. oaxacae
T. oolitica
T. petersi
T. planiceps
T. relicta
T. reticulata
T. robusta
T. rubra
T. schistosa
T. semicincta
T. sertula
T. shawi
T. slavensi
T. striata
T. supracincta
T. taeniata
T. tayrae
T. tecta
T. trilineata
T. triseriata
T. tritaeniata
T. vermiformis
T. vulcani
T. wilcoxi
T. yaquia
Tantillita [Genus]
    T. brevissima [Species]
    T. canula
    T. lintoni
Telescopus [Genus]
    T. beetzi [Species]
    T. dhara
    T. fallax
    T. gezirae
    T. hoogstraali
    T. nigriceps
    T. obtusus
    T. pulcher
    T. rhinopoma
    T. semiannulatus
    T. tessellatus
    T. variegata
Tetralepis [Genus]
    T. fruhstorferi [Species]
Thamnodynastes [Genus]
    T. chaquensis [Species]
    T. chimanta
    T. corocoroensis
    T. duida
    T. gambotensis
    T. hypoconia
    T. marahuaquensis
    T. pallidus
    T. rutilus
    T. strigatus
    T. strigilis
    T. yavi
Thamnophis [Genus]
    T. angustirostris [Species]
    T. atratus

T. brachystoma
T. butleri
T. chrysocephalus
T. couchii
T. cyrtopsis
T. elegans
T. eques
T. exsul
T. fulvus
T. gigas
T. godmani
T. hammondii
T. marcianus
T. melanogaster
T. mendax
T. ordinoides
T. proximus
T. radix
T. rossmani
T. rufipunctatus
T. sauritus
T. scalaris
T. scaliger
T. sirtalis
T. sumichrasti
T. valida
Thelotornis [Genus]
    T. capensis [Species]
    T. kirtlandii
    T. usambaricus
Thermophis [Genus]
    T. baileyi [Species]
Thrasops [Genus]
    T. flavigularis [Species]
    T. jacksonii
    T. occidentalis
Tomodon [Genus]
    T. degener [Species]
    T. dorsatus
    T. ocellatus
Trachischium [Genus]
    T. fuscum [Species]
    T. guentheri
    T. laeve
    T. monticola
    T. tenuiceps
Tretanorhinus [Genus]
    T. mocquardi [Species]
    T. nigroluteus
    T. taeniatus
    T. variabilis
Trimetopon [Genus]
    T. barbouri [Species]
    T. gracile
    T. pliolepis
    T. simile
    T. slevini
    T. viquezi
Trimorphodon [Genus]
    T. biscutatus [Species]
    T. tau

Tropidoclonion [Genus]
    T. lineatum [Species]
Tropidodipsas [Genus]
    T. fasciata [Species]
    T. philippii
    T. zweifeli
Tropidodryas [Genus]
    T. serra [Species]
    T. striaticeps
Tropidonophis [Genus]
    T. aenigmaticus [Species]
    T. dahlii
    T. dendrophiops
    T. doriae
    T. elongatus
    T. halmahericus
    T. hypomelas
    T. mairii
    T. mcdowelli
    T. montanus
    T. multiscutellatus
    T. negrosensis
    T. novaeguineae
    T. parkeri
    T. picturatus
    T. punctiventris
    T. statistictus
    T. truncatus
Umbrivaga [Genus]
    U. mertensi [Species]
    U. pyburni
    U. pygmaea
Uromacer [Genus]
    U. catesbyi [Species]
    U. frenatus
    U. oxyrhynchus
    U. ricardinii
Uromacerina [Genus]
    U. ricardinii [Species]
Urotheca [Genus]
    U. decipiens [Species]
    U. dumerilli
    U. elapoides
    U. fulviceps
    U. guentheri
    U. lateristriga
    U. multilineata
    U. myersi
    U. pachyura
Virginia [Genus]
    V. pulchra [Species]
    V. striatula
    V. valeriae
Waglerophis [Genus]
    W. merremi [Species]
Xenelaphis [Genus]
    X. ellipsifer [Species]
    X. hexagonotus
Xenochrophis [Genus]
    X. asperrimus [Species]

X. bellula
X. cerasogaster
X. flavipunctatum
X. maculatus
X. piscator
X. punctulatus
X. sanctijohannis
X. trianguligerus
X. vittatus
Xenodermus [Genus]
  X. javanicus [Species]
Xenodon [Genus]
  X. guentheri [Species]
  X. neuwiedii
  X. rabdocephalus
  X. severus
  X. werneri
Xenopholis [Genus]
  X. scalaris [Species]
  X. undulatus
Xenoxybelis [Genus]
  X. argenteus [Species]
Xylophis [Genus]
  X. perroteti [Species]
  X. stenorhynchus

Elapidae [Family]
  Acalyptophis [Genus]
    A. peronii [Species]
  Acanthophis [Genus]
    A. antarcticus [Species]
    A. barnetti
    A. crotalusei
    A. cummingi
    A. hawkei
    A. praelongus
    A. pyrrhus
    A. wellsei
    A. woolfi
  Aipysurus [Genus]
    A. apraefrontalis [Species]
    A. duboisii
    A. eydouxii
    A. foliosquama
    A. fuscus
    A. laevis
    A. pooleorum
    A. tenuis
  Aspidelaps [Genus]
    A. lubricus [Species]
    A. scutatus
  Aspidomorphus [Genus]
    A. lineaticollis [Species]
    A. muelleri
    A. schlegeli
  Astrotia [Genus]
    A. stokesii [Species]
  Austrelaps [Genus]
    A. labialis [Species]
    A. ramsayi
    A. superbus

Boulengerina [Genus]
  B. annulata [Species]
  B. christyi
Bungarus [Genus]
  B. andamanensis [Species]
  B. bungaroides
  B. caeruleus
  B. candidus
  B. ceylonicus
  B. fasciatus
  B. flaviceps
  B. lividus
  B. magnimaculatus
  B. multicinctus
  B. niger
  B. sindanus
Cacophis [Genus]
  C. churchilli
  C. harriettae
  C. krefftii
  C. squamulosus
Calliophis [Genus]
  C. beddomei [Species]
  C. bibroni
  C. calligaster
  C. gracilis
  C. kelloggi
  C. macclellandi
  C. maculiceps
  C. melanurus
  C. nigrescens
Demansia [Genus]
  D. atra
  D. calodera
  D. olivacea
  D. papuensis
  D. psammophis
  D. rufescens
  D. simplex
  D. torquata
Dendroaspis [Genus]
  D. angusticeps [Species]
  D. jamesoni
  D. polylepis
  D. viridis
Denisonia [Genus]
  D. devisi [Species]
  D. maculata
Disteira [Genus]
  D. kingii [Species]
  D. major
  D. nigrocincta
  D. walli
Drysdalia [Genus]
  D. coronoides [Species]
  D. mastersii
  D. rhodogaster
Echiopsis [Genus]
  E. atriceps [Species]
  E. curta

Elapognathus [Genus]
  E. coronata [Species]
  E. minor
Elapsoidea [Genus]
  E. broadleyi [Species]
  E. chelazzii
  E. guentherii
  E. laticincta
  E. loveridgei
  E. nigra
  E. semiannulata
  E. sundevallii
  E. trapei
Emydocephalus [Genus]
  E. annulatus [Species]
  E. ijimae
Enhydrina [Genus]
  E. schistosa [Species]
  E. zweifeli
Ephalophis [Genus]
  E. greyi [Species]
Furina [Genus]
  F. barnardi [Species]
  F. diadema
  F. dunmalli
  F. ornata
  F. tristis
Hemachatus [Genus]
  H. haemachatus [Species]
Hemiaspis [Genus]
  H. damelii [Species]
  H. signata
Homoroselaps [Genus]
  H. dorsalis [Species]
  H. lacteus
Hoplocephalus [Genus]
  H. bitorquatus [Species]
  H. bungaroides
  H. stephensii
Hydrelaps [Genus]
  H. darwiniensis [Species]
Hydrophis [Genus]
  H. atriceps [Species]
  H. belcheri
  H. bituberculatus
  H. brooki
  H. caerulescens
  H. coggeri
  H. cyanocinctus
  H. czeblukovi
  H. elegans
  H. fasciatus
  H. inornatus
  H. klossi
  H. laboutei
  H. lamberti
  H. lapemoides
  H. mamillaris
  H. mcdowelli
  H. melanocephalus

H. melanosoma
H. obscurus
H. ornatus
H. pacificus
H. parviceps
H. semperi
H. sibauensis
H. spiralis
H. stricticollis
H. torquatus
H. vorisi
Kerilia [Genus]
 K. jerdonii [Species]
Kolpophis [Genus]
 K. annandalei [Species]
Lapemis [Genus]
 L. curtus [Species]
 L. hardwickii
Laticauda [Genus]
 L. colubrina [Species]
 L. crockeri
 L. laticaudata
 L. schistorhynchus
 L. semifasciata
Loveridgelaps [Genus]
 L elapoides [Species]
Maticora [Genus]
 M. bivirgata [Species]
 M. intestinalis
Micropechis [Genus]
 M. ikaheka [Species]
Micruroides [Genus]
 M. euryxanthus [Species]
Micrurus [Genus]
 M. alleni [Species]
 M. altirostris
 M. ancoralis
 M. annellatus
 M. averyi
 M. bernadi
 M. bocourti
 M. bogerti
 M. browni
 M. catamayensis
 M. clarki
 M. collaris
 M. corallinus
 M. decoratus
 M. diana
 M. diastema
 M. dissoleucus
 M. distans
 M. dumerilii
 M. elegans
 M. ephippifer
 M. filiformis
 M. frontalis
 M. frontifasciatus
 M. fulvius
 M. hemprichii

M. hippocrepis
M. ibiboboca
M. isozonus
M. langsdorffi
M. laticollaris
M. latifasciatus
M. lemniscatus
M. limbatus
M. margaritiferus
M. medemi
M. mertensi
M. mipartitus
M. multifasciatus
M. multiscutatus
M. narduccii
M. nebularis
M. nigrocinctus
M. pacaraimae
M. pachecogili
M. paraensis
M. peruvianus
M. petersi
M. proximans
M. psyches
M. putumayensis
M. pyrrhocryptus
M. remotus
M. ruatanus
M. sangilensis
M. scutiventris
M. spixii
M. spurelli
M. steindachneri
M. stewarti
M. stuarti
M. surinamensis
M. tener
M. tricolor
M. tschudii
Naja [Genus]
 N. annulifera [Species]
 N. atra
 N. haje
 N. kaouthia
 N. katiensis
 N. mandalayensis
 N. melanoleuca
 N. mossambica
 N. naja
 N. nigricollis
 N. nivea
 N. oxiana
 N. pallida
 N. philippinensis
 N. sagittifera
 N. samarensis
 N. siamensis
 N. sputatrix
 N. sumatrana
Notechis [Genus]

N. ater [Species]
N. scutatus
Ogmodon [Genus]
 O. vitianus [Species]
Ophiophagus [Genus]
 O. hannah [Species]
Oxyuranus [Genus]
 O. microlepidotus [Species]
 O. scutellatus
Pailsus [Genus]
 P. pailsei [Species]
Parahydrophis [Genus]
 P. mertoni [Species]
Paranaja [Genus]
 P. multifasciata [Species]
Parapistocalamus [Genus]
 P. hedigeri [Species]
Pelamis [Genus]
 P. platurus [Species]
Pseudechis [Genus]
 P. australis [Species]
 P. butleri
 P. colletti
 P. guttatus
 P. papuanus
 P. porphyriacus
Pseudohaje [Genus]
 P. goldii [Species]
 P. nigra
Pseudonaja [Genus]
 P. affinis [Species]
 P. guttata
 P. inframacula
 P. ingrami
 P. modesta
 P. nuchalis
 P. textilis
Rhinoplocephalus [Genus]
 R. bicolor [Species]
 R. boschmai
 R. incredibilis
 R. nigrescens
 R. nigrostriatus
 R. pallidiceps
Salomonelaps [Genus]
 S. par [Species]
Simoselaps [Genus]
 S. anomalus [Species]
 S. approximans
 S. australis
 S. bertholdi
 S. bimaculatus
 S. calonotus
 S. fasciolatus
 S. incinctus
 S. littoralis
 S. minimus
 S. morrisi
 S. roperi
 S. semifasciatus

S. warro
Suta [Genus]
  S. dwyeri [Species]
  S. fasciata
  S. flagellum
  S. gouldii
  S. monachus
  S. nigriceps
  S. ordensis
  S. punctata
  S. spectabilis
  S. suta

Thalassophina [Genus]
  T. viperina [Species]
Thalassophis [Genus]
  T. anomalus [Species]
Toxicocalamus [Genus]
  T. buergersi [Species]
  T. grandis
  T. holopelturus
  T. longissimus
  T. loriae
  T. misimae
  T. preussi

T. spilolepidotus
T. stanleyanus
Tropidechis [Genus]
  T. carinatus [Species]
Vermicella [Genus]
  V. annulata [Species]
  V. intermedia
  V. multifasciata
  V. snelli
  V. vermiformis
Walterinnesia [Genus]
  W. aegyptia [Species]

· · · · ·

A brief geologic history of animal life

A note about geologic time scales: A cursory look will reveal that the timing of various geological periods differs among textbooks. Is one right and the others wrong? Not necessarily. Scientists use different methods to estimate geological time—methods with a precision sometimes measured in tens of millions of years. There is, however, a general agreement on the magnitude and relative timing associated with modern time scales. The closer in geological time one comes to the present, the more accurate science can be—and sometimes the more disagreement there seems to be. The following account was compiled using the more widely accepted boundaries from a diverse selection of reputable scientific resources.

Geologic time scale

Era	Period	Epoch	Dates	Life forms
Proterozoic			2,500-544 mya*	First single-celled organisms, simple plants, and invertebrates (such as algae, amoebas, and jellyfish)
Paleozoic	Cambrian		544-490 mya	First crustaceans, mollusks, sponges, nautiloids, and annelids (worms)
	Ordovician		490-438 mya	Trilobites dominant. Also first fungi, jawless vertebrates, starfish, sea scorpions, and urchins
	Silurian		438-408 mya	First terrestrial plants, sharks, and bony fish
	Devonian		408-360 mya	First insects, arachnids (scorpions), and tetrapods
	Carboniferous	Mississippian	360-325 mya	Amphibians abundant. Also first spiders, land snails
		Pennsylvanian	325-286 mya	First reptiles and synapsids
	Permian		286-248 mya	Reptiles abundant. Extinction of trilobytes
Mesozoic	Triassic		248-205 mya	Diversification of reptiles: turtles, crocodiles, therapsids (mammal-like reptiles), first dinosaurs
	Jurassic		205-145 mya	Insects abundant, dinosaurs dominant in later stage. First mammals, lizards, frogs, and birds
	Cretaceous		145-65 mya	First snakes and modern fish. Extinction of dinosaurs, rise and fall of toothed birds
Cenozoic	Tertiary	Paleocene	65-55.5 mya	Diversification of mammals
		Eocene	55.5-33.7 mya	First horses, whales, and monkeys
		Oligocene	33.7-23.8 mya	Diversification of birds. First anthropoids (higher primates)
		Miocene	23.8-5.6 mya	First hominids
		Pliocene	5.6-1.8 mya	First australopithecines
	Quaternary	Pleistocene	1.8 mya-8,000 ya	Mammoths, mastodons, and Neanderthals
		Holocene	8,000 ya-present	First modern humans

*Millions of years ago (mya)

Index

Bold page numbers indicate the primary discussion of a topic; page numbers in italics indicate illustrations; "t" indicates table.

A

Abaco boas, 7:410
Abronia spp. *See* Alligator lizards
Abronia aurita. See Coban alligator ligzards
Acanthodactylus spp., 7:297
Acanthophis spp. *See* Death adders
Acanthophis antarcticus. See Death adders
Acontias spp., 7:327, 7:328
Acontinae, 7:327
Acontophiops spp., 7:327
Acridophaga spp., 7:445, 7:447
Acrochordidae. *See* File snakes
Acrochordus spp. *See* File snakes
Acrochordus arafurae. See Arafura file snakes
Acrochordus granulatus. See Little file snakes
Acrochordus javanicus. See Java file snakes
Acutotyphlops spp., 7:380, 7:381
Acutotyphlops subocularis. See Bismarck
  blindsnakes
Adders
  African puff, 7:27, 7:447
  common, 7:447, 7:448, 7:449, 7:459–460
  death, 7:39, **7:483–488**, 7:489, **7:493–494**
  dwarf puff, 7:446
  Gaboon, 7:446, 7:447, 7:449, 7:451, 7:457
  night, 7:445–448, 7:486
  Rhombic night, 7:449, 7:451–452
  See also Eastern hog-nosed snakes
Adenorhinos spp., 7:445
Adolfus spp., 7:297
Aesculapius, 7:54–55
Aetosaurs, 7:17–18
African burrowing snakes, **7:461–464**
African egg-eaters. *See* Common egg-eaters
African pancake tortoises. *See* Pancake
  tortoises
African puff adders, 7:27, 7:447
African religions, snakes in, 7:55
African rock pythons, 7:55, 7:420, 7:421
African sideneck turtles, **7:129–134**, 7:131
African slender-snouted crocodiles, 7:180
African watersnakes, 7:467, 7:468
Afro-American river turtles, **7:137–142**, 7:139
Agama hispida. See Spiny agamas
Agamidae, 7:204, **7:209–222**, 7:212–213
  behavior, 7:211
  conservation status, 7:211
  distribution, 7:209, 7:210
  evolution, 7:209
  feeding ecology, 7:211
  habitats, 7:210
  humans and, 7:211
  physical characteristics, 7:209–210

  reproduction, 7:211
  species of, 7:214–221
  taxonomy, 7:209
Agaminae, 7:209
Agamodon spp., 7:287, 7:288
Agamodon anguliceps, 7:289
Agkistrodon spp., 7:445, 7:446, 7:448
Agkistrodon bilineatus, 7:448
Agkistrodon contortrix. See Southern
  copperheads
Agkistrodon piscivorus. See Cottonmouths
Ahaetulla spp., 7:466
Aida-Wedo (Loa god), 7:57
Aipysurus laevis. See Olive seasnakes
Aldabra tortoises, 7:47, 7:70, 7:72
Alethinophidia, 7:198, 7:201
Alligator lizards, **7:339–344**, 7:341
Alligator mississippiensis. See American
  alligators
Alligator sinensis. See Chinese alligators
Alligator snappers. *See* Alligator snapping
  turtles
Alligator snapping turtles, 7:94, 7:96
  behavior, 7:35, 7:39
  as food, 7:47
  physical characteristics, 7:70
  See also Snapping turtles
Alligatoridae, 7:19, **7:171–178**, 7:174
  See also Alligators; Caimans
Alligatorinae. *See* Alligators
Alligators, 7:32, 7:60, 7:157–165, **7:171–178**,
  7:174
  See also American alligators
Alopoglossus spp., 7:304
Alopoglossus angulatus, 7:304
Alsophis sanctaecrucis, 7:470
Amazon tree boas, 7:410, 7:411
Amazonian snail-eaters, 7:471, 7:479, 7:480
Amba Mata (Indian goddess), 7:56
Amblyodipsas spp. *See* Purple-glossed snakes
Amblyrhynchus cristatus. See Marine iguanas
Ameiva spp., 7:309, 7:310, 7:311, 7:312
Ameiva ameiva. See Giant ameivas
Ameiva polops. See St. Croix ground lizards
American alligators, 7:106, 7:172, 7:174,
  7:175
  distribution, 7:175
  evolution, 7:171
  farming, 7:48–49
  as food, 7:48
  humans and, 7:52
  physical characteristics, 7:171
  reproduction, 7:4, 7:36, 7:42, 7:157, 7:161,
    7:163, 7:164, 7:173

American chameleons. *See* Green anoles
American crocodiles, 7:162, 7:179, 7:183,
  7:184
American Medical Association, icon of, 7:55
American mud turtles, **7:121–127**, 7:124
American saltwater crocodiles. *See* American
  crocodiles
Ammonia toxicity, 7:3
Amniotes, 7:12–13, 7:14, 7:15
Amphibians, *vs.* reptiles, 7:3
Amphiglossus astrolabi, 7:329
Amphisbaena alba. See White-bellied
  wormlizards
Amphisbaenidae. *See* Wormlizards
Anacondas, 7:202, 7:409, 7:410, 7:411
  See also Green anacondas
Anadia spp., 7:303
Anapsida, 7:4–5, 7:12–14
Anelytropsis spp., 7:272
Angleheads, **7:209–222**
Angolan pythons, 7:420
Anguidae, 7:206, 7:207, **7:339–345**, 7:341
Anguimorpha, 7:196, 7:198
Anguinae, 7:339, 7:340
  See also Glass lizards
Anguis spp. *See* Slowworms
Anguis fragilis, 7:339, 7:340
Aniliidae. *See* False coral snakes
Anilioidea, 7:198
Anilius scytale. See False coral snakes
Annam leaf turtles, 7:119
Anniella spp. *See* Legless lizards
Anniella pulchra. See California legless lizards
Anniellinae. *See* Legless lizards
Anoles, **7:243–257**, 7:248
  behavior, 7:244–245
  conservation status, 7:246–247
  distribution, 7:243, 7:244
  evolution, 7:243
  feeding ecology, 7:245–246
  habitats, 7:244
  humans and, 7:247
  physical characteristics, 7:243–244
  reproduction, 7:207, 7:246
  species of, 7:250–257
  taxonomy, 7:243
  See also Green anoles
Anolis spp. *See* Anoles
Anolis carolinensis. See Green anoles
Anomalepididae. *See* Early blindsnakes
Anomalepis spp., 7:369, 7:370–371
Anomalepis aspinosus, 7:371
Anomalopus spp., 7:328
Anomochilidae. *See* False blindsnakes

Anomochilus spp. *See* False blindsnakes
Anomochilus leonardi, 7:*387*, 7:*388*
Anomochilus weberi, 7:387, 7:*388*
Ant hill pythons. *See* Pygmy pythons
Antaresia spp., 7:420, 7:421
Antaresia perthensis. See Pygmy pythons
Antsingy leaf chameleons. *See* Armored
  chameleons
Apalone spp. *See* Snapping turtles
Apalone spinifera. See Spiny softshells
Aparallactinae, 7:461
Aparallactus spp. *See* Centipede eaters
Aparallactus jacksonii. See Jackson's centipede
  eaters
Aphrodite (Greek goddess), 7:54
Apodora spp., 7:420, 7:421
Apodora papuana. See Papuan pythons
Apostolepis spp., 7:467, 7:468
Arabian sandboas, 7:409–411
Arafura file snakes, 7:439–*442*, 7:*443*
Archosauria, 7:16, 7:17–19
Arizona brush lizards. *See* Common sagebrush
  lizards
Armadillo lizards, 7:*320–321*
Armored chameleons, 7:232, 7:*234*,
  7:*235–236*
Arraus. *See* South American river turtles
Arthrosaura spp., 7:303
Artrellias. *See* Crocodile monitors
Asian giant softshell turtles, 7:152
Asian giant tortoises, 7:71
Asian grass lizards, 7:298
Asian kukrisnakes, 7:467, 7:469
Asian medicine, reptiles in, 7:50–51
Asian narrow-headed softshell turtles, 7:70,
  7:*152*
Asian river turtles, 7:66
Aspidites spp., 7:419, 7:420, 7:421
Aspidites melanocephalus. See Black-headed
  pythons
Aspidites ramsayi. See Womas
Athena (Greek goddess), 7:54
Atheris spp., 7:445, 7:447
Atheris squamigera. See Green bush vipers
Atlantic ridleys. *See* Kemp's ridley turtles
Atractaspididae. *See* African burrowing snakes
Atractaspidinae. *See Atractaspis* spp.
Atractaspis spp., 7:461, 7:483
Atractaspis bibronii. See Southern burrowing
  asps
Atractaspis microlepidota. See Small-scaled
  burrowing asps
Atropoides spp., 7:445
Atropoides nummifer, 7:447, 7:448
Audition. *See* Ears
Auffenberg, Walter, 7:53, 7:361
Australian freshwater crocodiles. *See*
  Johnstone's crocodiles
Australian pygmy monitors. *See* Short-tailed
  monitors
Australian sand goannas, 7:37, 7:*360*, 7:361,
  7:*362*
Australian skinks, 7:202
Australo-American sideneck turtles, **7:77–84**
Australo-Papuan elapids, 7:483, 7:484, 7:486
Austrelaps labialis, 7:486
Autarchoglossa, 7:197–198, 7:200, 7:202,
  7:203–204, 7:206, 7:207
Avila-Pires, T. C. S., 7:307
Azemiopinae. *See* Fea's vipers

Azemiops feae. See Fea's vipers
Aztecs, reptiles and, 7:55

B

Bachia spp., 7:303
Bachia bresslaui, 7:*305*, 7:*306*
Balanophis spp., 7:468
Ball pythons, 7:*23*, 7:420, 7:*423*, 7:427
Banana boas. *See* Southern bromeliad
  woodsnakes
Banded dwarf boas. *See* Banded woodsnakes
Banded ratsnakes, 7:466, 7:470
Banded seasnakes. *See* Sea kraits
Banded woodsnakes, 7:434, 7:*436*
Bandtailed earless lizards. *See* Common lesser
  earless lizards
Bandy-bandy snakes, 7:486, 7:*489*, 7:*494*,
  7:498
Barisia spp. *See* Alligator lizards
Barkudia spp., 7:328
Barrington iguanas, 7:*244*
Basiliscus basiliscus. See Common basilisks
Basiliscus plumifrons. See Green basilisks
Batrachemys spp., 7:77
Batrachemys dahli, 7:77
Beaded lizards, Mexican. *See* Mexican beaded
  lizards
Bearded dragons, 7:*41*, 7:*205*, 7:*212*, 7:*218*,
  7:219
Bearded pygmy chameleons. *See* Short-tailed
  chameleons
Bearded toad heads. *See* Toad-headed agamas
Beauty lizards. *See* Brown garden lizards
Behavior, **7:34–46**
  activity patterns, 7:43–45
  African burrowing snakes, 7:462–464
  African sideneck turtles, 7:129, 7:132–133
  Afro-American river turtles, 7:137, 7:140–141
  Agamidae, 7:211, 7:214–221
  Alligatoridae, 7:172–173, 7:175–177
  Anguidae, 7:340, 7:342–344
  Australo-American sideneck turtles, 7:78,
    7:81–83
  big-headed turtles, 7:135
  blindskinks, 7:272
  blindsnakes, 7:381–385
  boas, 7:411, 7:413–417
  Central American river turtles, 7:100
  chameleons, 7:228–230, 7:235–241
  chemosensory systems and, 7:34–38
  colubrids, 7:468–469, 7:473–481
  Cordylidae, 7:320–321, 7:324–325
  Crocodylidae, 7:180, 7:184–187
  early blindsnakes, 7:371
  Elapidae, 7:485, 7:491–498
  false blindsnakes, 7:388
  false coral snakes, 7:400
  file snakes, 7:441
  Florida wormlizards, 7:284
  Gekkonidae, 7:261, 7:265–269
  Geoemydidae, 7:116, 7:118–119
  gharials, 7:168
  Helodermatidae, 7:354–355
  Iguanidae, 7:244–*245*, 7:250–257
  Kinosternidae, 7:121–122, 7:125–126
  knob-scaled lizards, 7:348, 7:351
  Lacertidae, 7:298, 7:301–302

  learning, 7:39–40
  leatherback seaturtles, 7:101
  Microteiids, 7:304, 7:306–307
  mole-limbed wormlizards, 7:280, 7:281
  Neotropical sunbeam snakes, 7:406
  New World pond turtles, 7:105–106,
    7:109–113
  night lizards, 7:292, 7:294–295
  pig-nose turtles, 7:76
  pipe snakes, 7:396
  play, 7:39
  pythons, 7:420–421, 7:424–428
  seaturtles, 7:*86*, 7:89–91
  shieldtail snakes, 7:392–394
  skinks, 7:329, 7:332–337
  slender blindsnakes, 7:375–377
  snapping turtles, 7:93, 7:95–96
  softshell turtles, 7:152, 7:154–155
  spade-headed wormlizards, 7:288, 7:289
  splitjaw snakes, 7:430
  Squamata, 7:204–206
  sunbeam snakes, 7:402
  tactile cues, 7:40–42
  Teiidae, 7:311, 7:314–316
  tortoises, 7:143, 7:146–148
  Tropidophiidae, 7:434, 7:436–437
  tuatara, 7:190
  Varanidae, 7:360–361, 7:365–368
  Viperidae, 7:447, 7:451–459
  vision and, 7:38–39
  wormlizards, 7:274, 7:276
Behavioral fevers, 7:45
Bell geckos, 7:262
Bell's dab lizards. *See* Spiny-tailed agamas
Bengal monitor lizards, 7:52, 7:*360*, 7:361
Bern Convention of European Wildlife and
  Natural Habitats. *See* Convention on the
  Conservation of European Wildlife and
  Natural Habitats
Bibron, G., 7:379
Bibron's burrowing asps. *See* Southern
  burrowing asps
Bibron's geckos, 7:261
Big-headed turtles, 7:32, **7:135–136**, 7:*139*,
  7:*141*–142
Biogeography, Squamata, 7:203–204
Bipedidae. *See* Mole-limbed wormlizards
Bipes spp. *See* Mole-limbed wormlizards
Bipes biporus. See Two-legged wormlizards
Bipes canaliculatus. See Tropical wormlizards
Bipes tridactylus, 7:280
Bird snakes, 7:466
Bismarck blindsnakes, 7:380, 7:382,
  7:*384–385*
Bitia spp. *See* Keel-bellied watersnakes
Bitia hydroides. See Keel-bellied watersnakes
Bitis spp., 7:445, 7:446, 7:448
Bitis arietans. See African puff adders
Bitis gabonica. See Gaboon adders
Bitis nasicornis, 7:447
Bitis peringueyi, 7:447
Black and white tegus, 7:*311*, 7:312
Black-breasted leaf turtles, 7:7, 7:*116*
Black caimans, 7:157, 7:171, 7:173
Black-headed bushmasters, 7:*450*, 7:*452*,
  7:456
Black-headed pythons, 7:*423*, 7:*424*, 7:425
Black-headed snakes, 7:466
Black mambas, 7:*490*, 7:*491*
Black marsh turtles, 7:72

Black-necked spitting cobras, 7:*490*,
7:*492*–493
Black pythons, 7:420
Black ratsnakes, 7:43
Black snakes, 7:462, 7:485
Black threadsnakes, 7:*374*, 7:*485*
Black turtles. *See* Pacific green turtles
Blackish blindsnakes, 7:382, 7:*383*–*384*
Blacktails. *See* Indigo snakes
Blanding's turtles, 7:106
Blanus cinereus, 7:274
Bleached earless lizards. *See* Common lesser
earless lizards
Blindskinks, **7:271–272**
Blindsnakes, **7:379–385**, 7:*380*, 7:*381*
early, 7:198, 7:201, 7:202, **7:369–372**
false, **7:387–388**
skulls of, 7:201
slender, 7:371, **7:373–377**
spotted, 7:*197*
taxonomy, 7:198
Blood, crocodilian, 7:161
Blood pythons, 7:*423*, 7:*426*–427
Bloodsuckers. *See* Brown garden lizards
Blotched pipe snakes, 7:*395*, 7:*396*
Blue-tongued skinks, 7:*328*
Blunt-headed vinesnakes, 7:468
Blunt-nosed vipers. *See* Levantine vipers
Boa spp., 7:410, 7:411
Boa constrictors, 7:*409–411*, 7:*412*, 7:*413*
Boa dumerili. *See* Dumeril's boas
Boa madagascariensis. *See* Madagascar boas
Boa mandrita. *See* Madagascar tree boas
Boas, 7:26, 7:40, 7:42, 7:*197*, 7:205,
7:409–417, 7:*412*
See also specific types of boas
Bobtails, 7:*329*, 7:*331*, 7:*332*, 7:336
Bockadams. *See* Dog-faced watersnakes
Bog turtles, 7:105
Bogert, Charles M., 7:43
Boggi. *See* Bobtails
Böhme, Wolfgang, 7:223–224
Boidae. *See* Boas
Boiga irregularis. *See* Brown treesnakes
Boinae, 7:206, 7:409–411
Bolyeria multocarinata. *See* Smooth-scaled
splitjaws
Bolyeriidae. *See* Splitjaw snakes
Bon Dieu Loa, 7:57
Booidea, 7:198
Boomslangs, 7:466, 7:467, 7:*472*, 7:*474*–475
Borneo crocodiles, 7:179
Bothriechis spp., 7:445, 7:447
Bothriechis aurifer. *See* Yellow-blotched palm-
pitvipers
Bothriechis schlegelii. *See* Gold morph of
eyelash vipers
Bothriopsis spp., 7:445, 7:447
Bothriopsis bilineata, 7:447
Bothrochilus spp., 7:420, 7:421
Bothrochilus boa. *See* Ringed pythons
Bothrocophias spp., 7:445
Bothrocophias hyoprora, 7:447
Bothrops spp., 7:445
Bothrops alcatraz, 7:448
Bothrops ammodytoides. *See* Patagonian
lanceheads
Bothrops asper, 7:448
Bothrops insularis, 7:448
Box turtles, 7:*106*

Chinese three-striped, 7:119
eastern, 7:*68*, 7:106, 7:*108*, 7:112–*113*
Malayan, 7:*12*, 7:*115*
ornate, 7:72
yellow-margined, 7:*117*, 7:*118*–119
Brachylophus spp., 7:244
Brachymeles spp., 7:327, 7:328
Brachyophidium rhodogaster, 7:391
Bradypodion spp., 7:37, 7:224, 7:227, 7:232
Bradypodion thamnobates. *See* KwaZulu-Natal
Midlands dwarf chameleons
Brain, 7:8
See also Physical characteristics
Braswell, A. L., 7:315
Broad-banded sand swimmers, 7:*331*, 7:*333*,
7:334–335
See also Half-girdled snakes
Broad-headed skinks, 7:*328*, 7:*330*, 7:*336*–337
Broad-headed snakes, 7:485, 7:486
Broad-snouted caimans, 7:164, 7:171
Bromeliad boas. *See* Southern bromeliad
woodsnakes
Bromeliad dwarf boas. *See* Southern
bromeliad woodsnakes
Bromeliad woodsnakes, 7:*436*–437
Brookesia spp., 7:224–225, 7:227
Brookesia perarmata. *See* Armored chameleons
Brown caimans. *See* Common caimans
Brown garden lizards, 7:*213*, 7:*214*–215
Brown roofed turtles, 7:72
Brown sandboas, 7:410
Brown snakes, 7:483, 7:485, 7:*489*, 7:*496*,
7:497–498
Brown treesnakes, 7:470, 7:*472*, 7:*473*
Brush lizards. *See* Common sagebrush lizards
Brygoo, E. R., 7:235
Bufocephala spp., 7:77
Bullsnakes, in Hopi snake dance, 7:57
Bungarus fasciatus, 7:51
Bunker's earless lizards. *See* Common lesser
earless lizards
Burmese pythons, 7:420
Burrowing asps, 7:461
Burrowing boas. *See* Neotropical sunbeam
snakes
Burrowing pythons. *See* Calabar boas
Burrowing snakes
African, **7:461–464**
colubrids, 7:466
Neotropical, 7:468
Nilgiri, 7:*393*
yellow and black, 7:462
Burrowing techniques, 7:274, 7:280, 7:288
See also Behavior
Burton's legless lizards. *See* Burton's snake
lizards
Burton's snake lizards, 7:262, 7:*264*, 7:*265*,
7:269
Bushmasters, 7:*450*, 7:*452*, 7:456
Butterfly agamas, 7:*213*, 7:*215*, 7:220

C

Caduceus, 7:54–55
Caiman crocodilus. *See* Common caimans
Caiman latirostris. *See* Broad-snouted caimans
Caiman lizards, 7:201, 7:309, 7:310, 7:311,
7:*312*, 7:*313*, 7:*316*

Caiman yacare. *See* Yacaré caimans
Caimaninae. *See* Caimans
Caimans, 7:157–165, **7:171–178**, 7:*174*
Calabar boas, 7:409, 7:411, 7:*412*, 7:*415*–416
Calabar ground pythons. *See* Calabar boas
Calamaria spp. *See* Reedsnakes
Caldon turtles. *See* Leatherback seaturtles
Calechidna spp., 7:445
California horned lizards. *See* Texas horned
lizards
California legless lizards, 7:*341*, 7:*342*–*343*
California whiptail lizards, 7:*310*
Callagur borneoensis. *See* Painted terrapins
Callisaurus draconoides. *See* Zebra-tailed lizards
Callopistes spp., 7:309, 7:312
Callopistes flavipunctatus, 7:310
Callopistes maculatus, 7:310
Calloselasma spp., 7:445
Calloselasma rhodostoma. *See* Malaysian
pitvipers
Calodactylodes aureus. *See* Indian golden geckos
Calotes, **7:209–211**, 7:214–215
Calotes liocephalus, 7:211
Calotes versicolor. *See* Brown garden lizards
Calumma spp., 7:37, 7:224, 7:227, 7:232
Calumma boettgeri, 7:224
Calumma nasuta, 7:224
Calumma parsonii. *See* Parson's chameleons
Calyptommatus spp., 7:303
Campan's chameleons. *See* Jeweled
chameleons
Campbell, J., 7:340
Canal turtles. *See* Leatherback seaturtles
Canary Islands giant lizards, 7:59
Candoia spp., 7:409, 7:410
See also Viper boas
Candoia aspera. *See* Viper boas
Candoia bibroni. *See* Fiji Island boas
Canebrakes. *See* Timber rattlesnakes
Caouana turtles. *See* Leatherback seaturtles
Cape filesnakes, 7:*471*, 7:*477*, 7:*478*
Cape flat lizards, 7:*321*, 7:*323*, 7:*324*–325
Cape rough-scaled lizards, 7:298
Cape spinytail iguanas, 7:47–48, 7:51, 7:*248*,
7:*251*–252
Cape terrapins. *See* Helmeted turtles
Captivity, reptiles in, 7:40
See also Humans
Cardon turtles. *See* Leatherback seaturtles
Caretta caretta. *See* Loggerhead turtles
Carettochelyidae. *See* Pig-nose turtles
Carettochelys insculpta. *See* Pig-nose turtles
Caribbean land iguanas, 7:207
Carolina anoles. *See* Green anoles
Carpet vipers. *See* Saw-scaled vipers
Carphophis spp. *See* Wormsnakes
Carroll, Robert
on amniote eggs, 7:12–13
on archosauria, 7:17
on pterodactyloids, 7:19
Casarea dussumieri. *See* Splitjaw snakes
Cat-eyed snakes, 7:469, 7:481
Caucasilacerta spp., 7:297
Caudal autonomy. *See* Tails, regeneration of
Causinae. *See* Night adders
Causus spp. *See* Night adders
Causus rhombeatus. *See* Rhombic night adders
Celestus spp., 7:339
Celestus hylaius, 7:*341*, 7:*343*
Celestus macrotus, 7:339

INDEX

Cemophora spp. *See* Scarletsnakes
Centipede eaters, 7:461
Central American river turtles, **7:99–100**
Central netted dragons, 7:*212*, 7:216
Cerastes spp., 7:445
Cerastes cerastes. See Horned vipers
Ceratophora tennentii. See Leaf-horned agamas
Cerberus rynchops. See Dog-faced watersnakes
Cerrophidion spp., 7:445
Chain vipers. *See* Russel's vipers
Chalcides spp., 7:327, 7:329
Chamaeleo spp., 7:*37*, 7:224, 7:227, 7:232
Chamaeleo calyptratus. See Veiled chameleons
Chamaeleo caroliquarti, 7:223
Chamaeleo chamaeleon. See Common chameleons
Chamaeleo dilepis. See Flap-necked chameleons
Chamaeleo jacksonii. See Jackson's chameleons
Chamaeleo namaquensis, 7:228
Chamaeleo parsonii. See Parson's chameleons
Chamaeleonidae. *See* Chameleons
Chamaesaura spp. *See* Grass lizards
Chamaesaura anguina. See Common grass lizards
Chameleons, **7:223–242**, 7:*233–234*
  behavior, 7:38, 7:228–230
  conservation status, 7:232
  distribution, 7:223, 7:227–228
  evolution, 7:223–224
  eyes, 7:32
  feeding ecology, 7:230
  habitats, 7:228
  humans and, 7:232
  physical characteristics, 7:198, 7:224–227, 7:*225*, 7:*226*, 7:*231*
  reproduction, 7:*230–232*
  species of, 7:235–241
  superstitions and, 7:57
  taxonomy, 7:223–224
  See also specific types of chameleons
Chappell Island tiger snakes, 7:496
Charina spp., 7:410–411
Charina reinhardtii. See Calabar boas
Charina trivirgata. See Rosy boas
Chelidae. *See* Australo-American sideneck turtles
Chelidinae, 7:77
Chelodina longicollis. See Common snakeneck turtles
Chelodina rugosa. See Northern snakenecks
Chelodina siebenrocki. See New Guinea snakeneck turtles
Chelodina steindachneri. See Steindachner's turtles
Chelodininae, 7:77
Chelonia mydas. See Green seaturtles
Chelonians, 7:47, 7:53
  See also Tortoises; Turtles
Cheloniidae. *See* Seaturtles
Chelus fimbriatus. See Matamatas
Chelydra spp. *See* Snapping turtles
Chelydra serpentina. See Snapping turtles
Chelydridae. *See* Snapping turtles
Chemical pollution, 7:60–62
Chemosensory systems, 7:31
  behavior and, 7:34–38
  Squamata, 7:195, 7:198, 7:200, 7:204–205
  See also Physical characteristics
Chicken turtles, 7:106

Chihuahuan collared lizards. *See* Common collared lizards
Chilorhinophis spp. *See* Yellow and black burrowing snakes
China alligators. *See* Chinese alligators
Chinese Alligator Fund, 7:173
Chinese alligators, 7:59, 7:161, 7:171–176, 7:*174*
Chinese Materia Medica, 7:50
Chinese medicine. *See* Asian medicine
Chinese softshell turtles, 7:48, 7:151, 7:152, 7:*153*, 7:*154*, 7:155
Chinese stripe-necked turtles, 7:*117*, 7:*118*, 7:119
Chinese three-striped box turtles, 7:119
Chionactis spp. *See* Shovel-nosed snakes
Chirindia spp., 7:273
Chitra chitra. *See* Asian narrow-headed softshell turtles
Chitra indica. *See* Asian narrow-headed softshell turtles
Chlamydosaurus kingi. See Frilled lizards
Christianity, snakes in, 7:57
Christinus marmoratus. See Marbled geckos
Christmas Island blindsnakes, 7:381, 7:382
Chromatophores, 7:*226*
Chrysemys picta. See Painted turtles
Chrysemys picta picta. See Eastern painted turtles
Chrysopelea spp. *See* Flyingsnakes
Chuckawallas. *See* Common chuckwallas
Chucks. *See* Common chuckwallas
Cihuacoatl (Mayan goddess), 7:55
Circulatory system, 7:28–29, 7:67, 7:69–70
  See also Physical characteristics
Cistoclemmys flavomarginata. See Yellow-margined box turtles
CITES. *See* Convention on International Trade in Endangered Species
Clark's lacerta, 7:299
Claudius spp., 7:121
Clelia spp. *See* Mussaranas
Clelia clelia. See Mussuranas
Clemmys guttata. See Spotted turtles
Climatic change, reptiles and, 7:61
Cloaked lizards. *See* Frilled lizards
Cnemidophorus spp., 7:309, 7:310, 7:311, 7:312
Cnemidophorus gularis. See Texas spotted whiptails
Cnemidophorus inornatus. See Little striped whiptails
Cnemidophorus laredoensis. See Laredo striped whiptails
Cnemidophorus sexlineatus. See Six-lined racerunners
Cnemidophorus tigris, 7:312
Cnemidophorus tigris mundus. See California whiptail lizards
Cnemidophorus uniparens. See Desert grassland whiptails
Cnemidophorus vanzoi. See St. Lucia whiptails
Coachella sand-lizards. *See* Coachella Valley fringe-toed lizards
Coachella uma. *See* Coachella Valley fringe-toed lizards
Coachella Valley fringe-toed lizards, 7:207, 7:*248*, 7:*253*, 7:255
Coban alligator lizards, 7:*341*, 7:*343*–344
Cobras, 7:42, 7:56, 7:205–206, **7:483–488**, 7:*490*, **7:492–493**

Coeranoscincus spp., 7:328
Coffeesnakes, 7:468
Cogger, H. G., 7:216
Cole, Charles J., 7:306
Coleonyx variegatus. See Western banded geckos
Coleonyx variegatus abbotti. See San Diego banded geckos
Collared lizards, 7:243, 7:244, 7:*249*, 7:*250*
Colli, Guarino R., 7:306
Coloptychon spp. *See* Alligator lizards
Coloration, 7:30
  chameleons, 7:224–225, 7:*226*, 7:227
  crocodilians, 7:158–159
  Squamata, 7:203
  See also Physical characteristics
Coluber spp. *See* Racers
Coluber constrictor. See Eastern racers
Colubridae. *See* Colubrids
Colubrids, **7:465–482**, 7:*471*, 7:*472*
  behavior, 7:468–469
  conservation status, 7:470
  distribution, 7:465, 7:467
  evolution, 7:465–467
  feeding ecology, 7:469
  habitats, 7:468
  humans and, 7:470
  physical characteristics, 7:467
  reproduction, 7:469–470
  species of, 7:*473–481*
  taxonomy, 7:198, 7:465–467
Colubrinae, 7:466, 7:468
Colubroidea. *See* Colubrids
Combat behaviors. *See* Behavior
Common adders, 7:447, 7:448, 7:*449*, 7:*459–460*
Common basilisks, 7:25
Common bush vipers. *See* Green bush vipers
Common caimans, 7:161, 7:163, 7:164, 7:171, 7:173, 7:*174*, 7:*176–177*
Common chameleons, 7:228, 7:229, 7:*233*, 7:*237*
Common chuckwallas, 7:245, 7:248, 7:*250*, 7:*252–253*
Common collared lizards, 7:*249*, 7:*250*
Common egg-eaters, 7:*469*, 7:*472*, 7:*474*
Common garter snakes, 7:469, 7:*471*, 7:*478–479*
Common grass lizards, 7:*320*
Common iguanas. *See* Green iguanas
Common kingsnakes, 7:470
Common lesser earless lizards, 7:*248*, 7:*252*, 7:*253–254*
Common musk turtles. *See* Stinkpots
Common night adders. *See* Rhombic night adders
Common plate-tailed geckos, 7:*264*, 7:*267*, 7:*268*
Common rough-scaled lizards, 7:298
Common sagebrush lizards, 7:*249*, 7:*255–256*
Common side-blotched lizards, 7:246, 7:*249*, 7:*255*, 7:*256*
Common slug-eaters, 7:*471*, 7:*477*
Common snakeneck turtles, 7:78, 7:*80*, 7:*82–83*
Common snapping turtles. *See* Snapping turtles
Common sunbeam snakes, 7:*401*, 7:*402*
Conant, Roger, 7:455
Concertina locomotion, 7:26, 7:29

See also Behavior
Cone-headed chameleons. *See* Veiled chameleons
Conolophus pallidus. See Barrington iguanas
Conservation status, 7:10–11, **7:59–63**
  African burrowing snakes, 7:462–464
  African sideneck turtles, 7:130, 7:132–133
  Afro-American river turtles, 7:138, 7:140–142
  Agamidae, 7:211, 7:214–221
  Alligatoridae, 7:173, 7:175–177
  American alligators, 7:48–49
  Anguidae, 7:340, 7:342–344
  Australo-American sideneck turtles, 7:78, 7:81–83
  big-headed turtles, 7:136
  blindskinks, 7:272
  blindsnakes, 7:382–385
  boas, 7:411, 7:413–417
  Central American river turtles, 7:100
  chameleons, 7:232, 7:235–241
  colubrids, 7:470, 7:473–481
  Cordylidae, 7:322, 7:324–325
  crocodilians, 7:10–11
  Crocodylidae, 7:181–182, 7:184–187
  early blindsnakes, 7:371
  Elapidae, 7:486–487, 7:491–498
  false blindsnakes, 7:388
  false coral snakes, 7:400
  file snakes, 7:442
  Florida wormlizards, 7:285
  Gekkonidae, 7:262–263, 7:265–269
  Geoemydidae, 7:116, 7:118–119
  gharials, 7:169
  Helodermatidae, 7:357
  Iguanidae, 7:246–247, 7:250–257
  Kinosternidae, 7:122–123, 7:125–126
  knob-scaled lizards, 7:349, 7:351
  Lacertidae, 7:298–299, 7:301–302
  leatherback seaturtles, 7:102
  Microteiids, 7:304, 7:306–307
  mole-limbed wormlizards, 7:280, 7:281
  Neotropical sunbeam snakes, 7:406
  New World pond turtles, 7:106–107, 7:109–113
  night lizards, 7:292, 7:294–295
  pig-nose turtles, 7:76
  pipe snakes, 7:397
  pythons, 7:421–422, 7:424–428
  seaturtles, 7:87, 7:89–91
  shieldtail snakes, 7:392–394
  skinks, 7:329, 7:332–337
  slender blindsnakes, 7:375–377
  snapping turtles, 7:94, 7:95–96
  softshell turtles, 7:152, 7:154–155
  spade-headed wormlizards, 7:288, 7:289
  splitjaw snakes, 7:430–431
  Squamata, 7:207
  sunbeam snakes, 7:403
  Teiidae, 7:312, 7:314–316
  Testudines, 7:72–73
  tortoises, 7:10, 7:72–73, 7:144, 7:146–149
  Tropidophiidae, 7:435–437
  tuatara, 7:191
  turtles, 7:10–11, 7:72–73
  Varanidae, 7:363, 7:365–368
  Viperidae, 7:448, 7:451–460
  wormlizards, 7:274, 7:276
  See also Extinct species; Humans
Continental drift, Squamata and, 7:196

Convention on International Trade in Endangered Species
  American crocodiles, 7:184
  Central American river turtles, 7:100
  chameleons, 7:232
  colubrids, 7:470
  common chameleons, 7:238
  Cordylidae, 7:322
  gharials, 7:169
  Helodermatidae, 7:357
  Jackson's chameleons, 7:239
  jeweled chameleons, 7:239
  Johnstone's crocodiles, 7:185
  leatherback seaturtles, 7:102
  Madagascar day geckos, 7:268
  minor chameleons, 7:240
  mugger crocodiles, 7:185
  mussuranas, 7:480
  Neotropical sunbeam snakes, 7:406
  Nile crocodiles, 7:186
  panther chameleons, 7:240
  Parson's chameleons, 7:236–237
  Phelsuma spp., 7:263
  prehensile-tailed skinks, 7:333
  pythons, 7:421
  Russel's vipers, 7:458
  saltwater crocodiles, 7:187
  Teiidae, 7:312
  Varanidae, 7:363
  veiled chameleons, 7:237
  Viperidae, 7:448
Convention on the Conservation of European Wildlife and Natural Habitats
  Lacertidae, 7:299
  smooth snakes, 7:474
Cooper Jr., William E., 7:348
Cope, Edward Drinker
  on blindsnakes, 7:379
  on Gila monsters, 7:353
Cope's lizards. *See* Long-nosed leopard lizards
Cophotis spp., 7:211
Copperheads
  in religions, 7:57
  reproduction, 7:*36*
  southern, 7:52, 7:447, 7:448
Coral snakes
  behavior, 7:468
  coloration, 7:203
  distribution, 7:484
  feeding ecology, 7:486
  North American, 7:*490*, 7:*491*–492
  taxonomy, 7:483
  venom and, 7:206
Corallus spp., 7:410, 7:411
Corallus caninus. See Emerald tree boas
Corallus cropanii, 7:411
Corallus hortulanus. See Amazon tree boas
Cordylidae, 7:204, 7:206, **7:319–325**, 7:*323*
Cordylinae. *See* Girdled lizards
Cordylosaurus subtessellatus. See Dwarf plated lizards
Cordylus spp. *See* Girdled lizards
Cordylus cataphractus. See Armadillo lizards
Cordylus giganteus. See Giant girdled lizards
Cornsnakes, 7:*39*, 7:470
Coronella austriaca. See Smooth snakes
Corucia spp., 7:329
Corucia zebrata. See Prehensile-tailed skinks
Corytophanes percarinatus, 7:246

Corytophaninae, 7:243–244
Cottonmouths, 7:*35*, 7:37, 7:57, 7:447, 7:*450*, 7:*452*–453
Cotylosauria, 7:4–5
Courtship. *See* Reproduction
Crag lizards, 7:319
Crested lizards. *See* Desert iguanas
Cribos. *See* Indigo snakes
Cricosaura spp., 7:291
Cricosaura typica. See Cuban night lizards
Crocodile monitors, 7:*364*, 7:*365*, 7:368
Crocodile tegus, 7:309, 7:311, 7:*313*, 7:*315*–*316*
Crocodiles, 7:59, 7:157–165, **7:179–188**, 7:*183*
  See also Nile crocodiles; Saltwater crocodiles
Crocodilians, **7:157–165**
  conservation status, 7:10–11, 7:59, 7:61, 7:164
  evolution, 7:18–19, 7:157
  farming, 7:48
  as food, 7:48
  heart, 7:10
  humans and, 7:52–53
  integumentary system, 7:29
  limbs, 7:25
  in mythology, 7:55
  reproduction, 7:6–7, 7:42, 7:163–164
  salt glands, 7:30
  skeleton, 7:23–24
  skulls, 7:5, 7:26
  superstitions and, 7:57
  teeth, 7:9, 7:26
  territoriality, 7:44
  water balance, 7:30
Crocodilurus sp. *See* Crocodile tegus
Crocodilurus lacertinus. See Crocodile tegus
Crocodylidae, **7:179–188**, 7:*183*
  See also Crocodiles
Crocodylinae. *See* Crocodiles
Crocodylomorpha, 7:157
Crocodylus acutus. See American crocodiles
Crocodylus cataphractus. See African slender-snouted crocodiles
Crocodylus johnstonii. See Johnstone's crocodiles
Crocodylus niloticus. See Nile crocodiles
Crocodylus novaeguineae. See New Guinea crocodiles
Crocodylus palustris. See Mugger crocodiles
Crocodylus porosus. See Saltwater crocodiles
Crocodylus raninus. See Borneo crocodiles
Crocodylus rhombifer. See Cuban crocodiles
Crotalinae. *See* Pitvipers
Crotalus spp., 7:445, 7:446, 7:448
Crotalus adamanteus. See Eastern diamondback rattlesnakes
Crotalus atrox. See Western diamondback rattlesnakes
Crotalus cerastes. See Sidewinding
Crotalus durissus. See Sidewinding, 7:55, 7:448
Crotalus horridus. See Timber rattlesnakes
Crotalus viridis. See Prairie rattlesnakes
Crotalus willardi, 7:447
Crotaphytinae, 7:243, 7:244
Crotaphytus spp., 7:246
Crotaphytus collaris. See Common collared lizards
Cryptodira, 7:13–14, 7:65
  See also Turtles
Ctenophorus inermis. See Central netted dragons

INDEX

Ctenophorus isolepis. See Military dragons
Ctenosaura hemilopha. See Cape spinytail
  iguanas
Ctenotus spp., 7:329
Ctenotus quattuordecimlineatus. See Fourteen-
  lined comb eared skinks
Cuban boas, 7:*412,* 7:*413,* 7:415
Cuban crocodiles, 7:160
Cuban ground boas, 7:*434*–435
Cuban night lizards, 7:291, 7:*293,* 7:*294,*
  7:295
Cuora amboinensis. See Malayan box turtles
Cuora trifasciata. See Chinese three-striped
  box turtles
Cuvier's dwarf caimans, 7:157, 7:171
Cyclanorbinae, 7:151
Cyclodomorphus branchialis, 7:*331,* 7:*333*–334
Cyclotyphlops spp. *See Cyclotyphlops deharvengi*
Cyclotyphlops deharvengi, 7:380, 7:381
Cyclura spp. *See* Caribbean land iguanas
Cyclura carinata, 7:245
Cyclura carinata carinata. See Turks and
  Caicos iguanas
Cyclura collei. See Jamaican iguanas
Cyclura cornuta cornuta. See Rhinoceros
  iguanas
Cylindrophis aruensis, 7:396
Cylindrophis engkariensis, 7:395–396
Cylindrophis lineatus, 7:395
Cylindrophis maculatus. See Blotched pipe
  snakes
Cylindrophis ruffus. See Red-tailed pipe snakes

D

Daboia spp., 7:445
Daboia russelii. See Russel's vipers
Damballah-Wed (Loa god), 7:57
Danh-gbi (African god), 7:55
Darevskia spp., 7:297, 7:298
Darevskia clarkorum. See Clark's lacerta
Dasypeltis spp. *See* Egg-eaters
Dasypeltis scabra. See Common egg-eaters
"Dear enemy" phenomenon, 7:36–37
Death adders, 7:39, **7:483–488,** 7:*489,*
  7:493–494
Defense mechanisms
  colubrids, 7:468
  Cordylidae, 7:*320*–321
  Gekkonidae, 7:261
  Iguanidae, 7:244–245
  Lacertidae, 7:298
  pythons, 7:420
  slender blindsnakes, 7:375
  white-bellied wormlizards, 7:276
  See also Behavior
Deinagkistrodon spp., 7:445, 7:448
Deinagkistrodon acutus. See Hundred-pace
  pitvipers
Deinosuchus spp., 7:157, 7:158
Deirochelyinae, 7:105
Deirochelys reticularia. See Chicken turtles
DeKay's brown snakes. *See* Brown snakes
Delcourt's giant geckos, 7:259, 7:262–263
Demansia spp. *See* Whipsnakes
Demon adders. *See* Rhombic night adders
Dendroaspis spp. *See* Mambas
Dendroaspis polylepis. See Black mambas

Denisonia maculata, 7:486
Dermatemys mawii. See Central American river
  turtles
Dermochelys coriacea. See Leatherback
  seaturtles
Desert agamas. *See* Spiny agamas
Desert grassland whiptails, 7:312, 7:*313,*
  7:*314,* 7:315
Desert horned vipers. *See* Horned vipers
Desert iguanas, 7:36, 7:*248,* 7:*252*
Desert lizards. *See* Desert iguanas
Desert night lizards, 7:291, 7:*293,* 7:*294*
Desert plated lizards, 7:319–321
Desert tortoises, 7:*145,* 7:*146,* 7:147
Desmatosuchus spp., 7:157
Dhabb lizards, as food, 7:48
Diadophis spp. *See* Ring-necked snakes
Diadophis punctatus. See Ring-necked snakes
Diamond-backed watersnakes, 7:470
Diamond pythons, 7:*420*–421
Diamondback rattlesnakes
  eastern, 7:40, 7:50, 7:475
  western, 7:38, 7:40, 7:42, 7:44, 7:50, 7:*197*
Diamondback terrapins, 7:107, 7:*108,*
  7:109–*110*
Diapsida, 7:*5,* 7:15–17
Dibamidae. *See* Blindskinks
Dibamus spp., 7:271, 7:272
Dibamus bourreti, 7:271
Dicrodon spp., 7:309, 7:310
Dicromatism, sexual, 7:41
  See also Reproduction
Digestion, 7:9–10, 7:26–28
  See also Physical characteristics
Dimorphism, sexual, 7:41
  See also Reproduction
Dinilysia spp., 7:16
Dinosauria, 7:19–22
Diplodactylinae, 7:259, 7:261
Diploglossinae, 7:339, 7:340
Diploglossus spp., 7:339, 7:340
Diploglossus fasciatus, 7:340
Diplometopon spp., 7:287
Dipsas spp. *See* Snail-suckers
Dipsas indica. See Amazonian snail-eaters
Dipsosaurus spp., 7:246
Dipsosaurus dorsalis. See Desert iguanas
Diseases, reptilian, 7:61, 7:62
Dispholidus spp. *See* Boomslangs
Dispholidus typus. See Boomslangs
Distribution
  African burrowing snakes, 7:*461,* 7:*462,* 7:*463*
  African sideneck turtles, 7:*129,* 7:*132*–*133*
  Afro-American river turtles, 7:*137,*
    7:*140*—142
  Agamidae, 7:*209,* 7:210, 7:*214*–221
  Alligatoridae, 7:*171,* 7:*172,* 7:*175*–177
  Anguidae, 7:*339,* 7:340, 7:*342*–344
  Australo-American sideneck turtles, 7:77,
    7:*81*–83, 7:*82*
  big-headed turtles, 7:*135*
  blindskinks, 7:*271,* 7:272
  blindsnakes, 7:*379,* 7:381, 7:*383*–385
  boas, 7:*409,* 7:410–411, 7:*413*–417
  Central American river turtles, 7:*99,* 7:100
  chameleons, 7:*223,* 7:227–228, 7:*235*–240
  colubrids, 7:*465,* 7:467, 7:*473*–481
  Cordylidae, 7:*319,* 7:320, 7:*324*–325
  crocodilians, 7:161–162
  Crocodylidae, 7:*179,* 7:*184*–187

early blindsnakes, 7:*369,* 7:*370*–*371*
Elapidae, 7:*483,* 7:*484,* 7:*491*–498
false blindsnakes, 7:*387*–388
false coral snakes, 7:*399,* 7:400
file snakes, 7:*439,* 7:*441,* 7:*442*
Florida wormlizards, 7:*283,* 7:284
Gekkonidae, 7:*259,* 7:260, 7:*265*–269
Geoemydidae, 7:*115,* 7:*118*–119
gharials, 7:*167,* 7:168
Helodermatidae, 7:*353,* 7:*354*
Iguanidae, 7:*243,* 7:244, 7:*250*–256
Kinosternidae, 7:*121,* 7:*125*–126
knob-scaled lizards, 7:*347,* 7:348, 7:*351*
Lacertidae, 7:*297,* 7:298, 7:*301*–302
leatherback seaturtles, 7:*101*
Microteiids, 7:*303,* 7:*306*–307
mole-limbed wormlizards, 7:*279,* 7:280,
  7:*281*
Neotropical sunbeam snakes, 7:*405*–406
New World pond turtles, 7:*105,* 7:*109*–*113*
night lizards, 7:*291,* 7:*294*–295
pig-nose turtles, 7:75
pipe snakes, 7:*395,* 7:*396*
pythons, 7:*419,* 7:420, 7:*424*–428
seaturtles, 7:*85,* 7:*89*–91
shieldtail snakes, 7:*391,* 7:*392,* 7:*393*
skinks, 7:*327,* 7:*328,* 7:*332*–337
slender blindsnakes, 7:*373,* 7:*374,* 7:*376*–377
snapping turtles, 7:*93,* 7:*95*–96
softshell turtles, 7:*151,* 7:*154*–155
spade-headed wormlizards, 7:*287,* 7:*288,*
  7:*289*
splitjaw snakes, 7:*429,* 7:*430*
Squamata, 7:203–204
sunbeam snakes, 7:*401*–402
Teiidae, 7:*309,* 7:310, 7:*314*–316
tortoises, 7:*143,* 7:*146*–148
Tropidophiidae, 7:*433,* 7:*434,* 7:*436*–437
tuatara, 7:*189*–*190*
turtles, 7:70
Varanidae, 7:*359,* 7:360, 7:*365*–368
Viperidae, 7:*445,* 7:*447,* 7:*451*–459
wormlizards, 7:*273,* 7:*274,* 7:*276*
Dob lizards. *See* Spiny-tailed agamas
Dog-faced watersnakes, 7:465–466, 7:*472,*
  7:*473,* 7:476–477
Dogania subplana. See Malayan softshell turtles
Dome pressure receptors (DPRs), 7:160
DPRs. *See* Dome pressure receptors
Dracaena spp. *See* Caiman lizards
Dracaena guianensis, 7:310
Dracaena paraguayensis. See Paraguayan
  caiman lizards
Draco volans. See Flying lizards
Dragon lizards, **7:209–211,** 7:*212,* 7:*213,*
  7:216–220
Drymarchon spp. *See* Indigo snakes
Drymarchon corais. See Indigo snakes
Drymarchon corais couperi. See Eastern indigo
  snakes
Duberria spp. *See* Slug-eaters
Duberria lutrix. See Common slug-eaters
Duméril, A., 7:379
Dumeril's boas, 7:411
Durrell, Gerald, 7:430
Duvernoy's glands, 7:467
Dwarf boas, 7:*434*
Dwarf crocodiles, 7:157, 7:179
Dwarf plated lizards, 7:320
Dwarf puff adders, 7:446

E

Earless dragons, 7:*213*, 7:*218*, 7:220
Earless monitors, **7:359–368**, 7:*364*, 7:*365*
Early blindsnakes, 7:198, 7:201, 7:202, **7:369–372**
Ears, 7:9, 7:32, 7:159–160
  See also Physical characteristics
East African black mud turtles, 7:*130*, 7:*131*, 7:*133*
East African sandboas, 7:411, 7:*412*, 7:*416*–417
East African serrated mud turtles, 7:*131*, 7:*132*–*133*
Eastern blindsnakes. *See* Blackish blindsnakes
Eastern box turtles, 7:*68*, 7:106, 7:*108*, 7:*112*–*113*
Eastern collared lizards. *See* Common collared lizards
Eastern diamondback rattlesnakes, 7:40, 7:50, 7:475
Eastern earless lizards. *See* Common lesser earless lizards
Eastern hog-nosed snakes, 7:*197*, 7:*470*, 7:*471*, 7:*475*, 7:480–481
Eastern indigo snakes, 7:470, 7:475
Eastern mud turtles, 7:121, 7:*122*
Eastern painted turtles, 7:*65*
Eastern racers, 7:470
Eastern side-blotched lizards. *See* Common side-blotched lizards
Eastern spiny softshell turtles, 7:*152*
Eastern zebra-tailed lizards. *See* Zebra-tailed lizards
Eastwood's seps, 7:322
Echiopsis atriceps, 7:486
Echiopsis curta, 7:486
Echis spp., 7:445
Echis carinatus. See Saw-scaled vipers
Ectothermy, 7:28, 7:198
  See also Physical characteristics
Egernia spp., 7:327, 7:329
Egernia striata. See Night skinks
Egg-eaters, common, 7:*469*, 7:*472*, 7:*474*
Eggs, 7:6–7
  See also Reproduction
Eggs (food), 7:47–48
Egyptian banded cobras, 7:*484*
Egyptian mythology, reptiles in, 7:55
Elaphe spp. *See* Ratsnakes
Elaphe guttata. See Cornsnakes
Elaphe guttata guttata. See Cornsnakes
Elaphe obsoleta. See Black ratsnakes
Elapidae, 7:206, **7:483–499**, 7:*489*, 7:*490*
  behavior, 7:485
  conservation status, 7:486–487
  distribution, 7:*483*, 7:484
  evolution, 7:483–484
  feeding ecology, 7:485–486
  habitats, 7:484–485
  humans and, 7:487–488
  physical characteristics, 7:484
  reproduction, 7:486
  species accounts, 7:*491*–498
  taxonomy, 7:483–484
Elapinae, 7:484
Elapognathus minor, 7:486
Elapomorphus spp., 7:467
Elapotinus spp., 7:461
Elegant-eyed lizards, 7:*203*
Elephant trunk snakes. *See* Java file snakes

Elgaria spp. *See* Alligator lizards
Elgaria kingii, 7:340
Elgaria parva, 7:339
Elseya dentata. See Victoria river snappers
Elseya novaeguineae. See New Guinea snapping turtles
Elusor spp., 7:78
Embryos
  crocodilians, 7:163–164
  development of, 7:6–7, 7:13
  retention of, 7:7
  See also Reproduction
Emerald tree boas, 7:410, 7:*412*, 7:*414*
Emotional fevers, 7:45
Emydidae. *See* New World pond turtles
Emydinae, 7:105
Emydocephalus annulatus. See Turtle-headed seasnakes
Emydura spp., 7:78
Emys orbicularis. See European pond turtles
Endangered Species Act (U.S.)
  American alligators, 7:48–49
  pythons, 7:421
Energy, 7:10, 7:28, 7:43–45
Enhydris spp., 7:468
Enulius spp., 7:469
Eosuchia, 7:5
Epicrates spp., 7:410, 7:411
Epicrates angulifer. See Cuban boas
Epicrates cenchria. See Rainbow boas
Epicrates exsul. See Abaco boas
Epicrates gracilis. See Haitian vine boas
Epicrates inornatus. See Puerto Rican boas
Epicrates monensis. See Mona boas
Epicrates subflavus. See Jamaican boas
Eremiainae, 7:297
Eremiascincus spp., 7:329
Eremiascincus richardsonii. See Broad-banded sand swimmers
Eretmochelys imbricata. See Hawksbills
Eristicophis spp., 7:445
Ermia spp., 7:445
Erpeton tentaculatus. See Tentacled snakes
Erycinae, 7:409–411
Erymnochelys madagascariensis. See Madagascan big-headed turtles
Eryx spp., 7:411
Eryx colubrinus. See East African sandboas
Eryx jayakari. See Arabian sandboas
Eryx johnii. See Brown sandboas
Eryx muelleri. See Sahara sandboas
Eryx somalicus, 7:411
Eryx tataricus, 7:411
Escorpíon. *See* Mexican beaded lizards
Estuarine crocodiles. *See* Saltwater crocodiles
Eublepharinae, 7:259, 7:262
Eublepharis macularius. See Leopard geckos
Eudimorphodon spp., 7:*18*
Eugongylus spp., 7:327
Eulamprus quoyi, 7:329
Eumeces spp. *See* Skinks
Eumeces laticeps. See Broad-headed skinks
Eumeces obsoletus. See Great Plains skinks
Eunectes spp., 7:410
Eunectes murinus. See Green anacondas
Eurasian pond turtles, **7:115–120**
Eurasian river turtles, **7:115–120**
Eurasian watersnakes, 7:468
European chameleons. *See* Common chameleons

"European Community Habitats Directive," 7:238
European pond turtles, 7:71, 7:*107*, 7:*108*, 7:*112*
European Union Habitat and Species Directive, on smooth snakes, 7:474
Euryapsida, 7:*5*, 7:14–15
Eurydactylodes spp., 7:261
Evolution, **7:12–22**, 7:*13*
  African burrowing snakes, 7:461
  African sideneck turtles, 7:129
  Afro-American river turtles, 7:137
  Agamidae, 7:209
  Alligatoridae, 7:171
  Anguidae, 7:339
  Australo-American sideneck turtles, 7:77, 7:81–83
  big-headed turtles, 7:135
  blindskinks, 7:271
  blindsnakes, 7:379
  boas, 7:409
  Central American river turtles, 7:99
  chameleons, 7:223–224
  colubrids, 7:465–467
  Cordylidae, 7:319
  crocodilians, 7:18–19, 7:157
  Crocodylidae, 7:179
  early blindsnakes, 7:369
  Elapidae, 7:483–484
  false blindsnakes, 7:387
  false coral snakes, 7:399
  file snakes, 7:439
  Florida wormlizards, 7:283–284
  Gekkonidae, 7:259
  Geoemydidae, 7:115
  gharials, 7:167
  Helodermatidae, 7:353
  Iguanidae, 7:243
  Kinosternidae, 7:121
  knob-scaled lizards, 7:347
  Lacertidae, 7:297
  leatherback seaturtles, 7:101
  lizards, 7:16
  mesosaurs, 7:14
  Microteiids, 7:303
  mole-limbed wormlizards, 7:279
  Neotropical sunbeam snakes, 7:405
  New World pond turtles, 7:105
  night lizards, 7:291
  pareiasaurs, 7:14
  pig-nose turtles, 7:75
  pipe snakes, 7:395
  pythons, 7:419
  seaturtles, 7:14, 7:85, 7:89–91
  shieldtail snakes, 7:391
  skinks, 7:327–328
  slender blindsnakes, 7:373
  snakes, 7:16–17
  snapping turtles, 7:93
  softshell turtles, 7:151
  spade-headed wormlizards, 7:287
  splitjaw snakes, 7:429–430
  squamata, 7:195–198
  sunbeam snakes, 7:401
  Teiidae, 7:309
  tortoises, 7:143
  Tropidophiidae, 7:433
  tuatara, 7:189, 7:*190*
  turtles, 7:13–14, 7:65
  Varanidae, 7:359

Evolution *(continued)*
  Viperidae, 7:445–446
  wormlizards, 7:273
Excited coloration, 7:9
  See also Physical characteristics
Exercise, 7:10
Exiliboa spp., 7:434
Exiliboa placata, 7:434
Exports. *See* Imports and exports
Extinct species
  Alsophis sanctaecrucis, 7:470
  Ameiva spp., 7:312
  Delcourt's giant geckos, 7:259, 7:262–263
  Eastwood's seps, 7:322
  file snakes, 7:439
  Galápagos tortoises, 7:147
  Leiocephalus eremitus, 7:246–247
  Leiocephalus herminieri, 7:246–247
  Megalania prisca, 7:359
  Navassa woodsnakes, 7:435
  Phelsuma edwardnewtoni, 7:262
  polyglyphanodontines, 7:309
  reptilian skulls and, 7:4–5
  smooth-scaled splitjaws, 7:429
  Typhlops cariei, 7:382
Eyed lizards, 7:298
Eyes, 7:8–9
  chameleons, 7:32, 7:225–226
  crocodilians, 7:159
  See also Physical characteristics; Vision

F

False blindsnakes, **7:387–388**
False coral snakes, **7:399–400,** 7:*399,* 7:*400*
False gharials, 7:59, 7:*157,* **7:179–188,** 7:*183*
False pitvipers, 7:467, 7:468
False water-cobras, 7:468
Fangs, 7:27
  See also Venom
Farancia spp. *See* Mudsnakes
Farancia abacura. See Red-bellied mudsnakes
Farancia erytrogramma. See Rainbow snakes
Farming and ranching, reptiles, 7:48–*49*
Fea's vipers, 7:445–448, 7:*449,* 7:*451*
Feeding ecology, 7:3–4, 7:26–28
  African burrowing snakes, 7:462–464
  African sideneck turtles, 7:130, 7:132–133
  Afro-American river turtles, 7:137,
    7:140–141
  Agamidae, 7:211, 7:214–221
  Alligatoridae, 7:173, 7:175–177
  Anguidae, 7:340, 7:342–344
  Australo-American sideneck turtles, 7:78,
    7:81–83
  big-headed turtles, 7:135–136
  blindskinks, 7:272
  blindsnakes, 7:382–385
  boas, 7:411, 7:413–417
  Central American river turtles, 7:100
  chameleons, 7:230, 7:235–241
  colubrids, 7:469, 7:473–481
  Cordylidae, 7:321, 7:324–325
  crocodilians, 7:162–163
  Crocodylidae, 7:180, 7:184–187
  early blindsnakes, 7:371
  Elapidae, 7:485–486, 7:491–498
  false blindsnakes, 7:388

false coral snakes, 7:400
file snakes, 7:441–442
Florida wormlizards, 7:284
Gekkonidae, 7:261–262, 7:265–269
Geoemydidae, 7:116, 7:118–119
gharials, 7:168
Helodermatidae, 7:355, 7:357
Iguanidae, 7:245–246, 7:250–257
Kinosternidae, 7:122, 7:125–126
knob-scaled lizards, 7:348, 7:351
Lacertidae, 7:298, 7:301–302
leatherback seaturtles, 7:101–102
Microteiids, 7:304, 7:306–307
mole-limbed wormlizards, 7:280, 7:281
Neotropical sunbeam snakes, 7:406
New World pond turtles, 7:106, 7:109–113
night lizards, 7:292, 7:294–295
pig-nose turtles, 7:76
pipe snakes, 7:396
plasticity and, 7:39–40
preferences and plasticity, 7:39–40
pythons, 7:421, 7:424–428
reptiles, 7:3–4
seaturtles, 7:86, 7:89–91
shieldtail snakes, 7:392–394
skinks, 7:329, 7:332–337
slender blindsnakes, 7:375–377
snapping turtles, 7:94, 7:95–96
softshell turtles, 7:152, 7:154–155
spade-headed wormlizards, 7:288, 7:289
splitjaw snakes, 7:430
Squamata, 7:204–*205*
suction, 7:3–4
sunbeam snakes, 7:402–403
Teiidae, 7:311–312, 7:314–316
tortoises, 7:143, 7:146–148
Tropidophiidae, 7:434–437
tuatara, 7:190
turtles, 7:67
Varanidae, 7:361–362, 7:365–368
Viperidae, 7:447–448, 7:451–459
wormlizards, 7:274, 7:276
Feick's dwarf boas. *See* Banded woodsnakes
Fertilization, of reptilian eggs, 7:6
  See also Reproduction
Feylinia spp., 7:327, 7:328
Feylininae. *See Feylinia* spp.
Fierce snakes. *See* Taipans
Fiji Island boas, 7:410, 7:411
File snakes, 7:198, **7:439–444,** 7:*443*
  See also Cape filesnakes
Fitzroy River turtles, 7:69
Flap-necked chameleons, 7:*230*
Flat lizards, 7:319, 7:320, 7:321, 7:322
Flat-tailed horned lizards, 7:207
Florida alligators. *See* American alligators
Florida wormlizards, **7:283–285,** 7:*283,*
  7:*284*
Flying geckos, 7:*261*
Flying lizards, 7:*209,* 7:*211,* 7:*213,* 7:*217*
Flyingsnakes, 7:468
Folk medicine, reptiles in, 7:50–52, 7:*54*
  See also Humans; specific species
Folklore, reptiles in, 7:54
  See also Humans; specific species
Food, reptiles as, 7:47–48
  See also Feeding ecology; Humans
Fordonia leucobalia. See White-bellied
  mangrove snakes
Forest cobras, 7:*490,* 7:*492*

Fourteen-lined comb eared skinks, 7:*331,*
  7:*332,* 7:333
Freshies. *See* Johnstone's crocodiles
Frill-necked lizards. *See* Frilled lizards
Frilled lizards, 7:*35,* 7:*210,* 7:*212,* 7:215–216,
  7:*217*
Fringe-toed lizards. *See* Coachella Valley
  fringe-toed lizards
Frog-eyed geckos. *See* Common plate-tailed
  geckos
Frost, D., 7:340
Furcifer spp., 7:*37,* 7:224, 7:227, 7:232
Furcifer campani. See Jeweled chameleons
Furcifer labordi, 7:232
Furcifer minor. See Minor chameleons
Furcifer pardalis. See Panther chameleons
Furina dunmalli, 7:486

G

Gaboon adders, 7:446, 7:447, 7:*449,* 7:*451,*
  7:457
Gaboon vipers. *See* Gaboon adders
Galápagos tortoises, 7:47, 7:59, 7:*71,* 7:*145,*
  7:*146*–147
Galliwasps, **7:339–340,** 7:*341,* **7:343**
Gallotia spp. *See* Giant lizards
Gallotia gomerana. See Gomeran giant lizards
Gallotia simonyi. See Simony's giant lizards
Gallotiinae, 7:297
Gambelia spp., 7:246
Gambelia wislizenii. See Long-nosed leopard
  lizards
Gans, Carl, 7:391, 7:392
Garter snakes
  behavior, 7:44
  chemosensory behavior, 7:34–36
  common, 7:469, 7:*471,* 7:478–*479*
  food ecology, 7:39–40
  reproduction, 4:470, 7:42
  tactile cues, 7:40, 7:42
Gastropholis spp., 7:297
Gators. *See* American alligators
Gavialidae. *See* Gharials
Gavialis gangeticus. See Gharials
Geckos, **7:259–269,** 7:*264*
  behavior, 7:261
  conservation status, 7:262–263
  distribution, 7:259, 7:260
  evolution, 7:259
  feeding ecology, 7:261–262
  in folk medicine, 7:51
  habitats, 7:260–261
  hearing, 7:32
  humans and, 7:263
  in mythology and religion, 7:55
  physical characteristics, 7:259–*260*
  reproduction, 7:262
  species of, 7:265–269
  taxonomy, 7:259
Gehyra vorax. See Voracious geckos
Gekko gecko. See Tokay geckos
Gekko gecko azhari. See Tokay geckos
Gekko gecko gecko. See Tokay geckos
Gekko vittatus. See White-striped geckos
Gekkonidae, **7:259–269,** 7:*264*
  behavior, 7:261
  conservation status, 7:262–263

distribution, 7:*259*, 7:260
evolution, 7:259
feeding ecology, 7:261–262
habitats, 7:260–261
humans and, 7:263
physical characteristics, 7:259–260
reproduction, 7:262
species of, 7:*265–269*
Gekkoninae, 7:259, 7:262
Gekkota, 7:196, 7:197–198, 7:200, 7:203–204, 7:207
Genetic sex determination (GSD), 7:72
    See also Reproduction
Geochelone denticulata. See South American yellow-footed tortoises
Geochelone gigantea. See Aldabra tortoises
Geochelone nigra. See Galápagos tortoises
Geochelone nigra vandenburghi. See Giant tortoises
Geochelone radiata. See Radiated tortoises
Geochelone sulcata. See Great African tortoises
Geoemyda japonica, 7:119
Geoemyda spengleri. See Black-breasted leaf turtles
Geoemydidae, **7:115–120,** 7:*117*
Geophis spp. *See* Neotropical burrowing snakes
Gerrhonotinae. *See* Alligator lizards
Gerrhonotus spp. *See* Alligator lizards
Gerrhonotus liocephalus. See Texas alligator lizards
Gerrhosaurinae. *See* Plated lizards
Gerrhosaurus spp. *See* Plated lizards
Gerrhosaurus skoogi. See Desert plated lizards
Gharials, 7:19, 7:59, 7:158, 7:161, **7:167–170,** 7:*167*, 7:*168*, 7:*169*, 7:179, 7:180
    See also False gharials
Giant ameivas, 7:*313*, 7:*314*
Giant blindsnakes. *See* Schlegel's blindsnakes
Giant geckos, 7:262
Giant girdled lizards, 7:*323*, 7:*324*
Giant land iguanas, 7:59
Giant lizards, 7:59, 7:297, 7:298–299
Giant musk turtles, 7:72
Giant South American river turtles. *See* South American river turtles
Giant tortoises, 7:71, 7:*144*
Gibba turtles, 7:78, 7:*80*, 7:*81*
Gila monsters, 7:10, 7:37, **7:353–358,** 7:*354*, 7:*355*, 7:*356*
Girdled lizards, **7:319–325,** 7:*323*
Glands, 7:29–30
    See also Physical characteristics
Glass lizards, **7:339–342,** 7:*341*
Glen canyon chuckwallas. *See* Common chuckwallas
Glossy wormsnakes. *See* Peters' wormsnakes
Gloyd, Howard, 7:455
Gloydius spp., 7:445, 7:446
Gloydius caliginosus, 7:447
Gloydius halys, 7:447
Gloydius monticola, 7:447
Gloydius strauchi. See Tibetan pitvipers
Glyptemys muhlenbergii. See Bog turtles
Gnypetoscincus queenslandiae. See Prickly forest skinks
Goannas, 7:57, **7:359–368**
Gold morph of eyelash vipers, 7:*447*, 7:448
Golden arboreal alligator lizards. *See* Coban alligator lizards

Gomeran giant lizards, 7:299
Gonatodes albogularis. See Yellow-headed geckos
Gonatodes albogularis albogularis. See Yellow-headed geckos
Gonatodes albogularis bodinii. See Yellow-headed geckos
Gonatodes albogularis fuscus. See Yellow-headed geckos
Gonatodes albogularis notatus. See Yellow-headed geckos
Goose-neck turtles. *See* Spiny softshells
Gopherus agassizii. See Desert tortoises
Gorakhnath, 7:56
Gracilisuchus spp., 7:18
Graptemys spp. *See* Map turtles
Grass lizards, 7:298, 7:*300*, 7:*301*, 7:*302*, 7:319–322
Grass snakes, 7:466
Gray, John E., 7:189, 7:223
Grayia spp., 7:467
Gray's sliders, 7:105
Great African tortoises, 7:66
Great Plains skinks, 7:42
Greater five-lined skinks. *See* Broad-headed skinks
Greek mythology, reptiles in, 7:54–55
Greek tortoises, 7:*67*, 7:*143*
Green anacondas, 7:202; 7:410, 7:411, 7:*412*, 7:*414*, 7:415
Green anoles, 7:*35*, 7:38–39, 7:*244*, 7:246, 7:*248*, 7:250, 7:256–257
Green basilisks, 7:*245*
Green bush vipers, 7:*449*, 7:456–457
Green iguanas, 7:47, 7:243, 7:245, 7:*246*
Green pythons, 7:*206*, 7:420, 7:*421*, 7:*423*, 7:*425*, 7:426
Green seaturtles, 7:*38*, 7:43, 7:*85*, 7:86, 7:*88*, 7:90–91
Green tree pythons. *See* Green pythons
Green turtles. *See* Green seaturtles
Gridiron-tailed lizards. *See* Zebra-tailed lizards
Ground agamas. *See* Spiny agamas
Ground pythons. *See* Neotropical sunbeam snakes
Groundsnakes, 7:466, 7:468, 7:469
GSD. *See* Genetic sex determination
Guadalupe, Virgin of, 7:55
Gulf ridleys. *See* Kemp's ridley turtles
Günther, Albert, 7:189
Gymnophthalmidae. See Microteiids
Gymnophthalmus spp., 7:303, 7:304
Gymnophthalmus cryptus, 7:306
Gymnophthalmus underwoodi, 7:*305*, 7:*306*–307

H

Habit destruction and alteration, 7:60
    See also Conservation status
Habitats
    African burrowing snakes, 7:462, 7:463
    African sideneck turtles, 7:129, 7:132–133
    Afro-American river turtles, 7:137, 7:140—141
    Agamidae, 7:210, 7:214–221
    Alligatoridae, 7:172, 7:175–177

Anguidae, 7:340, 7:342–344
Australo-American sideneck turtles, 7:77, 7:81–83
big-headed turtles, 7:135
blindskinks, 7:272
blindsnakes, 7:381, 7:383–385
boas, 7:411, 7:413–417
Central American river turtles, 7:100
chameleons, 7:228, 7:235–241
colubrids, 7:468, 7:473–481
Cordylidae, 7:320, 7:324–325
Crocodylidae, 7:179–180, 7:184–187
early blindsnakes, 7:371
Elapidae, 7:484–485, 7:491–498
false blindsnakes, 7:388
false coral snakes, 7:400
file snakes, 7:441
Florida wormlizards, 7:284
Gekkonidae, 7:260–261, 7:265–269
Geoemydidae, 7:115, 7:118–119
gharials, 7:168
Helodermatidae, 7:354
Iguanidae, 7:244, 7:250–256
Kinosternidae, 7:121, 7:125–126
knob-scaled lizards, 7:348, 7:351
Lacertidae, 7:298, 7:301–302
leatherback seaturtles, 7:101
Microteiids, 7:304, 7:306–307
mole-limbed wormlizards, 7:280, 7:281
Neotropical sunbeam snakes, 7:406
New World pond turtles, 7:105, 7:109–112
night lizards, 7:292, 7:294–295
pig-nose turtles, 7:76
pipe snakes, 7:396
pythons, 7:420, 7:424–428
seaturtles, 7:86, 7:89–91
shieldtail snakes, 7:392–394
skinks, 7:329, 7:332–337
slender blindsnakes, 7:374–377
snapping turtles, 7:93, 7:95–96
softshell turtles, 7:152, 7:154–155
spade-headed wormlizards, 7:288, 7:289
splitjaw snakes, 7:430
Squamata, 7:204
sunbeam snakes, 7:402
Teiidae, 7:310–311, 7:314–316
tortoises, 7:143, 7:146–148
Tropidophiidae, 7:434, 7:436–437
tuatara, 7:190
Varanidae, 7:360, 7:365–368
Viperidae, 7:447, 7:451–459
wormlizards, 7:274, 7:276
Haitian vine boas, 7:410
Half-girdled snakes, 7:486, 7:*489*, 7:*497*, 7:498
    See also Broad-banded sand swimmers
Half-toed geckos. *See* House geckos
Halmahera pythons, 7:*423*, 7:*425*–426
Hambly, Wilfred, 7:55
Hardy, Laurence M., 7:306
Hawksbills, 7:*66*, 7:*87*
Hearing. *See* Ears
Heart, crocodilian, 7:160–161
    See also Physical characteristics
Hecht, Max K., 7:16
Hedonic receptors, in reptiles, 7:42
Helicops spp. *See* Neotropical watersnakes
Heliobolus spp., 7:297
Heliobolus lugubris. See Kalahari sand lizards
Helmeted terrapins. *See* Helmeted turtles
Helmeted turtles, 7:*131*, 7:*132*

Helminthophis spp., 7:369, 7:370, 7:371
Helminthophis flavoterminatus, 7:369–370, 7:371
Heloderma spp., 7:353
Heloderma horridum. *See* Mexican beaded lizards
Heloderma suspectum. *See* Gila monsters
Helodermatidae, 7:206, **7:353–358**, *7:356*
Hemachatus haemachatus. *See* Rinkhal's cobras
Hemidactylus frenatus. *See* House geckos
Hemidactylus turcicus. *See* Mediterranean geckos
Hemotoxins, 7:52
Henderson, Robert W., 7:410
Herbivory, 7:27, 7:204
See also Feeding ecology
Hermann's tortoises, *7:145*, *7:148*–149
Herrel, Anthony, 7:347
Heterodon spp. *See* Hog-nosed snakes
Heterodon platirhinos. *See* Eastern hog-nosed snakes
Hickety turtles. *See* Central American river turtles
Hidden-necked turtles. *See* Cryptodira
Hieremys annandalii. *See* Yellow-headed temple turtles
Hinduism, snakes in, 7:56
Hissing adders. *See* Eastern hog-nosed snakes
Hoburogecko suchanovi, 7:259
Hog-nosed snakes, 7:205, 7:467
See also Eastern hog-nosed snakes
Hoge's sideneck turtles, 7:79
Holaspis spp., 7:297
Holbrookia maculata. *See* Common lesser earless lizards
Homalopsinae, 7:465–466, 7:468, 7:470
Homalopsis spp. *See* Puff-faced watersnakes
Homing behavior, 7:43
See also Behavior
Homoroselaps spp., 7:483
Homoroselaps lacteus. *See* Spotted harlequin snakes
Hopi snake dance, 7:56–57
Hoplocephalus bungaroides. *See* Broad-headed snakes
Hoplocercinae, 7:243, 7:244
Hoplodactylus delcourti. *See* Delcourt's giant geckos
Horned lizards, 7:204, 7:207, 7:243–247, *7:249*, *7:251*, *7:254*
Horned toads. *See* Texas horned lizards
Horned vipers, *7:449*, *7:457*–458
Horny toads. *See* Texas horned lizards
House geckos, 7:260, 7:263, *7:264*, *7:267*
Hox genes, 7:24
Huachuca earless lizards. *See* Common lesser earless lizards
Humans
African burrowing snakes and, 7:462–464
African sideneck turtles and, 7:130, 7:132–133
Afro-American river turtles and, 7:138, 7:140–142
Agamidae and, 7:211, 7:214–221
Alligatoridae and, 7:173, 7:175–177
Anguidae and, 7:340, 7:342–344
Australo-American sideneck turtles and, 7:79, 7:81–83
big-headed turtles and, 7:136
blindskinks and, 7:272
blindsnakes and, 7:382–385
boas and, 7:411, 7:413–417

Central American river turtles and, 7:100
colubrids and, 7:470, 7:473–481
Cordylidae and, 7:322, 7:324–325
Crocodylidae and, 7:182, 7:184–187
early blindsnakes and, 7:371
Elapidae and, 7:487–488, 7:491–498
false blindsnakes and, 7:388
false coral snakes and, 7:400
feeding ecology and, 7:232, 7:235–241
file snakes and, 7:442
Florida wormlizards and, 7:285
Gekkonidae and, 7:263, 7:265–269
Geoemydidae and, 7:116, 7:118–119
gharials and, 7:170
Helodermatidae and, 7:357
Iguanidae and, 7:247, 7:250–257
Kinosternidae and, 7:123, 7:125–126
knob-scaled lizards and, 7:349, 7:351
Lacertidae and, 7:299, 7:301–302
leatherback seaturtles and, 7:102
Microteiids and, 7:304, 7:306–307
mole-limbed wormlizards and, 7:280, 7:281
mythology and religions, 7:54–57
Neotropical sunbeam snakes and, 7:406
New World pond turtles and, 7:107, 7:109–113
night lizards and, 7:292, 7:294–295
pig-nose turtles and, 7:76
pipe snakes and, 7:397
pythons and, 7:421, 7:422, 7:424–428
reptile farming and ranching, 7:48–*49*
reptiles and, **7:47–58**
seaturtles and, 7:87, 7:89–91
shieldtail snakes and, 7:392–394
skinks and, 7:329, 7:332–337
slender blindsnakes and, 7:375–377
snapping turtles and, 7:94, 7:95–96
softshell turtles and, 7:152, 7:154–155
spade-headed wormlizards and, 7:288, 7:289
splitjaw snakes and, 7:431
sunbeam snakes and, 7:403
Teiidae and, 7:312, 7:314–316
tortoises and, 7:144, 7:146–149
Tropidophiidae and, 7:435–437
tuatara and, 7:191
Varanidae and, 7:363, 7:365–368
Viperidae and, 7:448, 7:451–460
wormlizards and, 7:275, 7:276
See also Conservation status
Hundred-pace pitvipers, *7:450*, *7:454*–455
Hundred-pacers. *See* Hundred-pace pitvipers
Hydrodynastes spp. *See* False water-cobras
Hydromedusinae, 7:77
Hydrophiidae, 7:483
Hydrophiinae, 7:196–197, 7:484
Hydrops spp., 7:467, 7:468
Hydrosaurus amboinensis. *See* Sailfin lizards
Hylonomus lyelli, 7:14
Hypnale spp., 7:445
Hypnale hypnale, 7:447
Hypnale nepa. *See* Sri Lankan hump-nosed pitvipers

I

Ichnotropis spp., 7:297
Ichnotropis capensis. *See* Cape rough-scaled lizards

Ichnotropis squamulosa. *See* Common rough-scaled lizards
Ichthyosauria, 7:5, 7:15
Iguana spp. *See* Iguanas
Iguana iguana. *See* Green iguanas
Iguanas, **7:243–257**, *7:248*
behavior, 7:244–245
conservation status, 7:246–247
distribution, 7:243, 7:244
evolution, 7:243
farming, 7:48
feeding ecology, 7:245–246
in folk medicine, 7:51
as food, 7:48, *7:51*
habitats, 7:244
humans and, 7:247
physical characteristics, 7:243–244
reproduction, 7:246
species of, 7:250–257
taxonomy, 7:243
See also specific types of iguanas
Iguanidae, **7:243–257**, *7:248*, *7:249*
behavior, 7:204, 7:244–*245*
conservation status, 7:246–247
distribution, 7:203–204, *7:243*, 7:244
evolution, 7:195, 7:196, 7:243
feeding ecology, 7:245–246
habitats, 7:244
humans and, 7:247
physical characteristics, 7:200, 7:243–244
reproduction, 7:207, 7:246
species of, 7:250–257
taxonomy, 7:197–198, 7:243
Iguaninae, 7:204, 7:243–246
Illinois mud turtles. *See* Yellow mud turtles
Imantodes spp. *See* Blunt-headed vinesnakes
Imports and exports, of reptiles, 7:53t
Indian cobras, 7:56
Indian flapshell turtles, *7:153*, *7:154*
Indian garden lizards. *See* Brown garden lizards
Indian gharials. *See* Gharials
Indian golden geckos, 7:261
Indian mythology, snakes in, 7:52, 7:55–56
Indian pythons, 7:42, 7:56, 7:420–422
Indian tent turtles, 7:115
Indigo snakes, 7:466, 7:470, *7:472*, *7:475*–476
Indo-Pacific crocodiles. *See* Saltwater crocodiles
Indonesian Komodo dragons. *See* Komodo dragons
Insectivores, Iguanidae as, 7:246
See also Feeding ecology
Integumentary sense organs (ISOs), 7:160
Integumentary system, 7:29–30
chameleons, 7:224–225, *7:226*
impermeability of, 7:3
skin shedding, in Squamata, 7:198
turtles, 7:67
See also Physical characteristics
Intelligence, of Varanidae, 7:360
Internal organs
crocodilian, 7:160–161
lizards, 7:199
snakes, 7:200
Testudines, 7:68
tuatara, 7:191
See also Physical characteristics
International Union for Conservation of Nature. *See* IUCN Red List of Threatened Species

Island night lizards, 7:292
ISOs. *See* Integumentary sense organs
IUCN Red List of Threatened Species
African sideneck turtles, 7:130
Afro-American river turtles, 7:138
Agamidae, 7:211
alligator snapping turtles, 7:94, 7:96
Alligatoridae, 7:173
American crocodiles, 7:184
Anguidae, 7:340
armored chameleons, 7:236
Australo-American sideneck turtles, 7:79, 7:83
big-headed turtles, 7:136
blindsnakes, 7:382
boas, 7:411
Central American river turtles, 7:100
chameleons, 7:232
Chinese alligators, 7:173, 7:176
Chinese softshell turtles, 7:155
Chinese stripe-necked turtles, 7:119
Coachella Valley fringe-toed lizards, 7:255
colubrids, 7:470
common chameleons, 7:238
Cordylidae, 7:322
crocodiles, 7:182
desert tortoises, 7:147
diamondback terrapins, 7:109
Eastern box turtles, 7:113
Elapidae, 7:486
European pond turtles, 7:112
false blindsnakes, 7:388
Galápagos tortoises, 7:147
Geoemydidae, 7:116
gharials, 7:169
giant girdled lizards, 7:324
green seaturtles, 7:90
Helodermatidae, 7:357
Hermann's tortoises, 7:148–149
Iguanidae, 7:246–247
jeweled chameleons, 7:239
Kemp's ridley turtles, 7:91
Kinosternidae, 7:122
Komodo dragons, 7:368
KwaZulu-Natal Midlands dwarf chameleons, 7:235
Lacertidae, 7:298–299
leaf-horned agamas, 7:215
leatherback seaturtles, 7:102
loggerhead turtles, 7:89
Madagascan big-headed turtles, 7:142
minor chameleons, 7:240
mugger crocodiles, 7:185
New World pond turtles, 7:106
painted terrapins, 7:118
pancake tortoises, 7:148
pig-nose turtles, 7:76
pond sliders, 7:111
pythons, 7:421
skinks, 7:329
softshell turtles, 7:152
South American river turtles, 7:140
South American yellow-footed tortoises, 7:146
Sphenodon guntheri, 7:192
splitjaw snakes, 7:430
spotted turtles, 7:111
Teiidae, 7:312
tortoises, 7:144
Varanidae, 7:363

Viperidae, 7:448
yellow-margined box turtles, 7:119

J

Jackson's centipede eaters, 7:462
Jackson's chameleons, 7:228, 7:*229*, 7:*233*, 7:*235*, 7:238–239
Jacobson's organ, 7:*31*, 7:195
See also Physical characteristics
Jamaican boas, 7:411
Jamaican Iguana Research and Conservation Group, 7:247
Jamaican iguanas, 7:246, 7:247
Jaragua sphaero, 7:259
Java file snakes, 7:439–*440*, 7:441, 7:*442*, 7:*443*
Jaw prehension, 7:4, 7:198, 7:200, 7:204
See also Physical characteristics
Jersey Wildlife Preservation Trust, on splitjaw snakes, 7:430
Jesus Christ lizards. *See* Common basilisks; Green basilisks
Jeweled chameleons, 7:232, 7:*234*, 7:*239*
Jicotea turtles. *See* Central American river turtles
Johnstone's crocodiles, 7:163, 7:164, 7:180, 7:*183*, 7:184–*185*
Judeo-Christian religions, snakes in, 7:57

K

Kachuga tentoria. See Indian tent turtles
Kalahari sand lizards, 7:298
Keel-bellied watersnakes, 7:466, 7:468
Keel-scaled boas. *See* Splitjaw snakes
Keel-scaled splitjaws. *See* Splitjaw snakes
Kemp's ridley turtles, 7:*88*, 7:*89*, 7:91
Keniabitis spp., 7:445
Kentropyx spp., 7:309, 7:310, 7:312
King cobras, 7:42, 7:56, 7:484, 7:486, 7:*490*, 7:*493*
Kingsnakes, 7:42, 7:44, 7:466, 7:470
Kinosternidae, **7:121–127**, 7:*124*
Kinosterninae. *See* American mud turtles
Kinosternon spp. *See* Eastern mud turtles
Kinosternon angustipons, 7:123
Kinosternon baurii. See Striped mud turtles
Kinosternon dunni, 7:123
Kinosternon flavescens. See Yellow mud turtles
Kinosternon leucostomum. See White-lipped mud turtles
Kinosternon sonoriense, 7:123
Klaver, Charles, 7:223–224
Knob-scaled lizards, **7:347–352**, 7:*350*
Knob-tailed geckos, 7:*262*
Koelliker's glass lizards. *See* Moroccan glass lizards
Komodo dragons, 7:202, 7:359, 7:360, 7:*361*, 7:*363*, 7:*364*, 7:*365*, 7:367–368
Komodo monitors. *See* Komodo dragons
Kraits, **7:483–487**, 7:*490*, **7:495**
Kronosaurus spp., 7:15
Kukrisnakes, 7:467, 7:469
KwaZulu-Natal Midlands dwarf chameleons, 7:*234*, 7:*235*
KwaZulu-Natal Nature Conservation Service, 7:*61*

L

Lacerta spp., 7:297
Lacerta agilis. See Sand lizards
Lacerta lepida. See Eyed lizards
Lacertidae, 7:207, **7:297–302**, 7:*300*
Lachesis spp., 7:445, 7:446, 7:448
Lachesis melanocephala. See Black-headed bushmasters
Lachesis muta, 7:456
Lagosuchus spp., 7:19
Lahontan Basin lizards. *See* Long-nosed leopard lizards
Lampropeltis spp. *See* Kingsnakes
Lampropeltis getula. See Common kingsnakes
Lampropeltis getula floridana. See Kingsnakes
Lampropeltis triangulum. See Milksnakes
Lamprophiinae, 7:467, 7:468
Landscape with Reptile (Palmer), 7:454
Langaha spp. *See* Madagascan vinesnakes
Lanthanotines, 7:359
Lanthanotus spp., 7:202, 7:359, 7:360
Lanthanotus borneensis. See Earless monitors
Lapparentophis spp., 7:16
Laredo striped whiptails, 7:312
Large-scaled shieldtails, 7:*392*
Larutia spp., 7:328
Lateral undulation, 7:25–26
Laticauda spp. *See* Sea kraits
Laticauda colubrina. See Sea kraits
Laúd turtles. *See* Leatherback seaturtles
Leach's giant geckos. *See* New Caledonian giant geckos
Leaf-horned agamas, 7:211, 7:*213*, 7:*215*
Leaf vipers. *See* Green bush vipers
Learning, in reptiles, 7:39–40
Leatherback seaturtles, **7:101–103**, 7:*102*
conservation status, 7:62
endothermic, 7:28
physical characteristics, 7:67, 7:70
reproduction, 7:71
See also Seaturtles
Leatherback turtles. *See* Spiny softshells
Legend of White Snake, 7:56
Legends. *See* Folklore
Legless lizards, 7:339–740
Leiocephalus eremitus, 7:246–247
Leiocephalus herminieri, 7:246–247
Leiolamus spp., 7:246
Leiolepidinae, 7:204, 7:209
Leiolepis belliana. See Butterfly agamas
Leiopython spp., 7:420
Leiopython albertisii. See White-lipped pythons
Lemos-Espinal, Julio, 7:348
Leopard geckos, 7:262
Leopard lizards. *See* Long-nosed leopard lizards
Lepidochelys kempii. See Kemp's ridley turtles
Lepidodactylus lugubris. See Mourning geckos
Lepidophyma spp., 7:291, 7:292
Lepidophyma flavimaculatum. See Yellow-spotted night lizards
Lepidosauria, 7:16, 7:23–24
Leposoma spp., 7:303, 7:304
Leposoma percarinatum, 7:304
Leptodeira spp. *See* Cat-eyed snakes
Leptodeira septentrionalis. See Northern cat-eyed snakes
Leptotyphlopidae. *See* Slender blindsnakes
Leptotyphlops spp., 7:373–375

INDEX

Leptotyphlops alfredschmidti, 7:373–374
Leptotyphlops blanfordi, 7:374
Leptotyphlops borrichianus, 7:374
Leptotyphlops boulengeri, 7:373
Leptotyphlops broadleyi, 7:373
Leptotyphlops cairi, 7:374
Leptotyphlops conjunctus, 7:375
Leptotyphlops dulcis. See Texas blindsnakes
Leptotyphlops filiformis, 7:374
Leptotyphlops goudotii, 7:375
Leptotyphlops humilis, 7:373–375
Leptotyphlops macrops, 7:373
Leptotyphlops macrorhynchus, 7:373, 7:374
Leptotyphlops macrurus, 7:374
Leptotyphlops melanotermus, 7:373
Leptotyphlops natatrix, 7:375
Leptotyphlops nigricans. See Black threadsnakes
Leptotyphlops nursii, 7:374
Leptotyphlops occidentalis, 7:373, 7:374
Leptotyphlops parkeri, 7:374
Leptotyphlops rostratus, 7:374
Leptotyphlops scutifrons. See Peters' wormsnakes
Leptotyphlops septemstriatus, 7:374
Leptotyphlops teaguei, 7:373–374
Leptotyphlops tricolor, 7:373–374
Leptotyphlops unguirostris, 7:374
Leptotyphlops weyrauchi, 7:373
Leptotyphlops wilsoni, 7:374
Lerista spp., 7:328, 7:329
Lerista bougainvillii, 7:329
Lesser chameleons. *See* Minor chameleons
Lesser Sundas pythons, 7:420
Levantine vipers, 7:449, 7:459
Lialis spp., 7:205
Lialis burtonis. See Burton's snake lizards
Liasis spp., 7:420
Liasis fuscus. See Water pythons
Limbs, 7:24–25
  crocodilians, 7:160
  Lepidosauria, 7:23–24
  Squamata, 7:23–24, 7:202
  Testudines, 7:66, 7:67
  See also Physical characteristics
Line-tailed pygmy monitors. *See* Stripe-tailed monitors
Lined snakes, 7:468
Lioheterodon spp. *See* Madagascan hog-nosed snakes
Liolaeminae, 7:204
Liophidium spp., 7:205
Liophis spp., 7:467
Liotyphlops spp., 7:369–371
Liotyphlops anops, 7:369–370
Liotyphlops beui, 7:369–371
Liotyphlops schubarti, 7:369–370
Liotyphlops ternetzii, 7:369–370, 7:371
Lissemys punctata. See Indian flapshell turtles
Little file snakes, 7:439–440, 7:441, 7:443
Little striped whiptails, 7:310, 7:312
Lizards, **7:195–208**
  alligator, **7:339–344,** 7:341
  Anguidae, **7:339–345,** 7:341
  behavior, 7:204–206
  blindskinks, **7:271–272**
  conservation status, 7:59, 7:62, 7:207
  Cordylidae, **7:319–325,** 7:323
  distribution, 7:203–204
  dragon, **7:209–211,** 7:212, 7:213,
    7:216–220
  ears, 7:9

evolution, 7:16, 7:195–198
eyes, 7:9, 7:32
Florida wormlizards, **7:283–285,** 7:284
in folk medicine, 7:51
as food, 7:48
Gekkonidae, **7:259–269,** 7:264
girdled, **7:319–325,** 7:323
glass, **7:339–342,** 7:341
habitats, 7:204
Helodermatidae, **7:353–358,** 7:356
herbivorous, 7:27
homing, 7:43
humans and, 7:52–53
Iguanidae, **7:243–257,** 7:248, 7:249
integumentary system, 7:29
knob-scaled lizards, **7:347–352,** 7:350
Lacertidae, **7:297–302,** 7:300
limbs, 7:25
locomotion, 7:7–8, 7:25
Mexican beaded, 7:10, **7:353–358,** 7:354,
  7:356
mole-limbed wormlizards, **7:279–282**
in mythology and religions, 7:54, 7:55
night, 7:207
night lizards, **7:291–296,** 7:293
physical characteristics, 7:198–203, 7:199
plated, **7:319–325,** 7:323
reproduction, 7:6–7, 7:42, 7:206–207
rock, **7:297–302**
salt glands, 7:30
skinks, **7:327–338,** 7:330, 7:331
spade-headed wormlizards, **7:287–290**
taxonomy, 7:195–198, 7:196
teeth, 7:9
Teiidae, **7:309–317,** 7:313
territoriality, 7:43–44
thermoregulation, 7:28
Varanidae, **7:359–368,** 7:364
venom, 7:10
wall, **7:297–302**
whiptail, **7:309–317,** 7:313
wormlizards, **7:273–277,** 7:274
See also specific types of lizards
Loas (voodoo gods), 7:57
Lobulia spp., 7:329
Loch Ness monster, 7:54
Locomotion, 7:7–8, 7:24–26, 7:202
  See also Physical characteristics
Loggerhead turtles, 7:61, 7:85, 7:87, 7:88,
  7:89–90
Loggerheads. *See* Loggerhead turtles
Long-neck turtles. *See* Common snakeneck turtles
Long-nosed leopard lizards, 7:244, 7:249,
  7:250–251
Long-nosed pitvipers. *See* Hundred-pace pitvipers
Long-snouted dragons, 7:212, 7:218
Long-snouted lashtails. *See* Long-snouted dragons
Long-tailed utas. *See* Common sagebrush lizards
Lophognathus longirostris. See Long-snouted dragons
Louisiana alligators. *See* American alligators
Loveridgea spp., 7:274
Loxocemidae. *See* Neotropical sunbeam snakes
Loxocemus spp. *See* Neotropical sunbeam snakes

Loxocemus bicolor. See Neotropical sunbeam snakes
Lungs, 7:10, 7:28–29
  See also Respiration
Luth turtles. *See* Leatherback seaturtles
Lycodonomorphus spp. *See* African watersnakes
Lygosoma spp., 7:327–328
Lygosominae, 7:327–328
Lystrophis spp. *See* Neotropical hog-nosed snakes

M

Mabuya spp., 7:327–329
Mabuya heathi. See South American skinks
Mabuya striata. See Striped skinks
Macey, J. Robert, 7:340, 7:347
Macrelaps microlepidotus. See Natal black snakes
Macrocerastes spp., 7:445
Macrochelys spp., 7:93
Macrochelys temminckii. See Alligator snapping turtles
Macrostomata, 7:198
Macroteiids, 7:309, 7:311, 7:312
Macrovipera spp., 7:445, 7:447
Macrovipera lebetina. See Levantine vipers
Madagascan big-headed turtles, 7:139,
  7:141–142
Madagascan hog-nosed snakes, 7:467
Madagascan plated lizards, 7:320, 7:323,
  7:324, 7:325
Madagascan vinesnakes, 7:467
Madagascar boas, 7:411
Madagascar day geckos, 7:264, 7:267,
  7:268
Madagascar iguanas, 7:243
Madagascar tree boas, 7:411
Madras Crocodile Bank, 7:185
Mainland Holdings (Papua, New Guinea),
  7:182
Malaclemys terrapin. See Diamondback terrapins
Malacochersus tornieri. See Pancake tortoises
Malagasy day geckos, 7:260, 7:263
Malayan box turtles, 7:12, 7:115
Malayan gharials, 7:179, 7:180
Malayan softshell turtles, 7:152
Malaysian pitvipers, 7:52, 7:447, 7:448
Malpolon monspessulanus. See Montpellier snakes
Mambas, 7:483–486, 7:490, 7:491
Mansadevi (Indian goddess), 7:56
Map turtles, 7:106
Marbled geckos, 7:261
Marine crocodiles, 7:157
Marine iguanas, 7:27–28, 7:243, 7:244
Marine reptiles, 7:14–15, 7:30
  See also specific species
Marsh crocodiles. *See* Mugger crocodiles
Massasauga snakes, 7:39, 7:447
Masticophis spp. *See* Whipsnakes
Masticophis taeniatus. See Striped whipsnakes
Matamatas, 7:78, 7:80, 7:81–82
Maternal behavior, reptilian, 7:42
  See also Reproduction
Mating. *See* Reproduction

Mauremys annamensis. See Annam leaf turtles
Mayans, reptiles and, 7:55
Mechanoreceptors, 7:9, 7:32
See also Physical characteristics
Medicine, reptiles in, 7:50–52
See also Humans
Mediterranean chameleons. *See* Common chameleons
Mediterranean geckos, 7:260
Megalania spp., 7:16, 7:17
Megalania prisca, 7:359
Mebelya capensis. See Cape filesnakes
Melanophidium spp., 7:391
Melanoseps spp., 7:328
Melanosuchus niger. See Black caimans
Menetia spp. *See* Australian skinks
Menetia greyii, 7:330, 7:334, 7:335–336
Meroles spp., 7:297
Meroles anchietae. See Shovel-snouted lizards
Mesalina spp., 7:297
Mesaspis spp. *See* Alligator lizards
Mesaspis monticola. See Montane alligator lizards
Mesoamerican pythons. *See* Neotropical sunbeam snakes
Mesoclemmys spp., 7:77
Mesoclemmys gibba. See Gibba turtles
Mesopotamians, snakes and, 7:55
Mesosauria, 7:5, 7:14
Metabolism. *See* Energy
Mexican beaded lizards, 7:10, **7:353–358,** 7:354, 7:356
Mexican burrowing pythons. *See* Neotropical sunbeam snakes
Mexican cloud forest anguid lizards, 7:207
Mexican giant musk turtles, 7:121, 7:122
Mexican ridleys. *See* Kemp's ridley turtles
Meylan, Peter A., 7:151
Micrelaps muelleri, 7:462
Microteiids, 7:204, **7:303–308,** 7:305
Micrurus spp. *See* Coral snakes
Micrurus fulvius. See North American coral snakes
Migration, 7:43, 7:86, 7:89
See also Behavior
Military dragons, 7:212, 7:216–217
Milksnakes, 7:466, 7:472, 7:476
Minor chameleons, 7:234, 7:239–240
Mississippi alligators. *See* American alligators
Mojave zebratails. *See* Zebra-tailed lizards
Mole-limbed wormlizards, **7:279–282**
Mole lizards. *See* Two-legged wormlizards
Moloch horridus. See Thorny devils
Mona boas, 7:411
Mona Island blindsnakes, 7:382
Monitors, 7:206, **7:359–368,** 7:364
in folk medicine, 7:51
as food, 7:48
humans and, 7:52, 7:53
Monocled cobras, 7:485
Monopeltis spp., 7:274
Montane alligator lizards, 7:341, 7:343, 7:344
Montatheris spp., 7:445
Montivipera spp., 7:445, 7:447
Montpellier snakes, 7:466
Morelia spp., 7:420
Morelia amethistina, 7:426
Morelia boeleni. See Black pythons
Morelia carinata. See Rough-scaled pythons
Morelia oenpelliensis, 7:420

Morelia tracyae. See Halmahera pythons
Morelia viridis. See Green pythons
Moroccan glass lizards, 7:341, 7:342
Morrocoy amarillos. *See* South American yellow-footed tortoises
Mosasaurs, 7:16
Mountain boomers. *See* Common collared lizards
Mountain devils. *See* Thorny devils
Mountain earless lizards. *See* Common lesser earless lizards
Mountain spiny lizards, 7:249, 7:251, 7:254–255
Mt. Meru chameleons. *See* Jackson's chameleons
Mourning geckos, 7:260, 7:262
Mozambique spitting cobras, 7:195
Mud turtles
American, **7:121–127,** 7:124
East African black, 7:131, 7:133
East African serrated, 7:131, 7:132–133
eastern, 7:121, 7:122
striped, 7:122
white-lipped, 7:124, 7:125, 7:126
yellow, 7:72, 7:124, 7:125
Mudsnakes, 7:467–470
Mugger crocodiles, 7:163, 7:179, 7:182, 7:183, 7:185–186
Murphy, John C., 7:410
Muscles, 7:23–26
See also Physical characteristics
Musk turtles, 7:72, 7:121–127, 7:124
Mussuranas, 7:205, 7:467, 7:471, 7:479–480
Mythology, reptiles in, 7:54–57
See also Humans; specific species

N

Naga Panchami, 7:56
Naja melanoleuca. See Forest cobras
Naja mossambica. See Mozambique spitting cobras
Naja naja. See Indian cobras
Naja naja annulifera. See Egyptian banded cobras
Naja naja kaouthia. See Monocled cobras
Naja nigricollis. See Black-necked spitting cobras
Namib day geckos, 7:261
Natal black snakes, 7:462
National Trust for the Turks and Caicos Islands, 7:247
Native American mythology, snakes in, 7:56–57
Natricinae, 7:466, 7:468
Natrix spp. *See* Eurasian watersnakes
Natrix natrix. See Grass snakes
Natural Resources Conservation Authority, 7:247
Navassa woodsnakes, 7:435
Neck-banded snakes, 7:205, 7:468, 7:469
Neoseps spp., 7:328
Neotropical burrowing snakes, 7:468
Neotropical hog-nosed snakes, 7:467
Neotropical sunbeam snakes, **7:405–407,** 7:405, 7:406
Neotropical swampsnakes, 7:468
Neotropical watersnakes, 7:467, 7:468

Neotropical wood turtles, **7:115–120**
Nephrurus stellatus. See Knob-tailed geckos
Nerodia spp. *See* North American watersnakes
Nerodia rhombifera. See Diamond-backed watersnakes
Nervous system, 7:31
See also Physical characteristics
Nesting, crocodilians, 7:163–164, 7:181
See also Reproduction
Neurotoxins, 7:52
Neusticurus spp., 7:303, 7:304
Neusticurus ecpleopus, 7:305, 7:306, 7:307
Nevada side-blotched lizards. *See* Common side-blotched lizards
Nevada zebratails. *See* Zebra-tailed lizards
New Caledonian geckos, 7:262
New Caledonian giant geckos, 7:259, 7:264, 7:265–266
New Guinea crocodiles, 7:181
New Guinea ground boas. *See* Viper boas
New Guinea olive pythons. *See* Papuan pythons
New Guinea snakeneck turtles, 7:77
New Guinea snapping turtles, 7:68
New World natricine snakes, 7:206
New World pond turtles, **7:105–113,** 7:108
New World pythons. *See* Neotropical sunbeam snakes
Newman's knob-scaled lizards, 7:347–349, 7:350, 7:351
Nidhogger, 7:54
Night adders, 7:445–448, 7:449, 7:451–452, 7:486
Night lizards, 7:207, **7:291–296,** 7:293
Night skinks, 7:331, 7:334
Nile crocodiles, 7:48, 7:161–164, 7:183, 7:186
conservation status, 7:181
feeding ecology, 7:180
as food, 7:48
habitats, 7:179
reproduction, 7:182
Nile monitors, 7:48, 7:360, 7:362–363
Nilgiri burrowing snakes, 7:393
Ninia spp. *See* Coffeesnakes
Nocturnal desert skinks. *See* Night skinks
North American coral snakes, 7:490, 7:491–492
North American earthsnakes, 7:470
North American garter snakes, 7:466
North American watersnakes, 7:466, 7:468
Northern Australian snapping turtles. *See* Victoria river snappers
Northern brown-shouldered lizards. *See* Common side-blotched lizards
Northern cat-eyed snakes, 7:471, 7:479, 7:481
Northern coral snakes. *See* North American coral snakes
Northern crested lizards. *See* Desert iguanas
Northern earless lizards. *See* Common lesser earless lizards
Northern false iguanas. *See* Cape spinytail iguanas
Northern ground utas. *See* Common side-blotched lizards
Northern sagebrush lizards. *See* Common sagebrush lizards
Northern side-blotched lizards. *See* Common side-blotched lizards
Northern snakenecks, 7:78
Northern zebra-tailed lizards. *See* Zebra-tailed lizards

INDEX

Northwest snapping turtles. *See* Victoria river snappers
Notechis scutatus. See Tiger snakes
Nothosaurs, 7:14–15
Nucras tessellata. See Western sandveld lizards

O

Ocadia sinensis. See Chinese stripe-necked turtles
Odatria spp., 7:360
Ogmodon vitianus, 7:484, 7:486
Olfactory senses, 7:31, 7:34–36
    See also Physical characteristics
Oligodon spp. *See* Asian kukrisnakes
Olive ridley seaturtles, 7:71
Olive seasnakes, 7:*489*, 7:*494*
Omanosaura spp., 7:297
Opheodrys aestivus. See Rough greensnakes
Ophiodes spp., 7:339
Ophiomorus spp., 7:328
Ophiophagus hannah. See King cobras
Ophisaurus spp. *See* Glass lizards
Ophisaurus apodus, 7:339
Ophisaurus attenuatus, 7:340
Ophisaurus attenuatus attenuatus. See Western slender glass lizards
Ophisaurus koellikeri. See Moroccan glass lizards
Ophisops spp., 7:297
Ophryacus spp., 7:445, 7:446
Oplurinae, 7:243, 7:244
Oras. *See* Komodo dragons
Oriental grass lizards, 7:297
Oriental six-lined runners. *See* Six-lined grass lizards
Orinoco crocodiles, 7:164
Ornate box turtles, 7:72
Ornithischia, 7:5
Ornithodira, 7:19
Osteolaemus tetraspis. See Dwarf crocodiles
Otocryptis spp., 7:210
Oviparous reproduction, 7:6–7
    See also Reproduction; specific species
Ovophis spp., 7:445
Owenetta spp., 7:13
Oxybelis spp., 7:466
Oxyuranus spp. *See* Taipans
Oxyuranus microlepidotus. See Taipans
Oxyuranus scutellatus. See Taipans

P

Pachydactylus bibronii. See Bibron's geckos
Pachyrhachis spp., 7:16
Pacific green turtles, 7:90
Painted batagurs. *See* Painted terrapins
Painted terrapins, 7:*117*, 7:*118*
Painted turtles, 7:*65*, 7:72, 7:106, 7:*108*, 7:*109*, 7:*110*
Pale leopard lizards. *See* Long-nosed leopard lizards
Palea steindachneri. See Chinese softshell turtles
Paleosuchus palpebrosus. See Cuvier's dwarf caimans

Paleosuchus trigonatus. See Smooth-fronted caimans
Paliguana spp., 7:16
Palmatogecko rangei. See Web-footed geckos
Palmer, Thomas, 7:454
Palmer, W. M., 7:315
Pan hinged terrapins. *See* East African black mud turtles
Pan terrapins. *See* East African black mud turtles
Pancake tortoises, 7:66, 7:*145*, 7:*147–148*
Panther chameleons, 7:*37*, 7:224, 7:225, 7:*227*, 7:*230*, 7:*233*, 7:*239*, 7:240
Papuan ground boas. *See* Viper boas
Papuan pythons, 7:420, 7:*423*, 7:*424–425*
Parachute geckos, 7:*261*
Paraguayan caiman lizards, 7:310, 7:*313*, 7:*316*
Parapsid reptiles, 7:*5*
Pareatinae, 7:465, 7:468
Pareiasaurs, 7:14
Parietal organs, 7:32
Parson's chameleons, 7:*8*, 7:227, 7:*233*, 7:*236–237*
Parthenogenesis, 7:6, 7:207, 7:312
    See also Reproduction
Parvalacerta spp., 7:297
Patagonian lanceheads, 7:*450*, 7:*453–454*
Paulissen, Mark, 7:315
Pedioplanis spp., 7:297
Pelamis platurus. See Yellow-bellied seasnakes
Pelias spp., 7:445, 7:447
Pelochelys cantorii. See Asian giant softshell turtles
Pelodiscus sinensis. See Chinese softshell turtles
Pelomedusa spp., 7:129
Pelomedusa subrufa. See Helmeted turtles
Pelomedusidae. *See* African sideneck turtles
Pelusios spp., 7:129
Pelusios broadleyi, 7:130
Pelusios seychellensis, 7:130
Pelusios sinuatus. See East African serrated mud turtles
Pelusios subniger. See East African black mud turtles
Pelycosauria, 7:5
Peninsula cooter turtles, 7:*36*
Perenties, 7:361
Peters' earthsnakes. *See* Peters' wormsnakes
Peters' threadsnakes. *See* Peters' wormsnakes
Peters' wormsnakes, 7:375, 7:*376–377*
Petrolacosaurus spp., 7:15
Pets, reptilian, 7:53, 7:53t
    See also Humans; specific species
Petzold, H.-G., 7:448
Phelsuma spp. *See* Malagasy day geckos
Phelsuma edwardnewtoni. See Giant geckos
Phelsuma madagascariensis. See Madagascar day geckos
Philippine crocodiles, 7:59
Philippine Gray's monitors, 7:363
Phillips, John, 7:360
Philodryas spp., 7:467
Phipson's shieldtail snakes, 7:392, 7:*393–394*
Phrynocephalus spp., 7:211
Phrynocephalus mystaceus. See Toad-headed agamas
Phrynops spp., 7:77
Phrynops hilarii. See Toad head turtles
Phrynosoma spp. *See* Horned lizards

Phrynosoma cornutum. See Texas horned lizards
Phrynosoma mcalli. See Flat-tailed horned lizards
Phrynosomatinae, 7:243, 7:244
Physical characteristics, 7:4–*5*, 7:8–*9*, **7:23–33**
    African burrowing snakes, 7:461, 7:463
    African sideneck turtles, 7:129, 7:132–133
    Afro-American river turtles, 7:137, 7:140–141
    Agamidae, 7:209–210, 7:214–221
    Alligatoridae, 7:171–172, 7:175–177
    Anguidae, 7:339–340, 7:342–344
    Australo-American sideneck turtles, 7:77, 7:81–83
    big-headed turtles, 7:135
    blindskinks, 7:271
    blindsnakes, 7:379–380, 7:383–385
    boas, 7:409–410, 7:413–417
    Central American river turtles, 7:99
    chameleons, 7:224–227, 7:*225*, 7:*226*, 7:*231*, 7:235–241
    colubrids, 7:467, 7:473–481
    Cordylidae, 7:319–320, 7:324–325
    crocodilians, 7:157–*161*, 7:*159*
    Crocodylidae, 7:179, 7:184–187
    early blindsnakes, 7:369–370
    Elapidae, 7:484, 7:491–498
    false blindsnakes, 7:387
    false coral snakes, 7:400
    file snakes, 7:439–440
    Florida wormlizards, 7:284
    Gekkonidae, 7:259–*260*, 7:265–269
    Geoemydidae, 7:115, 7:118–119
    gharials, 7:167
    Helodermatidae, 7:353–354
    Iguanidae, 7:243–244, 7:250–256
    Kinosternidae, 7:121, 7:125–126
    knob-scaled lizards, 7:347–348, 7:351
    Lacertidae, 7:297, 7:301–302
    leatherback seaturtles, 7:101
    Microteiids, 7:303, 7:306–307
    mole-limbed wormlizards, 7:279–280, 7:281
    Neotropical sunbeam snakes, 7:405–*406*
    New World pond turtles, 7:105, 7:109–112
    night lizards, 7:291, 7:294–295
    Parson's chameleons, 7:*8*
    pig-nose turtles, 7:75
    pipe snakes, 7:395–396
    pythons, 7:420, 7:424–428
    seaturtles, 7:85, 7:89–91
    shieldtail snakes, 7:391–393
    skinks, 7:328, 7:332–337
    slender blindsnakes, 7:373–374, 7:376–377
    snapping turtles, 7:93, 7:95–96
    softshell turtles, 7:151, 7:154–155
    spade-headed wormlizards, 7:287–288, 7:289
    splitjaw snakes, 7:430
    Squamata, 7:*197*, 7:198–203, 7:*199–201*
    sunbeam snakes, 7:401–402
    Teiidae, 7:310, 7:314–316
    tortoises, 7:143, 7:146–148
    Tropidophiidae, 7:433–434, 7:436–437
    tuatara, 7:189–*190*, 7:*191*
    turtles, 7:65–70
    Varanidae, 7:359–360, 7:365–368
    Viperidae, 7:446–447, 7:451–459
    wormlizards, 7:273–274
Phytosaurs, 7:17
Pig-nose turtles, **7:75–76**

Pike-headed alligators. *See* American alligators
Pipe snakes, **7:395–397**
Pitted-shelled turtles. *See* Pig-nose turtles
Pituophis spp., 7:42
Pituophis catenifer. See Bullsnakes
Pitvipers, 7:9, 7:32–33, 7:41, 7:42, 7:206, **7:445–456,** 7:*450*
Placentation, 7:7
    See also Reproduction
Placodontia, 7:5, 7:15
Placodus spp., 7:15
Plains garter snakes, 7:44, 7:470
    See also Garter snakes
Plateau side-blotched lizards. *See* Common side-blotched lizards
Plated lizards, **7:319–325,** 7:*323*
Platyplectrurus spp., 7:391
Platyplectrurus madurensis, 7:392
Platysaurus spp. *See* Flat lizards
Platysaurus capensis. See Cape flat lizards
Platysternidae. *See* Big-headed turtles
Platysternon megacephalum. See Big-headed turtles
Play, reptiles and, 7:39
    See also Behavior
Plectrurus perrotetii. See Nilgiri burrowing snakes
Plesiosaurs, 7:15
Pleurodira, 7:13, 7:65
Podarcis lilfordi, 7:298
Podocnemididae. *See* Afro-American river turtles
Podocnemis expansa. See South American river turtles
Podocnemis unifilis. See Yellow-spotted river turtles
Podocnemis vogli, 7:137
Podophis spp., 7:16
Pogona barbata. See Bearded dragons
Pogona minor. See Bearded dragons
Pogona vitticeps. See Bearded dragons
Polemon spp. *See* Snake eaters
Pollution, chemical, 7:60–62
Polychrotinae. *See* Anoles
Polyglyphanodontines, 7:309
Pond sliders, 7:*108,* 7:*110*–111
Pope's pitvipers, 7:*446*
Poromera spp., 7:297
Porthidium spp., 7:445
Porthidium nasutum, 7:448
Porthidium ophryomegas, 7:447, 7:448
Prairie rattlesnakes, 7:34, 7:38, 7:44, 7:50
Prasinohaema spp., 7:329
Predation, 7:39, 7:162–163
    See also Behavior
Prehensile-tailed skinks, 7:329, 7:*331,* 7:*332*–333
Prickly forest skinks, 7:*330,* 7:*333,* 7:335
Prionodactylus argulus. See Elegant-eyed lizards
Proatheris spp., 7:445
Proctoporus spp., 7:303
Proganochelys spp., 7:13
Proprioceptors, 7:9
Prosymna spp. *See* Shovel-snouted snakes
Protobothrops spp., 7:445
Psammodromas spp., 7:297
Psammophis spp. *See* Sandsnakes
Pseudechis spp. *See* Black snakes
Pseudechis porphyriacus. See Red-bellied black snakes

Pseudemydura spp., 7:78
Pseudemydura umbrina. See Western swamp turtles
Pseudemys concinna. See River cooters
Pseudemys nelsoni. See Red-bellied turtles
Pseudoboines, 7:467
Pseudocerastes spp., 7:445
Pseudocordylus spp. *See* Crag lizards
Pseudoeryx spp., 7:468
Pseudonaja spp. *See* Brown snakes
Pseudonaja textilis. See Brown snakes
Pseudosuchia, 7:17
Pseustes spp. *See* Bird snakes
Ptenopus spp. *See* Bell geckos
Pteranodon spp., 7:19
Pterodactyloids, 7:19
Pterodaustro spp., 7:19
Pterosauria, 7:5, 7:19
Ptyas spp. *See* Banded ratsnakes
Ptyas korros, 7:51
Ptychoglossus spp., 7:303
Ptychozoon spp., 7:*261*
Puerto Rican boas, 7:411
Puff adders. *See* Eastern hog-nosed snakes
Puff-faced watersnakes, 7:468
Purple-glossed snakes, 7:461, 7:462
Pygmy goannas, 7:361
Pygmy pythons, 7:421, 7:*423,* 7:*424*
Pygopodinae. *See* Pygopods
Pygopods, 7:205, **7:259–269,** 7:*264*
  behavior, 7:261
  conservation status, 7:262–263
  distribution, 7:*259,* 7:260
  evolution, 7:259
  feeding ecology, 7:261–262
  habitats, 7:260–261
  humans and, 7:263
  physical characteristics, 7:260
  reproduction, 7:262
  species of, 7:*265–269*
Python spp., 7:419, 7:420
Python anchietae. See Angolan pythons
Python brongersmai. See Blood pythons
Python molurus. See Indian pythons
Python molurus bivattatus. See Burmese pythons
Python regius. See Ball pythons
Python reticulatus. See Reticulated pythons
Python sebae. See African rock pythons
Python timoriensis. See Lesser Sundas pythons
Pythonidae. *See* Pythons
Pythons, **7:419–428,** 7:*423*
  behavior, 7:205, 7:420–421
  boas *vs.,* 7:409, 7:419
  conservation status, 7:421–422
  distribution, 7:*419,* 7:420
  endothermic, 7:28
  evolution, 7:419
  feeding ecology, 7:421
  habitats, 7:420
  humans and, 7:421, 7:422
  muscles, 7:26
  in mythology and religions, 7:55–57
  physical characteristics, 7:202
  reproduction, 7:36, 7:42, 7:207, 7:421
  skeleton, 7:24
  snake charming and, 7:56
  species of, 7:*424–428*
  taxonomy, 7:419
  See also specific types of pythons

Q

Quetzalcoatl (Mayan deity), 7:55
Quetzalcoatlas, 7:19
Quill-snouted snakes, 7:461, 7:462, 7:*463*

R

R-complex. *See* Reptilian brain
Racers, 7:26, 7:205–206, 7:466, 7:469, 7:470
Radiated tortoises, 7:*30*
Rainbow boas, 7:410
Rainbow snakes, 7:469
Ramphotyphlops spp., 7:380–382
Ramphotyphlops bituberculatus, 7:380
Ramphotyphlops braminus, 7:379, 7:381, 7:382
Ramphotyphlops cumingii, 7:380, 7:381
Ramphotyphlops exocoeti. See Christmas Island blindsnakes
Ramphotyphlops grypus, 7:380
Ramphotyphlops nigrescens. See Blackish blindsnakes
Ramphotyphlops proximus, 7:379
Ramphotyphlops unguirostris, 7:379
Ramphotyphlops wiedii, 7:382
Ranacephala spp., 7:77
Ranacephala hogei. See Hoge's sideneck turtles
Ranching, reptiles, 7:48–*49*
Ratsnakes, 7:41, 7:42, 7:43, 7:466, 7:468, 7:470
Rattlesnakes
  behavior, 7:44
  chemosensory system, 7:36–37
  eastern diamondback, 7:40, 7:50, 7:475
  as food, 7:48
  homing, 7:43
  prairie, 7:34, 7:38, 7:44, 7:50
  in religions, 7:56–57
  reproduction, 7:42
  roundups, 7:49–50
  timber, 7:*50,* 7:448, 7:*450,* 7:*452,* 7:454
  venom, 7:10
  western diamondback, 7:38, 7:40, 7:42, 7:44, 7:50, 7:*197*
  See also Snakes
Rauisuchids, 7:17
Rectilinear locomotion, 7:26
Red-bellied black snakes, 7:486, 7:*489,* 7:*497*
Red-bellied mudsnakes, 7:469
Red-bellied turtles, 7:*106*
Red Data Book of Threatened Vertebrates (Greek), 7:238
Red-eared sliders, 7:47, 7:48, 7:66, 7:107
  See also Pond sliders
Red-sided garter snakes, 7:479
Red-tailed pipe snakes, 7:*396,* 7:*397*
Redtail boas. *See* Boa constrictors
Reedsnakes, 7:466
Regal pythons. *See* Reticulated pythons
Regeneration, of tails, 7:*24*
Regina alleni. See Striped crayfish snakes
Religion, reptiles and, 7:*50,* 7:55–57
  See also Humans; specific species
Renenutet (Egyptian goddess), 7:55
Reproduction, **7:6–7,** 7:*36*
  activity patterns and, 7:44
  African burrowing snakes, 7:462–464
  African sideneck turtles, 7:130, 7:132–133
  Afro-American river turtles, 7:137–138, 7:140–142

Reproduction *(continued)*
Agamidae, 7:211, 7:214–221
Alligatoridae, 7:*173*, 7:175–177
American alligators, 7:*4*, 7:*36*
Anguidae, 7:340, 7:342–344
Australo-American sideneck turtles, 7:78, 7:81–83
big-headed turtles, 7:136
blindskinks, 7:272
blindsnakes, 7:382–385
boas, 7:411, 7:413–417
Central American river turtles, 7:100
chameleons, 7:*230–232*, 7:235–241
colubrids, 7:469–470, 7:473–481
Cordylidae, 7:321–322, 7:324–325
crocodilians, 7:6–7, 7:162–163, 7:163–164
Crocodylidae, 7:180–*181*, 7:184–187
early blindsnakes, 7:371
Elapidae, 7:486, 7:491–498
false blindsnakes, 7:388
false coral snakes, 7:400
file snakes, 7:442
Florida wormlizards, 7:284–285
Gekkonidae, 7:262, 7:265–269
Geoemydidae, 7:116, 7:118–119
gharials, 7:168–169
Helodermatidae, 7:*357*
Iguanidae, 7:246, 7:250–257
Kinosternidae, 7:122, 7:125–126
knob-scaled lizards, 7:348, 7:351
Lacertidae, 7:298, 7:301–302
leatherback seaturtles, 7:*102*
lizards, 7:6–7
Microteiids, 7:304, 7:306–307
mole-limbed wormlizards, 7:280, 7:281
Neotropical sunbeam snakes, 7:406
New World pond turtles, 7:*106*, 7:109–113
night lizards, 7:292, 7:294–295
oviparous, 7:6–7
pig-nose turtles, 7:76
pipe snakes, 7:397
pythons, 7:421, 7:424–428
seaturtles, 7:86–87, 7:89–91
shieldtail snakes, 7:392–394
skinks, 7:329, 7:332–337
slender blindsnakes, 7:375–377
snakes, 7:6–7
snapping turtles, 7:6, 7:*94*, 7:95–96
softshell turtles, 7:152, 7:154–155
spade-headed wormlizards, 7:288, 7:289
splitjaw snakes, 7:430
Squamata, 7:*206–207*
sunbeam snakes, 7:403
tactile cues, 7:40–42
Teiidae, 7:312, 7:314–316
tortoises, 7:7, 7:70–72, 7:*71*, 7:143–144, 7:146–148
Tropidophiidae, 7:435–437
tuatara, 7:6–7, 7:190–191
turtles, 7:6–7, 7:70–72, 7:*71*
Varanidae, 7:*360*, 7:362–363, 7:365–368
Viperidae, 7:448, 7:451–459
viviparous, 7:6–7
vomeronasal system and, 7:35–36
wormlizards, 7:274, 7:276
Reptiles
vs. amphibians, 7:3
behavior, **7:34–46**
brains, 7:8
conservation status, 7:10–11, **7:59–63**

definition and description, **7:3–11**
diseases, 7:61, 7:62
diversity, 7:7–8
ears, 7:9
evolution, **7:12–22**, 7:*13*
eyes, 7:8–9
feeding ecology, 7:3–4, 7:26–28
in folk medicine, 7:50–52, 7:*54*
humans and, **7:47–58**
locomotion, 7:8, 7:24–26
as pets, 7:53
physical characteristics, 7:4–*5*, 7:8–9, **7:23–33**
reproduction, **7:6–7**, 7:*36*
See also specific reptiles
Reptilian brain, 7:8
Respiration, 7:28–29, 7:*67*, 7:160–161
See also Physical characteristics
Reticulated pythons, 7:56, 7:202, 7:420–422, 7:*423*, 7:*426*, 7:427–428
Rhabdophis tigrinus. See Yamakagashis
Rhacodactylus auriculatus. See New Caledonian geckos
Rhacodactylus leachianus. See New Caledonian giant geckos
Rhampholeon spp., 7:224–225, 7:227
Rhampholeon brevicaudatus. See Short-tailed chameleons
Rhamphorhyncoids, 7:19
Rheodytes leukops. See Fitzroy River turtles
Rhinemys spp., 7:77
Rhineura floridana. See Florida wormlizards
Rhineuridae. *See* Florida wormlizards
Rhinoceros iguanas, 7:246
Rhinoleptus spp., 7:373, 7:374
Rhinoleptus koniagui, 7:373, 7:374
Rhinophis spp., 7:391, 7:392
Rhinophis oxyrhynchus, 7:391
Rhinotyphlops spp., 7:380–382
Rhinotyphlops acutus, 7:379, 7:381
Rhinotyphlops feae, 7:381
Rhinotyphlops newtoni, 7:381
Rhinotyphlops schlegelii. See Schlegel's blindsnakes
Rhinotyphlops simoni, 7:381
Rhinotyphlops unitaeniatus, 7:380
Rhombic egg-eaters. *See* Common egg-eaters
Rhombic night adders, 7:*449*, 7:*451*–452
Rhoptropus afer. See Namib day geckos
Rhynchocephalia. *See* Tuataras
Ring-necked snakes, 7:*35*, 7:467–469
Ringed pythons, 7:420
Rinkhal's cobras, 7:485
Riopa spp., 7:328
River cooters, 7:*107*
Rock lizards, **7:297–302**
Roman mythology, reptiles in, 7:54–55
Romer, Alfred S., 7:13
Rosy boas, 7:409–411, 7:*412*, 7:*413*, 7:416
Rough earthsnakes, 7:468
Rough greensnakes, 7:468
Rough-scaled lizards, 7:298
Rough-scaled pythons, 7:420
Round Island boas. *See* Splitjaw snakes
Roundups, rattlesnakes, 7:49–50
Royal pythons. *See* Ball pythons
Rubber boas, 7:409
Russel's vipers, 7:56, 7:448, 7:*449*, 7:*451*, 7:458

S
Sagebrush lizards, 7:*249*, 7:*255–256*
Sahara sandboas, 7:409, 7:411
Saharan horned vipers. *See* Horned vipers
Sailfin lizards, 7:*213*, 7:*215*, 7:218
Saint Patrick, 7:54
St. Croix ground lizards, 7:312
St. Lucia whiptails, 7:312
Saiphos equalis, 7:329
Salivary glands, 7:9–10
See also Physical characteristics
Salt glands, 7:30
See also Physical characteristics
Saltwater crocodiles, 7:157–158, 7:161–163, 7:*183*, 7:*185*, 7:186–187
conservation status, 7:181–182
farming, 7:48
as food, 7:48
humans and, 7:182
muscular limbs of, 7:23
physical characteristics, 7:179
Samut Prakan Crocodile Farm (Thailand), 7:182
San Diego banded geckos, 7:265
San Diego horned lizards. *See* Texas horned lizards
San Francisco garter snakes, 7:479
Sand lizards, 7:298, 7:*300*, 7:*301*–302
Sand monitors, 7:362
Sand swimmers. *See* Half-girdled snakes
Sandboas, 7:409–411
Sandfish, 7:*330*, 7:*332*, 7:337
Sandsnakes, 7:466
Sarcosuchus spp., 7:157, 7:158
Sauresia spp., 7:339
Saurischia, 7:5
Sauromalus obesus. See Common chuckwallas
Sauropterygia, 7:5
Saw-jawed turtles. *See* Painted terrapins
Saw-scaled vipers, 7:447, 7:448, 7:*449*, 7:*451*, 7:458
Scaly-fronted wormsnakes. *See* Peters' wormsnakes
Scaphiodontophis spp. *See* Neck-banded snakes
Scarlet kingsnakes. *See* Milksnakes
Scarletsnakes, 7:469
Sceloporus spp. *See* Spiny lizards
Sceloporus jarrovii. See Mountain spiny lizards
Scelotes spp., 7:328
Schlegel's beaked snakes. *See* Schlegel's blindsnakes
Schlegel's blindsnakes, 7:379–380, 7:382, 7:*383*
Schmidt, Karl P., 7:475
Schneider's smooth-fronted caimans. *See* Smooth-fronted caimans
Scincella spp., 7:328, 7:329
Scincidae. *See* Skinks
Scincinae, 7:327
Scincomorpha, 7:196, 7:198
Scincus scincus. See Sandfish
Scleroglossa, 7:195–196, 7:197–198, 7:200, 7:202–204, 7:207
Scolecophidia. *See* Blindsnakes
Scolecoseps spp., 7:328
Scorpion. *See* Broad-headed skinks
Sea kraits, 7:483–488, 7:*486*, 7:*490*, 7:*495*
Seasnakes, 7:30, **7:483–488**, 7:*489*, 7:*490*, **7:494–497**

Seaturtles, 7:*38*, 7:43, **7:85–92**, 7:*87*, 7:*88*
  behavior, 7:*86*
  conservation status, 7:59–61, 7:*60*, 7:87
  evolution, 7:14, 7:85
  feeding ecology, 7:86
  as food, 7:47
  habitats, 7:86
  humans and, 7:87
  locomotion, 7:24
  physical characteristics, 7:85
  reproduction, 7:71, 7:86–87
  species of, 7:*89*–91
  taxonomy, 7:85
  See also Leatherback seaturtles
Semites, snakes and, 7:55
Sensory system, 7:8–10, 7:159–160
  See also Physical characteristics
Sepsophis spp., 7:328
Serpent Worship in Africa (Hambly), 7:55
Serpentine locomotion, 7:8
Serrated hinged terrapins. *See* East African
  serrated mud turtles
Serrated turtles. *See* East African serrated
  mud turtles
Seychelles tortoises, 7:59
Sharp-nosed pitvipers. *See* Hundred-pace
  pitvipers
Shesh Nag, 7:56
Shielded blindsnakes. *See* Peters' wormsnakes
Shieldtail snakes, **7:391–394**
Shinglebacks. *See* Bobtails
Short pythons. *See* Blood pythons
Short-tailed chameleons, 7:*234*, 7:*235*,
  7:240–241
Short-tailed monitors, 7:359, 7:361, 7:362,
  7:*364*, 7:*366*
Short-tailed pygmy monitors. *See* Short-tailed
  monitors
Short-tailed pythons. *See* Blood pythons
Shorthead wormlizards. *See* Spade-headed
  wormlizards
Shovel-nosed snakes, 7:466, 7:468
Shovel-snouted lizards, 7:298
Shovel-snouted snakes, 7:468
Siamese crocodiles, 7:59, 7:164
Sibon spp. *See* Snail-suckers
Sibynophis spp., 7:205
Side-blotched lizards. *See* Common side-
  blotched lizards
Side-stabbing snakes. *See* Southern burrowing
  asps
Sidewinding, 7:*25*–26, 7:39, 7:*40*, 7:447
Simbi (Loa god), 7:57
Simony's giant lizards, 7:298–299
Simoselaps spp. *See* Half-girdled snakes
Simoselaps calonotus, 7:486
Simoselaps semifasciatus. *See* Half-girdled snakes
Sistrurus spp., 7:445, 7:446
Six-lined grass lizards, 7:*300*, 7:*301*, 7:302
Six-lined racerunners, 7:*313*, 7:*314*–315
Skeleton, 7:23–26
  crocodilians, 7:23
  lepidosauria, 7:23–24
  turtles, 7:23, 7:*70*
  See also Physical characteristics
Skin. *See* Integumentary system
Skinks, 7:327, **7:327–338**, 7:328, 7:329,
  7:*330*, 7:*331*
  Australian, 7:202
  behavior, 7:204, 7:205, 7:329

blindskinks, **7:271–272**
chemosensory systems, 7:36
conservation status, 7:329
distribution, 7:*327*, 7:328
evolution, 7:196, 7:327–328
feeding ecology, 7:329
Great Plains, 7:42
habitats, 7:329
humans and, 7:329
physical characteristics, 7:328
reproduction, 7:206–207, 7:329
species of, 7:*332–337*
taxonomy, 7:327–328
Skins, human uses of, 52–53
Skulls, 7:4–*5*, 7:26
  crocodilian, 7:26, 7:*159*
  lizards, 7:*199*
  snakes, 7:26, 7:*201*
  Squamata, 7:198, 7:200
  tuatara, 7:*191*
  turtle, 7:26, 7:*65*, 7:*67*
  See also Physical characteristics
Sleepy lizards. *See* Bobtails
Slender blindsnakes, 7:198, 7:201, 7:202,
  7:371, **7:373–377**
Slowworms, 7:339, 7:340
Slug-eaters, 7:467, 7:*471*, 7:477
Small-scaled burrowing asps, 7:462
Small-spotted lizards. *See* Long-nosed leopard
  lizards
Smell. *See* Olfactory senses
Smooth-fronted caimans, 7:164, 7:171, 7:*174*,
  7:*175*, 7:177
Smooth-scaled splitjaws, 7:429, 7:*430*
Smooth snakes, 7:472, 7:*473*–474
Snail-suckers, 7:201, 7:467
Snake charming, cobras, 7:485
Snake eaters, 7:461
Snake lizards, 7:320, 7:321
  See also Burton's snake lizards
Snake oil, 7:50
Snakeneck turtles, 7:78, 7:*80*, 7:*82*–83
Snakes, **7:195–208**
  African burrowing, **7:461–464**
  behavior, 7:204–206
  blindsnakes, 7:198, 7:201, **7:379–385**, 7:*380*
  boas, **7:409–417**, 7:*412*
  colubrids, **7:465–482**, 7:*471*, 7:*472*
  conservation status, 7:59, 7:207
  corn, 7:*39*, 7:470
  digestion, 7:27
  distribution, 7:203–204
  early blindsnakes, 7:198, 7:201, 7:202,
    7:369–372
  ears, 7:9, 7:32
  Elapidae, **7:483–499**, 7:*489*, 7:*490*
  evolution, 7:16–17, 7:195–198
  eyes, 7:9, 7:32
  false blindsnakes, **7:387–388**
  false coral snakes, **7:399–400**
  fear of, 7:53–54
  file snakes, 7:198, **7:439–444**, 7:*443*
  in folk medicine, 7:50–52
  as food, 7:48
  habitats, 7:204
  homing, 7:43
  humans and, 7:49–50, 7:52–53
  integumentary system, 7:29
  locomotion, 7:8, 7:25–26
  lungs, 7:10

migrations, 7:43
in mythology and religion, 7:50, 7:52,
  7:54–57
Neotropical sunbeam, **7:405–407**
olfactory system, 7:34–36
physical characteristics, 7:*197*, 7:198–203,
  7:*200*, 7:*201*
pipe snakes, **7:395–397**
pythons, **7:419–428**, 7:*423*
reproduction, 7:6–7, 7:40–42, 7:206–207
salt glands, 7:30
shieldtail, **7:391–394**
skulls, 7:26, 7:*201*
slender blindsnakes, **7:373–377**
splitjaw, **7:429–431**, 7:*430*
sunbeam, **7:401–403**
taxonomy, 7:195–198, 7:*196*
teeth, 7:9, 7:26
territoriality, 7:44
Tropidophiidae, **7:433–438**
Viperidae, **7:445–460**, 7:*449*, 7:*450*
See also specific types of snakes
Snappers. *See* Snapping turtles
Snapping turtles, **7:93–97**, 7:*95*
  farming, 7:48
  as food, 7:47
  New Guinea, 7:68
  reproduction, 7:6, 7:*72*, 7:94, 7:*94*, 7:*95*–96
  See also Alligator snapping turtles
Soa soas. *See* Sailfin lizards
Sobek (Egyptian idol), 7:55
Softshell turtles, 7:48, 7:65–72, **7:151–155**,
  7:*153*
Sonora spp. *See* Groundsnakes
Sonoran collared lizards. *See* Common
  collared lizards
South American river turtles, 7:70, 7:*139*,
  7:*140*
South American skinks, 7:206, 7:329
South American tropical lizards, 7:243
South American yellow-footed tortoises,
  7:*145*, 7:*146*
Southern bromeliad woodsnakes, 7:*436–437*
Southern burrowing asps, 7:463–464
Southern copperheads, 7:52, 7:447, 7:448
Southern sagebrush lizards. *See* Common
  sagebrush lizards
Spade-headed wormlizards, **7:287–290**
Speckled earless lizards. *See* Common lesser
  earless lizards
Spectacled caimans. *See* Common caimans
Sphaerodactylus ariasae. *See* Jaragua sphaero
Sphenodon spp. *See* Tuataras
Sphenodon guntheri, 7:189–192
Sphenodon punctatus, 7:189–192
  See also Tuataras
Sphenodontidae. *See* Tuataras
Sphenomorphus spp., 7:327–329
Spilotes spp. *See* Tiger ratsnakes
Spinal cord, 7:31
  See also Physical characteristics; Skeleton
Spinejaw snakes, **7:433–438**
Spiny agamas, 7:*213*, 7:*214*
Spiny lizards, 7:41, 7:246, 7:*249*, 7:*251*,
  7:254–255
Spiny softshells, 7:*153*, 7:*154*–155
Spiny-tailed agamas, 7:*213*, 7:*214*, 7:221
Spiny-tailed iguanas. *See* Cape spinytail
  iguanas
Spitting cobras, 7:*487*, 7:*490*, 7:*492*–493

INDEX

Splitjaw boas. *See* Splitjaw snakes
Splitjaw snakes, **7:429–431**, *7:430*
Spotted blindsnakes, 7:*197*, 7:379
Spotted harlequin snakes, 7:*485*
Spotted lizards. *See* Common lesser earless
  lizards
Spotted turtles, 7:*108*, 7:*110*, 7:111
Spreading adders. *See* Eastern hog-nosed
  snakes
Squamata, **7:195–208**
  behavior, 7:204–206
  conservation status, 7:59, 7:207
  distribution, 7:203–204
  evolution, 7:195–198
  feeding ecology, 7:204–*205*
  habitats, 7:204
  hearing, 7:32
  integumentary system, 7:*30*
  physical characteristics, 7:23–24, 7:*197*,
    7:*199–201*
  reproduction, 7:*205*, 7:206–207
  skulls of, 7:5
  taxonomy, 7:195–198, 7:*196*
  teeth, 7:26
  See also Lizards; Snakes
Sri Lankan hump-nosed pitvipers, 7:*450*,
  7:*454*, 7:455–456
Stagnolepus spp., 7:17–18
Stansbury's swift. *See* Common side-blotched
  lizards
Staurotypinae. *See* Musk turtles
Staurotypus spp., 7:121, 7:122
Staurotypus triporcatus. *See* Mexican giant
  musk turtles
Steindachner's turtles, 7:78
Stem reptiles. *See* Cotylosauria
Stenorrhina spp., 7:469
Sternotherus spp., 7:121
Sternotherus depressus, 7:123
Sternotherus odoratus. *See* Stinkpots
Stiletto snakes. *See* African burrowing snakes
Stinkpots, 7:*122*, 7:*124*, 7:*125*, 7:126
Streptostyly, 7:195
Stripe-tailed monitors, 7:361, 7:362, 7:*364*,
  7:*366–367*
Striped blind legless skinks, 7:*330*, 7:*332*
Striped crayfish snakes, 7:469
Striped mud turtles, 7:*122*
Striped quill-snouted snakes. *See* Variable
  quill-snouted snakes
Striped sand lizards. *See* Western sandveld
  lizards
Striped skinks, 7:*330*, 7:*332*, 7:335
Striped whipsnakes, 7:43
Strophurus spp., 7:261
Stupendemys spp., 7:14
Stupendemys geographicus, 7:137
Suction feeding, 7:3–4
Suffusio predatrix, 7:369
Sunbeam snakes, **7:401–403**
Sungazers. *See* Giant girdled lizards
Sungei tuntong. *See* Painted terrapins
Superstitions, reptiles and, 7:57
Synapsid reptiles, 7:4–*5*

T

Tactile behavioral cues, 7:40–42
  See also Behavior

Tails
  autotomy, 7:261
  regeneration of, 7:*24*
  See also Physical characteristics
Taipans, 7:483, 7:485, 7:488, 7:*490*, 7:*496*
Takydromus spp. *See* Oriental grass lizards
Takydromus sexlineatus. *See* Six-lined grass
  lizards
Tantilla spp. *See* Black-headed snakes
Taste, 7:30
  See also Physical characteristics
Taxonomy
  African burrowing snakes, 7:461, 7:463
  African sideneck turtles, 7:129, 7:132–133
  Afro-American river turtles, 7:137,
    7:140–141
  Agamidae, 7:209, 7:214–221
  Alligatoridae, 7:171, 7:175–177
  Anguidae, 7:339, 7:342–344
  Australo-American sideneck turtles, 7:77,
    7:81–83
  big-headed turtles, 7:135
  blindsnakes, 7:379, 7:383–384
  boas, 7:409, 7:413–416
  Central American river turtles, 7:99
  chameleons, 7:223–224, 7:235–240
  colubrids, 7:465–467, 7:473–481
  Cordylidae, 7:319, 7:324–325
  crocodilians, 7:157, 7:*158*
  Crocodylidae, 7:179, 7:184–186
  early blindsnakes, 7:369
  Elapidae, 7:483–484, 7:491–498
  false blindsnakes, 7:387
  false coral snakes, 7:399
  file snakes, 7:439
  Florida wormlizards, 7:283
  Gekkonidae, 7:259, 7:265–269
  Geoemydidae, 7:115, 7:118–119
  gharials, 7:167
  Iguanidae, 7:243, 7:250–256
  Kinosternidae, 7:121, 7:125–126
  knob-scaled lizards, 7:347, 7:351
  Lacertidae, 7:297, 7:301–302
  leatherback seaturtles, 7:101
  Microteiids, 7:303, 7:306–307
  mole-limbed wormlizards, 7:279, 7:281
  Neotropical sunbeam snakes, 7:405
  New World pond turtles, 7:105, 7:109–112
  night lizards, 7:291, 7:294–295
  pig-nose turtles, 7:75
  pipe snakes, 7:395
  pythons, 7:419, 7:424
  seaturtles, 7:85, 7:89–91
  shieldtail snakes, 7:393
  skinks, 7:327–328, 7:332–337
  slender blindsnakes, 7:373, 7:376
  snapping turtles, 7:93, 7:95–96
  softshell turtles, 7:151, 7:154–155
  spade-headed wormlizards, 7:287
  splitjaw snakes, 7:430
  Squamata, 7:195–198, 7:*196*
  sunbeam snakes, 7:401
  Teiidae, 7:309, 7:314–316
  Testudines, 7:*66*
  tortoises, 7:143, 7:146–148
  Tropidophiidae, 7:433, 7:436
  tuatara, 7:189
  Varanidae, 7:359, 7:365–368
  Viperidae, 7:445, 7:451–459
  wormlizards, 7:273, 7:276

Taylor, Edward H., 7:369
Teeth, 7:9, 7:26
  See also Physical characteristics
Tegus, 7:207, **7:309–317**, 7:*313*
Teiidae, 7:204, 7:206, 7:207, **7:309–317**,
  7:*313*
Teiinae, 7:309
Teius spp., 7:309, 7:311, 7:312
Temperature-dependent sex determination
  (TSD), 7:7, 7:13, 7:72, 7:164
  See also Reproduction
Tentacled snakes, 7:465, 7:468, 7:*472*, 7:*473*,
  7:477
Teoidea, 7:307
Teratoscincus scincus. *See* Common plate-tailed
  geckos
Terecays. *See* Yellow-spotted river turtles
Teretrurus sanguineus, 7:391
Terrapene spp. *See* Box turtles
Terrapene carolina. *See* Eastern box turtles
Terrapins
  diamondback, 7:107, 7:*108*, 7:*109–110*
  painted, 7:*117*, 7:*118*
  See also Helmeted turtles
Terrestrial kraits, 7:483, 7:484, 7:486
Terrestrisuchus spp., 7:18
Territoriality, 7:43–44, 7:245
  See also Behavior
Testudines. *See* Turtles
Testudinidae. *See* Tortoises
Testudo graeca. *See* Greek tortoises
Testudo hermanni. *See* Hermann's tortoises
Tetradactylus eastwoodae. *See* Eastwood's seps
Texas alligator lizards, 7:*341*, 7:*343*, 7:344
Texas blindsnakes, 7:374–376
Texas horned lizards, 7:246, 7:*249*, 7:*251*,
  7:254
Texas Parks and Wildlife Department, on
  northern cat-eyed snakes, 7:481
Texas spotted whiptails, 7:312
Texas threadsnakes. *See* Texas blindsnakes
Texas wormsnakes. *See* Texas blindsnakes
Thalattosuchians. *See* Marine crocodiles
Thamnophiine natricines, 7:470
Thamnophiini, 7:466
Thamnophis spp. *See* North American garter
  snakes
Thamnophis butleri. *See* Garter snakes
Thamnophis radix. *See* Plains garter snakes
Thamnophis sauritus. *See* Garter snakes
Thamnophis sirtalis. *See* Common garter snakes
Thamnophis sirtalis infernalis. *See* San
  Francisco garter snakes
Thamnophis sirtalis parietalis. *See* Red-sided
  garter snakes
Thamnophis sirtalis tetrataenia. *See* San
  Francisco garter snakes
Thecodontia, 7:5
Theletornis spp. *See* Twigsnakes
Therapsida, 7:5
Thermoreception, 7:32–33
Thermoregulation, 7:28, 7:44–45
  Iguanidae, 7:244
  knob-scaled lizards, 7:348
  Teiidae, 7:310–311
  See also Behavior
Third eyes, 7:32
Thorny devils, 7:*212*, 7:*217*, 7:218–219
Threadsnakes, 7:*374*, 7:*485*
  See also Slender blindsnakes

Three-horned chameleons. *See* Jackson's chameleons
Three-stripe batagurs. *See* Painted terrapins
Tibetan pitvipers, 7:*450*, 7:*454*, 7:455
Ticinosuchus spp., 7:17
Tiger ratsnakes, 7:466, 7:468
Tiger snakes, 7:490, 7:495–496
Tiger watersnakes. *See* Yamakagashis
Tiliqua spp., 7:329
Tiliqua gerrardii, 7:329
Tiliqua rugosa. *See* Bobtails
Tiliqua scincoides. *See* Blue-tongued skinks
Timber rattlesnakes, 7:*50*, 7:448, 7:*450*, 7:*452*, 7:454
Timon spp., 7:297
T'o. *See* Chinese alligators
Toad head turtles, 7:78
Toad-headed agamas, 7:*213*, 7:*215*, 7:220–221
Toes, Squamata, 7:198
See also Physical characteristics
Tokay geckos, 7:*261–262*, 7:263, 7:*264*, 7:*265*, 7:266
Toltecs, reptiles and, 7:55
Tomistoma schlegelii. *See* False gharials; Malayan gharials
Tomistominae. *See* False gharials
Tongues, 7:9
    chameleons, 7:*224*, 7:*225*, 7:226–227
    crocodilians, 7:160
    Squamata, 7:198
    See also Physical characteristics
Tortoises, **7:65–73, 7:143–149**, 7:*145*
    Aldabra, 7:47, 7:70, 7:72
    conservation status, 7:10, 7:59, 7:61–62, 7:72–73, 7:144
    distribution, 7:70, 7:*143*
    in folk medicine, 7:51
    as food, 7:47
    hearing, 7:32
    in mythology and religions, 7:55
    physical characteristics, 7:65–70, 7:143
    radiated, 7:*30*
    reproduction, 7:7, 7:70–72, 7:*71*, 7:143–144
    See also specific types of tortoises
Tortuga aplanada. *See* Central American river turtles
Tortuga blanca. *See* Central American river turtles
Tortuga plana. *See* Central American river turtles
Tou lung. *See* Chinese alligators
Tracheloptychus spp. *See* Madagascan plated lizards
Trachemys scripta. *See* Pond sliders
Trachemys scripta elegans. *See* Red-eared sliders
Trachemys venusta grayi. *See* Gray's sliders
Trachyboa spp., 7:434
Trachyboa boulengeri, 7:434
Trachyboa gularis, 7:434
Tretanorhinus spp. *See* Neotropical swampsnakes
Tretioscincus bifasciatus, 7:303
Triceratolepidophis spp., 7:445
Trimeresurus spp., 7:445, 7:447
Trimeresurus popeiorum. *See* Pope's pitvipers
Trioceros spp., 7:224
Trionychidae. *See* Softshell turtles
Trionychinae, 7:151
Trionyx spiniferus. *See* Eastern spiny softshell turtles

Trogonophidae. *See* Spade-headed wormlizards
Trogonophis spp., 7:287, 7:288
Trogonophis wiegmanni. *See* Spade-headed wormlizards
Tropical ratsnakes. *See* Tiger snakes
Tropical wormlizards, 7:*280*
Tropidoclonion spp. *See* Lined snakes
Tropidolaemus spp., 7:445, 7:447
Tropidophiidae, **7:433–438**
Tropidophis spp., 7:433, 7:434, 7:435
Tropidophis bucculentus. *See* Navassa woodsnakes
Tropidophis celiae, 7:433
Tropidophis feicki. *See* Banded woodsnakes
Tropidophis fuscus, 7:433
Tropidophis melanurus. *See* Cuban ground boas
Tropidophis paucisquamis, 7:434
Tropidophis semicinctus, 7:433
Tropidophis spiritus, 7:433
Tropidophis taczanowskyi, 7:434
Tropidophorus grayi, 7:329
Tropidosaura spp., 7:297
Tropidurinae, 7:243, 7:244, 7:246–247
Trunk turtles. *See* Leatherback seaturtles
TSD. *See* Temperature-dependent sex determination
Tuatara, **7:189–193**, 7:195
    conservation status, 7:59, 7:192
    skeleton, 7:23–24
    skulls of, 7:5
    third eye, 7:32
    visual displays, 7:*35*
Tupinambinae, 7:309
Tupinambis spp. *See* Tegus
Tupinambis duseni, 7:312
Tupinambis merianae, 7:312
Tupinambis rufescens, 7:310, 7:312
Tupinambis teguixin. *See* Black and white tegus
Turks and Caicos iguanas, 7:247
Turtle-headed seasnakes, 7:490, 7:494–495
Turtles, **7:65–73**
    African sideneck, **7:129–134**, 7:*130*, 7:*131*
    Afro-American river, **7:137–142**, 7:*139*
    American mud, **7:121–127**, 7:*124*
    Australo-American sideneck, **7:77–84**
    big-headed, 7:32, **7:135–136**
    black-breasted leaf, 7:7, 7:*116*
    black marsh, 7:72
    Blanding's, 7:106
    bog, 7:105
    box, 7:*106*
    brown roofed, 7:72
    Central American river, **7:99–100**
    chicken, 7:106
    Chinese softshell, 7:48, 7:151, 7:152, 7:*153*, 7:*154*, 7:155
    Chinese stripe-necked, 7:*117*, 7:*118*, 7:119
    Chinese three-striped box, 7:119
    common snakeneck, 7:78, 7:*80*, 7:*82–83*
    conservation status, 7:10–11, 7:59–62, 7:72–73
    distribution, 7:70
    eastern box, 7:*68*, 7:106, 7:*108*, 7:112–*113*
    Eurasian pond, **7:115–120**
    Eurasian river, **7:115–120**
    evolution, 7:13–14, 7:65
    farming, 7:48
    in folk medicine, 7:51
    as food, 7:47
    food ecology, 7:67

homing, 7:43
    integumentary system, 7:29
    locomotion, 7:24
    in mythology and religion, 7:54–55
    Wood, **7:115–120**
    New World pond, **7:105–113**, 7:*108*
    physical characteristics, 7:65, 7:65–70
    pig-nose, **7:75–76**
    reproduction, 7:6–7, 7:42, 7:70–72, 7:*71*
    salt glands, 7:30
    shells, 7:13, 7:65–67, 7:*69*, 7:72
    skeleton, 7:23, 7:*70*
    skulls, 7:4–5, 7:26
    softshell, 7:48, 7:65–72, **7:151–155**, 7:*153*
    territoriality, 7:44
    See also specific types of turtles
Twigsnakes, 7:466, 7:467
Two-legged wormlizards, 7:280, 7:*281*
Tympanocryptis spp., 7:210
Tympanocryptis cephalus. *See* Earless dragons
Typhlacontias spp., 7:328
Typhlophis spp., 7:369–371
Typhlophis ayarzaguenai, 7:370, 7:*371*
Typhlophis squamosus, 7:*370*, 7:*371*
Typhlopidae. *See* Blindsnakes
Typhlops spp., 7:379–382
Typhlops andamanesis, 7:381
Typhlops angolensis, 7:379
Typhlops bibronii, 7:382
Typhlops cariei, 7:379
Typhlops comorensis, 7:381
Typhlops congestus, 7:380
Typhlops depressiceps, 7:380
Typhlops diardii, 7:382
Typhlops elegans, 7:381
Typhlops grivensis, 7:379
Typhlops lineolatus, 7:379
Typhlops meszoelyi, 7:381
Typhlops monensis. *See* Mona Island blindsnakes
Typhlops oatesii, 7:381
Typhlops paucisquamus, 7:380
Typhlops punctatus. *See* Spotted blindsnakes
Typhlops reticulatus, 7:380
Typhlops socotranus, 7:381
Typhlops vermicularis, 7:382
Typhlosaurus spp., 7:327–329
Typhlosaurus gariepensis, 7:329
Typhlosaurus lineatus. *See* Striped blind legless skinks

U

Uma spp., 7:245
Uma inornata. *See* Coachella Valley fringe-toed lizards
Umbrivaga spp., 7:469
Ungaliophis spp., 7:434, 7:435
Ungaliophis continentalis. *See* Dwarf boas
Ungaliophis panamensis. *See* Southern bromeliad woodsnakes
United States Fish and Wildlife Service
    American alligators, 7:49
    Central American river turtles, 7:100
    eastern indigo snakes, 7:475
    Gila monsters, 7:357
    leatherback seaturtles, 7:102
    reptilian imports/exports, 7:53*t*
    San Francisco garter snakes, 7:479

Urea, reptiles and, 7:3
Uric acid, reptiles and, 7:3
Uromacer spp., 7:467
Uromastyx spp. *See* Dhabb lizards
Uromastyx acanthinurus. See Spiny-tailed agamas
Uropeltidae. *See* Shieldtail snakes
Uropeltis spp., 7:391, 7:392
Uropeltis macrolepis macrolepis. See Large-scaled shieldtails
Uropeltis maculatus, 7:392
Uropeltis myhendrae, 7:391
Uropeltis ocellatus, 7:391
Uropeltis phipsonii. See Phipson's shieldtail snakes
Uropeltis rubromaculata, 7:392
Urosaurus graciosus. See Common sagebrush lizards
U.S. Endangered Species Act. *See* Endangered Species Act (U.S.)
U.S. Fish and Wildlife Service. *See* United States Fish and Wildlife Service
Uta stansburiana. See Common side-blotched lizards
Utas. *See* Common side-blotched lizards

V

Vanzosaura rubricauda, 7:304
Varanidae, 7:9, 7:196, 7:204, 7:206, 7:207, **7:359–368**, 7:364
Varanines, 7:359
Varanus spp., 7:207, 7:359–362
Varanus albigularis, 7:52, 7:360, 7:362
Varanus bengalensis. See Bengal monitor lizards
Varanus brevicauda. See Short-tailed monitors
Varanus caudolineatus. See Stripe-tailed monitors
Varanus dumerilii, 7:363
Varanus eremius, 7:361
Varanus exanthematicus, 7:52
Varanus giganteus. See Perenties
Varanus gilleni, 7:361, 7:362
Varanus gouldii. See Australian sand goannas; Sand monitors
Varanus griseus, 7:48, 7:363
Varanus indicus, 7:360–361
Varanus komodoensis. See Komodo dragons
Varanus mertensi, 7:360
Varanus mitchelli, 7:360–361
Varanus niloticus. See Nile monitors
Varanus olivaceus, 7:361
Varanus prasinus, 7:361–363
Varanus rosenbergi, 7:362–363
Varanus salvadorii. See Crocodile monitors
Varanus salvator. See Water monitors
Varanus tristis, 7:361, 7:363
Varanus varius, 7:362–363
Variable bush vipers. *See* Green bush vipers
Variable quill-snouted snakes, 7:463
Veiled chameleons, 7:224, 7:228, 7:233, 7:237
Venom, 7:9–10, 7:37–38
  folk medicine and, 7:50–52
  teeth and, 7:9, 7:26, 7:206
  See also Elapidae; Viperidae
Venomous snakes. *See* Elapidae; Viperidae
Venus (Roman goddess), 7:54

Vermicella spp. *See* Bandy-bandy snakes
Vermicella annulata. See Bandy-bandy snakes
Vestigial limbs, 7:23, 7:24
Victoria river snappers, 7:78, 7:80, 7:82, 7:83
Vinesnakes, 7:466–468
Viper boas, 7:40, 7:411, 7:412, 7:413–414
Vipera spp., 7:445, 7:446, 7:448
Vipera aspis, 7:448
Vipera berus. See Common adders
Vipera darevskii, 7:447
Vipera dinniki, 7:447
Vipera latifii, 7:448
Vipera ursinii, 7:448
Vipera wagneri, 7:448
Viperidae, **7:445–460**, 7:449, 7:450
  behavior, 7:447
  conservation status, 7:448
  distribution, 7:445, 7:447
  evolution, 7:445–446
  feeding ecology, 7:447–448
  habitats, 7:447
  humans and, 7:448
  physical characteristics, 7:446–447
  reproduction, 7:448
  species of, 7:451–460
  taxonomy, 7:445
Viperinae. *See* Vipers
Vipers, **7:445–460**, 7:449
  behavior, 7:205, 7:206, 7:447
  conservation status, 7:448
  distribution, 7:445, 7:447
  Elapidae *vs.*, 7:483
  evolution, 7:445–446
  feeding ecology, 7:447–448
  habitats, 7:447
  humans and, 7:448
  physical characteristics, 7:446–447
  reproduction, 7:41, 7:42, 7:448
  species of, 7:451–460
  taxonomy, 7:445
Virgin of Guadalupe, 7:55
Virginia spp. *See* North American earthsnakes
Virginia striatula. See Rough earthsnakes
Vishnu (Indian god), 7:55–56
Vision, 7:32, 7:38–39
  See also Eyes
Visuki (Indian god), 7:56
Vitt, Laurie J., 7:304, 7:307, 7:340
Viviparous reproduction, 7:6–7, 7:206–207, 7:298
  See also Reproduction
Vomeronasal system, 7:34–37, 7:195
  See also Physical characteristics
Voodoo, snakes in, 7:57
Voracious geckos, 7:263

W

Wager, Vincent A., 7:223
Wall lizards, **7:297–302**
Wart snakes. *See* File snakes
Water conservation, reptilian, 7:3
Water exchange, 7:30
Water moccasins. *See* Cottonmouths
Water monitors, 7:48, 7:52, 7:360–361, 7:363
Water pythons, 7:420
Web-footed geckos, 7:261, 7:264, 7:267–268
Werner, Franz, 7:223

West African pythons. *See* Ball pythons
Western banded geckos, 7:264, 7:265
Western chuckwallas. *See* Common chuckwallas
Western collared lizards. *See* Common collared lizards
Western diamondback rattlesnakes, 7:38, 7:40, 7:42, 7:44, 7:50, 7:197
Western earless lizards. *See* Common lesser earless lizards
Western ground uta. *See* Common side-blotched lizards
Western pond turtles, 7:106
Western rattlesnakes. *See* Prairie rattlesnakes
Western sagebrush lizards. *See* Common sagebrush lizards
Western sandveld lizards, 7:300, 7:301
Western side-blotched lizards. *See* Common side-blotched lizards
Western slender glass lizards, 7:340
Western swamp turtles, 7:79, 7:80, 7:82, 7:83
Western zebra-tailed lizards. *See* Zebra-tailed lizards
Wetmorena spp., 7:339
Whipsnakes, 7:26, 7:57, 7:486
Whiptail lizards, **7:309–317**, 7:313
White-bellied mangrove snakes, 7:465–466
White-bellied wormlizards, 7:273, 7:276
White-lipped mud turtles, 7:124, 7:125, 7:126
White-lipped pythons, 7:420
White-striped geckos, 7:260, 7:262
Wildlife Protection Fund (India), 7:56
Wilson, E. O., 7:54
Womas, 7:420
Wonambi spp., 7:17
Wonder geckos. *See* Common plate-tailed geckos
Wood turtles, 7:72
Woodsnakes, **7:433–438**
Wormlizards, **7:273–277**, 7:274
  Florida, **7:283–285**, 7:284
  mole-limbed, **7:279–282**
  spade-headed, **7:287–290**
Wormsnakes, 7:467, 7:470
  See also Slender blindsnakes

X

Xantusia spp., 7:291, 7:292
Xantusia henshawi. See Granite night lizards
Xantusia riversiana. See Island night lizards
Xantusia vigilis. See Desert night lizards
Xantusiidae. *See* Night lizards
Xenocalamus spp. *See* Quill-snouted snakes
Xenocalamus bicolor. See Variable quill-snouted snakes
Xenodermatinae, 7:465
Xenodon spp. *See* False pitvipers
Xenodontinae, 7:466–468
Xenopeltidae. *See* Sunbeam snakes
Xenopeltis spp. *See* Sunbeam snakes
Xenopeltis unicolor. See Common sunbeam snakes
Xenophidion spp., 7:433, 7:434
Xenosauridae. *See* Knob-scaled lizards
Xenosaurus spp. *See* Knob-scaled lizards
Xenosaurus grandis, 7:348, 7:350, 7:351
Xenosaurus grandis grandis, 7:347

Xenosaurus newmanorum. See Newman's knob-scaled lizards
Xenosaurus penai, 7:347
Xenosaurus phalaroantheron, 7:347
Xenosaurus platyceps, 7:347, *7:348*
Xenosaurus rectocollaris, 7:347
Xenotyphlops spp., 7:380, 7:381
Xenotyphlops grandidieri, 7:380, 7:381
Xenoxybelis spp., 7:467

Y

Yacaré caimans, 7:171, *7:172*
Yamakagashis, 7:467, 7:468, *7:471,* *7:473,* 7:478
Yangtze alligators. *See* Chinese alligators
Yarrow's spiny lizards. *See* Mountain spiny lizards

Yellow and black burrowing snakes, 7:462
Yellow-bellied seasnakes, 7:30, 7:484, *7:489,* 7:496–*497*
Yellow-bellied sliders. *See* Pond sliders
Yellow-blotched palm-pitvipers, *7:450, 7:453*
Yellow-footed tortoises. *See* South American yellow-footed tortoises
Yellow-headed geckos, *7:264, 7:265,* 7:266–267
Yellow-headed sidenecks. *See* Yellow-spotted river turtles
Yellow-headed temple turtles, *7:116*
Yellow-margined box turtles, *7:117,* *7:118*–119
Yellow mud turtles, *7:72, 7:124, 7:125*
Yellow-spotted Amazon turtles. *See* Yellow-spotted river turtles
Yellow-spotted night lizards, 7:291, *7:293,* *7:294*

Yellow-spotted river turtles, 7:*138,* 7:*139,* 7:*140,* 7:141
Yellowhead collared lizards. *See* Common collared lizards
Yemen chameleons. *See* Veiled chameleons
Yemeni chameleons. *See* Veiled chameleons
Young, J. Z., 7:19
Yow lung. *See* Chinese alligators
Yucca night lizards. *See* Desert night lizards

Z

Zani, P. A., 7:304, 7:307
Zebra-tailed lizards, 7:245, *7:248,* *7:253*
Zonosaurus spp. *See* Madagascan plated lizards
Zonosaurus madagascariensis. See Madagascan plated lizards
Zoos, 7:62